Handbook of Natural Exotoxins: Bacterial Exotoxins

Handbook of Natural Exotoxins: Bacterial Exotoxins

Editor: Robert Jones

MURPHY & MOORE
www.murphy-moorepublishing.com

www.murphy-moorepublishing.com

MURPHY & MOORE

Cataloging-in-Publication Data

Handbook of natural exotoxins : bacterial exotoxins / edited by Robert Jones.
p. cm.
Includes bibliographical references and index.
ISBN 978-1-63987-290-9
1. Bacterial toxins. 2. Microbial toxins. 3. Bacterial antigens. I. Jones, Robert.
QP632.B3 H36 2022
615.952 99--dc23

Murphy & Moore Publishing,
1 Rockefeller Plaza,
New York City, NY 10020, USA

ISBN 978-1-63987-290-9 (Hardback)

Contents

Preface

Bacteria are prokaryotic microorganisms, which are generally a few micrometers long. They are found in the soil, acidic hot springs, water and even radioactive waste. The toxin secreted by the bacteria is termed as an exotoxin. It disrupts the normal cellular mechanism of the host, causing major damage. Exotoxins can be classified into various types such as cell surface-active, membrane damaging and intracellular. They are used for the production of vaccines and cancer treatment. The treatment of cancer involves the elimination of cancer cells without causing harm to normal cells. This is achieved by attaching an antibody or receptor ligand to the exotoxin. This results in the creation of a recombinant toxin which harms specified cells. This book traces the progress in the field of bacteriology and highlights some of its key concepts and applications. It consists of contributions made by international experts. Those in search of information to further their knowledge will be greatly assisted by this book.

This book is a comprehensive compilation of works of different researchers from varied parts of the world. It includes valuable experiences of the researchers with the sole objective of providing the readers (learners) with a proper knowledge of the concerned field. This book will be beneficial in evoking inspiration and enhancing the knowledge of the interested readers.

In the end, I would like to extend my heartiest thanks to the authors who worked with great determination on their chapters. I also appreciate the publisher's support in the course of the book. I would also like to deeply acknowledge my family who stood by me as a source of inspiration during the project.

Editor

1

Bacterial Toxins as Pathogen Weapons Against Phagocytes

Ana do Vale[1,2]*, *Didier Cabanes*[1,3] *and Sandra Sousa*[1,3]*

[1] *Host Interaction and Response, Instituto de Investigação e Inovação em Saúde, Universidade do Porto, Porto, Portugal,* [2] *Group of Fish Immunology and Vaccinology, Instituto de Biologia Molecular e Celular, Universidade do Porto, Porto, Portugal,* [3] *Group of Molecular Microbiology, Instituto de Biologia Molecular e Celular, Universidade do Porto, Porto, Portugal*

Edited by:
Katharina F. Kubatzky,
University Hospital Heidelberg, Germany

Reviewed by:
Charles C. Caldwell,
University of Cincinnati, USA
Rance E. Berg,
University of North Texas Health Science Center, USA
Asparouh Iliev,
University of Bern, Switzerland

***Correspondence:**
Ana do Vale
avale@ibmc.up.pt;
Sandra Sousa
srsousa@ibmc.up.pt

Bacterial toxins are virulence factors that manipulate host cell functions and take over the control of vital processes of living organisms to favor microbial infection. Some toxins directly target innate immune cells, thereby annihilating a major branch of the host immune response. In this review we will focus on bacterial toxins that act from the extracellular milieu and hinder the function of macrophages and neutrophils. In particular, we will concentrate on toxins from Gram-positive and Gram-negative bacteria that manipulate cell signaling or induce cell death by either imposing direct damage to the host cells cytoplasmic membrane or enzymatically modifying key eukaryotic targets. Outcomes regarding pathogen dissemination, host damage and disease progression will be discussed.

Keywords: bacterial exotoxin, macrophage, neutrophil, phagocyte, immunomodulation, innate immune response

INTRODUCTION

Macrophages and neutrophils are central mediators of the innate immune system that act at early stages of bacterial infection and have the ability to clear the pathogen through phagocytosis and subsequent digestion (Mantovani et al., 2011; Wynn et al., 2013). Besides their overlapping functional properties, macrophages, and neutrophils also display distinct and specialized features allowing their concerted and cooperative action against pathogens (Appelberg, 2007; Silva, 2010; Silva and Correia-Neves, 2012). Accordingly to their antimicrobial capacity, toxicity, and lifespan, macrophages and neutrophils have different localizations. Long-lived macrophages are distributed in tissues throughout the body where they perform local immune surveillance activities, including recognition, phagocytosis, and rapid signaling of invading pathogens. The number of resident macrophages in resting tissues is rather low. Upon infection resident macrophages secrete chemokines, rapidly recruiting monocytes, and quiescent neutrophils from blood and bone marrow pools to the infectious foci. Monocytes differentiate to short-lived mature macrophages and neutrophils are activated to produce powerful antimicrobial molecules and release proinflammatory cytokines and chemokines amplifying the initial chemotactic role of resident macrophages and sustaining antimicrobial activities (Mantovani et al., 2011; Wynn et al., 2013).

The outcome of an infection is dictated by the nature of host-pathogen interactions and greatly depends on the efficacy of phagocytes. When, through the action of virulence determinants, the pathogen takes the control of the interaction in detriment of the host, the infection establishes. By the contrary, if the host immune defenses dominate over the pathogen, the infection is controlled and the pathogen is eliminated. In this scenario, several human pathogens evolved an arsenal of sophisticated mechanisms to evade host defenses. In particular, given the central role of macrophages and neutrophils as primary phagocytes, several pathogens deploy strategies to either survive or annihilate phagocytes antimicrobial functions (Flannagan et al., 2009). To avoid macrophage antimicrobial properties, some intracellular pathogens interfere with the classical maturation of the phagosome, blocking fusion with endosomes, and lysosomes thereby preventing destruction. Others escape the phagosome to multiply in the cytosol and modulate host gene expression to limit production of cytokines that would be deleterious for pathogen progression. Despite the mechanisms of evasion of neutrophil-killing are yet poorly defined, it is known that some pathogens survive and multiply inside neutrophils which may act as Trojan horses for microbial dissemination (Appelberg, 2007). From the extracellular milieu, some pathogens target macrophage and neutrophils mainly through the secretion of toxins that: (1) cause irreversible damage leading to phagocyte death or (2) heavily perturb intracellular signaling pathways, blocking phagocytosis or modulating inflammation (e.g., through the control of expression of chemokines and cytokines; Lemichez and Barbieri, 2013).

Toxins are potent molecules produced by a large variety of bacterial pathogens that target host cells and play key roles in the host–pathogen dialog. They are major virulence factors often sufficient to determine the outcome of the infection. Indeed, attesting their importance in pathogenesis, the injection of small amounts of some purified toxins can recapitulate many key symptoms of the disease. Bacterial toxins can be divided in several groups regarding their nature and mode of action (Lemichez and Barbieri, 2013). In this review we focus on bacterial exotoxins, which are secreted by the pathogen and act on the host cells from the extracellular milieu. Although exotoxins may target different cell types, some specifically target macrophages and neutrophils thus taking the control of innate immune response, providing the pathogen a suitable environment for active proliferation. Interestingly, while the initial steps of phagocytes intoxication are specific for different exotoxins, the ultimate cellular effects leading to the loss of host cell function are often the same. Taking this into account we review here the mechanisms of phagocyte targeting by archetypal exotoxins such as pertussis toxin (PT) and adenylate cyclase toxin (ACT) secreted by *Bordetella pertussis*, anthrax toxin from *Bacillus anthracis* and *Staphylococcus aureus* leukotoxins. We also include here AIP56, a recently described toxin from *Photobacterium damselae piscicida* (Phdp). In addition, the mode of action of mycolactone, a polyketide molecule produced by *Mycobacterium ulcerans*, and other bacterial secreted products not formally termed as toxins, such as *S. aureus* superantigens-like proteins (SSLs) and phenol-soluble modulins (PSMs), are also reviewed here. Clostridial C3 toxins, which target and modulate macrophage functions, are the focus of another review in this Topic (Barth et al., 2015).

Bordetella pertussis: TWO TOXINS ARE BETTER THAN ONE

Bordetella pertussis is a Gram-negative pathogen that infects the human respiratory tract causing whooping cough, an acute and highly contagious infection (Mattoo and Cherry, 2005; Melvin et al., 2014). Initially thought to be a toxin-mediated disease (Pittman, 1984), such as cholera and diphtheria, pertussis disease is instead the result of the coordinated action of a variety of bacterial factors that allow bacterial adherence to ciliated respiratory epithelium, survival to host innate immune defense, multiplication, and resistance to inflammatory cells (Carbonetti, 2007). Two of the major virulence factors of *B. pertussis* are the secreted toxins, PT and ACT, which emerged as key elements for suppression/modulation of the host immune and inflammatory responses (Carbonetti, 2010; Higgs et al., 2012; Melvin et al., 2014). Interestingly, mouse infections with different mutants suggested that these two toxins have complementary functions in pathogenesis assaulting the innate immune cells at different times and from different angles. PT would act at early stages of infection mainly inhibiting the recruitment of immune cells, while ACT would later intoxicate macrophages and neutrophils blocking bacterial engulfment and destruction (Carbonetti et al., 2005).

First Round: PT Inflicts First Blow to the Host

Pertussis toxin is a multisubunit AB-toxin exclusively produced by *B. pertussis*. Through B-subunits, PT binds to any sialic acid-containing glycoprotein at the cellular surface (Witvliet et al., 1989; Saukkonen et al., 1992; Stein et al., 1994), is internalized by endocytosis and follows a retrograde transport pathway through the Golgi complex to the endoplasmic reticulum (Plaut and Carbonetti, 2008; **Figure 1**), from which the A-subunit translocate to the cytoplasm of host cells (Locht et al., 2011). In the host cell cytosol, the A-subunit exhibits ADP-ribosyltransferase activity, hydrolysing NAD and ADP-ribosylating heteromeric G-proteins of the G_i family (Katada et al., 1983). This modification prevents the interaction of G_i proteins with their cognate G-protein coupled receptors (GPCRs) causing the disruption of downstream cell signaling transduction (Mangmool and Kurose, 2011). Besides the activity of the A domain, B-subunits also display signaling function by activating intracellular signaling cascades in a G_i-protein independent manner (Mangmool and Kurose, 2011).

Pertussis toxin has long been considered as a major *B. pertussis* virulence factor. Indeed, PT was reported to be the cause of systemic symptoms of pertussis disease such as lymphocytosis and leukocytosis (Morse and Morse, 1976) and was associated with lethal infection by *B. pertussis* in a neonatal mouse model (Goodwin and Weiss, 1990). However, evidences supporting its role in respiratory infection have only emerged in the last decade. Experiments in the mouse model revealed that the lack of PT confers a defect in *B. pertussis* colonization at the early stages of infection (Carbonetti et al., 2003, 2005). Interestingly, a PT-deficient strain reaches wild type levels of colonization whenever co-infections with both strains are performed or intranasal inoculation of purified PT precedes infection (Carbonetti et al., 2003). Further studies have shown that depletion of resident airway macrophages leads to exacerbated *B. pertussis* infection in a PT-independent manner (Carbonetti et al., 2007), indicating that PT targets

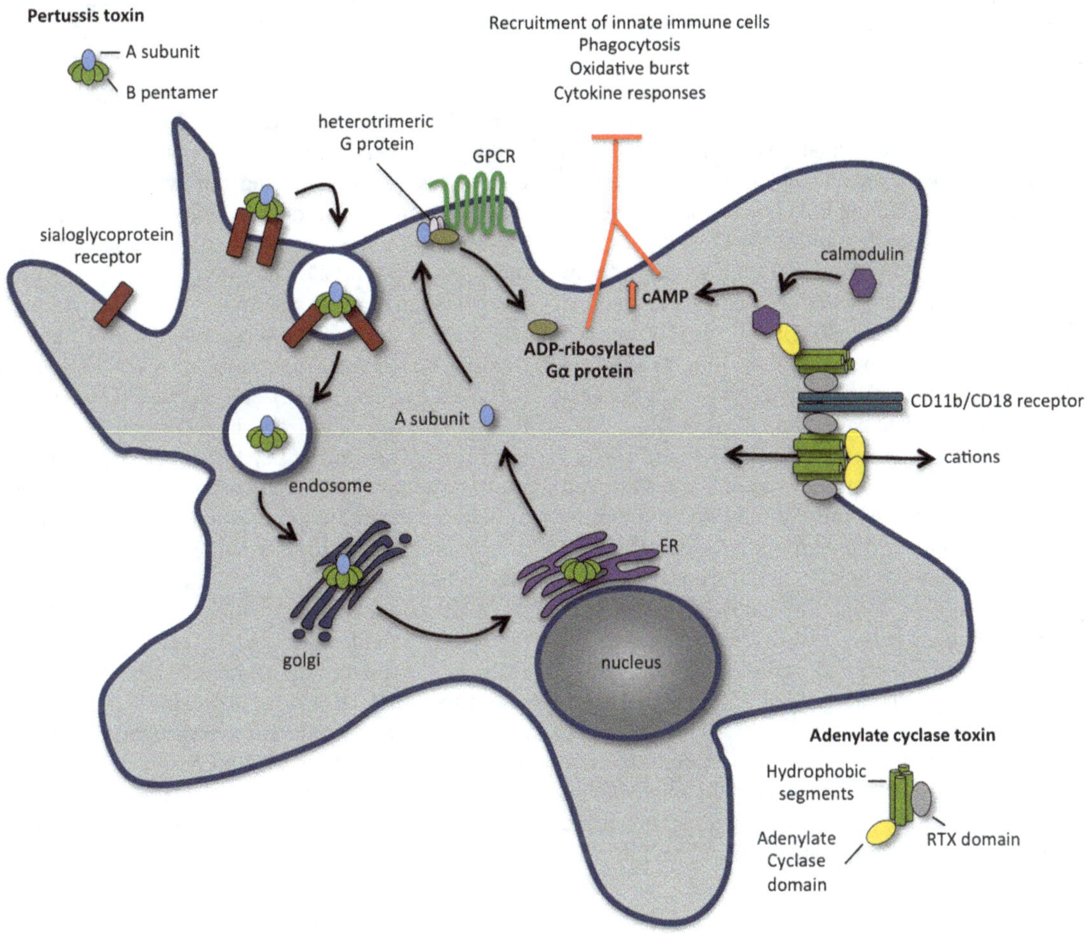

FIGURE 1 | The concerted action of pertussis toxin (PT) and adenylate cyclase toxin (ACT) annihilates recruitment and function of innate immune cells. Following binding to a sialoglyprotein receptor, PT is endocytosed and retrogradely transported to the endoplasmic reticulum (ER). From the ER, the A subunit is delivered into the cytosol and travels to the plasma membrane, where it ADP-ribosylates the alpha-subunit of heterotrimeric G proteins, perturbing their regulatory functions and leading to an increase in the cAMP concentration that contributes to the early suppression of inflammatory cytokine production and inhibits the recruitment of immune cells to the site of infection. ACT binds with high affinity to CD11b/CD18 receptor [also known as complement receptor 3 (CR3) or macrophage-1 antigen (Mac-1)] present at the surface of macrophages, neutrophils, and dendritic cells. Upon binding, ACT integrates the membrane of target cell in two different conformations: a translocation precursor that relocalises at lipid raft domains from where the adenylate cyclase activity translocates directly to the cell cytoplasm; and a pore precursor that oligomerises and permeabilises the cells causing ion concentration imbalance. Binding of calmodulin stimulates the adenylate cyclase, leading to an increase in intracellular levels of cAMP. The activity of ACT inhibits complement-mediated phagocytosis, inhibits the production of pro-inflammatory cytokines and interferes with immune cell recruitment.

airway macrophages disrupting their protective activity at early steps of infection (Carbonetti et al., 2007). Intranasal administration of PT resulted in ADP-ribosylation of airway macrophages G_i-proteins (Carbonetti et al., 2007), suggesting that its inhibitory function on macrophages *in vivo* results from the immunosuppressive activities ascribed to PT *in vitro*. In particular, *in vitro* PT was shown to inhibit macrophage and neutrophil migration (Meade et al., 1984), phagocytosis (Mork and Hancock, 1993) and cytokine response (He et al., 1988; Hume and Denkins, 1989).

In vivo, PT plays a dual role in the establishment of the disease. Whereas it has immunosuppressive functions at early stages of infection, later it potentiates inflammatory responses, likely prolonging healing and promoting bacterial dissemination (Eby et al., 2015). Early in infection, PT targets airway macrophages and inhibits neutrophil recruitment to the infection site (Carbonetti et al., 2003, 2005; Kirimanjeswara et al., 2005; Andreasen and Carbonetti, 2008). Neutrophils were reported to play a protective role against *B. pertussis* only in previously infected mice and in the presence of anti-*B. pertussis* antibodies (Kirimanjeswara et al., 2005; Andreasen and Carbonetti, 2009), which may suggest that PT delays neutrophil recruitment to the airways avoiding rapid antibody-mediated clearance of the pathogen. Later, at the peak of infection, high numbers of neutrophils are recruited to the lungs of mice infected with wild type strain, but not to those infected with PT-deficient strain (Carbonetti et al., 2005; Andreasen et al., 2009). This recruitment of neutrophils to the lungs correlates with an increase of proinflammatory cytokines such as IL-17, TNFα and IFNγ that appears to be dependent on PT activity (Andreasen et al., 2009). Recently, global transcriptional profiles of mice lungs infected with wild type or PT-deficient *B. pertussis* revealed that, at the peak of infection, the ADP-ribosylation activity of PT correlates with upregulation of immune and inflammatory response genes (Connelly et al., 2012). *In vitro* studies suggested that PT directly impairs neutrophil migration through ADP-ribosylation the G_i-proteins associated with surface chemokine receptors (Spangrude et al., 1985; Scott et al., 1988; **Figure 1**). However, *in vivo* data showed that besides acting directly on chemokine receptors signaling (Kirimanjeswara et al., 2005), PT also suppresses early neutrophil recruitment by inhibiting the production of neutrophil-attracting chemokines by airway macrophages and lung epithelial cells (Andreasen and Carbonetti, 2008). Transcription of genes expressing CXCL1, CXCL2, and CXCL5 is inhibited in the lungs of mice intranasally infected with wild type *B. pertussis*, as compared to animals infected with strains deficient for PT production or producing a PT variant devoid of ADP-ribosylation activity (Andreasen and Carbonetti, 2008). Interestingly, PT also blocks chemokine gene expression and early neutrophil recruitment to the airways following intranasal administration of LPS (Andreasen and Carbonetti, 2008), which occurs presumably through modulation of TLR-4 signaling, a G_i-protein independent pathway.

Altogether, these studies show that PT suppresses early inflammation in the respiratory tract and inhibit microbicidal function of inflammatory cells, potentially providing advantage to the pathogen by allowing its rapid growth and establishment within the host at early phases of infection.

Second Round: ACT Perpetuates the Pathogen Gain

Adenylate cyclase toxin is a bi-functional toxin produced by all species of *Bordetella* that infect mammals. While its N-terminal domain contains an adenylate cyclase activity that converts ATP in cAMP, the C-terminus possesses RTX motifs that bind mammalian cells and form cation-selective pores in the host plasma membrane. Through its combined adenylate cyclase and pore-forming activities, ACT manipulates host cell physiology in two different ways: it interferes with intracellular signaling by increasing the levels of cAMP, and disturbs ion homeostasis by disrupting the permeability barrier of the plasma membrane (**Figure 1**). ACT specifically binds with high affinity CD11b/CD18 (also known as CR3 or Mac-1) present at the surface of macrophages, neutrophils and dendritic cells (DCs; Guermonprez et al., 2001). Upon binding, ACT integrates the membrane of target cells in two different conformations: a translocation precursor that re-localizes at lipid raft domains from where the adenylate cyclase activity translocates directly to the cell cytoplasm; and a pore precursor that oligomerises and permeabilises the cells causing ion concentration imbalance (Fiser et al., 2007, 2012; Bumba et al., 2010; **Figure 1**). While the rapid translocation of adenylate cyclase activity across the host plasma membrane does not dependent on endocytosis (Gordon et al., 1988), the clathrin-dependent endocytosis of ACT together with its receptor CD11b/CD18 was reported (Khelef et al., 2001; Martin et al., 2011). Classical mechanisms of membrane repair upon toxin-induced pore formation include the removal of the pores through endocytosis. Interestingly, ACT-translocated molecules control the rate of ACT-pore removal delaying their endocytic uptake, thus exacerbating the permeabilisation of phagocytes and maximizing the cytotoxic action (Fiser et al., 2012). Despite its interaction with a specific receptor, ACT was reported to promiscuously bind and intoxicate many cell types, including CD11b/CD18-negative cells (Ladant and Ullmann, 1999; Paccani et al., 2008; Eby et al., 2012), although the biological relevance of these findings remains unclear. Indeed, while ACT triggers macrophage apoptosis *in vitro* (Khelef et al., 1993; Khelef and Guiso, 1995) and *in vivo* (Gueirard et al., 1998), the viability of cells from non-haematopietic origin remains unaffected (Gueirard et al., 1998; Bassinet et al., 2000). *In vivo* studies support that ACT primarily targets phagocytes such as alveolar macrophages and neutrophils, disrupting the early innate antibacterial host immune response (Harvill et al., 1999; **Figure 1**).

cAMP is a key second messenger with pleiotropic effects. Increased levels of cAMP severely compromise cellular functions such as the migration of neutrophils, the plasticity of DC responses, the release of cytokines by macrophages and the homeostasis of actin cytoskeleton. Thus, through the uncontrolled production of intracellular cAMP, ACT subverts phagocytic, and bactericidal function of macrophages and neutrophils in a variety of ways (**Figure 1**). In human alveolar macrophages and neutrophils, as in mouse macrophages, the ACT-mediated cAMP production blocks phagocytosis, chemotaxis, and oxidative burst (Confer and Eaton, 1982; Friedman et al., 1987; Weingart and Weiss, 2000; Kamanova et al., 2008; Eby et al., 2014). High levels of cAMP cause a transient inactivation of RhoA inducing massive actin rearrangements that dramatically decrease macropinocytosis, block complement-mediated phagocytosis (Kamanova et al., 2008) and possibly impair the chemotactic properties of primary monocytes. Recently, ACT-induced cAMP synthesis was shown to trigger pro-apoptotic signaling in phagocytes through the activity of tyrosine phosphatase SHP-1, the accumulation of cytosolic BimEL and the consequent activation of Bax, permeabilisation of the outer mitochondrial membrane and activation of programmed cell death (Ahmad et al., 2015; Cerny et al., 2015). Besides the immediate action of ACT on the ablation of bactericidal functions of phagocytes, ACT activity was also reported to block the release of TNFα and the production of ROS in human monocytes (Njamkepo et al., 2000), to promote incomplete or aberrant maturation of DCs (Skinner et al., 2004; Boyd et al., 2005) and to impair T-cell activation by interfering with immunological synapse signaling (Paccani et al., 2011). ACT was described to suppress the secretion of pro-inflammatory cytokines such as IL-12 and TNFα and favor the production of anti-inflammatory IL-10 molecules (Spensieri et al., 2006; Hickey et al., 2008).

Despite the fact that studies in mouse models established ACT as an important virulence factor for *B. pertussis* infection (Khelef et al., 1992), the effects described above were obtained from *in vitro* studies, and their significance during *in vivo* infection requires further investigation. Importantly, recent studies using baboon infection model and clinical samples from humans showed that the concentration of ACT in tissues is much lower than the amount used for *in vitro* experiments (Eby et al., 2013), which may compromise the relevance of some effects reported *in vitro*.

Pertussis toxin and ACT have undeniable roles during *B. pertussis* infection and certainly play key functions in the pathophysiology of pertussis disease. Numerous and wide-ranging effects of the purified toxins on cultured cell lines have been reported, however, establishing the correlation of such effects with the human pathology appeared as an incredible difficult task. In addition, the mouse model only provides limited possibilities to address this issue. Together with analysis of clinical samples from humans, the use of baboons as non-human primate model is expected to shed new light on the mechanisms of action of PT and ACT and on pertussis disease.

Bacillus anthracis: ARMED TO ANNIHILATE THE HOST INNATE IMMUNE DEFENSES

Bacillus anthracis is a Gram-positive, spore-forming rod that causes anthrax, an acute and fast progressing disease that affects humans and other animals, and results from a combination of bacterial infection and toxemia (Moayeri et al., 2015). The spores, which are the infectious form of the pathogen, are able to resist harsh environmental conditions and to infect new hosts when inhaled, ingested, or exposed to skin breaks. Upon entering a potential host, spores germinate into vegetative bacilli that replicate and disseminate through the bloodstream, leading to a systemic infection. Whereas in experimental inhalational anthrax, spore germination requires engulfment by macrophages/DCs (Guidi-Rontani et al., 1999; Hanna and Ireland, 1999), in experimental cutaneous infections it has been shown that most of the spore germination occurs extracellularly (Bischof et al., 2007; Corre et al., 2013). Soon after spore germination, vegetative *B. anthracis* start producing two potent exotoxins – anthrax lethal toxin (LT) and edema toxin (ET) – that along with a poly-D-glutamic acid capsule, are its major virulence factors (Moayeri et al., 2015). *B. anthracis* also secretes anthrolysin O (ALO), a member of the cholesterol-dependent cytolysin family (Shannon et al., 2003). ALO is cytotoxic for several mono- and polymorphonuclear cells (Shannon et al., 2003; Cocklin et al., 2006; Mosser and Rest, 2006) and together with LT, induces apoptosis of macrophages (Shannon et al., 2003). However, ALO-deficient and wild-type strains revealed no differences in virulence in an inhalation infection model (Heffernan et al., 2007). Consistent with these observations, ALO-based vaccines, although conferring protection against lethal intravenous challenge with ALO, do not protect mice against peritoneal infection by *B. anthracis* (Cowan et al., 2007). Altogether, available data indicate that ALO is not essential for *B. anthracis* virulence and thus further studies are needed to determine if ALO contributes to *B. anthracis* pathogenicity.

Anthrax toxins originate from the association of three different protein components: a host cell receptor binding protein named protective antigen (PA) and two enzymatic proteins, edema factor (EF) and lethal factor (LF) (Young and Collier, 2007; **Figure 2**). EF or LF associated to PA is referred as ET or LT, respectively. Host cell intoxication begins when PA binds to either tumor endothelium marker 8 (TEM8, also known as anthrax toxin receptor 1, ANTXR1) or capillary morphogenesis protein 2 (CMG2, also known as anthrax toxin receptor 2, ANTXR2), which are expressed by different cell types, including macrophages and neutrophils (Bradley et al., 2001; Scobie et al., 2003). Although both TEM8 and CMG2 can function as anthrax toxin receptors, studies with TEM8- and CMG2-null mice have shown that CMG2 is the major anthrax toxin receptor mediating toxin lethality *in vivo* and that TEM8 plays only a minor role in anthrax toxin-associated pathogenesis (Liu et al., 2009). Upon binding to its cell surface receptor, PA is proteolytically processed at its N-terminus by a furin-like protease and self-assembles forming an oligomeric prepore able to bind EF and LF. The EF and/or LF-prepore-receptor complex undergoes receptor-mediated endocytosis (Abrami et al., 2003) and the acidic conditions in endosomes induce conversion of the prepore to a pore, allowing translocation of EF and LF into the cell cytosol (Collier, 2009) to exert their cytotoxic effects (**Figure 2**). In endosomes, the toxin complex can also be sorted into intraluminal vesicles that undergo back fusion with the endosomal membrane allowing sequestered EF and LF to reach the cytosol (Abrami et al., 2004).

The toxicity of LT and ET results from the catalytic activities of LF and EF. LF is a zinc-dependent metalloprotease that cleaves mitogen activate protein kinase (MAPK) kinases (MKK) 1-4 and 6-7, preventing the activation of the ERK1/2, p38 and JNK-pathways (Duesbery et al., 1998; Pellizzari et al., 1999; Vitale et al., 2000; Tournier et al., 2005) and thus interfering with critical signaling pathways involved in host defense (**Figure 2**). EF is a calmodulin- and Ca^{2+}-dependent adenylate cyclase that causes a sustained increase of cAMP (Leppla, 1982) by converting ATP to cAMP. As discussed above in the context of the ACT from *B. pertussis*, increased cAMP levels perturb key cellular functions, leading to severe consequences for the host (Leppla, 1982; Firoved et al., 2005).

Anthrax toxins: Weapons to Cripple Host Macrophages and Neutrophils

Anthrax has for long been described as a toxin-mediated disease, mainly because LT and ET can be lethal for experimental animals and anti-toxin immunization is effective in protecting against infection (Kaur et al., 2013). However, it has recently been proposed that extreme bacteraemia and severe sepsis rather than the anthrax toxins *per se* are the cause of anthrax-induced lethality (Coggeshall et al., 2013). Indeed, the fact that anthrax toxins act on multiple tissues simultaneously, due to the ubiquitous expression of the anthrax toxin receptors, complicate untangling their effects on the host and delayed the identification of the key tissue targets responsible for its lethal effects. Recently, using cell-type specific CMG2-null mice and the correspondent

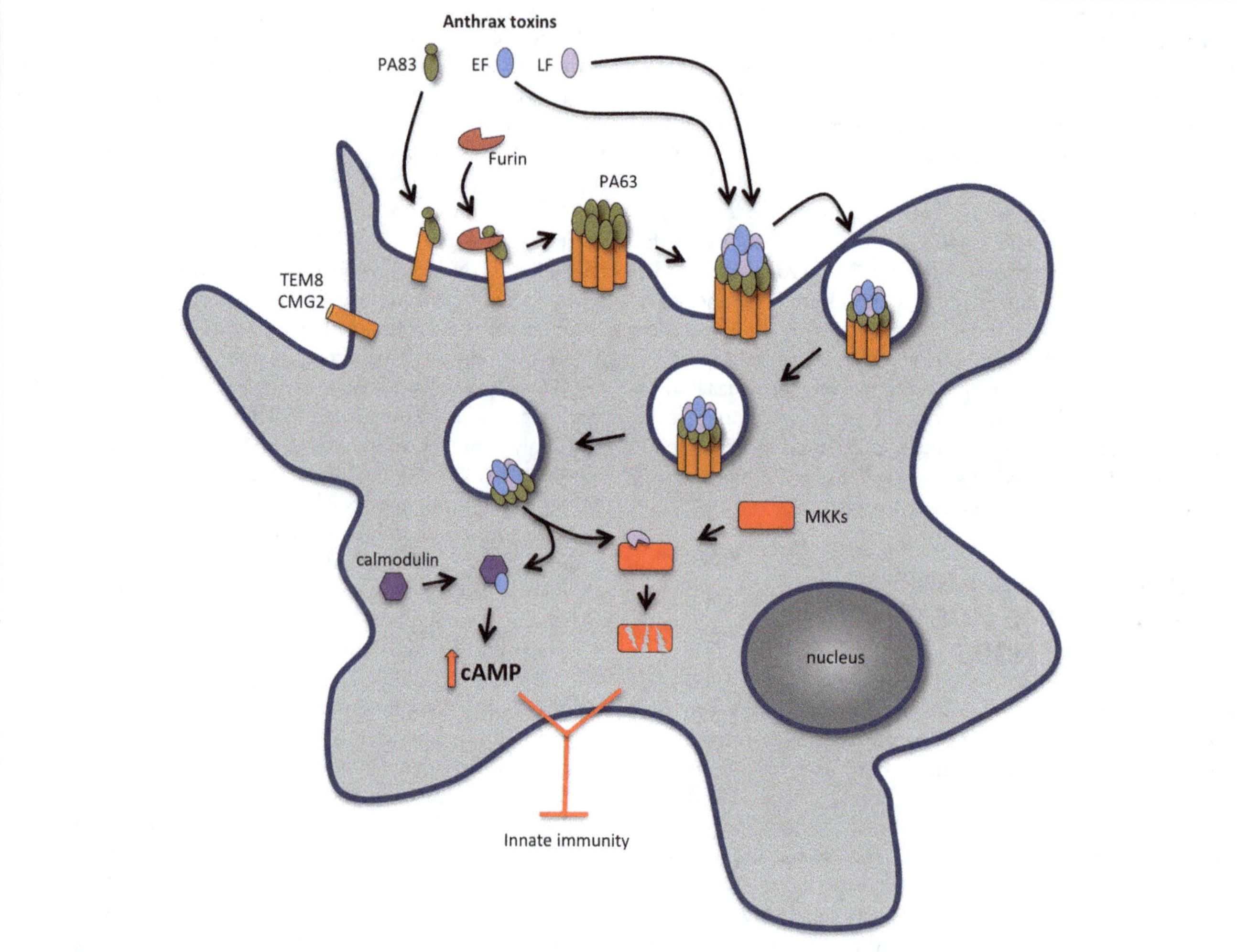

FIGURE 2 | Anthrax toxins cooperatively disable host innate immune response. Host cell intoxication by anthrax toxins involves interaction of protective antigen (PA, 83 kDa) with two cellular receptors [tumor endothelium marker 8 (TEM8), also known as anthrax toxin receptor 1 (ANTXR1) and capillary morphogenesis protein 2 (CMG2) also known as anthrax toxin receptor 2 (ANTXR2)], which are expressed by different cell types, including macrophages and neutrophils. Upon binding to the cell surface receptor, PA83 is proteolytically processed at its N-terminus by a furin-like protease yielding the C-terminal fragment PA63 that oligomerises into a heptameric prepore able to bind edema factor (EF) and lethal factor (LF). The EF and/or LF-prepore-receptor complex undergoes receptor-mediated endocytosis and the acidic conditions in endosomes induce conversion of the prepore to a pore, allowing translocation of EF and LF into the cell cytosol to exert their cytotoxic effects. LT is a zinc-dependent metalloprotease that inhibits activation of neutrophils and macrophages, expression of inflammatory cytokines and cell motility by disrupting mitogen activated protein kinase kinases (MAPKKs)-regulated pathways. LT activity also promotes macrophage apoptosis by interfering with pro-survival MAPKK dependent pathways. ET is a calcium and calmodulin-dependent adenylate cyclase that increases intracellular cAMP concentration, leading to the suppression of the expression of inflammatory cytokines and cell chemotaxis through protein kinase A (PKA)-dependent pathways. The concerted action of LT and ET blocks the function of phagocytic cells.

cell-type specific CMG2-expressing mice, Liu and colleagues have shown that LT-induced mortality requires targeting of cardiomyocytes and vascular smooth muscle cells, whereas ET-induced lethality relies mainly on targeting hepatocytes (Liu et al., 2013). Using myeloid-specific CMG2-null mice, in which both macrophages and neutrophils are insensitive to LT and ET due to their inability to bind and internalize the toxins, the same authors have also clarified the role of macrophages and other myeloid cells in anthrax toxins induced lethality and in *B. anthracis* infection (Liu et al., 2010). Myeloid-specific CMG2-null mice are fully sensitive to both LT and ET, indicating that lethality does not depend on the targeting of macrophages, neutrophils, and other myeloid cells (Liu et al., 2010). However, they are completely resistant to infection by *B. anthracis*, indicating that the targeting of myeloid cells by anthrax toxins is required for the establishment of a successful infection (Liu et al., 2010). In what concerns the relative contributions of ET and LT to the establishment of *B. anthracis* infection, trials with Sterne strains deleted of PA, LF, or EF suggest that LT plays a more prominent role (Liu et al., 2010).

Available data indicate that LT and ET act in concert to inhibit macrophage activation as well as the activation and recruitment of other immune cells, such as neutrophils, early in infection (Baldari et al., 2006; Tournier et al., 2009; Liu et al., 2014). This favors bacterial escape and multiplication, and contributes to the severe bacteraemia observed in terminal disease. Macrophage activation requires signaling through MAPK cascades, including JNK and p38 pathways, which are central for induction of inflammatory molecules, including cytokines and chemoattractants, as well as Cox-2 and iNOS. LT interrupts MAPK cascades by directly cleaving MAPKK, whereas ET inhibits MAPK-dependent gene expression by interfering with multiple PKA-related pathways (Baldari et al., 2006). Several reports show that LT inhibits the secretion of pro-inflammatory cytokines by macrophages as well as by DCs *in vitro* and *in vivo* (Pellizzari et al., 1999; Erwin et al., 2001; Agrawal et al., 2003; Alileche et al., 2005; Bergman et al., 2005; Brittingham et al., 2005; Tournier et al., 2005; Ribot et al., 2006). ET has been shown to suppress secretion of inflammatory mediators by DCs (Tournier et al., 2005). Whereas LT inhibits IL-10 secretion by these cells, ET inhibits IL12p70 production (Tournier et al., 2005). Interestingly, LT and ET have a cumulative suppressive effect upon TNFα secretion (Tournier et al., 2005). In addition to suppress pro-inflammatory cytokine secretion by macrophages, LT is also able to trigger programmed cell death in these cells *in vitro*. Indeed, it has been reported that LT induced apoptosis of RAW264.7 cells (Popov et al., 2002). Furthermore, although treatment of three different human monocytic cell lines (HL-60, THP-1, and U937) did not result in cell death, upon differentiation into macrophage-like phenotypes, the cells become susceptible to a cell death program that has features of apoptosis but apparently does not requires the activity of effector caspases (Kassam et al., 2005). The apoptogenic activity of LT toward macrophages likely relates to LT-dependent disruption of survival signals triggered by TLR4 and mediated by p38 MAPK (Park et al., 2002, 2005; Hsu et al., 2004) that activates the NF-kB-dependent expression of pro-survival genes. By cleaving the upstream MAPKK MKK3, LT blocks p38 MAPK and NF-kB activation, leading to macrophage apoptosis (Huang et al., 2004; Park et al., 2005).

Lethal toxin has also been shown to induce a rapid and lytic form of caspase-1-dependent cell death, called pyroptosis, in macrophages from specific rat and mice strains. The susceptibility of macrophages to pyroptosis has been linked to polymorphisms of the *Nlrp1b* gene in mice (Boyden and Dietrich, 2006) and of the orthologous *Nlrp1* gene in rat (Newman et al., 2010). The ability of LT to induce macrophage pyroptosis was initially interpreted as a virulence mechanism of *B. anthracis* (Muehlbauer et al., 2007). It was speculated that the destruction of macrophages by LT compromised their role in restricting *B. anthracis* infection and that the cytokine burst associated to LT-induced macrophage lysis contributed to LT-dependent pathological effects by aggravating the vascular damage occurring in anthrax (Muehlbauer et al., 2007). More recently, it has been shown that LT is able to directly cleave mouse Nlrp1b and rat Nlrp1 close to their N-terminus (Hellmich et al., 2012; Levinsohn et al., 2012; Chavarria-Smith and Vance, 2013) resulting in the activation of Nlrp1 inflammasomes in rat and in mice with LT-sensitive macrophages. Ultimately, this leads to caspase-1 activation and pyroptosis accompanied by the release of the inflammatory cytokines IL-1β and IL-18, which induces a strong innate immune response that is protective against *B. anthracis* infection (Moayeri et al., 2010; Terra et al., 2010). Therefore, the current view is that LT-mediated activation of Nlrp1 that leads to inflammasome activation and macrophage pyroptosis is not a virulence mechanism used by *B. anthracis* to promote infection, but rather a protective host-response against anthrax (Chavarria-Smith and Vance, 2015).

Neutrophils play a major role in controlling *B. anthracis* infection and anthrax-toxin mediated neutralization of neutrophil functions is essential for successful infection (Liu et al., 2010). In *vitro* studies suggest that by reducing F-actin formation, LT and ET cooperate to inhibit neutrophil chemotaxis, chemokinesis, and ability to polarize (During et al., 2005, 2007; Szarowicz et al., 2009). It was also reported that neutrophils intoxicated with ET have reduced phagocytic activity (O'Brien et al., 1985). Additionally, LT has been shown to suppress cytokine production by neutrophils *in vitro* (Barson et al., 2008) and LT and ET were found to block neutrophil priming by LPS or muramyl dipeptide, thereby dampening the oxidative burst normally elicited by bacterial products and required for full antimicrobial activity (Wright and Mandell, 1986). In the case of ET, it has been shown that the inhibition of superoxide production results from an impairment of the activation of the neutrophil NADPH oxidase, an effect that likely results from the activity of ET as an adenylate cyclase (Crawford et al., 2006). Although it has been proposed that these effects are due to the ET-induced rise in cAMP leading to phosphorylation (activation) of PKA (Szarowicz et al., 2009), the downstream targets remain to be identified.

It is now unquestionable that anthrax toxins are crucial in anthrax pathogenesis, but their precise roles during human anthrax infections remain to be further clarified. Following the discovery of the anthrax toxins more than half a century ago, a myriad of studies were developed aiming at defining their role in anthrax disease and lethality. However, most of the infection studies were performed in animal models that may not completely reflect the events occurring during human infections. Concerning the data documenting the immunomodulatory effects of the anthrax toxins, most were obtained *in vitro*, often in experimental set-ups that involve the use of purified toxins and do not allow examining intoxication in the context of infection. Therefore, the available data need to be validated in the context of relevant animal models of infections before being extrapolated to the human disease scenario.

Photobacterium damselae piscicida: KILLING TWO BIRDS WITH ONE STONE

Photobacterium damselae piscicida is a Gram-negative extracellular bacterium that causes a systemic and deadly infection with a rapid course and very high mortalities in both wild and cultured marine fish (Romalde, 2002; Barnes et al., 2005). Phdp infections are characterized by the occurrence of generalized bacteraemia and extensive cytopathology with abundant tissue necrosis (do Vale et al., 2007). Infected fish often present whitish tubercle-like lesions of about 0.5 to

3.5 mm in diameter in several internal organs (Tung et al., 1985; Hawke et al., 1987; Noya et al., 1995; Magariños et al., 1996; do Vale et al., 2007), leading to the coining of the disease as fish pseudotuberculosis (Kubota et al., 1970). The lesions consist of accumulations of bacteria and apoptotic and necrotic cell debris (Kubota et al., 1970; Hawke et al., 1987; do Vale et al., 2007).

Photobacterium damselae piscicida-associated pathology is triggered by AIP56 (apoptosis inducing protein of 56 kDa), a plasmid-encoded toxin secreted by virulent Phdp strains (do Vale et al., 2005). The toxin is systemically disseminated in infected animals and induces selective apoptotic destruction of macrophages and neutrophils (do Vale et al., 2003, 2005, 2007). The simultaneous destruction of these cell types by AIP56 has two dramatic consequences for the host. On one hand, the drastic reduction of the number of phagocytes impairs the phagocytic defense, favoring pathogen dissemination (do Vale et al., 2007). On the other hand, it compromises the host capacity to clear apoptosing cells, leading to the lysis of the phagocytes by post-apoptotic secondary necrosis with consequent release of their highly cytotoxic tissue-damaging contents (do Vale et al., 2007; Silva et al., 2008).

AIP56: Spreading the Misery by Killing the Soldiers and Preventing their Burial

AIP56 is the founding and the only characterized member of a continuously growing family of bacterial proteins identified in different organisms, mainly marine *Vibrio* species and *Arsenophonus nasoniae*. It is an AB-type toxin, possessing a catalytic A domain at its N-terminal region and a B domain involved in binding/internalization into target cells at its C-terminal region (Silva et al., 2013; **Figure 3**). The catalytic domain of AIP56 is a zinc-dependent metalloprotease that cleaves the p65 subunit of NF-κB (Silva et al., 2013), an evolutionarily conserved transcription factor that regulates the expression of inflammatory and anti-apoptotic genes, playing a key role in host responses to microbial pathogen invasion. AIP56 has likely originated from a fusion of two components: its A domain is related to NleC, a type III secreted effector present in several enteric pathogenic bacteria (Yen et al., 2010; Baruch et al., 2011; Pearson et al., 2011; Sham et al., 2011; Hodgson et al., 2015) that are associated with severe human illness and death worldwide, whereas its B domain is related to a protein of unknown function from the lambda-like bacteriophage APSE2, a phage that infects *Hamiltonella defensa* (Degnan et al., 2009).

Although mammals are not susceptible to Phdp infection, likely due to temperature and osmolality restrictions, AIP56 is able to intoxicate mouse bone marrow derived macrophages (mBMDM; Pereira et al., 2014), through a mechanism similar to the one operating during intoxication of fish cells. Upon encountering susceptible cells, AIP56 binds to a still unidentified cell-surface receptor and is internalized through clathrin-mediated endocytosis (Pereira et al., 2014; **Figure 3**). Once in early endosomes, the toxin either follows the recycling pathway back to the extracellular medium or undergoes low pH-induced translocation across the endosomal membrane into the cytosol to display its toxic activity (Pereira et al., 2014). AIP56 cleaves an evolutionarily conserved peptide bond of the Rel homology domain of NF-κB p65, removing residues crucial for p65-DNA interaction and compromising NF-κB activity (Silva et al., 2013; **Figure 3**). During intoxication, the proteolytic activity of AIP56 results in a complete depletion of p65 and leads to the apoptotic death of cells (Silva et al., 2013; Pereira et al., 2014) through a process involving quick activation of caspases-8, -9 and -3, loss of mitochondrial membrane potential, translocation of cytochrome c to the cytosol and overproduction of ROS (do Vale et al., 2007; Reis et al., 2007a,b, 2010; Costa-Ramos et al., 2011).

AIP56 plays a pivotal role in the establishment of Phdp infection and in the development of the infection-associated pathology. In the initial phase of Phdp infection, when local multiplication of Phdp becomes detectable in infected tissues, extensive infiltration of macrophages and neutrophils occurs (do Vale et al., 2007). As the infection progresses, the pathogen extensively multiplies and disseminates systemically, which leads to a septicemic situation paralleled by the occurrence of AIP56 in the systemic circulation (do Vale et al., 2007). The presence of circulating toxin correlates with the appearance of high numbers of apoptotic macrophages and neutrophils in the peripheral blood, in the spleen, liver, and head–kidney vasculature, as well as in the splenic and head-kidney parenchyma and gut lamina propria (do Vale et al., 2007). This systemic apoptotic destruction of macrophages and neutrophils triggered by AIP56 explains the extensive phagocyte depletion observed in advanced Phdp infections (do Vale et al., 2007). The ability of the toxin to neutralize the main players responsible for the phagocytic defense of the host is a very effective pathogenicity strategy that contributes to the severity of Phdp infections by promoting survival of the pathogen and its unrestricted extracellular multiplication. Concomitantly, the AIP56-induced apoptosis of both professional phagocytes leads to tissue damage with deleterious consequences for the host. In fact, the destruction of macrophages, the cells with the crucial role of eliminating apoptotic cells (Parnaik et al., 2000; Wood et al., 2000), compromises the efficient clearance of apoptotic phagocytes and leads to their lysis by secondary necrosis (do Vale et al., 2003) with release of their cytotoxic intracellular contents (do Vale et al., 2007). This has particularly serious consequences in the case of neutrophils, due to their richness in highly cytotoxic molecules, which damage many cell types and produce tissue injury, thus contributing to the genesis of the Phdp-associated cytopathology.

Staphylococcus aureus: A BENCH OF SECRETED MOLECULES TO SHOOT NEUTROPHILS

Staphylococcus aureus is a Gram-positive bacterium that often colonizes the human nares and the skin. Besides being a commensal, *S. aureus* is also a redoubtable human pathogen that causes a variety of severe diseases. To set up a successful infection, *S. aureus* evolved an amazing variety of immune evasive strategies wiping out both innate and adaptive immune responses. *S. aureus* infections involve the invasion of host tissues, replication in

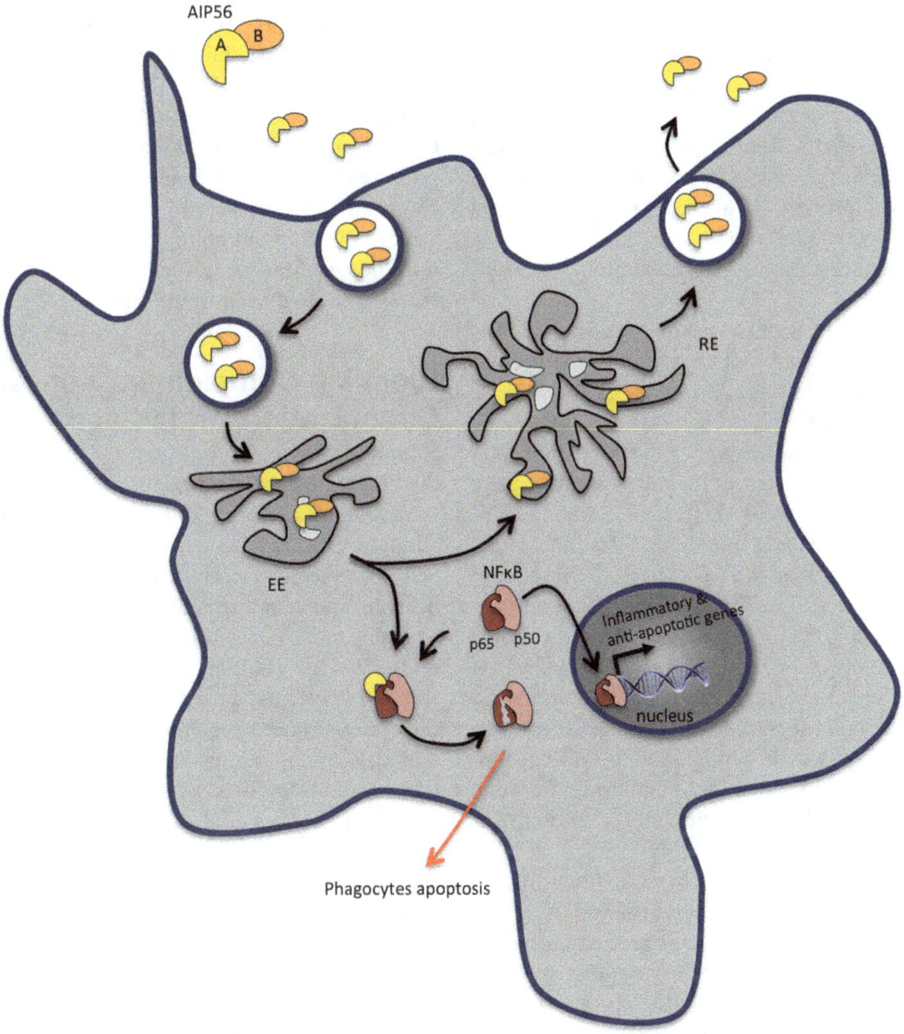

FIGURE 3 | AIP56 blocks innate immunity by inducing massive apoptosis of host macrophages and neutrophils. Upon encountering susceptible cells, apoptosis-inducing protein of 56 kDa (AIP56) binds to a still unidentified cell-surface receptor and undergoes clathrin-mediated endocytosis. Once in early endosomes, the toxin either follows the recycling pathway back to the extracellular medium or suffers low pH-induced translocation across the endosomal membrane into the cytosol to display its toxic activity. AIP56 is a zinc-dependent metalloprotease that cleaves the p65 subunit of nuclear factor-κB (NF-κB), an evolutionarily conserved transcription factor that regulates the expression of inflammatory and anti-apoptotic genes and plays a key role in host responses to microbial pathogen invasion. During infection, AIP56 disseminates systemically and its activity leads to depletion of macrophages and neutrophils by post-apoptotic secondary necrosis, thereby blocking the phagocytic defense of the host and contributing to the occurrence of tissue damage.

abscess lesions and dissemination through purulent drainage of these lesions and require the recruitment of immune cells to the site of infection. Such infiltrated immune cells would usually eliminate the bacteria. However, to counter their action, *S. aureus* secretes soluble molecules targeting multiple pathways to manipulate the capacity of neutrophils for chemotaxis, phagocytosis, and bacterial killing, thus enabling pathogen replication and ensuring the success of the infection. Several recent and comprehensive reviews highlight how *S. aureus* virulence factors manipulate the host immune response (Spaan et al., 2013b; Otto, 2014; Thammavongsa et al., 2015). Here we will focus on its secreted molecules mainly targeting neutrophils, modulating their function, or inducing cell killing.

Avoiding Neutrophil Extravasation, Chemotaxis, and Activation

In general, upon pathogen recognition, pro-inflammatory signals released by resident macrophages promote the adhesion of circulating neutrophils and further extravasation across capillary endothelium to the site of infection. This process relies on interactions between the endothelial surface receptors (e.g., selectins and ICAM1) and their respective ligands on the surface of neutrophils (e.g., PSGL1 and β2-integrin) (Spaan et al., 2013b). To inhibit neutrophil recruitment to the infected tissues, *S. aureus* secretes two anti-inflammatory factors that prevent neutrophil adhesion to the blood vessels and further transmigration (**Figure 4A**). The staphylococcal SSL5 binds

PSGL1 in a glycan dependent manner at the surface of neutrophils, blocking its interaction with P-selectin expressed by endothelial cells and abrogating the early steps of neutrophil attachment (Bestebroer et al., 2007). In addition, SSL5 was shown to inactivate matrix metalloproteinase from human neutrophils, accounting for the limited capacity of neutrophils to transmigrate into infected tissues (Itoh et al., 2010). The extracellular adherence protein (Eap) recognizes endothelial ICAM1, preventing its interaction with β2-integrins at the surface of neutrophils and further inhibiting extravasation (Chavakis et al., 2002).

Staphylococcus aureus also secretes a number of antagonists of neutrophil receptors interfering with chemokine signaling and limiting neutrophil recruitment (**Figure 4A**). In particular, SSL5 directly binds the N-terminus of G-protein coupled chemokine receptors (GPCRs) inhibiting calcium mobilization and actin polymerization, thus impairing neutrophil responses to a huge diversity of chemokines (e.g., CXCL8, CXCL1, CCL2, and CCL5) and to complement fragments C3a and C5a (Bestebroer et al., 2009). Similarly to the recognition of PSGL1 by SSL5, binding to GPCRs relies on the presence of sialic acid residues (Baker et al., 2007; Bestebroer et al., 2007, 2009). In addition, SSL5 was shown to bind to platelet glycoproteins, inducing platelet activation and aggregation, which could be important for colonization and immune evasion by *S. aureus* (de Haas et al., 2009). SSL10 inhibits CXCL12-mediated responses by targeting CXCR4 (Walenkamp et al., 2009) and SSL3 binds to TLR2, hindering immune recognition of staphylococcal lipoproteins and peptidoglycan (Yokoyama et al., 2012).

Phenol-soluble modulins are produced by all *S. aureus* strains and have multiple roles in staphylococcal pathogenesis (Cheung et al., 2014a). In particular, PSMs are potent pro-inflammatory molecules that interact with human formyl peptide receptor 2 (FPR2) (Kretschmer et al., 2010), a GPCR involved in the recognition of pathogens. At nanomolar concentrations, PSMs bind to and activate FPRs, with the strongest activation occurring through FPR2, stimulating several FPRs effector functions such as chemotaxis and pro-inflammatory cytokine production (e.g., CXCL8; Fu et al., 2006). However, human isolates of *S. aureus* evolved other strategies to counter neutrophil chemotaxis by directly interfering with FPRs signaling and limiting cytotoxicity while promoting bacterial replication (McCarthy and Lindsay, 2013; Cheung et al., 2014b). Inhibition of FPRs-mediated pro-inflammatory signaling occurs *via* the secretion of the chemotaxis-inhibitor protein of *S. aureus* (CHIPS) (de Haas et al., 2004) and the FPR2/ALS-inhibitory protein (FLIPr) and its homologue FLIPrL (Prat et al., 2006, 2009; **Figure 4A**). While CHIPS binds and inhibits FPR1 and C5aR, FLIPr and FLIPrL block FRP2-mediated signaling, thus avoiding recognition of PSMs secreted by *S. aureus*, impairing pro-inflammatory response and reducing neutrophil recruitment (Haas et al., 2004, 2005; Prat et al., 2006, 2009). Clearly, CHIPS, FLIPr and FLIPrL, and PSMs have opposite effects on FPRs activation. Thus the production/secretion of these bacterial molecules is likely to be under strict control to allow the establishment of the infection and to evade immune recognition.

The repertoire of secreted molecules by *S. aureus* to evade the early steps of immune response also include Stathopain A (ScpA), a cysteine protease that specifically cleaves the N-terminus of human CXCR2 (Laarman et al., 2012), a GPCR responding to several chemokines (e.g., CXCL1-3 and CXCL5-8). Stathopain A inhibits CXCR2-mediated calcium mobilization, migration, intracellular signaling and activation of neutrophils (Laarman et al., 2012), appearing as an important immunomodulatory molecule causing neutrophil unresponsiveness to several chemokines and blocking their recruitment to the site of infection.

Blocking Complement Activation, Opsonization, and Phagocytosis

Staphylococcus aureus display intrinsic physical features (e.g. thick peptidoglycan layer and capsule) that confer resistance to complement-mediated killing and neutrophil phagocytosis. In addition, *S. aureus* secretes a variety of highly specific proteins with complement-modulating functions (**Figure 4B**), thereby delaying the innate immune attack and generating a window of opportunity to replicate and establish within the host (Spaan et al., 2013b; Thammavongsa et al., 2015). Several of these secreted molecules target C3 or C3 convertases, both central molecules in the complement activation cascade. Aureolysin is a metalloproteinase that cleaves C3, generating a modified C3b fragment that is further degraded instead of being covalently linked to the bacterial surface where it would promote the generation of the chemoattractant molecule C5a (Laarman et al., 2011). In addition, this proteinase degrades human antimicrobial peptides highly potent against *S. aureus* (Sieprawska-Lupa et al., 2004). Thus, aureolysin activity promotes infection by blocking the complement cascade impairing bacterial C3b opsonization, phagocytosis, and neutrophil-mediated killing (Sieprawska-Lupa et al., 2004; Laarman et al., 2011). Staphylococcal component inhibitor (SCIN) specifically binds to human C3 convertase and blocks its activity thereby preventing the production of C3a, C3b, and C5a, thus interfering with complement activation and with neutrophil-mediated killing of *S. aureus* (Rooijakkers et al., 2005a, 2009; **Figure 4B**). In addition, *S. aureus* secreted extracellular fibrinogen-binding protein (Efb) together with its homologue extracellular complement-binding protein (Ecb), were shown to inhibit both C3 and C5 convertases (Jongerius et al., 2010a,b, 2007). Other *S. aureus* secreted proteins target both C3 and immunoglobulin binding, thus impairing complement activation and opsonization, this is the case of staphylococcal binder of immunoglobulin (Sbi; Haupt et al., 2008) and staphylokinase (Rooijakkers et al., 2005b).

Besides their role in blocking neutrophil extravasation and chemotaxis (described above), some SSLs also hinder complement activation and phagocytosis. SSL7 binds IgA and complement C5, inhibiting the production of C5a and further phagocytosis and bacterial clearance during *in vivo* infection (Bestebroer et al., 2010). SSL10 binds to IgG1 preventing recognition by Fc Receptors (FcR), thus impairing IgG1 opsonization and phagocytosis (Patel et al., 2010; **Figure 4B**).

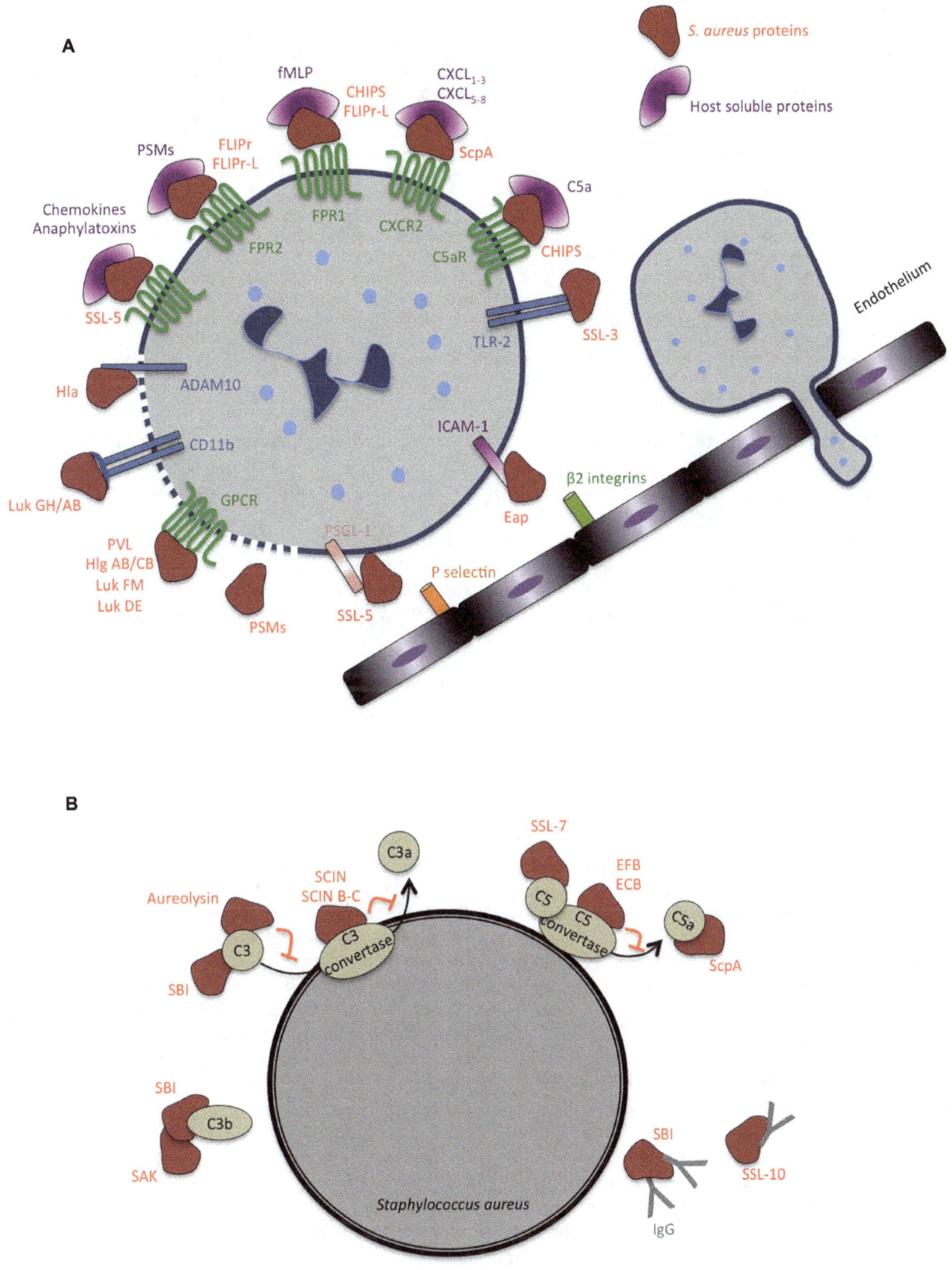

FIGURE 4 | Continued

FIGURE 4 | Continued

Strategies evolved by *Staphylococcus aureus* to counteract innate immune response. (A) Secreted bacterial factors that inhibit neutrophils extravasation, chemotaxis and activation. Neutrophil rolling is modulated by staphylococcal superantigen-like 5 (SSL5) that binds P-selectin glycoprotein ligand-1 (PSGL-1), blocking its interaction with P-selectin. The adhesion of neutrophils to the endothelium and consequent transmigration is inhibited by extracellular adherence protein (Eap), which binds to intercellular adhesion molecule 1 (ICAM-1). In addition to inhibiting PSGL-1, SSL5 inhibits neutrophil responses to chemokines and to anaphylatoxins, by binding to different chemokine receptors. Several staphylococcal molecules impair neutrophil chemotaxis and important co-signaling events during migration and phagocytosis: chemotaxis-inhibitor protein of *S. aureus* (CHIPS) binds and inhibits formyl peptide receptor 1 (FPR1) and C5a receptor (C5aR); formyl peptide receptor-like 1 inhibitor (FLIPr)-like inhibit FPR1; FLIPr and FLIPr-like inhibit FPR2; staphopain (ScpA) cleaves Chemokine (C–X–C Motif) Receptor 2 (CXCR2); staphylococcal SSL3 inhibits toll-like receptor 2 (TLR2)-mediated signaling, the bicomponent leukocidins Panton-Valentine leukocidin (PVL), gamma-hemolysin (Hlg) ABC, leukocidin (Luk) FM, Luk GH/AB, and Luk DE interact with chemoattractant receptors of the G-protein-coupled receptor (GPCR) family. Both Hla and Luk GH/AB induce cell lysis by binding ADAM metallopeptidase domain 10 (ADAM10) and CD11b, respectively. The cytolytic peptides phenol-soluble modulins (PSMs) have an amphipathic alpha-helical region that likely contributes to their lytic activity, presumably by membrane insertion and pore formation. **(B)** Secreted bacterial factors that inhibit opsonization and phagocytosis by neutrophils. The secreted metalloprotease aureolysin inhibits phagocytosis and killing of bacteria by neutrophils by cleaving C3. Staphylococcal complement inhibitor (SCIN), SCIN-B, and SCIN-C associate with and inhibits C3 convertase, thereby preventing the production of C3a, C3b, and further C5a and consequently interfering with complement activation. The extracellular fibrinogen binding protein (EFB) and the extracellular complement-binding protein (ECB) also inhibit complement activation by inactivating C5 convertase and staphylococcal SSL7 targets C5. Staphylococcal binder of immunoglobulin (SBI) affect both the function of complement and immunoglobulin binding, blocking the classical complement activation pathway, and associates with C3 inhibiting the alternative pathway. Staphylokinase (SAK) forms enzymatically active complexes with C3b blocking complement activation. Staphylococcal SSL10 binds IgG, affecting Fc receptor (FcR) recognition and complement activation.

Avoiding NET Bactericidal Activities

In addition to phagocytosis and intracellular killing, neutrophils evolved an alternative defense mechanism to trap extracellular pathogens and prevent their dissemination. This strategy relies on the release of nuclear content together with antimicrobial cytosolic and granular proteins to form neutrophil extracellular traps (NETs), which are scaffolds that act as physical barriers to pathogen progression protecting host tissues from damage (Papayannopoulos and Zychlinsky, 2009). A secreted staphylococcal nuclease (Nuc) has the capacity to degrade NETs thereby allowing *S. aureus* to resist their bactericidal activities *in vitro* as well as in a respiratory tract infection model (Berends et al., 2010). NET degradation by Nuc leads to the production of monophosphate nucleotides that are further converted into deoxyadenosine through the activity of adenosine synthase (AdsA), another *S. aureus* secreted protein. Interestingly, the accumulation of deoxyadenosine generated by AdsA activity promotes the autocleavage of pro-caspase-3, triggering caspase-3-induced apoptosis of infiltrating macrophages (Thammavongsa et al., 2013). Together, staphylococcal Nuc and AdsA act in a concerted mode to promote bacterial survival in *S. aureus* abscesses, by excluding macrophages from the infection foci.

Neutrophil Killing: The Ultimate Defense

In line with its ability to evade almost every step of the innate immune response, *S. aureus* induce the death of innate immune cells, trough the secretion of PSMs and several other toxins. As mentioned above, PSMs are *S. aureus* secreted molecules with multiples roles in infection (Wang et al., 2007; Cheung et al., 2014a). They trigger inflammatory responses by interacting with FPR2 and display, at higher concentrations, FPR2-independent cytolytic activity likely through membrane insertion and pore formation (Kretschmer et al., 2010; **Figure 4A**). In particular, PSMα peptides are able to trigger the lysis of the phagosome after neutrophil ingestion allowing intracellular bacterial replication, and ultimately are responsible for lysis of neutrophils promoting bacterial survival and escape to the extracellular milieu (Wang et al., 2007; Geiger et al., 2012; Chatterjee et al., 2013; Grosz et al., 2014). *In vitro,* expression levels of PSMs correlate with levels of cytotoxicity (Rasigade et al., 2013). Moreover, mutants deficient for PSMα production are perturbed in biofilm formation (Peschel and Otto, 2013) and attenuated in the mouse bloodstream infection model (Wang et al., 2007). Altogether, these observations strongly suggest that PSMα-triggered effects may play a key role in *in vivo S. aureus* infection.

In addition to PSMs whose cytolytic activity is receptor-independent, *S. aureus* secretes other cytolytic toxins that interact with specific receptors at the surface of eukaryotic cells, oligomerize and form pores inducing cell leakage and ultimately total lysis. *S. aureus*-produced toxins targeting white blood cells belong to the beta-barrel pore-forming toxins and comprise hemolysin-α (Hla, also called α-toxin) and bicompetent leukocidins (Otto, 2014; Thammavongsa et al., 2015; **Figure 4A**). *In vivo* infection studies have shown that Hla is required for several *S. aureus*-associated pathologies, such as pneumonia and severe skin infections (Bubeck Wardenburg et al., 2007; Bubeck Wardenburg and Schneewind, 2008; Kennedy et al., 2010). Hla interacts with high affinity with ADAM10 at the surface of host cells (Wilke and Bubeck Wardenburg, 2010) to damage epithelial, endothelial and immune cells (Berube and Bubeck Wardenburg, 2013; **Figure 4A**). Mice lacking ADAM10 expression in the lung epithelium resist to lethal pneumonia (Becker et al., 2014), whereas animals lacking ADAM10 specifically on myeloid lineage develop exacerbated skin infections (Inoshima et al., 2011). Although these results suggest that the outcome of Hla-mediated effects may dependent on the infected tissue, the role of Hla on tissue-specific innate immunity requires further analysis.

Leukocidins are composed by two distinct and independently secreted subunits that form heteromultimeric pores in the membrane of host myeloid cells (Otto, 2014). *S. aureus* produce different arrays of leukocidins with different species and cell type specificities, which are mainly dictated by their interaction with host GPCRs (Spaan et al., 2013b; Otto, 2014; Thammavongsa et al., 2015; **Figure 4A**). *In vitro* assays with purified proteins, as well as *ex-vivo* infections with *S. aureus*, have shown that leukocidin AB (LukAB, also called LukGH) kills human, but not mouse, neutrophils upon binding to CD11b (DuMont et al., 2013a). In addition, purified LukAB induce the release of NETs (Malachowa et al., 2013) and also promotes the escape from the phagosome in neutrophils, thus enabling *S. aureus* replication (DuMont et al., 2013b). Gamma-hemolysin HlgAB and HlgCB bind chemokine receptors (e.g., CXCR1, CXCR2, and CCR2) and complement receptors (e.g., C5aR), respectively (Spaan et al., 2014), and promote the lysis of human neutrophils and macrophages *in vitro*. LukED, produced by a large majority of clinical isolates of *S. aureus*, triggers the lysis of neutrophils and macrophages of different vertebrates by binding chemokine receptors such as CCR5, CXCR1 and CXCR2 (Alonzo et al., 2013; Reyes-Robles et al., 2013). Importantly, LukED was shown to play a critical role in *S. aureus* systemic infections in mice, by promoting bacterial replication *in vivo* through direct killing of neutrophils (Alonzo et al., 2012). Lastly, Panton-Valentine leukocidin (PVL), which is secreted by a small percentage of *S. aureus* isolates, binds C5aR on neutrophils and macrophages and has a restricted activity toward human and rabbit cells (Spaan et al., 2013a, 2015). Despite the several attempts to evaluate the exact contribution of each of these toxins to *S. aureus*-associated pathologies, their vast species and cell type specificities have rendered this analysis highly challenging. In this context, the data obtained from commonly used animal infection models (e.g., mice and rats) should be interpreted with caution.

Mycobacterium ulcerans: SABOTAGE OF THE HOST IMMUNE RESPONSE BY A POLYKETIDE TOXIN

Mycobacterium ulcerans is the causative agent of Buruli ulcer, a chronic ulcerative skin disease that usually starts as painless nodules on the limbs that then develop into large ulcers. These can lead to severe scars and local deformities, including disabling contractures, if not treated at early stages (WHO Buruli ulcer fact sheet N° 199, Updated July 2014). Buruli ulcer occurs most frequently in children living in tropical environments, near wetlands. The disease is more common in poor and rural areas of Africa but is also found in South America, Asia and Australia. *Mycobacterium ulcerans* is currently recognized as an environmental pathogen, but its reservoirs and mode of transmission remain doubtful (WHO Buruli ulcer fact sheet N° 199, Updated July 2014). Genetically very close to *M. tuberculosis* and *M. marinum*, *M. ulcerans* is unique among human pathogenic mycobacteria due to the secretion of a lipid toxin, the mycolactone (George et al., 1999; Hong et al., 2008). Mycolactone displays cytotoxic and immunosuppressive activities and is considered the major pathogenicity factor in Buruli ulcer, being essential for *M. ulcerans* virulence, immune modulation, and colonization (George et al., 1999; Coutanceau et al., 2005; Marsollier et al., 2005; Torrado et al., 2007a, 2010; Simmonds et al., 2009; Guenin-Mace et al., 2011).

In animal models, injection of mycolactone alone is sufficient to cause ulcers similar to those found in infected hosts (George et al., 1999, 2000). Concerning the mechanism of cellular intoxication, it has been proposed that, due to its hydrophobic nature, mycolactone passively diffuses through the plasma membrane (Snyder and Small, 2003; **Figure 5**). At micromolar concentrations, mycolactone is highly cytotoxic to a variety of mammalian cells, with variable susceptibility levels depending on the cell type (Hall and Simmonds, 2014). The mycolactone cytotoxicity has been linked to its apoptogenic activity. Apoptosis was observed in several cell types incubated *in vitro* with mycolactone (George et al., 2000; Gama et al., 2014), when primary mouse macrophages were incubated with toxigenic *M. ulcerans* strains (Oliveira et al., 2005), and in guinea pigs (George et al., 2000) and mice (Oliveira et al., 2005; Torrado et al., 2007b) infected with mycolactone-producing *M. ulcerans*. More importantly, massive apoptosis has been observed in Buruli ulcer lesions (Walsh et al., 2005). The mycolactone-induced cell death mechanisms appear to be complex and are not completely understood. It has been shown that mycolactone induces cell cycle arrest at the G1/G0-phase (George et al., 1999), but the connection between this effect of mycolactone and its cytotoxicity remains unclear. Early studies on mycolactone reported the occurrence of early actin cytoskeleton rearrangements, cell rounding, and detachment following incubation with the toxin (George et al., 1998, 1999). More recently, studies with HeLa and Jurkat T cells have shown that mycolactone induces increased actin polymerisation in intoxicated cells, as a consequence of its binding to the GTPase domain of the actin-cytoskeleton regulator Wiskott-Aldrich syndrome protein WASP (Guenin-Mace et al., 2013). This leads to hyper-activation of WASP and re-localization of the Arp2/3 complex and consequently, to major cytoskeletal rearrangements, including the formation of filopodia (Guenin-Mace et al., 2013). In epithelial cells, this causes loss of cell adhesion and E-cadherin-dependent tight junctions, ultimately leading to the death of detached cells by anoikis (Guenin-Mace et al., 2013). Recent studies performed *in vitro* with murine fibroblasts confirmed the cytoskeleton as a main target of mycolactone, by showing that mycolactone causes changes in microtubules and affects several regulators and structural components of microtubules and microfilaments (Gama et al., 2014). These deleterious effects inflicted by mycolactone upon the cytoskeleton likely contribute to the formation of the lesions characteristic of Burulli ulcer. Additionally, given the central role of the cytoskeleton in controlling key cellular functions, such as endocytosis, intracellular trafficking, cell adhesion and migration, it is reasonable to speculate that, by highjacking cytoskeleton functions, mycolatone perturbs the functions of phagocytic cells. Indeed, it is likely that the decreased phagocytic activity of macrophages exposed to mycolactone (Adusumilli et al., 2005; Coutanceau et al., 2005) results from the effect of the toxin upon the cytoskeleton of those cells. Further investigations are required to determine whether

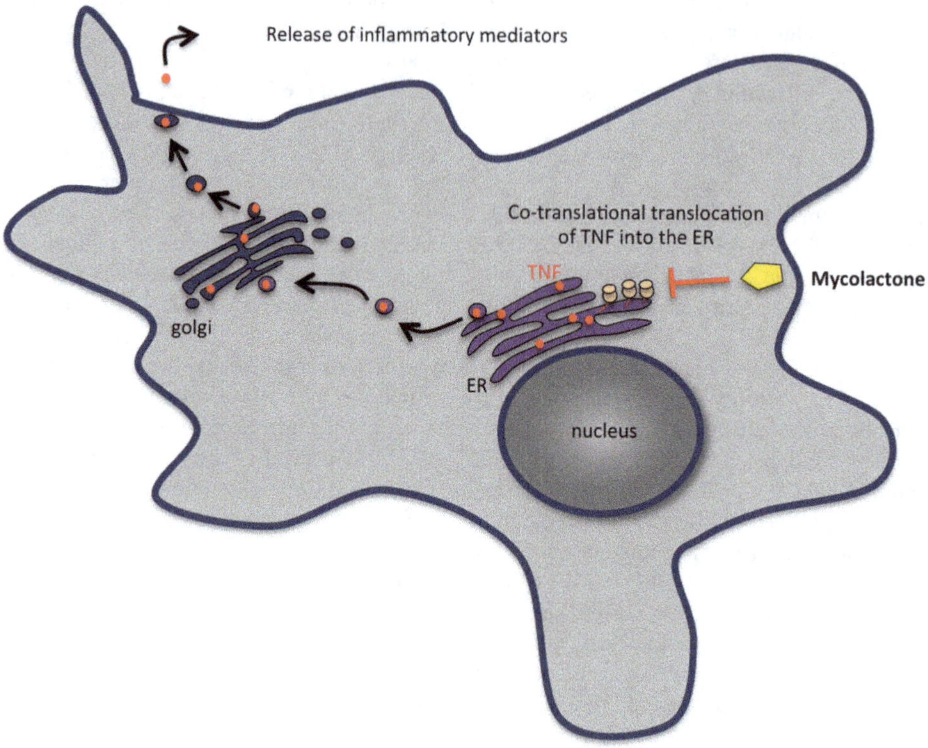

FIGURE 5 | Mycolactone inhibits the secretion of most cytokines, chemokines and other inflammatory mediators by macrophages. In eukaryotic cells, secretory proteins cross the ER membrane before being transported in vesicles to the Golgi complex and then to the plasma membrane. Mycolactone enters cells by passive diffusion through the plasma membrane and inhibits the production of inflammatory mediators by macrophages by blocking the translocation of nascent proteins into the ER. The proteins wrongly accumulated in the cytosol are then degraded by the proteasome.

cytoskeleton is manipulated by mycolactone in macrophages and neutrophils and what are the consequences of this manipulation *in vivo*.

In addition to its known cytotoxic effects toward distinct cell types, at non-cytotoxic concentrations, mycolactone interferes with important functions of immune cells, including monocytes, macrophages, and DCs (Arango Duque and Descoteaux, 2014; Hall and Simmonds, 2014). It is well recognized that macrophages play crucial roles in mycobacterial infections, including in Buruli ulcer. Although Buruli ulcer histopathology is characterized by extensive areas of necrosis with abundant extracellular bacteria, studies on infected humans, and experimental *M. ulcerans* infections revealed that *M. ulcerans* is phagocytosed by macrophages and neutrophils, similarly to the other pathogenic mycobacteria (Torrado et al., 2007b; Silva et al., 2009) and escape the microbicidal activity of the macrophages presumably by mycolactone-dependent interference with the IFNγ–dependent phagosome maturation and NO production required to control *M. ulcerans* infection (Torrado et al., 2010). After an initial phase of intracellular proliferation, varying according to the strain cytotoxicity/virulence, *M. ulcerans* causes apoptosis/necrosis of the host macrophage through a mycolactone-dependent mechanism and becomes extracellular (Torrado et al., 2007b).

Usually, phagocytosis of a microorganism triggers signaling events that rapidly culminate in a controlled inflammatory response involving the secretion of several cytokines and chemokines that recruit other inflammatory cells to the site of infection. However, available evidence suggests that this early response is heavily perturbed by mycolactone. Indeed, cells exposed to mycolactone-producing strains of *M. ulcerans* secrete much less TNF than those infected with mycolactone-negative strains (Torrado et al., 2007a) and purified mycolactone has been shown to supress the production of several cytokines, chemokines, and other inflammatory mediators by macrophages (Arango Duque and Descoteaux, 2014; Hall and Simmonds, 2014). It has been proposed that this suppression is associated with a mycolactone-induced blockade of co-translational protein translocation into the ER and subsequent degradation of the aberrantly located proteins in the cytosol (Hall et al., 2014; **Figure 5**). Failure to produce cytokines and chemokines may contribute to the absence of inflammatory infiltrate at the central necrotic areas of the lesion containing high numbers of extracellular bacilli, in addition to the lysis of recruited inflammatory cells induced by the build-up of mycolactone. The inflammatory infiltrates occupy a band at the periphery of the lesion that represents a front that is continuously advancing into healthy tissues in progressive *M. ulcerans* lesions (Silva et al., 2009).

CONCLUDING REMARKS

The clearance of infectious agents greatly depends on the host innate immune responses that take place at early stages of infection and in which macrophages and neutrophils are the central players. To counteract the host defense mechanisms, bacterial pathogens secrete a bench of different toxins that neutralize, at different levels, the host innate immune response and in particular, annihilate the function of macrophages and neutrophils. Despite having different features, secreted toxins targeting the function of innate immune cells often display similar and/or complementary activities and modulate the same central pathways of the host cell (e.g., inflammatory response, cytoskeleton dynamics, and cAMP signaling). Furthermore, a single pathogen may secrete several toxins that act differently to produce the same outcome (e.g., inhibit chemotaxis or induce phagocyte death). These apparently redundant strategies of bacterial attack ensure the multistep impairment of the early host immune responses mounted against the pathogen and guarantee the control of host–pathogen interaction providing a window of time and opportunity for bacterial growth and establishment within the host.

In the past, many studies aiming to uncover the molecular functions of bacterial toxins on host cells were performed *in vitro* in several cultured cell lines, and more recently in primary cells, using a wide range of concentrations of purified toxins. In addition, studies on animal models (mainly in rodents) using either purified toxins, wild type bacteria and toxin-deficient mutants, provided a number of important observations regarding the toxin-mediated pathologies. Together, these studies generated an incredible amount of data that paradoxically poorly contributed to the understanding of the role of toxins in human infections. Whereas in the perspective of using toxins as molecular tools to address cell biology topics there is great value in testing toxin effects in many *in vitro* cell systems, several issues render extremely difficult the interpretation of data from *in vitro* studies in the context of infection. In particular, the concentration of purified toxin used is often much higher than that produced by bacteria during infection and it is highly variable among different studies, and the cell lines tested are often non-relevant for the pathophysiology of the infection. Regarding studies performed in animal models, two major concerns have been pointed out: (1) many toxins display species-specificity and thus routinely used models, specially rodents, are non-relevant for the study of many toxin-mediated human pathologies compromising the extrapolation of data and (2) direct inoculation of purified toxins in the animals only provide limited information that do not necessarily recapitulate the effect of a given toxin in the context of human bacterial infection. Thus, the data generated so far needs to be cautiously analyzed whenever we aim to better understand the role of toxins in the *in vivo* infectious process. The accurate role of toxins in human infections needs to be analyzed in the context of bacterial infections in different animal species. In this perspective, future efforts should concentrate in the development and use of appropriate animal models, possibly non-human primates, in which available *in vitro* and *in vivo* data can be confirmed and possibly extrapolated to the human pathologies.

AUTHOR CONTRIBUTIONS

AdV and SS wrote the manuscript. DC contributed with figures design. The manuscript content was discussed and decided by all the authors.

FUNDING

This work was supported by FEDER funds through Programa Operacional Factores de Competitividade – COMPETE and by national funds through FCT – Fundação para a Ciência e a Tecnologia (project PTDC/BIA-MIC/3463/2012 FCOMP-01-0124-FEDER-028364; to AV). Research in the groups of Molecular Microbiology and Fish Immunology and Vaccinology is supported by national funds through FCT – Fundação para a Ciência e a Tecnologia/MEC – Ministério da Educação e Ciência and co-funded by FEDER within the partnership agreement PT2020 related with the research unit number 4293. The Group of Molecular Microbiology also receives support from a Research Grant 2014 by the European Society of Clinical Microbiology and Infectious Diseases (ESCMID) (to SS) and the PT2020 research project Infect-ERA/0001/2013 PROANTILIS. AdV received the FCT fellowship SFRH/BPD/95777/2013 by national funds through FCT – Fundação para a Ciência e a Tecnologia/MEC – Ministério da Educação e Ciência and co-funded by QREN and POPH (Programa Operacional Potencial Humano). SS is supported by FCT-Investigator program.

ACKNOWLEDGMENTS

We apologize to authors whose relevant work could not be cited owing to space limitations. We thank Nuno MS dos Santos (IBMC/I3S, Porto, Portugal) for critical reading of the manuscript and Jorge Pedrosa (ICVS/Universidade do Minho, Braga, Portugal) for reviewing the content concerning mycolactone.

REFERENCES

Abrami, L., Lindsay, M., Parton, R. G., Leppla, S. H., and van der Goot, F. G. (2004). Membrane insertion of anthrax protective antigen and cytoplasmic delivery of lethal factor occur at different stages of the endocytic pathway. *J. Cell Biol.* 166, 645–651. doi: 10.1083/jcb.200312072

Abrami, L., Liu, S., Cosson, P., Leppla, S. H., and van der Goot, F. G. (2003). Anthrax toxin triggers endocytosis of its receptor via a lipid raft-mediated clathrin-dependent process. *J. Cell. Biol.* 160, 321–328. doi: 10.1083/jcb.200211018

Adusumilli, S., Mve-Obiang, A., Sparer, T., Meyers, W., Hayman, J., and Small, P. L. (2005). *Mycobacterium ulcerans* toxic macrolide, mycolactone modulates the host immune response and cellular location of *M. ulcerans* in vitro and in vivo. *Cell. Microbiol.* 7, 1295–1304. doi: 10.1111/j.1462-5822.2005.00557.x

Agrawal, A., Lingappa, J., Leppla, S. H., Agrawal, S., Jabbar, A., Quinn, C., et al. (2003). Impairment of dendritic cells and adaptive immunity by anthrax lethal toxin. *Nature* 424, 329–334. doi: 10.1038/nature01794

Ahmad, J. N., Cerny, O., Linhartova, I., Masin, J., Osicka, R., and Sebo, P. (2015). cAMP signaling of *Bordetella* adenylate cyclase toxin through the SHP-1 phosphatase activates the BimEL-Bax pro-apoptotic cascade in phagocytes. *Cell Microbiol.* doi: 10.1111/cmi.12519 [Epub ahead of print].

Alileche, A., Serfass, E. R., Muehlbauer, S. M., Porcelli, S. A., and Brojatsch, J. (2005). Anthrax lethal toxin-mediated killing of human and murine dendritic cells impairs the adaptive immune response. *PLoS Pathog* 1:e19. doi: 10.1371/journal.ppat.0010019

Alonzo, F. III, Benson, M. A., Chen, J., Novick, R. P., Shopsin, B., and Torres, V. J. (2012). *Staphylococcus aureus* leucocidin ED contributes to systemic infection by targeting neutrophils and promoting bacterial growth in vivo. *Mol. Microbiol.* 83, 423–435. doi: 10.1111/j.1365-2958.2011.07942.x

Alonzo, F. III, Kozhaya, L., Rawlings, S. A., Reyes-Robles, T., DuMont, A. L., Myszka, D. G., et al. (2013). CCR5 is a receptor for *Staphylococcus aureus* leukotoxin ED. *Nature* 493, 51–55. doi: 10.1038/nature11724

Andreasen, C., and Carbonetti, N. H. (2008). Pertussis toxin inhibits early chemokine production to delay neutrophil recruitment in response to *Bordetella pertussis* respiratory tract infection in mice. *Infect. Immun.* 76, 5139–5148. doi: 10.1128/IAI.00895-08

Andreasen, C., and Carbonetti, N. H. (2009). Role of neutrophils in response to *Bordetella pertussis* infection in mice. *Infect. Immun.* 77, 1182–1188. doi: 10.1128/IAI.01150-08

Andreasen, C., Powell, D. A., and Carbonetti, N. H. (2009). Pertussis toxin stimulates IL-17 production in response to *Bordetella pertussis* infection in mice. *PLoS ONE* 4:e7079. doi: 10.1371/journal.pone.0007079

Appelberg, R. (2007). Neutrophils and intracellular pathogens: beyond phagocytosis and killing. *Trends Microbiol.* 15, 87–92. doi: 10.1016/j.tim.2006.11.009

Arango Duque, G., and Descoteaux, A. (2014). Macrophage cytokines: involvement in immunity and infectious diseases. *Front. Immunol.* 5:491. doi: 10.3389/fimmu.2014.00491

Baker, H. M., Basu, I., Chung, M. C., Caradoc-Davies, T., Fraser, J. D., and Baker, E. N. (2007). Crystal structures of the staphylococcal toxin SSL5 in complex with sialyl Lewis X reveal a conserved binding site that shares common features with viral and bacterial sialic acid binding proteins. *J. Mol. Biol.* 374, 1298–1308. doi: 10.1016/j.jmb.2007.09.091

Baldari, C. T., Tonello, F., Paccani, S. R., and Montecucco, C. (2006). Anthrax toxins: a paradigm of bacterial immune suppression. *Trends Immunol.* 27, 434–440. doi: 10.1016/j.it.2006.07.002

Barnes, A. C., dos Santos, N. M., and Ellis, A. E. (2005). Update on bacterial vaccines: *Photobacterium damselae* subsp. *piscicida*. *Dev. Biol. (Basel)* 121, 75–84.

Barson, H. V., Mollenkopf, H., Kaufmann, S. H., and Rijpkema, S. (2008). Anthrax lethal toxin suppresses chemokine production in human neutrophil NB-4 cells. *Biochem. Biophys. Res. Commun.* 374, 288–293. doi: 10.1016/j.bbrc.2008.07.018

Barth, H., Fischer, S., Moglich, A., and Fortsch, C. (2015). Clostridial C3 toxins target monocytes/macrophages and modulate their functions. *Front. Immunol.* 6:339. doi: 10.3389/fimmu.2015.00339

Baruch, K., Gur-Arie, L., Nadler, C., Koby, S., Yerushalmi, G., Ben-Neriah, Y., et al. (2011). Metalloprotease type III effectors that specifically cleave JNK and NF-kappaB. *EMBO J.* 30, 221–231. doi: 10.1038/emboj.2010.297

Bassinet, L., Gueirard, P., Maitre, B., Housset, B., Gounon, P., and Guiso, N. (2000). Role of adhesins and toxins in invasion of human tracheal epithelial cells by *Bordetella pertussis*. *Infect. Immun.* 68, 1934–1941. doi: 10.1128/IAI.68.4.1934-1941.2000

Becker, R. E., Berube, B. J., Sampedro, G. R., DeDent, A. C., and Bubeck Wardenburg, J. (2014). Tissue-specific patterning of host innate immune responses by *Staphylococcus aureus* alpha-toxin. *J. Innate Immun.* 6, 619–631. doi: 10.1159/000360006

Berends, E. T., Horswill, A. R., Haste, N. M., Monestier, M., Nizet, V., and von Kockritz-Blickwede, M. (2010). Nuclease expression by *Staphylococcus aureus* facilitates escape from neutrophil extracellular traps. *J. Innate Immun.* 2, 576–586. doi: 10.1159/000319909

Bergman, N. H., Passalacqua, K. D., Gaspard, R., Shetron-Rama, L. M., Quackenbush, J., and Hanna, P. C. (2005). Murine macrophage transcriptional responses to *Bacillus anthracis* infection and intoxication. *Infect. Immun.* 73, 1069–1080. doi: 10.1128/IAI.73.2.1069-1080.2005

Berube, B. J., and Bubeck Wardenburg, J. (2013). *Staphylococcus aureus* alpha-toxin: nearly a century of intrigue. *Toxins (Basel)* 5, 1140–1166. doi: 10.3390/toxins5061140

Bestebroer, J., Aerts, P. C., Rooijakkers, S. H., Pandey, M. K., Kohl, J., van Strijp, J. A., et al. (2010). Functional basis for complement evasion by staphylococcal superantigen-like 7. *Cell Microbiol.* 12, 1506–1516. doi: 10.1111/j.1462-5822.2010.01486.x

Bestebroer, J., Poppelier, M. J., Ulfman, L. H., Lenting, P. J., Denis, C. V., van Kessel, K. P., et al. (2007). Staphylococcal superantigen-like 5 binds PSGL-1 and inhibits P-selectin-mediated neutrophil rolling. *Blood* 109, 2936–2943.

Bestebroer, J., van Kessel, K. P., Azouagh, H., Walenkamp, A. M., Boer, I. G., Romijn, R. A., et al. (2009). Staphylococcal SSL5 inhibits leukocyte activation by chemokines and anaphylatoxins. *Blood* 113, 328–337. doi: 10.1182/blood-2008-04-153882

Bischof, T. S., Hahn, B. L., and Sohnle, P. G. (2007). Characteristics of spore germination in a mouse model of cutaneous anthrax. *J. Infect. Dis.* 195, 888–894. doi: 10.1086/511824

Boyd, A. P., Ross, P. J., Conroy, H., Mahon, N., Lavelle, E. C., and Mills, K. H. (2005). *Bordetella pertussis* adenylate cyclase toxin modulates innate and adaptive immune responses: distinct roles for acylation and enzymatic activity in immunomodulation and cell death. *J. Immunol.* 175, 730–738. doi: 10.4049/jimmunol.175.2.730

Boyden, E. D., and Dietrich, W. F. (2006). Nalp1b controls mouse macrophage susceptibility to anthrax lethal toxin. *Nat. Genet.* 38, 240–244. doi: 10.1038/ng1724

Bradley, K. A., Mogridge, J., Mourez, M., Collier, R. J., and Young, J. A. (2001). Identification of the cellular receptor for anthrax toxin. *Nature* 414, 225–229. doi: 10.1038/n35101999

Brittingham, K. C., Ruthel, G., Panchal, R. G., Fuller, C. L., Ribot, W. J., Hoover, T. A., et al. (2005). Dendritic cells endocytose *Bacillus anthracis* spores: implications for anthrax pathogenesis. *J. Immunol.* 174, 5545–5552. doi: 10.4049/jimmunol.174.9.5545

Bubeck Wardenburg, J., Patel, R. J., and Schneewind, O. (2007). Surface proteins and exotoxins are required for the pathogenesis of *Staphylococcus aureus* pneumonia. *Infect. Immun.* 75, 1040–1044. doi: 10.1128/IAI.01313-06

Bubeck Wardenburg, J., and Schneewind, O. (2008). Vaccine protection against *Staphylococcus aureus* pneumonia. *J. Exp. Med.* 205, 287–294. doi: 10.1084/jem.20072208

Bumba, L., Masin, J., Fiser, R., and Sebo, P. (2010). *Bordetella* adenylate cyclase toxin mobilizes its beta2 integrin receptor into lipid rafts to accomplish translocation across target cell membrane in two steps. *PLoS Pathog.* 6:e1000901. doi: 10.1371/journal.ppat.1000901

Carbonetti, N. H. (2007). Immunomodulation in the pathogenesis of *Bordetella pertussis* infection and disease. *Curr. Opin. Pharmacol.* 7, 272–278. doi: 10.1016/j.coph.2006.12.004

Carbonetti, N. H. (2010). Pertussis toxin and adenylate cyclase toxin: key virulence factors of *Bordetella pertussis* and cell biology tools. *Future Microbiol.* 5, 455–469. doi: 10.2217/fmb.09.133

Carbonetti, N. H., Artamonova, G. V., Andreasen, C., and Bushar, N. (2005). Pertussis toxin and adenylate cyclase toxin provide a one-two punch for establishment of *Bordetella pertussis* infection of the respiratory tract. *Infect. Immun.* 73, 2698–2703. doi: 10.1128/IAI.73.5.2698-2703.2005

Carbonetti, N. H., Artamonova, G. V., Mays, R. M., and Worthington, Z. E. (2003). Pertussis toxin plays an early role in respiratory tract colonization by *Bordetella pertussis*. *Infect. Immun.* 71, 6358–6366. doi: 10.1128/IAI.71.11.6358-63 66.2003

Carbonetti, N. H., Artamonova, G. V., Van Rooijen, N., and Ayala, V. I. (2007). Pertussis toxin targets airway macrophages to promote *Bordetella pertussis* infection of the respiratory tract. *Infect. Immun.* 75, 1713–1720. doi: 10.1128/IAI.01578-06

Cerny, O., Kamanova, J., Masin, J., Bibova, I., Skopova, K., and Sebo, P. (2015). *Bordetella pertussis* adenylate cyclase toxin blocks induction of bactericidal nitric oxide in macrophages through cAMP-dependent activation of the SHP-1 phosphatase. *J. Immunol.* 194, 4901–4913. doi: 10.4049/jimmunol.1402941

Chatterjee, S. S., Joo, H. S., Duong, A. C., Dieringer, T. D., Tan, V. Y., Song, Y., et al. (2013). Essential *Staphylococcus aureus* toxin export system. *Nat. Med.* 19, 364–367. doi: 10.1038/nm.3047

Chavakis, T., Hussain, M., Kanse, S. M., Peters, G., Bretzel, R. G., Flock, J. I., et al. (2002). *Staphylococcus aureus* extracellular adherence protein serves as anti-inflammatory factor by inhibiting the recruitment of host leukocytes. *Nat. Med.* 8, 687–693. doi: 10.1038/nm728

Chavarria-Smith, J., and Vance, R. E. (2013). Direct proteolytic cleavage of NLRP1B is necessary and sufficient for inflammasome activation by anthrax lethal factor. *PLoS Pathog.* 9:e1003452. doi: 10.1371/journal.ppat.1003452

Chavarria-Smith, J., and Vance, R. E. (2015). The NLRP1 inflammasomes. *Immunol. Rev.* 265, 22–34. doi: 10.1111/imr.12283

Cheung, G. Y., Joo, H. S., Chatterjee, S. S., and Otto, M. (2014a). Phenol-soluble modulins–critical determinants of staphylococcal virulence. *FEMS Microbiol. Rev.* 38, 698–719. doi: 10.1111/1574-6976.12057

Cheung, G. Y., Kretschmer, D., Duong, A. C., Yeh, A. J., Ho, T. V., Chen, Y., et al. (2014b). Production of an attenuated phenol-soluble modulin variant unique to the MRSA clonal complex 30 increases severity of bloodstream infection. *PLoS Pathog.* 10:e1004298. doi: 10.1371/journal.ppat.1004298

Cocklin, S., Jost, M., Robertson, N. M., Weeks, S. D., Weber, H. W., Young, E., et al. (2006). Real-time monitoring of the membrane-binding and insertion properties of the cholesterol-dependent cytolysin anthrolysin O from *Bacillus anthracis*. *J. Mol. Recogn.* 19, 354–362. doi: 10.1002/jmr.784

Coggeshall, K. M., Lupu, F., Ballard, J., Metcalf, J. P., James, J. A., Farris, D., et al. (2013). The sepsis model: an emerging hypothesis for the lethality of inhalation anthrax. *J. Cell Mol. Med.* 17, 914–920. doi: 10.1111/jcmm.12075

Collier, R. J. (2009). Membrane translocation by anthrax toxin. *Mol. Aspects Med.* 30, 413–422. doi: 10.1016/j.mam.2009.06.003

Confer, D. L., and Eaton, J. W. (1982). Phagocyte impotence caused by an invasive bacterial adenylate cyclase. *Science* 217, 948–950. doi: 10.1126/science.6287574

Connelly, C. E., Sun, Y., and Carbonetti, N. H. (2012). Pertussis toxin exacerbates and prolongs airway inflammatory responses during *Bordetella pertussis* infection. *Infect. Immun.* 80, 4317–4332. doi: 10.1128/IAI.00808-12

Corre, J. P., Piris-Gimenez, A., Moya-Nilges, M., Jouvion, G., Fouet, A., Glomski, I. J., et al. (2013). In vivo germination of *Bacillus anthracis* spores during murine cutaneous infection. *J. Infect. Dis.* 207, 450–457. doi: 10.1093/infdis/jis686

Costa-Ramos, C., Vale, A., Ludovico, P., Dos Santos, N. M., and Silva, M. T. (2011). The bacterial exotoxin AIP56 induces fish macrophage and neutrophil apoptosis using mechanisms of the extrinsic and intrinsic pathways. *Fish Shellfish Immunol.* 30, 173–181. doi: 10.1016/j.fsi.2010.10.007

Coutanceau, E., Marsollier, L., Brosch, R., Perret, E., Goossens, P., Tanguy, M., et al. (2005). Modulation of the host immune response by a transient intracellular stage of *Mycobacterium ulcerans*: the contribution of endogenous mycolactone toxin. *Cell. Microbiol.* 7, 1187–1196. doi: 10.1111/j.1462-5822.2005.00546.x

Cowan, G. J., Atkins, H. S., Johnson, L. K., Titball, R. W., and Mitchell, T. J. (2007). Immunisation with anthrolysin O or a genetic toxoid protects against challenge with the toxin but not against *Bacillus anthracis*. *Vaccine* 25, 7197–7205. doi: 10.1016/j.vaccine.2007.07.040

Crawford, M. A., Aylott, C. V., Bourdeau, R. W., and Bokoch, G. M. (2006). *Bacillus anthracis* toxins inhibit human neutrophil NADPH oxidase activity. *J. Immunol.* 176, 7557–7565. doi: 10.4049/jimmunol.176.12.7557

de Haas, C. J., Veldkamp, K. E., Peschel, A., Weerkamp, F., Van Wamel, W. J., Heezius, E. C., et al. (2004). Chemotaxis inhibitory protein of *Staphylococcus aureus*, a bacterial antiinflammatory agent. *J. Exp. Med.* 199, 687–695. doi: 10.1084/jem.20031636

de Haas, C. J., Weeterings, C., Vughs, M. M., de Groot, P. G., Van Strijp, J. A., and Lisman, T. (2009). Staphylococcal superantigen-like 5 activates platelets and supports platelet adhesion under flow conditions, which involves glycoprotein Ibalpha and alpha IIb beta 3. *J. Thromb. Haemost.* 7, 1867–1874. doi: 10.1111/j.1538-7836.2009.03564.x

Degnan, P. H., Yu, Y., Sisneros, N., Wing, R. A., and Moran, N. A. (2009). Hamiltonella defensa, genome evolution of protective bacterial endosymbiont from pathogenic ancestors. *Proc. Natl. Acad. Sci. U.S.A.* 106, 9063–9068. doi: 10.1073/pnas.0900194106

do Vale, A., Costa-Ramos, C., Silva, A., Silva, D. S., Gartner, F., dos Santos, N. M., et al. (2007). Systemic macrophage and neutrophil destruction by secondary necrosis induced by a bacterial exotoxin in a Gram-negative septicaemia. *Cell. Microbiol.* 9, 988–1003. doi: 10.1111/j.1462-5822.2006.00846.x

do Vale, A., Marques, F., and Silva, M. T. (2003). Apoptosis of sea bass (*Dicentrarchus labrax* L.) neutrophils and macrophages induced by experimental infection with *Photobacterium damselae* subsp. *piscicida*. *Fish Shellfish Immunol.* 15, 129–144. doi: 10.1016/S1050-4648(02)00144-4

do Vale, A., Silva, M. T., dos Santos, N. M., Nascimento, D. S., Reis-Rodrigues, P., Costa-Ramos, C., et al. (2005). AIP56, a novel plasmid-encoded virulence factor of *Photobacterium damselae* subsp. *piscicida* with apoptogenic activity against sea bass macrophages and neutrophils. *Mol. Microbiol.* 58, 1025–1038. doi: 10.1111/j.1365-2958.2005.04893.x

Duesbery, N. S., Webb, C. P., Leppla, S. H., Gordon, V. M., Klimpel, K. R., Copeland, T. D., et al. (1998). Proteolytic inactivation of MAP-kinase-kinase by anthrax lethal factor. *Science* 280, 734–737. doi: 10.1126/science.280.53 64.734

DuMont, A. L., Yoong, P., Day, C. J., Alonzo, F. III, McDonald, W. H., Jennings, M. P., et al. (2013a). *Staphylococcus aureus* LukAB cytotoxin kills human neutrophils by targeting the CD11b subunit of the integrin Mac-1. *Proc. Natl. Acad. Sci. U.S.A.* 110, 10794–10799. doi: 10.1073/pnas.1305121110

DuMont, A. L., Yoong, P., Surewaard, B. G., Benson, M. A., Nijland, R., van Strijp, J. A., et al. (2013b). *Staphylococcus aureus* elaborates leukocidin AB to mediate escape from within human neutrophils. *Infect. Immun.* 81, 1830–1841. doi: 10.1128/IAI.00095-13

During, R. L., Gibson, B. G., Li, W., Bishai, E. A., Sidhu, G. S., Landry, J., et al. (2007). Anthrax lethal toxin paralyzes actin-based motility by blocking Hsp27 phosphorylation. *EMBO J.* 26, 2240–2250. doi: 10.1038/sj.emboj.7601687

During, R. L., Li, W., Hao, B., Koenig, J. M., Stephens, D. S., Quinn, C. P., et al. (2005). Anthrax lethal toxin paralyzes neutrophil actin-based motility. *J. Infect. Dis.* 192, 837–845. doi: 10.1086/432516

Eby, J. C., Gray, M. C., and Hewlett, E. L. (2014). Cyclic AMP-mediated suppression of neutrophil extracellular trap formation and apoptosis by the *Bordetella pertussis* adenylate cyclase toxin. *Infect. Immun.* 82, 5256–5269. doi: 10.1128/IAI.02487-14

Eby, J. C., Gray, M. C., Mangan, A. R., Donato, G. M., and Hewlett, E. L. (2012). Role of CD11b/CD18 in the process of intoxication by the adenylate cyclase toxin of *Bordetella pertussis*. *Infect. Immun.* 80, 850–859. doi: 10.1128/IAI.05979-11

Eby, J. C., Gray, M. C., Warfel, J. M., Paddock, C. D., Jones, T. F., Day, S. R., et al. (2013). Quantification of the adenylate cyclase toxin of *Bordetella pertussis* in vitro and during respiratory infection. *Infect. Immun.* 81, 1390–1398. doi: 10.1128/IAI.00110-13

Eby, J. C., Hoffman, C. L., Gonyar, L. A., and Hewlett, E. L. (2015). Review of the neutrophil response to *Bordetella pertussis* infection. *Pathog. Dis.* 73, pii:ftv081. doi: 10.1093/femspd/ftv081

Erwin, J. L., DaSilva, L. M., Bavari, S., Little, S. F., Friedlander, A. M., and Chanh, T. C. (2001). Macrophage-derived cell lines do not express proinflammatory cytokines after exposure to *Bacillus anthracis* lethal toxin. *Infect. Immun.* 69, 1175–1177. doi: 10.1128/IAI.69.2.1175-1177.2001

Firoved, A. M., Miller, G. F., Moayeri, M., Kakkar, R., Shen, Y., Wiggins, J. F., et al. (2005). *Bacillus anthracis* edema toxin causes extensive tissue lesions and rapid lethality in mice. *Am. J. Pathol.* 167, 1309–1320. doi: 10.1016/S0002-9440(10)61218-7

Fiser, R., Masin, J., Basler, M., Krusek, J., Spulakova, V., Konopasek, I., et al. (2007). Third activity of *Bordetella* adenylate cyclase (AC) toxin-hemolysin. Membrane translocation of AC domain polypeptide promotes calcium influx into CD11b+ monocytes independently of the catalytic and hemolytic activities. *J. Biol. Chem.* 282, 2808–2820. doi: 10.1074/jbc.M609979200

Fiser, R., Masin, J., Bumba, L., Pospisilova, E., Fayolle, C., Basler, M., et al. (2012). Calcium influx rescues adenylate cyclase-hemolysin from rapid cell membrane removal and enables phagocyte permeabilization by toxin pores. *PLoS Pathog.* 8:e1002580. doi: 10.1371/journal.ppat.1002580

Flannagan, R. S., Cosio, G., and Grinstein, S. (2009). Antimicrobial mechanisms of phagocytes and bacterial evasion strategies. *Nat. Rev. Microbiol.* 7, 355–366. doi: 10.1038/nrmicro2128

Friedman, R. L., Fiederlein, R. L., Glasser, L., and Galgiani, J. N. (1987). *Bordetella pertussis* adenylate cyclase: effects of affinity-purified adenylate cyclase on human polymorphonuclear leukocyte functions. *Infect. Immun.* 55, 135–140.

Fu, H., Karlsson, J., Bylund, J., Movitz, C., Karlsson, A., and Dahlgren, C. (2006). Ligand recognition and activation of formyl peptide receptors in neutrophils. *J. Leukoc. Biol.* 79, 247–256. doi: 10.1189/jlb.0905498

Gama, J. B., Ohlmeier, S., Martins, T. G., Fraga, A. G., Sampaio-Marques, B., Carvalho, M. A., et al. (2014). Proteomic analysis of the action of the *Mycobacterium ulcerans* toxin mycolactone: targeting host cells cytoskeleton and collagen. *PLoS Negl. Trop. Dis.* 8:e3066. doi: 10.1371/journal.pntd.0003066

Geiger, T., Francois, P., Liebeke, M., Fraunholz, M., Goerke, C., Krismer, B., et al. (2012). The stringent response of *Staphylococcus aureus* and its impact on survival after phagocytosis through the induction of intracellular PSMs expression. *PLoS Pathog.* 8:e1003016. doi: 10.1371/journal.ppat.1003016

George, K. M., Barker, L. P., Welty, D. M., and Small, P. L. (1998). Partial purification and characterization of biological effects of a lipid toxin produced by *Mycobacterium ulcerans. Infect. Immun.* 66, 587–593.

George, K. M., Chatterjee, D., Gunawardana, G., Welty, D., Hayman, J., Lee, R., et al. (1999). Mycolactone: a polyketide toxin from *Mycobacterium ulcerans* required for virulence. *Science* 283, 854–857. doi: 10.1126/science.283.5403.854

George, K. M., Pascopella, L., Welty, D. M., and Small, P. L. (2000). A *Mycobacterium ulcerans* toxin, mycolactone, causes apoptosis in guinea pig ulcers and tissue culture cells. *Infect. Immun.* 68, 877–883. doi: 10.1128/IAI.68.2.877-883.2000

Goodwin, M. S., and Weiss, A. A. (1990). Adenylate cyclase toxin is critical for colonization and pertussis toxin is critical for lethal infection by *Bordetella pertussis* in infant mice. *Infect. Immun.* 58, 3445–3447.

Gordon, V. M., Leppla, S. H., and Hewlett, E. L. (1988). Inhibitors of receptor-mediated endocytosis block the entry of *Bacillus anthracis* adenylate cyclase toxin but not that of *Bordetella pertussis* adenylate cyclase toxin. *Infect. Immun.* 56, 1066–1069.

Grosz, M., Kolter, J., Paprotka, K., Winkler, A. C., Schafer, D., Chatterjee, S. S., et al. (2014). Cytoplasmic replication of *Staphylococcus aureus* upon phagosomal escape triggered by phenol-soluble modulin alpha. *Cell. Microbiol.* 16, 451–465. doi: 10.1111/cmi.12233

Gueirard, P., Druilhe, A., Pretolani, M., and Guiso, N. (1998). Role of adenylate cyclase-hemolysin in alveolar macrophage apoptosis during *Bordetella pertussis* infection in vivo. *Infect. Immun.* 66, 1718–1725.

Guenin-Mace, L., Carrette, F., Asperti-Boursin, F., Le Bon, A., Caleechurn, L., Di Bartolo, V., et al. (2011). Mycolactone impairs T cell homing by suppressing microRNA control of L-selectin expression. *Proc. Natl. Acad. Sci. U.S.A.* 108, 12833–12838. doi: 10.1073/pnas.1016496108

Guenin-Mace, L., Veyron-Churlet, R., Thoulouze, M. I., Romet-Lemonne, G., Hong, H., Leadlay, P. F., et al. (2013). Mycolactone activation of Wiskott-Aldrich syndrome proteins underpins Buruli ulcer formation. *J. Clin. Invest.* 123, 1501–1512. doi: 10.1172/JCI66576

Guermonprez, P., Khelef, N., Blouin, E., Rieu, P., Ricciardi-Castagnoli, P., Guiso, N., et al. (2001). The adenylate cyclase toxin of *Bordetella pertussis* binds to target cells via the alpha(M)beta(2) integrin (CD11b/CD18). *J. Exp. Med.* 193, 1035–1044. doi: 10.1084/jem.193.9.1035

Guidi-Rontani, C., Weber-Levy, M., Labruyere, E., and Mock, M. (1999). Germination of *Bacillus anthracis* spores within alveolar macrophages. *Mol. Microbiol.* 31, 9–17. doi: 10.1046/j.1365-2958.1999.01137.x

Haas, P. J., de Haas, C. J., Kleibeuker, W., Poppelier, M. J., van Kessel, K. P., Kruijtzer, J. A., et al. (2004). N-terminal residues of the chemotaxis inhibitory protein of *Staphylococcus aureus* are essential for blocking formylated peptide receptor but not C5a receptor. *J. Immunol.* 173, 5704–5711. doi: 10.4049/jimmunol.173.9.5704

Haas, P. J., de Haas, C. J., Poppelier, M. J., van Kessel, K. P., van Strijp, J. A., Dijkstra, K., et al. (2005). The structure of the C5a receptor-blocking domain of chemotaxis inhibitory protein of *Staphylococcus aureus* is related to a group of immune evasive molecules. *J. Mol. Biol.* 353, 859–872. doi: 10.1016/j.jmb.2005.09.014

Hall, B., and Simmonds, R. (2014). Pleiotropic molecular effects of the *Mycobacterium ulcerans* virulence factor mycolactone underlying the cell death and immunosuppression seen in Buruli ulcer. *Biochem. Soc. Trans.* 42, 177–183. doi: 10.1042/BST20130133

Hall, B. S., Hill, K., McKenna, M., Ogbechi, J., High, S., Willis, A. E., et al. (2014). The pathogenic mechanism of the *Mycobacterium ulcerans* virulence factor, mycolactone, depends on blockade of protein translocation into the ER. *PLoS Pathog.* 10:e1004061. doi: 10.1371/journal.ppat.1004061

Hanna, P. C., and Ireland, J. A. (1999). Understanding *Bacillus anthracis* pathogenesis. *Trends Microbiol.* 7, 180–182. doi: 10.1016/S0966-842X(99)01507-3

Harvill, E. T., Cotter, P. A., Yuk, M. H., and Miller, J. F. (1999). Probing the function of *Bordetella* bronchiseptica adenylate cyclase toxin by manipulating host immunity. *Infect. Immun.* 67, 1493–1500.

Haupt, K., Reuter, M., van den Elsen, J., Burman, J., Halbich, S., Richter, J., et al. (2008). The *Staphylococcus aureus* protein Sbi acts as a complement inhibitor and forms a tripartite complex with host complement Factor H and C3b. *PLoS Pathog.* 4:e1000250. doi: 10.1371/journal.ppat.1000250

Hawke, J. P., Plakas, S. M., Minton, R. V., McPhearson, R. M., Snider, T. G., and Guarino, A. M. (1987). Fish pasteurellosis of cultured striped bass (*Morone saxatilis*) in coastal Alabama. *Aquaculture* 65, 193–204. doi: 10.1016/0044-8486(87)90231-6

He, Y. X., Hewlett, E., Temeles, D., and Quesenberry, P. (1988). Inhibition of interleukin 3 and colony-stimulating factor 1-stimulated marrow cell proliferation by pertussis toxin. *Blood* 71, 1187–1195.

Heffernan, B. J., Thomason, B., Herring-Palmer, A., and Hanna, P. (2007). *Bacillus anthracis* anthrolysin O and three phospholipases C are functionally redundant in a murine model of inhalation anthrax. *FEMS Microbiol. Lett.* 271, 98–105. doi: 10.1111/j.1574-6968.2007.00713.x

Hellmich, K. A., Levinsohn, J. L., Fattah, R., Newman, Z. L., Maier, N., Sastalla, I., et al. (2012). Anthrax lethal factor cleaves mouse nlrp1b in both toxin-sensitive and toxin-resistant macrophages. *PLoS ONE* 7:e49741. doi: 10.1371/journal.pone.0049741

Hickey, F. B., Brereton, C. F., and Mills, K. H. (2008). Adenylate cycalse toxin of *Bordetella pertussis* inhibits TLR-induced IRF-1 and IRF-8 activation and IL-12 production and enhances IL-10 through MAPK activation in dendritic cells. *J. Leukoc. Biol.* 84, 234–243. doi: 10.1189/jlb.0208113

Higgs, R., Higgins, S. C., Ross, P. J., and Mills, K. H. (2012). Immunity to the respiratory pathogen *Bordetella pertussis. Mucosal Immunol.* 5, 485–500. doi: 10.1038/mi.2012.54

Hodgson, A., Wier, E. M., Fu, K., Sun, X., Yu, H., Zheng, W., et al. (2015). Metalloprotease NleC suppresses host NF-kappaB/inflammatory responses by cleaving p65 and interfering with the p65/RPS3 interaction. *PLoS Pathog* 11:e1004705. doi: 10.1371/journal.ppat.1004705

Hong, H., Coutanceau, E., Leclerc, M., Caleechurn, L., Leadlay, P. F., and Demangel, C. (2008). Mycolactone diffuses from *Mycobacterium ulcerans*-infected tissues and targets mononuclear cells in peripheral blood and lymphoid organs. *PLoS Negl. Trop. Dis.* 2:e325. doi: 10.1371/journal.pntd.0000325

Hsu, L. C., Park, J. M., Zhang, K., Luo, J. L., Maeda, S., Kaufman, R. J., et al. (2004). The protein kinase PKR is required for macrophage apoptosis after activation of Toll-like receptor 4. *Nature* 428, 341–345. doi: 10.1038/nature02405

Huang, C., Jacobson, K., and Schaller, M. D. (2004). MAP kinases and cell migration. *J. Cell Sci.* 117, 4619–4628. doi: 10.1242/jcs.01481

Hume, D. A., and Denkins, Y. M. (1989). Activation of macrophages to express cytocidal activity correlates with inhibition of their responsiveness to macrophage colony-stimulating factor (CSF-1): involvement of a pertussis toxin-sensitive reaction. *Immunol. Cell Biol.* 67(Pt 4), 243–249. doi: 10.1038/icb.1989.37

Inoshima, I., Inoshima, N., Wilke, G. A., Powers, M. E., Frank, K. M., Wang, Y., et al. (2011). A *Staphylococcus aureus* pore-forming toxin subverts the activity of ADAM10 to cause lethal infection in mice. *Nat. Med.* 17, 1310–1314. doi: 10.1038/nm.2451

Itoh, S., Hamada, E., Kamoshida, G., Takeshita, K., Oku, T., and Tsuji, T. (2010). Staphylococcal superantigen-like protein 5 inhibits matrix metalloproteinase 9 from human neutrophils. *Infect. Immun.* 78, 3298–3305. doi: 10.1128/IAI.00178-10

Jongerius, I., Garcia, B. L., Geisbrecht, B. V., van Strijp, J. A., and Rooijakkers, S. H. (2010a). Convertase inhibitory properties of Staphylococcal extracellular complement-binding protein. *J. Biol. Chem.* 285, 14973–14979. doi: 10.1074/jbc.M109.091975

Jongerius, I., Puister, M., Wu, J., Ruyken, M., van Strijp, J. A., and Rooijakkers, S. H. (2010b). Staphylococcal complement inhibitor modulates phagocyte responses by dimerization of convertases. *J. Immunol.* 184, 420–425. doi: 10.4049/jimmunol.0902865

Jongerius, I., Kohl, J., Pandey, M. K., Ruyken, M., van Kessel, K. P., van Strijp, J. A., et al. (2007). Staphylococcal complement evasion by various convertase-blocking molecules. *J. Exp. Med.* 204, 2461–2471. doi: 10.1084/jem.20070818

Kamanova, J., Kofronova, O., Masin, J., Genth, H., Vojtova, J., Linhartova, I., et al. (2008). Adenylate cyclase toxin subverts phagocyte function by RhoA inhibition and unproductive ruffling. *J. Immunol.* 181, 5587–5597. doi: 10.4049/jimmunol.181.8.5587

Kassam, A., Der, S. D., and Mogridge, J. (2005). Differentiation of human monocytic cell lines confers susceptibility to *Bacillus anthracis* lethal toxin. *Cell Microbiol.* 7, 281–292. doi: 10.1111/j.1462-5822.2004.00458.x

Katada, T., Tamura, M., and Ui, M. (1983). The A protomer of islet-activating protein, pertussis toxin, as an active peptide catalyzing ADP-ribosylation of a membrane protein. *Arch. Biochem. Biophys.* 224, 290–298. doi: 10.1016/0003-9861(83)90212-6

Kaur, M., Singh, S., and Bhatnagar, R. (2013). Anthrax vaccines: present status and future prospects. *Expert Rev. Vaccines* 12, 955–970. doi: 10.1586/14760584.2013.814860

Kennedy, A. D., Bubeck Wardenburg, J., Gardner, D. J., Long, D., Whitney, A. R., Braughton, K. R., et al. (2010). Targeting of alpha-hemolysin by active or passive immunization decreases severity of USA300 skin infection in a mouse model. *J. Infect. Dis.* 202, 1050–1058. doi: 10.1086/656043

Khelef, N., Gounon, P., and Guiso, N. (2001). Internalization of *Bordetella pertussis* adenylate cyclase-haemolysin into endocytic vesicles contributes to macrophage cytotoxicity. *Cell Microbiol.* 3, 721–730. doi: 10.1046/j.1462-5822.2001.00151.x

Khelef, N., and Guiso, N. (1995). Induction of macrophage apoptosis by *Bordetella pertussis* adenylate cyclase-hemolysin. *FEMS Microbiol. Lett.* 134, 27–32. doi: 10.1111/j.1574-6968.1995.tb07909.x

Khelef, N., Sakamoto, H., and Guiso, N. (1992). Both adenylate cyclase and hemolytic activities are required by *Bordetella pertussis* to initiate infection. *Microb. Pathog.* 12, 227–235. doi: 10.1016/0882-4010(92)90057-U

Khelef, N., Zychlinsky, A., and Guiso, N. (1993). *Bordetella pertussis* induces apoptosis in macrophages: role of adenylate cyclase-hemolysin. *Infect. Immun.* 61, 4064–4071.

Kirimanjeswara, G. S., Agosto, L. M., Kennett, M. J., Bjornstad, O. N., and Harvill, E. T. (2005). Pertussis toxin inhibits neutrophil recruitment to delay antibody-mediated clearance of *Bordetella pertussis*. *J. Clin. Invest.* 115, 3594–3601. doi: 10.1172/JCI24609

Kretschmer, D., Gleske, A. K., Rautenberg, M., Wang, R., Koberle, M., Bohn, E., et al. (2010). Human formyl peptide receptor 2 senses highly pathogenic *Staphylococcus aureus*. *Cell Host Microbe* 7, 463–473. doi: 10.1016/j.chom.2010.05.012

Kubota, S., Kimura, T., and Egusa, S. (1970). Studies of a bacterial tuberculosis of yellow-tail. I symptomology and histopathology. *Gyobyo Kenkyu* 4, 111–118. doi: 10.3147/jsfp.4.111

Laarman, A. J., Mijnheer, G., Mootz, J. M., van Rooijen, W. J., Ruyken, M., Malone, C. L., et al. (2012). *Staphylococcus aureus* Staphopain A inhibits CXCR2-dependent neutrophil activation and chemotaxis. *EMBO J.* 31, 3607–3619. doi: 10.1038/emboj.2012.212

Laarman, A. J., Ruyken, M., Malone, C. L., van Strijp, J. A., Horswill, A. R., and Rooijakkers, S. H. (2011). *Staphylococcus aureus* metalloprotease aureolysin cleaves complement C3 to mediate immune evasion. *J. Immunol.* 186, 6445–6453. doi: 10.4049/jimmunol.1002948

Ladant, D., and Ullmann, A. (1999). *Bordatella pertussis* adenylate cyclase: a toxin with multiple talents. *Trends Microbiol.* 7, 172–176. doi: 10.1016/S0966-842X(99)01468-7

Lemichez, E., and Barbieri, J. T. (2013). General aspects and recent advances on bacterial protein toxins. *Cold Spring Harb Perspect. Med.* 3:a013573. doi: 10.1101/cshperspect.a013573

Leppla, S. H. (1982). Anthrax toxin edema factor: a bacterial adenylate cyclase that increases cyclic AMP concentrations of eukaryotic cells. *Proc. Natl. Acad. Sci. U.S.A.* 79, 3162–3166. doi: 10.1073/pnas.79.10.3162

Levinsohn, J. L., Newman, Z. L., Hellmich, K. A., Fattah, R., Getz, M. A., Liu, S., et al. (2012). Anthrax lethal factor cleavage of Nlrp1 is required for activation of the inflammasome. *PLoS Pathog* 8:e1002638. doi: 10.1371/journal.ppat.1002638

Liu, S., Crown, D., Miller-Randolph, S., Moayeri, M., Wang, H., Hu, H., et al. (2009). Capillary morphogenesis protein-2 is the major receptor mediating lethality of anthrax toxin in vivo. *Proc. Natl. Acad. Sci. U.S.A.* 106, 12424–12429. doi: 10.1073/pnas.0905409106

Liu, S., Miller-Randolph, S., Crown, D., Moayeri, M., Sastalla, I., Okugawa, S., et al. (2010). Anthrax toxin targeting of myeloid cells through the CMG2 receptor is essential for establishment of *Bacillus anthracis* infections in mice. *Cell Host Microbe* 8, 455–462. doi: 10.1016/j.chom.2010.10.004

Liu, S., Moayeri, M., and Leppla, S. H. (2014). Anthrax lethal and edema toxins in anthrax pathogenesis. *Trends Microbiol.* 22, 317–325. doi: 10.1016/j.tim.2014.02.012

Liu, S., Zhang, Y., Moayeri, M., Liu, J., Crown, D., Fattah, R. J., et al. (2013). Key tissue targets responsible for anthrax-toxin-induced lethality. *Nature* 501, 63–68. doi: 10.1038/nature12510

Locht, C., Coutte, L., and Mielcarek, N. (2011). The ins and outs of pertussis toxin. *FEBS J.* 278, 4668–4682. doi: 10.1111/j.1742-4658.2011.08237.x

Magariños, B., Toranzo, A. E., and Romalde, J. L. (1996). Phenotypic and pathobiological characteristics of *Pasteurella piscicida*. *Annu. Rev. Fish Dis.* 6, 41–64. doi: 10.1016/S0959-8030(96)00004-X

Malachowa, N., Kobayashi, S. D., Freedman, B., Dorward, D. W., and DeLeo, F. R. (2013). *Staphylococcus aureus* leukotoxin GH promotes formation of neutrophil extracellular traps. *J. Immunol.* 191, 6022–6029. doi: 10.4049/jimmunol.1301821

Mangmool, S., and Kurose, H. (2011). G(i/o) protein-dependent and -independent actions of Pertussis Toxin (PTX). *Toxins (Basel.)* 3, 884–899. doi: 10.3390/toxins3070884

Mantovani, A., Cassatella, M. A., Costantini, C., and Jaillon, S. (2011). Neutrophils in the activation and regulation of innate and adaptive immunity. *Nat. Rev. Immunol.* 11, 519–531. doi: 10.1038/nri3024

Marsollier, L., Aubry, J., Coutanceau, E., Andre, J. P., Small, P. L., Milon, G., et al. (2005). Colonization of the salivary glands of *Naucoris* cimicoides by *Mycobacterium ulcerans* requires host plasmatocytes and a macrolide toxin, mycolactone. *Cell Microbiol.* 7, 935–943. doi: 10.1111/j.1462-5822.2005.00521.x

Martin, C., Uribe, K. B., Gomez-Bilbao, G., and Ostolaza, H. (2011). Adenylate cyclase toxin promotes internalisation of integrins and raft components and decreases macrophage adhesion capacity. *PLoS ONE* 6:e17383. doi: 10.1371/journal.pone.0017383

Mattoo, S., and Cherry, J. D. (2005). Molecular pathogenesis, epidemiology, and clinical manifestations of respiratory infections due to *Bordetella pertussis* and other *Bordetella* subspecies. *Clin. Microbiol. Rev.* 18, 326–382. doi: 10.1128/CMR.18.2.326-382.2005

McCarthy, A. J., and Lindsay, J. A. (2013). *Staphylococcus aureus* innate immune evasion is lineage-specific: a bioinfomatics study. *Infect. Genet. Evol.* 19, 7–14. doi: 10.1016/j.meegid.2013.06.012

Meade, B. D., Kind, P. D., Ewell, J. B., McGrath, P. P., and Manclark, C. R. (1984). In vitro inhibition of murine macrophage migration by *Bordetella pertussis* lymphocytosis-promoting factor. *Infect. Immun.* 45, 718–725.

Melvin, J. A., Scheller, E. V., Miller, J. F., and Cotter, P. A. (2014). *Bordetella pertussis* pathogenesis: current and future challenges. *Nat. Rev. Microbiol.* 12, 274–288. doi: 10.1038/nrmicro3235

Moayeri, M., Crown, D., Newman, Z. L., Okugawa, S., Eckhaus, M., Cataisson, C., et al. (2010). Inflammasome sensor Nlrp1b-dependent resistance to anthrax is mediated by caspase-1, IL-1 signaling and neutrophil recruitment. *PLoS Pathog* 6:e1001222. doi: 10.1371/journal.ppat.1001222

Moayeri, M., Leppla, S. H., Vrentas, C., Pomerantsev, A., and Liu, S. (2015). Anthrax pathogenesis. *Annu. Rev. Microbiol.* 69, 185–208.

Mork, T., and Hancock, R. E. (1993). Mechanisms of nonopsonic phagocytosis of *Pseudomonas aeruginosa*. *Infect. Immun.* 61, 3287–3293.

Morse, S. I., and Morse, J. H. (1976). Isolation and properties of the leukocytosis- and lymphocytosis-promoting factor of *Bordetella pertussis*. *J. Exp. Med.* 143, 1483–1502. doi: 10.1084/jem.143.6.1483

Mosser, E. M., and Rest, R. F. (2006). The *Bacillus anthracis* cholesterol-dependent cytolysin, Anthrolysin O, kills human neutrophils, monocytes and macrophages. *BMC Microbiol.* 6:56.

Muehlbauer, S. M., Evering, T. H., Bonuccelli, G., Squires, R. C., Ashton, A. W., Porcelli, S. A., et al. (2007). Anthrax lethal toxin kills macrophages in a strain-specific manner by apoptosis or caspase-1-mediated necrosis. *Cell Cycle* 6, 758–766. doi: 10.4161/cc.6.6.3991

Newman, Z. L., Printz, M. P., Liu, S., Crown, D., Breen, L., Miller-Randolph, S., et al. (2010). Susceptibility to anthrax lethal toxin-induced rat death is controlled by a single chromosome 10 locus that includes rNlrp1. *PLoS Pathog* 6:e1000906. doi: 10.1371/journal.ppat.1000906

Njamkepo, E., Pinot, F., Francois, D., Guiso, N., Polla, B. S., and Bachelet, M. (2000). Adaptive responses of human monocytes infected by *Bordetella pertussis*: the role of adenylate cyclase hemolysin. *J. Cell Physiol.* 183, 91–99. doi: 10.1002/(SICI)1097-4652(200004)183:1%3C91::AID-JCP11%3E3.0.CO;2-S

Noya, M., Magariños, B., Toranzo, A., and Lamas, J. (1995). Sequential pathology of experimental pasteurellosis in gilthead seabream *Sparus aurata*, A light- and electron-microscopic study. *Dis. Aquat. Organ.* 21, 177–186. doi: 10.1111/j.1365-2052.2010.02110.x

O'Brien, J., Friedlander, A., Dreier, T., Ezzell, J., and Leppla, S. (1985). Effects of anthrax toxin components on human neutrophils. *Infect. Immun.* 47, 306–310.

Oliveira, M. S., Fraga, A. G., Torrado, E., Castro, A. G., Pereira, J. P., Filho, A. L., et al. (2005). Infection with *Mycobacterium ulcerans* induces persistent inflammatory responses in mice. *Infect. Immun.* 73, 6299–6310. doi: 10.1128/IAI.73.10.6299-6310.2005

Otto, M. (2014). *Staphylococcus aureus* toxins. *Curr. Opin. Microbiol.* 17, 32–37. doi: 10.1016/j.mib.2013.11.004

Paccani, S. R., Dal Molin, F., Benagiano, M., Ladant, D., D'Elios, M. M., Montecucco, C., et al. (2008). Suppression of T-lymphocyte activation and chemotaxis by the adenylate cyclase toxin of *Bordetella pertussis*. *Infect. Immun.* 76, 2822–2832. doi: 10.1128/IAI.00200-08

Paccani, S. R., Finetti, F., Davi, M., Patrussi, L., D'Elios, M. M., Ladant, D., et al. (2011). The *Bordetella pertussis* adenylate cyclase toxin binds to T cells via LFA-1 and induces its disengagement from the immune synapse. *J. Exp. Med.* 208, 1317–1330. doi: 10.1084/jem.20101558

Papayannopoulos, V., and Zychlinsky, A. (2009). NETs: a new strategy for using old weapons. *Trends Immunol.* 30, 513–521. doi: 10.1016/j.it.2009.07.011

Park, J. M., Greten, F. R., Li, Z. W., and Karin, M. (2002). Macrophage apoptosis by anthrax lethal factor through p38 MAP kinase inhibition. *Science* 297, 2048–2051. doi: 10.1126/science.1073163

Park, J. M., Greten, F. R., Wong, A., Westrick, R. J., Arthur, J. S., Otsu, K., et al. (2005). Signaling pathways and genes that inhibit pathogen-induced macrophage apoptosis–CREB and NF-kappaB as key regulators. *Immunity* 23, 319–329. doi: 10.1016/j.immuni.2005.08.010

Parnaik, R., Raff, M. C., and Scholes, J. (2000). Differences between the clearance of apoptotic cells by professional and non-professional phagocytes. *Curr. Biol.* 10, 857–860. doi: 10.1016/S0960-9822(00)00598-4

Patel, D., Wines, B. D., Langley, R. J., and Fraser, J. D. (2010). Specificity of staphylococcal superantigen-like protein 10 toward the human IgG1 Fc domain. *J. Immunol.* 184, 6283–6292. doi: 10.4049/jimmunol.0903311

Pearson, J. S., Riedmaier, P., Marches, O., Frankel, G., and Hartland, E. L. (2011). A type III effector protease NleC from enteropathogenic *Escherichia coli* targets NF-kappaB for degradation. *Mol. Microbiol.* 80, 219–230. doi: 10.1111/j.1365-2958.2011.07568.x

Pellizzari, R., Guidi-Rontani, C., Vitale, G., Mock, M., and Montecucco, C. (1999). Anthrax lethal factor cleaves MKK3 in macrophages and inhibits the LPS/IFNgamma-induced release of NO and TNFalpha. *FEBS Lett.* 462, 199–204. doi: 10.1016/S0014-5793(99)01502-1

Pereira, L. M., Pinto, R. D., Silva, D. S., Moreira, A. R., Beitzinger, C., Oliveira, P., et al. (2014). Intracellular trafficking of AIP56, an NF-kappaB-cleaving toxin from *Photobacterium damselae* subsp. *piscicida*. *Infect Immun.* 82, 5270–5285. doi: 10.1128/IAI.02623-14

Peschel, A., and Otto, M. (2013). Phenol-soluble modulins and staphylococcal infection. *Nat. Rev. Microbiol.* 11, 667–673. doi: 10.1038/nrmicro3110

Pittman, M. (1984). The concept of pertussis as a toxin-mediated disease. *Pediatr. Infect. Dis.* 3, 467–486. doi: 10.1097/00006454-198409000-00019

Plaut, R. D., and Carbonetti, N. H. (2008). Retrograde transport of pertussis toxin in the mammalian cell. *Cell Microbiol.* 10, 1130–1139. doi: 10.1111/j.1462-5822.2007.01115.x

Popov, S. G., Villasmil, R., Bernardi, J., Grene, E., Cardwell, J., Wu, A., et al. (2002). Lethal toxin of *Bacillus anthracis* causes apoptosis of macrophages. *Biochem. Biophys. Res. Commun.* 293, 349–355. doi: 10.1016/S0006-291X(02)00227-9

Prat, C., Bestebroer, J., de Haas, C. J., van Strijp, J. A., and van Kessel, K. P. (2006). A new staphylococcal anti-inflammatory protein that antagonizes the formyl peptide receptor-like 1. *J. Immunol.* 177, 8017–8026. doi: 10.4049/jimmunol.177.11.8017

Prat, C., Haas, P. J., Bestebroer, J., de Haas, C. J., van Strijp, J. A., and van Kessel, K. P. (2009). A homolog of formyl peptide receptor-like 1 (FPRL1) inhibitor from *Staphylococcus aureus* (FPRL1 inhibitory protein) that inhibits FPRL1 and FPR. *J. Immunol.* 183, 6569–6578. doi: 10.4049/jimmunol.0801523

Rasigade, J. P., Trouillet-Assant, S., Ferry, T., Diep, B. A., Sapin, A., Lhoste, Y., et al. (2013). PSMs of hypervirulent *Staphylococcus aureus* act as intracellular toxins that kill infected osteoblasts. *PLoS ONE* 8:e63176. doi: 10.1371/journal.pone.0063176

Reis, M. I., Costa-Ramos, C., do Vale, A., and dos Santos, N. M. (2010). Molecular cloning of sea bass (*Dicentrarchus labrax* L.) caspase-8 gene and its involvement in *Photobacterium damselae* ssp. *piscicida* triggered apoptosis. *Fish Shellfish Immunol.* 29, 58–65. doi: 10.1016/j.fsi.2010.02.016

Reis, M. I., do Vale, A., Pinto, C., Nascimento, D. S., Costa-Ramos, C., Silva, D. S., et al. (2007a). First molecular cloning and characterisation of caspase-9 gene in fish and its involvement in a gram negative septicaemia. *Mol. Immunol.* 44, 1754–1764. doi: 10.1016/j.molimm.2006.07.293

Reis, M. I., Nascimento, D. S., do Vale, A., Silva, M. T., and dos Santos, N. M. (2007b). Molecular cloning and characterisation of sea bass (*Dicentrarchus labrax* L.) caspase-3 gene. *Mol. Immunol.* 44, 774–783. doi: 10.1016/j.molimm.2006.04.028

Reyes-Robles, T., Alonzo, F., III, Kozhaya, L., Lacy, D. B., Unutmaz, D., and Torres, V. J. (2013). *Staphylococcus aureus* leukotoxin ED targets the chemokine receptors CXCR1 and CXCR2 to kill leukocytes and promote infection. *Cell Host Microbe* 14, 453–459. doi: 10.1016/j.chom.2013.09.005

Ribot, W. J., Panchal, R. G., Brittingham, K. C., Ruthel, G., Kenny, T. A., Lane, D., et al. (2006). Anthrax lethal toxin impairs innate immune functions of alveolar macrophages and facilitates *Bacillus anthracis* survival. *Infect. Immun.* 74, 5029–5034. doi: 10.1128/IAI.00275-06

Romalde, J. L. (2002). *Photobacterium damselae* subsp. *piscicida*: an integrated view of a bacterial fish pathogen. *Int. Microbiol.* 5, 3–9. doi: 10.1007/s10123-002-0051-6

Rooijakkers, S. H., Ruyken, M., Roos, A., Daha, M. R., Presanis, J. S., Sim, R. B., et al. (2005a). Immune evasion by a staphylococcal complement inhibitor that acts on C3 convertases. *Nat. Immunol.* 6, 920–927. doi: 10.1038/ni1235

Rooijakkers, S. H., van Wamel, W. J., Ruyken, M., van Kessel, K. P., and van Strijp, J. A. (2005b). Anti-opsonic properties of staphylokinase. *Microbes Infect.* 7, 476–484. doi: 10.1016/j.micinf.2004.12.014

Rooijakkers, S. H., Wu, J., Ruyken, M., van Domselaar, R., Planken, K. L., Tzekou, A., et al. (2009). Structural and functional implications of the alternative complement pathway C3 convertase stabilized by a staphylococcal inhibitor. *Nat. Immunol.* 10, 721–727. doi: 10.1038/ni.1756

Saukkonen, K., Burnette, W. N., Mar, V. L., Masure, H. R., and Tuomanen, E. I. (1992). Pertussis toxin has eukaryotic-like carbohydrate recognition domains. *Proc. Natl. Acad. Sci. U.S.A.* 89, 118–122. doi: 10.1073/pnas.89.1.118

Scobie, H. M., Rainey, G. J., Bradley, K. A., and Young, J. A. (2003). Human capillary morphogenesis protein 2 functions as an anthrax toxin receptor. *Proc. Natl. Acad. Sci. U.S.A.* 100, 5170–5174. doi: 10.1073/pnas.0431098100

Scott, S. J., Olson, C. V., and Lad, P. M. (1988). Peptide mapping studies of the pertussis toxin substrate in human neutrophils, platelets and erythrocytes. *Mol. Immunol.* 25, 305–311. doi: 10.1016/0161-5890(88)90022-3

Sham, H. P., Shames, S. R., Croxen, M. A., Ma, C., Chan, J. M., Khan, M. A., et al. (2011). Attaching and effacing bacterial effector NleC suppresses epithelial inflammatory responses by inhibiting NF-kappaB and p38 mitogen-activated protein kinase activation. *Infect. Immun.* 79, 3552–3562. doi: 10.1128/IAI.05033-11

Shannon, J. G., Ross, C. L., Koehler, T. M., and Rest, R. F. (2003). Characterization of anthrolysin O, the *Bacillus anthracis* cholesterol-dependent cytolysin. *Infect. Immun.* 71, 3183–3189. doi: 10.1128/IAI.71.6.3183-3189.2003

Sieprawska-Lupa, M., Mydel, P., Krawczyk, K., Wojcik, K., Puklo, M., Lupa, B., et al. (2004). Degradation of human antimicrobial peptide LL-37 by *Staphylococcus aureus*-derived proteinases. *Antimicrob. Agents Chemother.* 48, 4673–4679. doi: 10.1128/AAC.48.12.4673-4679.2004

Silva, D. S., Pereira, L. M., Moreira, A. R., Ferreira-da-Silva, F., Brito, R. M., Faria, T. Q., et al. (2013). The apoptogenic toxin AIP56 is a metalloprotease A-B toxin that cleaves NF-kappab P65. *PLoS Pathog* 9:e1003128. doi: 10.1371/journal.ppat.1003128

Silva, M. T. (2010). When two is better than one: macrophages and neutrophils work in concert in innate immunity as complementary and cooperative partners of a myeloid phagocyte system. *J. Leukoc. Biol.* 87, 93–106. doi: 10.1189/jlb.0809549

Silva, M. T., and Correia-Neves, M. (2012). Neutrophils and macrophages: the main partners of phagocyte cell systems. *Front. Immunol.* 3:174. doi: 10.3389/fimmu.2012.00174

Silva, M. T., do Vale, A., and dos Santos, N. M. (2008). Secondary necrosis in multicellular animals: an outcome of apoptosis with pathogenic implications. *Apoptosis* 13, 463–482. doi: 10.1007/s10495-008-0187-8

Silva, M. T., Portaels, F., and Pedrosa, J. (2009). Pathogenetic mechanisms of the intracellular parasite *Mycobacterium ulcerans* leading to Buruli ulcer. *Lancet Infect. Dis.* 9, 699–710. doi: 10.1016/S1473-3099(09)70234-8

Simmonds, R. E., Lali, F. V., Smallie, T., Small, P. L., and Foxwell, B. M. (2009). Mycolactone inhibits monocyte cytokine production by a posttranscriptional mechanism. *J. Immunol.* 182, 2194–2202. doi: 10.4049/jimmunol.0802294

Skinner, J. A., Reissinger, A., Shen, H., and Yuk, M. H. (2004). *Bordetella* type III secretion and adenylate cyclase toxin synergize to drive dendritic cells into a semimature state. *J. Immunol.* 173, 1934–1940. doi: 10.4049/jimmunol.173.3.1934

Snyder, D. S., and Small, P. L. (2003). Uptake and cellular actions of mycolactone, a virulence determinant for *Mycobacterium ulcerans. Microb. Pathog.* 34, 91–101. doi: 10.1016/S0882-4010(02)00210-3

Spaan, A. N., Henry, T., van Rooijen, W. J., Perret, M., Badiou, C., Aerts, P. C., et al. (2013a). The staphylococcal toxin Panton-Valentine Leukocidin targets human C5a receptors. *Cell Host Microbe* 13, 584–594. doi: 10.1016/j.chom.2013.04.006

Spaan, A. N., Surewaard, B. G., Nijland, R., and van Strijp, J. A. (2013b). Neutrophils versus *Staphylococcus aureus*: a biological tug of war. *Annu. Rev. Microbiol.* 67, 629–650. doi: 10.1146/annurev-micro-092412-155746

Spaan, A. N., Schiepers, A., de Haas, C. J., van Hooijdonk, D. D., Badiou, C., Contamin, H., et al. (2015). Differential interaction of the staphylococcal toxins panton-valentine leukocidin and gamma-hemolysin CB with human C5a receptors. *J. Immunol.* 195, 1034–1043. doi: 10.4049/jimmunol.1500604

Spaan, A. N., Vrieling, M., Wallet, P., Badiou, C., Reyes-Robles, T., Ohneck, E. A., et al. (2014). The staphylococcal toxins gamma-haemolysin AB and CB differentially target phagocytes by employing specific chemokine receptors. *Nat. Commun.* 5:5438. doi: 10.1038/ncomms6438

Spangrude, G. J., Sacchi, F., Hill, H. R., Van Epps, D. E., and Daynes, R. A. (1985). Inhibition of lymphocyte and neutrophil chemotaxis by pertussis toxin. *J. Immunol.* 135, 4135–4143.

Spensieri, F., Fedele, G., Fazio, C., Nasso, M., Stefanelli, P., Mastrantonio, P., et al. (2006). *Bordetella pertussis* inhibition of interleukin-12 (IL-12) p70 in human monocyte-derived dendritic cells blocks IL-12 p35 through adenylate cyclase toxin-dependent cyclic AMP induction. *Infect. Immun.* 74, 2831–2838. doi: 10.1128/IAI.74.5.2831-2838.2006

Stein, P. E., Boodhoo, A., Armstrong, G. D., Heerze, L. D., Cockle, S. A., Klein, M. H., et al. (1994). Structure of a pertussis toxin-sugar complex as a model for receptor binding. *Nat. Struct. Biol.* 1, 591–596. doi: 10.1038/nsb0994-591

Szarowicz, S. E., During, R. L., Li, W., Quinn, C. P., Tang, W. J., and Southwick, F. S. (2009). *Bacillus anthracis* edema toxin impairs neutrophil actin-based motility. *Infect. Immun.* 77, 2455–2464. doi: 10.1128/IAI.00839-08

Terra, J. K., Cote, C. K., France, B., Jenkins, A. L., Bozue, J. A., Welkos, S. L., et al. (2010). Cutting edge: resistance to *Bacillus anthracis* infection mediated by a lethal toxin sensitive allele of Nalp1b/Nlrp1b. *J. Immunol.* 184, 17–20. doi: 10.4049/jimmunol.0903114

Thammavongsa, V., Kim, H. K., Missiakas, D., and Schneewind, O. (2015). Staphylococcal manipulation of host immune responses. *Nat. Rev. Microbiol.* 13, 529–543. doi: 10.1038/nrmicro3521

Thammavongsa, V., Missiakas, D. M., and Schneewind, O. (2013). *Staphylococcus aureus* degrades neutrophil extracellular traps to promote immune cell death. *Science* 342, 863–866. doi: 10.1126/science.1242255

Torrado, E., Adusumilli, S., Fraga, A. G., Small, P. L., Castro, A. G., and Pedrosa, J. (2007a). Mycolactone-mediated inhibition of tumor necrosis factor production by macrophages infected with *Mycobacterium ulcerans* has implications for the control of infection. *Infect. Immun.* 75, 3979–3988. doi: 10.1128/IAI.00290-07

Torrado, E., Fraga, A. G., Castro, A. G., Stragier, P., Meyers, W. M., Portaels, F., et al. (2007b). Evidence for an intramacrophage growth phase of *Mycobacterium ulcerans. Infect. Immun.* 75, 977–987. doi: 10.1128/IAI.00889-06

Torrado, E., Fraga, A. G., Logarinho, E., Martins, T. G., Carmona, J. A., Gama, J. B., et al. (2010). IFN-gamma-dependent activation of macrophages during experimental infections by *Mycobacterium ulcerans* is impaired by the toxin mycolactone. *J. Immunol.* 184, 947–955. doi: 10.4049/jimmunol.0902717

Tournier, J. N., Quesnel-Hellmann, A., Mathieu, J., Montecucco, C., Tang, W. J., Mock, M., et al. (2005). Anthrax edema toxin cooperates with lethal toxin to impair cytokine secretion during infection of dendritic cells. *J. Immunol.* 174, 4934–4941. doi: 10.4049/jimmunol.174.8.4934

Tournier, J. N., Rossi Paccani, S., Quesnel-Hellmann, A., and Baldari, C. T. (2009). Anthrax toxins: a weapon to systematically dismantle the host immune defenses. *Mol. Aspects Med.* 30, 456–466. doi: 10.1016/j.mam.2009.06.002

Tung, M.-C., Tsai, S.-S., Ho, L.-F., Huang, S.-T., and Chen, S.-C. (1985). An acute septicemic infection of Pasteurella organism in pond-cultured Formosa snake-head fish (Channa maculata Lacepeda) in Taiwan. *Fish Pathol.* 20, 143–148. doi: 10.3147/jsfp.20.143

Vitale, G., Bernardi, L., Napolitani, G., Mock, M., and Montecucco, C. (2000). Susceptibility of mitogen-activated protein kinase kinase family members to proteolysis by anthrax lethal factor. *Biochem. J.* 352(Pt 3), 739–745. doi: 10.1042/bj3520739

Walenkamp, A. M., Boer, I. G., Bestebroer, J., Rozeveld, D., Timmer-Bosscha, H., Hemrika, W., et al. (2009). Staphylococcal superantigen-like 10 inhibits CXCL12-induced human tumor cell migration. *Neoplasia* 11, 333–344. doi: 10.1593/neo.81508

Walsh, D. S., Meyers, W. M., Portaels, F., Lane, J. E., Mongkolsirichaikul, D., Hussem, K., et al. (2005). High rates of apoptosis in human *Mycobacterium ulcerans* culture-positive buruli ulcer skin lesions. *Am. J. Trop. Med. Hyg.* 73, 410–415.

Wang, R., Braughton, K. R., Kretschmer, D., Bach, T. H., Queck, S. Y., Li, M., et al. (2007). Identification of novel cytolytic peptides as key virulence determinants for community-associated MRSA. *Nat. Med.* 13, 1510–1514. doi: 10.1038/nm1656

Weingart, C. L., and Weiss, A. A. (2000). *Bordetella pertussis* virulence factors affect phagocytosis by human neutrophils. *Infect. Immun.* 68, 1735–1739. doi: 10.1128/IAI.68.3.1735-1739.2000

Wilke, G. A., and Bubeck Wardenburg, J. (2010). Role of a disintegrin and metalloprotease 10 in *Staphylococcus aureus* alpha-hemolysin-mediated cellular injury. *Proc. Natl. Acad. Sci. U.S.A.* 107, 13473–13478. doi: 10.1073/pnas.1001815107

Witvliet, M. H., Burns, D. L., Brennan, M. J., Poolman, J. T., and Manclark, C. R. (1989). Binding of pertussis toxin to eucaryotic cells and glycoproteins. *Infect. Immun.* 57, 3324–3330.

Wood, W., Turmaine, M., Weber, R., Camp, V., Maki, R. A., McKercher, S. R., et al. (2000). Mesenchymal cells engulf and clear apoptotic footplate cells in macrophageless PU.1 null mouse embryos. *Development* 127, 5245–5252.

Wright, G. G., and Mandell, G. L. (1986). Anthrax toxin blocks priming of neutrophils by lipopolysaccharide and by muramyl dipeptide. *J. Exp. Med.* 164, 1700–1709. doi: 10.1084/jem.164.5.1700

Wynn, T. A., Chawla, A., and Pollard, J. W. (2013). Macrophage biology in development, homeostasis and disease. *Nature* 496, 445–455. doi: 10.1038/nature12034

Yen, H., Ooka, T., Iguchi, A., Hayashi, T., Sugimoto, N., and Tobe, T. (2010). NleC, a type III secretion protease, compromises NF-kappaB activation by targeting p65/RelA. *PLoS Pathog* 6:e1001231. doi: 10.1371/journal.ppat.1001231

Yokoyama, R., Itoh, S., Kamoshida, G., Takii, T., Fujii, S., Tsuji, T., et al. (2012). Staphylococcal superantigen-like protein 3 binds to the Toll-like receptor 2 extracellular domain and inhibits cytokine production induced by *Staphylococcus aureus*, cell wall component, or lipopeptides in murine macrophages. *Infect. Immun.* 80, 2816–2825. doi: 10.1128/IAI.00399-12

Young, J. A., and Collier, R. J. (2007). Anthrax toxin: receptor binding, internalization, pore formation, and translocation. *Annu. Rev. Biochem.* 76, 243–265. doi: 10.1146/annurev.biochem.75.103004.142728

2

Underestimated Manipulative Roles of *Mycobacterium tuberculosis* Cell Envelope Glycolipids During Infection

Andreu Garcia-Vilanova[1], *John Chan*[2,3] *and Jordi B. Torrelles*[1*]

[1] Population Health Program, TB Group, Texas Biomedical Research Institute, San Antonio, TX, United States, [2] Department of Medicine (Infectious Diseases), Albert Einstein College of Medicine & Montefiore Medical Center, Bronx, NY, United States, [3] Department of Microbiology and Immunology, Albert Einstein College of Medicine & Montefiore Medical Center, Bronx, NY, United States

Edited by:
Andrea Cooper,
University of Leicester,
United Kingdom

Reviewed by:
Shigeto Yoshida,
Kanazawa University, Japan
Wayne Robert Thomas,
Telethon Kids Institute, University of Western Australia, Australia

***Correspondence:**
Jordi B. Torrelles
jtorrelles@txbiomed.org

The *Mycobacterium tuberculosis* cell envelope has been evolving over time to make the bacterium transmissible and adaptable to the human host. In this context, the *M. tuberculosis* cell envelope contains a peripheral barrier full of lipids, some of them unique, which confer *M. tuberculosis* with a unique shield against the different host environments that the bacterium will encounter at the different stages of infection. This lipid barrier is mainly composed of glycolipids that can be characterized by three different subsets: trehalose-containing, mannose-containing, and 6-deoxy-pyranose-containing glycolipids. In this review, we explore the roles of these cell envelope glycolipids in *M. tuberculosis* virulence and pathogenesis, drug resistance, and further, how these glycolipids may dictate the *M. tuberculosis* cell envelope evolution from ancient to modern strains. Finally, we address how these *M. tuberculosis* cell envelope glycolipids are impacted by the host lung alveolar environment, their role in vaccination and masking host immunity, and subsequently the impact of these glycolipids in shaping how *M. tuberculosis* interacts with host cells, manipulating their immune response to favor the establishment of an infection.

Keywords: tuberculosis, *Mycobacterium tuberculosis*, cell envelope glycolipids, infectious diseases, vaccine strategies, immunomodulatory lipids, immune responses

INTRODUCTION

Currently, in the twenty-first century, one person is infected with *Mycobacterium tuberculosis* (*M. tuberculosis*) every 6 s, and one person dies of tuberculosis (TB) every 21 s. Multi-, extensively, extremely (also known as total) drug-resistant (MDR/XDR/XXDR/TDR) *M. tuberculosis* strains are emerging worldwide as a threat to public health and TB control, raising concerns of a future epidemic of virtually untreatable TB (1–5). It is estimated that drug-resistant TB will kill 75 million people and cost the global economy $16.7 trillion over the next 35 years (6).

Most current and proposed therapies target the *M. tuberculosis* cell envelope, a complex and dynamic structure. Despite this, little is known regarding the role of *M. tuberculosis* cell envelope lipids, and specially glycolipids, in the establishment and maintenance of the infection and in the development of active TB disease. During its path to infection and disease outcome, *M. tuberculosis* gets in contact with different host environments (7), where a critical gap in our knowledge is defining the changes in the *M. tuberculosis* cell envelope structures during the course of pulmonary infection, as well as how these changes determine the *M. tuberculosis* infection outcome.

Little is also known about the cell envelope composition of MDR/XDR/XXDR/TDR *M. tuberculosis* strains. For these strains, it is still unknown what bacterial and host factors are involved in the induction of their overwhelming tissue damaging, driving to the collapse of the lung merely months after active TB is diagnosed. The *M. tuberculosis* cell envelope, in its majority (~80%), comprises carbohydrates and lipids readily exposed to the host immune system and thus being instrumental in determining the *M. tuberculosis* pathway of infection and the host inflammatory response that defines the TB disease outcome. In this context, structural-biological function relationships for the majority of the *M. tuberculosis* cell envelope glycolipids are still being elucidated. Mainly, *M. tuberculosis* glycolipids are composed of two carbohydrates, trehalose (a glucose disaccharide) and mannose (**Figure 1**). Related to the *M. tuberculosis* cell envelope trehalose glycolipid content, it is plausible that trehalose synthesis and degradation play a role in the process of reciprocal carbon mobilization across the *M. tuberculosis* cell envelope, especially during infection, where the regulation of trehalose containing lipids on the *M. tuberculosis* cell envelope may be modulated by many environmental and physiological factors that affect both *M. tuberculosis* and the host cell (8, 9). In regard to mannose-containing glycolipids and lipoglycans, studies have focused on showing how *M. tuberculosis* mimics the host by containing cell envelope components (e.g., mannose-containing molecules) whose terminal epitopes closely resemble mammalian glycoforms such as mannoproteins (**Figure 2**). Indeed, studies have shown that *M. tuberculosis* uses this host resemblance to its advantage, gaining entrance and establishing a unique intracellular niche within the host (10, 11).

Many *M. tuberculosis*-host factors are involved in this first encounter interface at the initial stages of the infection. First, the constitution of the *M. tuberculosis* cell envelope, which is strain-dependent, from the worldwide used laboratory strains (*M. tuberculosis* $H_{37}R_a$, *M. tuberculosis* $H_{37}R_v$, and *M. tuberculosis* Erdman), to the highly transmissible CDC1551, to the hypervirulent HN878, and to several relevant drug-resistant *M. tuberculosis* clinical isolates in endemic TB areas (12–16). We will discuss their cell envelope glycolipid constitution in relation to their infection outcome. Second, we will focus on the alveolar space and its predominant cells, the alveolar macrophage (AM) and the alveolar epithelial cell, and how these alveolar cells act as a niche for *M. tuberculosis* and discuss their contribution to control or favor the infection. We will also discuss our results in an area frequently bypassed, the host lung alveolar environment and its impact on the *M. tuberculosis* cell envelope, addressing how the lung environment can determine and change the nature of the *M. tuberculosis* cell envelope during infection and how these changes on the *M. tuberculosis* cell envelope may determine/contribute to the pathway of infection and disease outcome. Importantly, we will also discuss our recent findings on the potential role of the mycobacterial glycolipids in masking the generation of the host protective immune response against *M. tuberculosis* infection and TB disease.

THE *M. TUBERCULOSIS* CELL ENVELOPE AND ITS GLYCOLIPIDS

The *M. tuberculosis* complex includes *M. tuberculosis*, *M. africanum*, *M. bovis*, and the Bacillus Calmette–Guérin (BCG) strain, *M. canetti*, *M. caprae*, *M. microti*, *M. mungi*, *M. orygis*, *M. pinnipedii*, and *M. suricattae* (with *M. dassie* being not yet included). The *M. tuberculosis* complex cell envelope has a unique, complex, thick, and waxy cell envelope, with properties, such as low permeability and intrinsic resistance to entry of many hydrophobic antibiotics and resistance to harsh environments which, distinguish *M. tuberculosis* from other pathogenic bacteria (17, 18). This envelope comprises an array of lipids with diverse polarity, such as glycolipids, aminolipids, phospholipids, phosphoglycolipids, sulfolipids, phenolic-glycolipids, and other lipids; all of them exhibit important roles in *M. tuberculosis* pathogenicity and are an important target, due to their alleged essentiality and proved immunogenicity, for new drug treatments and diagnostics. Most, if not all, *M. tuberculosis* cell envelope components have been shown to provide *M. tuberculosis* with some advantage for establishing an infection or survival in the host, for inhibiting bacteriostatic components of the host lung mucosa, or by allowing it to survive harsh intracellular environments, adapting its metabolism to the host environment. Thus, the cell envelope of *M. tuberculosis*, *per se*, may contain all of the elements associated with active TB disease, including elements driving tissue caseation, hypersensitization, cytotoxicity, and the antigens triggering or dampening humoral and adaptive immunity (19). However, although obvious, cell envelope compositional differences exist among *M. tuberculosis* complex strains and non-pathogenic mycobacteria; a detailed electron microscopy study has not yet identified any special and unique features on the *M. tuberculosis* complex strains.

The *M. tuberculosis* cell envelope is defined in five distinct compartments, the plasma membrane, a periplasmic space, the cell envelope core, the peripheral lipid layer, and around it, the outer material. All these parts are thought to provide structural, mechanical, and metabolic support, as well as osmotic protection and transport and communication with host environments surrounding the bacillus during different stages of infection. Here, we will discuss *M. tuberculosis* cell envelope, focusing on the *M. tuberculosis* peripheral lipid layer constituents and their role in *M. tuberculosis* pathogenesis.

THE *M. TUBERCULOSIS* CELL ENVELOPE PERIPHERAL LIPID LAYER

The interest of studying the tubercle bacillus lipoids remotes to more than 90 years ago by Anderson et al. in 1929 stating: "*The crude wax obtained from the chloroform extract of the human type of tubercle bacillus, Strain H-37, yielded on purification, two fractions; viz., a white solid substance designated as purified wax and a soft yellowish brown salve-like mass which was called soft wax and it possessed a slight but agreeable perfume-like odor*" (20). Since then, a fascinating diversity of lipid

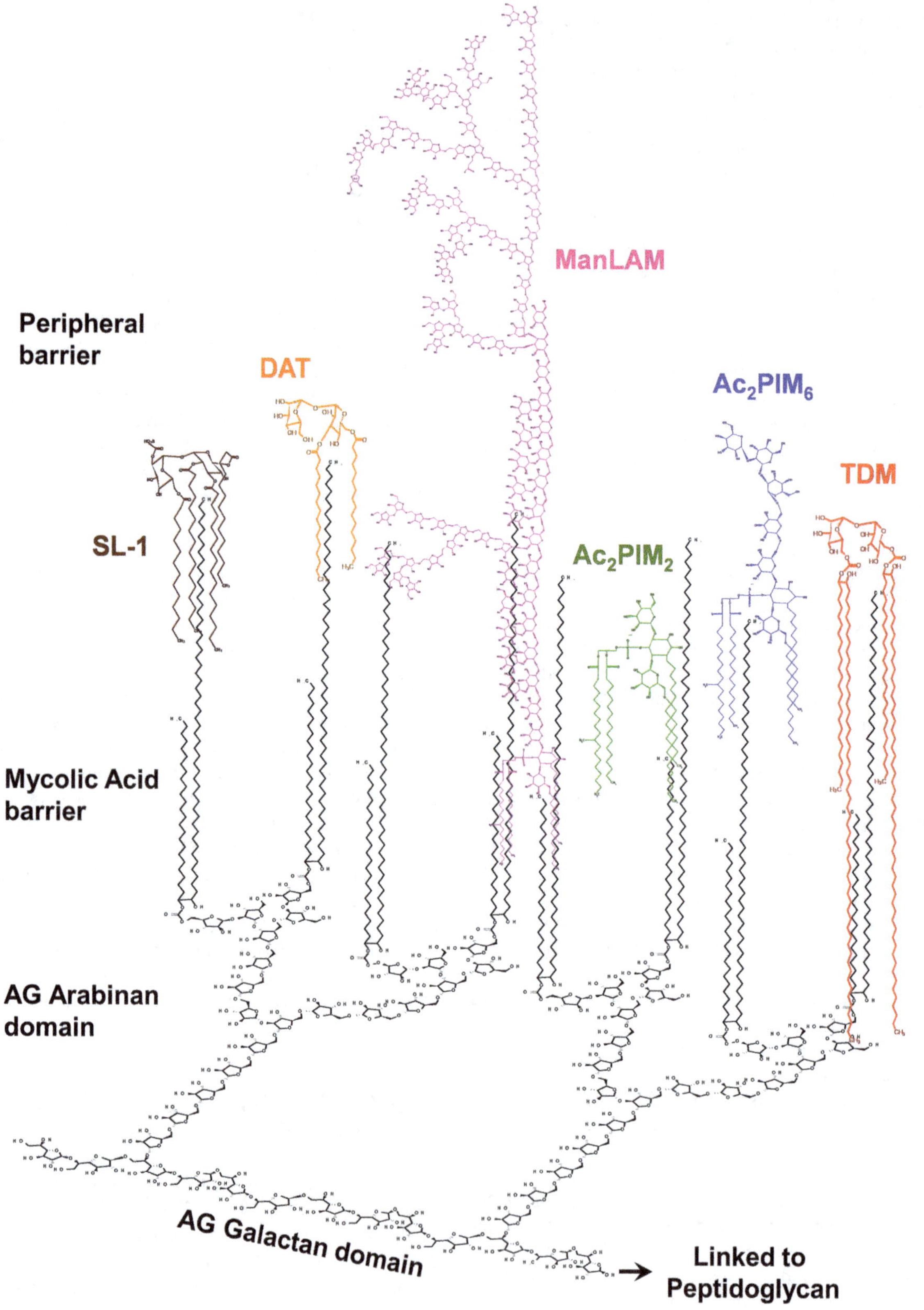

FIGURE 1 | The *M. tuberculosis* complex cell envelope with emphasis on surface glycolipids. The *M. tuberculosis* cell envelope is composed of a covalently linked core called mycolic-acids-arabinogalactan-peptidoglycan (mAGP) complex. On the top of this core is described to be a peripheral lipid barrier, where lipids intercalate among the mycolic acids of mAGP. Mycolic acids are perpendicular to the plasma membrane. Here the *M. tuberculosis* cell envelope surface location of

(Continued)

FIGURE 1 | mannose-containing glycolipids (phosphatidyl-*myo*-inositol mannosides, or PIMs) and lipoglycans (e.g., the mannose capped lipoarabinomannan, or ManLAM), as well as trehalose containing lipids [e.g., trehalose-dimycolate (TDM), sulfolipid-1 (SL-1), and diacyl-trehalose (DAT)], is depicted. The trehalose containing lipooligoscharides (LOSs), phthienoates and hydroxyphthienoates, 6-deoxy-pyranose-lipids (phenolic glycolipids, or PGLs), and mannose-containing mannosyl-β-1–phosphomycoketides (MPM) are not depicted. Outer material components (i.e., α-glucan, mannan, arabinomannan, and xylan) are also not shown. Relative number of molecules and size are not accurately depicted, reflecting published experimental data.

structures and their biological activities have been and are still being described. In this context, glycolipids are one of the major *M. tuberculosis* cell envelope constituents, defined by their polarity, cytotoxicity, and/or immunological properties (17) (**Figure 1**). These mainly comprise the acyl trehaloses family sharing a common α-D-Glc*p*(1→ 1′)-α-D-Glc*p* unit and are the class of *M. tuberculosis* lipids most extensively studied by lipidologists and mycobacteriologists. These specifically include di-, tri-, and penta-acyltrehaloses (DATs/TATs/PATs); mono- and di-mycolyl trehaloses (TMM/TDM); and tetraacyl-trehalose sulfo(glyco)lipids (SLs) (21), and the lesser studied to date, oligosaccharides containing lipids [lipooligosaccharides (LOSs)] (**Figure 3**). Another class of *M. tuberculosis* glycolipids of important relevance is the phosphatidyl-*myo*-inositol mannose derivatives (PIMs). Some of them are precursors of two major lipoglycans on the *M. tuberculosis* cell surface, the lipomannan, and the mannose capped lipoarabinomannan (ManLAM) (**Figure 3**). Lastly, and no less important, some *M. tuberculosis* strains also contain phenolic glycolipids (PGLs) (**Figure 3**). All these glycolipids are thought to perform filler roles in completing the outer leaflet of the asymmetric peripheral lipid layer on the *M. tuberculosis* cell envelope (17). The importance of these lipids in *M. tuberculosis* infection and pathogenesis is described in the sections below and summarized in **Table 1** and **Figure 4**.

THE DAT/TAT/PAT GLYCOLIPID FAMILY

Diacyl- (DAT; a 2, 3-diacyltrehalose), triacyl-trehalose (TAT; a 2, 3, 6-triacyltrehalose), and pentaacyl-trehalose (PAT; a 2, 3, 6, 2′, 4′-pentaacyltrehalose) contain polymethylated acyl groups (22). Recent studies unveiled that DAT family alone isolated from the *M. tuberculosis* cell envelope contains more than 30 molecular species, with each species containing up to six isomeric structures (23).

Their purpose and function in the *M. tuberculosis* cell envelope are still unclear. Some studies related DAT functions to its capacity to negatively regulate the host pro-inflammatory response induced by *M. tuberculosis* during infection in a dose-dependent manner (24). Indeed, DATs do not seem to participate in the receptor-mediated phagocytosis process of *M. tuberculosis*, do not induce host cytotoxicity, but trigger IL-10 production and are capable of inhibiting T-cell proliferation (25), supporting their role in potentially being used by *M. tuberculosis* to offset the host immune response during initial stages of the infection (24) (**Table 1**). Moreover, other studies indicate the capacity of DATs acting together with PAT (the major polyacyltrehalose of *M. tuberculosis* containing four mycolipenoyl and one fully saturated palmitoyl) in the maintenance of the *M. tuberculosis* cell envelope, as well as in pathogenesis promoting the *M. tuberculosis* survival within host cells (26) and being antigenic (27). In this context, several studies emphasize the potential primary role of DATs in regulating the composition of the *M. tuberculosis* cell envelope, serving as anchors of the *M. tuberculosis* outer material, which blocks *M. tuberculosis* recognition and phagocytosis by host cells (28, 29). In this regard, *M. tuberculosis* DAT mutants have altered cell envelope surface and have enhanced phagocytosis by host cells, but their absence from the *M. tuberculosis* cell envelope does not alter the bacterium capacity to replicate and survive *in vivo* using the mouse model (28, 30). Conversely, other studies have shown that *M. tuberculosis* mutants lacking DATs and sulfoglycolipids cannot block phagosome maturation, thus revealing the importance of these molecules in the *M. tuberculosis* pathogenesis (31). Currently, the contribution of *M. tuberculosis* acytrehaloses to pathogenesis is still understudied, and the main use of these *M. tuberculosis* glycolipids is still for serodiagnosis (32–36).

THE MYCOLIC ACID ESTERS OF TREHALOSE GLYCOLIPID FAMILY

Other trehalose-containing family of lipids characterized by their long acyl groups is the mycolyl-containing trehalose lipids (37). These are widely distributed and considered the most abundant, non-covalently associated *M. tuberculosis* cell envelope lipids, providing *M. tuberculosis* with a waxy, impermeable appearance. Both TDM (6,6′-dimycoloyl-α-D-trehalose) and TMM (6-monomycoloyl-α-D-trehalose) have their mycolates oriented toward the cell plasma membrane, presenting their trehalose moiety to the external environment.

TDM (or cord factor) was first obtained after a petroleum ether extraction from cells of a virulent strain of *M. tuberculosis* (38). After the extraction, *M. tuberculosis* presented a drastic disorganization of the typical cellular cords formed in a culture, and the resulting petroleum ether organic extract was toxic in mice. This toxicity was later shown to be related to TDM (39). Subsequently, further studies indicated that TDM toxicity is linked to the regulation of tissue-specific nicotinamide adenine dinuclease (NAD) activity by TDM-induced NADse, decreasing tissue-specific NAD levels by blocking the electron flow along the mitochondrial respiratory chain and thus oxidative phosphorylation (40–44). *M. tuberculosis* TDM has two mycolic acids (or mycolates). Each TDM mycolate has two acyl-chains, a long β-hydroxy chain consisting of 50 to 60 carbons, and a shorter α-alkyl chain typically with 24 carbons (45). Thus, each mycolic acid of TDM is reported to contain between 74 and 90 carbons (45–47). The length and ester-linkages

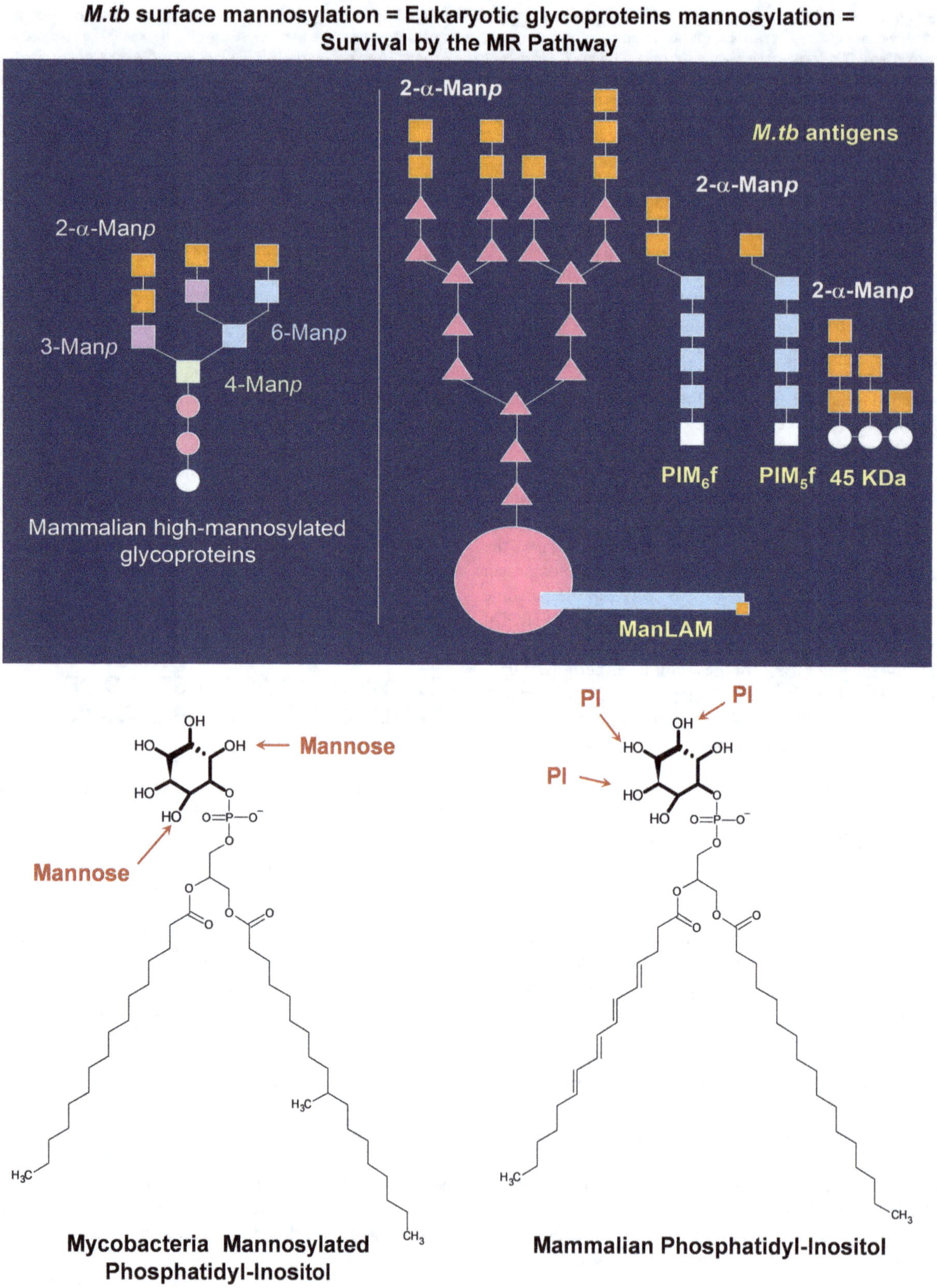

FIGURE 2 | *M. tuberculosis* complex mannose-containing lipids mimic mammalian structures allowing the bacterium to gain entry into host cells and thus potentially bypassing the host immune response. *M. tuberculosis* mannose-containing glycolipids (PIMs) and lipoglycans [ManLAM and lipomannan (LM), and some mannose containing proteins] contain α-2-mannose in their non-reducing termini, mimicking non-siliated mammalian glycoproteins in circulation.

(Continued)

FIGURE 2 | These non-siliated mammalian proteins can trigger inflammation, and thus, the host contains a series of homeostatic receptors, such as the mannose receptor (MR), to remove them out from circulation. *M. tuberculosis* can take advantage of this host clearance system by increasing its mannose surface, allowing the *M. tuberculosis* bacterium to interact with the homeostatic MR (among others), gaining entry and surviving within host cells. Moreover, these same glycolipids and lipoglycans contain a GPI-anchor of similar structure to mamalian phosphatidyl-inositol, which intercalates within vesicular membranes, regulating vesicular trafficking and fusion.

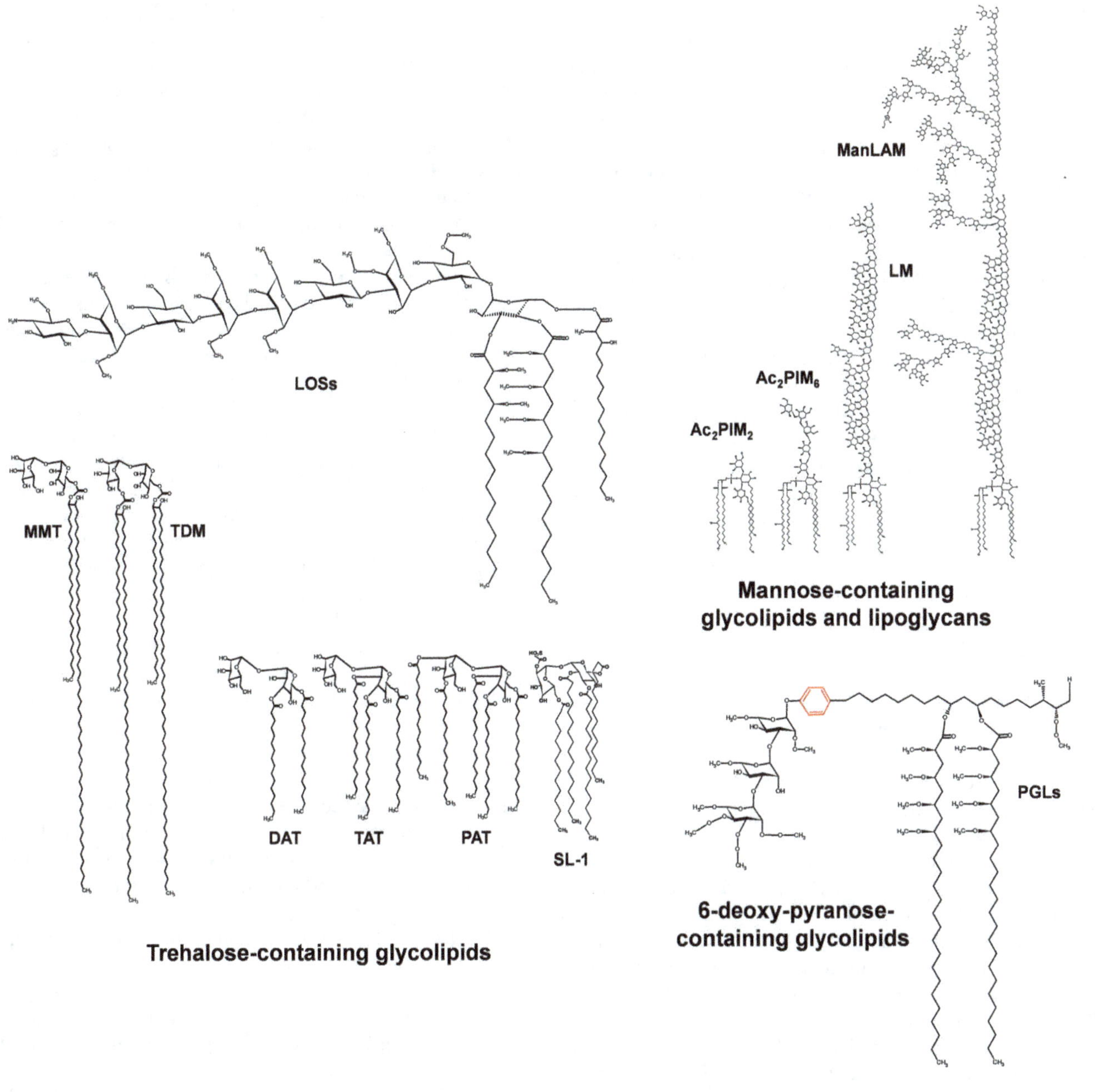

FIGURE 3 | Major glycolipids of the *M. tuberculosis* cell envelope surface. Three groups of glycolipids can be found on the *M. tuberculosis* cell envelope: (i) trehalose containing lipids, which involve di-, tri-, and penta-acylated trehaloses (DAT, TAT, PAT); trehalose mono- and di-mycolate (TMM, TDM); sulfolipids (SL-1); and lipooligosaccharides (LOSs); (ii) mannose-containing lipids, which involve the phosphatidyl-*myo*-inositol mannosides of different lengths (1 to 6 mannoses) and acylation patterns 1 to 4 fatty acid of different nature, being the most common palmitic (16:0) and tuberculostearic acid (TBST); and (iii) 6-deoxy-pyranose containing lipids, mainly phenolic glycolipids (PGLs). LOS and PGL types are restricted to a few strains of the *M. tuberculosis* complex.

of the mycolates on TDM molecules define their toxicity; a longer mycolate shows higher toxicity (48, 49). Related to the *M. tuberculosis* cell envelope, TDM seems to interlock with ManLAM, which is also found on the surface of *M. tuberculosis*, to form an asymmetrical bilayer and thus directly participates in the maintenance of the *M. tuberculosis* cell envelope integrity

(37). Studies removing TDM from the bacterium surface further indicated the importance of TDM in the *M. tuberculosis*-host relationship that takes place in infections *in vitro* and *in vivo* (50, 51). In this regard, *M. tuberculosis* is shown to produce large quantities of TDM during infection and when inside of the phagosome (52). TDM also seems to interlock host fibrinogen, the latter then acting as a cofactor for TDM biological effects (53).

M. tuberculosis TDM's biological properties have been described to rely on the whole molecule (54, 55), and these include (**Table 1**): interacting with Mincle (macrophage-inducible C-type lectin) (56) on the macrophage surface, driving *M. tuberculosis*-host recognition, and interacting with Mincle working together with the Fcγ-receptor transmembrane segment inducing Th1 (e.g., TNF, IL-12), Th2 (IL-6), and inflammasome (IL-1β) pro-inflammation mediators (57–59); actively participating in driving mycobacterial phagosome maturation arrest (55, 60) after phagocytosis of virulent strains of *M. tuberculosis*, allowing the bacillus to survive within the phagocyte; indirectly participating via its recognition by Mincle in the influx of neutrophils to the infection site (61); inducing host cell apoptosis (62–64); participating in the induction and vascularization of lung granulomas (65); being antigenic and having antitumoral activity (63, 66); having adjuvant properties being able to generate optimal antibody responses (67); inducing host cachexia, a typical symptom in active TB disease (68); and being able to trigger non-specific host immune responses against both parasitic and bacterial infections (63, 69–72). TDM is also shown to undergo cyclopropane modifications on its mycolates (73), driving a TDM-dependent reprogramming of the macrophage global gene expression during infection *in vitro* and *in vivo* (74).

TMM is known to be an essential precursor for the biosynthesis of TDM and the mycolic acid layer covalently linked to the arabinogalactan-peptidoglycan (mAGP) core of the *M. tuberculosis* cell envelope (75). Although TDM has been largely studied in its close relation with *M. tuberculosis* pathogenesis, this is not the case for TMM. TMM is thought to be involved in the maintenance of the structure and function of the *M. tuberculosis* cell envelope, working together with arabinogalactan mycolate, TDM (76), or phosphatidyl-ethanolamine, the latter shown to participate in the TMM translocation from the plasma membrane to *M. tuberculosis* cell envelope core (77). In this context, TMM is shown to be used by *M. tuberculosis* to transfer mycolic acids toward molecules like the envelope-linked arabinogalactan. In agreement with this fact, the known secreted immunogenic Ag 85 complex has been identified as a trehalose mycolyltransferase (78, 79).

As TDM, TMM has similar biological properties, also inducing lethal host cytotoxicity and host cell apoptosis; has granulomagenic activity; has adjuvant activity; and is capable of stimulating TNF via activation of the protein kinase C pathway (10, 80–84) (**Table 1**).

THE LOS GLYCOLIPID FAMILY

Another family of trehalose-based lipids present on the *M. tuberculosis* cell envelope is the LOSs. LOSs in the *M. tuberculosis* complex are understudied, produced by some strains (e.g., *M. canettii*) (85), and absent in most clinical isolates of *M. tuberculosis* as well as laboratory strains such as $H_{37}R_v$, $H_{37}R_a$, Erdman (86). The majority of their biological attributes are derived from studies performed in other mycobacterial species such as *Mycobacterium kansasii*, *Mycobacterium smegmatis*, and *Mycobacterium marinum* among others (85–89). LOSs are considered to be located on the surface of the mycobacterial cell envelope (90–92), but their contribution in shaping it is still unclear (93). However, the typical smooth and glossy colony formation observed in the *M. tuberculosis* complex strain *canettii* is speculated to be directly linked to the heavy presence of LOSs in this strain cell envelope (85, 94, 95).

All described LOSs share one trehalose with at least one or two long poly-methylated branched fatty acid chains that can be saturated or unsaturated, and located in both glucose residues of the trehalose end of the polymer (90). Depending on the mycobacterial species, a monosaccharide, or more frequently, an oligosaccharide (2 to 6 carbohydrate residues) is also present linked either at C3, C4, or C6 of the trehalose end (89, 96, 97). LOSs from different mycobacteria present different grade of variability in this oligosaccharide, the most frequent one being a tetraglucose structure [D-Glc*p*(β1→3)-D-Glc*p*-(β1→4)-**D-Glc*p*-(α1→ 1α)-D-Glc*p***], which includes the trehalose (in bold). Pyruvic acid residues (carboxyethylidene) have also been described giving an anionic character to these molecules.

Related to LOSs biological properties, these may play an important role in regulating bacterial transmission, motility, and biofilm formation; are immunogenic and also phage receptors; and may play a role in the establishment of the infection (86, 98) (**Table 1**). LOSs are mainly used for serotyping, as LOS-specific antibodies are raised against the immunodominant LOS oligosaccharide. LOSs are also shown to directly or indirectly participate in the mycobacterial infection of host macrophages (86). In this regard, their role in *M. tuberculosis* virulence is unclear. Some studies suggest LOSs being attenuated/masking factors such as DATs, where mycobacterial species containing LOS are readily cleared *in vivo*; conversely, mycobacterial strains lacking LOSs cause chronic and systemic infections *in vivo* (92, 99).

One role of LOSs is to participate in reducing the hydrophobicity of the mycobacterial cell envelope. Studies suggest that *M. tuberculosis* complex strain *canettii* is an ancestor strain, mainly isolated in the horn of Africa, containing abundant LOSs (100). However, as evolution goes, modern *M. tuberculosis* complex strains became more hydrophobic, replacing LOSs in their cell envelope in detriment of adding more apolar lipids. This change of polarity/hydrophobicity is thought to provide critical advantages to the *M. tuberculosis* complex, increasing their capability for aerosol transmission (101), affecting their virulence and pathogenicity (100, 102).

TABLE 1 | *M. tuberculosis* cell wall glycolipids and lipoglycans and their interaction with the host.

M. tuberculosis cell envelope lipid	Host cell receptor(s)	Phagosome maturation arrest	Host immune response	Sero activity	Cyto toxicity	Pathobiological context	Disease stage
Diacyl-, triacyl-, and pentacyl-trehalose (DAT, TAT, and PAT)	ND	Yes	Suppressor (anti-Inflammatory)	Yes	No	• Host cell recognition and activation Inhibitor (Mϕ, Nϕ, AT) • Phagocytosis inhibitor • B-cell priming • T-cell proliferation inhibitor	• Initial infection establishment • Increase transmissibility
Trehalose dimycolate (TDM)	Mincle-FcyR TLRs	Yes	Stimulator (pro-inflammatory)	Yes	Yes	• Host cell recognition and activation (Mϕ, Nϕ, AT) • Inflammasome activation • Apoptosis	• Initial infection establishment • Granuloma formation and maintenance • Cavitation • Increase transmissibility
Trehalose monomycolate (TMM)	ND	ND	Stimulator (pro-inflammatory)	Yes	Yes	• Apoptosis	• Initial infection establishment • Granuloma formation and maintenance • Cavitation • Increase transmissibility
Lipooligosaccharides (LOSs)	ND	ND	Suppressor (anti-inflammatory)	Yes	ND	• Host cell recognition (Mϕ, others?) • Motility	• Initial infection establishment • Decrease transmissibility
Sulfated trehalose glycolipid family (SL-1, Ac_2SGL)	ND	Yes (by SL-1)	Suppressor (anti-inflammatory)	Yes	Yes (Ac_2SGL only)	• Host cell recognition and activation Inhibitor (Mϕ, DCs, others?) • Phagocytosis inhibitor • Inflammasome inhibition • T-cell proliferation activator	• Initial infection establishment • Increase transmissibility
Phenolic glycolipid (PGL-TB)	CR3? TLRs	ND	Anti-/Pro-Inflammatory	Yes	Yes	• Host cell recognition and activation (Mϕ, others?)	• Initial infection establishment • Granuloma formation and maintenance (chronic infection)
Phthiocerol dimycocerosate (PDIM)	Direct insertion into host Mbrs	Yes	Suppressor (anti-inflammatory)	Unknown	Yes	• Host cell recognition and activation (Mϕ, DCs, others?)	• Initial infection establishment • Cavitation • Increase transmissibility
Lower-order phosphatidyl-*myo*-inositol mannosides (Ac_xPIM_{1-4})	CR3, TLRs, DC-SIGN	No	Stimulator (pro-inflammatory)	Yes	ND	• Host cell recognition and activation (Mϕ, Nϕ, others?) • Early endosome fusion stimulator • Oxidative response stimulator • Activate T-cell proliferation via CD1	• Initial infection establishment

(Continued)

TABLE 1 | Continued

M. tuberculosis cell envelope lipid	Host cell receptor(s)	Phagosome maturation arrest	Host immune response	Sero activity	Cyto toxicity	Pathobiological context	Disease stage
Higher-order phosphatidyl-*myo*-inositol mannosides (Ac_xPIM_{5-6})	MR, DC-SIGN	Yes (through the MR only)	Suppressor (anti-inflammatory)	Yes	ND	• Host cell recognition and activation (Mϕ, Nϕ, others?) • Apoptosis modulator • Activate T-cell proliferationvia CD1	• Initial infection establishment • Granuloma formation and maintenance (chronic infection)
Lipomannan (LM)	TLRs, DC-SIGN	No	Stimulator (pro-inflammatory)	Yes	ND	• Host cell recognition and activation (Mϕ, Nϕ, others?) • Apoptosis modulator • Activate T-cell proliferationvia CD1	• Initial infection establishment
Mannose-capped lipoarabinomannan (ManLAM)	MR, DC-SIGN	Yes (through the MR only)	Suppressor (anti-inflammatory)	Yes	ND	• Host cell recognition and activation (Mϕ, Nϕ, others?) • Apoptosis modulator • Activate T-cell proliferationvia CD1	• Initial infection establishment • Granuloma formation and maintenance (chronic infection)

THE SULFATED TREHALOSE GLYCOLIPID FAMILY

As in the case of the trehalose containing DATs, TATs, PATs, TDMs, and TMMs, sulfated trehaloses (or sulfolipids, designated by SL) (103) are also present in virulent strains of *M. tuberculosis*, specifically SL-1 (104, 105). Sulfate derivatives are rare in nature, where the degree of sulfation could directly change the binding properties directing extracellular interactions between biomolecules from both, *M. tuberculosis* and host cells (106). Concretely, *M. tuberculosis* SLs consists of a trehalose-2-sulfate disaccharide that can be acylated by two [defining the diacylated sulfoglycolipid or Ac_2SGL (107)] to four [defining SL-1 (108)] very long (up to C_{64}) saturated and unsaturated, highly branched fatty acids (109, 110).

Like TDM, SL-1 [and potentially Ac_2SGL (107)] is involved in the establishment of the infection. SL-1 seems to induce phagosome maturation arrest in macrophages (111); is cytotoxic *in vitro* (112, 113); inhibits macrophage priming/activation (114, 115), phagocytosis, and the inflammasome (115); induces superoxide release in activated neutrophils and monocytes; and are immunogenic with potential for serodiagnosis (33, 116–118) (**Table 1**). Specifically for Ac_2SGL (107), this can stimulate and proliferate CD1b-restricted T cells (119), and stimulate the production of Th1 cytokines by $CD8^+$ T cells (119). However, *in vitro* and *in vivo* studies, the latter using different animal models, dispute the role of SL-1 in *M. tuberculosis* virulence and pathogenesis (114–117, 120–122). SL-1 has been shown acting together with TDM, damaging the mitochondrial structures in the host cell *in vitro*; however, SL-1 alone intraperitoneally injected in mice did not induce cytotoxicity. *M. tuberculosis* mutants (*pks2*) lacking both SL-1 and its precursor Ac_2SGL indicate that both are not required for *M. tuberculosis* to grow and cause pathogenesis *in vivo*. Conversely, an *M. tuberculosis* mutant (*mmpl8*) only lacking SL-1 but accumulating Ac_2SGL were shown to be attenuated *in vivo*, indicating that the SL-1 precursor Ac_2SGL is indeed immunogenically participating in *M. tuberculosis* virulence (108). In this regard, it has been suggested that as Ac_2SGL is a precursor of SL-1 lacking two acyl groups (phthioceranic acid and hydroxyphthioceranic acid), it could be thus plausible that *M. tuberculosis* uses the ratio of SL-1/Ac_2SGL in its cell envelope to modulate the host immune response to its advantage (119, 123).

OTHER TREHALOSE GLYCOLIPID FAMILY

Lesser-studied *M. tuberculosis* trehalose-containing lipids are the phthienoates (or mycolipenates) and hydroxyphthienoates (or mycolipanolates). These consist of four species with different polarities (124). The less polar species have acyl substituents mainly straight-chain 16:0 and 18:0 acids and 2,4,6-trimethyl tetracos-2-enoic (C27 phthienoic or mycolipenic) acid, or a 3-hydroxy-2,4,6-trimethyltetracosanoic (C27-hydroxyphthienoic or mycolipanolic) acid. The polar pair species mainly have straight-chain 16:0 and 18:0, C27-mycolipanolic acid, minor proportions of C25- and C27-mycolipenic acids, and major proportions of 2,4-dimethyldocosanoic acid (124, 125). Their relative contribution to the *M. tuberculosis* pathogenesis remains

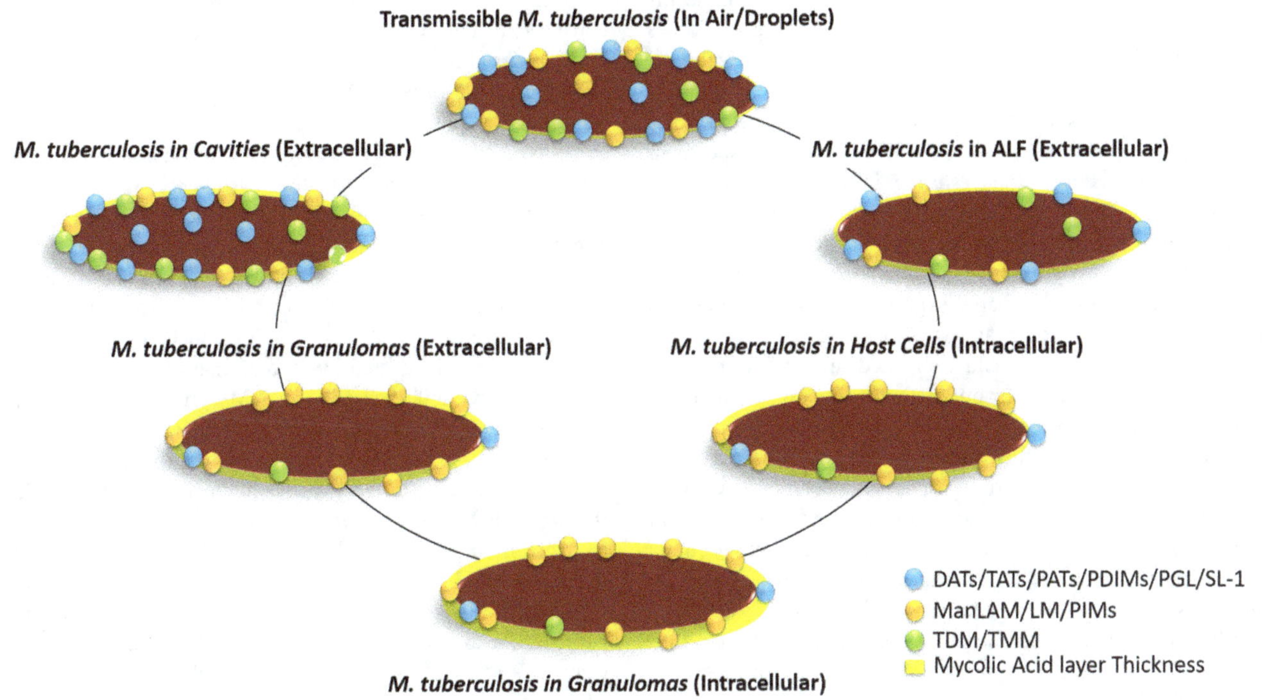

FIGURE 4 | Hypothetical distribution of glycolipids on the *M. tuberculosis* bacterium cell envelope surface at the different stages of infection. The transmissible bacterium is thought to contain a large number of glycolipids (DAT, TAT, PAT, SL-1, TMM, TDM, PDIMS, PIMs, etc.) on the cell envelope driving surface hydrophobicity, which in turn favors its transmission through the air. Upon infection and after reaching the alveolar space, *M. tuberculosis* gets in contact with alveolar lining fluid (ALF; hypophase of the lung mucosa). ALF contains an array of homeostatic hydrolases that significantly alter the *M. tuberculosis* cell envelope surface removing glycolipids and lipoglycans, and thus somehow altering its cell surface hydrophobicity and determining how the bacterium will interact with alveolar host cells. After phagocytosis by AMs, the bacterium is shown to hyperproduce PIMs, as well as switching to metabolic networks such as beta-oxidation, glyoxylate shunt, and the reverse methylcitrate cycle. This metabolic switch within host cells allows the bacterium to break down own and host captured long-chain fatty acids and cholesterol to generate acetyl-CoA and propionyl-CoA, increasing its production of mycolic acids in detriment of producing glycolipids. This bacterium metabolic state is thought to be maintained at latency stages when the bacterium remains intracellular within granulomas. Upon reactivation, the bacterium metabolic state changes again increasing glycolipid production in detriment of producing mycolic acids, restoring the bacterium surface hydrophobicity, while remaining extracellular within granulomas. This bacterial surface hydrophobicity gets further accentuated when *M. tuberculosis* escapes disrupted granulomas becoming extracellular within cavities, where host-driven hydrolytic destruction of tissues enhances *M. tuberculosis* escape to the blood stream and airways becoming transmissible again closing the cycle. Relative number of molecules and size are not accurately depicted reflecting published experimental data.

uncertain, but their function in maintaining the integrity of the cell envelope is stipulated (124).

THE PGL FAMILY

Other *M. tuberculosis* glycolipids containing oligosaccharides are the PGLs. PGLs are considered a glycosylated form of phthiocerol dimycocerosates (PDIM) (126). These glycolipids have been mainly studied in *Mycobacterium leprae* (triglycosylated PGL-I) (127–129) and *Mycobacterium marinum* (monoglycosylated PGL), being used for diagnostic testing in leprosy patients (130). The exact structure of PGL is species-specific, and its biological properties are starting to be elucidated. In this regard, while in leprosy, *M. leprae* PGL-1 subverts tissue macrophages to produce neurotoxic radical nitrogen species (e.g., nitric oxide) leading to nerve damage (131). Along this line, PGL-1 has also been shown to be effective oxygen radicals scavengers (132). In *M. marinum* infection, PGL drives phagosome maturation arrest (133), and allows the bacterium to escape from hostile tissue-resident macrophages during the initial stages of the infection (131, 134, 135). By doing so, *M. marinum* uses PGL to infect resident macrophages and induce the secretion of the chemokine (C-C motif) ligand 2 (CCL2) attracting non-bactericidal monocytes, which will be subsequently infected by *M. marinum* and used as Trojan horses to cause a systemic infection (134–136).

As LOSs, PGLs are not present in all *M. tuberculosis* complex strains. These have been found in abundance in the *M. tuberculosis* complex *canettii* strain (PGL-Tb) (137) and in same *M. tuberculosis* clinical isolates of the W-Beijing family. Due to the different amounts of PGLs produced by different *M. tuberculosis* strains (138), their use for *M. tuberculosis* serodiagnostic did not prove to be useful as TB patients presented significant variations in their response to this antigen (85). Interestingly, the hypervirulent phenotype observed in several strains of *M. tuberculosis* (e.g., the Beijing strain HN878) has been associated with the presence of PGL and triglycerides in their cell envelope (139, 140).

Contrary to the debatable PGL hypervirulence effects where *M. tuberculosis* PGL-deficient strains induce pro-inflammation (141), recent studies, using a necrotic *M. tuberculosis* mouse model, show that loss of *M. tuberculosis* PGLs decreases the

Th17 response (e.g., IL-17A), where IL-17A production seems critical to limit the development of hypoxic necrotic granulomas and reduces disease severity in TB (142). Thus, indirectly, *M. tuberculosis* PGL may favor *M. tuberculosis* infection reducing tissue damage. This concept is supported by other studies analyzing multiple clinical isolates, where the clinical spectrum of *M. tuberculosis* infection resulting in active TB or latent infection could be dictated not only by the host but also by the amounts and ratios of surface-exposed glycolipids defined by the strain genotype (134, 143, 144). Thus, serum opsonized *M. tuberculosis* clinical isolates expressing abundant PGL on their surface had reduced phagocytosis by macrophages, but the ingested *M. tuberculosis* bacilli replicated at faster rate (144). This was hypothetically attributed to the opsonization of *M. tuberculosis* surface-exposed PGL by host-soluble complement component 3 (C3) and the subsequent use by *M. tuberculosis* of the macrophage phagocytic complement receptor 3 (CR3) to infect host cells (144). Later studies confirmed that the sugar moiety of PGL is unique in its capacity to bind the lectin domain of CR3 for efficient infection of human macrophages, and *M. tuberculosis* PGLs are able to inhibit Toll-like receptor 2-triggered NF-κB activation and thus the pro-inflammatory response, thus assisting *M. tuberculosis* to subvert the host immune response (145) (**Table 1**). This capacity to dampen the host response is further corroborated by studies using *M. tuberculosis* PGL analogs, where these inhibited the production of Th1 (TNF), Th2 (IL-6), and the inflammasome (IL1-β)-related responses in a Toll-like receptor 2-mediated manner (143). Other supporting studies using recombinant BCG expressing PGL-1 show that this strain can exploit the host CR3, promoting bacterial uptake but at the same time inhibiting pro-inflammatory responses (e.g., TNF) (146). Overall, the direct interaction of PGL with host cell receptors is still unclear. Studies focused in determining which host receptors could recognize and bind *M. tuberculosis* complex BSA-conjugated glycans indicate that dendritic cell-specific ICAM-3-grabbing non-integrin (DC-SIGN or CD209), the DC-SIGN homolog DC-SIGNR (for DC-SIGN related), Langerin (CD207, C-type lectin receptor on Langerhans cells), the mannose receptor (MR or CD206), galactose receptor, dendritic cell C-type lectin 2 (Dectin-2), macrophage-inducible Ca^{2+}-dependent lectin receptor (Mincle or CLEC4E), and BDCA-2 (CLEC4C or CD303) had low affinity for the rhamnose (6-deoxy-L-mannose) and fucose (6-deoxy-L-galactose) found in PGLs, indicating that PGL could be targeted by other, yet to be discovered, host C-type lectins (147).

Interestingly, studies also highlighted the role of PGL in BCG efficacy, where a mutant BCG (deleting *fadD28*) lacking both PDIMs and PGLs not only made BCG more attenuated in mice but also reduced its efficacy as a vaccine against *M. tuberculosis* infection (148).

THE MANNOSE-CONTAINING GLYCOLIPID FAMILY

This family involves phosphatidyl-*myo*-inositol (PI) glycosylated derivatives known as PIMs and the predominant mannosyl-β-1-phosphomycoketides (MPM). PIMs are important lipids in the cell envelope of *M. tuberculosis*, both as key cell envelope constituents involved in metabolic processes and as essential participants in *M. tuberculosis*-host interactions. PI is an acidic (anionic) phospholipid consisting of a phosphatidic acid linked via a phosphate group to *myo*-D-inositol. PIMs are considered, together with phosphatidyl-ethanolamine, the major phospholipid components of the *M. tuberculosis* cell envelope (75). There are found on the *M. tuberculosis* cell envelope as a mixture of compounds differing from each other by their number of mannose and fatty acid residues. The basic structure of PIM_1 consists of a mannosyl unit attached to position C-2 of the *myo*-inositol of a PI anchor. Position C-6 of *myo*-inositol can be further substituted by an α-D-mannosyl or a linked di- or trimannosyl unit, giving PIM_2, PIM_3, and PIM_4, respectively. The terminal mannose residue of PIM_4 can be further substituted at position C-2 by a α-D-mannosyl leading to PIM_5, which can also be further substituted at the same position leading to PIM_6, the higher PIM encountered in mycobacteria to date (10). PIMs are shown to be located both in the *M. tuberculosis* plasma membrane and on the cell envelope surface (149). PIMs are intrinsically highly heterogeneous not only due to their mannose content but also due to the number and nature of their acyl groups. PIMs can be mono- to tetra-acylated, where the basic PIM is the diacylated form (150) esterified by palmitic (C16) and tuberculostearic (TBST or 10-methyl-octadecanoic) acids on C-1 and C-2 position of the glycerol, with a third fatty acyl substituent on the C-6 position of the (1→ 2) linked mannose to the inositol and with a fourth fatty acyl on the C-3 of the *myo*-inositol (150, 151). To simplify, PIMs can be grouped into lower- and higher-order PIMs depending on their number of mannoses, where lower-order PIMs contain one to four mannoses and higher-order PIMs contain five to six mannoses (152). The most common PIMs found in *M. tuberculosis* complex are $AcPIM_2$ and Ac_1PIM_2 (di- and triacylated PIM_2) and $AcPIM_6$ and Ac_1PIM_6 (di-and triacylated PIM_6) (150). All lower-order PIMs have a terminal α(1→ 6)-mannose, are exposed on the cell surface, and directly participate in the establishment of the infection, playing a role in macrophage recognition and phagocytosis process through association with the non-opsonic domain of CR3 (153). Lower-order PIMs also participate in facilitating trafficking processes within the phagocyte by promoting early endosomal fusion with *M. tuberculosis* containing phagosomes, thus providing nutrients to the bacterium (154) (**Table 1**).

In contrast, all higher-order PIMs have a terminal α(1→ 2)-mannose. In the case of Ac_xPIM_5, it is a single α(1→ 2)-mannose, and in the case of Ac_xPIM_6, it is an α(1→ 2)-dimannoside similar to the mannose caps of mannose-capped lipoarabinomannan [ManLAM, reviewed in Turner and Torrelles (155)]. Only tri-acylated forms of higher-order PIMs are shown to interact with the MR and interfering with trafficking pathways resulting in phagosome maturation arrest (152). Conversely, all *M. tuberculosis* PIMs seem to interact with DC-SIGN (152), although differences in DC-SIGN PIM recognition specificity may be species-dependent (156). Further studies have also shown that cytosolic soluble CD1e is involved in $AcPIM_6$ processing and presentation via CD1 with subsequent T-cell activation (157) (**Table 1**).

Overall, without discerning the type of PIMs used, these have been shown to participate in host cell recognition, phagocytosis, oxidant and cytokine production, regulating intracellular vesicular trafficking, regulating signaling pathways, apoptosis, and antigen presentation (158–160) (**Table 1**). However, as pointed out, these intrinsic PIM properties, apart of being species-specific and depending on their heterogeneous nature, also may be derived from differences in the protocols used for PIM isolation and for *in vitro* experiments including immune cell types and procedures to generate them (160, 161).

To date, the role and metabolism of PIMs during *M. tuberculosis* infection and pathogenesis is still unclear, as different PIMs are recognized by different receptors triggering different host responses and outcomes. PIMs are downregulated in stationary phase (162), but it has been observed that the production of PIMs by *M. tuberculosis* is enhanced during infection in primary human macrophages (163). The latter result indicates that *M. tuberculosis* can adapt to the environment during infection by modulating its cell envelope composition (164–168). That PIM expression levels during early stages of infection may be in part directly or indirectly involved in bacterial growth, virulence, and entry into persistence is still unclear; however, studies show that a PIM-hyperproducing *M. tuberculosis* strain (ΔRv2623) is hypervirulent *in vivo* being unable to establish a chronic infection (163), while a PIM-hypoproducing *M. tuberculosis* strain (ΔRv1747) is attenuated *in vivo* in mice (169) (**Table 1**).

Another *M. tuberculosis* mannose-containing glycolipid is MPM. This glycolipid consists of a mannosyl-β-1-phosphoisoprenoid with a C32 4,8,12,16,20-pentamethylpentacosyl chain (170) and has been shown to activate CD1c-restricted T cells (171).

THE ROLE OF CELL ENVELOPE GLYCOLIPIDS IN *M. TUBERCULOSIS* EVOLUTION, ADAPTATION TO THE HOST ENVIRONMENTS, DRUG RESISTANCE, AND PATHOGENESIS

There are no doubts that the cell envelope of *M. tuberculosis* has evolved over the years to become the perfect shield for this pathogen, from the perspective of its hydrophobicity, transmissibility, adaptation to the host, drug resistance, virulence, and pathogenesis.

Several studies have discussed at large how *M. tuberculosis* has evolved increasing the hydrophobicity of its cell envelope to its advantage. In this regard, the role of the cell envelope lipids in the transition of environmental mycobacteria becoming highly transmissible, pathogenic and host-dependent is pointed out (102). The possibility that ancient *M. tuberculosis* complex strains such as *M. canettii*, mainly isolated in the horn of Africa (172, 173), have evolved has been explored, that is, changing their metabolism and increasing their cell envelope hydrophobicity as they evolved to the modern *M. tuberculosis* complex strains, e.g., changing their cell envelope's outmost exposed lipid polarity and thus evolving into a more hydrophobic organism. This was achieved by stopping the production of polar lipids such as LOSs or PGLs (which currently are just described in a subset of *M. tuberculosis* complex modern strains, Beijing clinical isolates), reducing their polarity (e.g., by losing sugars or adding acyl chains in their cell envelope glycolipids, e.g., DAT, TAT, PAT, SLs, TDM, TMM), and increasing the presence of apolar lipids in their cell envelopes [e.g., dimycocerosates of the phthiocerol family (PDIMs)] (102). Thus, cell envelope hydrophobicity could be linked to a simplified cell envelope, efficiency in transmissibility, and increased pathogenesis [extended review in Jankute et al. (102)]. Conversely, other studies shuffle the possibility that modern *M. tuberculosis* complex strains are adapting to the host by decreasing the hydrophobicity of their cell envelopes by adding mannose-containing components on their cell envelope, mimicking mammalian glycoproteins (**Figure 2**), and thus deceiving the host immune response during infection (10).

Looking at both scenarios and at the properties described for the glycolipids present in the cell envelope of *M. tuberculosis* complex, these two possibilities seem compatible, but potentially with different outcomes. Increasing hydrophobicity allows *M. tuberculosis* to be highly transmissible, and this could be the case for CDC1551, an *M. tuberculosis* strain shown to have this highly transmissible property (174). Increasing the hydrophobicity of *M. tuberculosis* can also trigger increased interactions of *M. tuberculosis* bacilli with signaling receptors that drive pro-inflammation and cellular trafficking to the infection site, and depending on the metabolic status of the host and the bacterium at this initial encounter, could result in active TB disease. This is supported by the role of triglycerides in driving hypervirulence in place of the initially described PGL (139, 140), which seems now to be involved in attenuating the host immune response (142).

Conversely, the other scenario is decreasing hydrophobicity by *M. tuberculosis* adding mannose-containing biomolecules in its cell envelope, such as the glycolipids PIMs, and their biosynthetically related lipoglycans and their derivatives [lipomannan, ManLAM, mannan, and arabinomannan (10)] (**Table 1**). *M. tuberculosis* complex strains with high mannose content in their cell envelope could be less effectively transmissible; however, when infecting the host, these trigger an anti-inflammatory response by interacting their mannose-containing glycolipids with homeostatic receptors of phagocytes [e.g., the MR (10, 175)], confusing the host cell. This scenario can result in a permissive intracellular environment for *M. tuberculosis* to permanently stay, without being detected, allowing host cells to control the infection, ultimately driving to the latency state. This is supported by the role of other glycolipids present in modern *M. tuberculosis* complex strains, such as DATs, which are shown dampening the host immune response (24).

The fact is that *M. tuberculosis*-host cell interactions, infection, and potential outcomes may also depend on the *M. tuberculosis* metabolic status prior to and during infection (**Figure 4**). *M. tuberculosis* overproduces PIMs during *in vitro* infections in macrophages (163). *M. tuberculosis* Ac_1PIM_6, as well as ManLAM, interacts with the MR defining a pathway of intracellular survival for *M. tuberculosis* and at the same time dampens the host immune response potentially benefiting *M. tuberculosis* adaption to the host (152, 176). However, a third unexplored possibility is that the cell envelope of *M. tuberculosis* shaping directly depends on the host lung environment, during the early stages of infection (in the alveolar space) or in later stages of infection, in cavities, when transmissible bacilli are shaped to be transmitted. In both cases, *M. tuberculosis* remains extracellular. In the initial stages, when infection occurs, *M. tuberculosis* is deposited on the alveolar space of the lung, where it will find the human lung mucosa before encountering any host cell [e.g., AMs, alveolar epithelial cells (7)]. In this scenario, the alveolar epithelium is covered by lung mucosal fluid, which is composed of two phases, a lipid layer called surfactant (mainly composed of phospholipids) and an hypophase called alveolar lining fluid (or ALF). Host cells are thought to be immersed and operating in the ALF. ALF is thought to trap any particle that reaches the alveolar space. ALF has been described to contain an array of hydrolytic enzymes [hydrolases (164)], whose main function is to assist host cells in recycling ALF every 18 h (177). Studies have demonstrated that the nature of these host hydrolases impacts the *M. tuberculosis* cell envelope as it has been described, where 15-min exposure of *M. tuberculosis* to human ALF reduces the cell surface presence of ManLAM and TDM by ~70 and ~40%, respectively (164). Other affected lipids significantly reduced from the *M. tuberculosis* cell envelope upon contact with human ALF are triglycerides, PDIMs, SLs, PIMs, and at lower degree DATs (personal communication). The impact of these human ALF modifications on the *M. tuberculosis* cell envelope results in a better control of the infection by human phagocytes (macrophages and neutrophils) by increasing *M. tuberculosis*-containing phagosome-lysosome fusion and acidification events (164, 178). Thus, the repertoire of lipids exposed on the *M. tuberculosis* cell envelope surface during the infection initial stages and prior to the first contact with host cells may in fact depend on the host ALF status, e.g., activity of the ALF hydrolases shaping the cell envelope.

Nevertheless, these *M. tuberculosis* cell envelope glycolipid modifications also result in the release of *M. tuberculosis* cell envelope fragments to the milieu (164). Their implications in *M. tuberculosis* pathogenesis are starting to be elucidated, but studies have shown that these fragments further assist macrophages in controlling *M. tuberculosis* infection by further increasing phagosome-lysosome fusion events in an IL-10-dependent manner (179), as well as drive neutrophils to control *M. tuberculosis* infection better by increasing their oxidative response (180).

The fact is that the status of the host lung environment could somehow "*normalize the cell envelope of M. tuberculosis*," as well as redirecting *M. tuberculosis* adaptation to the host. Furthermore, *M. tuberculosis* exposure to ALF alters the cell envelope but also differentially drives how other innate soluble components described in the lung mucosa will interact with *M. tuberculosis* and drive infection and potentially disease outcome. In this context, TB comorbidities such as aging (181, 182), HIV, diabetes, smoking, etc., may impact the status of ALF, its hydrolases, and other soluble innate components functionality (182), and thus, the *M. tuberculosis* cell envelope may be differentially impacted depending on the status of the human ALF at the moment of the establishment of the infection.

This scenario has been tested in studies where the BCG vaccine has been exposed to human ALF and subsequently used to vaccinate mice showing that ALF-exposed BCG has an altered cell envelope and increased protection against *M. tuberculosis* challenge (183). Based on this study where ALF modified the BCG cell envelope and drove improved protection, studies delipidating BCG mostly removing apolar lipids showed the ability of this delipidated BCG to be delivered directly into the lung, minimizing tissue inflammation and to further increase protection against *M. tuberculosis* infection and significantly reducing tissue damage, the hallmark of TB disease (184). All these studies put in trial the role of the *M. tuberculosis* glycolipids during infection, how *M. tuberculosis* metabolically adapts to the lung environment, how *M. tuberculosis* uses its cell envelope glycolipids to mask its presence to host cells or to confuse them by mimicking mammalian structures, and how *M. tuberculosis* reconstitutes and reorganizes its cell envelope during infection and in the different environments that may encounter within the host. These environments could be extracellular at the earlier stages of infection, after escaping the infected cell via necrosis, or in cavities, where hypothetically *M. tuberculosis* bacilli could be exposed to lung tissue hydrolases causative of cavities, and thus, these hydrolases could shape bacteria that escape to the airways and are ready to be transmitted. These environments could also be intracellular within host cells in the alveolar space (e.g., AMs and alveolar epithelial cells) or within the interstitium in granuloma structures.

When it comes to drug resistance, limited information exists about the exact glycolipid composition and their role conferring the drug resistance phenotype in multi- (MDR), extensive- (XDR), and extreme- (XXDR) drug-resistant *M. tuberculosis* strains. In this regard, studies investigating *M. tuberculosis* strains containing rifampicin drug resistance-related mutations show that these strains change their cell envelope lipid composition by overexpressing PGL biosynthesis structurally related PDIMs influencing the macrophage metabolism and response during infection (16). Other studies looking at the role of tesA (a type II thioesterase) in *M. marinum* indicated a potential link between increased drug susceptibility and the lack of PDIM and PGLs on the bacterium cell envelope (185). Additional studies found that mycolic acid cyclopropanation is essential for *M. tuberculosis* viability, drug resistance, and cell envelope integrity (186). Overall, how glycolipids participate in the development of drug resistance directly or indirectly rearranging the *M. tuberculosis* cell envelope is still uncertain. Studies performed using transmission electron and atomic force microscopy techniques dig into this question showing that MDR-, XDR-, and XXDR-*M. tuberculosis* strains have thicker cell envelope and irregular cell surface with tubular extensions than susceptible strains (187, 188). As some XDR- and XXDR-*M. tuberculosis* strains are related to the Beijing family (189), which are shown to have their cell envelope overpopulated with triglycerides, it is plausible to question any relationship between the abundance of a specific hydrophobic lipid on the

M. tuberculosis cell envelope and drug resistance (190). Many of these questions remain unanswered.

At the end, studies have been focused on depicting the role of purified glycolipids on *M. tuberculosis*-host cell interactions and pathogenesis, using purified lipids, embedded into vehicles (coated in beads or nanoparticles), or using *M. tuberculosis* glycolipid(s)-depleted mutants. This conferred us an enormous knowledge about their role in *M. tuberculosis* cell envelope rearrangements, adaptation, metabolism, catabolism, virulence, and pathogenesis. However, it is still unknown how *M. tuberculosis* uses these lipids during the different stages of infection, how the host shapes and alters them, and how the status of the host with active TB and *M. tuberculosis* itself works together within the cavities to define the successful transmissible form of *M. tuberculosis*.

In this context, persistence of *M. tuberculosis* within host cells and in granulomas depends on the bacterial use of lipid metabolic networks such as beta-oxidation, glyoxylate shunt, and the reverse methylcitrate cycle, to break down own and host captured long-chain fatty acids and cholesterol to generate acetyl-CoA and propionyl-CoA (191, 192). *M. tuberculosis* could regulate acetyl-CoA and propionyl-CoA production depending of the status of the host. Where during initial infection stages and granuloma establishment and maintenance (chronic infection), *M. tuberculosis* converts acetyl-CoA into mycolic acids to thicken its cell envelope and to use them as energy storage units in detriment of producing SL-1, DAT, and PAT, among others (47, 193, 194). Conversely, during reactivation leading to cavity formation, *M. tuberculosis* could further convert acetyl-CoA into mycolic acids, further increasing their production (which in turn are used to produce TDM and TMM), and convert propionyl-CoA and methyl malonyl-CoA into Ac_2SG, SL-1, PAT, and PDIM (methyl-CoA-lipids) (14, 195, 196), increasing its cell envelope hydrophobicity, which will enhance its host to host transmissibility (**Table 1**, **Figure 4**).

THE ROLE OF *M. TUBERCULOSIS* CELL ENVELOPE GLYCOLIPIDS IN VACCINATION STRATEGIES

In vitro and *in vivo* studies using mycobacterial lipids and lipoglycans have shown that these induce both humoral and CD1-restricted T cell-mediated responses in the context of BCG vaccination (197–200). In this context, the CD1 locus encodes five related proteins (CD1a–e). CD1a–c are shown to present antigens to T cells. Concretely, CD1a is shown to present lipopeptides (201); CD1b presents mycobacterial cell envelope lipids and lipoglycans, such as mycolic acids, glucose monomycolate, glycerol monomycolate, SLs, LAM, and PIMs, among others (119, 202–205); and CD1c presents isoprenoid phosphoglycolipids (206). CD1d presents antigens to NKT cells, with the sponge-derived α-galactosyl-ceramide being the major antigen described (207). CD1e is located intracellularly and shown to present metabolized PIM_6 (157). Overall these data indicate that mycobacterial lipids could directly participate in generating long-standing protective immune responses via CD1-restricted T cells during *M. tuberculosis* infection. The role of these lipid-restricted CD1 T-cell populations in generating a protective immune response against *M. tuberculosis* infection has been debatable due to the lack of human homologs of CD1a, CD1b, and CD1c in mice, but studies using human CD1a–c transgenic (hCD1Tg) mice (208) concluded that, contrary to what has been observed in NK T cells, CD1-restricted T cells in reality exhibit delayed primary responses and more rapid secondary responses, similar to conventional T cells, indicating that these lipid-restricted CD1 T-cell populations in reality participate in adaptive immune responses upon mycobacterial infection and thus embrace the use of mycobacterial lipids to target these CD1-restricted T cells for vaccine strategies.

Conversely, for several years, the use of adjuvants in TB vaccination strategies relied on liposomes and oil extracts (209–211). Current vaccine strategies evolved the adjuvant field by targeting the stimulation of pattern recognition receptors, such as Toll-like receptors (212). These new generations of adjuvants include cationic lipids, micro-/nanoparticles, cytokines, toxin derivatives, CpG-containing DNA-based molecules, antimicrobial peptides/proteins, and mycobacterial proteins conjugates and lipids (209). These adjuvants in combination with the current BCG vaccine or with vaccine subunits are shown to be successful in boosting vaccine efficacies (209, 210, 213–215). In this context, there are ongoing vaccination strategies using mycobacterial lipids (or their analogs) as adjuvants driving better and prolonged protection. As an example, the combination of the Hybrid 1(H1) subunit vaccine with the liposomal adjuvant CAF01 [composed of a cationic liposome vehicle (dimethyldioctadecyl-ammonium) and the synthetic TDM analog, trehalose-6,6-dibehenate or TDB] (216–218). This novel vaccine seems to be well tolerated and able of generating long-lasting protective T-cell responses (216).

Having said this, we need to account that *M. tuberculosis* is a master manipulator of host immune responses, being able to passively (directed by the host) or actively (directed by the bacterium itself) change its metabolism and cell envelope lipid composition, adapting to the different adverse host environments (e.g., ALF, AMs, granuloma) during the course of infection (10, 219–222). Thus, mycobacterial lipids may generate or mask a protective host immune response against infection depending on the human lung environment and thus their debatable potential application in vaccine strategies, which has been reviewed by us in detail elsewhere (223, 224).

Overall, a comprehensive study of the *M. tuberculosis* cell envelope lipid constitution during infection could be critical to develop novel vaccine strategies. Our *in vitro* data demonstrate that *M. tuberculosis*, upon contact with the lung mucosa from healthy adult individuals, is temporally depleted of surface exposed lipids resulting in a better control of the infection by alveolar resident cells (164, 178–180, 225). Our *in vivo* data also show that this temporal depletion of lipids on the *M. tuberculosis* surface results in a better control of *M. tuberculosis* growth in the lung of infected mice over time (personal communication). However, this is not observed when *M. tuberculosis* is upon contact with the lung mucosa of healthy elderly individuals (182) or upon contact with the lung mucosa of HIV+ individuals (personal communication). Instead, in both of these cases, *M. tuberculosis* contact with the lung mucosa of the elderly and HIV+ populations accentuates its intracellular growth in AMs (*in vitro*) and in mice (*in vivo*) (182) (personal communication). These results point out the importance of the status of the human lung mucosa in altering the *M. tuberculosis* cell envelope lipid composition (164), resulting in a successful immune response initially protective against *M. tuberculosis* infection and intracellular growth. Subsequently, when inside human macrophages, *M. tuberculosis* seems to overproduce specific lipids, e.g. PIMs (163). Thus, the host-directed temporal removal of lipids and the bacterial overproduction of lipids within host cells may be important to understand which lipids (their presence and/or absence) are important to generate a prolonged and protective immune response, which may improve current vaccination strategies.

CONCLUSIONS

M. tuberculosis is proven to expend energy elaborating a great variety of glycolipids of rather unusual structure, both in the inner cell envelope and exposed on its cell surface. These include mainly three groups of glycolipids, trehalose-containing lipids [e.g., acylglucosides (DAT, TAT, PAT), sulfatides (SLs), and LOSs], the 6-deoxy-pyranose containing phenolic glycolipids (e.g., PGLs), and the ubiquitous mannose-containing glycolipids (e.g., PIMs). Some of these glycolipids are described as *M. tuberculosis* virulence factors being recognized by specific host cell receptors (or directly by their interaction with membranes) and thus helping the bacillus to survive within host cells. Therefore, enzymes involved in glycolipids biosynthesis have been and are extensively studied as these may represent potential drug targets (226). In this context, the role of the *M. tuberculosis* cell envelope lipids during dormancy, nutrient starvation, hypoxia, and oxidative stress has been reported using the *in vitro* Wayne's model, pointing out that under these conditions *M. tuberculosis* adapts its proteome and lipidome (222, 227–231).

Attempts to define glycolipids biological function by genetically manipulating *M. tuberculosis* to obtain glycolipid deficient mutants are somehow providing us if these glycolipids are important for *M. tuberculosis* infection and pathogenesis; however, these studies may bypass the impact of *M. tuberculosis* cell envelope rearrangements and the host environment impact on these glycolipids or new rearrangements due to the mutation. Thus, the production of mutants or conditional mutants that provide different levels of production of these glycolipids in the *M. tuberculosis* cell envelope at different stages of the infection may provide us more accurate information about the role of these glycolipids during infection and their real implication in pathogenesis, in place of studying mutants lacking a specific glycolipid.

AUTHOR CONTRIBUTIONS

AG-V, JC, and JT made substantial contributions to the conception, writing, and editing of this review.

FUNDING

This work has been partially supported by NIH/NIA (P01 AG-051428) and the Robert J. Kleberg and Helen C. Kleberg Foundation to JT, and by NIH/NIAID (R01 AI 146340) to JC and JT.

REFERENCES

1. (CDC) CfDCaP. *Emergence of Mycobacterium tuberculosis With Extensive Resistance to Second-Line Drugs–Worldwide, 2000-2004.* MMWR Morb Mortal Wkly Rep (2006). p. 301–5.
2. Gandhi NR, Moll A, Sturm AW, Pawinski R, Govender T, Lalloo U, et al. Extensively drug-resistant tuberculosis as a cause of death in patients co-infected with tuberculosis and HIV in a rural area of South Africa. *Lancet.* (2006) 368:1575–80. doi: 10.1016/S0140-6736(06)69573-1
3. Basu S, Friedland GH, Medlock J, Andrews JR, Shah NS, Gandhi NR, et al. Averting epidemics of extensively drug-resistant tuberculosis. *Proc Natl Acad Sci USA.* (2009) 106:7672–7. doi: 10.1073/pnas.0812472106
4. Andrews JR, Shah NS, Weissman D, Moll AP, Friedland G, Gandhi NR. Predictors of multidrug- and extensively drug-resistant tuberculosis in a high HIV prevalence community. *PLoS ONE.* (2010) 5:e15735. doi: 10.1371/journal.pone.0015735
5. Gandhi NR, Shah NS, Andrews JR, Vella V, Moll AP, Scott M, et al. HIV coinfection in multidrug- and extensively drug-resistant tuberculosis results in high early mortality. *Am J Respir Crit Care Med.* (2010) 181:80–6. doi: 10.1164/rccm.200907-0989OC
6. WHO. *Global Tuberculosis Report 2018.* Geneva: WHO Organization, (2018). Available online at: https://www.who.int/tb/publications/global_report/en/.
7. Torrelles JB, Schlesinger LS. Integrating lung physiology, immunology, and tuberculosis. *Trends Microbiol.* (2017) 25:688–97. doi: 10.1016/j.tim.2017.03.007
8. Avigad G, Dey PM. 4- carbohydrate metabolism: storage carbohydrates. In: Dey PM, Harborne JB, editors. *Plant Biochemistry.* London: Academic Press (1997). p. 143–204. doi: 10.1016/B978-012214674-9/50005-9
9. Kalscheuer R, Weinrick B, Veeraraghavan U, Besra GS, Jacobs WR Jr. Trehalose-recycling ABC transporter LpqY-SugA-SugB-SugC is essential for virulence of *Mycobacterium tuberculosis. Proc Natl Acad Sci USA.* (2010) 107:21761–6. doi: 10.1073/pnas.1014642108
10. Torrelles JB, Schlesinger LS. Diversity in *Mycobacterium tuberculosis* mannosylated cell wall determinants impacts adaptation to the host. *Tuberculosis.* (2010) 90:84–93. doi: 10.1016/j.tube.2010.02.003
11. Dobos KM, Khoo KH, Swiderek KM, Brennan PJ, Belisle JT. Definition of the full extent of glycosylation of the 45- kilodalton glycoprotein of *Mycobacterium tuberculosis. J Bacteriol.* (1996) 178:2498–506. doi: 10.1128/jb.178.9.2498-2506.1996
12. Gago G, Diacovich L, Gramajo H. Lipid metabolism and its implication in mycobacteria-host interaction. *Curr Opin Microbiol.* (2018) 41:36–42. doi: 10.1016/j.mib.2017.11.020
13. Velayati AA, Farnia P, Hoffner S. Drug-resistant *Mycobacterium tuberculosis*: Epidemiology and role of morphological alterations. *J Glob Antimicrob Resist.* (2018) 12:192–6. doi: 10.1016/j.jgar.2017.10.006
14. Ghazaei C. *Mycobacterium tuberculosis* and lipids: insights into molecular mechanisms from persistence to virulence. *J Res Med Sci.* (2018) 23:63. doi: 10.4103/jrms.JRMS_904_17
15. Chakraborty P, Kulkarni S, Rajan R, Sainis K. Drug resistant clinical isolates of *Mycobacterium tuberculosis* from different genotypes exhibit

differential host responses in THP-1 cells. *PLoS ONE.* (2013) 8:e62966. doi: 10.1371/journal.pone.0062966

16. Howard NC, Marin ND, Ahmed M, Rosa BA, Martin J, Bambouskova M, et al. *Mycobacterium tuberculosis* carrying a rifampicin drug resistance mutation reprograms macrophage metabolism through cell wall lipid changes. *Nat Microbiol.* (2018) 3:1099–108. doi: 10.1038/s41564-018-0245-0
17. Brennan PJ, Nikaido H. The envelope of mycobacteria. *Annu Rev Biochem.* (1995) 64:29–63. doi: 10.1146/annurev.bi.64.070195.000333
18. Jarlier V, Nikaido H. Mycobacterial cell wall: structure and role in natural resistance to antibiotics. *FEMS Microbiol Lett.* (1994) 123:11–8. doi: 10.1016/0378-1097(94)90267-4
19. Brennan PJ. *Mycobacterium* and other actinomycetes. In: Ratledge C, Wilkinson VL, editors. *Microbial Lipid.* London: Acad. Press. (1988). p. 203–98.
20. Anderson RJ. Chemical Investigation of biologically active lipoids of tubercle bacilli. *Proc Natl Acad Sci USA.* (1929) 15:628–33. doi: 10.1073/pnas.15.8.628
21. Rhoades ER, Streeter C, Turk J, Hsu FF. Characterization of sulfolipids of *Mycobacterium tuberculosis* H37Rv by multiple-stage linear ion-trap high-resolution mass spectrometry with electrospray ionization reveals that the family of sulfolipid II predominates. *Biochemistry.* (2011) 50:9135–47. doi: 10.1021/bi2012178
22. Gautier N, López Marín LM, Lanéelle M-A, Daffé M. Structure of mycoside F, a family of trehalose-containing glycolipids of *Mycobacterium fortuitum. FEMS Microbiol Lett.* (1992) 98:81–8. doi: 10.1016/0378-1097(92)90136-C
23. Frankfater C, Abramovitch RB, Purdy GE, Turk J, Legentil L, Lemiegre L, et al. Multiple-stage precursor ion separation and high resolution mass spectrometry toward structural characterization of 2,3-diacyltrehalose family from *Mycobacterium tuberculosis. Separations.* (2019) 6:4. doi: 10.3390/separations6010004
24. Lee KS, Dubey VS, Kolattukudy PE, Song CH, Shin AR, Jung SB, et al. Diacyltrehalose of *Mycobacterium tuberculosis* inhibits lipopolysaccharide- and mycobacteria-induced proinflammatory cytokine production in human monocytic cells. *FEMS Microbiol Lett.* (2007) 267:121–8. doi: 10.1111/j.1574-6968.2006.00553.x
25. Saavedra R, Segura E, Leyva R, Esparza LA, Lopez-Marin LM. Mycobacterial di-O-acyl-trehalose inhibits mitogen- and antigen-induced proliferation of murine T cells *in vitro. Clin Diagn Lab Immunol.* (2001) 8:1081–8. doi: 10.1128/CDLI.8.6.1-91-1088.2001
26. Bailo R, Bhatt A, Ainsa JA. Lipid transport in *Mycobacterium tuberculosis* and its implications in virulence and drug development. *Biochem Pharmacol.* (2015) 96:159–67. doi: 10.1016/j.bcp.2015.05.001
27. Papa F, Cruaud P, David HL. Antigenicity and specificity of selected glycolipid fractions from *Mycobacterium tuberculosis. Res Microbiol.* (1989) 140:569–78. doi: 10.1016/0923-2508(89)90089-2
28. Rousseau C, Neyrolles O, Bordat Y, Giroux S, Sirakova TD, Prevost MC, et al. Deficiency in mycolipenate- and mycosanoate-derived acyltrehaloses enhances early interactions of *Mycobacterium tuberculosis* with host cells. *Cell Microbiol.* (2003) 5:405–15. doi: 10.1046/j.1462-5822.2003.00289.x
29. Hatzios SK, Schelle MW, Holsclaw CM, Behrens CR, Botyanszki Z, Lin FL, et al. PapA3 is an acyltransferase required for polyacyltrehalose biosynthesis in *Mycobacterium tuberculosis. J Biol Chem.* (2009) 284:12745–51. doi: 10.1074/jbc.M809088200
30. Dubey VS, Sirakova TD, Kolattukudy PE. Disruption of msl3 abolishes the synthesis of mycolipanoic and mycolipenic acids required for polyacyltrehalose synthesis in *Mycobacterium tuberculosis* H37Rv and causes cell aggregation. *Mol Microbiol.* (2002) 45:1451–9. doi: 10.1046/j.1365-2958.2002.03119.x
31. Brodin P, Poquet Y, Levillain F, Peguillet I, Larrouy-Maumus G, Gilleron M, et al. High content phenotypic cell-based visual screen identifies *Mycobacterium tuberculosis* acyltrehalose-containing glycolipids involved in phagosome remodeling. *PLoS Pathog.* (2010) 6:e1001100. doi: 10.1371/journal.ppat.1001100
32. Muñoz M, Luquin M, García-Barcelo M, Julián E, Ausina V, Lanéelle MA. Distribution of surface-exposed antigenic glycolipids in recent clinical isolates off *Mycobacterium tuberculosis. Res Microbiol.* (1997) 148:405–12. doi: 10.1016/S0923-2508(97)83871-5
33. Julian E, Matas L, Perez A, Alcaide J, Laneelle MA, Luquin M. Serodiagnosis of tuberculosis: comparison of immunoglobulin A (IgA) response to sulfolipid I with IgG and IgM responses to 2,3-diacyltrehalose, 2,3,6-triacyltrehalose, and cord factor antigens. *J Clin Microbiol.* (2002) 40:3782–8. doi: 10.1128/JCM.40.10.3782-3788.2002
34. Sempere M, Ariza M, Sanchez-Monton T, Rodriguez-Gonzalez T, Martin-Luengo F, Valero-Guillen PL. Evaluacion de aciltrehalosas de *Mycobacterium fortuitum* en el diagnostico serologico de tuberculosis pulmonar. *Enferm Infecc Microbiol Clin.* (1995) 13:292–6.
35. Ridell M, Wallerstrom G, Minnikin DE, Bolton RC, Magnusson M. A comparative serological study of antigenic glycolipids from *Mycobacterium tuberculosis. Tubercle Lung Dis.* (1992) 73:101–5. doi: 10.1016/0962-8479(92)90063-P
36. Tortola MT, Laneelle MA, Martin-Casabona N. Comparison of two 2,3-diacyl trehalose antigens from *Mycobacterium tuberculosis* and *Mycobacterium fortuitum* for serology in tuberculosis patients. *Clin Diagn Lab Immunol.* (1996) 3:563–6.
37. Hunter RL, Olsen MR, Jagannath C, Actor JK. Multiple roles of cord factor in the pathogenesis of primary, secondary, and cavitary tuberculosis, including a revised description of the pathology of secondary disease. *Ann Clin Lab Sci.* (2006) 36:371–86. Available online at: http://www.annclinlabsci.org/content/36/4/371.full.pdf+html
38. Bloch H. Studies on the virulence of tubercle bacilli: isolation and biological properties of a constituent of virulent organisms. *J Exp Med.* (1950) 91:197–219. doi: 10.1084/jem.91.2.197
39. Noll H, Bloch H, Asselineau J, Lederer E. The chemical structure of the cord factor of Mycobacterium tuberculosis. *Biochim. Biophys. Acta.* (1956) 20:299–309. doi: 10.1016/0006-3002(56)90289-X
40. Mdluli K, Slayden RA, Zhu Y, Ramaswamy S, Pan X, Mead D, et al. Inhibition of a *Mycobacterium tuberculosis* β-ketoacyl ACP synthase by isoniazid. *Science.* (1998) 280:1607–10. doi: 10.1126/science.280.5369.1607
41. Artman M, Bekierkunst A, Goldenberg I. Tissue methabolism in infection: Biochemical changes in mice treated with cord factor. *Arch Biochem Biophys.* (1964) 105:80–5. doi: 10.1016/0003-9861(64)90237-1
42. Yuan Y, Zhu Y, Crane DD, Barry CEI. The effect of oxygenated mycolic acid composition on cell wall function and macrophage growth in *Mycobacterium tuberculosis. Mol Microbiol.* (1998) 29:1449–58. doi: 10.1046/j.1365-2958.1998.01026.x
43. Kordulakova J, Gilleron M, Puzo G, Brennan PJ, Gicquel B, Mikusova K, et al. Identification of the required acyltransferase step in the biosynthesis of the phosphatidylinositol mannosides of mycobacterium species. *J Biol Chem.* (2003) 278:36285–95. doi: 10.1074/jbc.M303639200
44. Rajni, Rao N, Meena LS. Biosynthesis and virulent behavior of lipids produced by *Mycobacterium tuberculosis*: LAM and cord factor: an overview. *Biotechnol Res Int.* (2011) 2011:274693. doi: 10.4061/2011/274693
45. Actor JK. Trehalose dimycolate (cord factor) as a contributing factor to tuberculosis pathogenesis. In: Cirillo JD, Kong Y, editors. *Tuberculosis Host-Pathogen Interactions.* Cham: Springer International Publishing. (2019). p. 43–61. doi: 10.1007/978-3-030-25381-3_3
46. Marrakchi H, Laneelle MA, Daffe M. Mycolic acids: structures, biosynthesis, and beyond. *Chem Biol.* (2014) 21:67–85. doi: 10.1016/j.chembiol.2013.11.011
47. Takayama KCW, Besra GS. Pathway to synthesis and processing of mycolic acids in *Mycobacterium tuberculosis. Clin Microbiol Rev.* (2005) 18:81–101. doi: 10.1128/CMR.18.1.81-101.2005
48. Fujita Y, Okamoto Y, Uenishi Y, Sunagawa M, Uchiyama T, Yano I. Molecular and supra-molecular structure related differences in toxicity and granulomatogenic activity of mycobacterial cord factor in mice. *Microb Pathog.* (2007) 43:10–21. doi: 10.1016/j.micpath.2007.02.006
49. Kato M. Action of a toxic glycolipid of Corynebacterium diphtheriae on mitochondrial structure and function. *J Bacteriol.* (1970) 101:709–16.
50. Silva CL, Ekizlerian SM, Fazioli RA. Role of cord factor in the modulation of infection caused by mycobacteria. *Am J Pathol.* (1985) 118:238–47.
51. Indrigo J, Hunter RL Jr, Actor JK. Influence of trehalose 6,6'-dimycolate (TDM) during mycobacterial infection of bone marrow macrophages. *Microbiology.* (2002) 148(Pt 7):1991-8. doi: 10.1099/00221287-148-7-1991
52. Fischer K, Chatterjee D, Torrelles J, Brennan PJ, Kaufmann SHE, Schaible UE. Mycobacterial lysocardiolipin is exported from phagosomes upon cleavage of cardiolipin by a macrophage-derived lysosomal phospholipase

A_2. *J Immunol.* (2001) 167:2187–92. doi: 10.4049/jimmunol.167.4.2187
53. Sakamoto K, Geisel RE, Kim MJ, Wyatt BT, Sellers LB, Smiley ST, et al. Fibrinogen regulates the cytotoxicity of mycobacterial trehalose dimycolate but is not required for cell recruitment, cytokine response, or control of mycobacterial infection. *Infect Immun.* (2010) 78:1004–11. doi: 10.1128/IAI.00451-09
54. Hunter RL, Olsen M, Jagannath C, Actor JK. Trehalose 6,6'-dimycolate and lipid in the pathogenesis of caseating granulomas of tuberculosis in mice. *Am J Pathol.* (2006) 168:1249–61. doi: 10.2353/ajpath.2006.050848
55. Spargo BJ, Crowe LM, Ioneda T, Beaman BL, Crowe JH. Cord factor (*a,a*-trehalose 6,6'-dimycolate) inhibits fusion between phospholipid vesicles. *Proc Natl Acad Sci USA.* (1991) 88:737–40. doi: 10.1073/pnas.88.3.737
56. Yamasaki S, Ishikawa E, Sakuma M, Hara H, Ogata K, Saito T. Mincle is an ITAM-coupled activating receptor that senses damaged cells. *Nat Immunol.* (2008) 9:1179–88. doi: 10.1038/ni.1651
57. Ishikawa E, Ishikawa T, Morita YS, Toyonaga K, Yamada H, Takeuchi O, et al. Direct recognition of the mycobacterial glycolipid, trehalose dimycolate, by C-type lectin Mincle. *J Exp Med.* (2009) 206:2879–88. doi: 10.1084/jem.20091750
58. Schoenen H, Bodendorfer B, Hitchens K, Manzanero S, Werninghaus K, Nimmerjahn F, et al. Cutting edge: mincle is essential for recognition and adjuvanticity of the mycobacterial cord factor and its synthetic analog trehalose-dibehenate. *J Immunol.* (2010) 184:2756–60. doi: 10.4049/jimmunol.0904013
59. Welsh KJ, Abbott AN, Hwang SA, Indrigo J, Armitige LY, Blackburn MR, et al. A role for tumour necrosis factor-alpha, complement C5 and interleukin-6 in the initiation and development of the mycobacterial cord factor trehalose 6,6′-dimycolate induced granulomatous response. *Microbiology.* (2008) 154:1813–24. doi: 10.1099/mic.0.2008/016923-0
60. Indrigo J, Hunter RL Jr, Actor JK. Cord factor trehalose 6,6′-dimycolate (TDM) mediates trafficking events during mycobacterial infection of murine macrophages. *Microbiology.* (2003) 149:2049–59. doi: 10.1099/mic.0.26226-0
61. Lee WB, Kang JS, Yan JJ, Lee MS, Jeon BY, Cho SN, et al. Neutrophils promote mycobacterial trehalose dimycolate-induced lung inflammation via the Mincle pathway. *PLoS Pathog.* (2012) 8:e1002614. doi: 10.1371/journal.ppat.1002614
62. Ozeki Y, Kaneda K, Fujiwara N, Morimoto M, Oka S, Yano I. *In vivo* induction of apoptosis in the thymus by administration of mycobacterial cord factor (trehalose 6,6′-dimycolate). *Infect Immun.* (1997) 65:1793–9.
63. Bekierkunst A. Acute granulomatous response produced in mice by trehalose-6,6-dimycolate. *J Bacteriol.* (1968) 96:958–61.
64. Sakaguchi I, Ikeda N, Nakayama M, Kato Y, Yano I, Kaneda K. Trehalose 6,6'-dimycolate (Cord factor) enhances neovascularization through vascular endothelial growth factor production by neutrophils and macrophages. *Infect Immun.* (2000) 68:2043–52. doi: 10.1128/IAI.68.4.2043-2052.2000
65. Werninghaus K, Babiak A, Gross O, Holscher C, Dietrich H, Agger EM, et al. Adjuvanticity of a synthetic cord factor analogue for subunit *Mycobacterium tuberculosis* vaccination requires FcRgamma-Syk-Card9-dependent innate immune activation. *J Exp Med.* (2009) 206:89–97. doi: 10.1084/jem.20081445
66. Bekierkunst A, Levij IS, Yarkoni E. Suppression of urethan-induced lung adenomas in mice treated with trehalose-6,6-dimycolate (cord factor) and living bacillus Calmette Guerin. *Science.* (1971) 174:1240–2. doi: 10.1126/science.174.4015.1240
67. Bekierkunst A, Yarkoni E, Flechner I, Morecki S, Vilkas E, Lederer E. Immune response to sheep red blood cells in mice pretreated with mycobacterial fractions. *Infect Immun.* (1971) 4:256–63.
68. Silva CL, Faccioli LH. Tumor necrosis factor (cachectin) mediates induction of cachexia by cord factor from mycobacteria. *Infect Immun.* (1988) 56:3067–71.
69. Parant M, Parant F, Chedid L, Drapier JC, Petit JF, Wietzerbin J, et al. Enhancement of nonspecific immunity to bacterial infection by cord factor (6,6'-trehalose dimycolate). *J Infect Dis.* (1977) 135:771–7. doi: 10.1093/infdis/135.5.771
70. Parant M, Audibert F, Parant F, Chedid L, Soler E, Polonsky J, et al. Nonspecific immunostimulant activities of synthetic trehalose-6,6'-diesters (lower homologs of cord factor). *Infect Immun.* (1978) 20:12–9.
71. Granger DL, Yamamoto KI, Ribi E. Delayed hypersensitivity and granulomatous response after immunization with protein antigens associated with a mycobacterial glycolipid and oil droplets. *J Immunol.* (1976) 116:482–8.
72. Yarkoni E, Bekierkunst A. Nonspecific resistance against infection with *Salmonella typhi* and *Salmonella typhimurium* induced in mice by cord factor (trehalose-6,6'-dimycolate) and its analogues. *Infect Immun.* (1976) 14:1125–9.
73. Rao V, Fujiwara N, Porcelli SA, Glickman MS. *Mycobacterium tuberculosis* controls host innate immune activation through cyclopropane modification of a glycolipid effector molecule. *J Exp Med.* (2005) 201:535–43. doi: 10.1084/jem.20041668
74. Sakamoto K, Kim MJ, Rhoades ER, Allavena RE, Ehrt S, Wainwright HC, et al. Mycobacterial trehalose dimycolate reprograms macrophage global gene expression and activates matrix metalloproteinases. *Infect Immun.* (2013) 81:764–76. doi: 10.1128/IAI.00906-12
75. Crick DC, Brennan PJ, McNeil MR. The cell wall of *Mycobacterium tuberculosis.* In: Rom WM, Garay SM, editors. *Tuberculosis,* 2nd ed. Philadelphia: Lippincott Williams and Wilkins (2003).
76. Walker RW, Prome JC, Lacave CS. Biosynthesis of mycolic acids. Formation of a C32 beta-keto ester from palmitic acid in a cell-free system of *Corynebacterium diphtheriae. Biochim Biophys Acta.* (1973) 326:52–62. doi: 10.1016/0005-2760(73)90027-1
77. Su C-C, Klenotic PA, Bolla JR, Purdy GE, Robinson CV, Yu EW. MmpL3 is a lipid transporter that binds trehalose monomycolate and phosphatidylethanolamine. *Proc Natl Acad Sci USA.* (2019) 116:11241–6. doi: 10.1073/pnas.1901346116
78. Sathyamoorthy N, Takayama K. Purification and characterization of a novel mycolic acid exchange enzyme from *Mycobacterium smegmatis. J Biol Chem.* (1987) 262:13417–23.
79. Belisle JT, Vissa VD, Sievert T, Takayama K, Brennan PJ, Besra GS. Role of the major antigen of *Mycobacterium tuberculosis* in cell wall biogenesis. *Science.* (1997) 276:1420–2. doi: 10.1126/science.276.5317.1420
80. Numata F, Nishimura K, Ishida H, Ukei S, Tone Y, Ishihara C, et al. Lethal and adjuvant activities of cord factor (trehalose-6,6'-dimycolate) and synthetic analogs in mice. *Chem Pharm Bull.* (1985) 33:4544–55. doi: 10.1248/cpb.33.4544
81. Sueoka E, Nishiwaki S, Okabe S, Iida N, Suganuma M, Yano I, et al. Activation of protein kinase C by mycobacterial cord factor, trehalose 6-monomycolate, resulting in tumor necrosis factor-α release in mouse lung tissues. *Jpn J Cancer Res.* (1995) 86:749–55. doi: 10.1111/j.1349-7006.1995.tb02464.x
82. Rhoades ER, Geisel RE, Butcher BA, McDonough S, Russell DG. Cell wall lipids from *Mycobacterium bovis* BCG are inflammatory when inoculated within a gel matrix: characterization of a new model of the granulomatous response to mycobacterial components. *Tuberculosis.* (2005) 85:159–76. doi: 10.1016/j.tube.2004.10.001
83. Lemaire G, Tenu JP, Petit JF, Lederer E. Natural and synthetic trehalose diesters as immunomodulators. *Med Res Rev.* (1986) 6:243–74. doi: 10.1002/med.2610060302
84. Fiolek TJ, Banahene N, Kavunja HW, Holmes NJ, Rylski AK, Pohane AA, et al. Engineering the mycomembrane of live mycobacteria with an expanded set of trehalose monomycolate analogues. *Chembiochem.* (2019) 20:1282–91. doi: 10.1002/cbic.201800687
85. Daffe M, McNeil M, Brennan PJ. Novel type-specific lipooligosaccharides from *Mycobacterium tuberculosis. Biochemistry.* (1991) 30:378–88. doi: 10.1021/bi00216a011
86. Ren H, Dover LG, Islam ST, Alexander DC, Chen JM, Besra GS, et al. Identification of the lipooligosaccharide biosynthetic gene cluster from *Mycobacterium marinum. Mol Microbiol.* (2007) 63:1345–59. doi: 10.1111/j.1365-2958.2007.05603.x
87. Hunter SW, Murphy RC, Clay K, Goren MB, Brennan PJ. Trehalose-containing lipooligosaccharides. A new class of species-specific antigens from Mycobacterium. *J Biol Chem.* (1983) 258:10481–7.
88. Saadat S, Ballou CE. Pyruvylated glycolipids from *Mycobacterium smegmatis.* Structures of two oligosaccharide components. *J Biol Chem.* (1983) 258:1813–8.

89. Daffé M. Lemassu A. Glycobiology of the mycobacterial surface: structures and biological activities of the cell envelope glycoconjugates. In: Doyle RJ, editor. *Glycomicrobiology*. Boston, MA: Springer. (2000).
90. Etienne G, Malaga W, Laval F, Lemassu A, Guilhot C, Daffe M. Identification of the polyketide synthase involved in the biosynthesis of the surface-exposed lipooligosaccharides in mycobacteria. *J Bacteriol.* (2009) 191:2613–21. doi: 10.1128/JB.01235-08
91. Ortalo-Magne A, Lemassu A, Laneelle M-A, Bardou F, Silve G, Gounon P, et al. Identification of the surface-exposed lipids on the cell envelopes of *Mycobacterium tuberculosis* and other mycobacterial species. *J Bacteriol.* (1996) 178:456–61. doi: 10.1128/jb.178.2.456-461.1996
92. Belisle JT, Brennan PJ. Chemical basis of rough and smooth variation in mycobacteria. *J Bacteriol.* (1989) 171:3465–70. doi: 10.1128/jb.171.6.3465-3470.1989
93. Lemassu A, Lévy-Frébault VV, Lanéelle M-A, Daffé M. Lack of correlation between colony morphology and lipooligosaccharide content in the *Mycobacterium tuberculosis* complex. *J Gen Microbiol.* (1992) 138:1535–41. doi: 10.1099/00221287-138-7-1535
94. van Soolingen D, Hoogenboezem T, de Haas PE, Hermans PW, Koedam MA, Teppema KS, et al. A novel pathogenic taxon of the *Mycobacterium tuberculosis* complex, Canetti: characterization of an exceptional isolate from Africa. *Int J Syst Bacteriol.* (1997) 47:1236–45. doi: 10.1099/00207713-47-4-1236
95. Pfyffer GE, Auckenthaler R, van Embden JD, van Soolingen D. *Mycobacterium canettii*, the smooth variant of *M. tuberculosis*, isolated from a Swiss patient exposed in Africa. *Emerg Infect Dis.* (1998) 4:631–4. doi: 10.3201/eid0404.980414
96. Gilleron M, Vercauteren J, Puzo G. Lipo-oligosaccharidic antigen from *Mycobacterium gastri*. Complete structure of a novel C4-branched 3,6-dideoxy-β- *xylo*-hexopyranose. *Biochemistry.* (1994) 33:1930–7. doi: 10.1021/bi00173a041
97. Hunter SW, Jardine I, Yanagihara DL, Brennan PJ. Trehalose-containing lipooligosaccharides from mycobacteria: structures of the oligosaccharide segments and recognition of a unique N-acylkanosamine-containing epitope. *Biochemistry.* (1985) 24:2798–805. doi: 10.1021/bi00332a030
98. Besra GS, Chatterjee D. Lipids and carbohydrates of *Mycobacterium tuberculosis*. In: Bloom BR, editor. *Tuberculosis: Pathogenesis, Protection and Control.* Washington, DC: American Society for Microbiology. (1994). p. 285–306. doi: 10.1128/9781555818357.ch20
99. Collins FM, Cunningham DS. Systemic *Mycobacterium kansasii* infection and regulation of the alloantigenic response. *Infect Immun.* (1981) 32:614–24.
100. Minnikin DE, Lee OYC, Wu HHT, Besra GS, Bhatt A, Nataraj V, et al. Ancient mycobacterial lipids: key reference biomarkers in charting the evolution of tuberculosis. *Tuberculosis.* (2015) 95(Suppl. 1):S133–9. doi: 10.1016/j.tube.2015.02.009
101. Falkinham JO, 3rd. Mycobacterial aerosols and respiratory disease. *Emerg Infect Dis.* (2003) 9:763–7. doi: 10.3201/eid0907.020415
102. Jankute M, Nataraj V, Lee OYC, Wu HHT, Ridell M, Garton NJ, et al. The role of hydrophobicity in tuberculosis evolution and pathogenicity. *Sci Rep.* (2017) 7:1315. doi: 10.1038/s41598-017-01501-0
103. Middlebrook G, Coleman CM, Schaefer WB. Sulfolipid from virulent *Tubercle bacilli. Proc Natl Acad Sci USA.* (1959) 45:1801–4. doi: 10.1073/pnas.45.12.1801
104. Goren MB. Sulfolipid I of *Mycobacterium tuberculosis*, strain H37Rv. II. Structural studies. *Biochim Biophys Acta.* (1970) 210:127–38. doi: 10.1016/0005-2760(70)90068-8
105. Goren MB. Sulfolipid I of *Mycobacterium tuberculosis*, strain H37Rv. I. Purification and properties. *Biochim Biophys Acta.* (1970) 210:116–26. doi: 10.1016/0005-2760(70)90067-6
106. Mougous JD, Petzold CJ, Senaratne RH, Lee DH, Akey DL, Lin FL, et al. Identification, function and structure of the mycobacterial sulfotransferase that initiates sulfolipid-1 biosynthesis. *Nat Struct Mol Biol.* (2004) 11:721–9. doi: 10.1038/nsmb802
107. Kumar P, Schelle MW, Jain M, Lin FL, Petzold CJ, Leavell MD, et al. PapA1 and PapA2 are acyltransferases essential for the biosynthesis of the *Mycobacterium tuberculosis* virulence factor sulfolipid-1. *Proc Natl Acad Sci USA.* (2007) 104:11221–6. doi: 10.1073/pnas.0611649104
108. Converse SE, Mougous JD, Leavell MD, Leary JA, Bertozzi CR, Cox JS. MmpL8 is required for sulfolipid-1 biosynthesis and *Mycobacterium tuberculosis* virulence. *Proc Natl Acad Sci USA.* (2003) 100:6121–6. doi: 10.1073/pnas.1030024100
109. Goren MB, Mor N. Influence of phagosomal contents on the apparent inhibition of phagosome-lysosome fusion mediated by polyanionic substances in mouse peritoneal macrophages. *Biochem Cell Biol.* (1990) 68:24–32. doi: 10.1139/o90-004
110. Leigh CD, Bertozzi CR. Synthetic studies toward *Mycobacterium tuberculosis* sulfolipid-I. *J Org Chem.* (2008) 73:1008–17. doi: 10.1021/jo702032c
111. Goren MB, Hart PD, Young MR, Armstrong JA. Prevention of phagosome-lysosome fusion in cultured macrophages by sulfatides of *Mycobacterium tuberculosis. Proc Natl Acad Sci USA.* (1976) 73:2510–4. doi: 10.1073/pnas.73.7.2510
112. Kato M, Goren MB. Enhancement of cord factor-toxicity by sulfolipid. *Jpn J Med Sci Biol.* (1974) 27:120–4.
113. Kato M, Goren MB. Synergistic action of cord factor and mycobacterial sulfatides on mitochondria. *Infect Immun.* (1974) 10:733–41.
114. Brozna JP, Horan M, Rademacher JM, Pabst KA, Pabst MJ. Monocyte responses to sulfatide from *Mycobacterium tuberculosis*: inhibition of priming for enhanced release of superoxide, associated with increased secretion of interleukin-1 and tumor necrosis factor alpha, and altered protein phosphorylation. *Infect Immun.* (1991) 59:2542–8.
115. Pabst JJ, Gross JM, Prozna JP, Goren MB. Inhibition of macrophage priming by sulfatide from *Mycobacterium tuberculosis. J Immunol.* (1988) 140:634–40.
116. Zhang L, English D, Andersen BR. Activation of human neutrophils by *Mycobacterium tuberculosis*-derived sulfolipid-1. *J Immunol.* (1991) 146:2730–6.
117. Zhang L, Goren MB, Holzer TJ, Andersen BR. Effect of *Mycobacterium tuberculosis*-derived sulfolipid I on human phagocytic cells. *Infect Immun.* (1988) 56:2876–83.
118. Julian E, Matas L, Alcaide J, Luquin M. Comparison of antibody responses to a potential combination of specific glycolipids and proteins for test sensitivity improvement in tuberculosis serodiagnosis. *Clin Diagnos Lab Immunol.* (2004) 11:70–6. doi: 10.1128/CDLI.11.1.70-76.2004
119. Gilleron M, Stenger S, Mazorra Z, Wittke F, Mariotti S, Bohmer G, et al. Diacylated sulfoglycolipids are novel mycobacterial antigens stimulating CD1-restricted T cells during infection with *Mycobacterium tuberculosis. J Exp Med.* (2004) 199:649–59. doi: 10.1084/jem.20031097
120. Gangadharam PR, Cohn ML, Middlebrook G. Infectivity, Pathogenicity and sulpholipid fraction of some Indian and British strains of *Tubercle bacilli. Tubercle.* (1963) 44:452–5. doi: 10.1016/S0041-3879(63)80087-2
121. Goren MB, Brokl O, Schaefer WB. Lipids of putative relevance to virulence in *Mycobacterium tuberculosis*: correlation of virulence with elaboration of sulfatides and strongly acidic lipids. *Infect Immun.* (1974) 9:142–9.
122. Rousseau C, Sirakova TD, Dubey VS, Bordat Y, Kolattukudy PE, Gicquel B, et al. Virulence attenuation of two Mas-like polyketide synthase mutants of *Mycobacterium tuberculosis. Microbiology.* (2003) 149:1837–47. doi: 10.1099/mic.0.26278-0
123. Queiroz A, Riley LW. Bacterial immunostat: *Mycobacterium tuberculosis* lipids and their role in the host immune response. *Rev Soc Bras Med Trop.* (2017) 50:9–18. doi: 10.1590/0037-8682-0230-2016
124. Minnikin DE, Dobson G, Sesardic D, Ridell M. Mycolipenates and mycolipanolates of trehalose from *Mycobacterium tuberculosis. J Gen Microbiol.* (1985) 131:1369–74. doi: 10.1099/00221287-131-6-1369
125. Daffe M, Lacave C, Laneelle MA, Gillois M, Laneelle G. Polyphthienoyl trehalose, glycolipids specific for virulent strains of the tubercle bacillus. *Eur J Biochem.* (1988) 172:579–84. doi: 10.1111/j.1432-1033.1988.tb13928.x
126. Arbues A, Lugo-Villarino G, Neyrolles O, Guilhot C, Astarie-Dequeker C. Playing hide-and-seek with host macrophages through the use of mycobacterial cell envelope phthiocerol dimycocerosates and phenolic glycolipids. *Front Cell Infect Microbiol.* (2014) 4:173. doi: 10.3389/fcimb.2014.00173
127. Hunter SW, Brennan PJ. Further specific extracellular phenolic glycolipid antigens and a related diacylphthiocerol from *Mycobacterium leprae. J Biol Chem.* (1983) 258:7556–62.

128. Hunter SW, Fujiwara T, Brennan PJ. Structure and antigenicity of the major specific glycolipid antigen of *Mycobacterium leprae*. *J Biol Chem.* (1983) 257:15072–8.
129. Hunter SW, Brennan PJ. A novel phenolic glycolipid from *Mycobacterium leprae* possibly involved in immunogenicity and pathogenicity. *J Bacteriol.* (1981) 147:728–35.
130. Cho S-N, Yanagihara DL, Hunter SW, Gelber RH, Brennan PJ. Serological specificity of phenolic glycolipid 1 from *Mycobacterium leprae* and use in serodiagnosis of leprosy. *Infect Immun.* (1983) 41:1077–83.
131. Madigan CA, Cambier CJ, Kelly-Scumpia KM, Scumpia PO, Cheng T-Y, Zailaa J, et al. A macrophage response to *Mycobacterium leprae* phenolic glycolipid initiates nerve damage in leprosy. *Cell.* (2017) 170:973–85.e10. doi: 10.1016/j.cell.2017.07.030
132. Chan J, Fujiwara T, Brennan P, McNeil M, Turco SJ, Sibille J-C, et al. Microbial glycolipids: possible virulence factors that scavenge oxygen radicals. *Proc Natl Acad Sci USA.* (1989) 86:2453–7. doi: 10.1073/pnas.86.7.2453
133. Robinson N, Kolter T, Wolke M, Rybniker J, Hartmann P, Plum G. Mycobacterial phenolic glycolipid inhibits phagosome maturation and subverts the pro-inflammatory cytokine response. *Traffic.* (2008) 9:1936–47. doi: 10.1111/j.1600-0854.2008.00804.x
134. Cambier CJ, O'Leary SM, O'Sullivan MP, Keane J, Ramakrishnan L. Phenolic glycolipid facilitates mycobacterial escape from microbicidal tissue-resident macrophages. *Immunity.* (2017) 47:552–65.e4. doi: 10.1016/j.immuni.2017.08.003
135. Singh AK, Bishai WR. Partners in crime: phenolic glycolipids and macrophages. *Trends Mol Med.* (2017) 23:981–3. doi: 10.1016/j.molmed.2017.09.003
136. Bolz M, Ernst JD. Fishing for answers in human mycobacterial infections. *Immunity.* (2017) 47:395–7. doi: 10.1016/j.immuni.2017.09.005
137. Daffe M, Lacave C, Laneelle M, Laneelle G. Structure of the major triglycosyl phenol-phthiocerol of *Mycobacterium tuberculosis* (strain Canetti). *Eur J Biochem.* (1987) 167:155–60. doi: 10.1111/j.1432-1033.1987.tb13317.x
138. Cho S-N, Shin J-S, Daffe M, Chong Y, Kim S-K, Kim J-D. Production of monoclonal antibody to a phenolic glycolipid of *Mycobacterium tuberculosis* and its use in detection of the antigen in clinical isolates. *J Clin Microbiol.* (1992) 30:3065–9.
139. Reed MB, Domenech P, Manca C, Su H, Barczak AK, Kreiswirth BN, et al. A glycolipid of hypervirulent tuberculosis strains that inhibits the innate immune response. *Nature.* (2004) 431:84–7. doi: 10.1038/nature02837
140. Reed MB, Gagneux S, Deriemer K, Small PM, Barry CE, III. The W-Beijing lineage of *Mycobacterium tuberculosis* overproduces triglycerides and has the DosR dormancy regulon constitutively upregulated. *J Bacteriol.* (2007) 189:2583–9. doi: 10.1128/JB.01670-06
141. Sinsimer D, Huet G, Manca C, Tsenova L, Koo MS, Kurepina N, et al. The phenolic glycolipid of *Mycobacterium tuberculosis* differentially modulates the early host cytokine response but does not in itself confer hypervirulence. *Infect Immun.* (2008) 76:3027–36. doi: 10.1128/IAI.01663-07
142. Domingo-Gonzalez R, Das S, Griffiths KL, Ahmed M, Bambouskova M, Gopal R, et al. Interleukin-17 limits hypoxia-inducible factor 1alpha and development of hypoxic granulomas during tuberculosis. *JCI Insight.* (2017) 2:92973. doi: 10.1172/jci.insight.92973
143. Elsaidi HRH, Lowary TL. Inhibition of cytokine release by *Mycobacterium tuberculosis* phenolic glycolipid analogues. *Chembiochem.* (2014) 15:1176–82. doi: 10.1002/cbic.201402001
144. Torrelles JB, Knaup R, Kolareth A, Slepushkina T, Kaufman TM, Kang PB, et al. Identification of *Mycobacterium tuberculosis* clinical isolates with altered phagocytosis by human macrophages due to a truncated lipoarabinomannan. *J Biol Chem.* (2008) 283:31417–28. doi: 10.1074/jbc.M806350200
145. Arbues A, Malaga W, Constant P, Guilhot C, Prandi J, Astarie-Dequeker C. Trisaccharides of phenolic glycolipids confer advantages to pathogenic mycobacteria through manipulation of host-cell pattern-recognition receptors. *ACS Chem Biol.* (2016) 11:2865–75. doi: 10.1021/acschembio.6b00568
146. Tabouret G, Astarie-Dequeker C, Demangel C, Malaga W, Constant P, Ray A, et al. *Mycobacterium leprae* phenolglycolipid-1 expressed by engineered *M. bovis* BCG modulates early interaction with human phagocytes. *PLoS Pathogens.* (2010) 6:e1001159. doi: 10.1371/journal.ppat.1001159
147. Zheng RB, Jegouzo SAF, Joe M, Bai Y, Tran H-A, Shen K, et al. Insights into interactions of mycobacteria with the host innate immune system from a novel array of synthetic *Mycobacterial glycans*. *ACS Chem Biol.* (2017) 12:2990–3002. doi: 10.1021/acschembio.7b00797
148. Tran V, Ahn SK, Ng M, Li M, Liu J. Loss of lipid virulence factors reduces the efficacy of the BCG vaccine. *Sci Rep.* (2016) 6:29076. doi: 10.1038/srep29076
149. Ortalo-Magne A, Andersen AB, Daffe M. The outermost capsular arabinomannans and other mannoconjugates of virulent and avirulent tubercle bacilli. *Microbiology.* (1996) 142:927–35. doi: 10.1099/00221287-142-4-927
150. Khoo K-H, Dell A, Morris HR, Brennan PJ, Chatterjee D. Structural definition of acylated phosphatidylinositol mannosides from *Mycobacterium tuberculosis*: definition of a common anchor for lipomannan and lipoarabinomannan. *Glycobiology.* (1995) 5:117–27. doi: 10.1093/glycob/5.1.117
151. Gilleron M, Nigou J, Cahuzac B, Puzo G. Structural study of the lipomannans from Mycobacterium bovis BCG: characterisation of multiacylated forms of the phosphatidyl-*myo*-inositol anchor. *J Mol Biol.* (1999) 285:2147–60. doi: 10.1006/jmbi.1998.2438
152. Torrelles JB, Azad AK, Schlesinger LS. Fine discrimination in the recognition of individual species of phosphatidyl-*myo*-inositol mannosides from *Mycobacterium tuberculosis* by C-type lectin pattern recognition receptors. *J Immunol.* (2006) 177:1805–16. doi: 10.4049/jimmunol.177.3.1805
153. Villeneuve C, Gilleron M, Maridonneau-Parini I, Daffe M, Astarie-Dequeker C, Etienne G. Mycobacteria use their surface-exposed glycolipids to infect human macrophages through a receptor-dependent process. *J Lipid Res.* (2005) 46:475–83. doi: 10.1194/jlr.M400308-JLR200
154. Vergne I, Fratti RA, Hill PJ, Chua J, Belisle J, Deretic V. *Mycobacterium tuberculosis* phagosome maturation arrest: mycobacterial phosphatidylinositol analog phosphatidylinositol mannoside stimulates early endosomal fusion. *Mol Biol Cell.* (2004) 15:751–60. doi: 10.1091/mbc.e03-05-0307
155. Turner J, Torrelles JB. Mannose-capped lipoarabinomannan in *Mycobacterium tuberculosis* pathogenesis. *Pathog Dis.* (2018) 76:fty026. doi: 10.1093/femspd/fty026
156. Driessen NN, Ummels R, Maaskant JJ, Gurcha SS, Besra GS, Ainge GD, et al. Role of phosphatidylinositol mannosides in the interaction between mycobacteria and DC-SIGN. *Infect Immun.* (2009) 77:4538–47. doi: 10.1128/IAI.01256-08
157. H. dlS, Mariotti S, Angenieux C, Gilleron M, Garcia-Alles LF, Malm D, et al. Assistance of microbial glycolipid antigen processing by CD1e. *Science.* (2005) 310:1321–4. doi: 10.1126/science.1115301
158. Gilleron M, Quesniaux VF, Puzo G. Acylation state of the phosphatidylinositol hexamannosides from *Mycobacterium bovis* bacillus Calmette Guerin and *Mycobacterium tuberculosis* H37Rv and its implication in Toll-like receptor response. *J Biol Chem.* (2003) 278:29880–9. doi: 10.1074/jbc.M303446200
159. Jones BW, Means TK, Heldwein KA, Keen MA, Hill PJ, Belisle JT, et al. Different toll-like receptor agonists induce distinct macrophage responses. *J Leukoc Biol.* (2001) 69:1036–44. doi: 10.1189/jlb.69.6.1036
160. Mazurek J, Ignatowicz L, Kallenius G, Svenson SB, Pawlowski A, Hamasur B. Divergent effects of mycobacterial cell wall glycolipids on maturation and function of human monocyte-derived dendritic cells. *PLoS ONE.* (2012) 7:e42515. doi: 10.1371/journal.pone.0042515
161. Kallenius G, Correia-Neves M, Buteme H, Hamasur B, Svenson SB. Lipoarabinomannan, and its related glycolipids, induce divergent and opposing immune responses to *Mycobacterium tuberculosis* depending on structural diversity and experimental variations. *Tuberculosis.* (2016) 96:120–30. doi: 10.1016/j.tube.2015.09.005
162. Guerin ME, Kordulakova J, Alzari PM, Brennan PJ, Jackson M. Molecular basis of phosphatidyl-myo-inositol mannoside biosynthesis and regulation in mycobacteria. *J Biol Chem.* (2010) 285:33577–83. doi: 10.1074/jbc.R110.168328
163. Glass LN, Swapna G, Chavadi SS, Tufariello JM, Mi K, Drumm JE, et al. *Mycobacterium tuberculosis* universal stress protein Rv2623 interacts with the putative ATP binding cassette (ABC) transporter Rv1747

to regulate mycobacterial growth. *PLoS Pathog.* (2017) 13:e1006515. doi: 10.1371/journal.ppat.1006515

164. Arcos J, Sasindran SJ, Fujiwara N, Turner J, Schlesinger LS, Torrelles JB. Human lung hydrolases delineate *Mycobacterium tuberculosis*-macrophage interactions and the capacity to control infection. *J Immunol.* (2011) 187:372–81. doi: 10.4049/jimmunol.1100823
165. Guirado E, Mbawuike U, Keiser TL, Arcos J, Azad AK, Wang SH, et al. Characterization of host and microbial determinants in individuals with latent tuberculosis infection using a human granuloma model. *MBio.* (2015) 6:e02537-14. doi: 10.1128/mBio.02537-14
166. Keiser TL, Azad AK, Guirado E, Bonacci R, Schlesinger LS. Comparative transcriptional study of the putative mannose donor biosynthesis genes in virulent *Mycobacterium tuberculosis* and attenuated Mycobacterium bovis BCG strains. *Infect Immun.* (2011) 79:4668–73. doi: 10.1128/IAI.05635-11
167. Lavollay M, Arthur M, Fourgeaud M, Dubost L, Marie A, Veziris N, et al. The peptidoglycan of stationary-phase *Mycobacterium tuberculosis* predominantly contains cross-links generated by L,D-transpeptidation. *J Bacteriol.* (2008) 190:4360–6. doi: 10.1128/JB.00239-08
168. Singh A, Crossman DK, Mai D, Guidry L, Voskuil MI, Renfrow MB, et al. *Mycobacterium tuberculosis* WhiB3 maintains redox homeostasis by regulating virulence lipid anabolism to modulate macrophage response. *PLoS Pathogens.* (2009) 5:e1000545. doi: 10.1371/journal.ppat.1000545
169. Curry JM, Whalan R, Hunt DM, Gohil K, Strom M, Rickman L, et al. An ABC transporter containing a forkhead-associated domain interacts with a serine-threonine protein kinase and is required for growth of *Mycobacterium tuberculosis* in mice. *Infect Immun.* (2005) 73:4471–7. doi: 10.1128/IAI.73.8.4471-4477.2005
170. Jackson M, Stadthagen G, Gicquel B. Long-chain multiple methyl-branched fatty acid-containing lipids of *Mycobacterium tuberculosis*: biosynthesis, transport, regulation and biological activities. *Tuberculosis.* (2007) 87:78–86. doi: 10.1016/j.tube.2006.05.003
171. Matsunaga I, Bhatt A, Young DC, Cheng TY, Eyles SJ, Besra GS, et al. *Mycobacterium tuberculosis* pks12 produces a novel polyketide presented by CD1c to T cells. *J Exp Med.* (2004) 200:1559–69. doi: 10.1084/jem.20041429
172. Fabre M, Koeck J-L, Le Fleche P, Simon F, Herve V, Vergnaud G, et al. High genetic diversity revealed by variable-number tandem repeat genotyping and analysis of hsp65 gene polymorphism in a large collection of "*Mycobacterium canettii*" strains indicates that the *M. tuberculosis* complex is a recently emerged clone of "*M. canettii*". *J Clin Microbiol.* (2004) 42:3248–55. doi: 10.1128/JCM.42.7.3248-3255.2004
173. Briquet A, Vong R, Roseau J-B, Javelle E, Cazes N, Riviere F, et al. Clinical features of *Mycobacterium canettii* infection: a retrospective study of 20 cases among French soldiers and relatives. *Clin Infect Dis.* (2019) 69:2003–10. doi: 10.1093/cid/ciz107
174. Valway SE, Sanchez MPC, Shinnick TF, Orme I, Agerton T, Hoy D, et al. An outbreak involving extensive transmission of a virulent strain of *Mycobacerium tuberculosis. N Engl J Med.* (1998) 338:633–9. doi: 10.1056/NEJM199803053381001
175. Schlesinger LS. Macrophage phagocytosis of virulent but not attenuated strains of *Mycobacterium tuberculosis* is mediated by mannose receptors in addition to complement receptors. *J Immunol.* (1993) 150:2920–30.
176. Kang PB, Azad AK, Torrelles JB, Kaufman TM, Beharka A, Tibesar E, et al. The human macrophage mannose receptor directs *Mycobacterium tuberculosis* lipoarabinomannan-mediated phagosome biogenesis. *J Exp Med.* (2005) 202:987–99. doi: 10.1084/jem.20051239
177. Notter RH. *Lung Surfactants: Basic Science and Clinical Applications.* New York, NY: Marcel Dekker (2000). p. 1–444. doi: 10.1201/9781482270426
178. Arcos J, Diangelo L, Scordo J, Sasindran J, Moliva J, Turner J, et al. Lung mucosa lining fluid modifies *Mycobacterium tuberculosis* to reprogram human neutrophil killing mechanisms. *J Infect Dis.* (2015) 212:948–58. doi: 10.1093/infdis/jiv146
179. Arcos J, Sasindran SJ, Moliva JI, Scordo JM, Sidiki S, Guo H, et al. *Mycobacterium tuberculosis* cell wall released fragments by the action of the human lung mucosa modulate macrophages to control infection in an IL-10-dependent manner. *Mucosal Immunol.* (2017) 10:1248–58. doi: 10.1038/mi.2016.115
180. Scordo JM, Arcos J, Kelley HV, Diangelo L, Sasindran SJ, Youngmin E, et al. *Mycobacterium tuberculosis* cell wall fragments released upon bacterial contact with the human lung mucosa alter the neutrophil response to infection. *Front Immunol.* (2017) 8:307. doi: 10.3389/fimmu.2017.00307
181. Moliva JI, Rajaram MV, Sidiki S, Sasindran SJ, Guirado E, Pan XJ, et al. Molecular composition of the alveolar lining fluid in the aging lung. *Age.* (2014) 36:9633. doi: 10.1007/s11357-014-9633-4
182. Moliva JI, Duncan MA, Olmo-Fontanez A, Akhter A, Arnett E, Scordo JM, et al. The lung mucosa environment in the elderly increases host susceptibility to *Mycobacterium tuberculosis* infection. *J Infect Dis.* (2019). doi: 10.1093/infdis/jiz138
183. Moliva JI, Hossfeld AP, Canan CH, Dwivedi V, Wewers MD, Beamer G, et al. Exposure to human alveolar lining fluid enhances *Mycobacterium bovis* BCG vaccine efficacy against *Mycobacterium tuberculosis* infection in a CD8(+) T-cell-dependent manner. *Mucosal Immunol.* (2018) 11:968–78. doi: 10.1038/mi.2017.80
184. Moliva JI, Hossfeld AP, Sidiki S, Canan CH, Dwivedi V, Beamer G, et al. Selective delipidation of *Mycobacterium bovis* BCG enables direct pulmonary vaccination and enhances protection against *Mycobacterium tuberculosis. Mucosal Immunol.* (2019) 12:805–15. doi: 10.1038/s41385-019-0148-2
185. Chavadi SS, Edupuganti UR, Vergnolle O, Fatima I, Singh SM, Soll CE, et al. Inactivation of tesA reduces cell wall lipid production and increases drug susceptibility in mycobacteria. *J Biol Chem.* (2011) 286:24616–25. doi: 10.1074/jbc.M111.247601
186. Barkan D, Liu Z, Sacchettini JC, Glickman MS. Mycolic acid cyclopropanation is essential for viability, drug resistance, and cell wall integrity of *Mycobacterium tuberculosis. Chem Biol.* (2009) 16:499–509. doi: 10.1016/j.chembiol.2009.04.001
187. Velayati AA, Farnia P, Ibrahim TA, Haroun RZ, Kuan HO, Ghanavi J, et al. Differences in cell wall thickness between resistant and nonresistant strains of *Mycobacterium tuberculosis*: using transmission electron microscopy. *Chemotherapy.* (2009) 55:303–7. doi: 10.1159/000226425
188. Velayati AA, Farnia P, Merza MA, Zhavnerko GK, Tabarsi P, Titov LP, et al. New insight into extremely drug-resistant tuberculosis: using atomic force microscopy. *Eur Respir J.* (2010) 36:1490–3. doi: 10.1183/09031936.00064510
189. Velayati AA, Masjedi MR, Farnia P, Tabarsi P, Ghanavi J, Ziazarifi AH, et al. Emergence of new forms of totally drug-resistant tuberculosis bacilli: super extensively drug-resistant tuberculosis or totally drug-resistant strains in iran. *Chest.* (2009) 136:420–5. doi: 10.1378/chest.08-2427
190. Torrelles JB. Broadening our view about the role of *Mycobacterium tuberculosis* cell envelope components during infection: a battle for survival. In: Cardona PJ, editor. *Understanding Tuberculosis—Analyzing the Origin of Mycobacterium tuberculosis Pathogenicity.* Rijeka, Croatia: Intech (2012). p. 1–46.
191. Jain M, Petzold CJ, Schelle MW, Leavell MD, Mougous JD, Bertozzi CR, et al. Lipidomics reveals control of *Mycobacterium tuberculosis* virulence lipids via metabolic coupling. *Proc Natl Acad Sci USA.* (2007) 104:5133–8. doi: 10.1073/pnas.0610634104
192. Serafini A, Tan L, Horswell S, Howell S, Greenwood DJ, Hunt DM, et al. *Mycobacterium tuberculosis* requires glyoxylate shunt and reverse methylcitrate cycle for lactate and pyruvate metabolism. *Mol Microbiol.* (2019) 112:1284–307. doi: 10.1111/mmi.14362
193. Sequeira PC, Senaratne RH, Riley LW. Inhibition of toll-like receptor 2 (TLR-2)-mediated response in human alveolar epithelial cells by mycolic acids and *Mycobacterium tuberculosis* mce1 operon mutant. *Pathog Dis.* (2014) 70:132–40. doi: 10.1111/2049-632X.12110
194. Dunphy KY, Senaratne RH, Masuzawa M, Kendall LV, Riley LW. Attenuation of *Mycobacterium tuberculosis* functionally disrupted in a fatty acyl-coenzyme A synthetase gene fadD5. *J Infect Dis.* (2010) 201:1232–9. doi: 10.1086/651452
195. Munoz-Elias EJ, Upton AM, Cherian J, McKinney JD. Role of the methylcitrate cycle in *Mycobacterium tuberculosis* metabolism, intracellular growth, and virulence. *Mol Microbiol.* (2006) 60:1109–22. doi: 10.1111/j.1365-2958.2006.05155.x
196. McKinney JD, Honer zu Bentrup K, Munoz-Elias EJ, Miczak A, Chen B, Chan WT, et al. Persistence of *Mycobacterium tuberculosis* in macrophages and mice requires the glyoxylate shunt enzyme isocitrate lyase. *Nature.* (2000) 406:735–8. doi: 10.1038/35021074
197. Moody DB, Reinhold BB, Guy MR, Beckman EM, Frederique DE, Furlong ST, et al. Structural requirements for glycolipid antigen

recognition by CD1b-restricted T cells. *Science.* (1997) 278:283–6. doi: 10.1126/science.278.5336.283
198. Porcelli SA, Modlin RL. The CD1 system: antigen-presenting molecules for T cell recognition of lipids and glycolipids. *Annu Rev Immunol.* (1999) 17:297–329. doi: 10.1146/annurev.immunol.17.1.297
199. Kawashima T, Norose Y, Watanabe Y, Enomoto Y, Narazaki H, Watari E, et al. Cutting edge: major CD8 T cell response to live bacillus Calmette-Guerin is mediated by CD1 molecules. *J Immunol.* (2003) 170:5345–8. doi: 10.4049/jimmunol.170.11.5345
200. Watanabe Y, Watari E, Matsunaga I, Hiromatsu K, Dascher CC, Kawashima T, et al. BCG vaccine elicits both T-cell mediated and humoral immune responses directed against mycobacterial lipid components. *Vaccine.* (2006) 24:5700–7. doi: 10.1016/j.vaccine.2006.04.049
201. Moody DB, Young DC, Cheng TY, Rosat JP, Roura-Mir C, O'Connor PB, et al. T cell activation by lipopeptide antigens. *Science.* (2004) 303:527–31. doi: 10.1126/science.1089353
202. Beckman EV, Porcelli SA, Morita CT, Behar SM, Furlong ST, Brenner MB. Recognition of a lipid antigen by CD1-restricted alpha beta$^+$ T cells. *Nature.* (1994) 372:691–4. doi: 10.1038/372691a0
203. Sieling PA, Chatterjee D, Porcelli SA, Prigozy TI, Mazzaccaro RJ, Soriano T, et al. CD1-restricted T cell recognition of microbial lipoglycan antigens. *Science.* (1995) 269:227–30. doi: 10.1126/science.7542404
204. Moody DB, Sugita M, Peters PJ, Brenner MB, Porcelli SA. The CD1-restricted T-cell response to mycobacteria. *Res Immunol.* (1996) 147:550–9. doi: 10.1016/S0923-2494(97)85221-2
205. Layre E, Collmann A, Bastian M, Mariotti S, Czaplicki J, Prandi J, et al. Mycolic acids constitute a scaffold for mycobacterial lipid antigens stimulating CD1-restricted T cells. *Chem Biol.* (2009) 16:82–92. doi: 10.1016/j.chembiol.2008.11.008
206. Moody DB, Ulrichs T, Muhlecker W, Young DC, Gurcha SS, Grant E, et al. CD1c-mediated T-cell recognition of isoprenoid glycolipids in *Mycobacterium tuberculosis* infection. *Nature.* (2000) 404:884–8. doi: 10.1038/35009119
207. Bendelac A, Savage PB, Teyton L. The biology of NKT cells. *Annu Rev Immunol.* (2007) 25:297–336. doi: 10.1146/annurev.immunol.25.022106.141711
208. Felio K, Nguyen H, Dascher CC, Choi HJ, Li S, Zimmer MI, et al. CD1-restricted adaptive immune responses to Mycobacteria in human group 1 CD1 transgenic mice. *J Exp Med.* (2009) 206:2497–509. doi: 10.1084/jem.20090898
209. Moreno-Mendieta SA, Rocha-Zavaleta L, Rodriguez-Sanoja R. Adjuvants in tuberculosis vaccine development. *FEMS Immunol Med Microbiol.* (2010) 58:75–84. doi: 10.1111/j.1574-695X.2009.00629.x
210. Ribi E, Granger DL, Milner KC, Yamamoto K, Strain SM, Parker R, et al. Induction of resistance to tuberculosis in mice with defined components of mycobacteria and with some unrelated materials. *Immunology.* (1982) 46:297–305.
211. Andersen P. Effective vaccination of mice against *Mycobacterium tuberculosis* infection with a soluble mixture of secreted mycobacterial proteins. *Infect Immun.* (1994) 62:2536–44.
212. Ishikawa E, Mori D, Yamasaki S. Recognition of mycobacterial lipids by immune receptors. *Trends Immunol.* (2017) 38:66–76. doi: 10.1016/j.it.2016.10.009
213. Agger EM. Novel adjuvant formulations for delivery of anti-tuberculosis vaccine candidates. *Adv Drug Deliv Rev.* (2016) 102:73–82. doi: 10.1016/j.addr.2015.11.012
214. Lindblad EB, Elhay MJ, Silva R, Appelberg R, Andersen P. Adjuvant modulation of immune responses to tuberculosis subunit vaccines. *Infect Immun.* (1997) 65:623–9.
215. Hwang SA, Arora R, Kruzel ML, Actor JK. Lactoferrin enhances efficacy of the BCG vaccine: comparison between two inbred mice strains (C57BL/6 and BALB/c). *Tuberculosis.* (2009) 89(Suppl. 1):S49–54. doi: 10.1016/S1472-9792(09)70012-5
216. van Dissel JT, Joosten SA, Hoff ST, Soonawala D, Prins C, Hokey DA, et al. A novel liposomal adjuvant system, CAF01, promotes long-lived *Mycobacterium tuberculosis*-specific T-cell responses in human. *Vaccine.* (2014) 32:7098–107. doi: 10.1016/j.vaccine.2014.10.036
217. van Dissel JT, Arend SM, Prins C, Bang P, Tingskov PN, Lingnau K, et al. Ag85B-ESAT-6 adjuvanted with IC31 promotes strong and long-lived *Mycobacterium tuberculosis* specific T cell responses in naive human volunteers. *Vaccine.* (2010) 28:3571–81. doi: 10.1016/j.vaccine.2010.02.094
218. van Dissel JT, Soonawala D, Joosten SA, Prins C, Arend SM, Bang P, et al. Ag85B-ESAT-6 adjuvanted with IC31(R) promotes strong and long-lived *Mycobacterium tuberculosis* specific T cell responses in volunteers with previous BCG vaccination or tuberculosis infection. *Vaccine.* (2011) 29:2100–9. doi: 10.1016/j.vaccine.2010.12.135
219. Fozo EM, Rucks EA. The Making and Taking of Lipids: The role of bacterial lipid synthesis and the harnessing of host lipids in bacterial pathogenesis. *Adv Microb Physiol.* (2016) 69:51–155. doi: 10.1016/bs.ampbs.2016.07.001
220. Kaufmann SH, Cole ST, Mizrahi V, Rubin E, Nathan C. *Mycobacterium tuberculosis* and the host response. *J Exp Med.* (2005) 201:1693–7. doi: 10.1084/jem.20050842
221. Peterson EJ, Bailo R, Rothchild AC, Arrieta-Ortiz ML, Kaur A, Pan M, et al. Path-seq identifies an essential mycolate remodeling program for mycobacterial host adaptation. *Mol Syst Biol.* (2019) 15:e8584. doi: 10.15252/msb.20188584
222. Eoh H, Wang Z, Layre E, Rath P, Morris R, Branch Moody D, et al. Metabolic anticipation in *Mycobacterium tuberculosis. Nat Microbiol.* (2017) 2:17084. doi: 10.1038/nmicrobiol.2017.84
223. Moliva JI, Turner J, Torrelles JB. Prospects in *Mycobacterium bovis* Bacille Calmette et Guerin (BCG) vaccine diversity and delivery: why does BCG fail to protect against tuberculosis? *Vaccine.* (2015) 33:5035–41. doi: 10.1016/j.vaccine.2015.08.033
224. Moliva JI, Turner J, Torrelles JB. Immune responses to BCG vaccination: why do they fail to protect against *Mycobacterium tuberculosis*? *Front Immunol.* (2017) 8:407. doi: 10.3389/fimmu.2017.00407
225. Scordo JM, Olmo-Fontanez AM, Kelley HV, Sidiki S, Arcos J, Akhter A, et al. The human lung mucosa drives differential *Mycobacterium tuberculosis* infection outcome in the alveolar epithelium. *Mucosal Immunol.* (2019) 12:795–804. doi: 10.1038/s41385-019-0156-2
226. Kaur D, Guerin ME, Skovierova H, Brennan PJ, Jackson M. Chapter 2: Biogenesis of the cell wall and other glycoconjugates of *Mycobacterium tuberculosis. Adv Appl Microbiol.* (2009) 69:23–78. doi: 10.1016/S0065-2164(09)69002-X
227. Raghunandanan S, Jose L, Gopinath V, Kumar RA. Comparative label-free lipidomic analysis of *Mycobacterium tuberculosis* during dormancy and reactivation. *Sci Rep.* (2019) 9:3660. doi: 10.1038/s41598-019-40051-5
228. Schubert OT, Ludwig C, Kogadeeva M, Zimmermann M, Rosenberger G, Gengenbacher M, et al. Absolute proteome composition and dynamics during dormancy and resuscitation of *Mycobacterium tuberculosis. Cell Host Microbe.* (2015) 18:96–108. doi: 10.1016/j.chom.2015.06.001
229. Rodriguez JG, Hernandez AC, Helguera-Repetto C, Aguilar Ayala D, Guadarrama-Medina R, Anzola JM, et al. Global adaptation to a lipid environment triggers the dormancy-related phenotype of *Mycobacterium tuberculosis. mBio.* (2014) 5:e01125-14. doi: 10.1128/mBio.01125-14
230. Deb C, Lee CM, Dubey VS, Daniel J, Abomoelak B, Sirakova TD, et al. A novel *in vitro* multiple-stress dormancy model for *Mycobacterium tuberculosis* generates a lipid-loaded, drug-tolerant, dormant pathogen. *PLoS ONE.* (2009) 4:e6077. doi: 10.1371/journal.pone.0006077
231. Galagan JE, Minch K, Peterson M, Lyubetskaya A, Azizi E, Sweet L, et al. The *Mycobacterium tuberculosis* regulatory network and hypoxia. *Nature.* (2013) 499:178–83. doi: 10.1038/nature12337

3

Modulation of Host Biology by *Pseudomonas aeruginosa* Quorum Sensing Signal Molecules: Messengers or Traitors

Yi-Chia Liu[1], *Kok-Gan Chan*[2] *and Chien-Yi Chang*[3,4]*

[1] *Division of Molecular Microbiology, School of Life Sciences, University of Dundee, Dundee, UK,* [2] *Division of Genetics and Molecular Biology, Institute of Biological Sciences, Faculty of Science, University of Malaya, Kuala Lumpur, Malaysia,* [3] *Centre for Bacterial Cell Biology, Medical School, Newcastle University, Newcastle upon Tyne, UK,* [4] *Interdisciplinary Computing and Complex BioSystems (ICOS) Research Group, School of Computing Science, Newcastle University, Newcastle upon Tyne, UK*

Edited by:
Katharina F. Kubatzky,
University Hospital Heidelberg, Germany

Reviewed by:
Rodolfo García-Contreras,
Instituto Nacional de Cardiologia, Mexico
Gertrud Maria Hänsch,
University of Heidelberg, Germany

***Correspondence:**
Chien-Yi Chang
chien-yi.chang@ncl.ac.uk

Bacterial cells sense their population density and respond accordingly by producing various signal molecules to the surrounding environments thereby trigger a plethora of gene expression. This regulatory pathway is termed quorum sensing (QS). Plenty of bacterial virulence factors are controlled by QS or QS-mediated regulatory systems and QS signal molecules (QSSMs) play crucial roles in bacterial signaling transduction. Moreover, bacterial QSSMs were shown to interfere with host cell signaling and modulate host immune responses. QSSMs not only regulate the expression of bacterial virulence factors but themselves act in the modulation of host biology that can be potential therapeutic targets.

Keywords: Quorum sensing, *N*-acyl homoserine lactones, *Pseudomonas* quinolone signal, *Pseudomonas aeruginosa*, immunomodulation

INTRODUCTION

Quorum sensing (QS) is coined to describe the phenomenon of an intercellular co-operative behavior of bacteria used to coordinate the activities of individual cells. Diffusible QS signal molecules (QSSMs) play crucial roles in signal transduction of which, when QSSMs reach a threshold concentration, can coordinate multiple gene expression and a change in the behavior of bacterial population through the activation of sensor regulatory proteins (Fuqua et al., 1994; Miller and Bassler, 2001; Williams and Cámara, 2009). Bacteria produce a broad-range of signal molecules. Different types of QSSMs have been identified and characterized (LaSarre and Federle, 2013). Besides prokaryote, bacterial QSSMs also affect the settlement and germination of eukaryotic seaweed zoospores (Joint et al., 2002; Twigg et al., 2013). In several pathogenic bacteria, QS control their virulence determinants and contribute to bacterial pathogenesis. Due to the fact that the population density-dependent regulatory systems used by many bacterial pathogens are not essential for survival under most conditions, the disruption/interference of QS is considered an alternative approach to attenuate bacterial virulence in infections (LaSarre and Federle, 2013). However, this point has recently been argued that the resistance mechanisms against QS inhibitors could be possible and have been identified (Defoirdt et al., 2010, 2013; García-Contreras et al., 2013, 2015a).

Pseudomonas aeruginosa is an ubiquitous Gram-negative bacterium with remarkably large and complex genome and is capable of adapting to versatile environments. In human cystic fibrosis (CF) lungs where *P. aeruginosa* has evolved the ability to form biofilms which are difficult to be eradicated by antibiotics (Heeb et al., 2011; Winsor et al., 2011). QS is responsible for the regulation of a large number of genes, for instance, around 10% of genes in the genome of *P. aeruginosa* are regulated by QS (Williams and Cámara, 2009). Here we review recent advances of *P. aeruginosa* QSSMs focusing on their roles in interference with host cells (**Table 1**) and the development of novel compounds that counteract the QSSMs activities.

N-ACYL HOMOSERINE LACTONES (AHLS) AND THEIR MODULATIONS IN HOST CELLS

Gram-negative bacteria, like *Aliivibrio fischeri* (previous *Vibrio fischeri*; Urbanczyk et al., 2007), have a conserved QS system with two central components, the LuxR-type and LuxI-type proteins, which serve as the signal receptor and signal synthase, respectively. LuxI catalyzes the synthesis of signaling molecules called *N*-acyl homoserine lactones (AHLs). When an AHLs concentration of 10 nM is reached, AHLs interact with LuxR and form a complex which promotes the expression of target genes, *luxICDABE* for bioluminescence production and also the LuxI production (Kaplan and Greenberg, 1985). This forms a positive loop to produce more signal molecules (Fuqua et al., 1994; Cámara et al., 2002; Fuqua and Greenberg, 2002). The *N*-acyl homoserine lactone consists of a homoserine lactone ring from *S*-adenosylmethionine (SAM) and acyl chain from acyl acyl-carrier-protein (acyl-ACP) linked by an amide bond (Parsek et al., 1999). Based on the acyl-ACP binding site, different LuxI homologs produce different AHLs with various acyl side chains (Watson et al., 2002; Gould et al., 2004). A broad range of AHLs is produced in Gram-negative bacteria and AHL-QS systems control various bacterial behaviors (LaSarre and Federle, 2013). In *A. fischeri* *N*-(3-oxohexanoyl) homoserine lactone (3-oxo-C6-HSL) is produced for controlling bioluminescence production. In *P. aeruginosa* two AHL synthases, RhlI and LasI, produce a wide spectrum of AHLs including *N*-butanoyl-homoserine lactone (C4-HSL), *N*-hexanoyl-homoserine lactone (C6-HSL) by RhlI and *N*-(3-oxooctanoyl)-homoserine lactone (3-oxo-C8-HSL), *N*-(3-oxodecanoyl)-homoserine lactone (3-oxo-C10-HSL), *N*-(3-oxododecanoyl)-homoserine lactone (3-oxo-C12-HSL) and *N*-(3-oxotetradecanoyl)-homoserine lactone (3-oxo-C14-HSL) by LasI (Ortori et al., 2011). An unusual *N*-(3-oxohexadecanoyl)-homoserine lactone (3-oxo-C16-HSL) secreted by an environmental *Pseudomonas* sp. from a diseased Tilapia fish suggests that 3-oxo-C16-HSL may contribute to the pathogenesis (Chang et al., 2012).

The abundant concentration of 3-oxo-C12-HSL in the culture of *P. aeruginosa* prompted investigations for its role in the pathogenesis with a mechanism potentially distinct from other pathogens. Indeed, 3-oxo-C12-HSL was found to activate mammalian cells through a mechanism independent of the toll-like receptor (TLR) pathways (Kravchenko et al., 2006). 3-oxo-C12-HSL was shown to activate pro-inflammatory responses in human epithelial and fibroblast cells through the induction the transcriptional factor, nuclear factor kappa-light-chain-enhancer of activated B cells (NF-κB) via the phosphorylation of ERK/MARK (Smith et al., 2001). However, this molecule selectively disrupts the NF-κB signaling pathway in activated macrophages (Kravchenko et al., 2008). Studies indicated that 3-oxo-C12-HSL not only induces apoptosis in haematopoietic cells but is cytototoxic to non-haematopoietic cells including airway epithelial cells, endothelial cells, fibroblasts, and mesenchymal stem cells (Tateda et al., 2003; Shiner et al., 2006; Schwarzer et al., 2010, 2012; Grabiner et al., 2014; Holban et al., 2014). *P. aeruginosa* 3-oxo-C12-HSL also impairs the epithelial barrier integrity through the alternations of calcium signaling and phosphorylation status of junctional proteins in the intestinal epithelial cells (Vikström et al., 2009, 2010).

In addition to its cytotoxicity, the role of 3-oxo-C12-HSL in immunomodulation has been intensively investigated (**Table 1**). Ritchie et al. (2005) reported that 3-oxo-C12-HSL inhibits the differentiation of Th1 and Th2 cells. Human polymorphonuclear neutrophils (PMNs) are attracted by 3-oxo-C12-HSL and increasingly express the adhesion proteins CD11b/CD18 and the immunoglobulin receptors CD16 and CD64 (Zimmermann et al., 2006; Wagner et al., 2007). The downregulation of the immune responses by 3-oxo-C12-HSL was demonstrated in human monocytes and murine macrophage-like cells in the presence of lipopolysaccharides (LPS) that 3-oxo-C12-HSL inhibits the production of pro-inflammatory cytokine tumor necrosis factor α (TNF-α) but promotes the production of anti-inflammatory cytokine interleukin-10 (IL-10; Hooi et al., 2004; Glucksam-Galnoy et al., 2013). Grabiner et al. (2014) noticed that despite the increasing transcriptional expression of the murine interleukin 8 (IL-8) homologs KC and interleukin 6 (IL-6) in murine embryonic fibroblasts (MEFs), KC and IL-6 protein secretion were inhibited by the treatment of 3-oxo-C12-HSL. It was shown that 3-oxo-C12-HSL acts upon the activation of endoplasmic reticulum (ER) stress transducer protein kinase RNA-like ER kinase (PERK) leading to the inhibition of protein synthesis. However, PERK is independent of 3-oxo-C12-HSL induced apoptosis indicating that 3-oxo-C12-HSL interferes with host cell biological activities through different mechanisms (Grabiner et al., 2014). Recent advances on the interactions between 3-oxo-C12-HSL and various types of host cells are highlighted in the review (Holm and Vikström, 2014).

Several host targets of 3-oxo-C12-HSL have been identified (**Figure 1**). In murine fibroblasts and human lung epithelial cells peroxisome proliferator-activated receptor beta/delta (PPARβ/δ) and PPARγ may be the 3-oxo-C12-HSL receptors for pro-inflammatory responses (Jahoor et al., 2008; Cooley et al., 2010). 3-oxo-C12-HSL interacts and co-localizes with the IQ-motif-containing GTPase-activating protein IQGAP1 in human intestinal epithelial cells that causes the alternation of cell migration in a Rac1 and Cdc42- dependent manner (Karlsson et al., 2012). MEFs in lack of a transcriptional factor X-box binding protein 1 transcription factor (XBP1s) are

TABLE 1 | Bacterial quorum sensing molecules and their roles in pathogenesis/immunomodulation.

Quorum sensing (QS) signal molecules (QSSMs)	Mechanism of virulence	Effect concentrations	Reference
3-oxo-C12-HSL	**Apoptosis and cytotoxicity**		
	3-oxo-C12-HSL is cytotoxic to murine bone-marine derived macrophages, neutrophils and monocytic cell lines.	12–50 μM	Tateda et al., 2003
	3-oxo-C12-HSL induces apoptosis in murine fibroblasts and human vascular endothelial cells (HUVEC)	100 μM	Shiner et al., 2006
	3-oxo-C12-HSL triggers intrinsic apoptotic pathway in airway epithelial cells including depolarization of mitochondrial membrane potential, release of cytochrome C and activation of caspases 3/7, 8, and 9. 3-oxo-C12-HSL-mediated apoptosis is independent of the presence of CFTR in airway epithelial cells	>10 μM	Schwarzer et al., 2010
	3-oxo-C12-HSL promotes human mesenchymal stem cells (MSCs) apoptosis	50 μM	Holban et al., 2014
	Disruption of barrier integrity		
	Pseudomonas aeruginosa 3O-C(12)-HSL causes the loss of epithelial barrier function via calcium signaling and further alteration in the phosphorylation status of junction proteins	20–200 and 10 μM with slower response	Vikström et al., 2009, 2010
	Immunomodulation and/or signaling		
	3-oxo-C12-HSL promotes the expression and production of IL-8 in human epithelial and fibroblast cells through the induction of NF-kB via the phosphorylation of ERK/MARK	100 μM	Smith et al., 2001
	3-oxo-C12-HSL inhibits ConA-activated PBMCs proliferation and IL-2 secretion	[a]IC_{50}: 18.24 μM	Hooi et al., 2004
	3-oxo-C12-HSL inhibits the proliferation of anti-CD3/anti-CD28 antibody activated T cells	IC_{50}: 44.47 μM	Hooi et al., 2004
	3-oxo-C12-HSL inhibits the differentiation of Th1 and Th2 cells	5 μM	Ritchie et al., 2005
	3-oxo-C12-HSL increases the cytosolic calcium levels and calcium release through inositol triphosphate (IP3) receptors in the ER.	1 mM	Shiner et al., 2006
	3-oxo-C12-HSL promotes neutrophil chemotaxis, phagocytosis and up-regulates the expression of CD11b/CD18 and CD16/CD64 receptors	100 μM	Zimmermann et al., 2006; Wagner et al., 2007
	3-oxo-C12-HSL selectively disrupts NF-κB signaling but not TLR-dependent pathways in activated macrophages	50 μM	Kravchenko et al., 2006, 2008
	3-oxo-C12-HSL binds to PPARγ ligand binding domain	25–50 μM	Jahoor et al., 2008; Cooley et al., 2010
	3-oxo-C12-HSL increases the secretion of IL-1β in human MSCs	50 μM	Holban et al., 2014
	3-oxo-C12-HSL activates NF-κB p65 by preventing the re-synthesis of IκB, increases transcription of KC and IL-6 but inhibits secretion of KC and IL-6 by MEFs. 3-oxo-C12-HSL activates PERK and inhibits protein synthesis	50 or 100 μM	Grabiner et al., 2014
Alkylquinolones	**Change of bacterial behaviors**		
	Bacterial autolysis	Spent culture supernatant 15 μg on filter discs	Williams and Cámara, 2009
	Iron chelation	50 μM	Diggle et al., 2007
	exDNA release	Genetic and phenotype study	Allesen-Holm et al., 2006
	Oxidative functions	>100 μM *in vitro*	Häussler and Becker, 2008

(Continued)

TABLE 1 | Continued

Quorum sensing (QS) signal molecules (QSSMs)	Mechanism of virulence	Effect concentrations	Reference
	Apoptosis		
	HHQ promotes human MSCs apoptosis	50 μM	Holban et al., 2014
	Immunomodulation and/or signaling		
	Pseudomonas quinolone signal (PQS) inhibits the proliferation of ConA-activated PBMCs	IC_{50}: 0.90 μM	Hooi et al., 2004
	PQS affects IL-2 secretion of ConA-stimulated PBMCs	IC_{50}: 2.03 μM	Hooi et al., 2004
	PQS promotes TNF-α production in LPS-treated monocytes	>25 uM	Hooi et al., 2004
	Inhibition of IL-12 production in dendritic cells resulting in reduction of T-cell proliferation	IC_{50}: 17.2 μM	Skindersoe et al., 2009
	Inhibition of NF-κB and HIP-1α pathways in murine epithelial cells and murine macrophages	Bacterial culture supernatants approx. PQS in PA14, 15 μM approx. PQS in *pqsL*, 45 μM	Kim et al., 2010; Legendre et al., 2012
	Stimulation of chemotaxis of neutrophils via the MAPK and p38 signaling pathways	10–100 μM	Hänsch et al., 2014
	PQS activates the secretion of IL-6 and HHQ induces IL-10 secretion by human MSCs	50 μM	Holban et al., 2014

[a]*50% inhibitory concentration (IC_{50}).*

protective from 3-oxo-C12-HSL and C14-HSL (*N*-tetradecanoyl-homoserine lactone) mediated apoptosis indicating that XBP1s is a critical host target in response of AHLs (Valentine et al., 2013). Paraoxonase 2, in response to 3-oxo-C12-HSL through its lactonase activity, leads to apoptosis in human and murine embryotic epithelial cells (Schwarzer et al., 2015). Interestingly, 3-oxo-C12-HSL activates the expression of a taste receptor T2R38 on the surface of primary human sinonasal cells (Lee et al., 2014) and neutrophils (Maurer et al., 2015). This recognition regulates calcium-dependent NO production thereby stimulates the mucociliary clearance and antibacterial effects suggesting an alternative innate immune defense mechanism distinct from the activation by canonical pattern recognition receptors (PRRs; Lee et al., 2012, 2014). Identification of the host compartments targeted by QSSMs could be the milestone for developing effective therapeutic methods against infections.

PATHOGENIC ROLES OF ALKYL-QUINOLONE SIGNALS

Pseudomonas aeruginosa also employs the alkyl-quinolone (AQ)-based QS system and the signal molecule was termed *Pseudomonas* quinolone signal (PQS; Pesci et al., 1999). The study of the AQs began from their intriguing structures similar to antimicrobial quinolones, although AQs were found no antimicrobial activities. Further studies unveiled that among more than 50 alkyl-quinolones found in *P. aeruginosa*, 2-heptyl-3-hydroxy-4-(1H)-quinolone (PQS) and its precursor molecular 2-heptyl-4(1*H*)-quinolone (HHQ) are major QSSMs that cooperates with the AHL-QS (Xiao et al., 2006; Heeb et al., 2011). Synthesis of PQS depends on the *pqsABCDE* operon. PqsA, the anthranilate co-enzyme A ligase, catalyzes anthranilate that is produced by PhnAB to anthraniloyl-coenzyme A. PqsD mediates the synthesis of 2-aminobenzoylacetate (2-ABA) from anthraniloyl-coenzyme A and malonyl-CoA, decarboxylating coupling of 2-ABA to an octanoate group of octanoic acid that linked to PqsBC to produce HHQ (Dulcey et al., 2013). A recent study suggested PqsE is involved in the HHQ synthesis through hydrolysing the 2-ABA-CoA to form 2-ABA (Drees and Fetzner, 2015). HHQ can be transformed to PQS by the mono-oxygenase PqsH encoded by *pqsH* located elsewhere on the chromosome (Pesci et al., 1999; Diggle et al., 2006). PqsR, also known as MvfR, is a LysR-type transcriptional regulator, with a conserved N-terminal DNA-binding helix-turn-helix and a C-terminal co-inducer-binding domain. PqsR activates the transcription of *pqsABCDE* and possibly the *phnAB* operon when binding to PQS or HHQ and triggers the typical QS autoinducing response enhancing AQ biosynthesis (Maddocks and Oyston, 2008; Heeb et al., 2011). PQS has been shown to reach the maximal production at late logarithmic phase (Diggle et al., 2003) and its production is promoted by the availability of the substrate anthranilate and the presence of aromatic amino acids (Palmer et al., 2005). AQ- and AHL-QS in *P. aeruginosa* are hierarchical and involved in the regulation of multiple virulence factors including rhamnolipids, pyocyanin, elastases, exotoxin A, and alkaline protease (Xiao

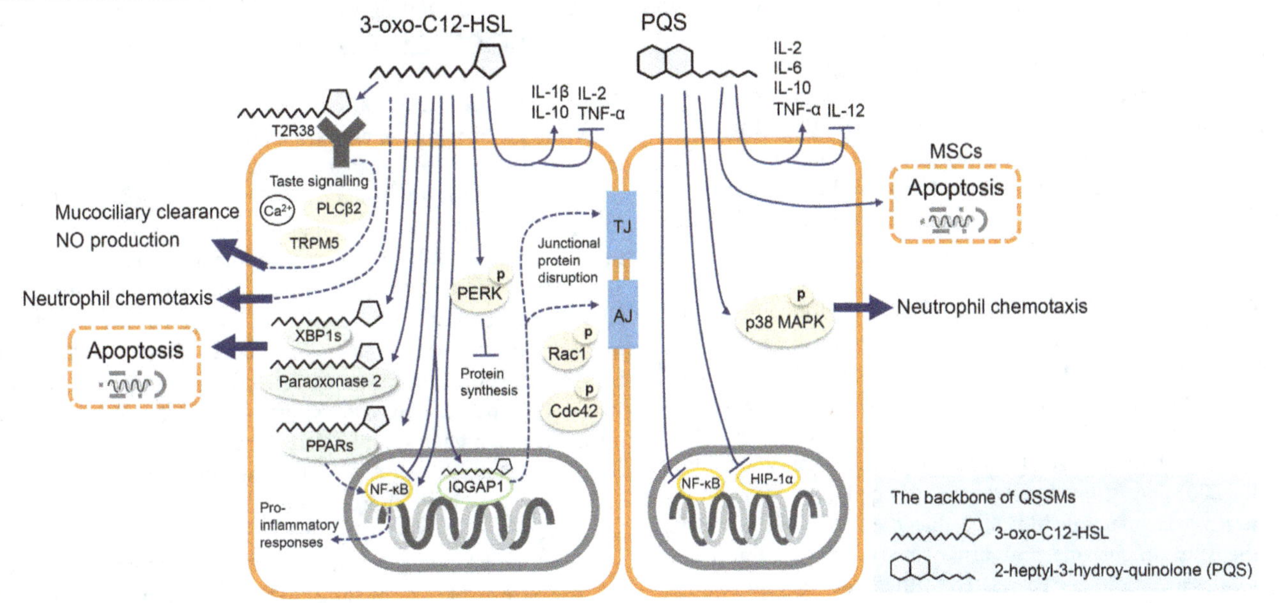

FIGURE 1 | Schematic illustration of the best characterized *Pseudomonas aeruginosa* quorum sensing signal molecules (QSSMs) that interfere with mammalian host biological functions. *P. aeruginosa* 3-oxo-C12-HSL targets XBP1 and paraoxonase 2 leading to host cell apoptosis, and the binding of 3-oxo-C12-HSL with IQGAP1 impairs the host cell integrity. 3-oxo-C12-HSL modulates innate immune responses via the activation of T2R38 receptor and inhibition of NF-κB pathways. *P. aeruginosa Pseudomonas* quinolone signal (PQS) molecule induces MSCs apoptosis, induces the secretion of inflammatory cytokines via the inhibition NF-κB and HIP-1α pathways. PQS molecule interferes with neutrophils chemotaxis potentially through the activation of p38 MAPK pathways. See text for details. PLCβ2, phospholipase C β2; TRPM5, transient receptor potential cation channel subfamily M member 5; PPARs, peroxisome proliferator-activated receptors; TJ, tight junction; AJ, adhesional junction; MAPK, mitogen-activated protein kinase; MSCs, mesenchymal stem cells.

et al., 2006; Dubern and Diggle, 2008; Nadal Jimenez et al., 2012).

Pseudomonas quinolone signal is considered a multifunctional molecule. PQS is involved in bacterial cell autolysis at high population densities in nutrient deprived conditions (Williams and Cámara, 2009). PQS also has iron-chelating properties that contribute to iron transport and facilitates siderophore-mediated iron delivery (Diggle et al., 2007). It has been demonstrated that there is far less extracellular DNA (exDNA) released by a *pqsA* mutant than its wild-type counterpart either in planktonic or biofilm cultures (Allesen-Holm et al., 2006). PQS has dual pro- and anti-oxidative functions for developing different levels of tolerance in *P. aeruginosa* cells to environmental stress (Häussler and Becker, 2008). This may shape the whole population structure, increase the fitness in hostile environments and lead to the development of resistance to host immune systems (García-Contreras et al., 2015a,b).

A new QS molecule, 2-(2-hydroxyphenyl)-thiazole-4-carbaldehyde (IQS) encoded by the *ambBCDE* operon was discovered recently (Lee et al., 2013). IQS is induced when *P. aeruginosa* is exposed to a phosphate-deprived environment. Under this unfavorable environment, expression of IQS overcomes the *las*-led QS circuit and promotes the expression of virulence factors. This finding may partially explain how *P. aeruginosa* clinical isolates persist in CF respiratory infections in the absence of a functional *las* system. Despite the fact that the AHL-QS of *P. aeruginosa* have been shown to play central roles in the regulation of virulence and immune modulation *in vitro*, this situation could differ *in vivo*. Around 50% of strains isolated from lungs of late stage CF patients are deficient in *lasR* function (Winstanley and Fothergill, 2009). Moreover, abolishing the whole AHL-QS by generating a quadruple mutation of *rhlIR* and *lasIR* exerted comparable infectivity to the wild-type strain in a mouse lung infection model (Lazenby et al., 2013), suggesting that the AHL-QS may not be required for full pathogenesis *in vivo* and other regulatory mechanism could be involved.

The contribution of the AQ-QS system to *P. aeruginosa* virulence was firstly described by Cao et al. (2001). PqsR positively regulates the expression of *phnAB* operon and the production of elastase, 3-oxo-C12-HSL and PQS that promotes the production of numerous virulence determinants. The *pqsR* mutant was attenuated up to 320-fold in the *Arabidopsis* plant infection model and caused a 65% reduction of mortality in a murine burn wound model (Cao et al., 2001). The interaction of PqsR with the AHL-QS was investigated by other groups showed that the effect of *pqsR* deficiency on pathogenesis is independent from LasRI/RhlRI (Déziel et al., 2004; Dubern and Diggle, 2008). Mutations in the multidrug efflux pump, such as *mexI* and *opmD* led to the inhibition of PQS production and the attenuation of *P. aeruginosa* in rat and plant infection models. Provision of exogenous AQs to these mutants restored the virulence on plants (Aendekerk, 2005). Rampioni et al. (2010) found that both *pqsA* and *pqsE* mutants in PAO1 were attenuated in plant, nematode and mouse burn wound infection models. In an acute urinary tract infection model, PQS molecules were present in the renal and bladder tissue of mice infected with wild-type *P. aeruginosa* but absent in the mice with PQS mutants infections (Bala et al., 2014). Wild-type *P. aeruginosa* caused more severe inflammation and tissue destruction and greater levels of inflammatory cytokines TNF-α, IL-6, and IL-10 at the site of infection in mice infected with wild-type strain than with PQS mutants. The virulence of PQS mutants can be restored by the addition of exogenous PQS molecules (Bala et al., 2014). These findings indicate that PQS participates in the pathogenesis of *P. aeruginosa*.

Pseudomonas quinolone signal has been identified in sputum, bronchoalveolar lavage fluid (BAL) and mucopurulent fluid from distal airways of end-stage CF lungs removed for transplant and at different stages from asymptotic early stage to late progression, suggesting a potential role of PQS in coordinating virulence factors during the course of infections (Collier et al., 2002; Guina et al., 2003). A study involving 60 CF patients with chronic *P. aeruginosa* infection indicated that the AQs were detectable in the sputum, plasma and urine and the concentrations of molecules are positively correlated to the *P. aeruginosa* bacterial cell density. 2-nonyl-4-hydroxy-quinoline (NHQ) in plasma was suggested to be the biomarker for *P. aeruginosa* infection in CF lungs (Barr et al., 2015). An *in vitro* transcriptomic study investigating the physiology of *P. aeruginosa* grown in CF sputum revealed that the genes associated with PQS metabolism, such as those coding for the aromatic amino acid aminotransferase, 4-hydroxyphenylpyruvate dioxygenase (*hpd*) and *pqsABCDE*, were expressed 10-fold greater than the expression when *P. aeruginosa* was cultured in media containing glucose alone as the carbon source (Palmer et al., 2005).

Pseudomonas aeruginosa AQ molecules have been implicated in the immuno-modulation on host cells. PQS was shown to modulate cell proliferation, the production of interleukin-2 (IL-2) and TNF-α in mitogen-stimulated human peripheral blood mononuclear cells (PBMCs; Hooi et al., 2004). PQS inhibited the production of IL-12 by LPS-stimulated bone marrow-derived dendritic cells which led to reduced T-cell proliferation (Skindersoe et al., 2009). Additionally AQ extracts derived from *P. aeruginosa* PA14 supernatants down-regulated host innate immune responses via inhibition of the NF-κB and hypoxia-inducible factor 1 alpha (HIF-1α) pathways in murine macrophages and cells obtained from BAL (Kim et al., 2010; Legendre et al., 2012). A recent study addressed the importance of timing in neutrophil infiltration in relation to the role of PQS in interference with neutrophil chemotaxis. Low levels of PQS stimulated the chemotaxis of neutrophils via the MAPK and p38 signaling pathways, whereas high levels of PQS, most likely produced by biofilm-like *P. aeruginosa*, did not interfere with neutrophils phagocytic capability and viability (Hänsch et al., 2014). Massive neutrophil accumulation is commonly seen in CF airways and high levels of neutrophil elastase correlate with poor pulmonary functions (Downey et al., 2008; Gifford and Chalmers, 2014). PQS may thus provide *P. aeruginosa* with another strategy for bacterial survival via the interference in multiple aspects of host biological activities.

THE DEVELOPMENT OF INHIBITORS AND VACCINES AGAINST QSSMS

Due to the fact that QSSMs have been implicated in the involvement of pathogenesis, the search for inhibitors and the development of vaccines that antagonize QSSMs are currently intensively investigated. Chang et al. (2014) suggested a strategy to screen novel anti-QS compounds from plant extraction that potentially could tackle the QS-mediated infections. The antibody 3-oxo-C12-HSL-BSA conjugate was also shown to alleviate the inflammatory responses by *P. aeruginosa* infections in an acute murine lung infection model (Miyairi et al., 2006). In a burn wound infection model, mice immunized with the vaccine 3-oxo-C12-HSL-r-PcrV conjugate before *P. aeruginosa* infection had higher survival than those without immunization (Golpasha et al., 2015). A high-throughput screening approach based on the inhibition of C12-mediated host responses identified triazolo[4,3-a]quinolines as 3-oxo-C12-HSL inhibitors with nanomolar potency that restore NF-κB activity in 3-oxo-C12-HSL treated cell lines and shown protective using an *in vivo* dermal infection model (Valentine et al., 2014). Since anthranilate (AA) being the precursor of AQs, halogenated AA analogs were found to inhibit the AQ biosynthesis and down-regulate the expression of PqsR controlled genes. Treatment with AA analogs prior to *P. aeruginosa* infection increased mice survival and lowered the bacterial dissemination to the organs (Lesic et al., 2007). MvfR-regulon inhibitors that bind QS transcriptional regulator MvfR (PqsR) were not only protective in murine acute and persistent infections against *P. aeruginosa* but also effectively reduced the formation of antibiotic-tolerant persisters (Starkey et al., 2014). These studies suggest the therapeutic potential of inhibitors and vaccines against QSSMs in both acute and chronic infections.

CONCLUSION

Quorum sensing-based bacterial communication links the individual bacterial cells to behave as multicellular organisms by employing signal molecules and to promote its population survival in the environment or hosts. QSSMs also interact with host cell signal pathways and the modulation of immune cell biology. For more than a decade strategies have been proposed from the use of inhibitors of QS for containing chronic infections (Hentzer et al., 2003) to the application of QSSMs for modulating immune responses to bacterial infections (Hancock et al., 2012). Understanding how QSSMs interact with host cells seems the promising land to tackle bacterial infections. Here we discussed recent advances on the interference of QSSMs with mammalian cells, the recently identified receptors on mammalian cells that target QSSMs and the QSSM inhibitors and their mechanisms. However, contradictory results suggested that many unknown mechanisms in complex bacteria-host interactions are remained. Also concern about bacterial resistance to QS inhibitors (García-Contreras et al., 2015a) and unexpected impact of QS inhibitors to environment (Decho et al., 2010) is increasing. Extending our understanding of the multiple roles of QSSMs would be valuable in the development of new therapeutic strategies against bacterial infections.

AUTHOR CONTRIBUTIONS

Y-CL, K-GC, C-YC wrote the paper. Y-CL made the figure. Y-CL and C-YC made the table.

ACKNOWLEDGMENTS

We are grateful for the Open Access Funding from the University of Dundee. This work was also supported by the University of Malaya High Impact Research Grants (UM C/625/1/HIR/MOHE/CHAN/01, A-000001-50001, and UM C/625/1/HIR/MOHE/CHAN/14/1, H-50001-A000027) awarded to K-GC.

REFERENCES

Aendekerk, S. (2005). The MexGHI-OpmD multidrug efflux pump controls growth, antibiotic susceptibility and virulence in *Pseudomonas aeruginosa* via 4-quinolone-dependent cell-to-cell communication. *Microbiology* 151, 1113–1125. doi: 10.1099/mic.0.27631-0

Allesen-Holm, M., Barken, K. B., Yang, L., Klausen, M., Webb, J. S., Kjelleberg, S., et al. (2006). A characterization of DNA release in *Pseudomonas aeruginosa* cultures and biofilms. *Mol. Microbiol.* 59, 1114–1128. doi: 10.1111/j.1365-2958.2005.05008.x

Bala, A., Chhibber, S., and Harjai, K. (2014). *Pseudomonas* quinolone signalling system: a component of quorum sensing cascade is a crucial player in the acute urinary tract infection caused by *Pseudomonas aeruginosa. Int. J. Med. Microbiol.* 304, 1199–1208. doi: 10.1016/j.ijmm.2014.08.013

Barr, H. L., Halliday, N., Cámara, M., Barrett, D. A., Williams, P., Forrester, D. L., et al. (2015). *Pseudomonas aeruginosa* quorum sensing molecules correlate with clinical status in cystic fibrosis. *Eur. Respir. J.* 46, 1046–1054. doi: 10.1183/09031936.00225214

Cámara, M., Williams, P., and Hardman, A. (2002). Controlling infection by tuning in and turning down the volume of bacterial small-talk. *Lancet Infect. Dis.* 2, 667–676. doi: 10.1016/S1473-3099(02)00447-4

Cao, H., Krishnan, G., Goumnerov, B., Tsongalis, J., Tompkins, R., and Rahme, L. G. (2001). A quorum sensing-associated virulence gene of *Pseudomonas aeruginosa* encodes a LysR-like transcription regulator with a unique self-regulatory mechanism. *Proc. Natl. Acad. Sci. U.S.A.* 98, 14613–14618. doi: 10.1073/pnas.251465298

Chang, C.-Y., Koh, C.-L., Sam, C.-K., Chan, X.-Y., Yin, W. F., and Chan, K. G. (2012). Unusual long-chain N-acyl homoserine lactone production by and presence of quorum quenching activity in bacterial isolates from diseased tilapia fish. *PLoS ONE* 7:e44034. doi: 10.1371/journal.pone.0044034.t002

Chang, C.-Y., Krishnan, T., Wang, H., Chen, Y., Yin, W. F., Chong, Y. M., et al. (2014). Non-antibiotic quorum sensing inhibitors acting against N-acyl homoserine lactone synthase as druggable target. *Sci. Rep.* 4:7245. doi: 10.1038/srep07245

Collier, D. N., Anderson, L., McKnight, S. L., Noah, T. L., Knowles, M., Boucher, R., et al. (2002). A bacterial cell to cell signal in the lungs of cystic fibrosis patients. *FEMS Microbiol. Lett.* 215, 41–46. doi: 10.1111/j.1574-6968.2002.tb11367.x

Cooley, M. A., Whittall, C., and Rolph, M. S. (2010). *Pseudomonas* signal molecule 3-oxo-C12-homoserine lactone interferes with binding of rosiglitazone to human PPARγ. *Microbes Infect.* 7, 231–237. doi: 10.1016/j.micinf.2009.12.009

Decho, A. W., Norman, R. S., and Visscher, P. T. (2010). Quorum sensing in natural environments: emerging views from microbial mats. *Trends Microbiol.* 18, 73–80. doi: 10.1016/j.tim.2009.12.008

Defoirdt, T., Boon, N., and Bossier, P. (2010). Can bacteria evolve resistance to quorum sensing disruption? *PLoS Pathogens* 6:e1000989. doi: 10.1371/journal.ppat.1000989.t001

Defoirdt, T., Brackman, G., and Coenye, T. (2013). Quorum sensing inhibitors: how strong is the evidence? *Trends Microbiol.* 21, 619–624. doi: 10.1016/j.tim.2013.09.006

Déziel, E., Gopalan, S., Tampakaki, A. P., Lépine, F., Padfield, K. E., Saucier, M., et al. (2004). The contribution of MvfR to *Pseudomonas aeruginosa* pathogenesis and quorum sensing circuitry regulation: multiple quorum sensing-regulated genes are modulated without affecting lasRI, rhlRI or the production of N-acyl- l-homoserine lactones. *Mol. Microbiol.* 55, 998–1014. doi: 10.1111/j.1365-2958.2004.04448.x

Diggle, S. P., Cornelis, P., Williams, P., and Cámara, M. (2006). 4-Quinolone signalling in *Pseudomonas aeruginosa*: old molecules, new perspectives. *Int. J. Med. Microbiol.* 296, 83–91. doi: 10.1016/j.ijmm.2006.01.038

Diggle, S. P., Matthijs, S., Wright, V. J., Fletcher, M. P., Chhabra, S. R., Lamont, I. L., et al. (2007). The *Pseudomonas aeruginosa* 4-quinolone signal molecules HHQ and PQS play multifunctional roles in quorum sensing and iron entrapment. *Chem. Biol.* 14, 87–96. doi: 10.1016/j.chembiol.2006.11.014

Diggle, S. P., Winzer, K., Chhabra, S. R., Worrall, K. E., Cámara, M., and Williams, P. (2003). The *Pseudomonas aeruginosa* quinolone signal molecule overcomes the cell density-dependency of the quorum sensing hierarchy, regulates rhl-dependent genes at the onset of stationary phase and can be produced in the absence of LasR. *Mol. Microbiol.* 50, 29–43. doi: 10.1046/j.1365-2958.2003.03672.x

Downey, D. G., Bell, S. C., and Elborn, J. S. (2008). Neutrophils in cystic fibrosis. *Thorax* 64, 81–88. doi: 10.1136/thx.2007.082388

Drees, S. L., and Fetzner, S. (2015). PqsE of *Pseudomonas aeruginosa* acts as pathway-specific thioesterase in the biosynthesis of alkylquinolone signaling molecules. *Chem. Biol.* 22, 611–618. doi: 10.1016/j.chembiol.2015.04.012

Dubern, J.-F., and Diggle, S. P. (2008). Quorum sensing by 2-alkyl-4-quinolones in *Pseudomonas aeruginosa* and other bacterial species. *Mol. BioSyst.* 4, 882–888. doi: 10.1039/b803796p

Dulcey, C. E., Dekimpe, V., Fauvelle, D.-A., Milot, S., Groleau, M.-C., Doucet, N., et al. (2013). The end of an old hypothesis: the *Pseudomonas* signaling molecules 4-hydroxy-2-alkylquinolines derive from Fatty Acids, not 3-ketofatty acids. *Chem. Biol.* 20, 1481–1491. doi: 10.1016/j.chembiol.2013.09.021

Fuqua, C., and Greenberg, E. P. (2002). Signalling: listening in on bacteria: acyl-homoserine lactone signalling. *Nat. Rev. Mol. Cell Biol.* 3, 685–695. doi: 10.1038/nrm907

Fuqua, W. C., Winans, S. C., and Greenberg, E. P. (1994). Quorum sensing in bacteria: the LuxR-LuxI family of cell density-responsive transcriptional regulators. *J. Bacteriol.* 176, 269–275.

García-Contreras, R., Maeda, T., and Wood, T. K. (2013). Resistance to quorum-quenching compounds. *Appl. Environ. Microbiol.* 79, 6840–6846. doi: 10.1128/AEM.02378-13

García-Contreras, R., Maeda, T., and Wood, T. K. (2015a). Can resistance against quorum-sensing interference be selected? *ISME J.* doi: 10.1038/ismej.2015.84

García-Contreras, R., Nuñez-López, L., Jasso-Chávez, R., Kwan, B. W., Belmont, J. A., Rangel-Vega, A., et al. (2015b). Quorum sensing enhancement of the stress response promotes resistance to quorum quenching and prevents social cheating. *ISME J.* 9, 115–125. doi: 10.1038/ismej.2014.98

Gifford, A. M., and Chalmers, J. D. (2014). The role of neutrophils in cystic fibrosis. *Curr. Opin. Hematol.* 21, 16–22. doi: 10.1097/MOH.0000000000000009

Glucksam-Galnoy, Y., Sananes, R., Silberstein, N., Krief, P., Kravchenko, V. V., Meijler, M. M., et al. (2013). The bacterial quorum-sensing signal molecule N-3-oxo-dodecanoyl-L-homoserine lactone reciprocally modulates pro- and anti-inflammatory cytokines in activated macrophages. *J. Immunol.* 191, 337–344. doi: 10.4049/jimmunol.1300368

Golpasha, I. D., Mousavi, S. F., Owlia, P., Siadat, S. D., and Irani, S. (2015). Immunization with 3-oxododecanoyl-L-homoserine lactone-r-PcrV conjugate enhances survival of mice against lethal burn infections caused by *Pseudomonas aeruginosa. Bosn. J. Basic Med. Sci.* 15, 15–24. doi: 10.17305/bjbms.2015.292

Gould, T. A., Schweizer, H. P., and Churchill, M. E. A. (2004). Structure of the *Pseudomonas aeruginosa* acyl-homoserinelactone synthase LasI. *Mol. Microbiol.* 53, 1135–1146. doi: 10.1111/j.1365-2958.2004.04211.x

Grabiner, M. A., Fu, Z., Wu, T., Barry, K. C., Schwarzer, C., and Machen, T. E. (2014). *Pseudomonas aeruginosa* quorum-sensing molecule homoserine lactone modulates inflammatory signaling through PERK and eI-F2. *J. Immunol.* 193, 1459–1467. doi: 10.4049/jimmunol.1303437

Guina, T., Purvine, S. O., Yi, E. C., Eng, J., Goodlett, D. R., Aebersold, R., et al. (2003). Quantitative proteomic analysis indicates increased synthesis of a quinolone by *Pseudomonas aeruginosa* isolates from cystic fibrosis airways. *Proc. Natl. Acad. Sci. U.S.A.* 100, 2771–2776. doi: 10.1073/pnas.0435846100

Hancock, R. E. W., Nijnik, A., and Philpott, D. J. (2012). Modulating immunity as a therapy for bacterial infections. *Nat. Rev. Microbiol.* 10, 243–254. doi: 10.1038/nrmicro2745

Hänsch, G. M., Prior, B., Brenner-Weiss, G., Obst, U., and Overhage, J. (2014). The *Pseudomonas* quinolone signal (PQS) stimulates chemotaxis of polymorphonuclear neutrophils. *JABFM* 12, 21–26. doi: 10.5301/jabfm.5000204

Häussler, S., and Becker, T. (2008). The *Pseudomonas* quinolone signal (PQS) balances life and death in *Pseudomonas aeruginosa* populations. *PLoS Pathogens* 4:e1000166. doi: 10.1371/journal.ppat.1000166.s002

Heeb, S., Fletcher, M. P., Chhabra, S. R., Diggle, S. P., Williams, P., and Cámara, M. (2011). Quinolones: from antibiotics to autoinducers. *FEMS Microbiol. Rev.* 35, 247–274. doi: 10.1111/j.1574-6976.2010.00247.x

Hentzer, M., Wu, H., Andersen, J. B., Riedel, K., Rasmussen, T. B., Bagge, N., et al. (2003). Attenuation of *Pseudomonas aeruginosa* virulence by quorum sensing inhibitors. *EMBO J.* 22, 3803–3815. doi: 10.1093/emboj/cdg366

Holban, A.-M., Bleotu, C., Chifiriuc, M. C., Bezirtzoglou, E., and Lazar, V. (2014). Role of *Pseudomonas aeruginosa* quorum sensing (QS) molecules on the viability and cytokine profile of human mesenchymal stem cells. *Virulence* 5, 303–310. doi: 10.4161/viru.27571

Holm, A., and Vikström, E. (2014). Quorum sensing communication between bacteria and human cells: signals, targets, and functions. *Front. Plant Sci.* 5:309. doi: 10.3389/fpls.2014.00309

Hooi, D. S. W., Bycroft, B. W., Chhabra, S. R., Williams, P., and Pritchard, D. I. (2004). Differential immune modulatory activity of *Pseudomonas aeruginosa* quorum-sensing signal molecules. *Infect. Immun.* 72, 6463–6470. doi: 10.1128/IAI.72.11.6463-6470.2004

Jahoor, A., Patel, R., Bryan, A., Do, C., Krier, J., Watters, C., et al. (2008). Peroxisome proliferator-activated receptors mediate host cell proinflammatory responses to *Pseudomonas aeruginosa* autoinducer. *J. Bacteriol.* 190, 4408–4415. doi: 10.1128/JB.01444-07

Joint, I., Tait, K., Callow, M. E., Callow, J. A., Milton, D., Williams, P., et al. (2002). Cell-to-cell communication across the prokaryote-eukaryote boundary. *Science* 298, 1207. doi: 10.1126/science.1077075

Kaplan, H. B., and Greenberg, E. P. (1985). Diffusion of autoinducer is involved in regulation of the Vibrio fischeri luminescence system. *J. Bacteriol.* 163, 1210–1214.

Karlsson, T., Turkina, M. V., Yakymenko, O., Magnusson, K.-E., and Vikström, E. (2012). The *Pseudomonas aeruginosa* N-acylhomoserine lactone quorum sensing molecules target IQGAP1 and modulate epithelial cell migration. *PLoS Pathogens* 8:e1002953. doi: 10.1371/journal.ppat.1002953

Kim, K., Kim, Y. U., Koh, B. H., Hwang, S. S., Kim, S.-H., Lépine, F., et al. (2010). HHQ and PQS, two *Pseudomonas aeruginosa* quorum-sensing molecules, down-regulate the innate immune responses through the nuclear factor-kappaB pathway. *Immunology* 129, 578–588. doi: 10.1111/j.1365-2567.2009.0 3160.x

Kravchenko, V. V., Kaufmann, G. F., Mathison, J. C., Scott, D. A., Katz, A. Z., Grauer, D. C., et al. (2008). Modulation of gene expression via disruption of NF-κB signaling by a bacterial small molecule. *Science* 321, 259–263. doi: 10.1126/science.1156499

Kravchenko, V. V., Kaufmann, G. F., Mathison, J. C., Scott, D. A., Katz, A. Z., Wood, M. R., et al. (2006). N-(3-Oxo-acyl) homoserine lactones signal cell activation through a mechanism distinct from the canonical pathogen-associated molecular pattern recognition receptor pathways. *J. Biol. Chem.* 281, 28822–28830. doi: 10.1074/jbc.M606613200

LaSarre, B., and Federle, M. J. (2013). Exploiting quorum sensing to confuse bacterial pathogens. *Microbiol. Mol. Biol. Rev.* 77, 73–111. doi: 10.1128/MMBR.00046-12

Lazenby, J. J., Griffin, P. E., Kyd, J., Whitchurch, C. B., and Cooley, M. A. (2013). A quadruple knockout of lasIR and rhlIR of *Pseudomonas aeruginosa* PAO1 that retains wild-type twitching motility has equivalent infectivity and persistence to PAO1 in a mouse model of lung infection. *PLoS ONE* 8:e60973. doi: 10.1371/journal.pone.0060973

Lee, J., Wu, J., Deng, Y., Wang, J., Wang, C., Wang, J., et al. (2013). A cell-cell communication signal integrates quorum sensing and stress response. *Nat. Chem. Biol.* 9, 339–343. doi: 10.1038/nchembio.1225

Lee, R. J., Chen, B., Redding, K. M., Margolskee, R. F., and Cohen, N. A. (2014). Mouse nasal epithelial innate immune responses to *Pseudomonas aeruginosa* quorum-sensing molecules require taste signaling components. *Innate Immun.* 20, 606–617. doi: 10.1177/1753425913503386

Lee, R. J., Xiong, G., Kofonow, J. M., Chen, B., Lysenko, A., Jiang, P., et al. (2012). T2R38 taste receptor polymorphisms underlie susceptibility to upper respiratory infection. *J. Clin. Invest.* 122, 4145–4159. doi: 10.1172/JCI 64240

Legendre, C., Reen, F. J., Mooij, M. J., McGlacken, G. P., Adams, C., and O'Gara, F. (2012). *Pseudomonas aeruginosa* alkyl quinolones repress hypoxia-inducible factor 1 (HIF-1) signaling through HIF-1α degradation. *Infect. Immun.* 80, 3985–3992. doi: 10.1128/IAI.00554-12

Lesic, B., Lépine, F., Déziel, E., Zhang, J., Zhang, Q., Padfield, K., et al. (2007). Inhibitors of pathogen intercellular signals as selective anti-infective compounds. *PLoS Pathogens* 3:e126. doi: 10.1371/journal.ppat.0030126.st003

Maddocks, S. E., and Oyston, P. C. F. (2008). Structure and function of the LysR-type transcriptional regulator (LTTR) family proteins. *Microbiology* 154, 3609–3623. doi: 10.1099/mic.0.2008/022772-0

Maurer, S., Wabnitz, G. H., Kahle, N. A., Stegmaier, S., Prior, B., Giese, T., et al. (2015). Tasting *Pseudomonas aeruginosa* biofilms: human neutrophils express the bitter receptor T2R38 as sensor for the quorum sensing molecule N-(3-oxododecanoyl)-l-Homoserine Lactone. *Front. Immunol.* 6:369. doi: 10.3389/fimmu.2015.00369

Miller, M. B., and Bassler, B. L. (2001). Quorum sensing in bacteria. *Ann. Rev. Microbiol.* 55, 165–199. doi: 10.1146/annurev.micro.55.1.165

Miyairi, S., Tateda, K., Fuse, E. T., Ueda, C., Saito, H., Takabatake, T., et al. (2006). Immunization with 3-oxododecanoyl-L-homoserine lactone-protein conjugate protects mice from lethal *Pseudomonas aeruginosa* lung infection. *J. Med. Microbiol.* 55, 1381–1387. doi: 10.1099/jmm.0.46658-0

Nadal Jimenez, P., Koch, G., Thompson, J. A., Xavier, K. B., Cool, R. H., and Quax, W. J. (2012). The multiple signaling systems regulating virulence in *Pseudomonas aeruginosa*. *Microbiol. Mol. Biol. Rev.* 76, 46–65. doi: 10.1128/MMBR.05007-11

Ortori, C. A., Dubern, J.-F., Chhabra, S. R., Cámara, M., Hardie, K., Williams, P., et al. (2011). Simultaneous quantitative profiling of N-acyl-L-homoserine lactone and 2-alkyl-4(1H)-quinolone families of quorum-sensing signaling molecules using LC-MS/MS. *Anal. Bioanal. Chem.* 399, 839–850. doi: 10.1007/s00216-010-4341-0

Palmer, K. L., Mashburn, L. M., Singh, P. K., and Whiteley, M. (2005). Cystic fibrosis sputum supports growth and cues key aspects of *Pseudomonas aeruginosa* physiology. *J. Bacteriol.* 187, 5267–5277. doi: 10.1128/JB.187.15.5267-5277.2005

Parsek, M. R., Val, D. L., Hanzelka, B. L., Cronan, J. E., and Greenberg, E. P. (1999). Acyl homoserine-lactone quorum-sensing signal generation. *Proc. Natl. Acad. Sci. U.S.A.* 96, 4360–4365. doi: 10.1073/pnas.96.8.4360

Pesci, E. C., Milbank, J. B., Pearson, J. P., McKnight, S., Kende, A. S., Greenberg, E. P., et al. (1999). Quinolone signaling in the cell-to-cell communication system of *Pseudomonas aeruginosa*. *Proc. Natl. Acad. Sci. U.S.A.* 96, 11229–11234. doi: 10.1073/pnas.96.20.11229

Rampioni, G., Pustelny, C., Fletcher, M. P., Wright, V. J., Bruce, M., Rumbaugh, K. P., et al. (2010). Transcriptomic analysis reveals a global alkyl-quinolone-independent regulatory role for PqsE in facilitating the environmental adaptation of *Pseudomonas aeruginosa* to plant and animal hosts. *Environ. Microbiol.* 12, 1659–1673. doi: 10.1111/j.1462-2920.2010.02214.x

Ritchie, A. J., Jansson, A., Stallberg, J., Nilsson, P., Lysaght, P., and Cooley, M. A. (2005). The *Pseudomonas aeruginosa* quorum-sensing molecule N-3-(oxododecanoyl)-L-homoserine lactone inhibits T-cell differentiation and cytokine production by a mechanism involving an early step in T-cell activation. *Infect. Immun.* 73, 1648–1655. doi: 10.1128/IAI.73.3.1648-165 5.2005

Schwarzer, C., Fu, Z., Morita, T., Whitt, A. G., Neely, A. M., Li, C., et al. (2015). Paraoxonase 2 serves a proapopotic function in mouse and human cells in response to the *Pseudomonas aeruginosa* quorum-sensing molecule N-(3-oxododecanoyl)-homoserine lactone. *J. Biol. Chem.* 290, 7247–7258. doi: 10.1074/jbc.M114.620039

4

Structural and Functional Characterization of the Type Three Secretion System (T3SS) Needle of *Pseudomonas aeruginosa*

Charlotte Lombardi[1†], James Tolchard[2†], Stephanie Bouillot[3], Luca Signor[1], Caroline Gebus[3], David Liebl[3‡], Daphna Fenel[1], Jean-Marie Teulon[1], Juliane Brock[1], Birgit Habenstein[2], Jean-Luc Pellequer[1], Eric Faudry[3], Antoine Loquet[2], Ina Attrée[3], Andréa Dessen[1,4] and Viviana Job[1,3]*

[1] Univ. Grenoble Alpes, CEA, CNRS, Institut de Biologie Structurale (IBS), Grenoble, France, [2] Institute of Chemistry and Biology of Membranes and Nanoobjects, Institut Européen de Chimie et Biologie (CBMN), UMR5248 CNRS, University of Bordeaux, Pessac, France, [3] Univ. Grenoble Alpes, Bacterial Pathogenesis and Cellular Responses Group, U1036 INSERM, ERL5261 CNRS, CEA, Grenoble, France, [4] Brazilian Biosciences National Laboratory (LNBio), Centro Nacional de Pesquisa em Energia e Materiais (CNPEM), Campinas, Brazil

Edited by:
Ignacio Arechaga,
University of Cantabria, Spain

Reviewed by:
Romé Voulhoux,
UMR 7255 Laboratoire d'Ingénierie des Systèmes Macromoléculaires (LISM), France
Paul Dean,
Teesside University, United Kingdom

***Correspondence:**
Viviana Job
viviana.job@cea.fr

†These authors have contributed equally to this work

‡Present address:
David Liebl,
*Agency for Science, Technology and Research (A*STAR), Singapore, Singapore*

The type three secretion system (T3SS) is a macromolecular protein nano-syringe used by different bacterial pathogens to inject effectors into host cells. The extracellular part of the syringe is a needle-like filament formed by the polymerization of a 9-kDa protein whose structure and proper localization on the bacterial surface are key determinants for efficient toxin injection. Here, we combined *in vivo*, *in vitro,* and *in silico* approaches to characterize the *Pseudomonas aeruginosa* T3SS needle and its major component PscF. Using a combination of mutagenesis, phenotypic analyses, immunofluorescence, proteolysis, mass spectrometry, atomic force microscopy, electron microscopy, and molecular modeling, we propose a model of the *P. aeruginosa* needle that exposes the N-terminal region of each PscF monomer toward the outside of the filament, while the core of the fiber is formed by the C-terminal helix. Among mutations introduced into the needle protein PscF, D76A, and P47A/Q54A caused a defect in the assembly of the needle on the bacterial surface, although the double mutant was still cytotoxic on macrophages in a T3SS-dependent manner and formed filamentous structures in *vitro*. These results suggest that the T3SS needle of *P. aeruginosa* displays an architecture that is similar to that of other bacterial needles studied to date and highlight the fact that small, targeted perturbations in needle assembly can inhibit T3SS function. Therefore, the T3SS needle represents an excellent drug target for small molecules acting as virulence blockers that could disrupt pathogenesis of a broad range of bacteria.

Keywords: type III secretion system, *Pseudomonas aeruginosa*, T3SS needle, structure, virulence, immunofluorescence microscopy, mutagenesis

INTRODUCTION

Bacterial pathogens have developed different strategies to colonize, invade and kill eukaryotic cells. One of the most successful mechanisms involves the use of the type III secretion system (T3SS), a macromolecular complex present on the surface of numerous pathogenic and colonizing species such as *Salmonella, Shigella, enteropathogenic Escherichia coli* (EPEC), *Yersinia,* and

Schwarzer, C., Fu, Z., Patanwala, M., Hum, L., Lopez-Guzman, M., Illek, B., et al. (2012). *Pseudomonas aeruginosa* biofilm-associated homoserine lactone C12 rapidly activates apoptosis in airway epithelia. *Cell. Microbiol.* 14, 698–709. doi: 10.1111/j.1462-5822.2012.01753.x

Schwarzer, C., Wong, S., Shi, J., Matthes, E., Illek, B., Ianowski, J. P., et al. (2010). *Pseudomonas aeruginosa* homoserine lactone activates store-operated cAMP and cystic fibrosis transmembrane regulator-dependent Cl-secretion by human airway epithelia. *J. Biol. Chem.* 285, 34850–34863. doi: 10.1074/jbc.M110.167668

Shiner, E. K., Terentyev, D., Bryan, A., Sennoune, S., Martinez-Zaguilan, R., Li, G., et al. (2006). *Pseudomonas aeruginosa* autoinducer modulates host cell responses through calcium signalling. *Cell. Microbiol.* 8, 1601–1610. doi: 10.1111/j.1462-5822.2006.00734.x

Skindersoe, M. E., Zeuthen, L. H., Brix, S., Fink, L. N., Lazenby, J., Whittall, C., et al. (2009). *Pseudomonas aeruginosa* quorum-sensing signal molecules interfere with dendritic cell-induced T-cell proliferation. *FEMS Immun. Med. Microbiol.* 55, 335–345. doi: 10.1111/j.1574-695X.2008.00533.x

Smith, R. S., Fedyk, E. R., Springer, T. A., Mukaida, N., Iglewski, B. H., and Phipps, R. P. (2001). IL-8 production in human lung fibroblasts and epithelial cells activated by the *Pseudomonas* autoinducer N-3-oxododecanoyl homoserine lactone is transcriptionally regulated by NF-kappa B and activator protein-2. *J. Immunol.* 167, 366–374. doi: 10.4049/jimmunol.167.1.366

Starkey, M., Lépine, F., Maura, D., Bandyopadhaya, A., Lesic, B., He, J., et al. (2014). Identification of anti-virulence compounds that disrupt quorum-sensing regulated acute and persistent pathogenicity. *PLoS Pathogens* 10:e1004321. doi: 10.1371/journal.ppat.1004321.s009

Tateda, K., Ishii, Y., Horikawa, M., Matsumoto, T., Miyairi, S., Pechere, J. C., et al. (2003). The *Pseudomonas aeruginosa* autoinducer N-3-oxododecanoyl homoserine lactone accelerates apoptosis in macrophages and neutrophils. *Infect. Immun.* 71, 5785–5793. doi: 10.1128/IAI.71.10.5785-5793.2003

Twigg, M. S., Tait, K., Williams, P., Atkinson, S., and Cámara, M. (2013). Interference with the germination and growth of Ulva zoospores by quorum-sensing molecules from Ulva-associated epiphytic bacteria. *Environ. Microbiol.* 16, 445–453. doi: 10.1111/1462-2920.12203

Urbanczyk, H., Ast, J. C., Higgins, M. J., Carson, J., and Dunlap, P. V. (2007). Reclassification of *Vibrio fischeri*, *Vibrio logei*, *Vibrio salmonicida* and *Vibrio wodanis* as *Aliivibrio fischeri* gen. nov., comb. nov., *Aliivibrio logei* comb. nov., *Aliivibrio salmonicida* comb. nov. and *Aliivibrio wodanis* comb. nov. *Int. J. Syst. Evol. Microbiol.* 57, 2823–2829. doi: 10.1099/ijs.0.65081-0

Valentine, C. D., Anderson, M. O., Papa, F. R., and Haggie, P. M. (2013). X-box binding protein 1 (XBP1s) is a critical determinant of *Pseudomonas aeruginosa* homoserine lactone-mediated apoptosis. *PLoS Pathog.* 9:e1003576. doi: 10.1371/journal.ppat.1003576

Valentine, C. D., Zhang, H., Phuan, P.-W., Nguyen, J., Verkman, A. S., and Haggie, P. M. (2014). Small molecule screen yields inhibitors of *Pseudomonas* homoserine lactone-induced host responses. *Cell. Microbiol.* 16, 1–14. doi: 10.1111/cmi.12176

Vikström, E., Bui, L., Konradsson, P., and Magnusson, K.-E. (2009). The junctional integrity of epithelial cells is modulated by *Pseudomonas aeruginosa* quorum sensing molecule through phosphorylation-dependent mechanisms. *Exp. Cell Res.* 315, 313–326. doi: 10.1016/j.yexcr.2008.10.044

Vikström, E., Bui, L., Konradsson, P., and Magnusson, K.-E. (2010). Role of calcium signalling and phosphorylations in disruption of the epithelial junctions by *Pseudomonas aeruginosa* quorum sensing molecule. *Eur. J. Cell Biol.* 89, 584–597. doi: 10.1016/j.ejcb.2010.03.002

Wagner, C., Zimmermann, S., Brenner-Weiss, G., Hug, F., Prior, B., Obst, U., et al. (2007). The quorum-sensing molecule N-3-oxododecanoyl homoserine lactone (3OC12-HSL) enhances the host defence by activating human polymorphonuclear neutrophils (PMN). *Anal. Bioanal. Chem.* 387, 481–487. doi: 10.1007/s00216-006-0698-5

Watson, W. T., Minogue, T. D., Val, D. L., von Bodman, S. B., and Churchill, M. E. A. (2002). Structural basis and specificity of acyl-homoserine lactone signal production in bacterial quorum sensing. *Mol. Cell* 9, 685–694. doi: 10.1016/S1097-2765(02)00480-X

Williams, P., and Cámara, M. (2009). Quorum sensing and environmental adaptation in *Pseudomonas aeruginosa*: a tale of regulatory networks and multifunctional signal molecules. *Curr. Opin. Microbiol.* 12, 182–191. doi: 10.1016/j.mib.2009.01.005

Winsor, G. L., Lam, D. K. W., Fleming, L., Lo, R., Whiteside, M. D., Yu, N. Y., et al. (2011). *Pseudomonas* genome database: improved comparative analysis and population genomics capability for *Pseudomonas* genomes. *Nucleic Acids Res.* 39, D596–D600. doi: 10.1093/nar/gkq869

Winstanley, C., and Fothergill, J. L. (2009). The role of quorum sensing in chronic cystic fibrosis *Pseudomonas aeruginosa* infections. *FEMS Microbiol. Lett.* 290, 1–9. doi: 10.1111/j.1574-6968.2008.01394.x

Xiao, G., He, J., and Rahme, L. G. (2006). Mutation analysis of the *Pseudomonas aeruginosa* mvfR and pqsABCDE gene promoters demonstrates complex quorum-sensing circuitry. *Microbiology* 152, 1679–1686. doi: 10.1099/mic.0.28605-0

Zimmermann, S., Wagner, C., Muller, W., Brenner-Weiss, G., Hug, F., Prior, B., et al. (2006). Induction of neutrophil chemotaxis by the quorum-sensing molecule N-(3-oxododecanoyl)-L-homoserine lactone. *Infect. Immun.* 74, 5687–5692. doi: 10.1128/IAI.01940-05

Pseudomonas spp. (Galán and Collmer, 1999; Galán et al., 2014; Wagner et al., 2018). The T3SS can be described as a multicomponent protein structure consisting of four major constituents: (i) a basal body that anchors the system to the bacterial membranes; (ii) an export apparatus which includes a cytoplasmic sorting platform that selects substrates and provides energy for the secretion process; (iii) a needle filament which protrudes toward the outside of the bacterial surface and serves as a passage for translocator proteins and effectors; and (iv) a translocation pore formed in the eukaryotic cell membrane that allows the entry of virulence effectors into the host cytoplasm.

The basal body and the needle form the so-called needle complex (NC), which represents the core structure of the T3SS. The number of NCs per bacterial cell varies between species. For instance, *Salmonella typhimurium*, assembles between 10 and 100 NCs per cell (Kubori et al., 1998), while only a few NCs were detected on the surface of *Pseudomonas aeruginosa* (Perdu et al., 2015). Notably, the T3SS needle is formed by a single polymerized protein that has been studied in the context of vaccine development against several pathogens (Charro and Mota, 2015; Jneid et al., 2016; Koroleva et al., 2017). It has been shown that immunization with needle proteins induces specific humoral and T cell responses and abrogates bacterial pathogenicity in animal models (Koroleva et al., 2017). In addition, small molecules such as phenoxyacetamide, that are directed against specific residues of the needle protein PscF can inhibit *P. aeruginosa* T3SS secretion (Bowlin et al., 2014) and the same compound also inhibits abscess formation in mice by directly blocking the T3SS (Berube et al., 2017). These results highlight the need for further characterization of molecular aspects and structures of NC assemblies to develop more specific anti-microbial molecules (Izoré et al., 2011).

Formation of the T3SS needle is a spatio-temporally regulated process. Protein monomers of the T3SS needle from different bacterial species have an approximate molecular weight of 9-kDa and display 13% sequence identity and 42% sequence similarity (considering the main human pathogens: *Salmonella, Shigella, P. aeruginosa,* and *Yersinia* sp.) (Kubori et al., 2000; Hoiczyk and Blobel, 2001; Pastor et al., 2005). Prior to secretion and assembly on the bacterial surface the spontaneous assembly of needle monomers in the bacteria cytosol is prevented by binding to two distinct chaperones concomitantly (Quinaud et al., 2005; Sun et al., 2008; Sal-Man et al., 2013), as is also the case for translocator proteins (Job et al., 2010; Discola et al., 2014; Nguyen et al., 2015). High resolution structures of T3SS needle proteins from different bacteria that have been solved to date (Deane et al., 2006; Quinaud et al., 2007; Wang et al., 2007; Sun et al., 2008; Poyraz et al., 2010) by crystallography and solution NMR reveal a similar protein fold composed of a helix-turn-helix motif with two long helices. Furthermore, one notable feature of T3SS needle proteins is an amphipathic C-terminal helix, which is 'trapped' within a hydrophobic concave region of the dual chaperone interface, presumably in order to block self-association (and early polymerization) (Quinaud et al., 2007; Sun et al., 2008). Upon T3SS activation, the needle protein dissociates from its chaperones, binding to the inner rod protein within the basal body (Kuhlen et al., 2018). Polymerization then occurs through addition of 100s of monomers to the distal end of growing needle. Interestingly, some bacteria possess accessory proteins that aid in assembly of the needle filament and accelerate the polymerization process *in vitro,* such as recently reported for *Salmonella typhimurium* (Kato et al., 2018). The final assembled needle in all bacteria is then capped by a tip protein (or soluble translocator), which is absolutely necessary to allow the formation of the translocation pore for the injection of effectors (Matteï et al., 2011; Harmon et al., 2013). During cell contact the specific needle length is essential for proper function yet the exact mechanism of length control still remains controversial (Cornelis, 2006). It has been suggested that a molecular ruler is involved (Kubori et al., 2000; Journet et al., 2003; Wagner et al., 2010; Bergeron et al., 2016), or that needle length is regulated by a precise timing of substrate switching, as suggested by recent mathematical models (Nariya et al., 2016).

The length of a fully assembled T3SS needle varies between bacterial species from 45 to 80 nm (Kubori et al., 1998; Tamano et al., 2000; Pastor et al., 2005; Radics et al., 2014; Park et al., 2018) while the outer and inner diameter range from 5–13 and 2–2.5 nm, respectively (Kubori et al., 2000; Pastor et al., 2005; Poyraz et al., 2010; Fujii et al., 2012; Loquet et al., 2012; Demers et al., 2013; Radics et al., 2014; Hu et al., 2018). Solid-state NMR structures of the T3SS needles of PrgI from *Salmonella* (Loquet et al., 2012) and MxiH from *Shigella* (Demers et al., 2014) indicated a super helical multimeric structure with 5.7 subunits per turn (Loquet et al., 2012) composed of three intermolecular subunit–subunit interfaces, one of which is axial (subunit i to i+11) and two are lateral (subunit i to i+5/6). Interestingly, the orientation of the two aforementioned helices within the needle structure has been the subject of debate (Fujii et al., 2012; Demers et al., 2013, 2014; Verasdonck et al., 2015). The first cryo-electron microscopy studies performed on the *Shigella* T3SS suggested that the C-terminus of each protomer points toward the outside (Deane et al., 2006; Fujii et al., 2012), while studies involving a combination of cryo-EM and solid state NMR indicated that at least in *Salmonella* and *Shigella* T3SS needles have the N-terminus that points toward the outside of the super-helical structure (Demers et al., 2013, 2014; Verasdonck et al., 2015).

In this work we address the architecture of the T3SS needle of *P. aeruginosa* using a combination of mutagenesis followed by phenotypic studies, proteolysis, mass spectrometry (MS), electron and atomic force microscopies (EM and AFM) and molecular modeling. We show that the *P. aeruginosa* T3SS needle monomer PscF has an N-terminal region oriented toward the outside of the filament, while the C-terminal helix of PscF forms the needle core. Using mutagenesis, we found that D76 is essential for needle polymerization as well as its assembly on the bacterial surface, while residues P47 and Q54 are concomitantly dispensable for polymerization but affect needle stability and effector secretion. Our results, analyzed in the context of other published studies, suggest that a common architecture exists between T3SS needles of different bacterial species, which could facilitate the rational design of

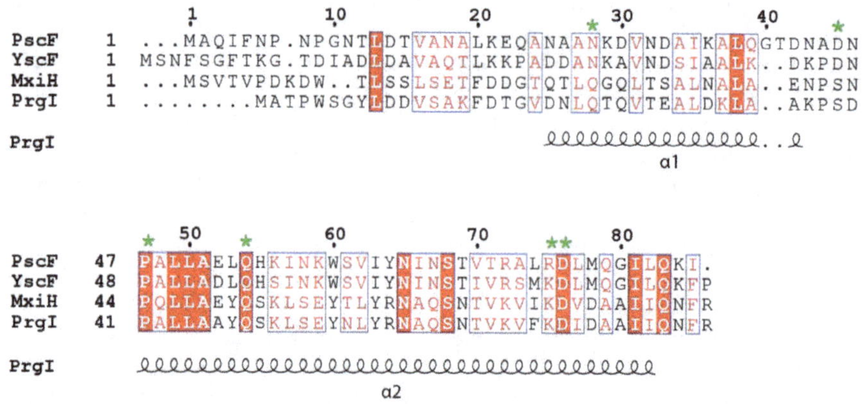

FIGURE 1 | Sequence alignment of needle proteins from different bacteria. PscF from *Pseudomonas aeruginosa*, YscF from *Yersinia* sp., MxiH from *Shigella* sp., PrgI from *Salmonella* sp. The residues mutated in this study are indicated with a green asterisk. Numbers refer to the PscF sequence. Secondary structure elements from the PrgI structure are indicated underneath the sequence alignment. The figure was generated with ESPript, using the new ENDscript server (Robert and Gouet, 2014). Conserved and similar residues are shown in red and blue boxes, respectively.

needle-specific anti-virulence molecules that block or destabilize the T3SS machinery.

RESULTS AND DISCUSSION

Identification of Key Functional Residues of PscF

To uncover the structural elements that are important for the functionality of the T3SS needle in *P. aeruginosa*, we first used a mutagenesis approach coupled to phenotypic studies. We designed seven different PscF point mutants on several bases: conservation of residues between different bacterial species, results obtained in *Shigella flexneri* (Kenjale et al., 2005; Veenendaal et al., 2007), *Yersinia pestis* (Davis and Mecsas, 2007), and *Yersinia pseudotuberculosis* (Torruellas et al., 2005) and predictions/models of PscF fold (**Figure 1**). We generated six single mutants: N28S, D45A, P47A, Q54A, R75A, D76A, and an additional double mutant (P47A/Q54A) whose homologous residues in the *Shigella* needle protein (MxiH-P44A/Q51A) had been shown to be critical for wild-type function (Kenjale et al., 2005; Veenendaal et al., 2007). The D76A mutant had already been studied by our groups and others (Kenjale et al., 2005; Quinaud et al., 2007) and it was further characterized in this study. All mutations were introduced into the pIApG-*pscF* plasmid and transformed into a *P. aeruginosa* clinical isolate CHA strain deleted for the PscF-encoding gene (Δ*pscF*) (Pastor et al., 2005). The cytotoxicity toward macrophages of the Δ*pscF/pscF*wt complemented strain had already been reported to be the same as the original CHA strain (Quinaud et al., 2007). We then tested the ability of all complemented strains to infect macrophages measuring the release of the host cytoplasmic enzyme lactate dehydrogenase (LDH) into the supernatant. Most mutant strains killed macrophages with comparable kinetics as the wild-type strain suggesting the existence of a robust core in the protein that can accept local perturbation. However, the two strains carrying PscF-D76A or the double mutant PscF-P47A/Q54A showed significantly attenuated cytotoxicity (with $P < 0.001$) toward macrophages. The *P. aeruginosa* PscF-D76A strain killed only 7.0 ± 1.8% of all cells present in the assay while *P. aeruginosa* PscF-P47A/Q54A strain killed 43.3 ± 6.2% of cells during 3 h of infection. As a comparison, the wild-type *P. aeruginosa* strain killed 87.5 ± 2.8% of all cells during the same period (**Figure 2A**). Both mutations P47A and Q54A alone had no effect on cytotoxicity after 3 h of infection, in agreement with previous observations of hemolysis and invasion in *Shigella* (Kenjale et al., 2005).

To understand the cause of decreased virulence observed with the selected mutants, we first analyzed the expression of the PscF mutants and the secretion profiles of translocator proteins PopB and PcrV in those strains (**Figure 2B**). The T3SS was induced *in vitro* by Ca^{2+} depletion in bacterial cultures, the pellet (containing the entire cell content) and supernatant (containing secreted proteins) were separated by centrifugation and analyzed by Western blotting. All *P. aeruginosa* mutants expressed the different PscF variants (**Figure 2B**), demonstrating that the mutations did not affect the expression of PscF. Analysis of PopB and PcrV in secreted fractions showed reduced levels for the P47A/Q54A strain and a thin band of PopB and no PcrV for the D76A strain, results that are consistent with the decreased cytotoxicity of these two mutants in our macrophage killing assay.

Our results point out that among the conserved residues of the needle protein PscF, D76, P47 and Q54 play essential roles in T3SS functionality, in agreement with other reports on *Shigella flexneri* and *Yersinia pestis* (Kenjale et al., 2005; Torruellas et al., 2005). Interestingly in *Shigella*, the corresponding mutants (MxiH-D73A and P44A/Q51A) secrete Ipa/Ipg proteins constitutively, and are unable to invade epithelial cells and to cause hemolysis (Kenjale et al., 2005; Veenendaal et al., 2007). Other mutations like PscF-R75A in *P. aeruginosa* have no effect on cytotoxicity toward macrophages, while the homologous mutation (MxiH-K72A) in *Shigella* has a significant impact in hemolysis, suggesting a potential different mode of regulation/activation

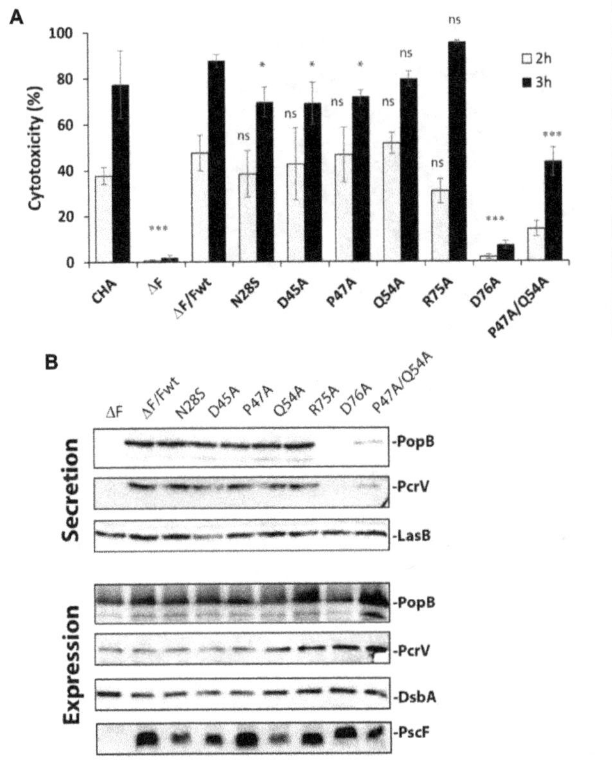

FIGURE 2 | T3SS activity in *Pseudomonas* PscF mutants. **(A)** The cytotoxicity of *P. aeruginosa* strains toward macrophage cells was measured by monitoring LDH release after 2 and 3 h of infection using a multiplicity of infection (MOI) of 5. The PscF deletion mutant (ΔF), D76A, and P47A/Q54A strains show a higher statistical difference at both 2 and 3 h of infection compared to wild-type complemented strain (ΔF/Fwt). LDH measurements were corrected with values corresponding to cells that were not infected; we considered as 100% of cytotoxicity a 1% Triton X-100 treated well. Pairwise differences relative to ΔF/Fwt based on the Tukey test are indicated: ns, non-significant, ***$P < 0.001$, *$P < 0.05$. Cytotoxicity of original CHA strain is also reported. **(B)** Western blot of the supernatant (secreted proteins) and total bacteria (expression) fractions after centrifugation were developed with anti-PopB and anti-PcrV antibodies. The expression of PscF variants was confirmed by loading total bacterial extracts on SDS- 15% PAGE. The Western blot was developed using anti-PscF antibodies obtained using a monomeric 6His-PscF (see section "Materials and Methods"). The D76A strain secretes almost no PopB or PcrV while P47A/Q54A secretes less of both translocators as compared to other mutants and the wild-type strain. LasB was used as loading control for the secreted protein, and DsbA as loading control for bacterial expression and as a lysis control for the supernatant (data not shown).

of needle protein polymerization. We thus set out to further understand the specific role of these residues in regards to: (i) proper folding of PscF, (ii) PscF's ability to polymerize and form the needle, (iii) interaction of PscF with the tip translocator protein PcrV, (iv) secretion, localization and abundance of the PscF needle on the bacterial surface.

In vitro Characterization of PscF Needle Mutants

In order to investigate the first two possibilities, i.e., that D76A and P47A/Q54A mutations could affect PscF folding and needle polymerization, we expressed and purified the two PscF mutants in *E. coli* in the absence of their two chaperones PscE and PscG. This allowed the *in vitro* production of filaments, as it was shown in the case of wild-type PscF (Quinaud et al., 2005). Purified PscF wild-type, PscF-D76A and PscF-P47A/Q54A filaments were then analyzed by negative staining electron microscopy (EM) and atomic force microscopy (AFM) (**Figure 3A** and **Supplementary Figure S1**). Both the wild-type and the P47A/Q54A PscF variants formed elongated fibers with outer diameters ranging from 2 to 10 nm indicating that the P47A/Q54A mutation does not prevent the needle polymerization process. The measurement of the diameters of both filaments on EM images suggested a tendency of P47A/Q54A PscF filaments to be slightly thinner than the wild-type ones with a statistically significant difference on diameter size distribution (*t*-test, $P < 0.001$) (**Figure 3B**). The diameter measurements by both EM and AFM thus indicated smaller values than the 8 nm-diameter determined by high resolution methods. This is probably due to a difficulty in identifying the borders of thin needles by EM, and the strong adsorption of the filaments on mica in the AFM experiments, as already found for cylindrical viruses (Godon et al., 2017). The D76A mutant protein could not form filaments suggesting a loss of capacity to polymerize (**Figure 3A**), thus explaining the lack of cytotoxicity of *P. aeruginosa* strains expressing PscF-D76A. On the other hand, in the PscF-P47A/Q54A mutant the 50% decrease in cytotoxicity could be linked to the formation of thinner fibers or could be correlated to a more subtle folding defect of the needle filament.

In order to investigate this last hypothesis we incubated both wild-type and P47A/Q54A PscF needle filaments with trypsin at room temperature and monitored protease digestion for up to 23 h. Samples were taken at different times and analyzed by SDS-PAGE and by liquid chromatography coupled to electrospray ionization mass spectrometry (LC/ESI-MS) (**Figure 3C** and **Table 1**). In our experiments, PscF filaments are a result of the polymerization of a 93-residue monomeric protein that carries a 6His-tag at the C-terminus with an expected mass of 10,218 Da (for the wild-type) and 10,135 Da (for the P47A/Q54A mutant). The SDS-PAGE patterns were apparently similar: both wild-type and P47A/Q54A proteins were surprisingly stable to trypsin digestion, and the full-length protein was also still apparent after 23 h. A major fragment at 6,500 Da (corresponding to fragment Ala37-His93) appeared on SDS-PAGE after 10 min (one asterisk in **Figure 3C**). The kinetics of trypsin cleavage were followed by LC/ESI-MS analysis at different time points (10, 30 min and 6 h of digestion) which allowed precise identification and relative quantification of proteolytic fragments (**Table 1**). All expected trypsin cleavage sites and the fragments identified by LC/ESI-MS are reported in **Figure 3C**. The first trypsin cleavage occurred after Lys21 and Lys36, generating the following three fragments: (Ala2-Lys21), (Ala2-Lys36) and (Ala37-His93). Subsequently, cleavage occurred after Lys56, leading to fragments (Lys37-Lys56) and (Ile57-His93). Moreover, LC/ESI-MS analyses of digested fragments highlighted interesting differences between PscF wild-type and PscF P47A/Q54A needle filaments. Indeed, an additional digested band at a lower molecular mass was visible on SDS-PAGE after 90 min for the wild-type filaments

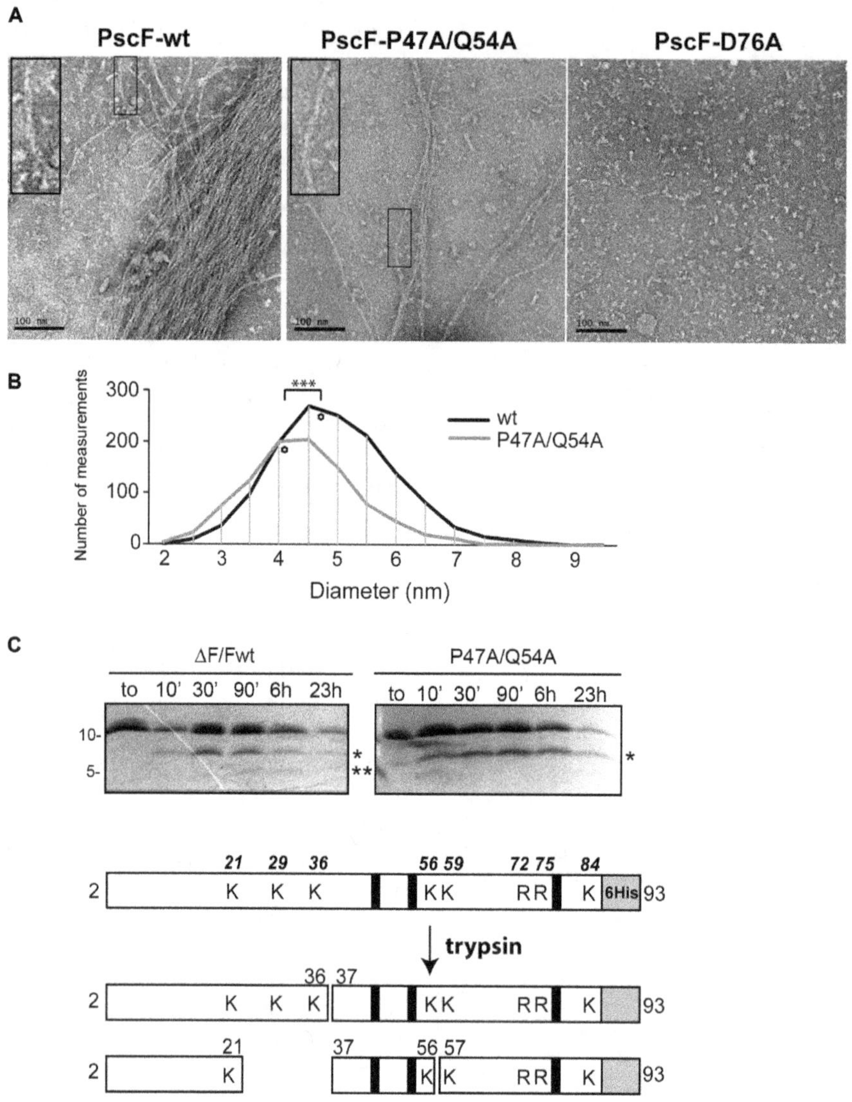

FIGURE 3 | Characterization of PscF wild-type and mutant filaments. **(A)** Negative-staining electron microscopy images of purified PscF wild-type (left panel), PscF P47A/Q54A (middle panel) and PscF D76A (right panel) visualized at the same protein concentration (0.1 mg/ml). Scale bar: 100 nm. **(B)** Diameter distribution of PscF filaments measured on EM images shown in **(A)**. The diamonds represent the average diameter. Wild-type PscF has an average diameter of 4.72 ± 1.02 nm (n = 1390) while PscF P47A/Q54A filaments are thinner with average diameters of 4.15 ± 0.97 nm (n = 986). Statistical analysis: ***P < 0.001. **(C)** Trypsin digestion profiles of purified wild-type PscF and P47A/Q54A filaments were analyzed by 16.5% Tris-Tricine and LC/ESI Mass spectrometry. Both proteins are stable with a major cleavage-band at 6.5 kDa (one asterisk) corresponding to fragments 37–93. Only for the wild-type protein is visible a second lower band of 4.5 kDa (residues 57–93) (two asterisks). A scheme of trypsin digestion kinetics determined by MS analysis is reported below, with the corresponding identified fragments. The gray box represents the 6His tag, black lines show the position of D76, P47, and Q54 residues mutated in this study. All trypsin-cleavable sites (lysines, K and arginines, R) are shown with their relative position on the top scheme.

only, corresponding to specific cleavage after Lys56 (two asterisks in **Figure 3C**). This fragment of 4,454 Da (residues Ile57-His93) for the PscF-P47A/Q54A mutant was not visible on SDS-PAGE and detectable only by MS analysis after 6 h in the proteolysis experiment representing 1.1% of the fragment population versus 10.5% in the wild-type PscF at the same time point (**Table 1**). These results suggest that Lys56 is less accessible to proteolysis by trypsin in the PscF-P47A/Q54A filament compared to the PscF wild-type filament. An opposite effect was observed for Lys36, that is more accessible to trypsin cleavage in the PscF-P47A/Q54A filament (with 10.7% of fragment population corresponding to Ala37-His93 with a mass of 6,472 Da) compared to the PscF wild-type filament (with 2.5% of fragment population) (**Table 1** and one asterisk in **Figure 3C**). Therefore, the two filaments are structurally different at least in the region around Lys36 and Lys56.

These results indicate that the P47A/Q54A and wild-type filaments only present a small difference on the outer surface structure (see below). Along with the fact that P47A/Q54A filaments could be slightly thinner, this could explain the decreased capacities of secretion and cytotoxicity of the strains carrying these mutations. However, one must also consider that the filaments purified form *E. coli* could be structurally distinct from needles assembled within the T3SS apparatus in *P. aeruginosa*.

Needle Visualization on the Surface of *P. aeruginosa* PscF Mutants

In order to further characterize the aforementioned mutations on PscF in a more physiological context, we looked at the localization and/or abundance of the needle on the *P. aeruginosa* surface as well as its interaction with the PcrV tip protein using fluorescence microscopy. We performed immune detection of PscF on fixed cells, using antibodies generated against the PscF native needles (produced during this study) (**Figure 4**). Fixed cells were prepared as described in Section "Materials and Methods" and then separated into two samples: one was incubated with anti-PcrV antibodies (shown in cyan), while the other one was incubated with anti-PscF antibodies (shown in green). In both samples, DNA was stained with SYTO24 (shown in red). The Δ*pscF* and Δ*pcrV* samples were used as negative controls and presented no unspecific labeling with the anti-PcrV antibodies and just weak background with the anti-PscF used at higher concentrations. ΔPscF/PscF-wt and ΔPcrV/PcrV-wt were used as positive controls and showed clearly distinct cell surface-associated fluorescent spots for both proteins all around the bacteria (**Figures 4A,B**). Of note, PcrV could not be detected on the surface of the ΔPscF strain because it only assembles at the tip of the PscF needle. All PscF mutants showed similar patterns of labeling as the positive controls except for the PscF D76A- and P47A/Q54A-expressing strains that showed markedly smaller amounts of PscF and PcrV spots all around their surface (**Figure 4B** and **Supplementary Figure S2**). To eliminate the possibility that the absence of PscF on the *P. aeruginosa* surface was due to a defect in expression of PscF D76A and P47A/Q54A, we repeated the experiment with cells made permeable with Triton X-100 in order to allow the entry of the antibody into the cell. The antibodies directed toward PscF detected the protein in both permeabilized mutant strains (**Supplementary Figure S3**). These data are in agreement with the Western blot results and indicate that PscF mutant proteins were produced within *P. aeruginosa*. Moreover, this experiment showed that D76A and P47A/Q54A mutants have a defect in PscF export and/or assembly onto the bacterial surface.

We then quantified the PscF and PcrV spots per cell using the MicrobeJ plugin (Ducret et al., 2016) of the ImageJ program (Schneider et al., 2012), considering that one spot corresponded to one needle. Between 30 and 40% of wild-type bacterial cells showed one or more PscF spots (up to 6) per cell, with a majority of cells with 1 or 2 spots/cell (**Figure 5B**). Immunolabeling of bacteria expressing PscF D76A showed that only 6.9 ± 1.6% of the counted cells (n = 1713) presented a single visible spot and the P47A/Q54A mutant displayed 8.9 ± 2.4% of the population with 1 or 2 spots per cell (n = 2307). For comparison, the complemented ΔPscF/PscF-wt stain presented 32.8 ± 4.0% of cells with at least one needle (n = 1829) (**Figures 5A,B**) while the negative control ΔPscF strain presented 2.3 ± 1.4% of spots near the bacteria. In all samples, except in D76A and P47A/Q54A, the percentage of cells with PscF and PcrV spots was very similar, indicating that the mutations do not affect binding between the tip protein and the needle. In the case of D76A and P47A/Q54A only 3.2 ± 1.2% (n = 2126) and 5.6 ± 2.5% (n = 2128) of cells presented PcrV spots, respectively, suggesting that the tip/needle interaction is present, but we could not exclude that some D76A and P47A/Q54A needles do not present PcrV at their tip.

These observations explain the absence of cytotoxicity toward macrophages in the case of D76A. Interestingly, although P47A/Q54A mutations affected the number of needles detected,

TABLE 1 | Identification and relative quantifation of PscF (*wt*/mutant) proteolytic fragments by LC/ESI MS.

Proteolysis time	10 min		30 min		6 h	
Theoretical mass (Da) for tryptic fragments of PscF (wt/mutant)	PscF-WT	PscF-P47A/Q54A	PscF-WT	PscF- P47A/Q54A	PscF-WT	PscF- P47A/Q54A
	Measured mass (Da) *(% Area of EIC peaks)*					
2097.09 [(1)] (2-21)	2097.08 *(9.4%)*	2097.09 *(5.6%)*	2097.09 *(18.1%)*	2097.09 *(10.8%)*	2097.08 *(28.1%)*	2097.09 *(30.8%)*
2118.07/2035.03 [(1)] (37-56)	2118.07 *(1.6%)*	2035.02 *(0.2%)*	2118.07 *(3.7%)*	2035.03 *(0.3%)*	2118.07 *(17.6%)*	2035.03 *(3.5%)*
3678.86 [(1)] (2-36)	3678.87 *(7.7%)*	3678.86 *(13.2%)*	3678.85 *(11.4%)*	3678.85 *(12.6%)*	3678.85 *(2.9%)*	3678.85 *(3.9%)*
4454.15 [(2)] (57-93)	4454.36 *(2.6%)*	n.d.	4454.52 *(3.4%)*	n.d.	4454.88 *(10.5%)*	4454.18 *(1.1%)*
6555.45/6472.36 [(2)] (37-93)	6555.63 *(9.4%)*	6472.50 *(9.1%)*	6555.72 *(9.4%)*	6472.55 *(8.5%)*	6555.60 *(2.5%)*	6472.55 *(10.7%)*
10218.47/10135.38 [(3)] (2-93)	10218.67 *(69.3%)*	10135.59 *(71.9%)*	10218.68 *(54.0%)*	10135.46 *(67.8%)*	10218.80 *(38.5%)*	10135.46 *(49.8%)*

Theoretical mass values for tryptic fragments of PscF (wild-type and P47A/Q54A mutant) are reported in the first column. Both PscFwt and P47A/Q54A mutant were digested with trypsin (trypsin:protein ratio 1:500 w/w) at room temperature. After 10 min, 30 min and 6 h of incubation the samples were analyzed by LC/ESI-MS. The area of the extracted ion chromatogram (EIC) peaks for each fragment were calculated and reported as a percentage of the sum of all fragment (total area of all the peaks). [(1)] monoisotopic mass, [(2)] average mass, [(3)] mass value for the intact undigested protein, missing the methionine at N-termini; n.d., not detected.

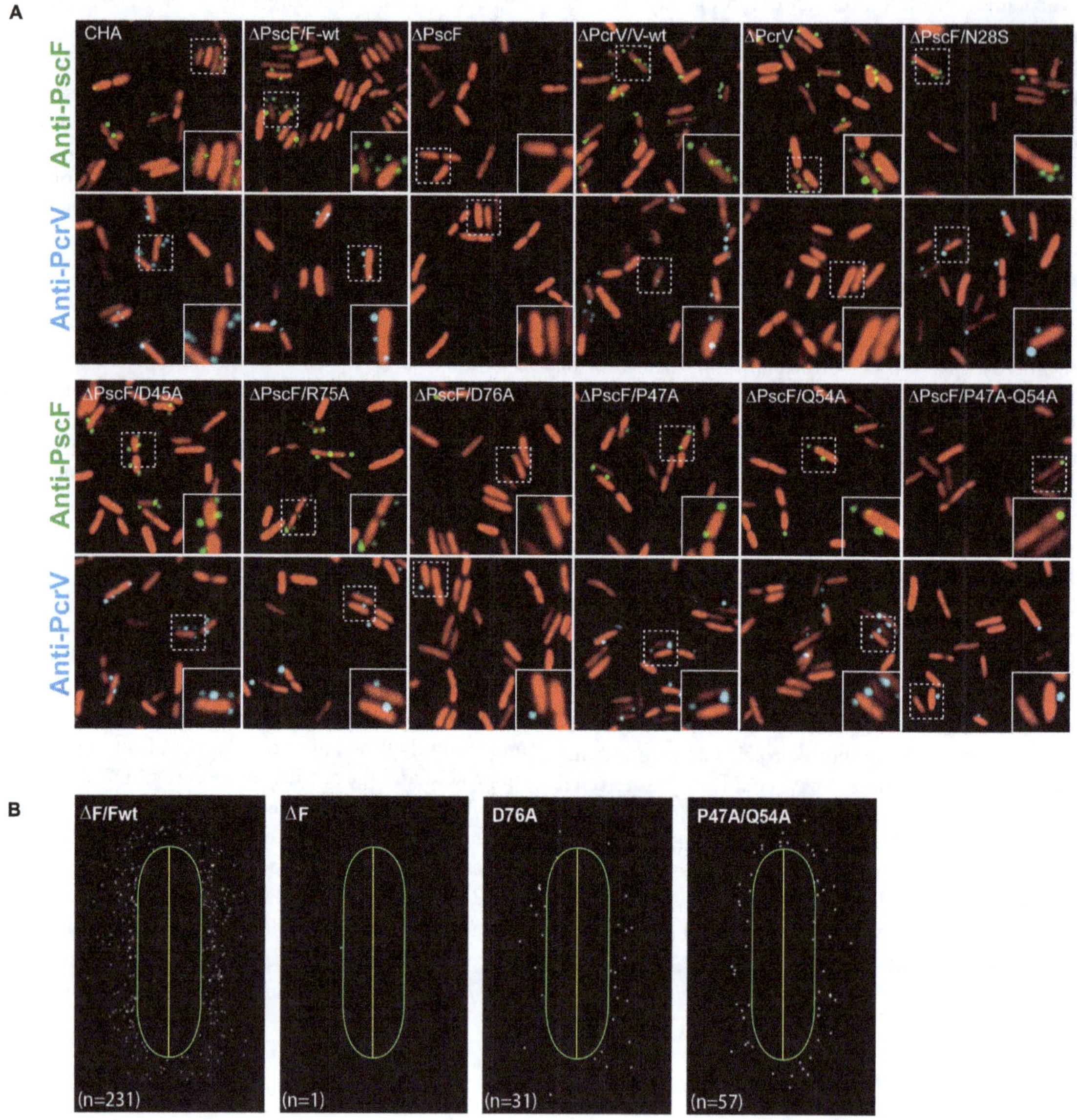

FIGURE 4 | Localization of PscF-needles on the surface of *Pseudomonas aeruginosa*. **(A)** *P. aeruginosa* CHA strain isolated on cystic fibrosis patient depleted for the *pscF* gene (Δ*pscF*) and complemented with pIApG-*pscF* constructs were grown in T3SS-inducing conditions. PscF needles were visualized by immuno fluorescence on fixed bacteria using anti-PscF (in green) or anti-PcrV (in cyan). SYTO24 was used to visualize bacteria (in red). As a negative controls we used *P. aeruginosa* ΔF and ΔV, while ΔF/Fwt and ΔV/Vwt were used as a positive control. Only few PscF and PcrV spots were visible on *P. aeruginosa* strains carrying a PscF-D76A or a PscF-P47A/Q54A. **(B)** MicrobeJ reconstitution of PscF distribution around the bacteria. n = number of spots in the image.

the synthesized needles still allowed injection of toxins, suggesting that the number of needles on the bacterial surface is less crucial than the structure of the needle *per se*. All these data strongly suggest that in *P. aeruginosa* D76A and P47A/Q54A mutants are defective in the *in vivo* needle assembly process while a concomitant instability or defect in the export of PscF through the basal body could not be excluded.

In *Yersinia pestis*, the YscF-D77A variant also shows a defect in the surface exposure of the needle whilst maintaining normal protein expression levels (Torruellas et al., 2005). In contrast, the MxiH-D73A and P44A/Q51A mutants in *Shigella flexneri* were both able to export the needle protein and assemble a NC almost normally (Kenjale et al., 2005; Veenendaal et al., 2007). Furthermore, MxiH-D73A was able to polymerize (Fujii et al., 2012) (unlike PscF-D76A) with no significant structural changes but it lacks the tip complex. Once again, this suggests that the same mutation will cause different effects on the assembly of *Shigella, Yersinia,* and *Pseudomonas* T3SS needles.

Modeling of the *P. aeruginosa* PscF Needle

In order to understand at the molecular level the impact of the D76A and P47A/Q54A mutations, we performed molecular modeling of the *P. aeruginosa* PscF needle, taking advantage of the solid-state NMR structure of the *Salmonella* PrgI needle

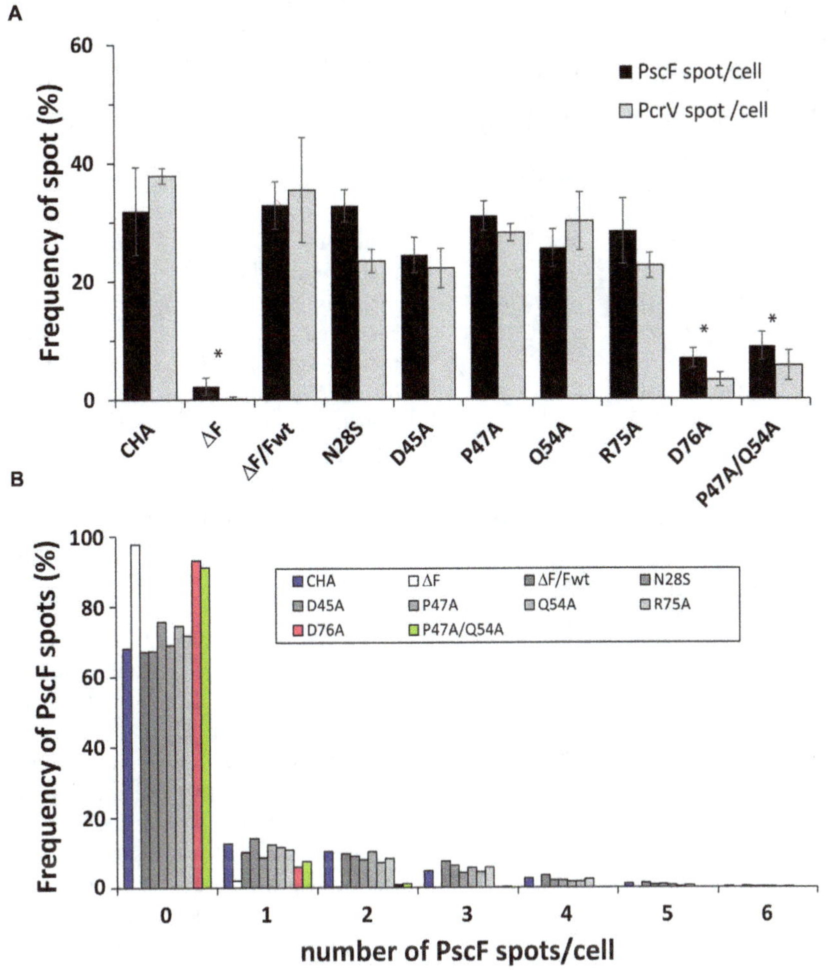

FIGURE 5 | Presence of PscF and PcrV on *Pseudomonas aeruginosa* PscF-mutant strains. **(A)** Images of immunolabeled fixed bacteria in **Figure 4** were analyzed with MicrobeJ to detect and associate each PscF or PcrV spot to only one bacterium. Frequency was considered as the percentage of bacteria with at least one PscF and PcrV spot on the total counted bacteria. About 30–40% of cells had needles with the PcrV tip protein on their surface, with the exception of the D76A strain with only 6.9 ± 1.6% (PscF) and 3.2 ± 1.2% (PcrV), and the P47A/Q54A strain with 8.9 ± 2.4% (PscF) and 5.6 ± 2.5% (PcrV) of cells with spots.
(B) Distribution of the number of PscF spots/needles per bacterial cell. Between 630 and 2300 individual bacteria were counted. *P. aeruginosa* wild-type and mutant strains had between 1 to 3 PscF spots per cell, except for D76A (in pink) and P47A/Q54A (in green) that presented only one or two spots per cell. Overall comparisons using the Kruskal–Wallis' test indicates significant differences between classes ($P < 0.001$). Pairwise differences relative to wild-type based on Dunn's *post hoc* test are shown: $^{*}P<0.05$.

(Loquet et al., 2012). We constructed two different models for the PscF needle: one using the protein in the same orientation as PrgI in the filament (i.e., with the C-termini inside and the N-termini outside the needle; that model will be named PscF-forward); and the other one supposing that the PscF protomer is inverted with the N-termini residue inside and the C-termini residue outside the needle structure (PscF-reverse) (**Figure 6**). Optimal models were subjected to a brief structural minimization protocol using the Chimera molecular modeling/visualization package (Pettersen et al., 2004) (default parameters) to remove atomic clashes (VDW overlap). Pre- and post-minimization structures were analyzed using the PSVS webserver (Bhattacharya et al., 2007) to evaluate the structural improvements in the monomeric models. After minimization, all homology models showed significant improvements in both clash scores and favorable Ramachandran angles. The energetically minimized structures were then used to reconstruct PrgI-like T3SS needles, and were built up from 29 monomers to form the needle filaments. To compare the two models of PscF to the needle architecture in *Salmonella* we also constructed a PrgI-inverted needle (PrgI-reverse).

The three resulting needle architectures: PscF-reverse, PscF-forward and PrgI-reverse are shown in **Figure 6**. We then tried to understand if one of the two PscF needle models (PscF-reverse or PscF-forward) was more favorable. To do so, we computed several parameters related to the global architecture: surface electrostatic potential (**Figure 6A**), position of conserved residues (**Figure 6B**) and hydrophobicity (**Figure 6C**) and compared them to the experimental solid-state NMR *Salmonella* PrgI needle structure. Surprisingly none of the PscF needle models showed

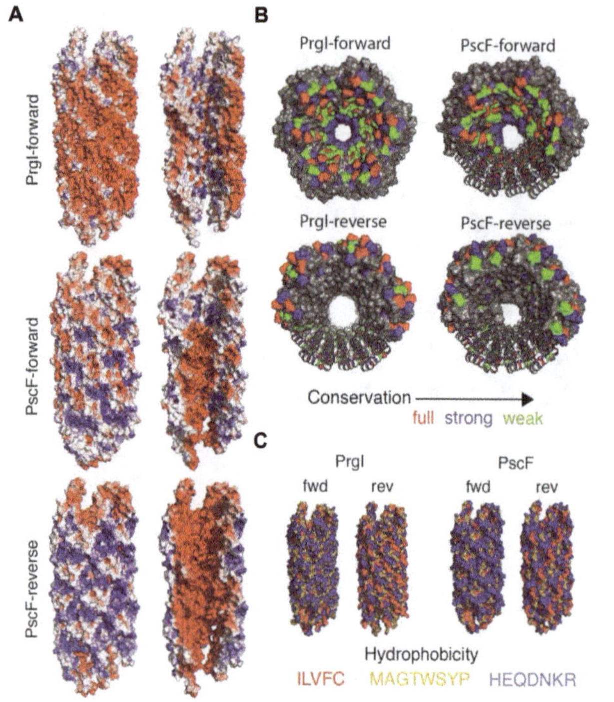

FIGURE 6 | Needle reconstruction for PscF wild-type and mutants. **(A)** The PrgI Solid-state NMR structure was used as a model for the construction of PscF in the Forward and Reverse orientations. Mutants D76A and P74A/Q54A were constructed on the PscF-forward model. The exterior view (left) and lumen view (right) of the T3SS needle models are colored by surface electrostatic potentials, at pH 7, using the APBS plugin for PyMol. **(B)** Distribution of conserved residues distribution on PrgI and PscF needle models in Forward and Reverse orientations. **(C)** Hydrophobicity distribution of the different models shown in **(B)**.

the same electrostatic distribution as the experimental PrgI needle structure (top in **Figure 6A**), displaying an astonishing pattern of circular succession of negative/positive charges at the lumen surface, while the external surface was predominantly negatively charged. Both PscF needle models showed a strong negative charge distribution at the inner surface, while the external surfaces were much more positively charged compared to PrgI needles. In addition, no particular succession of positively/negatively charged patterns could be seen for either PscF model. In PrgI needles, this lumen pattern arises mostly from D70 and Q77 (negative) and K66 and R80 (positive). Equivalent amino acid positions in PscF D76, Q83, and R72 are conserved (**Figures 1, 6B**); however, R80, the last residue in PrgI, is absent from PscF, leading to a pronounced change in the electrostatic charge distribution at the lumen surface.

Next, we investigated the hydrophobic surface of the models (**Figure 6C**) considering that the PrgI needle is characterized by a strong network of hydrophobic interactions at the lateral interfaces between subunit i and i+5/6 (Loquet et al., 2012). PscF-reverse and PrgI-reverse models both exhibit an unfavorable hydrophobic external surface while the PscF-forward model shows a favorable hydrophilic surface, strengthening the energetic relevance of the PscF-forward model. Finally, we examined the distribution of the conserved residues on the different needle models (**Figure 6B**). The PrgI needle structure has been defined by a typical distribution of conserved residues among the T3SS needle subunits, with almost all conserved residues pointing inside the needle pore while the external surface is made of non-conserved amino acids. Visualization of these conserved needle residues on the PscF models lead to two distinct situations: PscF-forward presents many conserved residues at the lumen surface, while this lumen surface in the PscF-reverse model is mostly composed of non-conserved residues. It has been suggested that the presence of non-conserved residues exposed to the external surface of PrgI could provide a way to evade immune response of the host cell (Loquet et al., 2012). The same distribution is observed for the PscF-forward model, and together with our observations on the hydrophobic and electrostatic surfaces it provides a substantial hint toward the relevance of the PscF-forward model compared to the PscF-reverse model.

Subsequently, we constructed PscF needle models for the two mutants: D76A and P47A/Q54A (**Figure 7B**), adopting the forward model orientation. In the D76A model, the distribution of the charges is drastically affected by the point mutation; D76 is located in the central hollow section of the needle and the mutation into alanine modifies the overall negative charge distribution, which could be an explanation to the lack of polymerization *in vitro* of this mutant. The P47A/Q54A model, however, suggests that the mutated residues are buried in the subunit–subunit interfaces and the overall charges are more similar to the wild-type PscF-forward model. According to the atomic depth measurements (Chen and Pellequer, 2013), the regions 45–47 is the deepest, farthest away from the solvent, in the whole needle. Any change in deeply buried residues is expected to impact proteins stability (Chakravarty and Varadarajan, 1999), that could explain the significant difference in the number of needle on bacteria surface. Considering the equivalent amino acids in PrgI (i.e., P41 and Q48), we examined the residues involved in the subunit–subunit interfaces by taking into account the solid-state NMR restraints used for the PrgI needle structure determination (see **Figure 2** in Loquet et al., 2012). P41 in PrgI is involved in strong contacts between subunit i to i+11 forming the so-called axial interface, while Q48 is weakly involved in the intramolecular fold. This suggests that the double P47A/Q54A mutation in PscF could lead just to a moderate perturbation of the needle arrangement that could explain why this mutant still polymerizes *in vitro*.

We then used the data obtained from trypsin proteolysis and MS analysis to explore the structural differences between wild-type and mutant P47A/Q54A filaments and to validate our model. As mentioned above, we monitored trypsin digestion of both fibers by LC/ESI-MS (**Figure 3C** and **Table 1**). This experiment allowed us to conclude that the residues that were the most accessible to trypsin digestion, and thus were solvent exposed are: Lys36, Lys21, and Lys56 (and probably Lys29, absent in MS digested-fragments). Interestingly, these experimental results are only in agreement with the PscF-forward wild-type model (with the N-termini exposed toward the outside), where Lys36, Lys29, and Lys21 are expected to be the most exposed

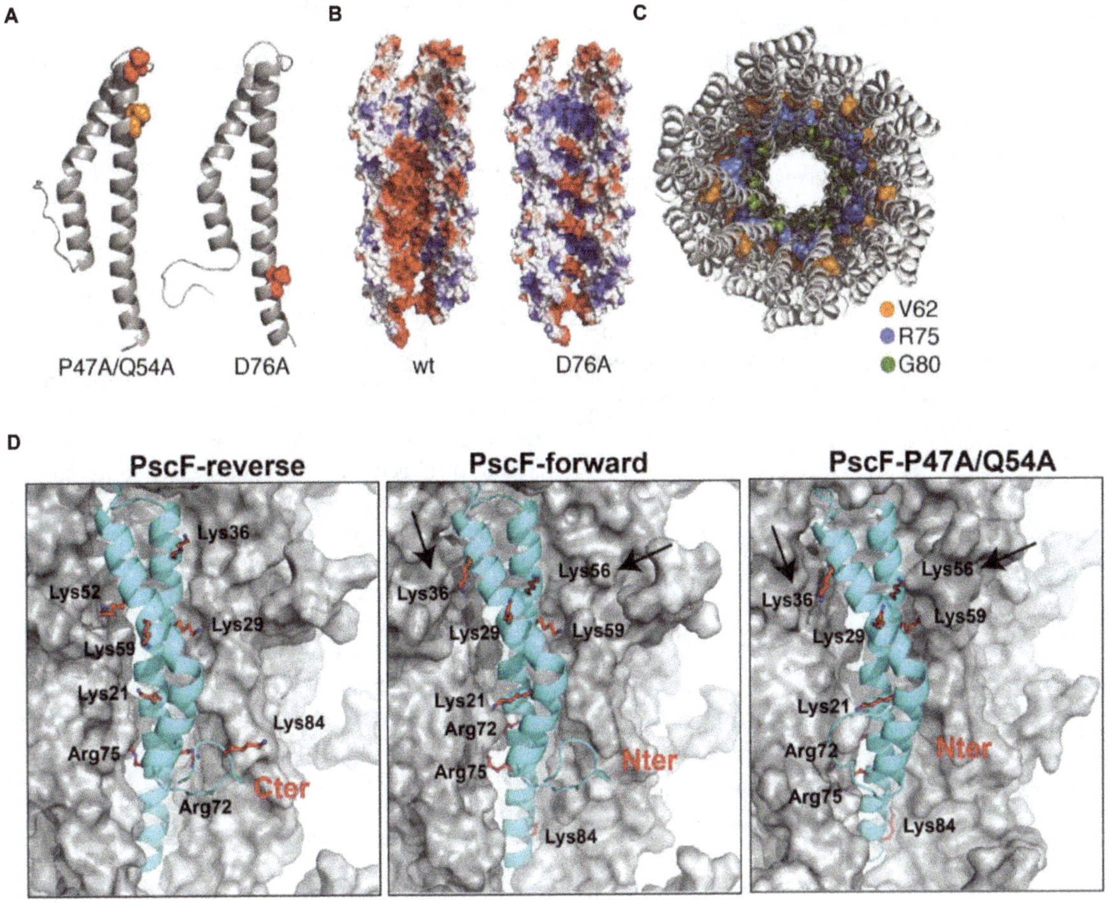

FIGURE 7 | Model of the PscF P47A/Q54A and PscF D76 assembled filament. **(A)** Ribbon representation of PscF monomer carrying the two mutations: D76A and P47A/Q54A. **(B)** Electrostatic surface representation of the two mutant reconstituted filaments with a channel view. **(C)** Top view of a PscF-forward wild-type model, colors show residues implicated in binding of the T3SS-inhibitor phenoxyacetamide (Bowlin et al., 2014). **(D)** Comparison between reverse and forward PscF and PscF-P47A/Q54A filaments. PscF-forward model (with the N-termini exposed toward the outside), with Lys36, Lys29, and Lys21 solvent-exposed, Lys56 and Lys59 less accessible, and Lys84, Arg72, and Arg75 completely buried. PscF-reverse model (with the C-termini exposed toward the outside), with solvent accessible residues: Lys84, Arg75, Lys59, and Lys36. The PscF-P47A/Q54A forward model showing accessibility of the same residues as compared to the PscF-forward wild-type model except for the Lys56 and Lys36 that are slightly less or more trypsin-accessible, respectively (indicated by two arrows). All needle models are represented as surfaces with one monomer of PscF represented as a cyan ribbon. All trypsin-cleavable sites are shown as sticks.

to the solvent, Lys56 and Lys59 are less accessible, and Lys84, Arg72 and Arg75 are completely buried (middle in **Figure 7D**). On the other hand, in the PscF-reverse wild-type model (with the C-termini exposed toward the outside), the residues that would be solvent accessible are: Lys84, Arg75, Lys59, and Lys36 (left in **Figure 7D**). Lys84, in particular, is located just before the C-terminus. In this configuration, the C-terminal 6His-tag would have been expected to be quickly cleaved by trypsin, but no fragment corresponding to this cleavage was observed for any of the peptides analyzed by LC/ESI-MS (**Table 1**). Moreover, the PscF-forward P47A/Q54A model with Lys56 less solvent-exposed and Lys36 more solvent-exposed (right in **Figure 7D**) is also in agreement with our trypsin digestion results. The digestion-fragment at 4,454 Da (corresponding to a cleavage after Lys56) was less abundant in the mutant compared to the wild-type (1.1 vs. 10.5%), while the digestion fragment corresponding to a cleavage after Lys36 was more abundant in the mutant compared to the wild-type (10.7 vs. 2.5%) as discussed above (**Figure 3C** and **Table 1**).

Thus, based on our biochemical, structural and molecular modeling data, we hypothesize that the T3SS needle of *P. aeruginosa* follows the PscF-forward model with the N-terminus pointing toward the outside of the fiber and the C-terminus consequently hidden inside the needle's hollow cavity. Our results imply that the overall architecture of the needle is therefore conserved between *Pseudomonas*, *Salmonella*, and *Shigella* spp., highlighting the importance of the orientation of the small 9-kDa protein inside the oligomeric structure. D76A and P47A/Q54A mutations are located at the C-terminal helix, which is well-conserved in the sequence among bacterial species. Their effects on needle polymerization and stability as well as their importance for functional regulation of needle assembly as shown by our analyses of electrostatic charges present at the lumen surface point to an important role of the C-terminal helix in the assembly mechanism. The impact of the mutations D76A and P47A/Q54A, and their homologs in T3SS needles of other bacteria, suggests that regions facing the inner cavity or deeply buried within the needle play a significant role in needle stability and/or function. It also suggests that local changes of the sequence of the subunit protein in crucial regions, as exemplified here for the two mutants and in previous studies, might lead to important perturbations on different aspects of the needle structure-function interplay, on both a structural (e.g., structural instability, surface electrostatic distribution) and functional basis. Moreover, the charge distribution in the lumen may be critical in needle assembly as suggest by the defect in polymerization in the D76A variant and the previous observation that the PscF-D76A protein overexpressed in *P. aeruginosa* CHA strain has a dominant-negative effect on assembly of the wild-type PscF needle (Quinaud et al., 2007).

Understanding the Mode of Action of T3SS Inhibitors

To go further we used our proposed *P. aeruginosa* T3SS needle model to understand the mode of action of T3SS-inhibitors, such as the phenoxyacetamide family, that was hypothesized to act inside the cavity of PscF (Bowlin et al., 2014). In this work, the authors suggested that PscF residues V62, R75 and G80 were involved in direct binding of phenoxyacetamide. In our model, only G80 is exposed to the lumen surface (**Figure 7C** in green), while V62 (in orange) and R75 (in blue) are located at subunit interfaces. These observations suggest that the three amino acid positions, upon binding with phenoxyacetamide, may decrease needle stability (for V62 and R75); in addition, it is also possible that binding could change the surface of the lumen (G80). These three residues appear to be close enough in our PscF needle model to speculate the presence of a single binding site, as suggested by the Moir group (Bowlin et al., 2014). We therefore suggest that it is possible to design other molecules that interact with multiple residues starting from the lumen structure model in order to block T3SS action.

The T3SS needle represents an excellent and promising molecular target in the fight against Gram-negative pathogens, since it is easily accessible to small molecules that do not need to cross bacterial membranes, and is non-essential for survival. Thus, small molecules affecting its function could block virulence without eliciting high levels of antibiotic resistance.

Our work provides a new model for the *P. aeruginosa* PscF-needle, that was validated *in vitro* by experimental data combining mutagenesis, proteolysis, and Mass Spectrometry experiments. Moreover, the phenotypic characterization of PscF mutants *in vivo* showed that the majority of the single mutations introduced in the needle have no effect on T3SS-toxicity and are not sufficient to disrupt needle polymerization, with the exception of D76A. This underlines the importance of the charge in the face of the lumen for assembly of a functional needle.

A detailed molecular comprehension of the structure–function relationship of the T3SS needle should advance future developments of anti-virulence molecules.

MATERIALS AND METHODS

Bacterial Strains and Culture Conditions for T3SS Expression

Strains and plasmids used in this study are presented in **Supplementary Table S2**. Cytotoxic *P. aeruginosa* cystic fibrosis isolate CHA carrying appropriate chromosomal deletions of *pscF* gene (Δ*pscF)* was transformed with plasmid pIApG-*pscF* wild-type or mutants that expressed *pscF* with a T3SS inducible promoter. Cultures were grown overnight in Luria-Bertani (LB) at 37°C at 300 rpm in the presence of 300 μg/ml of carbenicillin. Then next day, they were diluted to an optical density measured at 600 nm (OD_{600}) of 0.1 A.U. in LB in T3SS-inducing conditions by adding 5 mM EGTA and 20 mM $MgCl_2$ (Pastor et al., 2005; Dasgupta et al., 2006). When the OD_{600} reached 1.0 A.U., cells were centrifuged, and then the pellet (entire bacterial protein content) and the supernatant (the secreted protein fraction) was analyzed by Western blotting. Anti-PopB and anti-PcrV antibodies (Goure et al., 2004) were diluted to 1:10,000 and 1:3,000, respectively. Anti-PscF antibodies were raised in rabbits (Covalab, France) using monomeric 6His-PscF (Quinaud et al., 2005). The antibodies were further purified on Protein-A column and diluted 1:500 for Western blot analysis. As a loading control we used LasB (a secreted protein) and DsbA (a periplasmic protein) diluted to 1:2,000 and 1:10,000, respectively. Lysis control was done on secreted protein fractions using DsbA antibodies. All mutants were produced using the Quick-Change site-directed mutagenesis II kit (Stragatene). Primers are listed in **Supplementary Table S2**.

Cytotoxicity Assay

Lactate dehydrogenase release into the supernatant was measured using the Cytotoxicity Detection Kit by Roche Applied Science, following the recommended protocol. Briefly, J774 cells were seeded at 2×10^5 in 48-well plates the day prior to the experiment, and infected in DMEM medium at MOI of 5 with *P. aeruginosa*Δ*pscF*-pIApG-*pscF* constructs at OD_{600} = 1 A.U. After 2 and 3 h, 30 μL of cell supernatants were mixed with 100 μl of reaction mix and the OD was read at 490 nm. OD values were subtracted from that of uninfected cells. The 100% cell death value was quantified after addition of 1% Triton X-100, in duplicates. The experiment was performed in triplicate wells (except for the P47A/Q54A mutant for which six replicates were performed). Results represent the means and standard deviations of the triplicates. Statistics were calculated using SigmaPlot software. For multiple comparisons, a one-way analysis of variance (ANOVA) test was performed, followed by Tukey's test for pairwise comparisons.

Immunofluorescence

Two different *P. aeruginosa* genetic backgrounds were used: CHAΔ*pcrV* complemented or not with pIApG-*pcrV*-wt, and CHAΔ*pscF* complemented with all pIApG-*pscF* constructs or with the wild-type *pscF*. Cells were grown under T3SS-inducing conditions up to an optical density measured at 600 nm (OD_{600}) of 1 A.U. and then culture was fixed with 4% PFA in 25 mM

HEPES pH 7.4 overnight at 4°C. After centrifugation at 6,000 rpm for 15 min the pellet was washed three times with 1 ml PBS, then incubated for 30 min at RT in 1 ml PBS with 0.5% BSA (BS, Blocking Solution). Then cells were divided into two Eppendorf tubes and spin down, each pellet was resuspended and incubated for 30 min at room temperature in 50 μl of BS containing of anti-PcrV (1:200) (Goure et al., 2004) or 50 μl BS with anti-PscF (1:50). Anti-PscF antibodies were produced in rabbits (Biotem) from the PscF fibers purified under native conditions from *E. coli* BL21 (DE3). After three washing steps with 1 ml PBS, the pellet was resuspended into 50 μl of BS containing anti-rabbit coupled to Cy3 (Jackson ImmunoResearch Laboratories) with a final dilution of (1:500) and SYTO24-Green (Life Technologies) with a final dilution of (1:2000) and further incubated 30 min at RT. After three final washes, the cells were resuspended into BS. One drop of each sample was deposited on a 8-chambered labteck (Thermo Fisher Scientific) and liquid 1.5% low-melting agarose at 37°C or 1.5% agar pad was added on the top of the drop and visualized on a IX71 Olympus microscope controlled by the CellR Olympus system and driven by Xcellence software (Olympus). Images were captured with a Hammamatsu Orca-ER camera using a 100x (N.A. 1,30) oil objective.

To investigate for the presence of PscF inside the cells in PscF-mutants, cells were fixed with 4% PFA, washed as previously described, then permeabilized with 0.25% Triton X-100 for 5 min at room temperature to allow the anti-PscF to enter cells (method adapted from Cimino et al., 2006). Triton X-100 was then washed out three times with 1 ml PBS before continuing the protocol as described above from BS incubation.

Quantification of Fluorescent Spots Associated With Bacteria

Quantification was done using the MicrobeJ plugin of ImageJ (Ducret et al., 2016)[1] and [2]. Briefly, for bacterial identification we used the same setting (area, length, width, curvature of bacteria) for all images. For PcrV and PscF spot analyses a Tolerance of 100 was used, and only in the case of PscF a filter of intensity between 250-max was added in order to eliminate background noise and unrelated spots. For the final step of analyses by MicrobeJ, namely the association between each identified bacterium and the spots, we used two filters: inside and outside with "exclusive" settings, in order to associate each spot (that could be inside or outside the bacteria) with just one bacterium. Averages were calculated from the results of 5 to 11 independent images taken for each mutant. The frequency was considered as being the percentage of cells with at least one spot.

Statistics were calculated using SigmaPlot. For multiple comparisons, a one-way analysis of variance (ANOVA) test was performed, followed by Dunn's test.

PscF Expression and Purification and Mutant Construction

Escherichia coli BL21 (DE3) cells were transformed with pET22b-PscF (wild-type, D76A and P47A/Q54A) that generated a protein with a 6His-tag at the C terminus (Quinaud et al., 2005). Cells were grown in LB media supplemented with 100 μg/ml ampicillin at 37°C, under agitation (200 rpm) until they reached an OD_{600} = 0.7–0.8 A.U. Protein expression was induced by addition of 1 mM IPTG and growth was continued for up to 3 h, then cells were centrifuged for 30 min at 5,500 rpm at 4°C. Pellets were resuspended at 4°C in 25 ml of lysis buffer (25 mM Tris-HCl pH 8.0, 200 mM NaCl, 25 mM Imidazole) containing benzonase. Cells were lysed at 18 kPsi in a cell disrupter and centrifuged for 30 min at 15,000 rpm at 4°C. The soluble fraction was applied on a 1 ml resin Ni-NTA superflow (QIAGEN) in a batch column. The column was washed with lysis buffer and eluted the same buffer containing 200 mM imidazole. Sample purity was checked by SDS-PAGE and Mass Spectrometry. Mutants were produced using the Quick-Change site-directed mutagenesis II kit (Stragatene) and verified by DNA sequencing. Primers are listed in the **Supplementary Table S2**.

Trypsin Digestion

PscF-wt and PscF-P47A/Q54A proteins were purified as described above. Trypsin digestion was carried out according to the following procedure: trypsin was dissolved into a buffer containing 25 mM Tris-HCl pH 8.0, 200 mM NaCl and digestion was performed using a (trypsin:protein) ratio of 1:500 (w/w). The reaction was performed at RT and samples were collected at different time points (10 min, 30 min, 1 h 30 min, 6 h, ON) and analyzed by 16.5% Tris-Tricine gel and by LC/ESI-MS.

Electron Microscopy

Four μl of the protein samples were absorbed to the clean side of a carbon film on mica, stained with sodium silico tungstate and transferred to a 400-mesh copper grid. The images were taken under low dose conditions (<10 $e^-/Å^2$) at a magnification of 23,000 and 49,000 times with defocus values between 1.2 and 2.5 μm on a Tecnai 12 LaB6 electron microscope at 120 kV accelerating voltage using a CCD Camera Gatan Orius 1000. Size determination of the PscF filaments was performed in Gwyddion (Nečas and Klapetek, 2012) using the raw.dm3 EM files, measurements were done at different positions along the filaments using the "Measure distance" tool. Additional measurements were performed in ImageJ (Schneider et al., 2012). Statistics were calculated using SigmaPlot. For single comparisons, a Mann–Whitney Rank Sum Test was used.

Atomic Force Microscopy

A 2.5 μl drop of reconstituted filaments in deionized water (dilution 1/5000) was deposited on freshly cleaved mica, incubated for 3 min, washed with 1 ml of water with 80 μl drop steps to remove excessive salt crystals, and finally dried with nitrogen gas. Imaging was performed on a Multimode 8, Nanoscope V (Bruker) equipped with NanoScope software (Bruker, Santa Barbara, CA, United States). Imaging was done with peak force imaging mode at ~1 Hz rate, with 512 or 1024 pixel sampling and other parameters were adjusted automatically with ScanAsyst mode in air. A ScanAsyst-air (Bruker) cantilever with nominal 2 nm tip radius, 70 kHz frequency and 0.4 N/m spring constant was used. Images

[1] https://imagej.nih.gov/ij/

[2] http://www.microbej.com/

were processed with Gwyddion (Nečas and Klapetek, 2012), and if needed stripe noise was removed using DeStripe (Chen and Pellequer, 2011). Size determination of the filaments was performed in Gwyddion by using manual cross-sections (3px thick) and by measuring their maximum heights. Average values and standard deviations are reported. The computation of atomic depth was performed using the Adepth server (Chen and Pellequer, 2013).

LC/ESI Mass Spectrometry

Liquid Chromatography Electrospray Ionization Mass Spectrometry (LC/ESI-MS) was performed using a 6210 LC/ESI-TOF mass spectrometer interfaced with an HPLC binary pump system (Agilent Technologies). The mass spectrometer was calibrated in the mass-to-charge (*m/z*) 300–3200 range with standard calibrants (ESI-L, Low concentration tuning mix, Agilent Technologies) before measurements and mass spectra were recorded in the 300–3200 *m/z* range. MS acquisition was carried out in the positive ion mode with spectra in the profile mode. The MS instrument was operated with the following experimental settings: the ESI source temperature was set at 300°C; nitrogen was used as drying gas (7 l/min) and as nebulizer gas (10 psi); the capillary needle voltage was set at 4000 V. Spectra acquisition rate was of 1.03 spectra/s. All solvents used were HPLC grade (Chromasolv, Sigma-Aldrich), trifluoroacetic acid (TFA) was from Acros Organics (puriss., p.a.). Solvent A was 0.03% TFA in water, solvent B was 95% acetonitrile-5% water-0.03% TFA. The MS spectra were acquired and the data processed with MassHunter workstation software (*v*. B.02.00, Agilent Technologies) and with GPMAW software (v. 7.00b2, Lighthouse Data, Denmark).

Just before analysis each trypsin digested protein was diluted with solvent A to a final concentration of 5 μM and thermostatted at 10°C in the autosampler; 8 μl of each sample were injected. Samples were first trapped and desalted on a reverse phase-C8 cartridge (Zorbax 300SB-C8, 5 μm, 300 μm ID × 5 mm, Agilent Technologies) for 3 min at a flow rate of 50 μl/min with 100% solvent A and then eluted and separated onto a RP-C8 column (Jupiter, 5 μm, 300 Å, 1 mm ID × 50 mm, Phenomenex) at a flow rate of 50 μl/min using the following linear gradient: from 5 to 95% solvent B in 15 min, then remaining 2 min at 100% solvent B and finally re-equilibrating the column at 5% solvent B for 10 min.

Modeling

Optimal models of PscF-forward, PscF-reverse and PrgI-reverse protomers were built based on the PrgI protomer structure of the *Salmonella* needle structure by extracting chain A from the 2LPZ pdb file. MODELLER (Eswar et al., 2007), version 9.16, was used to individually align the forward (N-C:N-C) and reverse (N-C:C-N) PscF sequences to PrgI as well as a reverse (N-C:C-N) variant of PrgI for comparison (align2d.py). Initial single-chain models for the D76A and P47A/Q54A were prepared in the same way. The resulting alignment files and the PrgI subunit model were used as additional input to MODELLER to construct subunit models with secondary structure corresponding to that of PrgI (model-single.py). Ten models were calculated in each case and an overall assessment was manually carried out, judging energetic qualities and alignment to PrgI, to derive single optimal models for all cases. In the case of both forward and reverse PscF models, C-terminal helices were additionally enforced by alteration of the Phi and Psi backbone angles with PyMOL. Optimal models were subjected to a brief structural minimization protocol using the Chimera molecular modeling/visualization package (Pettersen et al., 2004) (default parameters) to remove atomic clashes (VDW overlap). PrgI-like T3SS needles were reconstructed in PyMol by aligning the constructed subunit models, by secondary structure, to the 29 chains of the PrgI (PDB code: 2LPZ) needle model.

The complete 29-chain needle structures, for wild-type (forward and reverse) and mutated subunits, were refined using a cluster-based installation of Rosetta (Das and Baker, 2008) (relax.mpi.linuxgccrelease - relax:fast), wherein the backbone heavy atoms were fixed in space but the positions of side-chains were allowed to evolve with respect to energetic minimization. The best resulting structures (as determined by overall Rosetta energy), calculated from a pool of 3 were then submitted to the MolProbity webserver (Davis et al., 2007) in a protonated form for validation (**Supplementary Table S1**).

AUTHOR CONTRIBUTIONS

VJ, IA, and AL designed the study and assembled the results. CL performed protein purifications and sample preparations for biochemical, EM and AFM experiments. DF and J-MT performed the EM and AFM experiments, respectively. J-LP analyzed and interpreted both images. All model building, computer minimizations and structure interpretations were performed by JT, with the supervision of AL and BH. CG constructed plasmids and mutants and perform initial cytotoxicity assays under the supervision of IA and EF. Macrophages cultures were prepared by SB. VJ performed cytotoxicity, florescence microscopy and biochemical experiments, with data analysis and interpretation. SB and DL contributed to the microscopy experiments and ImageJ/MicrobeJ analysis, while JB helped with Western blots. LS performed mass spectrometry experiments, data analysis, and interpretation. VJ, AD, and AL wrote the paper, with input from all authors.

FUNDING

This work used the platforms of the Grenoble Instruct-ERIC Center (ISBG: UMS 3518 CNRS-CEA-UGA-EMBL) with support from FRISBI (ANR-10 INBS-05-02) and GRAL (ANR-10-LABX-49-01) within the Grenoble Partnership for

Structural Biology (PSB). The electron microscope facility is supported by the Rhône-Alpes Region, the Fondation Recherche Médicale (FRM), the fonds FEDER, the Centre National de la Recherche Scientifique (CNRS), the CEA, the University of Grenoble, EMBL, and the GIS-Infrastrutures en Biologie Santé et Agronomie (IBISA). The authors wish to acknowledge ERC Starting Grant No. 639020, ANR-14-CE09-0020-01 and the Fondation pour la Recherche Médicale (FRM)—AJE20140630090 to AL, as well as grant VLM (RF20150501349) to EF, Fondation pour la Recherche Médicale (DEQ20170336705) to IA, and VLM (RIF20150501457) to AD.

ACKNOWLEDGMENTS

The authors thank Michel Ragno for antibody purification. This work acknowledges the AFM platform at the IBS.

REFERENCES

Bergeron, J. R. C., Fernández, L., Wasney, G. A., Vuckovic, M., Reffuveille, F., Hancock, R. E. W., et al. (2016). The structure of a type 3 secretion system (T3SS) ruler protein suggests a molecular mechanism for needle length sensing. *J. Biol. Chem.* 291, 1676–1691. doi: 10.1074/jbc.M115.684423

Berube, B. J., Murphy, K. R., Torhan, M. C., Bowlin, N. O., Williams, J. D., Bowlin, T. L., et al. (2017). Impact of type III secretion effectors and of phenoxyacetamide inhibitors of type III secretion on abscess formation in a mouse model of *Pseudomonas aeruginosa* infection. *Antimicrob. Agents Chemother.* 61, e1202–e1217. doi: 10.1128/AAC.01202-17

Bhattacharya, A., Tejero, R., and Montelione, G. T. (2007). Evaluating protein structures determined by structural genomics consortia. *Proteins* 66, 778–795. doi: 10.1002/prot.21165

Bowlin, N. O., Williams, J. D., Knoten, C. A., Torhan, M. C., Tashjian, T. F., Li, B., et al. (2014). Mutations in the *Pseudomonas aeruginosa* needle protein gene pscF confer resistance to phenoxyacetamide inhibitors of the type III secretion system. *Antimicrob. Agents Chemother.* 58, 2211–2220. doi: 10.1128/AAC.02795-13

Chakravarty, S., and Varadarajan, R. (1999). Residue depth: a novel parameter for the analysis of protein structure and stability. *Struct. Lond. Engl.* 7, 723–732. doi: 10.1016/S0969-2126(99)80097-5

Charro, N., and Mota, L. J. (2015). Approaches targeting the type III secretion system to treat or prevent bacterial infections. *Expert Opin. Drug Discov.* 10, 373–387. doi: 10.1517/17460441.2015.1019860

Chen, S. W., and Pellequer, J.-L. (2011). DeStripe: frequency-based algorithm for removing stripe noises from AFM images. *BMC Struct. Biol.* 11:7. doi: 10.1186/1472-6807-11-7

Chen, S. W., and Pellequer, J.-L. (2013). Adepth: New Representation and its implications for atomic depths of macromolecules. *Nucleic Acids Res.* 41, W412–W416. doi: 10.1093/nar/gkt299

Cimino, M., Alamo, L., and Salazar, L. (2006). Permeabilization of the mycobacterial envelope for protein cytolocalization studies by immunofluorescence microscopy. *BMC Microbiol.* 6:35. doi: 10.1186/1471-2180-6-35

Cornelis, G. R. (2006). The type III secretion injectisome. *Nat. Rev. Microbiol.* 4, 811–825. doi: 10.1038/nrmicro1526

Das, R., and Baker, D. (2008). Macromolecular modeling with rosetta. *Annu. Rev. Biochem.* 77, 363–382. doi: 10.1146/annurev.biochem.77.062906.171838

Dasgupta, N., Ashare, A., Hunninghake, G. W., and Yahr, T. L. (2006). Transcriptional induction of the *Pseudomonas aeruginosa* type III secretion system by low Ca2+ and host cell contact proceeds through two distinct signaling pathways. *Infect. Immun.* 74, 3334–3341. doi: 10.1128/IAI.00090-06

Davis, A. J., and Mecsas, J. (2007). Mutations in the *Yersinia pseudotuberculosis* type III secretion system needle protein, YscF, that specifically abrogate effector translocation into host cells. *J. Bacteriol.* 189, 83–97. doi: 10.1128/JB.01396-06

Davis, I. W., Leaver-Fay, A., Chen, V. B., Block, J. N., Kapral, G. J., Wang, X., et al. (2007). MolProbity: all-atom contacts and structure validation for proteins and nucleic acids. *Nucleic Acids Res.* 35, W375–W383. doi: 10.1093/nar/gkm216

Deane, J. E., Roversi, P., Cordes, F. S., Johnson, S., Kenjale, R., Daniell, S., et al. (2006). Molecular model of a type III secretion system needle: implications for host-cell sensing. *Proc. Natl. Acad. Sci. U.S.A.* 103, 12529–12533. doi: 10.1073/pnas.0602689103

Demers, J.-P., Habenstein, B., Loquet, A., Kumar Vasa, S., Giller, K., Becker, S., et al. (2014). High-resolution structure of the *Shigella* type-III secretion needle by solid-state NMR and cryo-electron microscopy. *Nat. Commun.* 5:4976. doi: 10.1038/ncomms5976

Demers, J.-P., Sgourakis, N. G., Gupta, R., Loquet, A., Giller, K., Riedel, D., et al. (2013). The common structural architecture of *Shigella flexneri* and *Salmonella typhimurium* type three secretion needles. *PLoS Pathog.* 9:e1003245. doi: 10.1371/journal.ppat.1003245

Discola, K. F., Förster, A., Boulay, F., Simorre, J.-P., Attree, I., Dessen, A., et al. (2014). Membrane and chaperone recognition by the major translocator protein PopB of the type III secretion system of *Pseudomonas aeruginosa*. *J. Biol. Chem.* 289, 3591–3601. doi: 10.1074/jbc.M113.517920

Ducret, A., Quardokus, E. M., and Brun, Y. V. (2016). MicrobeJ, a tool for high throughput bacterial cell detection and quantitative analysis. *Nat. Microbiol.* 1:16077. doi: 10.1038/nmicrobiol.2016.77

Eswar, N., Webb, B., Marti-Renom, M. A., Madhusudhan, M. S., Eramian, D., Shen, M.-Y., et al. (2007). Comparative protein structure modeling using MODELLER. *Curr. Protoc. Protein Sci.* 50, 2.9.1–2.9.31. doi: 10.1002/0471140864.ps0209s50 Unit 2.9,

Fujii, T., Cheung, M., Blanco, A., Kato, T., Blocker, A. J., and Namba, K. (2012). Structure of a type III secretion needle at 7-Å resolution provides insights into its assembly and signaling mechanisms. *Proc. Natl. Acad. Sci. U.S.A.* 109, 4461–4466. doi: 10.1073/pnas.1116126109

Galán, J. E., and Collmer, A. (1999). Type III secretion machines: bacterial devices for protein delivery into host cells. *Science* 284, 1322–1328. doi: 10.1126/science.284.5418.1322

Galán, J. E., Lara-Tejero, M., Marlovits, T. C., and Wagner, S. (2014). Bacterial type III secretion systems: specialized nanomachines for protein delivery into target cells. *Annu. Rev. Microbiol.* 68, 415–438. doi: 10.1146/annurev-micro-092412-155725

Godon, C., Teulon, J.-M., Odorico, M., Basset, C., Meillan, M., Vellutini, L., et al. (2017). Conditions to minimize soft single biomolecule deformation when imaging with atomic force microscopy. *J. Struct. Biol.* 197, 322–329. doi: 10.1016/j.jsb.2016.12.011

Goure, J., Pastor, A., Faudry, E., Chabert, J., Dessen, A., and Attree, I. (2004). The V antigen of *Pseudomonas aeruginosa* is required for assembly of the functional PopB/PopD translocation pore in host cell membranes. *Infect. Immun.* 72, 4741–4750. doi: 10.1128/IAI.72.8.4741-4750.2004

Harmon, D. E., Murphy, J. L., Davis, A. J., and Mecsas, J. (2013). A mutant with aberrant extracellular LcrV-YscF interactions fails to form pores and translocate Yop effector proteins but retains the ability to trigger Yop secretion in response to host cell contact. *J. Bacteriol.* 195, 2244–2254. doi: 10.1128/JB.02011-12

Hoiczyk, E., and Blobel, G. (2001). Polymerization of a single protein of the pathogen *Yersinia enterocolitica* into needles punctures eukaryotic cells. *Proc. Natl. Acad. Sci. U.S.A.* 98, 4669–4674. doi: 10.1073/pnas.071065798

Hu, J., Worrall, L. J., Hong, C., Vuckovic, M., Atkinson, C. E., Caveney, N., et al. (2018). Cryo-EM analysis of the T3S injectisome reveals the structure of the needle and open secretin. *Nat. Commun.* 9:3840. doi: 10.1038/s41467-018-06298-8

Izoré, T., Job, V., and Dessen, A. (2011). Biogenesis, regulation, and targeting of the type III secretion system. *Struct. Lond. Engl.* 1993, 603–612. doi: 10.1016/j.str.2011.03.015

Jneid, B., Moreau, K., Plaisance, M., Rouaix, A., Dano, J., and Simon, S. (2016). Role of T3SS-1 SipD protein in protecting mice against Non-typhoidal *Salmonella* typhimurium. *PLoS Negl. Trop. Dis.* 10:e0005207. doi: 10.1371/journal.pntd.0005207

Job, V., Matteï, P.-J., Lemaire, D., Attree, I., and Dessen, A. (2010). Structural basis of chaperone recognition of type III secretion system minor translocator proteins. *J. Biol. Chem.* 285, 23224–23232. doi: 10.1074/jbc.M110.111278

Journet, L., Agrain, C., Broz, P., and Cornelis, G. R. (2003). The needle length of bacterial injectisomes is determined by a molecular ruler. *Science* 302, 1757–1760. doi: 10.1126/science.1091422

Kato, J., Dey, S., Soto, J. E., Butan, C., Wilkinson, M. C., De Guzman, R. N., et al. (2018). A protein secreted by the *Salmonella* type III secretion system controls needle filament assembly. *eLife* 7:e35886. doi: 10.7554/eLife.35886

Kenjale, R., Wilson, J., Zenk, S. F., Saurya, S., Picking, W. L., Picking, W. D., et al. (2005). The needle component of the type III secreton of *Shigella* regulates the activity of the secretion apparatus. *J. Biol. Chem.* 280, 42929–42937. doi: 10.1074/jbc.M508377200

Koroleva, E. A., Kobets, N. V., Shcherbinin, D. N., Zigangirova, N. A., Shmarov, M. M., Tukhvatulin, A. I., et al. (2017). Chlamydial type III secretion system needle protein induces protective immunity against *Chlamydia muridarum* intravaginal infection. *Biomed Res. Int.* 2017:3865802. doi: 10.1155/2017/3865802

Kubori, T., Matsushima, Y., Nakamura, D., Uralil, J., Lara-Tejero, M., Sukhan, A., et al. (1998). Supramolecular structure of the *Salmonella* typhimurium type III protein secretion system. *Science* 280, 602–605. doi: 10.1126/science.280.5363.602

Kubori, T., Sukhan, A., Aizawa, S. I., and Galán, J. E. (2000). Molecular characterization and assembly of the needle complex of the *Salmonella* typhimurium type III protein secretion system. *Proc. Natl. Acad. Sci. U.S.A.* 97, 10225–10230. doi: 10.1073/pnas.170128997

Kuhlen, L., Abrusci, P., Johnson, S., Gault, J., Deme, J., Caesar, J., et al. (2018). Structure of the core of the type III secretion system export apparatus. *Nat. Struct. Mol. Biol.* 25, 583–590. doi: 10.1038/s41594-018-0086-9

Loquet, A., Sgourakis, N. G., Gupta, R., Giller, K., Riedel, D., Goosmann, C., et al. (2012). Atomic model of the type III secretion system needle. *Nature* 486, 276–279. doi: 10.1038/nature11079

Matteï, P.-J., Faudry, E., Job, V., Izoré, T., Attree, I., and Dessen, A. (2011). Membrane targeting and pore formation by the type III secretion system translocon. *FEBS J.* 278, 414–426. doi: 10.1111/j.1742-4658.2010.07974.x

Nariya, M. K., Israeli, J., Shi, J. J., and Deeds, E. J. (2016). Mathematical model for length control by the timing of substrate switching in the type III secretion system. *PLoS Comput. Biol.* 12:e1004851. doi: 10.1371/journal.pcbi.1004851

Nečas, D., and Klapetek, P. (2012). Gwyddion: an open-source software for SPM data analysis. *Open Phys.* 10, 181–188. doi: 10.2478/s11534-011-0096-2

Nguyen, V. S., Jobichen, C., Tan, K. W., Tan, Y. W., Chan, S. L., Ramesh, K., et al. (2015). Structure of AcrH-AopB chaperone-translocator complex reveals a role for membrane hairpins in type III secretion system translocon assembly. *Struct. Lond. Engl.* 23, 2022–2031. doi: 10.1016/j.str.2015.08.014

Park, D., Lara-Tejero, M., Waxham, M. N., Li, W., Hu, B., Galán, J. E., et al. (2018). Visualization of the type III secretion mediated *Salmonella*-host cell interface using cryo-electron tomography. *eLife* 7:e39514. doi: 10.7554/eLife.39514

Pastor, A., Chabert, J., Louwagie, M., Garin, J., and Attree, I. (2005). PscF is a major component of the *Pseudomonas aeruginosa* type III secretion needle. *FEMS Microbiol. Lett.* 253, 95–101. doi: 10.1016/j.femsle.2005.09.028

Perdu, C., Huber, P., Bouillot, S., Blocker, A., Elsen, S., Attrée, I., et al. (2015). ExsB is required for correct assembly of the *Pseudomonas aeruginosa* type III secretion apparatus in the bacterial membrane and full virulence in vivo. *Infect. Immun.* 83, 1789–1798. doi: 10.1128/IAI.00048-15

Pettersen, E. F., Goddard, T. D., Huang, C. C., Couch, G. S., Greenblatt, D. M., Meng, E. C., et al. (2004). UCSF Chimera–a visualization system for exploratory research and analysis. *J. Comput. Chem.* 25, 1605–1612. doi: 10.1002/jcc.20084

Poyraz, O., Schmidt, H., Seidel, K., Delissen, F., Ader, C., Tenenboim, H., et al. (2010). Protein refolding is required for assembly of the type three secretion needle. *Nat. Struct. Mol. Biol.* 17, 788–792. doi: 10.1038/nsmb.1822

Quinaud, M., Chabert, J., Faudry, E., Neumann, E., Lemaire, D., Pastor, A., et al. (2005). The PscE-PscF-PscG complex controls type III secretion needle biogenesis in *Pseudomonas aeruginosa. J. Biol. Chem.* 280, 36293–36300. doi: 10.1074/jbc.M508089200

Quinaud, M., Plé, S., Job, V., Contreras-Martel, C., Simorre, J.-P., Attree, I., et al. (2007). Structure of the heterotrimeric complex that regulates type III secretion needle formation. *Proc. Natl. Acad. Sci. U.S.A.* 104, 7803–7808. doi: 10.1073/pnas.0610098104

Radics, J., Königsmaier, L., and Marlovits, T. C. (2014). Structure of a pathogenic type 3 secretion system in action. *Nat. Struct. Mol. Biol.* 21, 82–87. doi: 10.1038/nsmb.2722

Robert, X., and Gouet, P. (2014). Deciphering key features in protein structures with the new ENDscript server. *Nucleic Acids Res.* 42, W320–W324. doi: 10.1093/nar/gku316

Sal-Man, N., Setiaputra, D., Scholz, R., Deng, W., Yu, A. C. Y., Strynadka, N. C. J., et al. (2013). EscE and EscG are cochaperones for the type III needle protein EscF of enteropathogenic *Escherichia coli. J. Bacteriol.* 195, 2481–2489. doi: 10.1128/JB.00118-13

Schneider, C. A., Rasband, W. S., and Eliceiri, K. W. (2012). NIH Image to ImageJ: 25 years of image analysis. *Nat. Methods* 9, 671–675. doi: 10.1038/nmeth.2089

Sun, P., Tropea, J. E., Austin, B. P., Cherry, S., and Waugh, D. S. (2008). Structural characterization of the *Yersinia pestis* type III secretion system needle protein YscF in complex with its heterodimeric chaperone YscE/YscG. *J. Mol. Biol.* 377, 819–830. doi: 10.1016/j.jmb.2007.12.067

Tamano, K., Aizawa, S., Katayama, E., Nonaka, T., Imajoh-Ohmi, S., Kuwae, A., et al. (2000). Supramolecular structure of the *Shigella* type III secretion machinery: the needle part is changeable in length and essential for delivery of effectors. *EMBO J.* 19, 3876–3887. doi: 10.1093/emboj/19.15.3876

Torruellas, J., Jackson, M. W., Pennock, J. W., and Plano, G. V. (2005). The *Yersinia pestis* type III secretion needle plays a role in the regulation of Yop secretion. *Mol. Microbiol.* 57, 1719–1733. doi: 10.1111/j.1365-2958.2005.04790.x

Veenendaal, A. K. J., Hodgkinson, J. L., Schwarzer, L., Stabat, D., Zenk, S. F., and Blocker, A. J. (2007). The type III secretion system needle tip complex mediates host cell sensing and translocon insertion. *Mol. Microbiol.* 63, 1719–1730. doi: 10.1111/j.1365-2958.2007.05620.x

Verasdonck, J., Shen, D.-K., Treadgold, A., Arthur, C., Böckmann, A., Meier, B. H., et al. (2015). Reassessment of MxiH subunit orientation and fold within native *Shigella* T3SS needles using surface labelling and solid-state NMR. *J. Struct. Biol.* 192, 441–448. doi: 10.1016/j.jsb.2015.10.005

Wagner, S., Grin, I., Malmsheimer, S., Singh, N., Torres-Vargas, C. E., and Westerhausen, S. (2018). Bacterial type III secretion systems: a complex device for the delivery of bacterial effector proteins into eukaryotic host cells. *FEMS Microbiol. Lett.* 365:fny201. doi: 10.1093/femsle/fny201

Wagner, S., Stenta, M., Metzger, L. C., Dal Peraro, M., and Cornelis, G. R. (2010). Length control of the injectisome needle requires only one molecule of Yop secretion protein P (YscP). *Proc. Natl. Acad. Sci. U.S.A.* 107, 13860–13865. doi: 10.1073/pnas.1006985107

Wang, Y., Ouellette, A. N., Egan, C. W., Rathinavelan, T., Im, W., and De Guzman, R. N. (2007). Differences in the electrostatic surfaces of the type III secretion needle proteins PrgI, BsaL, and MxiH. *J. Mol. Biol.* 371, 1304–1314. doi: 10.1016/j.jmb.2007.06.034

5

A Critical Role of Bacterioferritin in *Salmonella pullorum*-Induced IFN-β Expression in DF-1 Cells

*Zhichao Xu[1,2,3], Yao Qin[1,2,3], Yongqiang Wang[1,2,3], Xiaoqi Li[1,2,3], Hong Cao[1,2,3] and Shijun J. Zheng[1,2,3]**

[1] State Key Laboratory of Agrobiotechnology, China Agricultural University, Beijing, China, [2] Key Laboratory of Animal Epidemiology and Zoonosis, Ministry of Agriculture, China Agricultural University, Beijing, China, [3] College of Veterinary Medicine, China Agricultural University, Beijing, China

Edited by:
Inka Sastalla,
National Institutes of Health, USA

Reviewed by:
Dane Parker,
Columbia University, USA
Kenneth James Genovese,
United States Department of Agriculture, Agricultural Research Service, USA

***Correspondence:**
Shijun J. Zheng
sjzheng@cau.edu.cn

Salmonella enterica serovar Pullorum (*S. pullorum*) causes pullorum disease in poultry and results in great economic losses to the poultry industry. Although an eradication program has been successfully performed in some countries, it remains a major threat to countries with poor poultry disease surveillance. Currently there are no effective control measures for pullorum disease except eradication. In particular, the pathogenesis of *S. pullorum* infection is still largely unknown. Here we identified bacterioferritin (Bfr) as a major antigen of *S. pullorum* to elicit a humoral immune response. Furthermore, we demonstrate that Bfr induces activation of IFN-β promoter and mRNA expression in DF-1 cells, and that the amino acids 1–50 form a critical domain involved in IFN-β expression. Moreover, we found that the p38 MAPK signaling pathway was essential for Bfr-induced IFN-β expression. Importantly, *S. pullorum*-induced IFN-β expression was totally abolished by deficiency of Bfr in the bacteria, indicating that Bfr plays a critical role in *S. pullorum* induced IFN-β expression in DF-1 cells. Our findings provide new insights into the molecular mechanisms of the host response to *S. pullorum* infection.

Keywords: *Salmonella pullorum*, bacterioferritin, interferon, IFN-β

INTRODUCTION

Pullorum disease, an acute systemic disease commonly seen in young birds, is caused by *Salmonella enterica* serovar Pullorum. The clinical signs of pullorum disease are characterized by anorexia, diarrhea, dehydration, weakness and high mortality in young chicks, but this disease usually shows a persistent infection and causes decreased egg production and diarrhea in adult fowls (Shivaprasad, 2000). Pullorum disease is basically controlled in Europe and North America, but it still occurs in many countries such as Brazil, Argentina, India, and China, leading to severe economic losses (Barrow and Freitas, 2011; Barrow et al., 2012). *Salmonella* spp. belongs to the Enterobacteriaceae family. *Salmonella* is a Gram-negative and facultative intracellular pathogen which, depending on the serotype and host, can cause diseases ranging from gastroenteritis to typhoid fever (Marcus et al., 2000). *S. pullorum*, currently belonging to biovars of serovar Gallinarum within serogroup D, has identical somatic antigens (O1, O9, O12) and no flagella due to mutations in flagellar genes while its pathogenicity is restricted only to avian species (Barrow and Freitas, 2011). The relatively high rate of accumulation of mutations in the genome of *S. pullorum* suggests a rapid rate of evolution associated with the host adaptation, particularly in the development of *S. pullorum* (Barrow and Freitas, 2011). During *S. pullorum* infection, the interaction of this pathogen with the immune system occurs in three main phases, including invasion via the gastrointestinal tract, establishment of systemic infection and induction of cytokine expression (Chappell et al., 2009).

High titers of anti-*Salmonella* IgY were produced by birds infected with *S. pullorum* from 5 weeks onwards and *S. pullorum* was detected in splenic macrophages from 3 days to 10 weeks postinfection (Wigley et al., 2001). It was found that approximate 1 to 2% of macrophages contained fluorescent *Salmonella* bacteria in all birds examined, and dropped to less than 1% at 5 weeks postinfection and even further at 10 weeks to less than 0.5% of cells infected (Wigley et al., 2001), indicating that macrophage plays a critical role in clearance of *S. pullorum*. An antigen-specific T-cell response to *S. pullorum* was found in birds at 5 and 9 weeks postinfection, but dropped to negligible levels at 17 weeks postinfection (Wigley et al., 2005). The numbers of *S. pullorum* bacteria recovered from the spleen, liver, the reproductive tracts and developing eggs increased following the fall in T-cell proliferation activity at 18 weeks postinfection, while T-cells proliferation began to increase at 22 weeks postinfection (Wigley et al., 2005). In contrast to T-cell response, antibody response did not decline (Wigley et al., 2005). Like other pathogens, *Salmonella* infection stimulates cytokine production. The induction of cytokines such as IL-1β, IL-8, IL-12, IL-17, IL-18, TNF-α, and IFN-γ following *Salmonella* infection of chickens have been previously reported (Withanage et al., 2004; Berndt et al., 2007; Crhanova et al., 2011). One of the most remarkable features of *Salmonella* infection is that IFN-β was induced in fibroblasts and macrophages (Hess et al., 1989; Robinson et al., 2012). The role of IFN-β in the response to bacterial infection is variable, and it contributes to a variety of beneficial and detrimental immune functions (Monroe et al., 2010).

The iron that is acquired by the pathogenic bacterium is used for numerous biochemical activities and any surplus iron that is available is stored within the bacterial cell in the form of Bfr (Ratledge, 2007). Bfr belongs to an outer membrane protein in *S. hadar* as examined by a proteomic approach (Snoussi et al., 2012). Bfr is a major iron storage protein and protects against hydrogen peroxide toxicity, and the haeme-containing Bfr was found exclusively in bacteria (Velayudhan et al., 2007).

Currently it is known that Bfr is a T-cell antigen that induced a strong IFN-γ production and the proliferation of lymphocytes (Denoel et al., 1997; Al-Mariri et al., 2002; Lee et al., 2006). In addition, Bfr induced humoral immune response in mice immunized with DNA vaccine encoding the Bfr or recombinant Bfr proteins (Al-Mariri et al., 2001a,b. The antibodies against Bfr were detected from Crohn's disease, and 53% of Crohn's disease patients were positive, indicating that Bfr was a specific protein antigen of *Mycobacterium paratuberculosis* (Walmsley et al., 1996). However, little is known about the role of Bfr in innate immune responses. DF-1, an immortal chicken embryo fibroblast cell line, is commonly used for the research of *Salmonella* (Li et al., 2006; Szmolka et al., 2015) and type I interferon (Li et al., 2013). To gain a better understanding about the role of Bfr in innate immune responses, we set out to determine if Bfr induces humoral immune response in chickens and induces type I IFN expression in *S. pullorum* infected DF-1 cells.

In this study, we demonstrate that Bfr is a major antigen of *S. pullorum*, and Bfr induced IFN-β mRNA expression in DF-1 cells. In addition, we show that the amino acids 1–50 form a critical domain involved in activation of the IFN-β promoter. Furthermore, we found that Bfr induced IFN-β expression likely via the p38 mitogen-activated protein (MAP) Kinase signaling transduction pathway. Importantly, we found that *S. pullorum*-induced IFN-β expression was totally abolished by deficiency of Bfr in the bacteria, indicating that Bfr plays a critical role in *S. pullorum*-induced IFN-β expression in DF-1 cells.

MATERIALS AND METHODS

Bacteria and Cells

Salmonella pullorum strain 533 was obtained from China Institute of Veterinary Drug Control (Beijing, China). *Escherichia coli* DH5α and *E. coli* BL21 (DE3) strains were obtained from TransGen Biotech (Beijing, China). Bacteria were grown in LB medium. DF-1 cells were obtained from ATCC (USA), and were cultured in DMEM supplemented with 10% fetal bovine serum (FBS) in 5% CO_2 incubator.

Reagents

Protein A/G beads were purchased from GE Healthcare (USA). Protease inhibitor cocktail C was purchased from Yataihengxin Company (Beijing, China). The restriction enzymes *BamH* I, *Xho* I, and *Sph* I were purchased from TaKaRa (Dalian, China). The Mops, $FeCl_2$ and H_2O_2 were purchased from Solarbio Company (Beijing, China). The Large Amount Without Endo-Toxin Plasmid Preparation Kits and the monoclonal antibody against GAPDH were purchased from CWBio (Beijing, China). DMEM medium was purchased from Hyclone (USA). The jetPRIME reagent was purchased from Polyplus-transfection (France). The serum against *S. pullorum* was collected from chickens with pullorum disease and the control serum from SPF chickens from Beijing Agricultural University Animal Technology Company (Beijing, China). Monoclonal antibody against His-tag fusion protein was purchased from Abmart (Shanghai, China). Anti-GFP monoclonal antibody, anti-p38 monoclonal antibody, and anti-p-p38 monoclonal antibody were purchased from Santa Cruz Biotechnology (USA). Monoclonal antibody against Bfr (Clone ID: EU-0218), pGL3-chIFN-α-luc and pGL3-chIFN-β-luc were obtained from CAEU Biological Company (Beijing, China). HRP-conjugated goat-anti mouse polyclonal antibodies and HRP-conjugated goat-anti rabbit polyclonal antibodies were purchased from DingGuoShengWu Company (Beijing, China). Rabbit-anti chicken polyclonal antibodies were purchased from Bioss (Beijing, China). pET28a (+) vector was obtained from Novagen (USA). pEGFP-N1 vector was purchased from Clontech (USA). p38 (mitogen-activated protein kinase, MAPK) inhibitor SB203580 and JNK inhibitor SP600125 were purchased from Enzo Life Sciences (USA). Red homologous recombination using the plasmids pKD46, pKD3, and pCP20 were kindly provided by Professor Guo-Qiang Zhu (Yangzhou University, China).

Pull-Down Assay

Salmonella pullorum was grown in LB medium. The bacterial culture were centrifuged at 6000 × *g* for 5 min, and the pellet was resuspended and lysed in pH 7.4 PBS buffer by ultrasonic treatment. Then 50 μL of 25% protein A/G beads were preincubated with 60 μg rabbit-anti chicken polyclonal antibodies and 200 μL pullorum-positive serum or negative serum as controls for 8 h at 4°C. The mixture was washed three times with pH7.4 PBS by centrifugation at 825 × *g* for 3 min at 4°C and the supernatant was removed after the last wash. The rabbit-anti chicken polyclonal antibody-conjugated beads with anti-*S. pullorum* antibodies mixed with the cell extract from *S. pullorum* and incubated at 4°C for 8 h. The mixture was washed as above described. The immunoprecipitates were suspended with 40 μL 1x SDS-PAGE loading buffer and boiled for 10 min before resolved on 12% SDS-PAGE gel. Then the gel was stained with Coomassie blue dye for analysis of specific bands.

Mass Spectrometric Identification of Proteins

After separation of proteins on SDS-PAGE gel, the interesting bands were cut out and subjected to liquid chromatography-mass spectrometry. Briefly, the interesting peptides extracted from gel were dissolved in 0.1% formic acid, and then separated by a Nano-LC system (Micro-Tech Scientific, Vista, CA, USA) equipped with a C_{18} reverse phase column. The peptides were eluted using a 120 min gradients from 0 to 50% acetonitrile in 0.1% formic acid at a constant flow rate of 400 nL/min. Mass spectra were recorded on a 7-T Fourier transform ion-cyclotron resonance (FTICR) mass spectrometer, Apex-Qe (Bruker Daltonics, Bremen, Germany). Data were acquired in data-dependent mode using ApexControl 1.0 software (Bruker Daltonics, Bremen, Germany). Three strongest peaks of each MS acquisition were selected for the following MS/MS analysis. The MS/MS spectra were processed by DataAnalysis 3.4 (Bruker Daltonics, Bremen, Germany) with $S/N \geq 4.0$, and automatically searched against IPI.RAT database (version 3.41) using the Mascot 2.1.0 (Matrix Science, London, U.K.). The NCBI database was used in the search.

Construction of Plasmids

The *bfr* gene was amplified from *S. pullorum* genomic DNA by PCR using the specific primers containing *BamH* I in sense and *Xho* I in antisense (sense: 5′-CGCGGATCCATGA AAGGTGATGTTAAA-3′; antisense: 5′-CCG CTCGAGATC GGTAACCTTAATTTG-3′) that were designed with reference to the published sequence (GenBank, gene ID: 661554730), and the PCR products were cloned into the pET28a (+). The resulting plasmid was named pET28a-*bfr*. The *bfr* gene was then subcloned into pEGFP-N1 vector using primers with *Xho* I in sense and *BamH* I in antisense (sense: 5′-CCGCTCGAGATGAAAGGTGATGTTAAA-3′; antisense: 5′-C GCGGATCCCGATCGGTAACCTTAATT-3′) and this plasmid was named pEGFP-*bfr*. Different truncated *bfr* segments were subcloned into pEGFP-N1 vectors and were named Bfr (1–50aa), Bfr (51–158aa), Bfr (101–158aa) accordingly. The Bfr (1–50aa) plasmid was constructed using the same sense primer as pEGFP-*bfr* plasmid and the antisense primer is 5′-GGATCCCGATCAATGGATTCATGGTAC-3′ containing *BamH* I restriction site. The Bfr (51–158aa) and the Bfr (101–158aa) plasmids were constructed using the same antisense primer as pEGFP-*bfr* plasmid. For Bfr (51–158aa) plasmid, the sense primer is 5′-CCGCTCGAGGAGATGAAACACGCCGAT A-3′. For Bfr (101–158aa) plasmid, the sense primer is 5′-CTCGAGCTACGTGAGGCAATAGCC-3′. All the sense primers contained *Xho* I restriction site. All the primers were synthesized by Sangon Company (Shanghai, China), and all the constructs were confirmed with sequencing analysis by Huada Company (Beijing, China).

Iron Uptake Assays

Iron uptake by rBfr was examined using SpectraMax M5 according to the method described by Timoteo et al. (2012). Briefly, reaction of 0.5 μM rBfr with 12 μM Fe^{2+} ions and 72 μM H_2O_2 in 0.2 M Mops buffer (pH 7) and 0.2 M NaCl, and BSA was used as control. The iron storage capacity was determined by plotting the OD optical density at 200–400 nm in SpectraMax M5.

Transfection and Reporter Gene Assays

DF-1 cells (1×10^5) were seeded on 24-well plates and cultured overnight before transfection with pEGFP-*bfr* or pEGFP-N1, together with pGL3-chIFN-β-luc (or pGL3-chIFN-α-luc) and pRL-TK using jetPRIME reagents (Polyplus-transfection). Twenty four hours after transfection, cell extracts were harvested, and the luciferase activities were examined with a dual-specific luciferase assay kit (Promega). Firefly luciferase activities were normalized based on Renilla luciferase activities. All reporter gene assays were repeated at least three times. Data are represented as mean ± SD.

RNA Isolation and RT-PCR Analysis

Total RNA was prepared from DF-1 cells using a RNeasy kit (Aidlab, China) per the manufacturer' instruction, and was treated with DNase I. Two μg of total RNA was used for cDNA synthesis by reverse transcription using RT-PCR kit (TaKaRa). The specific primers for chicken IFN-α1 (sense: 5′-CCAGCACCTCGAGCAAT-3′; antisense: 5′-GGCGCTGTAATCGTTGTCT-3′), IFN-β (sense: 5′-GCCTCC AGCTCCTTCAGAATACG-3′; antisense: 5′-CTGGATCTGGTT GAGGAGGCTGT-3′), and glyceraldehyde-3-phosphate dehydrogenase (GAPDH; sense: 5′-TGCCCATCACAGCCACA CAGAAG-3′; antisense: 5′-ACTTTCCCCACAGCCTTAGCA G-3′) were designed with reference to previous publications (Li et al., 2007; Abdul-Careem et al., 2008; Liu et al., 2010) and synthesized by Sangon Company (Shanghai, China). The real-time PCR assay was carried out with a Light Cycler 480 (Roche, USA). The PCR was performed in a 20-μL volume containing 1 μL of cDNA, 10 μL of 2 × SYBR green Premix *Ex Taq* (TaKaRa), and a 0.4 μM of each gene-specific primer. The thermal cycling parameters were referred to the previous study (Li et al., 2013), and they were as follows: 94°C for 2 min; 45 cycles of 94°C for 20 s, 55°C for 20 s, and 72°C for 20 s;

and 1 cycle of 95°C for 30 s, 60°C for 30 s, and 95°C for 30 s. The final step was to obtain a melt curve for the PCR products to determine the specificity of the amplification. All sample reactions were carried out in triplicate on the same plate, and the GAPDH gene was utilized as the reference gene. Expression levels of genes were calculated relative to the expression of the GAPDH gene and expressed as fold increase or decrease relative to the control samples.

Inhibition of Signal Transduction Pathways

DF-1 cells (4 × 10^5) were seeded on a 6-well plates and cultured for 24 h before treatment with p38 inhibitor SB203580 (20 μM), JNK inhibitor SP600125 (20 μM), and dimethyl sulfoxide (DMSO) as control for 1 h, and then transfected with pEGFP-*bfr* or pEGFP-N1 as controls using jetPRIME reagents. Twenty four hours after transfection, total RNA was extracted and used for cDNA synthesis. Real-time PCR was performed to examine the expressions of chicken IFN-β and GAPDH at an mRNA level as above described.

Western Blot Analysis

The pET28a-*bfr* recombinant or empty vectors were transformed into *E. coli* BL21 (DE3), and Bfr-his recombinant proteins were expressed by 1 mM IPTG induction at 37°C for 6 h. One mL bacterial cells were centrifuged at 6000 × *g* for 5 min and the pellets were resuspended with 120 μL 1x SDS-PAGE loading buffer and boiled for 10 min before fractionated by electrophoresis on 12% SDS-PAGE gels, and the resolved proteins were transferred onto PVDF membranes (Millipore, USA). After blocking with 5% skim milk, the membranes were probed with primary antibodies [anti-His-tag (1:10000), pullorum-positive chicken serum (1:500), or pullorum-negative SPF chicken serum (1:500)], followed by incubation with HRP-conjugated goat anti-mouse IgG (1:20000; DingGuoShengWu, China) or HRP-conjugated rabbit anti-chicken IgY (1:5000) secondary antibodies (Bioss, China). The blots were visualized using the ECL reagent according to the manufacturer's instructions (CWBio, China). For GFP-Bfr expression analysis, cells were transfected with pEGFP-*bfr* or pEGFP-N1, cells lysates were prepared and examined with anti-GFP antibodies. For signal transduction pathways analysis, cells were transfected with pEGFP-*bfr* or pEGFP-N1. The whole cell extracts were lysed in lysis buffer (150 mM NaCl, 5 mM EDTA, 50 mM Tris·Cl, 10% glycerin, 1% Triton X-100) containing with 1% protease inhibitor cocktail C and 1% 20 mM phosphatase inhibitors (NaF) and examined with Western Blot using anti-p-p38, anti-p38, anti-GFP, and anti-GAPDH antibodies.

Generation of the Bfr-Deficient *S. pullorum*

The Bfr-deficient strain was constructed by λ-Red-mediated recombination system, according to the method described by Datsenko and Wanner (2000). Briefly, the specific primers (ΔBfr1: 5′-ATGAAAGGTGATGTTAAAATC ATAAATTATCTCAAT AAACTATTGGGAAATGTTAGGCTGG AGCTGCTTCG-3′; ΔBfr2: 5′-TTAATGGTAACCTTAATTTGT GATTGCAGATAATTTTGCATACCAAGTTCATATGA ATAT CCTCCTTAG-3′), including 50-bp homology extension from the 5′ and 3′ of the *bfr* gene, were designed to amplify the chloramphenicol cassette from the template plasmid pKD3 by PCR. The PCR products were purified and electroporated into *S. pullorum* containing the pKD46 plasmid. Recombinant bacteria *S. pullorum* Bfr::cat was screened and selected on both Cm and Amp resistance LB agar plates. Gene deletion was confirmed by PCR using the specific primers (Bfr1: 5′-ATGAAAGGTGATGTTAAA-3′; Bfr2: 5′-ATCGGTAACCTTAATTTG-3′). Then the Cm cassette gene of *S. pullorum* Bfr::cat was excised via introducing the Flp recombinase-expressing vector pCP20 by electroporation. The Bfr-deficiency in parental strain was confirmed by PCR, DNA sequencing and Western Blot.

To generate the ΔBfr-complemented strain, the Bfr open-reading frame was amplified by PCR using *S. pullorum* genomic DNA with the specific primers containing *BamH* I and *Sph* I (PBR-Bfr1: 5′-CGCGGATCCATGAAAGGTGATGTTAAA-3′; PBR-Bfr2: 5′-ACATGCATGCTTAATCGGTAACCTTAATTTG-3′). The expected 477 kb PCR product of *bfr* gene was confirmed by DNA sequencing, and further cloned into the plasmid pBR322. The constructed recombinant plasmid was electroporated into the Bfr-deficient bacteria to obtain the ΔBfr-complemented strain. The restoration of Bfr in parental strains was confirmed by PCR and Western Blot.

Infection of DF-1 Cells with Wild Type (WT), Bfr-Deficient and Complemented *S. pullorum*

DF-1 cells were infected with WT, Bfr-deficient (KO) and ΔBfr-complemented (RS) strains at an MOI of 500 or indicated doses. Eight hours after infection, total RNA was extracted from the infected cells and used for cDNA synthesis. Real-time PCR was performed to examine chicken IFN-β expression as above described.

Statistical Analysis

The significance of the differences between the treatment group and control in the activation of promoters and mRNA expressions (cytokine and transcription factor) was determined by the ANOVA and Mann-Whitney accordingly.

RESULTS

Screening for the Major Antigens of *S. pullorum*

Since *S. pullorum* infection elicits a robust humoral immune response in chickens, we wanted to determine the major antigens of *S. pullorum* responsible for the induction of specific antibodies. We proposed that major antigens of *S. pullorum* could be pulled down by specific IgY in the anti-*S. pullorum* Ab positive serum. To test our hypothesis, we performed a pull-down assay using anti-*S. pullorum* Ab positive serum of chickens and the bacterial cell extract of *S. pullorum* according to the published method (Cho et al., 2013). We found that there was an extra clear protein band in the immunoprecipitates of the mixture of anti-*S. pullorum* Ab-positive serum with bacterial cell lysate as compared to that of controls as demonstrated by SDS-PAGE (**Figure 1A**), indicating that the antigens of *S. pullorum* could be pulled down by anti-*S. pullorum* antibodies. To analyze the amino acid sequence of this major antigen, we cut-down the interesting protein band and performed a mass spectrometry. As a result, the arrow-pointed band in **Figure 1A** is a protein named bacterioferritin (Bfr) based on online information from GenBank (Gene ID: 661554730; **Figure 1B**). Although Bfr is known as a protein antigen of *M. paratuberculosis* (Walmsley et al., 1996), our data indicate that this protein might be an important major antigen of *S. pullorum*.

To determine the antigenicity of Bfr, we cloned *bfr* gene from genomic DNA of *S. pullorum* and expressed Bfr-his recombinant protein using an *E. coli* expression system. We found that the Bfr-his fusion protein could be detected by anti-*S. pullorum* Ab positive serum of chickens but not by negative chicken serum (**Figure 1C**), indicating that Bfr-his protein is of good antigenicity. These results suggest that Bfr may serve as a major antigen of *S. pullorum*.

To determine the basic function of Bfr-his(rBfr) fusion protein, we performed the iron uptake assays with rBfr according to the published method (Timoteo et al., 2012). We found that the color of Fe^{3+} in the BSA control could be clearly observed. In contrast, Fe^{3+} color was markedly reduced with rBfr treatment (**Figures 2A,B**), indicating that Bfr can rapidly uptake free Fe^{2+} for oxidation in the presence of H_2O_2 as the oxidant. These data suggest that rBfr is a functional iron storage protein.

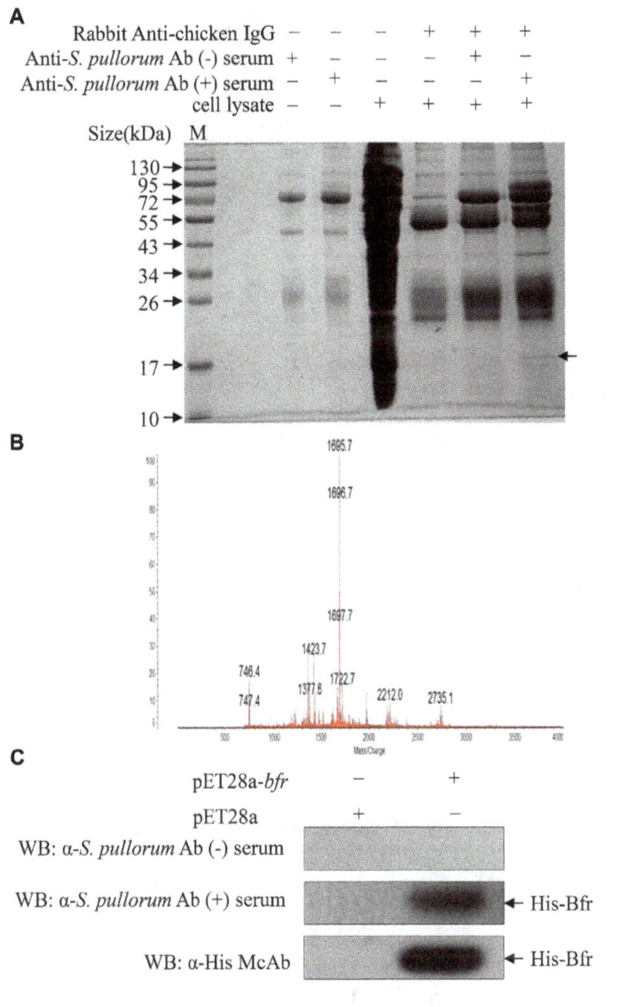

FIGURE 1 | Identification of major antigens of *Salmonella pullorum* by Pull-down assay. (A) Pull-down assay was performed to indentify major antigens using extracts of *S. pullorum* with anti-*S. pullorum* Ab positive serum of chickens and negative serum as controls. The pull-down pellets were examined by SDS-PAGE. **(B)** The arrow-pointed band was cut down and subjected to Mass Spectrometry analysis. **(C)** Recognition of recombinant His-bacterioferritin (Bfr) fusion protein by anti-*S. pullorum* Ab positive serum. Expressions of recombinant His-Bfr fusion protein were detected by Western Blot using anti-*S. pullorum* Ab positive serum, negative serum and anti-His McAb as control.

Bfr Induces IFN-β Expression in DF-1 Cells

Bfr is a major iron storage protein in bacteria (Velayudhan et al., 2007). It was reported that iron could affect innate immune response by influencing IFN-γ mediated pathways in macrophages (Nairz et al., 2014). This information prompted us to examine the effect of *S. pullorum* Bfr on the innate immune response in host cells. We made a pEGFP-*bfr* expression construct, and transfected DF-1 cells with pEGFP-*bfr* or pEGFP-N1 as control. As shown in **Figure 3A**, both GFP-Bfr and GFP were expressed well in the transfected cells as demonstrated by Western Blot using anti-GFP monoclonal antibody. Importantly, transfection of DF-1 cells with pEGFP-*bfr* markedly enhanced activation of IFN-β promoter, but not IFN-α, as compared to that of controls (**Figures 3B,C**). Consistent with this observation, the mRNA expressions of IFN-β in pEGFP-*bfr* transfected cells significantly increased as compared to that of pEGFP-N1 transfected controls, but the mRNA expression of IFN-α was unaffected (**Figures 3D,E**). These results suggest that intracellular Bfr induces IFN-β response in host cells.

Amino Acids 1–50 of Bfr are Responsible for Inducing IFN-β Expression

To determine the domain of Bfr that is responsible for inducing IFN-β expression, we constructed pEGFP-truncated *bfrs* encoding different lengths of Bfr as indicated in **Figure 4A**. We transfected DF-1 cells with the constructs and performed real-time PCR assay with specific primers. As shown in **Figures 4B,C**, transfection of cells with the vectors carrying the full-length and amino acids 1–50 portion of Bfr significantly induced IFN-β expression as compared to that of pEGFP-N1 transfected controls. In contrast, the construct without the gene encoding the amino acids 1–50 portion of Bfr failed to induce IFN-β

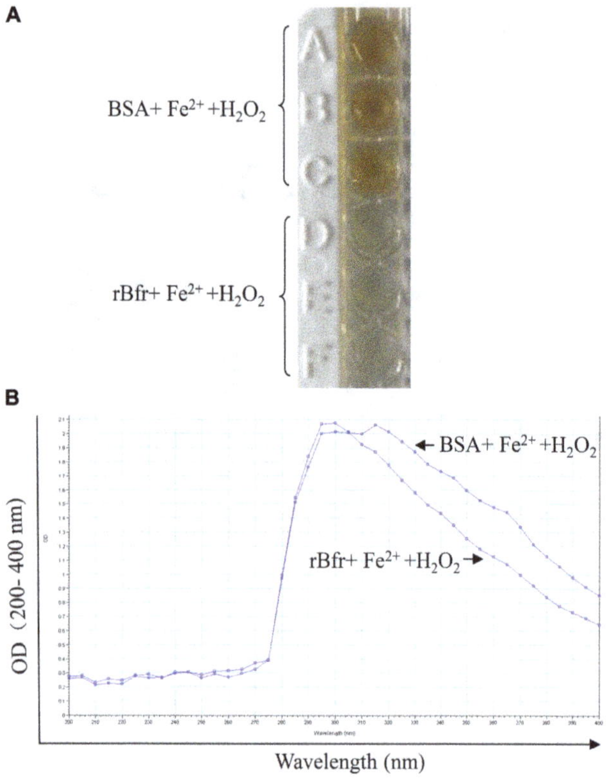

FIGURE 2 | Bacterioferritin uptakes iron *in vitro*. **(A)** Reaction of 0.5 μM rBfr with 12 μM Fe^{2+} ions and 72 μM H_2O_2 in 0.2 M Mops buffer (pH 7) and 0.2 M NaCl was carried out, and BSA was used as a control. **(B)** The OD optical density was examined at 200–400 nm in SpectraMax M5.

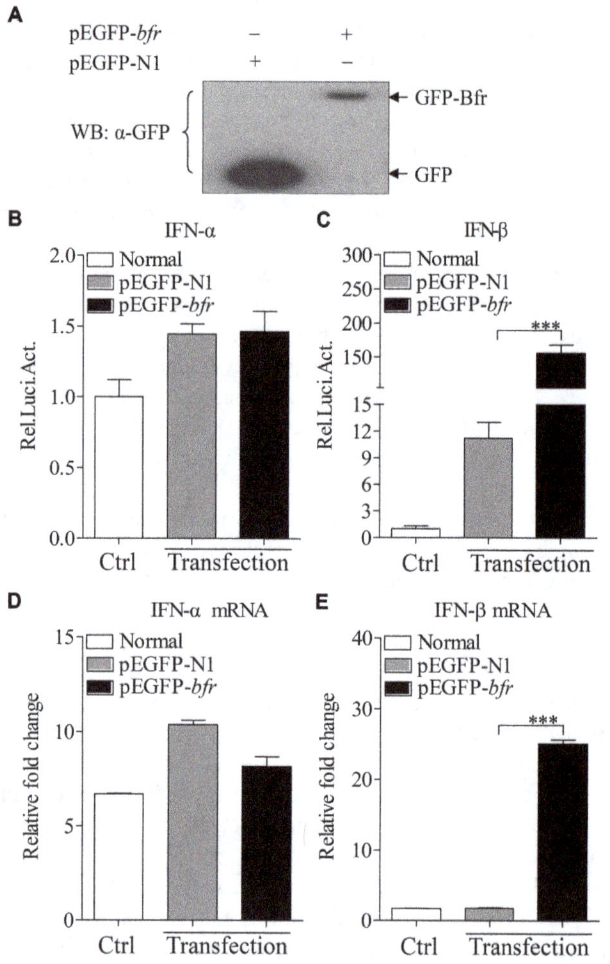

FIGURE 3 | Expression of Bfr induces activation of IFN-β promoter and enhances mRNA expressions of IFN-β in DF-1 cells. **(A)** DF-1 cells were transfected with pEGFP-*bfr* or pEGFP-N1. Twenty four hours after transfection, cell lysates were prepared and examined with Western Blot using anti-GFP monoclonal antibody. **(B,C)** DF-1 cells were transfected with pEGFP-*bfr* or pEGFP-N1 as controls. Twenty-four hours after transfection, cell lysates were prepared and activation of IFN-α and IFN-β promoters were examined with luciferase reporter gene assays. **(D,E)** DF-1 cells were transfected with pEGFP-*bfr* or pEGFP-N1 as controls. Twenty-four hours after transfection, cells were collected and mRNA expressions of IFN-α and IFN-β were examined with qRT-PCR using specific primers. The expression levels of mRNA were calculated in relation to the expression level of GAPDH. Results are representative of three independent experiments. Data are represented as mean ± SD, $n = 3$. *** $p < 0.001$.

expression. These results indicate that the amino acids 1–50 of Bfr is a critical domain responsible for inducing IFN-β response.

Bfr-Activated p38 MAP Kinase is Involved in IFN-β Expression

It was reported that activation of p38 MAP kinase was required for induction of *ifnb* gene expression in response to bacteria in the cytosol (O'Riordan et al., 2002). To dissect the signaling pathways involved in Bfr-induced IFN-β expression, we treated DF-1 cells for 1 h with the inhibitors of the key signaling molecules including p38 MAPK, JNK MAPK or DMSO as controls before pEGFP-*bfr* transfection. Twenty-four hours after transfection, real-time PCR was performed to examine the mRNA levels of IFN-β. As shown in **Figure 5A**, p38 MAPK inhibitor, but not JNK MAPK inhibitor, significantly inhibited IFN-β expression in cells with pEGFP-*bfr* transfection ($p < 0.001$). It is well known that the activation of p38 pathway requires phosphorylation of p38 (Wang et al., 1997). We therefore examined whether p38 is phosphorylated during Bfr expression. We transfected DF-1 cells with pEGFP-*bfr*, and examined the phosphorylation of p38 in pEGFP-*bfr* transfected cells with Western Blot using specific antibodies against p-p38, p38, and GAPDH. As a result, p38 phosphorylation was markedly enhanced in pEGFP-*bfr* transfected cells (**Figures 5B–D**). Taken together, these results clearly show that Bfr-induced phosphorylation of p38 is involved in induction of IFN-β expression. Thus the p38 MAPK pathway is essential for Bfr-induced IFN-β expression.

Bfr Plays a Critical Role in *S. pullorum*-Induced IFN-β Expression in Cells

The fact that Bfr induced expression of IFN-β in host cells prompted us to examine the role of Bfr in *S. pullorum*-induced IFN-β response in cells. We generated a Bfr-deficient *S. pullorum*

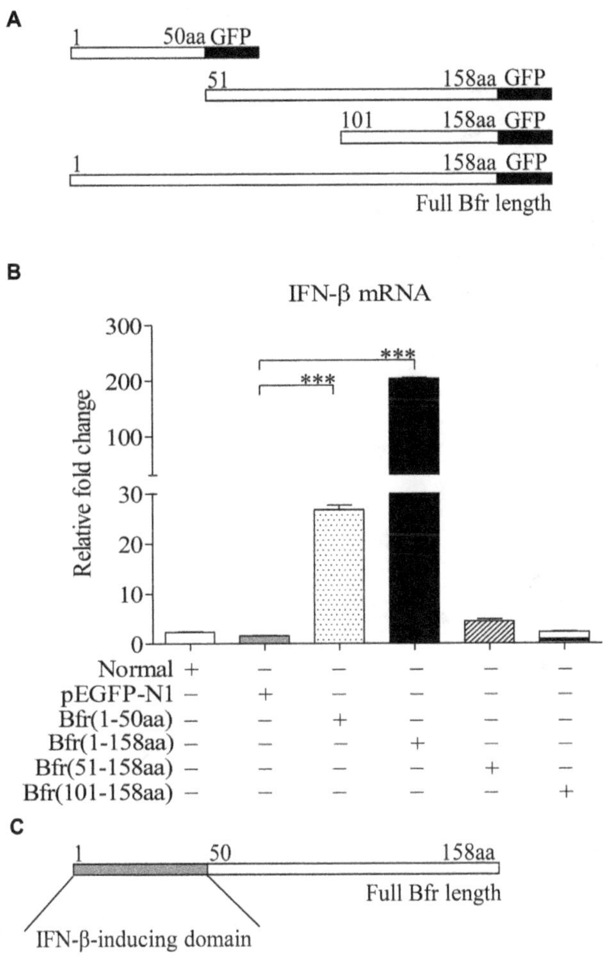

FIGURE 4 | Bacterioferritin 1–50aa is responsible for the induction of IFN-β expression in DF-1 cells. (A) Schematic diagrams showing the structure of full-length Bfr and truncated Bfrs. **(B)** DF-1 cells were transfected with indicated amounts of pEGFP-*bfr*, pEGFP-truncated *bfrs* or pEGFP-N1 as controls. Twenty four hours after transfection, cells were collected and mRNA expression of IFN-β was examined with qRT-PCR using specific primers. The expression levels of mRNA were calculated in relation to the expression level of GAPDH. **(C)** Schematic diagrams showing the IFN-β-inducing domain of Bfr. Data are represented as mean ± SD, $n = 3$. *** $p < 0.001$.

strain using λ-Red-mediated recombination system according to a simple gene disruption strategy (**Figure 6A**) and also generated the ΔBfr-complemented strain expressing Bfr by electroporation of Bfr-deficient *S. pullorum* with pBR322-*bfr*. As shown in **Figures 6B,C**, WT and the complemented *S. pullorum* strains expressed Bfr very well. In contrast, Bfr-deficient *S. pullorum* had no detectable Bfr as examined by PCR and Western Blot. These results indicate that Bfr-deficient *S. pullorum* strain and its complemented strain expressing Bfr were successfully generated.

We cultured WT, Bfr-deficient and the complemented *S. pullorum* strains in culture medium, and compared their growth at 0, 6, 12, and 24 h after culture. As a result, we did not find any difference between WT, Bfr-deficient and the complemented *S. pullorum* strains in terms of their growth (data not shown), suggesting that deficiency of Bfr in *S. pullorum* does not affect the bacterial replication. We infected DF-1 cells with WT *S. pullorum* at an MOI of 1, 5, 20, 100, or 500. Eight hour after infection, the mRNA expression of IFN-β was examined with real-time PCR assay. As shown in **Figure 7A**, infection of DF-1 cells with WT *S. pullorum* markedly induced mRNA expression of IFN-β in cells ($p < 0.001$). However, WT *S. pullorum*-induced IFN-β expression was completely abolished by deletion (knock-out) of *bfr* gene from the bacteria (**Figure 7B**), indicating that Bfr is required for *S. pullorum*-induced IFN-β expression. Thus, Bfr plays a critical role in *S. pullorum*-induced innate response in host cells.

DISCUSSION

Salmonella pullorum is a worldwide distributed poultry pathogen of considerable economic importance to the poultry industry, particularly to that of developing countries. An increasing number of *S. pullorum* strains were isolated in China, and many of these bacteria showed antimicrobial resistance (Pan et al., 2009). This pathogen causes high mortality in young chickens and persistent infection in adult chickens with clinical signs of decreased egg production and diarrhea. Although the humoral immune response against *S. pullorum* cannot clear the pathogens once the bacteria intrude host cells, the specific antibodies still play an important role in mediating phagocytosis of extracellular bacteria by phagocytes. In particular, the examination of anti-*S. pullorum* antibodies is of clinically diagnostic importance. Thus hemagglutination assay with the whole blood of chickens is generally performed to screen for the *S. pullorum*-infected chickens in flocks. Immunoprecipitation of whole bacterial cell lysate by antibodies against *S. pullorum* (Pull-down assay) is an efficient method that allows us to identify major antigens of the microorganism that elicit humoral immune response. The identified antigen would help to develop new diagnostic methods or vaccine in control of the disease (Britton et al., 1988; Sexton et al., 1991). High titers of anti-*Salmonella* IgY were produced by birds infected with *S. pullorum* after 5 weeks (Wigley et al., 2001). However, little information is available regarding the major antigen of *S. pullorum*. In this study, a major antigen Bfr that elicited antibody response was identified with a pull-down assay and Mass spectrometric method. Our results demonstrate that recombinant Bfr can be specifically recognized by pullorum-positive serum. Thus, Bfr-induced antibodies probably play an important role in host response against *S. pullorum* infection.

Bfr, a 17-kDa protein that was previously identified as an outer membrane protein of *S. hadar*, is composed of 24 identical subunits along with the usual ferroxidase sites that have 12 binding sites for heme iron (Wong et al., 2009; Snoussi et al., 2012). Bfr is also the major Fe storage protein in *Salmonella* (Velayudhan et al., 2007). We could observe the color of Fe^{3+} from the purified Bfr-his recombinant protein in our experiments. Inactivation of Bfr induces intracellular free Fe concentration and exhibits increased susceptibility to oxidative stress (Velayudhan et al., 2007). It was found that the

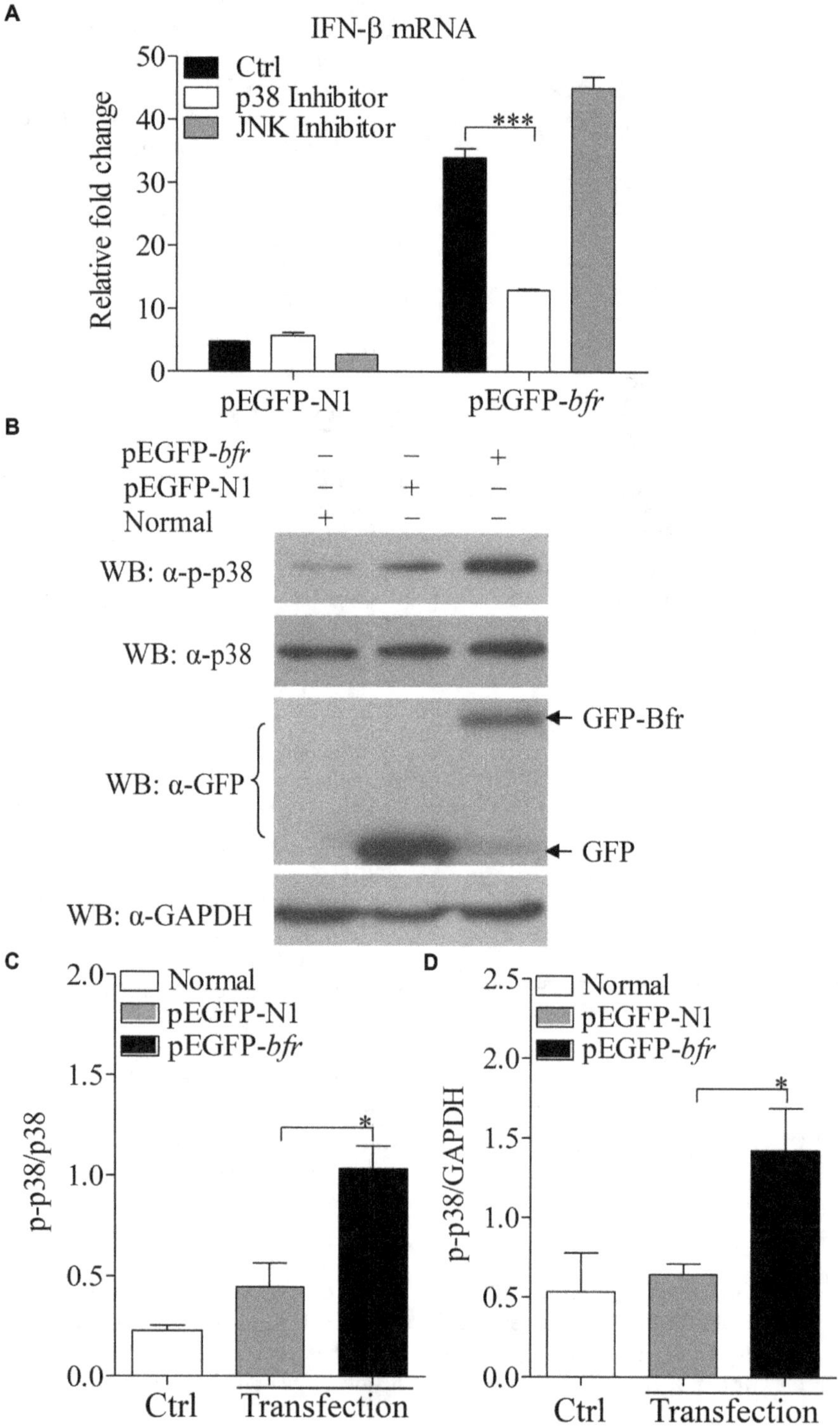

FIGURE 5 | Bacterioferritin induced IFN-β expression via p38 MAP Kinase signaling pathway. (A) Effects of p38 and JNK inhibitors on Bfr-induced IFN-β expression. DF-1 cells were transfected with indicated amounts of pEGFP-*bfr* or pEGFP-N1 as controls, and treated with inhibitors of p38, JNK, or DMSO as controls for 1 h. Twenty four hours after transfection, mRNA expression of IFN-β was examined with real-time PCR. **(B–D)** Effects of Bfr on phosphorylation of p38. DF-1 cells were transfected with pEGFP-*bfr* or pEGFP-N1 as controls, Twenty four hours after transfection, cell lysates were prepared and examined with Western Blot for the detection of p-p38, p38, GFP, GFP-Bfr, and GAPDH. The band density of p-p38, p38 and GAPDH in normal, pEGFP-*bfr* or pEGFP-N1 transfected cells in **(B)** was quantitated by densitometry, and the relative levels of p-p38 in **(B)** were calculated as follows: band density of p-p38/band density of p38 **(C)** or GAPDH **(D)**. Results are representative of three independent experiments. Data are represented as mean ± SD, $n = 3$. *** $p < 0.001$ and * $p < 0.05$.

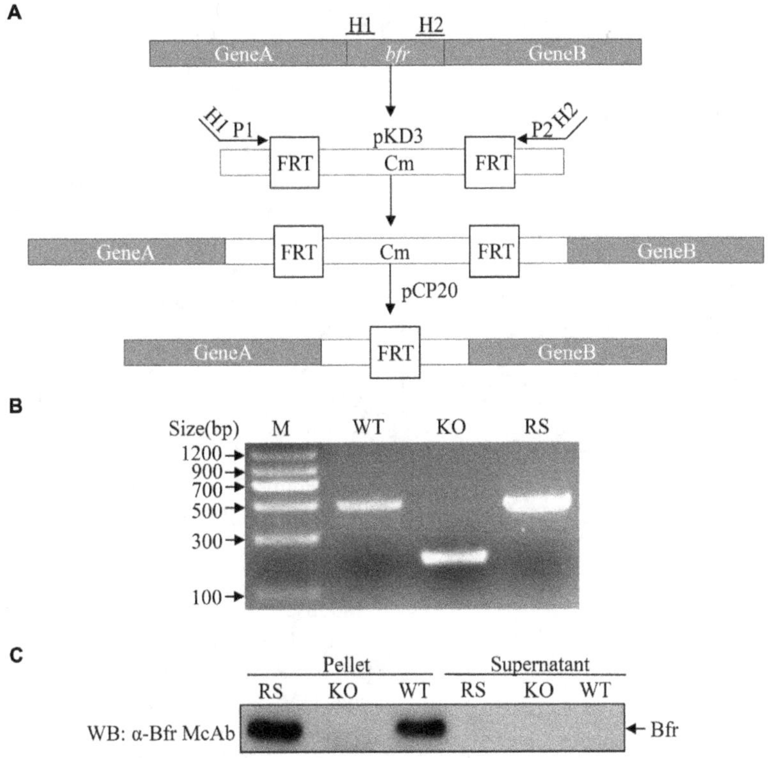

FIGURE 6 | Generation of Bfr-deficient *S. pullorum* strain by Red homologous recombination. (A) Schematic diagrams showing the strategy for the deletion of *bfr* gene. **(B)** Identification of recombinant (KO) and complemented (RS) strains by PCR. PCR was performed to examine *bfr* gene in WT, KO, and RS *S. pullorum* strains. **(C)** Examination of Bfr expressions in WT, KO, and RS *S. pullorum* strains. Cell lysates of WT, KO, and RS *S. pullorum* were prepared, and Bfr expressions were examined with Western Blot using anti-Bfr monoclonal antibody.

transcription factor SsrB controls resistance to reactive oxygen species through Bfr in *Salmonella* (Brown et al., 2014). This information suggests that Bfr is involved in regulation of iron homeostasis and protects against hydrogen peroxide toxicity in bacteria. H_2O_2 can react with Fe^{2+} ions via Fenton and Haber-Weiss reactions, producing ROS (reactive oxygen species) such as hydroxyl radical or superoxide, which are capable of damaging most cellular components, such as nucleic acids, protein or membrane lipids (Timoteo et al., 2012). And Bfr binds to DNA and reduces the damage of ROS by the rapid uptake and oxidation of free Fe^{2+}, using H_2O_2 as the oxidant (Timoteo et al., 2012). Our results demonstrate that recombinant Bfr can mediate the rapid uptake and oxidation of free Fe^{2+}, using H_2O_2 as the oxidant. Interestingly, we also found that Bfr could induce self-activation in a yeast two-hybrid screening and apoptosis in DF-1 cells in this study (data not shown). This information suggests that Bfr might be a muti-functional protein.

Bacterioferritin is an important factor in bacteria, but few reports are available regarding the cell response to Bfr. Currently it is known that Bfr is a T-cells antigen that induces a strong IFN-γ production and the proliferation of lymphocytes (Denoel et al., 1997; Al-Mariri et al., 2002; Lee et al., 2006). In addition, Bfr induced humoral immune response in mice immunized with DNA vaccine encoding the Bfr or recombinant Bfr proteins (Al-Mariri et al., 2001a,b). The patient sera of *M. leprae* could react with Bfr, and *M. paratuberculosi* antigen D was identified as Bfr (Brooks et al., 1991; Spencer et al., 2011). These findings suggest that Bfr plays an important role in acquired immunity. However, little is known about the role of Bfr in the innate immune response. Our data show that Bfr not only acts as a potent antigen inducing humoral immune response but also as an inducer for innate immune responses (inducing IFN-β expression), indicating that Bfr is not merely a protein for iron storage and detoxification (Bou-Abdallah et al., 2002). Furthermore, we found that the amino acids 1–50 of Bfr were responsible for induction of IFN-β expression. However, Bfr did not affect the expression of IFN-α in cells. It seems that the role of Bfr is specific, only for induction of IFN-β expression.

Recognition of bacterial products by host surveillance system results in transcription of the *ifnb* gene, and the activation of cytosol-specific signaling is associated with phosphorylation of the p38 mitogen-activated protein (MAP) kinase (O'Riordan et al., 2002). In this study, when DF-1 cells were treated with p38 MAP Kinase inhibitor, Bfr-induced IFN-β expression was markedly inhibited, indicating that Bfr might activate cytosol-specific signaling. In contrast, JNK MAP Kinase inhibitor had no effects on Bfr-induced IFN-β response. These results suggest that Bfr induces IFN-β expression via the p38 signal transduction pathway. As p38 MAP kinases are major players during inflammatory responses, they can be activated by

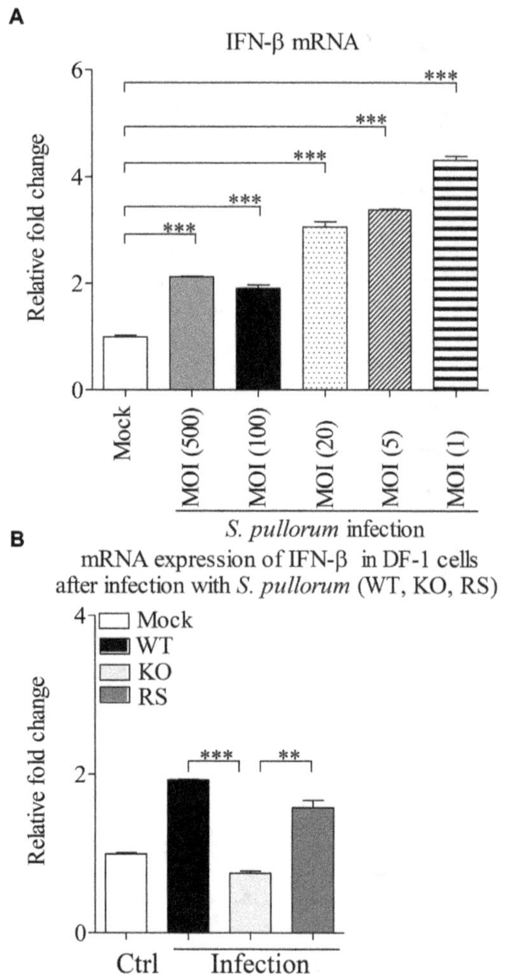

FIGURE 7 | A critical role of Bfr in *S. pullorum*-induced IFN-β expressions. (A) DF-1 cells were infected with WT *S. pullorum* at an indicated MOI, 8 h after infection, mRNA expressions of IFN-β in cells were examined with qRT-PCR using specific primers. The expression levels of mRNA were calculated in relation to the expression level of GAPDH. **(B)** DF-1 cells were infected with WT, KO, and RS *S. pullorum* strains at an MOI of 500. Eight hours after infection, mRNA expressions of IFN-β in cells were examined with qRT-PCR using specific primers. The mRNA expression levels of IFN-β in cells were calculated in relation to the expression level of GAPDH. Results are representative of three independent experiments. Data are represented as mean ± SD, $n = 3$. *** $p < 0.001$ and ** $p < 0.01$.

environmental and cellular stresses including pathogens, heat shock, growth factors, osmotic shock, ultraviolet irradiation and cytokines (Yang et al., 2014). p38 kinases have two domains: a 135 amino acid N-terminal domain and a 225 amino acid C-terminal domain. The phosphorylation lip of p38 consists of 13 residues, Leu-171-Val-183, and the protein is activated by phosphorylation of a signal threonine (Thr-180) and a single tyrosine residue (Tyr-182) in the lip (Wang et al., 1997). Furthermore, we found that p38 phosphorylation was induced by Bfr, indicating that p38 signal transduction pathway is essential for IFN-β expression in cells.

DF-1, an immortal chicken embryo fibroblast cell line, is commonly used for the research of *Salmonella* (Li et al., 2006; Szmolka et al., 2015) and type I interferon (Li et al., 2013). Our data show that *S. pullorum* infection significantly induces activation of the IFN-β promoter in DF-1 cells, supporting the previous publication by Hess et al. (1989). In contrast, *S. pullorum*-induced IFN-β expression was completely abolished by deficiency of Bfr in the bacteria, indicating that Bfr is required for *S. pullorum*-induced IFN-β expression in cells.

IFN-β is a key cytokine in the innate immune response, mediating expression of hundreds of IFN-stimulated genes (ISGs) that are responsible for the establishment of an antimicrobial state in the infected tissue (Schmolke et al., 2014). It was reported that IFN-β potently represses *S. typhimurum*-dependent induction of IL-1 family cytokines and neutrophil chemokines and IFN-$\beta^{-/-}$ mice exhibit greater resistance to oral *S. typhimurum* infection and a slower spread of *S. typhimurum* to distal sterile sites (Perkins et al., 2015). *S. typhimurum* induces the production of IFN-β, which drives necroptosis of macrophages and allows *Salmonella* to evade the immune response that is detrimental to the survival of mice (Robinson et al., 2012). Since Bfr induced IFN-β expression through the p38 MAP Kinase signaling pathway in cells, several important questions need to be addressed. For example, what is the host protein directly targeted by Bfr? What is the role of IFN-β induced by Bfr in host cell response to *Salmonella* infection? Elucidation of these questions will further our understandings of the mechanisms underlying pathogenesis of *Salmonella* infection.

CONCLUSION

Our results demonstrate that Bfr is a major antigen of *S. pullorum*. Our data also show that Bfr induced IFN-β expression via its amino acids 1–50 portion. Furthermore, we found that the p38 MAPK signaling pathway was essential for Bfr-induced IFN-β expression. Importantly, *S. pullorum* induced-IFN-β was totally abolished by deficiency of Bfr in the bacteria, indicating that Bfr plays a critical role in *S. pullorum* induced-IFN-β expression in DF-1 cells. Our findings provide new insights into the molecular mechanisms of the host response to *S. pullorum* infection.

AUTHOR CONTRIBUTIONS

SZ and ZX conceived and designed the experiments; ZX performed the experiments; SZ and ZX analyzed the data; SZ, YQ, YW, XL and HC contributed reagents/materials/analysis tools; SZ and ZX wrote the paper.

ACKNOWLEDGMENTS

We thank Drs. Guoqiang Zhu and Fuyong Chen for their generous assistance. This work was supported by grants from Earmarked Fund for Modern Agro-industry Technology Research System (#NYCYTX-41).

REFERENCES

Abdul-Careem, M. F., Hunter, B. D., Lee, L. F., Fairbrother, J. H., Haghighi, H. R., Read, L., et al. (2008). Host responses in the bursa of Fabricius of chickens infected with virulent Marek's disease virus. *Virology* 379, 256–265. doi: 10.1016/j.virol.2008.06.027

Al-Mariri, A., Tibor, A., Lestrate, P., Mertens, P., De Bolle, X., and Letesson, J. J. (2002). *Yersinia enterocolitica* as a vehicle for a naked DNA vaccine encoding *Brucella abortus* bacterioferritin or P39 antigen. *Infect. Immun.* 70, 1915–1923. doi: 10.1128/IAI.70.4.1915-1923.2002

Al-Mariri, A., Tibor, A., Mertens, P., De Bolle, X., Michel, P., Godefroid, J., et al. (2001a). Protection of BALB/c mice against *Brucella abortus* 544 challenge by vaccination with bacterioferritin or P39 recombinant proteins with CpG oligodeoxynucleotides as adjuvant. *Infect. Immun.* 69, 4816–4822. doi: 10.1128/IAI.69.8.4816-4822.2001

Al-Mariri, A., Tibor, A., Mertens, P., De Bolle, X., Michel, P., Godfroid, J., et al. (2001b). Induction of immune response in BALB/c mice with a DNA vaccine encoding bacterioferritin or P39 of *Brucella* spp. *Infect. Immun.* 69, 6264–6270. doi: 10.1128/IAI.69.10.6264-6270.2001

Barrow, P. A., and Freitas, N. O. (2011). Pullorum disease and fowl typhoid–new thoughts on old diseases: a review. *Avian Pathol.* 40, 1–13. doi: 10.1080/03079457.2010.542575

Barrow, P. A., Jones, M. A., Smith, A. L., and Wigley, P. (2012). The long view: *Salmonella*-the last forty years. *Avian Pathol.* 41, 413–420. doi: 10.1080/03079457.2012.718071

Berndt, A., Wilhelm, A., Jugert, C., Pieper, J., Sachse, K., and Methner, U. (2007). Chicken cecum immune response to *Salmonella enterica* serovars of different levels of invasiveness. *Infect. Immun.* 75, 5993–6007. doi: 10.1128/IAI.00695-07

Bou-Abdallah, F., Lewin, A. C., Le Brun, N. E., Moore, G. R., and Chasteen, N. D. (2002). Iron detoxification properties of *Escherichia coli* bacterioferritin, Attenuation of oxyradical chemistry. *J. Biol. Chem.* 277, 37064–37069. doi: 10.1074/jbc.M205712200

Britton, W. J., Hellqvist, L., Garsia, R. J., and Basten, A. (1988). Antigens of *Mycobacterium leprae* identified by immunoprecipitation with sera from leprosy and tuberculosis patients. *Clin. Exp. Immunol.* 71, 394–398.

Brooks, B. W., Young, N. M., Watson, D. C., Robertson, R. H., Sugden, E. A., Nielsen, K. H., et al. (1991). *Mycobacterium* paratuberculosis antigen D: characterization and evidence that it is a bacterioferritin. *J. Clin. Microbiol.* 29, 1652–1658.

Brown, N. F., Rogers, L. D., Sanderson, K. L., Gouw, J. W., Hartland, E. L., and Foster, L. J. (2014). A horizontally acquired transcription factor coordinates *Salmonella* adaptations to host microenvironments. *mBio* 5:e1727-14. doi: 10.1128/mBio.01727-14

Chappell, L., Kaiser, P., Barrow, P., Jones, M. A., Johnston, C., and Wigley, P. (2009). The immunobiology of avian systemic salmonellosis. *Vet. Immunol. Immunopathol.* 128, 53–59. doi: 10.1016/j.vetimm.2008.10.295

Cho, S. B., Zheng, Z., Ahn, K. J., Choi, M. J., Cho, S., Kim, D. Y., et al. (2013). Serum IgA reactivity against GroEL of *Streptococcus sanguinis* and human heterogeneous nuclear ribonucleoprotein A2/B1 in patients with Behcet disease. *Br. J. Dermatol.* 168, 977–983. doi: 10.1111/bjd.12128

Crhanova, M., Hradecka, H., Faldynova, M., Matulova, M., Havlickova, H., Sisak, F., et al. (2011). Immune response of chicken gut to natural colonization by gut microflora and to *Salmonella enterica* serovar enteritidis infection. *Infect. Immun.* 79, 2755–2763. doi: 10.1128/IAI.01375-10

Datsenko, K. A., and Wanner, B. L. (2000). One-step inactivation of chromosomal genes in *Escherichia coli* K-12 using PCR products. *Proc. Natl. Acad. Sci. U.S.A.* 97, 6640–6645. doi: 10.1073/pnas.120163297

Denoel, P. A., Vo, T. K., Weynants, V. E., Tibor, A., Gilson, D., Zygmunt, M. S., et al. (1997). Identification of the major T-cell antigens present in the *Brucella melitensis* B115 protein preparation, Brucellergene OCB. *J. Med. Microbiol.* 46, 801–806. doi: 10.1099/00222615-46-9-801

Hess, C. B., Niesel, D. W., and Klimpel, G. R. (1989). The induction of interferon production in fibroblasts by invasive bacteria: a comparison of *Salmonella* and *Shigella* species. *Microb. Pathog.* 7, 111–120. doi: 10.1016/0882-4010(89)90030-2

Lee, B. Y., Horwitz, M. A., and Clemens, D. L. (2006). Identification, recombinant expression, immunolocalization in macrophages, and T-cell responsiveness of the major extracellular proteins of *Francisella tularensis*. *Infect. Immun.* 74, 4002–4013. doi: 10.1128/IAI.00257-06

Li, L., Fang, W., Li, J., Huang, Y., and Yu, L. (2006). Oral DNA vaccination with the polyprotein gene of infectious bursal disease virus (IBDV) delivered by the attenuated *Salmonella* elicits protective immune responses in chickens. *Vaccine* 24, 5919–5927. doi: 10.1016/j.vaccine.2006.04.057

Li, Y. P., Handberg, K. J., Juul-Madsen, H. R., Zhang, M. F., and Jorgensen, P. H. (2007). Transcriptional profiles of chicken embryo cell cultures following infection with infectious bursal disease virus. *Arch. Virol.* 152, 463–478. doi: 10.1007/s00705-006-0878-9

Li, Z., Wang, Y., Li, X., Li, X., Cao, H., and Zheng, S. J. (2013). Critical roles of glucocorticoid-induced leucine zipper in infectious bursal disease virus (IBDV)-induced suppression of type I Interferon expression and enhancement of IBDV growth in host cells via interaction with VP4. *J. Virol.* 87, 1221–1231. doi: 10.1128/JVI.02421-12

Liu, H., Zhang, M., Han, H., Yuan, J., and Li, Z. (2010). Comparison of the expression of cytokine genes in the bursal tissues of the chickens following challenge with infectious bursal disease viruses of varying virulence. *Virol. J.* 7, 364. doi: 10.1186/1743-422X-7-364

Marcus, S. L., Brumell, J. H., Pfeifer, C. G., and Finlay, B. B. (2000). *Salmonella* pathogenicity islands: big virulence in small packages. *Microbes Infect.* 2, 145–156. doi: 10.1016/S1286-4579(00)00273-2

Monroe, K. M., McWhirter, S. M., and Vance, R. E. (2010). Induction of type I interferons by bacteria. *Cell. Microbiol.* 12, 881–890. doi: 10.1111/j.1462-5822.2010.01478.x

Nairz, M., Haschka, D., Demetz, E., and Weiss, G. (2014). Iron at the interface of immunity and infection. *Front. Pharmacol.* 5:152. doi: 10.3389/fphar.2014.00152

O'Riordan, M., Yi, C. H., Gonzales, R., Lee, K. D., and Portnoy, D. A. (2002). Innate recognition of bacteria by a macrophage cytosolic surveillance pathway. *Proc. Natl. Acad. Sci. U.S.A.* 99, 13861–13866. doi: 10.1073/pnas.202476699

Pan, Z., Wang, X., Zhang, X., Geng, S., Chen, X., Pan, W., et al. (2009). Changes in antimicrobial resistance among *Salmonella enterica* subspecies enterica serovar Pullorum isolates in China from 1962 to 2007. *Vet. Microbiol.* 136, 387–392. doi: 10.1016/j.vetmic.2008.11.015

Perkins, D. J., Rajaiah, R., Tennant, S. M., Ramachandran, G., Higginson, E. E., Dyson, T. N., et al. (2015). *Salmonella* typhimurium co-opts the host type I IFN system to restrict macrophage innate immune transcriptional responses selectively. *J. Immun.* 195, 2461–2471. doi: 10.4049/jimmunol.1500105

Ratledge, C. (2007). Iron metabolism and infection. *Food Nutr. Bull.* 28, S515–S523. doi: 10.1177/15648265070284S405

Robinson, N., McComb, S., Mulligan, R., Dudani, R., Krishnan, L., and Sad, S. (2012). Type I interferon induces necroptosis in macrophages during infection with *Salmonella enterica* serovar Typhimurium. *Nat. Immunol.* 13, 954–962. doi: 10.1038/ni.2397

Schmolke, M., Patel, J. R., de Castro, E., Sanchez-Aparicio, M. T., Uccellini, M. B., Miller, J. C., et al. (2014). RIG-I detects mRNA of intracellular *Salmonella enterica* serovar Typhimurium during bacterial infection. *mBio* 5:e1006-14. doi: 10.1128/mBio.01006-14

Sexton, J. L., Milner, A. R., and Campbell, N. J. (1991). Fasciola hepatica: immunoprecipitation analysis of biosynthetically labelled antigens using sera from infected sheep. *Parasite Immunol.* 13, 105–108. doi: 10.1111/j.1365-3024.1991.tb00267.x

Shivaprasad, H. L. (2000). Fowl typhoid and pullorum disease. *Rev. Sci. Tech.* 19, 405–424.

Snoussi, S., May, A. E., Coquet, L., Chan, P., Jouenne, T., Landoulsi, A., et al. (2012). Adaptation of *Salmonella enterica* Hadar under static magnetic field: effects on outer membrane protein pattern. *Proteome Sci.* 10:6. doi: 10.1186/1477-5956-10-6

Spencer, J. S., Kim, H. J., Wheat, W. H., Chatterjee, D., Balagon, M. V., Cellona, R. V., et al. (2011). Analysis of antibody responses to *Mycobacterium leprae* phenolic glycolipid I, lipoarabinomannan, and recombinant proteins to define disease subtype-specific antigenic profiles in leprosy. *Clin. Vaccine Immunol.* 18, 260–267. doi: 10.1128/CVI.00472-10

Szmolka, A., Wiener, Z., Matulova, M. E., Varmuzova, K., and Rychlik, I. (2015). Gene expression profiles of chicken embryo fibroblasts in response to *Salmonella* enteritidis infection. *PLoS ONE* 10:e0127708. doi: 10.1371/journal.pone.0127708

Timoteo, C. G., Guilherme, M., Penas, D., Folgosa, F., Tavares, P., and Pereira, A. S. (2012). Desulfovibrio vulgaris bacterioferritin uses H(2)O(2) as a co-substrate for iron oxidation and reveals DPS-like DNA protection and binding activities. *Biochem. J.* 446, 125–133. doi: 10.1042/BJ20111439

Velayudhan, J., Castor, M., Richardson, A., Main-Hester, K. L., and Fang, F. C. (2007). The role of ferritins in the physiology of *Salmonella enterica* sv. Typhimurium: a unique role for ferritin B in iron-sulphur cluster repair and virulence. *Mol. Microbiol.* 63, 1495–1507. doi: 10.1111/j.1365-2958.2007.05600.x

Walmsley, R. S., Ibbotson, J. P., Chahal, H., and Allan, R. N. (1996). Antibodies against *Mycobacterium* paratuberculosis in Crohn's disease. *QJM* 89, 217–221. doi: 10.1093/qjmed/89.3.217

Wang, Z., Harkins, P. C., Ulevitch, R. J., Han, J., Cobb, M. H., and Goldsmith, E. J. (1997). The structure of mitogen-activated protein kinase p38 at 2.1-A resolution. *Proc. Natl. Acad. Sci. U.S.A.* 94, 2327–2332. doi: 10.1073/pnas.94.6.2327

Wigley, P., Berchieri, A. J., Page, K. L., Smith, A. L., and Barrow, P. A. (2001). *Salmonella enterica* serovar Pullorum persists in splenic macrophages and in the reproductive tract during persistent, disease-free carriage in chickens. *Infect. Immun.* 69, 7873–7879. doi: 10.1128/IAI.69.12.7873-7879.2001

Wigley, P., Hulme, S. D., Powers, C., Beal, R. K., Berchieri, A. J., Smith, A., et al. (2005). Infection of the reproductive tract and eggs with *Salmonella enterica* serovar pullorum in the chicken is associated with suppression of cellular immunity at sexual maturity. *Infect. Immun.* 73, 2986–2990. doi: 10.1128/IAI.73.5.2986-2990.2005

Withanage, G. S., Kaiser, P., Wigley, P., Powers, C., Mastroeni, P., Brooks, H., et al. (2004). Rapid expression of chemokines and proinflammatory cytokines in newly hatched chickens infected with *Salmonella enterica* serovar typhimurium. *Infect. Immunol.* 72, 2152–2159. doi: 10.1128/IAI.72.4.2152-2159.2004

Wong, S. G., Tom-Yew, S. A., Lewin, A., Le Brun, N. E., Moore, G. R., Murphy, M. E., et al. (2009). Structural and mechanistic studies of a stabilized subunit dimer variant of *Escherichia coli* bacterioferritin identify residues required for core formation. *J. Biol. Chem.* 284, 18873–18881. doi: 10.1074/jbc.M901747200

Yang, Y., Kim, S. C., Yu, T., Yi, Y. S., Rhee, M. H., Sung, G. H., et al. (2014). Functional roles of p38 mitogen-activated protein kinase in macrophage-mediated inflammatory responses. *Mediators Inflamm.* 2014:352371. doi: 10.1155/2014/352371

Conflict of Interest Statement: The authors declare that the research was conducted in the absence of any commercial or financial relationships that could be construed as a potential conflict of interest.

Bacterial Exotoxins and the Inflammasome

*Allison J. Greaney, Stephen H. Leppla and Mahtab Moayeri**

Microbial Pathogenesis Section, Laboratory of Parasitic Diseases, National Institute of Allergy and Infectious Diseases, National Institutes of Health, Bethesda, MD, USA

Edited by:
Denise Monack, Stanford University School of Medicine, USA

Reviewed by:
Laurel L. Lenz, University of Colorado School of Medicine, USA
Sunny Shin, University of Pennsylvania, USA
Prosper N. Boyaka, The Ohio State University, USA

***Correspondence:**
Mahtab Moayeri
mmoayeri@niaid.nih.gov

The inflammasomes are intracellular protein complexes that play an important role in innate immune sensing. Activation of inflammasomes leads to activation of caspase-1 and maturation and secretion of the pro-inflammatory cytokines interleukin (IL)-1β and IL-18. In certain myeloid cells, this activation can also lead to an inflammatory cell death (pyroptosis). Inflammasome sensor proteins have evolved to detect a range of microbial ligands and bacterial exotoxins either through direct interaction or by detection of host cell changes elicited by these effectors. Bacterial exotoxins activate the inflammasomes through diverse processes, including direct sensor cleavage, modulation of ion fluxes through plasma membrane pore formation, and perturbation of various host cell functions. In this review, we summarize the findings on some of the bacterial exotoxins that activate the inflammasomes.

Keywords: inflammasome, caspase-1, interleukin-1, pyroptosis, exotoxins, NLRP3, Nod-like receptors, anthrax lethal toxin

INTRODUCTION

Inflammasomes are multiprotein complexes that form in response to microbial effectors, metabolites, nucleic acids, and other danger signals. These signals are sensed by cytosolic pattern recognition receptors [reviewed in Ref. (1)]. The most well-known inflammasome sensors, the nucleotide-binding domain/leucine rich repeat (NLR) proteins, contain common domains, including a leucine rich repeat (LRR), a nucleotide-binding domain (NACHT), a caspase activation and recruitment domain (CARD), and in some, but not all cases, a pyrin domain (PYD). Other inflammasome sensors are the absent in melanoma 2 domain (AIM2) protein, which will not be further discussed in this review, and a sensor called Pyrin, whose function remains controversial and has been described as both an inflammasome activator and inhibitor (2, 3). Upon activation, the inflammasome sensors initiate assembly of a complex that often includes an adaptor protein [usually the apoptosis-associated speck-like protein containing a CARD (ASC)], and pro-caspase-1. Proximity-based autoproteolysis of pro-caspase-1 then leads to cleavage of its substrates – the pro-inflammatory cytokines interleukin (IL)-1β and IL-18, and initiation of a rapid lytic cell death called pyroptosis that requires caspase-1 activity and targeting of unknown death substrates [reviewed in Ref. (1)]. IL-1β and IL-18 are well-studied for their roles in recruiting innate immune cells and promoting adaptive and humoral immunity. Pyroptosis and the accompanying release of cellular contents also act as danger signals, resulting in effects on bystander cells that can impact both innate and adaptive immune responses.

Different sensors activate inflammasome assembly in response to seemingly disparate stimuli. The mechanism of activation of some inflammasome sensors, such as rodent NLRP1, NAIP/NLRC4, and AIM2, are now known [for review see Ref. (1)]. Exactly how the promiscuous NLRP3 inflammasome, however, is activated by a diversity of seemingly disparate stimuli is still a matter of much debate [reviewed in Ref. (1, 4)].

This review focuses on effector bacterial exotoxins and how they activate different inflammasome sensors (**Table 1**; **Figure 1**). Most toxins that have been described to activate the inflammasomes are pore-forming toxins that activate NLRP3. However, toxins with unique enzymatic actions also activate the inflammasomes. There are also many other bacterial proteins that activate inflammasomes, such as flagellin and needle/rod components of the bacterial type III secretion system (T3SS), which activate NAIP/NLRC4 by direct binding. These are reviewed elsewhere (5–7). Furthermore, the "non-canonical" caspase-11 inflammasome, which directly senses bacterial endotoxin Lipid A, has been examined in a series of elegant studies (8–12) and is also reviewed elsewhere (1, 13, 14). We will discuss bacterial exotoxins that directly activate the inflammasome by sensor modification, namely, *Bacillus anthracis* lethal toxin (LT), and other bacterial exotoxins that activate the NLRP3 and Pyrin inflammasomes indirectly by altering host cell function in ways that include plasma membrane pore formation, co-option of the host actin cytoskeleton, and as-of-yet unknown mechanisms.

ANTHRAX LETHAL TOXIN AND DIRECT ACTIVATION OF THE NLRP1 INFLAMMASOME

Anthrax LT is a major virulence factor of *B. anthracis*, inducing vascular collapse during anthrax infection of animals ranging from rodents to monkeys (15–17). LT is a bipartite toxin made of a receptor-binding moiety, protective antigen (PA), and a zinc-dependent metalloprotease, lethal factor (LF). Upon endosome acidification, PA delivers LF to the host cell cytosol [reviewed in Ref. (18)]. LF activates the NLRP1 inflammasome in macrophages and dendritic cells from certain inbred rodent strains by cleavage of NLRP1 (19, 20). Cleavage occurs in an N-terminal region of unknown function, located in a position corresponding to the PYD of human NLRP1 (19, 21). This cleavage is necessary and sufficient for inflammasome activation (19, 22). In most rat strains, only one of the two NLRP1 paralogs is expressed, and susceptibility to pyroptosis is perfectly correlated with the ability of LT to cleave the expressed NLRP1 sensor (19, 21, 23). Rat strains, such as Fischer and Brown Norway, express NLRP1 variants that are cleaved by LF and have macrophages that pyroptose in response to the toxin. Furthermore, when these rats are challenged with LT, they undergo death in <1 h. Strains such as SHR, Lewis, and Copenhagen have an altered sequence within the LF cleavage site, rendering NLRP1 resistant to cleavage, macrophages resistant to pyroptosis, and the animals highly resistant to toxin challenge (19, 21, 23).

Rat NLRP1 proteins from all tested rat strains are 98% identical (21), pointing to the evolutionary pressure conserving the sensor's sequence outside of the few polymorphisms in the LF cleavage site. In mice, three *Nlrp1* alleles (*Nlrp1a*, *Nlrp1b*, *Nlrp1c*) exist (24, 25). The one most homologous to the expressed rat NLRP1 is NLRP1a, which is, in a manner similar to rat NLRP1, highly

TABLE 1 | Inflammasome-activating bacterial exotoxins.

Toxin	Source	Mechanism	Sensor	Effect sensed by NLR (direct or indirect)	Reference
Lethal toxin (LT)	*B. anthracis*	Protease	NLRP1	NLRP1 cleavage	(19–24)
Nigericin	*S. hygroscopicus*	Ionophore	NLRP3	K^+ efflux	(40)
α-Hemolysin	*S. aureus*	Pore former	NLRP3	K^+ efflux	(60–62)
Panton-valentine leukocidin	*S. aureus*	Pore former	NLRP3	K^+ efflux	(68)
Leukocidin A/B	*S. aureus*	Pore former	NLRP3	K^+ efflux	(69)
Listeriolysin O	*L. monocytogenes*	Pore former	NLRP3	K^+ efflux	(42, 43, 50)
Aerolysin	*A. hydrophila*	Pore former	NLRP3	K^+ efflux	(44)
Tetanolysin O	*C. tetani*	Pore former	NLRP3	K^+ efflux	(49)
Pneumolysin	*S. pneumonia*	Pore former	NLRP3	K^+ efflux	(51–53)
β-Hemolysin	Group B *Streptococcus*	Pore former	NLRP3	K^+ efflux	(55–57)
Streptolysin O	*S. pyogenes*	Pore former	NLRP3	K^+ efflux	(58, 59)
α-Hemolysin	*E. coli*	Pore former	NLRP3	K^+ efflux	(77)
Enterohemolysin	*E. coli* O157:H7	Pore former	NLRP3	K^+ efflux	(76)
Various hemolysins	*Vibrio* species	Pore former	NLRP3	K^+ efflux	(78–80)
TcdB	*C. difficile*	Glucosylase	Pyrin	Rho GTPase inactivation	(87, 90)
VopS	*V. parahaemolyticus*	Adenylyltransferase	Pyrin	Rho GTPase inactivation	(87)
IbpA	*H. somni*	Adenylyltransferase	Pyrin	Rho GTPase inactivation	(87)
C3 toxin	*C. botulinum*	ADP-ribosyltransferase	Pyrin	Rho GTPase inactivation	(87)
Unknown	*B. cenocepacia*	Deamidase	Pyrin	Rho GTPase inactivation	(87)
Pertussis toxin (PTX)	*B. pertussis*	ADP-ribosyltransferase	Pyrin	Unknown	(87)
CdtB	*Aggregatibacter actinomycetemcomitans*	Lipid phosphatase	NLRP3	GSK3β-induced P2X7 activation	(92, 93)
IpaH7.8	*S. flexneri*	E3 ubiquitin ligase	NLRP3, NLRC4	GLMN degradation	(94)
Pertussis adenylate cyclase toxin	*B. pertussis*	Adenylate cyclase and pore former	NLRP3	Pore formation; K^+ efflux	(98)
Heat-labile enterotoxin (LT)	*E. coli*	ADP-ribosyltransferase	NLRP3	cAMP increase	(100, 101)
Cholera toxin (CT)	*V. cholerae*	ADP-ribosyltransferase	Unknown	cAMP increase	(101)

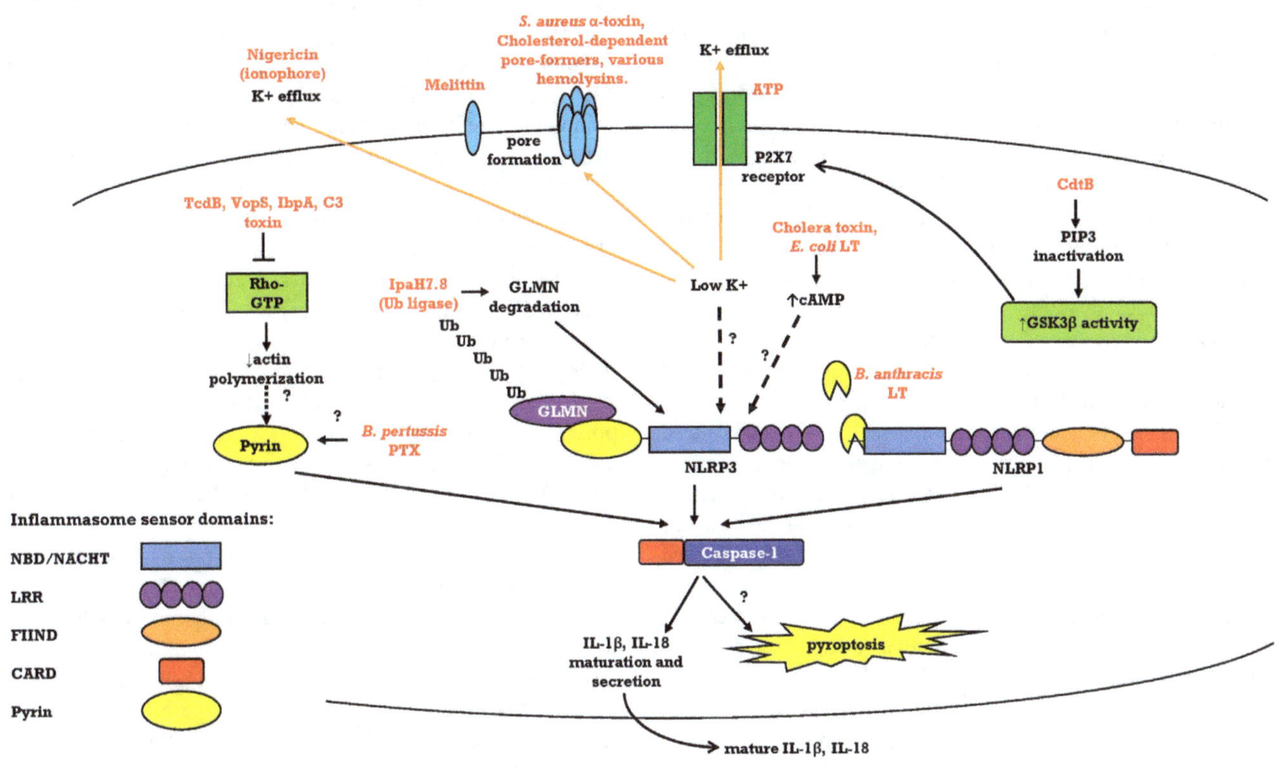

FIGURE 1 | Overview of mechanisms of toxin-induced activation of the Pyrin, NLRP3, and rodent NLRP1 inflammasomes. Bacterial exotoxins can activate the inflammasome sensors through diverse direct or indirect mechanisms. After inflammasome sensor activation, these different pathways converge on caspase-1 recruitment to the inflammasome platform and autoproteolytic activation. Caspase-1 cleaves its substrates pro-IL-1β and pro-IL-18 to their mature forms for secretion. Caspase-1 activation is also accompanied by pyroptosis through an unknown mechanism.

conserved among all mouse strains in which it is expressed1 (25). Murine NLRP1a, however, is not cleaved by LF and its activator is currently unknown.

Rather, the highly polymorphic mouse NLRP1b, which has up to 200 polymorphisms between different inbred mouse strains, is cleaved by the toxin (24, 25). NLRP1 activation in rodent macrophages and dendritic cells and the resulting pyroptosis requires proteasome activity (26–28) and can occur independently of ASC (29–31). Activation results in an inflammatory cytokine (16, 32, 33) and eicosanoid response (34) that confers an increased resistance to *B. anthracis* spore infection (32, 33). Thus, among inbred mouse strains, there is an inverse correlation between sensitivity of macrophages and dendritic cells to LT-induced pyroptosis and animal susceptibility to *B. anthracis* infection.

The unique, rapid death induced by LT in rats, which is not replicated in toxin-sensitive mice at even 10-fold higher doses, does not display the inverse relationship observed in mice. Instead, rat death following both toxin and spore challenge is positively correlated to NLRP1 sensitivity to LT cleavage (35). Human NLRP1 proteins sequenced thus far do not contain an LT cleavage site and instead have an N-terminal PYD. Human NLRP1 is not activated by LT, and human macrophages and dendritic cells are resistant to this toxin (35).

In addition, the obligate intracellular parasite, *Toxoplasma gondii*, has also been shown to activate the NLRP1 inflammasome in select inbred rat strains by an unknown mechanism (36–39). However, it is unknown if *T. gondii* activates NLRP1 in a manner similar to LT, through actions of a protease or toxin. We speculate that NLRP1 has evolved to sense diverse pathogen proteases, and polymorphisms present in different NLRP1 alleles may define responsiveness to different pathogens. Future work is needed to identify other NLRP1 agonists and the pressures driving the evolution of its conserved and polymorphic sequences.

PORE-FORMING TOXINS AND INDIRECT NLRP3 ACTIVATION

Unlike the toll-like receptors (TLRs) and the NAIP/NLRC4 and AIM2 inflammasomes, which directly sense microbial products, some inflammasome sensors indirectly sense the effects of bacterial toxins on host cell function. For example, NLRP3 is believed to be activated by an indirect mechanism. While the precise signals that activate NLRP3 remain unknown, it has been proposed that NLRP3 may be indirectly activated by K^+ efflux, lysosomal damage and cathepsin B release, mitochondrial damage, or reactive oxygen species production (1). The best-studied example of indirect inflammasome activation by bacterial toxins is the impact of pore formation on cellular potassium levels and subsequent NLRP3 activation.

Early studies linking IL-1β responses to addition of exogenous ATP or the *Streptomyces hygroscopicus*-derived potassium ionophore, nigericin (40), occurred before inflammasome components were identified and the term introduced. Bhakdi and colleagues confirmed that additional agents that deplete potassium in cells, including the pore-forming *Staphylococcus aureus* alpha toxin, the ionophore valinomycin, and the Na^+/K^+ ATPase inhibitor ouabain, can trigger processing of IL-1β (41). These investigators hypothesized that cellular K^+ concentration changes could control the function of caspase-1. Ten years later, studies with the calcium channel activator maitotoxin, nigericin, and ATP showed that induction of IL-1β and IL-18 secretion in TLR-primed macrophages treated with these K^+ efflux-inducing agents occurred in a manner dependent on the inflammasome adaptor ASC and the NLRP3 sensor (42). It was found that IL-1β secretion induced by *Listeria monocytogenes* infection of macrophages required listeriolysin expression (42, 43), suggesting that pore formation and perturbation of cellular K^+ levels could also be the basis for inflammasome activation by this toxin, although the purified toxin itself was not tested.

The first demonstration of NLRP3- and ASC-dependent inflammasome formation in response to a purified pore-forming bacterial toxin was reported with studies using aerolysin purified from *Aeromonas hydrophila* (44). Interestingly, this study identified a novel caspase-1-dependent protective response in cells treated with either aerolysin or *S. aureus* α-hemolysin. In contrast to the pyroptosis usually observed following caspase-1 activation, NLRP3-mediated activation of caspase-1 by sublytic doses of the pore-forming toxins resulted in induction of sterol regulatory element-binding proteins that altered membrane biogenesis and promoted cell survival (44). The link between pore-based lowered intracellular K^+ and NLRP3 activation was later confirmed by multiple other groups (45, 46). These early studies led to the testing of a large range of purified bacterial lysins and pore formers for their ability to activate the NLRP3 inflammasome.

Other cholesterol-dependent pore-forming toxins similar to aerolysin were also previously linked to IL-1β production (47, 48) and have since been tested in cells with inflammasome component deficiencies to link this response directly to caspase-1 activation. Tetanolysin O, from *Clostridium tetani* induces IL-1β maturation and release from bone marrow-derived macrophages (BMDMs) at low, non-lytic doses in an NLRP3-, cathepsin B-, and caspase-1-dependent manner (49). Listeriolysin, originally suggested to play a role in IL-1β responses during *Listeria* infection via pore formation (42), directly activates the inflammasome by K^+ efflux induction (50).

Streptococcus pneumoniae pneumolysin induces NLRP3-dependent IL-1β secretion that is linked to a pro-inflammatory cytokine cascade, which includes IL-17 and IFN-γ responses (51). Pneumolysin mutants and bacterial serotypes associated with variable toxin production confirm the requirement for this lysin in NLRP3-, ASC- and caspase-1-dependent IL-1β and IL-18 production (52). The host cytokine response to pneumolysin has also been shown to increase protection against pneumonia in a mouse model (52). In a murine model of pneumococcal meningitis, the extent of caspase-1 activation via the NLRP3 inflammasome has been linked to clinical disease severity (53). In contrast to the study described above, in this model, inflammasome activation was associated with pathology rather than bacterial clearance. Pneumolysin was the key inducer of the IL-1β response associated with disease, and inhibitors of IL-1β and IL-18 signaling-altered pathological responses. Pneumolysin activation of the inflammasome was suggested to require ATP release and lysosomal destabilization associated with cathepsin B cytosolic activity (53). Interestingly, pneumolysin has also been linked to the sensing of *S. pneumonia* by the cytosolic DNA-sensing molecule STING and the downstream type I interferon responses, likely in an inflammasome-independent manner (54). Other *Streptococcal* proteins can also mediate inflammasome activation through their pore-forming ability. The pigment associated with group B streptococci was recently shown to be hemolytic and capable of pore formation and induction of a K^+ efflux linked to NLRP3 activation (55). Previous studies demonstrated that β-hemolysin (56, 57) and other streptococcal lysins (58, 59) also activate the NLRP3 inflammasome.

Staphylococcus aureus is another gram-positive bacterium that has evolved to activate the NLRP3 inflammasome through the actions of multiple pore-forming toxins and hemolysins. Following up on Bhakdi's early observation that *S. aureus* α-hemolysin is associated with the IL-1β response, this toxin was verified to induce NLRP3-dependent activation of caspase-1 in both human and mouse monocytic cells (60). Activation of the inflammasome by α-hemolysin has been reported to have both protective and detrimental consequences *in vivo*. For example, α-hemolysin-mediated activation of NLRP3 in a mouse pneumonia model has been linked to IL-1β-independent necrosis, pulmonary damage, and severe pneumonia (61). By contrast, inflammasome activation is protective in other models. Neutrophil-derived IL-1β in response to α-hemolysin was protective in an abscess model of infection in mice (62). Studies of patients with atopic dermatitis have linked α-hemolysin upregulation of NLRP3 expression and *S. aureus*-mediated caspase-1 activation in monocytes to the control of Th2 responses, which may ameliorate this disease (63). IL-1β has also been linked to protection against *S. aureus*-induced CNS disease and brain abscesses (64). Hemolysin-dependent induction of IL-1β was attenuated in NLRP3- and ASC-deficient microglia following bacterial infection. Interestingly, in these studies, IL-18 secretion from microglia occurred in a manner independent of NLRP3, suggesting a different cleavage mechanism for IL-18 (65). Recently, *S. aureus* phenol soluble modulins (PSMs), another class of pore-forming substances that are important virulence factors [reviewed in Ref. (66)], were shown to induce IL-18 secretion from human keratinocytes in a caspase-1-independent manner (67). In a mouse model, PSMs were also shown to be important for *S. aureus*-induced neutrophil recruitment and systemic inflammation. These authors suggest that IL-18 release and neutrophil recruitment induced by PSMs is inflammasome independent, again suggesting novel mechanisms for IL-18 activation (67).

Other pore-forming toxins from *S. aureus* have also been shown to activate the inflammasome. The Panton–Valentine leukocidin associated with tissue necrosis has been shown to induce release of Il-1β and IL-18 in an NLRP3-dependent manner (68). Leukocidin A/B is another pore-forming toxin that can activate the NLRP3 inflammasome in human monocytes (69). Interestingly, it has also been suggested that *S. aureus* hemolysins do not activate caspase-1 alone and require *S. aureus* lipoproteins for inflammasome activation, albeit independently of the lipoprotein actions on TLR-mediated inflammasome priming (70).

Certain *S. aureus* toxins have also been described to have effects on necroptosis in a manner that may link to their activation of the inflammasome. For example, it was found that the virulent methicillin-resistant (MRSA) strain USA300 induces receptor-interacting serine–threonine kinase (RIP)1/RIP3/mixed lineage kinase domain-like (MLKL)-dependent necroptosis, thereby contributing to inflammatory lung pathology (71). *S. aureus* treatment of human and murine macrophages leads to phosphorylation of MLKL- and necroptosis-induced cell death. *Rip3*$^{-/-}$ macrophages are protected from *S. aureus*-induced cytotoxicity. Live bacteria or supernatants from mutants lacking either the α-hemolysin, leukocidin A/B, or PSM toxins display reduced induction of macrophage cytotoxicity. Interestingly, α-hemolysin-induced inflammasome activation has been shown to be dependent on MLKL pore formation, supporting the link between the NLRP3 inflammasome and necrosome components RIP1, RIP3, and MLKL (71–75). Necroptosis contributes to lung pathology and reduced bacterial clearance *in vivo* by depleting immunoregulatory resident alveolar macrophages that are essential for bacterial clearance (71).

Gram-negative bacteria also produce a variety of hemolysins that have been implicated in activation of the NLRP3 inflammasome. Enterohemorrhagic *Escherichia coli* (EHEC) O157:H7, the causative agent of hemorrhagic uremic syndrome, produces a hemolysin that induces IL-1β from the THP-1 macrophage/monocyte cell line, and RNA-interference experiments indicate a role for NLRP3, ASC, and caspase-1 in the initiation of the cytokine release (76). Uropathogenic *E. coli* also induces IL-1β, and *E. coli* α-hemolysin was found to be responsible for NLRP3-dependent responses in mouse macrophages infected with certain bacterial strains. Other strains activated an NLRP3-independent cell death pathway and hemolysin-independent IL-1β secretion from human macrophages (77). *Vibrio vulnificus* and *Vibrio cholerae* were first shown to activate the NLRP3 inflammasome in macrophages through the action of hemolysins (78). *Vibrio parahaemolyticus* also produces thermostable direct hemolysins known as TDHs that can activate the NLRP3 inflammasome (79). *Vibrio fluvialis*, which induces a diarrheal disease in humans, produces a hemolysin that can activate the inflammasome in mouse and human macrophages. Importantly, in a mouse model, the toxin was associated with IL-1β production, and toxin-containing cell-free culture supernatants induced higher levels of cytokine production. Hemolysin-deficient bacteria and their supernatants had lower levels of response *in vivo* (80).

The study of pore formation-mediated inflammasome activation first started with analysis of marine and fungal ionphores (42) and has now again been extended beyond bacterial toxins. Mold pore-forming mycotoxins (81), viral viraporins (82), melittin, the small cationic pore-forming peptide found in bee venom, which can form a single alpha helix spanning the plasma membrane (83), and the *Bombina maxima* frog derived aerolysin-like protein (84) are examples of unique non-bacterial toxin pore-forming agents that activate the NLRP3 inflammasome and caspase-1. The utility of these and many of the toxins described in this section as potential adjuvants for amplification of the immune response, or pro-inflammatory therapeutics is an area that awaits investigation.

INDIRECT ACTIVATION OF THE INFLAMMASOMES BY OTHER BACTERIAL EXOTOXINS HAVING ENZYMATIC ACTIONS

In addition to NLRP3 activation by pore-forming bacterial toxins, inflammasome activation can also occur by other mechanisms of indirect sensing of bacterial effectors.

Rho family of GTPases is molecular switches that control the dynamics of the actin cytoskeleton. The actin cytoskeleton can be co-opted by bacterial pathogen effectors targeting Rho GTPases for either hyper-activation or inactivation (85). Several cytosolic innate immune-sensing pathways have evolved to recognize the sequelae of Rho GTPase actions. For example, activation of Rho GTPases, including Rac1 and Cdc42, by bacterial toxins activates the cytosolic NOD1 sensor and leads to NFκB-dependent expression of pro-inflammatory genes (86). Recently, Shao and colleagues identified Pyrin as an innate immune sensor of bacterial toxin inhibition of Rho GTPases (87). They demonstrated that the *Clostridium difficile* toxin TcdB, which glucosylates and inactivates Rho proteins, activates the Pyrin inflammasome. This discovery provided a link between the previous findings that Pyrin controlled inflammatory familial Mediterranean fever pathogenesis (88, 89) through inflammasome activation (2), and that *C. difficile* toxins TcdA/B activate the ASC-dependent inflammasome (90). Interestingly, Pyrin specifically recognizes Rho subfamily inhibition, as cytochalasin D, an inhibitor of actin polymerization, and the *Clostridium sordellii* lethal toxin TcsL, which modifies Rac/Cdc42 and some Ras-related GTPases, do not activate the Pyrin inflammasome. Furthermore, a variety of bacterial toxins with different enzymatic activities can activate the Pyrin inflammasome. Other Pyrin activators include *Parahaemolyticus* VopS, *Histophilus somni* IbpA, *Clostridium botulinum* C3 toxin, and *Burkholderia cenocepacia*. These toxins inactivate Rho subfamily GTPases by modifying I-switch residues through glucosylation, adenylation, ADP-ribosylation, and deamidation. Together, the diversity of chemical modifications of Rho GTPases, and the inability to co-IP RHOA/B/C and Pryin suggest that Pyrin does not directly interact with Rho GTPases and detect modifications; rather, it is proposed that Pyrin senses downstream effects on the actin cytoskeleton (87).

Interestingly, the *Bordetella pertussis* toxin PTX, which is an ADP-ribosyltransferase (similar to the botulinum C3 toxin, but targeting a different substrate), has been shown to activate the Pyrin inflammasome *in vivo* (91), as well as to upregulate IL-1β expression in a TLR4-dependent manner. Both events require on the toxin's ADP-ribosyltransferase activity. While the mechanism of PTX activation of the Pyrin inflammasome is unknown, the intriguing possibility remains that PTX, which ADP-ribosylates the α_i subunit of heterotrimeric G proteins, may activate the

Pyrin inflammasome via an indirect mechanism similar to that of the other bacterial toxins targeting the GTPases, discussed above.

The *Aggregatibacter actinomycetemcomitans* cytolethal distending toxin (Cdt), can act as both signal 1 and signal 2 for inflammasome activation, up-regulating pro-inflammatory cytokine expression (signal 1) and also activating the NLRP3 sensor (signal 2) (92, 93). Cdt is a heterotrimeric protein found in several bacterial species, also including *Campylobacter jejuni*, *Shigella* species, and some *E. coli* isolates. The CdtA and CdtC subunits comprise the receptor-binding moiety, and the CdtB subunit is an lipid phosphatase. CdtB dephosphorylates the signaling lipid phosphatidylinositol-3,4,5-triphosphate (PIP3) and leads to its degradation, decreased phosphorylation of Akt and glycogen synthase kinase 3β (GSK3β), decreased Akt kinase activity but increased GSK3β kinase activity, and inhibition of the PI-3K signaling pathway. GSK3β activation induced by CdtB in human macrophages and monocytes is believed to lead to NFκB activation and the expression of pro-inflammatory cytokines, including IL-1β, dependent on CtdB lipid phosphatase activity (92). Cdt-induced GSK3β activation leads to the generation of extracellular ATP, activation of the P2X7 purinergic receptor, K^+ efflux, activation of the NLRP3 inflammasome, and IL-1β release (93). Inflammasome activation is NLRP3, ASC, and caspase-1 dependent (93). This is a case where a single bacterial toxin, once in the macrophage cytosol, may act as both signal 1 and signal 2 for inflammasome activation. Interestingly, it should be noted that Cdt toxins are also considered to act as DNases that induce DNA damage and cause cell cycle arrest. Which of these activities is more important *in vivo* is not clear, and the possibility remains that the DNase activity could also result in activation of other sensors.

The NAIP/NLRC4 inflammasome has been demonstrated to be activated by bacterial flagellin and T3SS needle and rod proteins [reviewed in Ref. (7)]. In contrast to previously known NAIP/NLRC4 agonists, which activate the inflammasome by direct binding to the NAIP sensors, a novel, indirect activator of the NLRC4 inflammasome has recently been identified (94). A *Shigella flexneri* E3 ubiquitin ligase effector protein secreted into the host cell cytosol via the type III secretion system (T3SS), invasion plasmid antigen H7.8 (IpaH7.8), was found to activate the NLRP3 and NLRC4 inflammasomes (94). This protein was found to be an important virulence factor, as enzymatically inactive mutants were defective in lung colonization following intranasal infection. Although in most cases inflammasome activation is important for controlling bacterial infection, this is a case were the activation is used by the bacterium to promote dissemination. The LRR domain of IpaH7.8 was found to directly interact with the host protein glomulin/flagellar-associated protein 68 (GLMN) and target both GLMN and itself for degradation by the proteasome. GLMN was suggested to inhibit the NLRP3 inflammasome through an unknown mechanism. Future studies may reveal whether GLMN inhibits the inflammasome, or if IpaH7.8 activates this sensor through a different mechanism (94).

Interestingly, ricin, a highly poisonous toxin found in the seeds of the *Ricinus communis* plant, and an inhibitor of protein syntheses, has also been shown to induce a macrophage IL-1β-mediated pro-inflammatory response in the airways, contributing to lethality (95). Upregulation of IL-1β transcription and the pro-inflammatory response is believed to be mediated by activation of the stress-activated protein kinases (SAPKs), including JNK and the p38 MAPK [reviewed in Ref. (96)]. Ricin was also demonstrated to activate the NLRP3 inflammasome in a proteasome-dependent manner (97). Inflammasome activation was independent of the ribotoxic stress response and phosphorylation of p38 and JNK. It was proposed that inflammasome activation was the result of breakdown of an unidentified, labile NLRP3 inhibitor whose synthesis is blocked by ricin. JNK- and p38-independent NLRP3 inflammasome activation may contribute to the previous finding that IL-1β plays an important role in ricin-induced severe lung inflammation and lethality (95).

BACTERIAL TOXIN INFLAMMASOME ACTIVATORS AND ADAPTIVE IMMUNITY

Some studies demonstrated that toxin-induced production of IL-1β also influences adaptive immunity, and it is likely that many toxins that activate the inflammasomes will have similar effects. The pertussis RTX adenylate cyclase toxin previously mentioned contains, in addition to its enzymatic domain, a separate pore-forming domain, which activates the NLRP3 inflammasome. This activation leads to IL-1β production by dendritic cells and induction of antigen-specific Th17 cells that require the cytokine for expansion. Th17 differentiation induced by inflammasome activation is required for control of infection and clearance of bacteria from the lungs in a mouse model (98). Similarly, the IL-17 response associated with NLRP3 inflammasome activation by *S. pneumonia* pneumolysin is also associated with protective immunity against intranasal infection, and IL-1β is required for promoting the IL-17 response (51). Interestingly, the non-toxic trehalose-6,6-dibehenate (TBD) adjuvant, which promotes Th1/Th17 responses, also requires inflammasome adaptor ASC-dependent activation of IL-1β production (99).

The adjuvant role of non-pore-forming bacterial enterotoxins and their enzymatically inactive mutants in inducing inflammasome dependent, Th17-polarized protective immunity has also been studied (100, 101). The *E. coli* heat-labile enterotoxin (HL-LT, commonly termed LT) and cholera toxin (CT) are AB toxins that ADP-ribosylate the Gs component of adenylate cyclase, leading to an increase in cAMP. They have been shown to be potent adjuvants, but their enterotoxicity precludes their use in human oral vaccines. HL-LT along with an enzymatically highly attenuated mutant can both activate the NLRP3/caspase-1 inflammasome and induce mature IL-1β secretion from LPS-primed dendritic cells (100). Inflammasome activation and the associated IL-1β response are required for promoting Th17 responses *in vivo*, which in turn protect against *B. pertussis* challenge in a model where the toxin is the primary adjuvant in the pertussis vaccine (100). In related studies, the heat-killed mycobacteria components in Complete Freund's Adjuvant (CFA) have been shown to drive Th17 differentiation through a mechanism that requires NLRP3 inflammasome activation (102).

A recent study examined the ability for double-mutant HL-LT and multiple-mutated CT to induce a Th17-mediated adjuvant response in human peripheral blood mononuclear cells (PBMCs) (101). These variants have highly reduced enzymatic activity and, unlike the wild type toxins, are non-toxic but still induce very low levels of cAMP and can stimulate IL-17A production from human PBMCs when administered to cells with a polyclonal T cell superantigen (101). This effect is dependent on cAMP/protein kinase A (PKA) signaling, as a PKA inhibitor inhibits IL-17A production, and a cAMP analog recapitulates the toxin effects. Monocytes were demonstrated to have a modest increase in cAMP in response to the mutant toxins, and monocytes pre-incubated with the toxins and then co-cultured with $CD4^{+}$ T cells led to an increase in IL-17A production in a caspase-1-, IL-1β-dependent manner (101). These studies demonstrate the important role that toxin-induced inflammasome activation may play in vaccine development.

CONCLUDING COMMENTS

The recognition of self from non-self is the foundation of the innate immune response. Pattern recognition receptors such as the TLRs and some of the inflammasome sensors detect highly conserved microbial molecular patterns, including LPS, peptidoglycan, flagellin, and CpG DNA. The inflammasome receptors discussed in this review, however, are able to detect highly diverse bacterial effectors because they sense these toxins through their functional effects rather than their molecular patterns. In the case of NLRP1, the sensor is activated by direct cleavage by a bacterial protease that also cleaves the mitogen-activated protein kinase kinases (MEKs) and results in cell – and animal – death. The NLRP3 and Pyrin inflammasomes, on the other hand, detect indirect effects on host cell state including ion fluxes and perturbation of the actin cytoskeleton. In this way, the inflammasome sensors have evolved to respond to different bacterial toxins with diverse mechanisms of action to converge on caspase-1 activation and initiation of the immune response.

Toxin activation of the inflammasome in the first-responder cells of the innate immune system plays an important role in pro-inflammatory and pyroptosis events that can have protective or pathogenic consequences in the host. Interestingly, bacterial effectors that inhibit the inflammasome or suppress its activation have also been described in recent years [reviewed in Ref. (103)], suggesting that microbes have also evolved to evade inflammasome detection by the host. With the threat of antimicrobial-resistant bacteria, an understanding of the mechanisms by which microbes modulate the innate immune responses is essential to studies of microbial pathogenesis and therapeutic development.

ACKNOWLEDGMENTS

This work was supported by the Intramural Research Program of the National Institutes of Health, National Institute of Allergy and Infectious Diseases. The content is solely the responsibility of the authors and does not necessarily represent the official views of the National Institute of Allergy and Infectious Diseases or the National Institutes of Health.

REFERENCES

1. Lamkanfi M, Dixit VM. Mechanisms and functions of inflammasomes. *Cell* (2014) **157**(5):1013–22. doi:10.1016/j.cell.2014.04.007
2. Chae JJ, Cho YH, Lee GS, Cheng J, Liu PP, Feigenbaum L, et al. Gain-of-function Pyrin mutations induce NLRP3 protein-independent interleukin-1beta activation and severe autoinflammation in mice. *Immunity* (2011) **34**(5):755–68. doi:10.1016/j.immuni.2011.02.020
3. Papin S, Cuenin S, Agostini L, Martinon F, Werner S, Beer HD, et al. The SPRY domain of Pyrin, mutated in familial Mediterranean fever patients, interacts with inflammasome components and inhibits proIL-1beta processing. *Cell Death Differ* (2007) **14**(8):1457–66. doi:10.1038/sj.cdd.4402142
4. Elliott EI, Sutterwala FS. Initiation and perpetuation of NLRP3 inflammasome activation and assembly. *Immunol Rev* (2015) **265**(1):35–52. doi:10.1111/imr.12286
5. Lage SL, Longo C, Branco LM, da Costa TB, Buzzo Cde L, Bortoluci KR. Emerging concepts about NAIP/NLRC4 inflammasomes. *Front Immunol* (2014) **5**:309. doi:10.3389/fimmu.2014.00309
6. Vance RE. The NAIP/NLRC4 inflammasomes. *Curr Opin Immunol* (2015) **32**:84–9. doi:10.1016/j.coi.2015.01.010
7. Zhao Y, Shao F. The NAIP-NLRC4 inflammasome in innate immune detection of bacterial flagellin and type III secretion apparatus. *Immunol Rev* (2015) **265**(1):85–102. doi:10.1111/imr.12293
8. Kayagaki N, Warming S, Lamkanfi M, Vande Walle L, Louie S, Dong J, et al. Non-canonical inflammasome activation targets caspase-11. *Nature* (2011) **479**(7371):117–21. doi:10.1038/nature10558
9. Kayagaki N, Wong MT, Stowe IB, Ramani SR, Gonzalez LC, Akashi-Takamura S, et al. Noncanonical inflammasome activation by intracellular LPS independent of TLR4. *Science* (2013) **341**(6151):1246–9. doi:10.1126/science.1240248
10. Aachoui Y, Leaf IA, Hagar JA, Fontana MF, Campos CG, Zak DE, et al. Caspase-11 protects against bacteria that escape the vacuole. *Science* (2013) **339**(6122):975–8. doi:10.1126/science.1230751
11. Hagar JA, Powell DA, Aachoui Y, Ernst RK, Miao EA. Cytoplasmic LPS activates caspase-11: implications in TLR4-independent endotoxic shock. *Science* (2013) **341**(6151):1250–3. doi:10.1126/science.1240988
12. Shi J, Zhao Y, Wang Y, Gao W, Ding J, Li P, et al. Inflammatory caspases are innate immune receptors for intracellular LPS. *Nature* (2014) **514**(7521):187–92. doi:10.1038/nature13683
13. Yang J, Zhao Y, Shao F. Non-canonical activation of inflammatory caspases by cytosolic LPS in innate immunity. *Curr Opin Immunol* (2015) **32**:78–83. doi:10.1016/j.coi.2015.01.007
14. Stowe I, Lee B, Kayagaki N. Caspase-11: arming the guards against bacterial infection. *Immunol Rev* (2015) **265**(1):75–84. doi:10.1111/imr.12292
15. Beall FA, Dalldorf FG. The pathogenesis of the lethal effect of anthrax toxin in the rat. *J Infect Dis* (1966) **116**(3):377–89. doi:10.1093/infdis/116.3.377
16. Moayeri M, Haines D, Young HA, Leppla SH. *Bacillus anthracis* lethal toxin induces TNF-alpha-independent hypoxia-mediated toxicity in mice. *J Clin Invest* (2003) **112**(5):670–82. doi:10.1172/JCI17991
17. Shoop WL, Xiong Y, Wiltsie J, Woods A, Guo J, Pivnichny JV, et al. Anthrax lethal factor inhibition. *Proc Natl Acad Sci U S A* (2005) **102**(22):7958–63. doi:10.1073/pnas.0502159102
18. Liu S, Moayeri M, Leppla SH. Anthrax lethal and edema toxins in anthrax pathogenesis. *Trends Microbiol* (2014) **22**(6):317–25. doi:10.1016/j.tim.2014.02.012
19. Levinsohn JL, Newman ZL, Hellmich KA, Fattah R, Getz MA, Liu S, et al. Anthrax lethal factor cleavage of NLRP1 is required for activation of the inflammasome. *PLoS Pathog* (2012) **8**(3):e1002638. doi:10.1371/journal.ppat.1002638

20. Hellmich KA, Levinsohn JL, Fattah R, Newman ZL, Maier N, Sastalla I, et al. Anthrax lethal factor cleaves mouse NLRP1B in both toxin-sensitive and toxin-resistant macrophages. *PLoS One* (2012) **7**(11):e49741. doi:10.1371/journal.pone.0049741
21. Newman ZL, Printz MP, Liu S, Crown D, Breen L, Miller-Randolph S, et al. Susceptibility to anthrax lethal toxin-induced rat death is controlled by a single chromosome 10 locus that includes rNlrp1. *PLoS Pathog* (2010) **6**(5):e1000906. doi:10.1371/journal.ppat.1000906
22. Chavarria-Smith J, Vance RE. Direct proteolytic cleavage of NLRP1B Is necessary and sufficient for inflammasome activation by anthrax lethal factor. *PLoS Pathog* (2013) **9**(6):e1003452. doi:10.1371/journal.ppat.1003452
23. Newman ZL, Crown D, Leppla SH, Moayeri M. Anthrax lethal toxin activates the inflammasome in sensitive rat macrophages. *Biochem Biophys Res Commun* (2010) **398**(4):785–9. doi:10.1016/j.bbrc.2010.07.039
24. Boyden ED, Dietrich WF. Nalp1b controls mouse macrophage susceptibility to anthrax lethal toxin. *Nat Genet* (2006) **38**(2):240–4. doi:10.1038/ng1724
25. Sastalla I, Crown D, Masters SL, McKenzie A, Leppla SH, Moayeri M. Transcriptional analysis of the three NLRP1 paralogs in mice. *BMC Genomics* (2013) **14**:188. doi:10.1186/1471-2164-14-188
26. Wickliffe KE, Leppla SH, Moayeri M. Anthrax lethal toxin-induced inflammasome formation and caspase-1 activation are late events dependent on ion fluxes and the proteasome. *Cell Microbiol* (2008) **10**(2):332–43. doi:10.1111/j.1462-5822.2007.01044.x
27. Squires RC, Muehlbauer SM, Brojatsch J. Proteasomes control caspase-1 activation in anthrax lethal toxin-mediated cell killing. *J Biol Chem* (2007) **282**(47):34260–7. doi:10.1074/jbc.M705687200
28. Fink SL, Bergsbaken T, Cookson BT. Anthrax lethal toxin and salmonella elicit the common cell death pathway of caspase-1-dependent pyroptosis via distinct mechanisms. *Proc Natl Acad Sci U S A* (2008) **105**(11):4312–7. doi:10.1073/pnas.0707370105
29. Broz P, von Moltke J, Jones JW, Vance RE, Monack DM. Differential requirement for caspase-1 autoproteolysis in pathogen-induced cell death and cytokine processing. *Cell Host Microbe* (2010) **8**(6):471–83. doi:10.1016/j.chom.2010.11.007
30. Van Opdenbosch N, Gurung P, Vande Walle L, Fossoul A, Kanneganti TD, Lamkanfi M. Activation of the NLRP1B inflammasome independently of ASC-mediated caspase-1 autoproteolysis and speck formation. *Nat Commun* (2014) **5**:3209. doi:10.1038/ncomms4209
31. Guey B, Bodnar M, Manie SN, Tardivel A, Petrilli V. Caspase-1 autoproteolysis is differentially required for NLRP1B and NLRP3 inflammasome function. *Proc Natl Acad Sci U S A* (2014) **111**(48):17254–9. doi:10.1073/pnas.1415756111
32. Moayeri M, Crown D, Newman ZL, Okugawa S, Eckhaus M, Cataisson C, et al. Inflammasome sensor NLRP1B-dependent resistance to anthrax is mediated by caspase-1, IL-1 signaling and neutrophil recruitment. *PLoS Pathog* (2010) **6**(12):e1001222. doi:10.1371/journal.ppat.1001222
33. Terra JK, Cote CK, France B, Jenkins AL, Bozue JA, Welkos SL, et al. Cutting edge: resistance to *Bacillus anthracis* infection mediated by a lethal toxin sensitive allele of NALP1B/NLRP1B. *J Immunol* (2010) **184**(1):17–20. doi:10.4049/jimmunol.0903114
34. von Moltke J, Trinidad NJ, Moayeri M, Kintzer AF, Wang SB, van Rooijen N, et al. Rapid induction of inflammatory lipid mediators by the inflammasome in vivo. *Nature* (2012) **490**(7418):107–11. doi:10.1038/nature11351
35. Moayeri M, Sastalla I, Leppla SH. Anthrax and the inflammasome. *Microbes Infect* (2012) **14**(5):392–400. doi:10.1016/j.micinf.2011.12.005
36. Cirelli KM, Gorfu G, Hassan MA, Printz M, Crown D, Leppla SH, et al. Inflammasome sensor NLRP1 controls rat macrophage susceptibility to *Toxoplasma gondii*. *PLoS Pathog* (2014) **10**(3):e1003927. doi:10.1371/journal.ppat.1003927
37. Ewald SE, Chavarria-Smith J, Boothroyd JC. NLRP1 is an inflammasome sensor for *Toxoplasma gondii*. *Infect Immun* (2014) **82**(1):460–8. doi:10.1128/IAI.01170-13
38. Gorfu G, Cirelli KM, Melo MB, Mayer-Barber K, Crown D, Koller BH, et al. Dual role for inflammasome sensors NLRP1 and NLRP3 in murine resistance to *Toxoplasma gondii*. *mBio* (2014) **5**(1):e1113–7. doi:10.1128/mBio.01117-13
39. Cavailles P, Flori P, Papapietro O, Bisanz C, Lagrange D, Pilloux L, et al. A highly conserved toxo1 haplotype directs resistance to toxoplasmosis and its associated caspase-1 dependent killing of parasite and host macrophage. *PLoS Pathog* (2014) **10**(4):e1004005. doi:10.1371/journal.ppat.1004005
40. Perregaux D, Gabel CA. Interleukin-1 beta maturation and release in response to ATP and nigericin. Evidence that potassium depletion mediated by these agents is a necessary and common feature of their activity. *J Biol Chem* (1994) **269**(21):15195–203.
41. Walev I, Reske K, Palmer M, Valeva A, Bhakdi S. Potassium-inhibited processing of IL-1 beta in human monocytes. *EMBO J* (1995) **14**(8):1607–14.
42. Mariathasan S, Weiss DS, Newton K, McBride J, O'Rourke K, Roose-Girma M, et al. Cryopyrin activates the inflammasome in response to toxins and ATP. *Nature* (2006) **440**(7081):228–32. doi:10.1038/nature04515
43. Ozoren N, Masumoto J, Franchi L, Kanneganti TD, Body-Malapel M, Erturk I, et al. Distinct roles of TLR2 and the adaptor ASC in IL-1beta/IL-18 secretion in response to *Listeria monocytogenes*. *J Immunol* (2006) **176**(7):4337–42. doi:10.4049/jimmunol.176.7.4337
44. Gurcel L, Abrami L, Girardin S, Tschopp J, van der Goot FG. Caspase-1 activation of lipid metabolic pathways in response to bacterial pore-forming toxins promotes cell survival. *Cell* (2006) **126**(6):1135–45. doi:10.1016/j.cell.2006.07.033
45. Petrilli V, Papin S, Dostert C, Mayor A, Martinon F, Tschopp J. Activation of the NALP3 inflammasome is triggered by low intracellular potassium concentration. *Cell Death Differ* (2007) **14**(9):1583–9. doi:10.1038/sj.cdd.4402195
46. Munoz-Planillo R, Kuffa P, Martinez-Colon G, Smith BL, Rajendiran TM, Nunez G. K(+) efflux is the common trigger of NLRP3 inflammasome activation by bacterial toxins and particulate matter. *Immunity* (2013) **38**(6):1142–53. doi:10.1016/j.immuni.2013.05.016
47. Hackett SP, Stevens DL. Streptococcal toxic shock syndrome: synthesis of tumor necrosis factor and interleukin-1 by monocytes stimulated with pyrogenic exotoxin A and streptolysin O. *J Infect Dis* (1992) **165**(5):879–85. doi:10.1093/infdis/165.5.879
48. Houldsworth S, Andrew PW, Mitchell TJ. Pneumolysin stimulates production of tumor necrosis factor alpha and interleukin-1 beta by human mononuclear phagocytes. *Infect Immun* (1994) **62**(4):1501–3.
49. Chu J, Thomas LM, Watkins SC, Franchi L, Nunez G, Salter RD. Cholesterol-dependent cytolysins induce rapid release of mature IL-1beta from murine macrophages in a NLRP3 inflammasome and cathepsin B-dependent manner. *J Leukoc Biol* (2009) **86**(5):1227–38. doi:10.1189/jlb.0309164
50. Hamon MA, Cossart P. K+ efflux is required for histone H3 dephosphorylation by *Listeria monocytogenes* listeriolysin O and other pore-forming toxins. *Infect Immun* (2011) **79**(7):2839–46. doi:10.1128/IAI.01243-10
51. McNeela EA, Burke A, Neill DR, Baxter C, Fernandes VE, Ferreira D, et al. Pneumolysin activates the NLRP3 inflammasome and promotes proinflammatory cytokines independently of TLR4. *PLoS Pathog* (2010) **6**(11):e1001191. doi:10.1371/journal.ppat.1001191
52. Witzenrath M, Pache F, Lorenz D, Koppe U, Gutbier B, Tabeling C, et al. The NLRP3 inflammasome is differentially activated by pneumolysin variants and contributes to host defense in pneumococcal pneumonia. *J Immunol* (2011) **187**(1):434–40. doi:10.4049/jimmunol.1003143
53. Hoegen T, Tremel N, Klein M, Angele B, Wagner H, Kirschning C, et al. The NLRP3 inflammasome contributes to brain injury in pneumococcal meningitis and is activated through ATP-dependent lysosomal cathepsin B release. *J Immunol* (2011) **187**(10):5440–51. doi:10.4049/jimmunol.1100790
54. Koppe U, Hogner K, Doehn JM, Muller HC, Witzenrath M, Gutbier B, et al. *Streptococcus pneumoniae* stimulates a STING- and IFN regulatory factor 3-dependent type I IFN production in macrophages, which regulates RANTES production in macrophages, cocultured alveolar epithelial cells, and mouse lungs. *J Immunol* (2012) **188**(2):811–7. doi:10.4049/jimmunol.1004143
55. Whidbey C, Vornhagen J, Gendrin C, Boldenow E, Samson JM, Doering K, et al. A streptococcal lipid toxin induces membrane permeabilization and pyroptosis leading to fetal injury. *EMBO Mol Med* (2015) **7**(4):488–505. doi:10.15252/emmm.201404883
56. Costa A, Gupta R, Signorino G, Malara A, Cardile F, Biondo C, et al. Activation of the NLRP3 inflammasome by group B streptococci. *J Immunol* (2012) **188**(4):1953–60. doi:10.4049/jimmunol.1102543
57. Gupta R, Ghosh S, Monks B, DeOliveira RB, Tzeng TC, Kalantari P, et al. RNA and beta-hemolysin of group B *Streptococcus* induce interleukin-1beta (IL-1beta) by activating NLRP3 inflammasomes in mouse macrophages. *J Biol Chem* (2014) **289**(20):13701–5. doi:10.1074/jbc.C114.548982

58. Harder J, Franchi L, Munoz-Planillo R, Park JH, Reimer T, Nunez G. Activation of the Nlrp3 inflammasome by *Streptococcus pyogenes* requires streptolysin O and NF-kappa B activation but proceeds independently of TLR signaling and P2x7 receptor. *J Immunol* (2009) **183**(9):5823–9. doi:10.4049/jimmunol.0900444
59. Keyel PA, Roth R, Yokoyama WM, Heuser JE, Salter RD. Reduction of streptolysin O (SLO) pore-forming activity enhances inflammasome activation. *Toxins (Basel)* (2013) **5**(6):1105–18. doi:10.3390/toxins5061105
60. Craven RR, Gao X, Allen IC, Gris D, Bubeck Wardenburg J, McElvania-Tekippe E, et al. *Staphylococcus aureus* alpha-hemolysin activates the NLRP3-inflammasome in human and mouse monocytic cells. *PLoS One* (2009) **4**(10):e7446. doi:10.1371/journal.pone.0007446
61. Kebaier C, Chamberland RR, Allen IC, Gao X, Broglie PM, Hall JD, et al. *Staphylococcus aureus* alpha-hemolysin mediates virulence in a murine model of severe pneumonia through activation of the NLRP3 inflammasome. *J Infect Dis* (2012) **205**(5):807–17. doi:10.1093/infdis/jir846
62. Cho JS, Guo Y, Ramos RI, Hebroni F, Plaisier SB, Xuan C, et al. Neutrophil-derived IL-1beta is sufficient for abscess formation in immunity against *Staphylococcus aureus* in mice. *PLoS Pathog* (2012) **8**(11):e1003047. doi:10.1371/journal.ppat.1003047
63. Niebuhr M, Baumert K, Heratizadeh A, Satzger I, Werfel T. Impaired NLRP3 inflammasome expression and function in atopic dermatitis due to Th2 milieu. *Allergy* (2014) **69**(8):1058–67. doi:10.1111/all.12428
64. Kielian T, Bearden ED, Baldwin AC, Esen N. IL-1 and TNF-alpha play a pivotal role in the host immune response in a mouse model of *Staphylococcus aureus*-induced experimental brain abscess. *J Neuropathol Exp Neurol* (2004) **63**(4):381–96.
65. Hanamsagar R, Torres V, Kielian T. Inflammasome activation and IL-1beta/IL-18 processing are influenced by distinct pathways in microglia. *J Neurochem* (2011) **119**(4):736–48. doi:10.1111/j.1471-4159.2011.07481.x
66. Cheung GY, Joo HS, Chatterjee SS, Otto M. Phenol-soluble modulins – critical determinants of staphylococcal virulence. *FEMS Microbiol Rev* (2014) **38**(4):698–719. doi:10.1111/1574-6976.12057
67. Syed AK, Reed TJ, Clark KL, Boles BR, Kahlenberg JM. *Staphylococcus aureus* phenol soluble modulins stimulate the release of pro-inflammatory cytokines from keratinocytes and are required for induction of skin inflammation. *Infect Immun* (2015) **83**(9):3428–37. doi:10.1128/IAI.00401-15
68. Holzinger D, Gieldon L, Mysore V, Nippe N, Taxman DJ, Duncan JA, et al. *Staphylococcus aureus* panton-valentine leukocidin induces an inflammatory response in human phagocytes via the NLRP3 inflammasome. *J Leukoc Biol* (2012) **92**(5):1069–81. doi:10.1189/jlb.0112014
69. Melehani JH, James DB, DuMont AL, Torres VJ, Duncan JA. *Staphylococcus aureus* leukocidin A/B (LukaB) kills human monocytes via host NLRP3 and ASC when extracellular, but not intracellular. *PLoS Pathog* (2015) **11**(6):e1004970. doi:10.1371/journal.ppat.1004970
70. Munoz-Planillo R, Franchi L, Miller LS, Nunez G. A critical role for hemolysins and bacterial lipoproteins in *Staphylococcus aureus*-induced activation of the NLRP3 inflammasome. *J Immunol* (2009) **183**(6):3942–8. doi:10.4049/jimmunol.0900729
71. Kitur K, Parker D, Nieto P, Ahn DS, Cohen TS, Chung S, et al. Toxin-induced necroptosis is a major mechanism of *Staphylococcus aureus* lung damage. *PLoS Pathog* (2015) **11**(4):e1004820. doi:10.1371/journal.ppat.1004820
72. Vince JE, Wong WW, Gentle I, Lawlor KE, Allam R, O'Reilly L, et al. Inhibitor of apoptosis proteins limit RIP3 kinase-dependent interleukin-1 activation. *Immunity* (2012) **36**(2):215–27. doi:10.1016/j.immuni.2012.01.012
73. Wang X, Jiang W, Yan Y, Gong T, Han J, Tian Z, et al. RNA viruses promote activation of the NLRP3 inflammasome through a RIP1-RIP3-DRP1 signaling pathway. *Nat Immunol* (2014) **15**(12):1126–33. doi:10.1038/ni.3015
74. Yabal M, Muller N, Adler H, Knies N, Gross CJ, Damgaard RB, et al. XIAP restricts TNF- and RIP3-dependent cell death and inflammasome activation. *Cell Rep* (2014) **7**(6):1796–808. doi:10.1016/j.celrep.2014.05.008
75. Weng D, Marty-Roix R, Ganesan S, Proulx MK, Vladimer GI, Kaiser WJ, et al. Caspase-8 and RIP kinases regulate bacteria-induced innate immune responses and cell death. *Proc Natl Acad Sci U S A* (2014) **111**(20):7391–6. doi:10.1073/pnas.1403477111
76. Zhang X, Cheng Y, Xiong Y, Ye C, Zheng H, Sun H, et al. Enterohemorrhagic *Escherichia coli* specific enterohemolysin induced IL-1beta in human macrophages and EHEC-induced IL-1beta required activation of NLRP3 inflammasome. *PLoS One* (2012) **7**(11):e50288. doi:10.1371/journal.pone.0050288
77. Schaale K, Peters KM, Murthy AM, Fritzsche AK, Phan MD, Totsika M, et al. Strain- and host species-specific inflammasome activation, IL-1beta release, and cell death in macrophages infected with uropathogenic *Escherichia coli*. *Mucosal Immunol* (2015). doi:10.1038/mi.2015.44
78. Toma C, Higa N, Koizumi Y, Nakasone N, Ogura Y, McCoy AJ, et al. Pathogenic *Vibrio* activate NLRP3 inflammasome via cytotoxins and TLR/nucleotide-binding oligomerization domain-mediated NF-Kappa B signaling. *J Immunol* (2010) **184**(9):5287–97. doi:10.4049/jimmunol.0903536
79. Higa N, Toma C, Koizumi Y, Nakasone N, Nohara T, Masumoto J, et al. *Vibrio parahaemolyticus* effector proteins suppress inflammasome activation by interfering with host autophagy signaling. *PLoS Pathog* (2013) **9**(1):e1003142. doi:10.1371/journal.ppat.1003142
80. Song L, Huang Y, Zhao M, Wang Z, Wang S, Sun H, et al. A critical role for hemolysin in *Vibrio fluvialis*-induced IL-1beta secretion mediated by the NLRP3 inflammasome in macrophages. *Front Microbiol* (2015) **6**:510. doi:10.3389/fmicb.2015.00510
81. Kankkunen P, Rintahaka J, Aalto A, Leino M, Majuri ML, Alenius H, et al. Trichothecene mycotoxins activate inflammatory response in human macrophages. *J Immunol* (2009) **182**(10):6418–25. doi:10.4049/jimmunol.0803309
82. Ito M, Yanagi Y, Ichinohe T. Encephalomyocarditis virus viroporin 2B activates NLRP3 inflammasome. *PLoS Pathog* (2012) **8**(8):e1002857. doi:10.1371/journal.ppat.1002857
83. Palm NW, Medzhitov R. Role of the inflammasome in defense against venoms. *Proc Natl Acad Sci U S A* (2013) **110**(5):1809–14. doi:10.1073/pnas.1221476110
84. Xiang Y, Yan C, Guo X, Zhou K, Li S, Gao Q, et al. Host-derived, pore-forming toxin-like protein and trefoil factor complex protects the host against microbial infection. *Proc Natl Acad Sci U S A* (2014) **111**(18):6702–7. doi:10.1073/pnas.1321317111
85. Aktories K. Bacterial protein toxins that modify host regulatory GTPases. *Nat Rev Microbiol* (2011) **9**(7):487–98. doi:10.1038/nrmicro2592
86. Keestra AM, Winter MG, Auburger JJ, Frassle SP, Xavier MN, Winter SE, et al. Manipulation of small Rho GTPases is a pathogen-induced process detected by NOD1. *Nature* (2013) **496**(7444):233–7. doi:10.1038/nature12025
87. Xu H, Yang J, Gao W, Li L, Li P, Zhang L, et al. Innate immune sensing of bacterial modifications of Rho GtPases by the Pyrin inflammasome. *Nature* (2014) **513**(7517):237–41. doi:10.1038/nature13449
88. The International Fmf Consortium. Ancient missense mutations in a new member of the RoRet gene family are likely to cause familial Mediterranean fever. *Cell* (1997) **90**(4):797–807. doi:10.1016/S0092-8674(00)80539-5
89. French FMF Consortium. A candidate gene for familial Mediterranean fever. *Nat Genet* (1997) **17**(1):25–31. doi:10.1038/ng0997-25
90. Ng J, Hirota SA, Gross O, Li Y, Ulke-Lemee A, Potentier MS, et al. *Clostridium difficile* toxin-induced inflammation and intestinal injury are mediated by the inflammasome. *Gastroenterology* (2010) **139**(2):.e1–3. doi:10.1053/j.gastro.2010.04.005
91. Dumas A, Amiable N, de Rivero Vaccari JP, Chae JJ, Keane RW, Lacroix S, et al. The inflammasome Pyrin contributes to pertussis toxin-induced IL-1beta synthesis, neutrophil intravascular crawling and autoimmune encephalomyelitis. *PLoS Pathog* (2014) **10**(5):e1004150. doi:10.1371/journal.ppat.1004150
92. Shenker BJ, Walker LP, Zekavat A, Dlakic M, Boesze-Battaglia K. Blockade of the PI-3K signalling pathway by the *Aggregatibacter actinomycetemcomitans* cytolethal distending toxin induces macrophages to synthesize and secrete pro-inflammatory cytokines. *Cell Microbiol* (2014) **16**(9):1391–404. doi:10.1111/cmi.12299
93. Shenker BJ, Ojcius DM, Walker LP, Zekavat A, Scuron MD, Boesze-Battaglia K. *Aggregatibacter actinomycetemcomitans* cytolethal distending toxin activates the NLRP3 inflammasome in human macrophages, leading to the release of proinflammatory cytokines. *Infect Immun* (2015) **83**(4):1487–96. doi:10.1128/IAI.03132-14
94. Suzuki S, Mimuro H, Kim M, Ogawa M, Ashida H, Toyotome T, et al. *Shigella* Ipah7.8 E3 ubiquitin ligase targets glomulin and activates inflammasomes to demolish macrophages. *Proc Natl Acad Sci U S A* (2014) **111**(40):E4254–63. doi:10.1073/pnas.1324021111

95. Lindauer ML, Wong J, Iwakura Y, Magun BE. Pulmonary inflammation triggered by ricin toxin requires macrophages and IL-1 signaling. *J Immunol* (2009) **183**(2):1419–26. doi:10.4049/jimmunol.0901119
96. Jandhyala DM, Thorpe CM, Magun B. Ricin and Shiga toxins: effects on host cell signal transduction. *Curr Top Microbiol Immunol* (2012) **357**:41–65. doi:10.1007/82_2011_181
97. Lindauer M, Wong J, Magun B. Ricin toxin activates the NALP3 inflammasome. *Toxins (Basel)* (2010) **2**(6):1500–14. doi:10.3390/toxins2061500
98. Dunne A, Ross PJ, Pospisilova E, Masin J, Meaney A, Sutton CE, et al. Inflammasome activation by adenylate cyclase toxin directs Th17 responses and protection against *Bordetella pertussis*. *J Immunol* (2010) **185**(3):1711–9. doi:10.4049/jimmunol.1000105
99. Desel C, Werninghaus K, Ritter M, Jozefowski K, Wenzel J, Russkamp N, et al. The mincle-activating adjuvant TDB induces MyD88-dependent Th1 and Th17 responses through IL-1R signaling. *PLoS One* (2013) **8**(1):e53531. doi:10.1371/journal.pone.0053531
100. Brereton CF, Sutton CE, Ross PJ, Iwakura Y, Pizza M, Rappuoli R, et al. *Escherichia coli* heat-labile enterotoxin promotes protective Th17 responses against infection by driving innate IL-1 and IL-23 production. *J Immunol* (2011) **186**(10):5896–906. doi:10.4049/jimmunol.1003789
101. Larena M, Holmgren J, Lebens M, Terrinoni M, Lundgren A. Cholera toxin, and the related nontoxic adjuvants mmCT and dmLT, promote human Th17 responses via cyclic AMP-protein kinase A and inflammasome-dependent IL-1 signaling. *J Immunol* (2015) **194**(8):3829–39. doi:10.4049/jimmunol.1401633
102. Shenderov K, Barber DL, Mayer-Barber KD, Gurcha SS, Jankovic D, Feng CG, et al. Cord factor and peptidoglycan recapitulate the Th17-promoting adjuvant activity of mycobacteria through mincle/CARD9 signaling and the inflammasome. *J Immunol* (2013) **190**(11):5722–30. doi:10.4049/jimmunol.1203343
103. Shin S, Brodsky IE. The inflammasome: learning from bacterial evasion strategies. *Semin Immunol* (2015) **27**(2):102–10. doi:10.1016/j.smim.2015.03.006

Conflict of Interest Statement: The authors declare no commercial or financial relationships that could be construed as a potential conflict of interest.

7

Tools and Approaches for Dissecting Protein Bacteriocin Import in Gram-Negative Bacteria

*Iva Atanaskovic and Colin Kleanthous**

Department of Biochemistry, University of Oxford, Oxford, United Kingdom

Edited by:
Ignacio Arechaga,
University of Cantabria, Spain

Reviewed by:
Vasvi Chaudhry,
University of Tübingen, Germany
Gerd M. Seibold,
University of Ulm, Germany

***Correspondence:**
Colin Kleanthous
colin.kleanthous@bioch.ox.ac.uk

Bacteriocins of Gram-negative bacteria are typically multi-domain proteins that target and kill bacteria of the same or closely related species. There is increasing interest in protein bacteriocin import; from a fundamental perspective to understand how folded proteins are imported into bacteria and from an applications perspective as species-specific antibiotics to combat multidrug resistant bacteria. In order to translocate across the cell envelope and cause cell death, protein bacteriocins hijack nutrient uptake pathways. Their import is energized by parasitizing intermembrane protein complexes coupled to the proton motive force, which delivers a toxic domain into the cell. A plethora of genetic, structural, biochemical, and biophysical methods have been applied to find cell envelope components involved in bacteriocin import since their discovery almost a century ago. Here, we review the various approaches that now exist for investigating how protein bacteriocins translocate into Gram-negative bacteria and highlight areas of research that will need methodological innovations to fully understand this process. We also highlight recent studies demonstrating how bacteriocins can be used to probe organization and architecture of the Gram-negative cell envelope itself.

Keywords: bacteriocin, import, Gram-negative bacteria, cell envelope, methods

INTRODUCTION

When exposed to environmental or competition stress bacteria often release proteinaceous toxins called bacteriocins that target and kill neighboring bacteria (Cascales et al., 2007; Cornforth and Foster, 2013). Bacteriocins of Gram-positive bacteria are mostly post-translationally modified peptides with a broad species target range, the best known example being the widely used food preservative nisin. Peptidic bacteriocins have been reviewed extensively (Héchard and Sahl, 2002; Cotter, 2014). The focus of the present review is on protein bacteriocins from Gram-negative bacteria. These are large folded proteins (40–80 kDa) that are composed of multiple domains and tend to have a narrow killing spectrum because of the numerous specific interactions at the cell surface involved in their import. A bacteriocin producer cell is protected from its own bacteriocin by an immunity protein, while sensitive strains lack such protection. These toxins can form pores in the inner membrane, act as nucleases to degrade DNA or RNA in the target cell, or interfere with cell

Abbreviations: CCCP, carbonyl cyanide 3-chlorophenylhydrazone; ColE9, colicin E9; CPA, common polysaccharide antigen; EK, enterokinase; FRAP, Forster recovery after photobleaching; FRET, Forster resonance energy transfer; Im9, ColE9 immunity protein; IM, inner membrane; ITC, isothermal titration calorimetry; LPS, lipopolysaccharide; MMBL, monocot mannose-binding lectin; NMR, nuclear magnetic resonance; NTD, N-terminal domain; OBS, OmpF-binding site; OM, outer membrane; OMP, outer membrane protein; PMF, proton motive force; PPI, protein–protein interactions; SAXS, small-angle X-ray scattering; TIRF, total internal reflection fluorescence.

wall biosynthesis. To do so, protein bacteriocins (hereafter referred to merely as bacteriocins) have to cross a multi-layered cell envelope which is accomplished by parasitizing host proteins involved in nutrient and metabolite trafficking (**Figure 1**).

Bacteriocins have the potential to be developed as much-needed therapeutics to treat multidrug resistant bacterial infections (Cotter, 2014; Behrens et al., 2017). A prerequisite for successfully applying bacteriocins as antimicrobials, however, is to understand how they are imported. Moreover, these import pathways might reveal further processes that could be exploited for newly designed drugs or chimeric bacteriocins with more potent toxicity (Lukacik et al., 2012).

The discovery of new bacteriocins has been accelerated by whole-genome sequencing technologies and implementation of gene mining tools (Jamet and Nassif, 2015; Sharp et al., 2017). While peptide Gram-negative bacteriocins, such as microcins, have been reviewed elsewhere (Duquesne et al., 2007), here we focus on methods used for the study of multidomain protein bacteriocins. The two most explored groups of protein bacteriocins are colicins produced by *Escherichia coli* (Cascales et al., 2007) and pyocins produced by *Pseudomonas aeruginosa* (Ghequire and De Mot, 2014). Nuclease colicins contain an N-terminal translocation domain, a central receptor binding domain, and a C-terminal cytotoxic domain that binds a cognate immunity protein, while in pyocins the location of the translocation and receptor binding domains appears reversed (Michel-Briand and Baysse, 2002). Bacteriocins can be divided into two groups based on the periplasmic energy transducing system they exploit for import. Group A use the Tol–Pal system, which is composed of periplasmic and IM proteins, Pal, TolA, TolB, TolR, TolQ. All but the outer membrane lipoprotein Pal have been documented to be involved in bacteriocin import. Group B use the Ton system, composed of TonB, ExbB, and ExbD proteins (**Figure 1**). It is likely that all protein bacteriocins fall into these two such groups (Kleanthous, 2010).

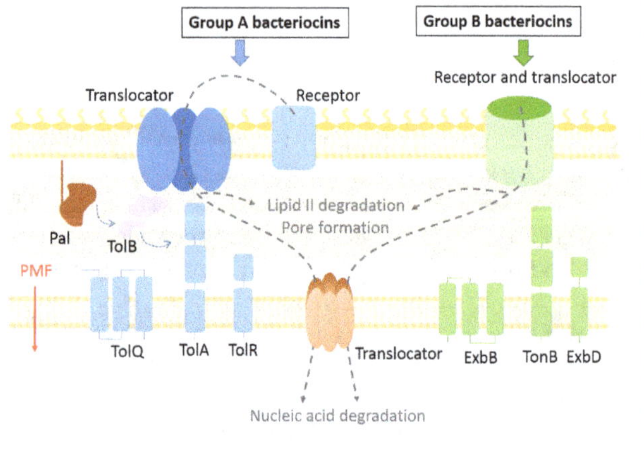

FIGURE 1 | Bacteriocin import pathways. Bacteriocins bind to outer membrane receptors to get imported into the cell. Some bacteriocins (group B) use the receptor protein also as a translocator to cross the outer membrane, while exploiting the TonB system and PMF as an energy source. In case of group A bacteriocins the translocator differs from the receptor protein and the Tol system is used to enter the periplasm. Bacteriocins that degrade lipid II and prevent peptidoglycan recycling remain in the periplasm, while pore forming bacteriocins are inserted in the inner membrane. Nuclease bacteriocins use a distinct protein translocator to cross the inner membrane.

Colicin E9 is one of the best understood of the Group A bacteriocins in terms of its translocation mechanism. Import of ColE9 involves assembly of a translocon complex. The OM portion of the translocon includes BtuB, its receptor, its porin translocator OmpF or OmpC, TolB, its periplasmic target, and Im9, its immunity protein. In order to form this OM translocon, ColE9 uses its intrinsically unstructured N-terminus to pass through a porin channel to engage the PMF-coupled Tol–Pal system in the periplasm (Housden et al., 2005). How ColE9 translocates across the OM is not understood, but it is known that the colicin exploits FtsH, the AAA^+ ATPase/protease, to cross the IM. Once inside the cell, the ColE9 DNase causes nonspecific cleavage of double-stranded DNA which results in cell death (Walker et al., 2007). Group B colicins and pyocins pass through Ton-dependent receptors without the involvement of porins. For instance, pyocin S2 binds to a TonB-coupled siderophore receptor FpvAI, and passes through the FpvAI lumen mimicking its cognate ligand (White et al., 2017).

Gram-negative bacteriocins use a variety of pathways and distinct combinations of cell envelope proteins to kill cells, making it challenging to dissect their import mechanisms. Since their discovery, genetic, structural, biochemical, and biophysical approaches have all been deployed to define these pathways often in combination (**Figure 2**). Here, we give an overview of these approaches, pointing out the most recent advancements in the toolkit used for dissecting bacteriocin translocation in Gram-negative bacteria.

PAIRING UP BACTERIOCINS WITH THEIR RECEPTORS

Every bacteriocin begins its journey through the cell envelope by binding to a specific receptor on the bacterial surface. While most bacteriocins bind OM proteins, some use LPS as their primary receptor (Kim et al., 2014; McCaughey et al., 2014, 2016). The specificity of a bacteriocin–receptor interaction narrows down the target range of these toxins. Therefore, pairing up bacteriocins with their receptors is an important first step in applying them as therapeutics (Cotter, 2014). If a sufficient number of bacteriocin receptors are known, a screen for receptor genes in a genome of a pathogen isolated from the site of infection could guide the design of bacteriocin cocktails to specifically eradicate the cause of infection. Additionally, linking a bacteriocin to a receptor of known biological function is the starting point for understanding which import pathway is being hijacked by the toxin.

When searching for a bacteriocin receptor, the first issue to address is its chemical nature. Bacteriocin *neutralization experiments* have often been used to determine if the receptor is within the protein or LPS fraction of the outer membrane (Weltzien and Jesaitis, 1971). If a certain cell fraction contains the receptor, it will bind to a bacteriocin and inhibit its toxic activity, which can be assessed by a plate killing assay (**Figure 2**). Neutralization experiments can also be used to test if a specific

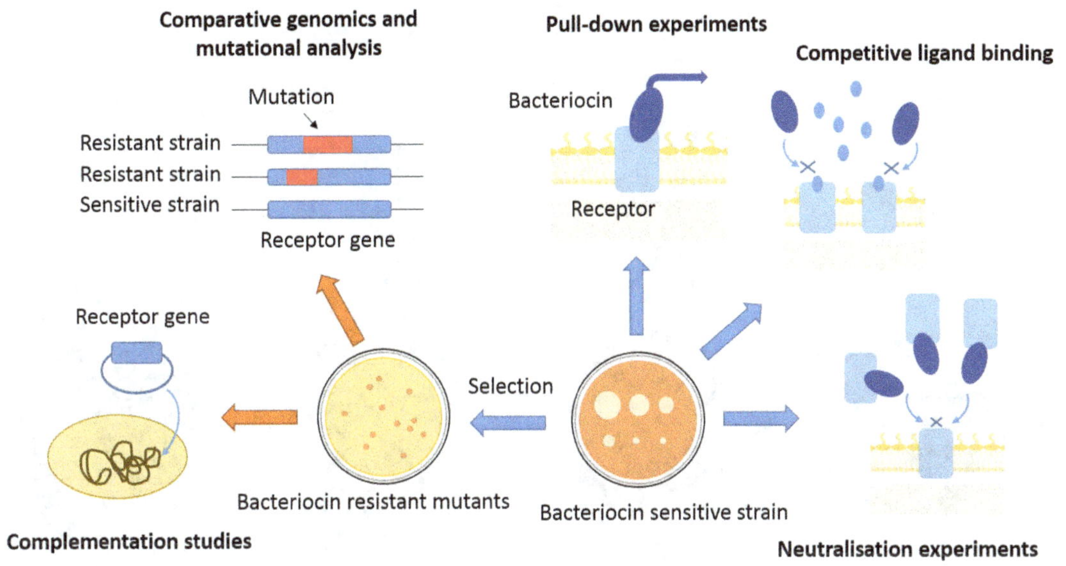

FIGURE 2 | Approaches to finding bacteriocin translocon components. Translocon components can be isolated from membranes of sensitive strains by use of affinity tagged bacteriocins as bait. Competitive ligand binding or bacteriocin binding to a purified translocon component can inhibit bacteriocin activity and give indications about proteins engaged in bacteriocin entry. Bacteriocin-resistant mutants can be selected by use of high bacteriocin concentrations, or generated by transposon mutagenesis. Genomes of resistant mutants can further be analyzed for mutations underlying the resistance phenotype, which can give gene candidates for translocon components. A component of bacteriocin entry has to be further confirmed by complementation studies, where bacteriocin sensitivity is re-established once a resistant mutant is transformed with a gene of interest.

nutrient import pathway is being hijacked by a bacteriocin. If cells are exposed to a bacteriocin in the presence of a nutrient with which it shares its import pathway, competitive binding of the ligand should either inhibit bacteriocin activity or nutrient import. Such experiments were successful in early attempts at receptor discovery. For instance, competitive binding of cyanocobalamin and E-type colicins gave early indications that they all share BtuB as a receptor (Masi et al., 1973). Similarly, such experiments suggested that pyocin S3 and pyoverdine both bind to the ferripyoverdine type II receptor in *P. aeruginosa* (Baysse et al., 1999), and ferredoxin and pectocin M bind to the same ferredoxin receptor in *Pectobacterium artrosepticum* (Grinter et al., 2012).

An alternative approach is one that combines *neutralization assays with cell wall fractionation and protein purification.* OM fractions can be tested for bacteriocin neutralization activity. After singling out the OM fraction with neutralization activity, further analysis by mass spectrometry can identify the receptor. Use of protease inhibitors is important for preventing OM proteases from degrading the bacteriocin, which can be misinterpreted as bacteriocin neutralization. Misinterpretations can also be avoided by co-fractionating OM proteins isolated from a bacteriocin resistant and a bacteriocin sensitive strain, where all OM proteins from each strain have been labeled with two different fluorescent or radioactive labels. Since the resistant mutants should lack the receptor, a fraction with receptor activity should contain only proteins that originate from the sensitive strain (Sabet and Schnaitman, 1973). A limitation of this approach is that receptor concentrations obtained by OM fractionation can often be insufficient to achieve neutralization. This approach is therefore limited to cases where bacteriocin receptors are expressed to high levels or bind the bacteriocin with high affinity.

Instead of testing for neutralization activity, bacteriocin OM receptors can be identified directly by *pull-down experiments*, using a bacteriocin of interest as a bait (**Figure 2**). A bacteriocin should co-elute from the chromatography column with its receptor, where further analysis of the co-eluent by proteolysis and peptide fingerprinting by mass spectrometry can give a receptor candidate. For example, Housden et al. (2005) used a hexahistidine-tagged immunity protein complexed to ColE9 and immobilized on a nickel affinity column to purify components of the ColE9 translocon from OM extracts. These pull-down experiments can be challenging in cases of low-abundance receptors. One way to circumvent this problem is to find receptor overproducers, marked by an increased sensitivity to a bacteriocin of interest (Bowles et al., 1983). Another option is to cultivate cells in conditions that increase bacteriocin sensitivity by inducing receptor expression (Bindereif et al., 1982). Defining growth conditions where a bacteriocin receptor is overexpressed can also give valuable hints about the nature of the receptor. This was the case for several S-type pyocin receptors where it was observed that killing of *P. aeruginosa* was more effective if cultivated in iron-limited conditions (Smith et al., 1992; Sano et al., 1993). This observation indicated that siderophore receptors, which are overexpressed when cells are starved of iron, are involved in import of S-type pyocins. Receptors for pyocins S3, S2, S4, and S5 (Baysse et al., 1999; Denayer et al., 2007; Elfarash et al., 2012, 2014) were all discovered by this route.

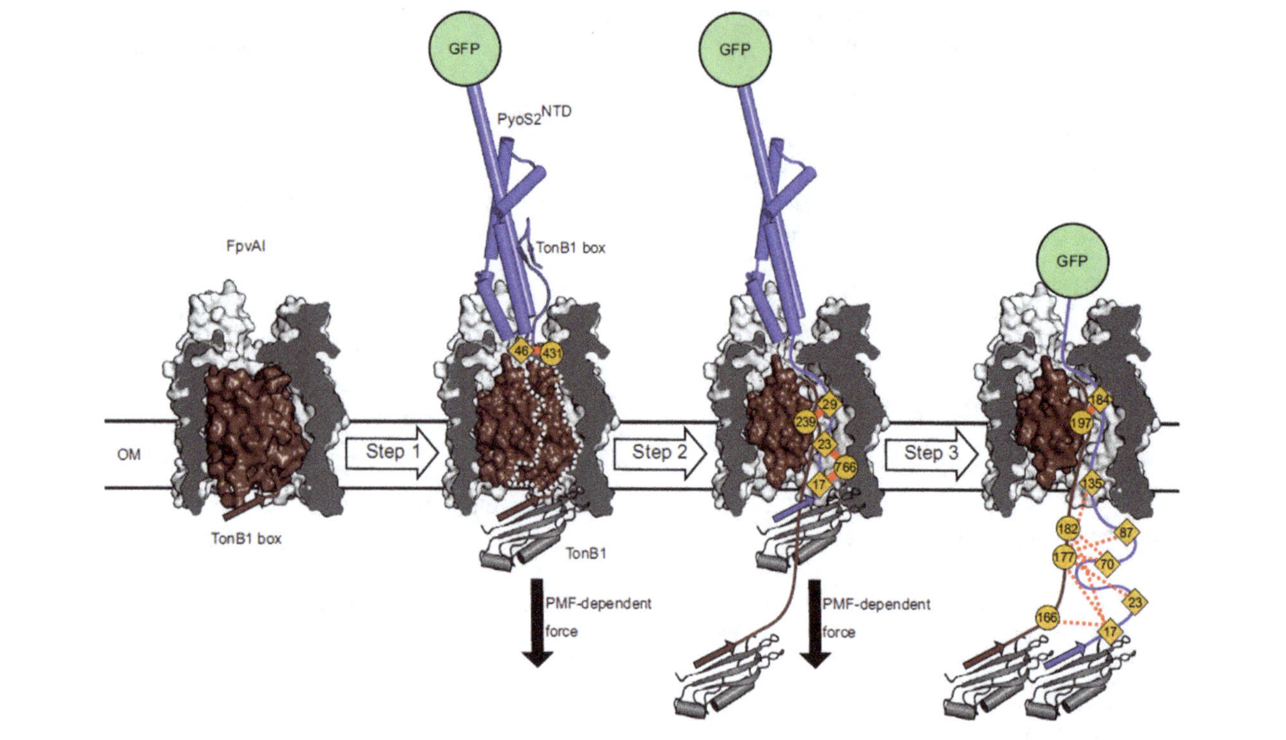

FIGURE 3 | *In vivo* cross-linking strategy developed by White et al. for dissecting the import pathway of pyocin S2 through the FpvAI receptor. Photoactivatable crosslinking data were used to establish a three-step model of translocation. In step 1, pyocin S2 binds to FpvAI and mimics its cognate ligand, ferro-pyoverdine. In step 2, a PMF-dependent mechanical force, applied via the ExbB–ExbD–TonB1 complex in the IM (not shown), drives unfolding of the labile half of the receptor plug domain and the N terminus of pyocin S2 enters the receptor lumen. In step 3, pyocin S2 binds to TonB1 and this interaction drives further passage of the pyocin through the receptor lumen. Complete translocation is blocked by the force-resistant GFP, which enables identification of translocation intermediates and mapping of interactions that govern pyocin S2 cell entry [image taken and used with permission from White et al. (2017), CC BY-NC-ND 4.0].

Another approach for receptor discovery is based on the isolation and characterization of *bacteriocin-resistant mutants*. A resistant mutant can lack a bacteriocin receptor; hence, picking up genetic differences between resistant and sensitive strains can pinpoint the receptor gene (**Figure 2**). When selecting for resistant mutants by use of lethal bacteriocin doses a major challenge is to distinguish between resistant and tolerant strains; only resistant strains bare mutations that specifically alter import machinery components. Therefore, it is always necessary to disregard tolerance by exposure to even higher bacteriocin doses than those used for selection and by checking if a mutant's growth kinetics is affected by the bacteriocin. Finally, resistance can be confirmed by checking if the mutant's OM has lost its bacteriocin binding properties by a neutralization assay or use of a fluorescently labeled bacteriocin to test for cell surface association (**Figure 3**; Rassam et al., 2015; White et al., 2017).

After isolating a resistant mutant, it is possible to compare its OM protein composition with the parental strain. A combination of electrophoresis and mass spectrometry can point out proteins absent in the OM of the resistant strain, but again, this has only been successful in cases where receptor genes are highly expressed (Ohkawa et al., 1980; Baysse et al., 1999). The advent of *whole-genome sequencing and comparative genomics* is now the preferred method of choice. Comparing a mutant's genome with a bacteriocin-sensitive reference strain can pinpoint potential receptor genes (McCaughey et al., 2014). This strategy has been successful in identifying bacteriocin receptors in Gram-positive bacteria (Cotter, 2014). The decreasing cost of DNA sequencing and continuous refinement of bioinformatics tools makes it likely that researchers in the field of Gram-negative bacteriocin import will increasingly turn to whole-genome sequencing. However, comparative genomics still has its limitations especially in the case of bacterial strains with a high rate of spontaneous mutations, where it can be difficult to filter out a mutation associated with bacteriocin resistance.

Receptor genes have also been identified by use of *cosmid libraries*. Genomic fragments from a bacteriocin-sensitive strain can be transformed into a bacteriocin-resistant background. The goal is to identify a fragment that can restore bacteriocin sensitivity and potentially carries a receptor gene (Pilsl et al., 1999; Smajs and Weinstock, 2001). An alternative approach is to construct *a library of transposon mutants* in a strain that is sensitive to the bacteriocin. Sequencing of genomic regions around the transposon insertion site in a resistant mutant can pinpoint genes linked to bacteriocin import (Baysse et al., 1999; de Chial et al., 2003; Elfarash et al., 2014; Ghequire et al., 2017). The power of this approach can be increased with the implementation of high-throughput transposon insertion sequencing techniques, such as TraDIS (Barquist et al., 2016), since the employment of a dense transposon library can give better genomic coverage and increase the probability of a resistance phenotype being detected. TraDIS has recently been used to show that LPSs bearing O-antigens shield bacteriocin receptors in uropathogenic *E. coli* but that this effect is modulated by growth conditions (Sharp et al., 2019). Nevertheless, if a receptor is an essential gene no transposon mutants will be obtained and a library search will fail to show the receptor. In this case, pull-down experiments or linkage analysis (see below) can be used.

Bacteriocin receptors can also be found by *linkage analysis*, as in the case of S-type pyocin receptors. S pyocins kill *P. aeruginosa* better under iron limiting conditions, which gave indications that their activity might be linked to the pyoverdine import system (Ohkawa et al., 1980; Smith et al., 1992). A collection of *P. aeruginosa* strains was screened for pyocin S2 sensitivity and ferripyoverdine receptor genes typed by multiplex PCR. All S2 sensitive strains had the type I ferripyoverdine *fpvA*I gene, indicating this was the S2 receptor (Denayer et al., 2007). The same approach was used to link pyocin S4 sensitivity to the *fpvAI* receptor gene (Elfarash et al., 2012). Another type of linkage analysis that was successful in bacteriocin receptor discovery was *metabolite analysis*, where pyoverdine production was compared between pyocin S3 resistant and sensitive strains. It was found that pyocin S3 kills only type II pyoverdine producers of *P. aeruginosa*, while type I and III producers were resistant to S3 (Govan, 1978; Baysse et al., 1999). This gave strong indications that pyocin S3 binds to the ferripyoverdine type II receptor, which was confirmed by subsequent studies (de Chial et al., 2003). A future challenge will be implementation of high-throughput linkage approaches, which can be used when there are no initial clues about the import mechanism. This could be achieved through *genome wide association studies*, if both a genome sequence database and a physical strain collection are available. One could then test bacteriocin sensitivity throughout the collection and conduct a gene linkage analysis for a collection of corresponding annotated genomic sequences (Brynildsrud et al., 2016). Genes that are present in a large number of sensitive, but absent in a large number of bacteriocin resistant strains, are then further tested for receptor coding activity. On the other hand, the development of new mass spectrometric approaches in high-throughput metabolomics (Zampieri et al., 2017) could enable full metabolome comparison between bacteriocin resistant and sensitive strains, where metabolites lacking in resistant strains could give hints about import mechanisms being hijacked by the bacteriocin.

The majority of bacteriocin receptors identified to-date are proteins; however, non-proteinaceous receptors have also been identified. Pyocin L1 is a lectin-like bacteriocin produced by *P. aeruginosa*. It consists of tandem MMBL domains and kills cells by targeting the CPA of *P. aeruginosa* LPS, which is predominantly a homopolymer of D-rhamnose. The widespread inclusion of D-rhamnose in the LPS of pseudomonads explains the unusual genus-specific activity of this lectin-like bacteriocin. The discovery of the pyocin L1 saccharide receptor was achieved through a combination of genetics, structural, and biophysical approaches (McCaughey et al., 2014). Alignment of the pyocin L1 protein sequence with other lectin-like bacteriocins revealed the presence of three conserved MMBL sugar-binding domains, giving the first indications that pyocin L1 might bind to polysaccharide rather than to a protein receptor. A *P. aeruginosa* strain sensitive to this pyocin was used to recover resistant mutants, the genome sequences of which showed a deletion in the *wbpZ* gene, which encodes a glycosyltransferase involved in the synthesis of the CPA component of LPS (Rocchetta et al., 1998). Subsequent immunoblotting with a CPA-specific antibody along with transposon insertions in genes *wzt* and *wzm*, which encode the ATP-binding and the membrane components of a CPA dedicated ABC transporter (Lam et al., 2011), confirmed that CPA on the cell surface is required for pyocin L1 killing. Direct binding of pyocin L1 to CPA and D-rhamnose was shown by ITC and NMR spectroscopy. Finally, X-ray crystallography defined the mode of binding of the D-rhamnose receptor to the C-terminal MMBL domain of pyocin L1. This study therefore provides an excellent example how a combination of genomics and mutational analysis combined with biophysics and structural data can identify non-proteinaceous bacteriocin receptors.

In summary, bacteriocin receptors come in many different types and so an equally varied toolkit is needed for their identification (**Figure 2**). Receptors with high expression levels can be pulled-down using a bacteriocin as bait. This might be the only viable approach if the receptor is essential. If non-essential, receptor coding genes can be discovered through isolation of bacteriocin-resistant mutants, either spontaneously generated or induced through transposon mutagenesis. Finally, the future development of high-throughput approaches based on whole-genome sequencing and comparative genomics will alleviate receptor discovery for an ever-growing number of newly identified bacteriocins (Sharp et al., 2017).

IDENTIFYING TRANSLOCON COMPONENTS DOWNSTREAM OF THE RECEPTOR

After binding to a specific receptor, bacteriocins translocate across the OM. The receptor itself can serve as a translocation pore, as for the group B pyocin S2 (White et al., 2017), or another protein can be recruited to the complex to serve as a translocator, as for the group A colicin ColE9 (Housden et al., 2013). Before establishing a translocation model, it is necessary to find all OM and periplasmic components of a bacteriocin's translocon.

Bacteriocin insensitive mutants can be used to find translocon components other than the receptor. Mutant library screens have been particularly useful in this regard. For example, a screen of the Keio collection library for mutants insensitive but still capable of binding colicin S4 showed that *ompF*, *tolA*, *tolB*, *tolQ*, and *tolR* genes are all linked to its translocation. In the same study, it was found that after binding to the receptor OmpW, colicin S4 recruits OmpF and the Tol–Pal system to translocate across the OM (Arnold et al., 2009). A complication of this approach, however, is that *tol* gene knock-outs have altered membrane stability and permeability; deletion of *tolA* in *E. coli* leads to a pleiotropic phenotype characterized by outer membrane blebbing, release of the periplasmic content, increased sensitivity to cholic acid and SDS, and defective O-antigen polymerization (Germon et al., 2001). Indeed, *P. aeruginosa tolQRA* knock-outs are lethal (Wei et al., 2009). Therefore, using *tol* deletion mutants to check if bacteriocin entry is Tol–Pal dependent is not always feasible.

Chimeric bacteriocins have been effectively used to separate the receptor binding and translocation phases of the import process, which can be important for finding translocon components downstream of the receptor. This approach was successfully used to show that colicin Ia uses two copies of the Cir protein for OM translocation - one copy is used as the receptor and the other copy is used for translocation. A chimeric colicin Ia, in which the receptor-binding domain was replaced by that from colicin E3, required BtuB, the colicin E3 receptor, but also a copy of Cir and TonB for its killing activity. This experiment gave indications that one copy of Cir interacts with the receptor-binding domain of colicin Ia to concentrate the protein on the cell surface, while the other copy of Cir interacts with the translocation domain of this colicin so it can pass through the OM and enter the periplasm (Buchanan et al., 2007; Jakes and Finkelstein, 2010). However, direct binding of the Cir translocator with colicin Ia has yet to be demonstrated biochemically. *Receptor bypass* experiments have also been used to find translocon components other than the receptor. In such experiments, the OM is first permeabilized, usually by osmotic shock, so the receptor binding step is bypassed (Thomas and Valvano, 1993). This can be useful when a receptor for a bacteriocin under investigation is not known or when suitable chimeras are not available.

Very little is known about *inner membrane translocation* of Gram-negative bacteriocins. However, mutational analysis successfully identified some IM proteins necessary for transport of nuclease colicins. Walker et al. (2007) used a Δ*ftsH E. coli* strain to show that the toxicity of all nuclease colicins (regardless of their Tol–Pal/Ton dependence) is dependent on FtsH, an inner membrane AAA+ ATPase/protease. This protease cleaves off the DNase domain during import to the cytoplasm (Chauleau et al., 2011; Mora and de Zamaroczy, 2014). In other studies, whole-genome sequencing of an *E. coli* mutant that is insensitive to colicin D gave indications that an inner membrane peptidase LepB is required for cell entry (de Zamaroczy et al., 2001). LepB is a key membrane component of the cellular secretion machinery, which releases secreted proteins into the periplasm by cleaving the inner membrane-bound leader. It was further shown that this protein binds to colicin D and probably directs it to the FtsH protease for cell entry (Mora et al., 2015).

Since IM translocation components are a part of protein translocation systems that are well conserved across different Gram-negative bacteria, it is possible to study the import of bacteriocins from other species using *E. coli* as a model system. The only limitation here is to bypass all the species-specific translocation steps that are mostly localized to the outer membrane. Hence, Mora et al. (2015) used a colicin D/klebicin D hybrid in which the N-terminal import domain of klebicin D was replaced with that of colicin D. Klebicin D targets *Klebsiella* species, but when fused to the receptor binding and the translocation domain of colicin D it can also kill *E. coli* obviating the need for a *Klebsiella* knock-out library. In this way, it was found that klebicin D, like colicin D, uses LepB for import (Mora et al., 2015).

DECONSTRUCTING THE PROTEIN–PROTEIN INTERACTIONS OF BACTERIOCIN TRANSLOCONS

Several approaches can be taken to define binding epitopes and binding induced conformational changes within bacteriocin translocons. PPIs between translocon components have to be first confirmed *in vitro* and *in vivo*. A common approach is *gene complementation* using a bacteriocin-resistant mutant or a bacteriocin-resistant species (Kjos et al., 2014), where the establishment of bacteriocin sensitivity can confirm the involvement of components in bacteriocin import (**Figure 2**). Apart from gene complementation assays, interactions between components can be assessed *in vivo* by *pull-down* experiments. A way to "freeze" the translocon in the assembly phase is by *in vivo* cross-linking (Masi et al., 2007; White et al., 2017) or by a disulphide locked bacteriocin (Housden et al., 2005, 2013). In both cases, a bacteriocin must be able to bind to its receptor and trigger translocon assembly without fully traversing the cell envelope and killing the target cell. Bacteriocin-bound protein complexes can further be extracted from the outer membrane by affinity chromatography using a bacteriocin or its immunity protein as bait. Pulldowns followed by limited proteolysis of the complex, and mass spectrometry of recovered protein fragments, can indicate which binding epitopes are involved in the translocon. For example, Housden et al. (2013) designed a disulphide lock that forms between the TolB-binding epitope of ColE9 and periplasmic TolB following recruitment of OmpF in the OM. A histidine tag on TolB enabled a heptameric assembly of the ColE9–Im9 complex, BtuB, OmpF trimer, and TolB to be purified and analyzed by mass spectrometry. Finally, limited proteolysis, in combination with planar lipid bilayer experiments and native mass spectrometry, demonstrated that within the translocon, ColE9's unstructured N-terminal region passes twice through its bound porin thereby presenting its TolB-binding epitope in a conformationally constrained orientation in the periplasm (Housden et al., 2013).

Calorimetric measurements have also been used extensively in colicin import studies (Housden and Kleanthous, 2011), in particular ITC. ITC parameters can provide evidence of conformational changes within the translocon (Housden et al., 2005). ITC has also been used to determine how a bacteriocin affects existing PPIs within the cell envelope. Changes in the heats of binding in presence of a bacteriocin can indicate if it abolishes or induces a certain PPI. In this way, it has been shown that ColE9 interacts with TolB when entering the periplasm, disrupting the TolB–Pal complex and stimulating formation of a TolA–TolB complex that traverses the periplasm (Bonsor et al., 2009). ITC combined with site-directed mutagenesis can map translocon-binding sites, as in the case of the ColE9 TolB-binding region. The favorable enthalpy and unfavorable entropy changes associated with ColE9-binding TolB correspond to a disorder-to-order transition that occurs when the intrinsically unstructured region of ColE9 folds and binds TolB (Loftus et al., 2006). In other words, ITC parameters can indicate not only which regions of bacteriocins and their translocon components interact, but also which regions undergo conformational changes and in which phases of the translocation process these conformational changes occur.

Stopped-flow FRET experiments demonstrated how a bacteriocin can remodel PPIs within the cell envelope during import. Papadakos et al. (2012a) used a series of pre-steady-state kinetic experiments utilizing FRET pairs of ColE9 TolB-binding epitope, TolB and Pal, to establish the kinetic basis for competitive recruitment of TolB by ColE9 during which Pal gets displaced from its TolB–Pal complex. Interactions between translocon components have also been investigated using *planar lipid bilayers*. For example, it has been shown that ColE1 occludes TolC channels and that ColE9 occludes OmpF channels in planar lipid bilayers, confirming previous findings that these OMPs are involved in bacteriocin translocation (Zakharov et al., 2004). This approach was also used to identify a TolC box in ColE1 (Zakharov et al., 2016).

ASSEMBLING THE TRANSLOCON

A major challenge in understanding protein bacteriocin import is to assemble their translocons for structural studies where, for the most part, only general import mechanisms have been described (Cascales et al., 2007; Ghequire and De Mot, 2014; McCaughey et al., 2016; White et al., 2017).

Since Gram-negative bacteriocins are large multi-domain proteins, it is thought they must unfold, either partially or completely, in order to translocate across the outer membrane. Introducing *disulphide bonds* to prevent conformational changes in certain regions of a bacteriocin has been a useful approach in delineating such structural changes (Penfold et al., 2004). Similarly, *protease cleavage sites* have been used to probe the accessibility of bacteriocin sequences within a translocon. Zhang et al. (2008) introduced unique EK cleavage sites in a disulphide-locked ColE9 at a number of locations to study the surface accessibility of colicin subdomains shortly after receptor binding. In this experiment, a disulphide-lock within the colicin was used to synchronize translocon assembly; disulfide bond reduction simultaneously triggers initiation of translocation in all bacterial cells. This enabled determination of EK cleavage site accessibility for different regions of the colicin, which gave important insight into the position of ColE9 in the assembled translocon (Zhang et al., 2008).

Studies are beginning to unravel the molecular mechanism(s) by which bacteriocins translocate across the OM. White et al. (2017) developed an *in vivo cross-linking strategy* to map the import of the pyocin S2 NTD through the FpvAI receptor (**Figure 3**). This approach depended on first blocking import of the NTD using a C-terminal GFP. GFP is able to withstand ~200 pN of force whereas the PMF can only deliver ~20 pN (Saeger et al., 2012). This strategy allowed the accumulation of translocation intermediates that would otherwise be undetectable by crosslinking. Variants of this GFP fusion were then generated in which benzoylphenylalanine was incorporated at different positions of the NTD using amber suppression and crosslinked following transport into *P. aeruginosa* cells. Detailed mass spectrometric analysis of crosslinked peptides demonstrated that the pyocin not only translocates through FpvAI but that it does so by a process which likely mimics that used by pyoverdine, the natural ligand for FpvAI (White et al., 2017).

An important aspect of nuclease bacteriocin import is the release of the tightly bound immunity protein. All nuclease bacteriocins are produced bound tightly ($K_d \sim 10^{-14}$ M at pH 7 and 25°C) to their immunity protein Papadakos et al. (2012b). The half-life for dissociation for this complex is several days yet killing occurs within minutes. Hence, it has been postulated that an energy transduction path exists that jettisons the immunity protein at the cell surface during import. Zhang et al. (2008) developed a sensitive *fluorescence assay* to investigate immunity release. The assay was based on release of a fluorescently labeled immunity protein into the cell supernatant. Bacteriocin import was synchronized using a disulfide-lock. In this way, fluorescently labeled Im9 was detected in the cell supernatant after the addition of a reducing agent and was dependent on the PMF across the inner membrane (Zhang et al., 2008; Vankemmelbeke et al., 2012), shown subsequently to be linked to global conformational changes within the colicin (Vankemmelbeke et al., 2013). How force might cause immunity dissociation has been investigated by *single molecule atomic force spectroscopy* (Farrance et al., 2013). These studies demonstrated that the ColE9 DNase–Im9 complex is exquisitely sensitive to mechanical deformation at the N-terminus of the nuclease, which could represent pulling into a cell, causing rapid dissociation of the immunity protein.

STRUCTURAL BIOLOGY APPROACHES FOR STUDYING BACTERIOCIN IMPORT

Structural studies are important in gaining a molecular understanding of the bacteriocin translocation process. To-date, there are only a few structures of intact Gram-negative bacteriocins (Wiener et al., 1997; Soelaiman et al., 2001; Helbig et al., 2011; Klein et al., 2016). More informative, in terms of translocation mechanism, are structures of bacteriocins or bacteriocin domains bound to their receptor or a component of the translocon assembly (Kurisu et al., 2003; Loftus et al., 2006; Buchanan et al., 2007; Sharma et al., 2007; Zhang et al., 2009; Housden et al., 2010; White et al., 2017). These structures give information about conformational changes that a bacteriocin induces and shed light on the import mechanism.

Crystallization of bacteriocins can often be challenging due to intrinsically disordered and flexible regions in these proteins. Flexibility can be reduced by deletion of these sequences or by the introduction of intramolecular disulfide bonds (Klein et al., 2016). To complete a structure of a bacteriocin it is often necessary to substitute missing X-ray diffraction data with results from other experiments. Modeling of data from *analytical ultracentrifugation* (Manon and Ebel, 2010) and *NMR experiments* (Collins et al., 2002; Hecht et al., 2009) can help in filling-in missing parts of a bacteriocin structure. Results from *SAXS* experiments can also be combined with diffraction data; a bacteriocin can be treated as a flexible system, namely by the ensemble optimization method, which enables a distribution of conformations to be included in the final model (Johnson et al., 2017).

Structures of *bacteriocin–receptor and bacteriocin–translocator complexes* are essential to understand bacteriocin import mechanisms. When compared to receptor–cognate ligand complexes, these structures can show to what extent a bacteriocin mimics the ligand (if at all) when traversing the cell envelope. For example, co-crystal structures for E-type colicins bound to the BtuB receptor (ColE2, ColE3) show that the colicins do not mimic the interactions of the ligand and do not induce conformational changes within the globular N-terminal plug domain of BtuB indicative of transport. This contrasts the situation of pyocin S2-NTD bound to its receptor FpvAI where the structure clearly supports a model (validated by crosslinking data) that the pyocin mimics the endogenous ligand, pyoverdine, and transports through the receptor in a TonB-dependent manner (White et al., 2017). For the E colicins (all of which require the Tol complex), entry to the periplasm requires OmpF or OmpC. These porins acts as translocators, which is supported by structures of OmpF in which fragments of E colicins are bound within its lumen (Yamashita et al., 2008; Housden et al., 2010) and by planar lipid bilayer experiments where colicin fragments block ion conductance (Zakharov et al., 2004).

The development of structure-prediction algorithms could broaden the understanding of bacteriocin translocons (Delarue and Koehl, 2018). For instance, protein fold predictions have already been employed in structure determination of a group A colicin translocon component, TolQ (Ovchinnikov et al., 2015). Additionally, future developments in translocon structures will undoubtedly involve cryo-EM methods. It may even be possible to eventually capture a bacteriocin in-transit and use cryo-tomography to map out its interactions within the cell envelope.

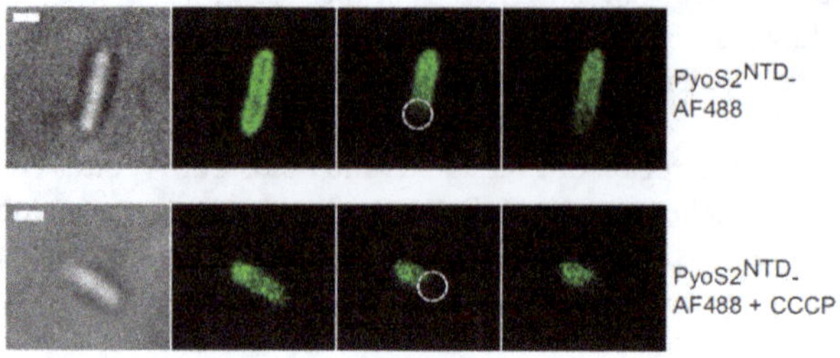

FIGURE 4 | FRAP experiments can be used to show that bacteriocin import is a PMF-dependent process. *P. aeruginosa* PAO1 cells are labeled with pyocin S2 AF488. The bleached region is highlighted (dashed circle). FRAP suggests pyoS2NTD-AF488 has translocated to the periplasm, where it can diffuse laterally. Absence of FRAP observed when cells are treated with 100 μM CCCP indicates that the pyocin remains bound to FpvAI in the OM and that the PMF is necessary for pyocin translocation. Scale bars, 1 μm [image taken and used with permission from White et al. (2017), CC BY-NC-ND 4.0].

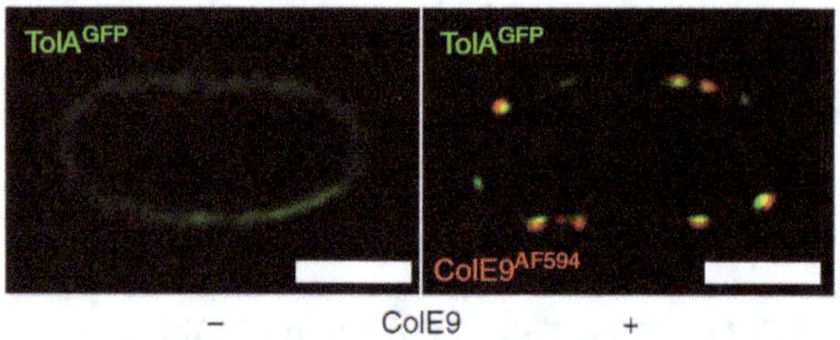

FIGURE 5 | Fluorescently labeled bacteriocins can be used to track outer and inner membrane protein clusters; used by Rassam et al. (2018) to show that bacteriocin-induced clustering of TolA in the IM mirrors that of OMPs in the OM. 2D-SIM z-slice showing significant co-clustering (yellow fluorescence) of GFP-TolA and ColE9AF594 in the IM and OM, respectively. Scale bars, 1 μm [image taken and used with permission from Rassam et al. (2018), CC BY-NC-ND 4.0].

VIEWING THE IMPORT PROCESS IN THE CONTEXT OF OUTER MEMBRANE ORGANIZATION

Bacteriocins have become valuable tools with which to investigate spatiotemporal organization in the cell envelope by fluorescence microscopy (Kleanthous et al., 2015). Fluorescently labeled bacteriocins have been used in conjunction with *single particle tracking TIRF microscopy* to show that OMPs display highly restricted mobility in the outer membrane due to the formation of supramolecular assemblies called OMP islands, which also explains the lack of fluorescence recovery for labeled OMPs in FRAP experiments (**Figure 4**). TIRF microscopy of bacteriocin-labeled OMPs has also shown that OMP islands move to the poles as new islands appear in the membrane (Rassam et al., 2015). Even more remarkably, the restricted mobility that is characteristic of OMPs within OMP islands becomes imprinted on inner membrane proteins when the bacteriocin forms its translocon across the two membranes (**Figure 5**; Rassam et al., 2018).

Fluorescently labeled bacteriocins have also been used for dissecting their import mechanism. GFP was deployed by White et al. (2017) to visualize association of pyocin S2 with *P. aeruginosa* cells, block translocation, and trap the pyocin within its receptor, FpvAI, for subsequent crosslinking studies. By switching to an organic dye (AF488), White et al. (2017) demonstrated import of the pyocin S2 domain since in contrast to GFP-labeled protein, fluorescence recovery was observed for AF488-pyocin S2-labeled cells in FRAP experiments. Imported fluorescent protein was also protected against exogenously added protease.

Live cell imaging can also be used to dissect the directionality of PPIs formed between bacteriocins and their translocators. Housden et al. (2018) used a combination of molecular dynamics simulations, fluorescence microscopy, and single channel recording planar lipid bilayer measurements to unambiguously demonstrate from which side of the OM different OBSs of

ColE9 associated with the lumen of the porin. The intrinsically unstructured N-terminal region of ColE9 houses two OBSs (OBS1 and OBS2) that reside within the pores of OmpF and that flank an epitope that binds periplasmic TolB. The studies revealed that OBS2 binds OmpF from the extracellular side, while the interaction of OBS1 occurs from the periplasmic face of OmpF, which ensures constrained presentation of the TolB epitope within the bacterial periplasm (Housden et al., 2018).

THE ENERGETICS AND KINETICS OF BACTERIOCIN IMPORT

The energetics of bacteriocin translocation are still controversial. Indeed, little is known about the energy dependence of individual translocation steps. Since bacteriocins are folded proteins, import is likely to rely on the input of energy, the main source being the PMF generated across the IM. However, some studies suggest that OM translocation is energy independent, as in the case of colicin A (Bourdineaud et al., 1990). Nevertheless, live cell imaging studies are beginning to show the link between PMF, Ton, Tol, and bacteriocin import (White et al., 2017; Rassam et al., 2018).

A way of testing if energy is necessary for a certain phase of bacteriocin import is to disrupt the PMF and then assess import with a suitable assay. Protonic ionophores have been used for this; for example, CCCP was used in FRAP experiments to show that PMF is necessary for import of pyocin S2 (**Figure 4**; White et al., 2017). Disulphide locked colicins and fluorescently labeled immunity proteins have been used to study the energy dependence of immunity protein release (Vankemmelbeke et al., 2012). A sensitive assay for detecting immunity protein release (described above) was used to confirm that both TolB and TolA are necessary for this process in ColE9 import and to define which regions of TolA are engaged. TolA is anchored in the cytoplasmic membrane via a single transmembrane region that interacts to TolQ and TolR and drives a TolA conformational change in response to PMF (Germon et al., 2001). Vankemmelbeke et al. (2009) focused on residue H22 of the TolA IM region that has been previously linked to energy-dependent conformational change of this protein (Larsen et al., 2007). They showed that an alanine mutation of this residue disrupts immunity protein release, which indicated that an energy-dependent conformational change of TolA is essential for nuclease colicin Im9 release at the cell surface. This points to a role for TolA in transducing cellular energy in a manner similar to that described for TonB (Postle and Larsen, 2007), but the mechanism remains unknown. Bonsor et al. (2009) went on to show that it is the contact between TolA and TolB within the periplasm that is the necessary energy-transduction step for nuclease colicin Im release. One model proposes that the rotation of Tol complex transmembrane helices is transduced into a conformational change in TolA that affects its interaction with the TolB–bacteriocin complex and triggers Im protein dissociation (Papadakos et al., 2012). New methodological approaches will be necessary to dissect how the electrochemical potential of the PMF is transduced by the Tol and Ton systems into the mechanical action of bacteriocin import.

Little is known of the kinetics of bacteriocin import beyond the onset of cellular changes these protein toxins induce. Early studies showed that conformational flexibility is a prominent bacteriocin trait, determining the kinetics of import (Duche et al., 1994). Additionally, bacteriocin charge (Walker et al., 2007) and cell envelope properties related to lipid composition are factors that influence the speed of import (Bourdineaud et al., 1990; Walker et al., 2007). In the case of pore forming colicins, measuring cytoplasmic potassium efflux caused by colicin A inner membrane insertion was used to show how inner membrane fluidity influences the kinetics of translocation (Bourdineaud et al., 1990). In the case of nuclease colicins, Walker et al. (2007) observed that inner membrane charge impacts the rate of import. In this study, a strain of *E. coli* in which the level of anionic phospholipids was regulated by isopropyl β-D-thiogalactopyranoside induction was used to show that both inner membrane and bacteriocin charge influence the rate of cell entry. Import kinetics were measured indirectly, through the efficiency of cell killing or DNA damage. Still, little is known about the kinetics of individual translocation steps and how PPIs between translocon components influence the rate of import. Therefore, to understand import kinetics in more detail, methods for direct measurement of individual translocation steps will have to be developed, most likely based on single molecule methods.

FUTURE PERSPECTIVES

As protein bacteriocins begin to be exploited as therapeutics for multidrug-resistant bacteria so our ability to understand their import mechanisms needs to develop (Behrens et al., 2017). Various approaches, ranging from classical biochemical methods to advanced molecular genetics, have been used over the years for translocon component discovery. The development of high-throughput approaches based on whole-genome sequencing and comparative genomics can speed up this process in the future. Calorimetric measurements have been used extensively to identify binding epitopes and to study the nature of interactions between translocon components. Moreover, implementation of various microscopy techniques, channel conductivity measurements, *in vivo* cross-linking, and disulphide locking of bacteriocins have been exploited for determining the directionality of PPIs within the translocon and for identifying entry paths that bacteriocins use when crossing the OM. Finally, *in vivo* imaging of cells bound to fluorescently labeled bacteriocins has enabled visualization of the import process for the first time at single cell level and provided new tools for probing the spatiotemporal organization of the cell envelope.

While substantial knowledge about OM translocation of bacteriocins has been gained over recent years, the IM component of translocation remains to be elucidated. Many questions remain unanswered – which proteins are involved in bacteriocin IM translocation, how is this process energized, which parts of a bacteriocin are necessary for IM translocation, and which parts enter the cytoplasm. Additionally, a clear link between the OM and IM components of bacteriocin translocons still has to be revealed. For example, are nuclease bacteriocins brought into the periplasm in their entirety before import across the cytoplasmic membrane begins? Finally, the study of bacteriocins in the context of cell envelope organization has breathed new life into bacteriocin research.

AUTHOR CONTRIBUTIONS

IA and CK contributed to the original manuscript and the editorial changes.

FUNDING

IA was funded by the Wellcome Trust Infection, Immunology and Translational Medicine Doctoral Training Centre. CK acknowledges funding from the ERC (OMPorg - 742555), Wellcome Trust (201505/Z/16/Z), MRC (MR/R009937/1), and BBSRC (BB/P009948/1).

ACKNOWLEDGMENTS

The authors would like to thank Ms. Hannah Behrens, Dr. Joanna Szczepaniak, Dr. Nathalie Reichmann, and Dr. Nick Housden for useful discussions and feedback on this manuscript.

REFERENCES

Arnold, T., Zeth, K., and Linke, D. (2009). Structure and function of colicin S4, a colicin with a duplicated receptor-binding domain. *J. Biol. Chem.* 284, 6403–6413. doi: 10.1074/jbc.M808504200

Barquist, L., Mayho, M., Cummins, C., Cain, A. K., Boinett, C. J., Page, A. J., et al. (2016). The TraDIS toolkit: sequencing and analysis for dense transposon mutant libraries. *Bioinformatics* 32, 1109–1111. doi: 10.1093/bioinformatics/btw022

Baysse, C., Meyer, J. M., Plesiat, P., Geoffroy, V., Michel-Briand, Y., and Cornelis, P. (1999). Uptake of pyocin S3 occurs through the outer membrane ferripyoverdine type II receptor of *Pseudomonas aeruginosa*. *J. Bacteriol.* 181, 3849–3851.

Behrens, H. M., Six, A., Walker, D., and Kleanthous, C. (2017). The therapeutic potential of bacteriocins as protein antibiotics. *Emerg. Top. Life Sci.* 1, 65–74. doi: 10.1042/ETLS20160016

Bindereif, A., Volkmar, B., and Klaus, H. (1982). The cloacin receptor of ColV-bearing *Escherichia coli* is part of the Fe3+-Aerobactin transport system. *J. Bacteriol.* 150, 1472–1475.

Bonsor, D. A., Hecht, O., Vankemmelbeke, M., Sharma, A., Krachler, A. M., Housden, N. G., et al. (2009). Allosteric β-propeller signalling in TolB and its manipulation by translocating colicins. *EMBO J.* 28, 2846–2857. doi: 10.1038/emboj.2009.224

Bourdineaud, J. P., Boulanger, P., Lazdunski, C., and Letellier, L. (1990). In vivo properties of colicin A: channel activity is voltage dependent but translocation may be voltage independent. *Proc. Natl. Acad. Sci. U.S.A.* 87, 1037–1041. doi: 10.1073/pnas.87.3.1037

Bowles, L. K., Miguel, A. G., and Konisky, J. (1983). Purification of the colicin I receptor. *J. Biol. Chem.* 258, 1215–1220.

Brynildsrud, O., Bohlin, J., Scheffer, L., and Eldholm, V. (2016). Rapid scoring of genes in microbial pan-genome-wide association studies with Scoary. *Genome Biol.* 17:238. doi: 10.1186/s13059-016-1108

Buchanan, S. K., Lukacik, P., Grizot, S., Ghirlando, R., Ali, M. M., Barnard, T. J., et al. (2007). Structure of colicin I receptor bound to the R-domain of colicin Ia: implications for protein import. *EMBO J.* 26, 2594–2604. doi: 10.1038/sj.emboj.7601693

Cascales, E., Buchanan, S. K., Duché, D., Kleanthous, C., Lloubès, R., Postle, K., et al. (2007). Colicin biology. *Microbiol. Mol. Biol. Rev.* 71, 158–229. doi: 10.1128/MMBR.00036-06

Chauleau, M., Mora, L., Serba, J., and de Zamaroczy, M. (2011). FtsH-dependent processing of RNase colicins D and E3 means that only the cytotoxic domains are imported into the cytoplasm. *J. Biol. Chem.* 286, 29397–29407. doi: 10.1074/jbc.M111.242354

Collins, E. S., Whittaker, S. B., Tozawa, K., MacDonald, C., Boetzel, R., Penfold, C. N., et al. (2002). Structural dynamics of the membrane translocation domain of colicin E9 and its interaction with TolB. *J. Mol. Biol.* 318, 787–804. doi: 10.1016/S0022-2836(02)00036-0

Cornforth, D. M., and Foster, K. R. (2013). Competition sensing: the social side of bacterial stress responses. *Nat. Rev. Microbiol.* 11, 285–293. doi: 10.1038/nrmicro2977

Cotter, P. D. (2014). An 'Upp'-turn in bacteriocin receptor identification. *Mol. Microbiol.* 92, 1159–1163. doi: 10.1111/mmi.12645

de Chial, M., Ghysels, B., Beatson, S. A., Geoffroy, V., Meyer, J. M., Pattery, T., et al. (2003). Identification of type II and type III pyoverdine receptors from *Pseudomonas aeruginosa*. *Microbiology* 149, 821–831. doi: 10.1099/mic.0.26136-0

de Zamaroczy, M., Mora, L., Lecuyer, A., Géli, V., and Buckingham, R. H. (2001). Cleavage of colicin D is necessary for cell killing and requires the inner membrane peptidase LepB. *Mol. Cell* 8, 159–168. doi: 10.1016/S1097-2765(01)00276-3

Delarue, M., and Koehl, P. (2018). Combined approaches from physics, statistics, and computer science for ab initio protein structure prediction: ex unitate vires (unity is strength)? *F1000Research* 7:F1000FacultyRev-1125. doi: 10.12688/f1000research.14870.1

Denayer, S., Matthijs, S., and Cornelis, P. (2007). Pyocin S2 (Sa) kills *Pseudomonas aeruginosa* strains via the FpvA type I ferripyoverdine receptor. *J. Bacteriol.* 189, 7663–7668. doi: 10.1128/JB.00992-07

Duche, D., Baty, D., Chartier, M., and Letellier, L. (1994). Unfolding of colicin A during its translocation through the *Escherichia coli* envelope as demonstrated by disulfide bond engineering. *J. Biol. Chem.* 269, 24820–24825.

Duquesne, S., Destoumieux-Garzon, D., Peduzzi, J., and Rebuffat, S. (2007). Microcins, gene-encoded antibacterial peptides from enterobacteria. *Nat. Prod. Rep.* 24, 708–734. doi: 10.1039/b516237h

Elfarash, A., Dingemans, J., Ye, L., Hassan, A. A., Craggs, M., Reimmann, C., et al. (2014). Pore-forming pyocin S5 utilizes the FptA ferripyochelin receptor to kill *Pseudomonas aeruginosa*. *Microbiology* 160, 261–269. doi: 10.1099/mic.0.070672-0

Elfarash, A., Wei, Q., and Cornelis, P. (2012). The soluble pyocins S2 and S4 from *Pseudomonas aeruginosa* bind to the same FpvAI receptor. *Microbiologyopen.* 1, 268–275. doi: 10.1002/mbo3.27

Farrance, O. E., Hann, E., Kaminska, R., Housden, N. G., Derrington, S. R., Kleanthous, C., et al. (2013). A force-activated trip switch triggers rapid dissociation of a colicin from its immunity protein. *PLoS Biol.* 11:e1001489. doi: 10.1371/journal.pbio.1001489

Germon, P., Ray, M. C., Vianney, A., and Lazzaroni, J. C. (2001). Energy-dependent conformational change in the TolA protein of *Escherichia coli* involves its N-terminal domain, TolQ, and TolR. *J. Bacteriol.* 183, 4110–4114. doi: 10.1128/JB.183.14.4110-4114.2001

Ghequire, M. G., and De Mot, R. (2014). Ribosomally encoded antibacterial proteins and peptides from *Pseudomonas*. *FEMS Microbiol. Rev.* 38, 523–568. doi: 10.1111/1574-6976.12079

Ghequire, M. G. K., Kemland, L., Anoz-Carbonell, E., Buchanan, S. K., and De Mot, R. (2017). A natural chimeric *Pseudomonas* bacteriocin with novel pore-forming activity parasitizes the ferrichrome transporter. *mBio* 8:e1961-16. doi: 10.1128/mBio.01961-16

Govan, J. R. W. (1978). Pyocin typing of *Pseudomonas aeruginosa*. *Methods Microbiol.* 10, 61–91. doi: 10.1016/S0580-9517(08)70660-3

Grinter, R., Milner, J., and Walker, D. (2012). Ferredoxin containing bacteriocins suggest a novel mechanism of iron uptake in *Pectobacterium* spp. *PLoS One* 7:e33033. doi: 10.1371/journal.pone.0033033

Héchard, Y., and Sahl, H. G. (2002). Mode of action of modified and unmodified bacteriocins from Gram-positive bacteria. *Biochimie* 84, 545–557. doi: 10.1016/S0300-9084(02)01417-7

Hecht, O., Ridley, H., Lakey, J. H., and Moore, G. R. (2009). A common interaction for the entry of colicin N and filamentous phage into *Escherichia coli*. *J. Mol. Biol.* 388, 880–893. doi: 10.1016/j.jmb.2009.03.035

Helbig, S., Patzer, S. I., Schiene-Fischer, C., Zeth, K., and Braun, V. (2011). Activation of colicin M by the FkpA prolyl cis-trans isomerase/chaperone. *J. Biol. Chem.* 286, 6280–6290. doi: 10.1074/jbc.M110.165274

Housden, N. G., Hopper, J. T., Lukoyanova, N., Rodriguez-Larrea, D., Wojdyla, J. A., Klein, A., et al. (2013). Intrinsically disordered protein threads through the bacterial outer-membrane porin OmpF. *Science* 340, 1570–1574. doi: 10.1126/science.1237864

Housden, N. G., and Kleanthous, C. (2011). Thermodynamic dissection of colicin interactions. *Methods Enzymol.* 488, 123–145. doi: 10.1016/B978-0-12-381268-1.00006-9

Housden, N. G., Loftus, S. R., Moore, G. R., James, R., and Kleanthous, C. (2005). Cell entry mechanism of enzymatic bacterial colicins: porin recruitment and the thermodynamics of receptor binding. *Proc. Natl. Acad. Sci. U.S.A.* 102, 13849–13854. doi: 10.1073/pnas.0503567102

Housden, N. G., Rassam, P., Lee, S., Samsudin, F., Kaminska, R., Sharp, C., et al. (2018). Directional porin binding of intrinsically disordered protein sequences promotes colicin epitope display in the bacterial periplasm. *Biochemistry* 57, 4374–4381. doi: 10.1021/acs.biochem.8b00621

Housden, N. G., Wojdyla, J. A., Korczynska, J., Grishkovskaya, I., Kirkpatrick, N., Brzozowski, A. M., et al. (2010). Directed epitope delivery across the *Escherichia coli* outer membrane through the porin OmpF. *Proc. Natl. Acad. Sci. U.S.A.* 107, 21412–21417. doi: 10.1073/pnas.1010780107

Jakes, K. S., and Finkelstein, A. (2010). The colicin Ia receptor, Cir, is also the translocator for colicin Ia. *Mol. Microbiol.* 75, 567–578. doi: 10.1111/j.1365-2958.2009.06966.x

Jamet, A., and Nassif, X. (2015). New players in the toxin field: polymorphic toxin systems in bacteria. *mBio* 6:e285-15. doi: 10.1128/mBio.00285-15

Johnson, C. L., Solovyova, A. S., Hecht, O., Macdonald, C., Waller, H., Grossmann, J. G., et al. (2017). The two-state prehensile tail of the antibacterial toxin colicin N. *Biophys. J.* 113, 1673–1684. doi: 10.1016/j.bpj.2017.08.030

Kim, Y. C., Tarr, A. W., and Penfold, C. N. (2014). Colicin import into *E. coli* cells: a model system for insights into the import mechanisms of bacteriocins. *Biochim. Biophys. Acta.* 1843, 1717–1731. doi: 10.1016/j.bpj.2017.08.030

Kjos, M., Oppegård, C., Diep, D. B., Nes, I. F., Veening, J.-W., Nissen-Meyer, J., et al. (2014). Sensitivity to the two-peptide bacteriocin lactococcin G is dependent on UppP, an enzyme involved in cell-wall synthesis. *Mol. Microbiol.* 92, 1177–1187. doi: 10.1111/mmi.12632

Kleanthous, C. (2010). Swimming against the tide: progress and challenges in our understanding of colicin translocation. *Nat. Rev. Microbiol.* 8, 843–848. doi: 10.1038/nrmicro2454

Kleanthous, C., Rassam, P., and Baumann, C. G. (2015). Protein-protein interactions and the spatiotemporal dynamics of bacterial outer membrane proteins. *Curr. Opin. Struct. Biol.* 35, 109–115. doi: 10.1016/j.sbi.2015.10.007

Klein, A., Wojdyla, J. A., Joshi, A., Josts, I., McCaughey, L. C., Housden, N. G., et al. (2016). Structural and biophysical analysis of nuclease protein antibiotics. *Biochem. J.* 473, 2799–2812. doi: 10.1042/BCJ20160544

Kurisu, G., Zakharov, S. D., Zhalnina, M. V., Bano, S., Eroukova, V. Y., Rokitskaya, T. I., et al. (2003). The structure of BtuB with bound colicin E3 R-domain implies a translocon. *Nat. Struct. Biol.* 10, 948–954. doi: 10.1038/nsb997

Lam, J. S., Taylor, V. L., Islam, S. T., Hao, Y., and Kocincova, D. (2011). Genetic and functional diversity of *Pseudomonas aeruginosa* lipopolysaccharide. *Front. Microbiol.* 2:118. doi: 10.3389/fmicb.2011.00118

Larsen, R. A., Deckert, G. E., Kastead, K. A., Devanathan, S., Keller, K. L., and Postle, K. (2007). His(20) provides the sole functionally significant side chain in the essential TonB transmembrane domain. *J. Bacteriol.* 189, 2825–2833. doi: 10.1128/JB.01925-26

Loftus, S. R., Walker, D., Maté, M. J., Bonsor, D. A., James, R., Moore, G. R., et al. (2006). Competitive recruitment of the periplasmic translocation portal TolB by a natively disordered domain of colicin E9. *Proc. Natl. Acad. Sci. U.S.A.* 103, 12353–12358. doi: 10.1073/pnas.0603433103

Lukacik, P., Barnard, T. J., Keller, P. W., Chaturvedi, K. S., Seddiki, N., Fairman, J. W., et al. (2012). Structural engineering of a phage lysin that targets gram-negative pathogens. *Proc. Natl. Acad. Sci. U.S.A.* 109, 9857–9862. doi: 10.1073/pnas.1203472109

Manon, F., and Ebel, C. (2010). "Analytical ultracentrifugation, a useful tool to probe intrinsically disordered proteins," in *Instrumental Analysis of Intrinsically Disordered Proteins*, eds V. Uversky and S. Longhi (Hoboken, NJ: John Wiley), 433–449.

Masi, D. R., White, J. C., Schnaitman, C. A., and Bradbeer, C. (1973). Transport of vitamin B12 in *Escherichia coli*: common receptor sites for vitamin B12 and the E colicins on the outer membrane of the cell envelope. *J. Bacteriol.* 115, 506–513.

Masi, M., Vuong, P., Humbard, M., Malone, K., and Misra, R. (2007). Initial steps of colicin E1 import across the outer membrane of *Escherichia coli*. *J. Bacteriol.* 189, 2667–2676. doi: 10.1128/JB.01448-06

McCaughey, L. C., Grinter, R., Josts, I., Roszak, A. W., Waløen, K. I., Cogdell, R. J., et al. (2014). Lectin-like bacteriocins from *Pseudomonas* spp. utilise D-rhamnose containing lipopolysaccharide as a cellular receptor. *PLoS Pathog.* 10:e1003898. doi: 10.1371/journal.ppat.1003898

McCaughey, L. C., Josts, I., Grinter, R., White, P., Byron, O., Tucker, N. P., et al. (2016). Discovery, characterization and in vivo activity of pyocin SD2, a protein antibiotic from *Pseudomonas aeruginosa*. *Biochem. J.* 473, 2345–2358. doi: 10.1042/BCJ20160470

Michel-Briand, Y., and Baysse, C. (2002). The pyocins of *Pseudomonas aeruginosa*. *Biochimie* 84, 499–510. doi: 10.1016/S0300-9084(02)01422-0

Mora, L., and de Zamaroczy, M. (2014). In vivo processing of DNase colicins E2 and E7 is required for their import into the cytoplasm of target cells. *PLoS One* 9:e96549. doi: 10.1371/journal.pone.0096549

Mora, L., Moncoq, K., England, P., Oberto, J., and de Zamaroczy, M. (2015). The stable interaction between signal peptidase LepB of *Escherichia coli* and nuclease bacteriocins promotes toxin entry into the cytoplasm. *J. Biol. Chem.* 290, 30783–30796. doi: 10.1074/jbc.M115.691907

Ohkawa, I., Shiga, S., and Kageyama, M. (1980). Effect of iron concentration in the growth medium on the sensitivity of *Pseudomonas aeruginosa* to pyocin S2. *J. Biochem.* 87, 323–331. doi: 10.1093/oxfordjournals.jbchem.a132740

Ovchinnikov, S., Kinch, L., Park, H., Liao, Y., Pei, J., Kim, D. E., et al. (2015). Large-scale determination of previously unsolved protein structures using evolutionary information. *eLife* 4:e09248. doi: 10.7554/eLife.09248

Papadakos, G., Housden, N. G., Lilly, K. J., Kaminska, R., and Kleanthous, C. (2012a). Kinetic basis for the competitive recruitment of TolB by the intrinsically disordered translocation domain of colicin E9. *J. Mol. Biol.* 418, 269–280. doi: 10.1016/j.jmb.2012.01.039

Papadakos, G., Wojdyla, J. A., and Kleanthous, C. (2012b). Nuclease colicins and their immunity proteins. *Q. Rev. Biophys.* 45, 57–103. doi: 10.1017/S0033583511000114

Penfold, C. N., Healy, B., Housden, N. G., Boetzel, R., Vankemmelbeke, M., Moore, G. R., et al. (2004). Flexibility in the receptor-binding domain of the enzymatic colicin E9 is required for toxicity against *Escherichia coli* cells. *J. Bacteriol.* 186, 4520–4527. doi: 10.1128/JB.186.14.4520-4527.2004

Pilsl, H., Smajs, D., and Braun, V. (1999). Characterization of colicin S4 and its receptor, OmpW, a minor protein of the *Escherichia coli* outer membrane. *J. Bacteriol.* 181, 3578–3581.

Postle, K., and Larsen, R. A. (2007). TonB-dependent energy transduction between outer and cytoplasmic membranes. *Biometals* 20, 453–465. doi: 10.1007/s10534-006-9071-76

Rassam, P., Copeland, N. A., Birkholz, O., Tóth, C., Chavent, M., Duncan, A. L., et al. (2015). Supramolecular assemblies underpin turnover of outer membrane proteins in bacteria. *Nature* 523, 333–336. doi: 10.1038/nature14461

Rassam, P., Long, K. R., Kaminska, R., Williams, D. J., Papadakos, G., Baumann, C. G., et al. (2018). Intermembrane crosstalk drives inner-membrane protein organization in *Escherichia coli*. *Nat. Commun.* 9:1082. doi: 10.1038/s41467-018-03521-24

Rocchetta, H. L., Burrows, L. L., Pacan, J. C., and Lam, J. S. (1998). Three rhamnosyltransferases responsible for assembly of the A-band D-rhamnan polysaccharide in *Pseudomonas aeruginosa*: a fourth transferase, WbpL, is required for the initiation of both A-band and B-band lipopolysaccharide synthesis. *Mol. Microbiol.* 28, 1103–1119. doi: 10.1046/j.1365-2958.1998.00871.x

Sabet, S. F., and Schnaitman, C. A. (1973). Purification and properties of the colicin E3 receptor of *Escherichia coli*. *J. Biol. Chem.* 248, 1797–1806.

Saeger, J., Hytönen, V. P., Klotzsch, E., and Vogel, V. (2012). GFP's mechanical intermediate states. *PLoS One* 7:e46962. doi: 10.1371/journal.pone.0046962

Sano, Y., Kobayashi, M., and Kageyama, M. (1993). Functional domains of S-type pyocins deduced from chimeric molecules. *J. Bacteriol.* 175, 6179–6185. doi: 10.1128/jb.175.19.6179-6185.1993

Sharma, O., Yamashita, E., Zhalnina, M. V., Zakharov, S. D., Datsenko, K. A., Wanner, B. L., et al. (2007). Structure of the complex of the colicin E2 R-domain and its BtuB receptor. The outer membrane colicin translocon. *J. Biol. Chem.* 282, 23163–23170. doi: 10.1074/jbc.M703004200

Sharp, C., Boinett, C., Cain, A., Housden, N., Kumar, S., Turner, K., et al. (2019). O-antigen dependent colicin insensitivity of uropathogenic *Escherichia coli*. *J. Bacteriol.* 201:e00545-18. doi: 10.1128/JB.00545-18

Sharp, C., Bray, J., Housden, N. G., Maiden, M. C. J., and Kleanthous, C. (2017). Diversity and distribution of nuclease bacteriocins in bacterial genomes revealed using hidden markov models. *PLoS Comput. Biol.* 13:e1005652. doi: 10.1371/journal.pcbi.1005652

Smajs, D., and Weinstock, G. M. (2001). The iron- and temperature-regulated cjrBC genes of *Shigella* and enteroinvasive *Escherichia coli* strains code for colicin Js uptake. *J. Bacteriol.* 183, 3958–3966. doi: 10.1128/JB.183.13.3958-3966.2001

Smith, A. W., Hirst, P. H., Hughes, K., Gensberg, K., and Govan, J. R. (1992). The pyocin Sa receptor of *Pseudomonas aeruginosa* is associated with ferripyoverdin uptake. *J. Bacteriol.* 174, 4847–4849. doi: 10.1128/jb.174.14.4847-4849.1992

Soelaiman, S., Jakes, K., Wu, N., Li, C., and Shoham, M. (2001). Crystal structure of colicin E3: implications for cell entry and ribosome inactivation. *Mol. Cell* 8, 1053–1062. doi: 10.1016/S1097-2765(01)00396-3

Thomas, J. A., and Valvano, M. A. (1993). Role of tol genes in cloacin DF13 susceptibility of *Escherichia coli* K-12 strains expressing the cloacin DF13-aerobactin receptor IutA. *J. Bacteriol.* 175, 548–552. doi: 10.1128/jb.175.2.548-552.1993

Vankemmelbeke, M., Housden, N. G., James, R., Kleanthous, C., and Penfold, C. N. (2013). Immunity protein release from a cell-bound nuclease colicin complex requires global conformational rearrangement. *Microbiologyopen* 2, 853–861. doi: 10.1002/mbo3.122

Vankemmelbeke, M., Shea, O. P., James, R., and Penfold, C. N. (2012). Interaction of nuclease colicins with membranes: insertion depth correlates with bilayer perturbation. *PLoS One* 7:e46656. doi: 10.1371/journal.pone.0046656

Vankemmelbeke, M., Zhang, Y., Moore, G. R., Kleanthous, C., Penfold, C. N., and James, R. (2009). Energy-dependent immunity protein release during *tol*-dependent Nuclease Colicin Translocation. *J. Biol. Chem.* 284, 18932–41. doi: 10.1074/jbc.M806149200

Walker, D., Mosbahi, K., Vankemmelbeke, M., James, R., and Kleanthous, C. (2007). The role of electrostatics in colicin nuclease domain translocation into bacterial cells. *J. Biol. Chem.* 282, 31389–31397. doi: 10.1074/jbc.M705883200

Wei, Y., Li, Z., Chen, B., Liang, H., and Duan, K. (2009). Characterization of the orf1-tolQRA operon in *Pseudomonas aeruginosa*. *Microbiol. Immunol.* 53, 309–318. doi: 10.1111/j.1348-0421.2009.00130.x

Weltzien, H. U., and Jesaitis, M. A. (1971). The nature of the cilicin K receptor of *Escherichia coli* Cullen. *J. Exp. Med.* 133, 534–553. doi: 10.1084/jem.133.3.534

White, P., Joshi, A., Rassam, P., Housden, N. G., Kaminska, R., Goult, J. D., et al. (2017). Exploitation of an iron transporter for bacterial protein antibiotic import. *Proc. Natl. Acad. Sci. U.S.A.* 114, 12051–12056. doi: 10.1073/pnas.1713741114

Wiener, M., Freymann, D., Ghosh, P., and Stroud, R. M. (1997). Crystal structure of colicin Ia. *Nature* 385, 461–464. doi: 10.1038/385461a0

Yamashita, E., Zhalnina, M. V., Zakharov, S. D., Sharma, O., and Cramer, W. A. (2008). Crystal structures of the OmpF porin: function in a colicin translocon. *EMBO J.* 27, 2171–2180. doi: 10.1038/emboj.2008.137

Zakharov, S. D., Eroukova, V. Y., Rokitskaya, T. I., Zhalnina, M. V., Sharma, O., Loll, P. J., et al. (2004). Colicin occlusion of OmpF and TolC channels: outer membrane translocons for colicin import. *Biophys. J.* 87, 3901–3911. doi: 10.1529/biophysj.104.046151

Zakharov, S. D., Wang, X. S., and Cramer, W. A. (2016). The colicin E1 TolC-binding conformer: pillar or pore function of TolC in colicin import? *Biochemistry* 55, 5084–5094. doi: 10.1021/acs.biochem.6b00621

Zampieri, M., Sekar, K., Zamboni, N., and Sauer, U. (2017). Frontiers of high-throughput metabolomics. *Curr. Opin. Chem. Biol.* 36, 15–23. doi: 10.1016/j.cbpa.2016.12.006

Zhang, Y., Li, C., Vankemmelbeke, M. N., Bardelang, P., Paoli, M., Penfold, C. N., et al. (2009). The crystal structure of the TolB box of colicin A in complex with TolB reveals important differences in the recruitment of the common TolB translocation portal used by group A colicins. *Mol. Microbiol.* 75, 623–636. doi: 10.1111/j.1365-2958.2009.06808.x

Zhang, Y., Vankemmelbeke, M. N., Holland, L. E., Walker, D. C., James, R., and Penfold, C. N. (2008). Investigating early events in receptor binding and translocation of colicin E9 using synchronized cell killing and proteolytic cleavage. *J. Bacteriol.* 190, 4342–4350. doi: 10.1128/JB.00047-8

8

eIF2α Confers Cellular Tolerance to *S. aureus* α-toxin

Gisela von Hoven[1†], Claudia Neukirch[1†], Martina Meyenburg[1], Sabine Füser[1], Maria Bidna Petrivna[1], Amable J. Rivas[1], Alexey Ryazanov[2], Randal J. Kaufman[3], Raffi V. Aroian[4] and Matthias Husmann[1*]*

[1] University Medical Center, Institute of Medical Microbiology and Hygiene, Johannes Gutenberg-University, Mainz, Germany, [2] Department of Pharmacology, Rutgers Robert Wood Johnson Medical School, Piscataway, NJ, USA, [3] Degenerative Diseases Program, Sanford-Burnham Medical Research Institute, La Jolla, CA, USA, [4] University of Massachusetts Medical School, Worcester, MA, USA

Edited by:
Katharina F. Kubatzky,
University Hospital Heidelberg,
Germany

Reviewed by:
Falko Hochgräfe,
University Greifswald, Germany
Ken Cadwell,
NYU School of Medicine, USA

***Correspondence:**
Matthias Husmann,
Mainz, Germany
husmann@uni-mainz.de;
Raffi V. Aroian,
Worcester, MA, USA
raffi.aroian@umassmed.edu

†Gisela von Hoven and Claudia Neukirch have contributed equally to this work.

We report on the role of conserved stress–response pathways for cellular tolerance to a pore forming toxin. First, we observed that small molecular weight inhibitors including of eIF2α-phosphatase, jun-*N*-terminal kinase (JNK), and PI3-kinase sensitized normal mouse embryonal fibroblasts (MEFs) to the small pore forming *S. aureus* α-toxin. Sensitization depended on expression of mADAM10, the murine ortholog of a proposed high-affinity receptor for α-toxin in human cells. Similarly, eIF2α$^{S51A/S51A}$ MEFs, which harbor an Ala knock-in mutation at the regulated Ser51 phosphorylation site of eukaryotic translation initiation factor 2α, were hyper-sensitive to α-toxin. Inhibition of translation with cycloheximide did not mimic the tolerogenic effect of eIF2α-phosphorylation. Notably, eIF2α-dependent tolerance of MEFs was toxin-selective, as wild-type MEFs and eIF2α$^{S51A/S51A}$ MEFs exhibited virtually equal sensitivity to *Vibrio cholerae* cytolysin. Binding of *S. aureus* α-toxin to eIF2α$^{S51A/S51A}$ MEFs and toxicity in these cells were enhanced as compared to wild-type cells. This led to the unexpected finding that the mutant cells carried more ADAM10. Because basal phosphorylation of eIF2α in MEFs required amino acid deprivation-activated eIF2α-kinase 4/GCN2, the data reveal that basal activity of this kinase mediates tolerance of MEFs to α-toxin. Further, they suggest that modulation of ADAM10 is involved. During infection, bacterial growth may cause nutrient shortage in tissues, which might activate this response. Tolerance to α-toxin was robust in macrophages and did not depend on GCN2. However, JNKs appeared to play a role, suggesting differential cell type and toxin selectivity of tolerogenic stress responses. Understanding their function or failure will be important to comprehend anti-bacterial immune responses.

Keywords: pore forming toxins, *S. aureus* α-toxin, cellular tolerance, EIF2AK4, MAPK

Introduction

Membrane perforation by pore forming toxins (PFT) is an ancient mode of attack employed by many bacteria, which helps them to establish or sustain infection (1–3). PFT represent a large group of bacterial toxins, which can be divided into various structural families (2). Many PFT have been discovered based on their ability not only to lyse red blood cells but they also affect nucleated cells, with effects ranging from induction of cell death to proliferation, time scales of occurrence from seconds to days after attack (4). Of particular relevance in the present context, cell autonomous defenses are in place to limit or reverse damage of nucleated target cells of PFT (4–13); they have been discussed as an integral part of the innate immune system, defending against bacteria (14). Work in *C. elegans* identified MAPK as master regulators of defense against PFT (7, 15). Whereas the importance of p38 MAPK is well established (3, 4, 9, 16–18), data on the role of jun-*N*-terminal kinases (JNKs) are somewhat conflicting (19–21). Large scale analyses of perforated cells have identified multiple additional changes taking place in response to PFT (15, 20, 22, 23), many of which appear to be triggered by the drop of cytosolic potassium (11, 20, 24, 25). Although the contribution of the various pathways to cell autonomous defense against PFT remains to be established in most cases, a basic concept emerges according to which removal of membrane pores (3, 10, 13, 26–32) and metabolic homeostasis (10, 12, 13, 20, 33, 34) are cornerstones of early cell autonomous defense against PFT. Importantly, mechanisms involved in pore removal depend on PFT and cell type (4, 8, 29). It appears that MAPK p38 and autophagy are required if the recovery process is prolonged as with *S. aureus* α-toxin and aerolysin (8, 20).

Phosphorylation of eukaryotic translation initiation factor 2 α (eIF2α) is a conserved stress–response activated by various PFT (12, 13, 20, 33–35). How this pathway impacts survival of target cells remains incompletely understood. Eukaryotic translation initiation requires assembly of a 43S ternary pre-initiation complex, consisting of met-t-RNAi(Met), eIF2, and GTP. In mammalian cells, this step is controlled through phosphorylation of eIF2α at serine 51 by 1 of 4 eIF2α-kinases (GCN2, PERK, PKR, and HRI), which respond to different types of stress (36). GCN2 serves as nutritional sensor, which is activated by uncharged t-RNAs (37, 38). Several lines of evidence indicate that membrane stress triggers this pathway: first, mutations that affect vesicular transport in yeast trigger phosphorylation of eIF2α (39). Second, in mammalian cells, plasma membrane perforation by bacterial PFT leads to activation of GCN2 (12, 33), phosphorylation of eIF2α, transient attenuation of translation, and activation of autophagy (12, 13, 20). Also, membrane damage by chlorpromazine or detergent triggers GCN2 (40). *S. aureus* α-toxin inhibited uptake of leucine by cells, providing an explanation for activation of GCN2 in target cells of PFT (12). In human epithelial cells, eIF2α, eIF2α-kinases and the regulatory eIF2α-phosphatase subunit CReP/Ppp1r15B are all required for efficient recovery from α-toxin-dependent loss of ATP (13). Surprisingly, these proteins served to remove membrane pores, thus, linking control of translation initiation and membrane traffic (13).

Conspicuously, many rodent cell types are not affected even by comparably high concentrations (micromolar range) of α-toxin; but the cause is not known. Receptor density on murine cells might be low, or murine ADAM10 might be an inefficient receptor as compared to its human counterpart. Alternatively, murine cells might be particularly tolerant to the consequences of successful attack. At any rate, to better understand results of *in vivo* experiments with *S. aureus* or α-toxin in mice, it is important to comprehend the mechanisms underlying tolerance of murine cells.

In ecoimmunology, "tolerance" denotes the ability of an organism to cope with high-pathogen load and resulting damage (41–43). To elucidate mechanisms of tolerance, it will be important to investigate the phenomenon at the cellular level using defined noxious agents. Here, we focus on cellular tolerance to PFT. A PFT may fail to cause overt damage to a cell if it cannot bind to, or attack, membranes in the first place. Alternatively, target cells may be able to cope with membrane perforation. In both cases, we consider the target cell "tolerant to the PFT." The term "susceptibility" shall denote responsiveness of a cell to a PFT as measured by loss of ATP or loss of potassium ions from the cytosol.

In the present work, we have investigated tolerance of mouse cells to *S. aureus* α-toxin. The results support a broader protective function of MAPK and document a cell type- and toxin-selective role of eIF2α.

Results

Normal Mouse Embryonal Fibroblasts are Tolerant to *S. aureus* α-Toxin

Exposure of human keratinocytes (HaCaT) to nanomolar concentrations of α-toxin for 2 h leads to significant loss of ATP. In contrast, murine keratinocytes (PDV) or mouse embryonal fibroblasts (MEFs) appeared to be not susceptible to α-toxin (**Figure 1A**). In contrast, MEFs were exquisitely susceptible to pVCC (**Figure 1B**), another PFT of the small β-barrel family (44).

Inhibitors of Various Signaling Pathways Sensitize MEFs for α-Toxin

One potential explanation for selective tolerance of MEF to α-toxin could be the lack of an appropriate receptor, or lower expression levels of receptor. Amino acid sequences of human and murine ADAM10, the proposed high-affinity receptor of α-toxin, are not identical. Western blots with an antibody against ADAM10 yielded weaker bands with the murine cells (**Figure 1C**). Although it cannot be excluded that the antibody binds human and murine cells with different efficiency, it is therefore possible that qualitative or quantitative differences in receptor expression could explain lower susceptibility of murine vs. human cells to α-toxin. Alternatively, however, efficient ongoing repair of membrane damage and fitness to cope with metabolic stress could also play a role. Therefore, we tested a panel of small molecular weight compounds that we knew to inhibit recovery from α-toxin-dependent ATP-depletion in HaCaT cells. Several (combinations of) inhibitors sensitized MEFs for α-toxin (**Figure 2A**). The effect depended on the concentration of inhibitor, as exemplified for JNK3XIISR3576 (**Figure 2B**). Although JNK3XIISR3576 is selective for JNK3 if applied at nanomolar concentrations, MAPK other than JNK3 seem to be involved here because MEFs do not express JNK3 (45). As shown for Sal/Dyn, inhibitors sensitized wild-type MEF for α-toxin, provided they expressed ADAM10 (**Figure 2C**). Thus, inhibitors did not sensitize cells by increasing unspecific toxicity of α-toxin.

Lack of Phosphorylatable eIF2α, GCN2, or Ppp1r15B Sensitizes MEFs for α-Toxin

That Salubrinal, an inhibitor of eIF2α-phosphatases (46) sensitized MEFs to α-toxin, and similar observations in

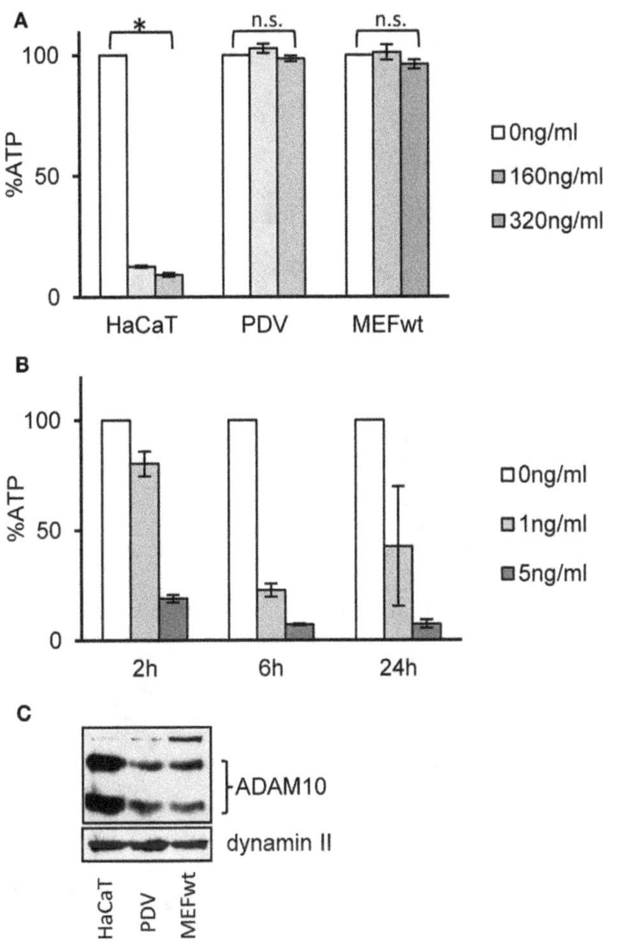

FIGURE 1 | Wild-type MEFs are tolerant to α-toxin. (A) HaCaT cells (human keratinocytes), PDV cells (murine keratinocytes), or MEFs wt cells (mouse embryonal fibroblasts) were treated or not with α-toxin. Cellular ATP levels were determined after 2 h; data are percent of untreated controls; mean values ±SE; $n \geq 4$; asterisk denotes $p \leq 0.05$. **(B)** MEFs wt cells were treated with indicated doses of pVCC and cellular ATP levels were determined after 2, 6, or 24 h; shown are percent of untreated controls; mean values ±SE; $n = 3$. **(C)** Cell lysates of wt, PDV, and HaCaT were analyzed by Western blot for ADAM10 and dynamin II (loading control).

keratinocytes (13), prompted us to investigate the response of MEFs with defined genetic modifications affecting expression or function of proteins involved in regulation of translation: first, in eIF2α$^{S51A/S51A}$ cells, the eIF2α locus is replaced with a non-phosphorylatable version, thus precluding regulated attenuation of translation via P-eIF2α. Second, GCN2$^{-/-}$ MEFs lack nutrient sensitive eIF2α-kinase GCN2/EIF2AK4, thereby blunting phosphorylation of eIF2α in response to amino acid deprivation. Third, Ppp1r15B$^{-/-}$ MEFs are devoid of the sole constitutive regulatory subunits of eIF2α-phosphatase (47); this defect leads to constitutively higher eIF2α-phosphorylation levels. Fourth, EeF2K$^{-/-}$ MEFs lack eukaryotic elongation factor 2 kinase (EeF2K), which functions downstream of mTOR to control protein synthesis (48).

First, we compared ATP levels in these MEFs to assess metabolic perturbation after treatment with α-toxin. EeF2K$^{-/-}$ cells were not susceptible to 10 μg/ml of α-toxin and thus behaved like wild-type MEFs. In contrast, lack of GCN2 or Ppp1r15B both sensitized cells to α-toxin. The strongest effect was observed with eIF2α$^{S51A/S51A}$ MEFs, which lost up to ~80% of ATP (**Figure 3A**). Although α-toxin led to dose-dependent loss of ATP in eIF2α$^{S51A/S51A}$ cells, there was barely any effect on wild-type cells (**Figure 3B**). Notably, pVCC-dependent loss of ATP was similar in wild-type or mutant MEFs (**Figure 3C**). Next, we determined the frequency of sub-G1 events, a measure of DNA fragmentation, in cells treated with α-toxin for 48 h. No significant difference was observed between wild-type and eIF2α$^{S51A/S51A}$ cells in the presence of β-ME and non-essential amino acids (NEAA). However, without these supplements, the number of sub-G1 events was doubled in eIF2α$^{S51A/S51A}$ cells (**Figure 3D**), although toxin-dependent loss of ATP and phosphorylation of eIF2α were equal in media with or without additives (data not shown). Additives thus appear to protect cells from secondary damage.

GCN2/EIF2AK4 Contains α-Toxin-Dependent Stress

Next, we assessed basal or toxin-dependent eIF2α-phosphorylation in the various MEFs lines. In wild-type cells, some basal phosphorylation of eIF2α was noted, which was moderately increased by α-toxin; after normalization for eIF2α, the effect was, however, statistically insignificant. As expected, P-eIF2α was not detected in eIF2α$^{S51A/S51A}$ MEFs (**Figure 4A**). No P-eIF2α was also discerned with samples of untreated GCN2$^{-/-}$ cells; this indicated that basal phosphorylation of eIF2α in cultured MEFs depends on GCN2. Paradoxically, however, α-toxin-dependent phosphorylation of eIF2α was enhanced in cells lacking this eIF2α-kinase. This showed that increased susceptibility of GCN2$^{-/-}$ MEFs to α-toxin (**Figure 3A**) cannot be accounted for by diminished toxin-dependent eIF2α-phosphorylation. Membranes were re-probed with antibodies against p38, which becomes phosphorylated in response to PFT. Strikingly, α-toxin caused robust phosphorylation of stress activated protein kinase p38 in GCN2$^{-/-}$ MEFs, Ppp1r15B MEFs, and eIF2α$^{S51A/S51A}$ MEFs (**Figure 4A**), but not in wild-type MEFs, supporting the notion that cells with imbalanced eIF2α-phosphorylation experienced more severe toxin-dependent stress. p70S6K, substrate of mTORC1 became de-phosphorylated (**Figure 4A**), indicating that α-toxin inhibits mTORC1, master regulator of translation and autophagy.

CHX Does Not Tolerize eIF2α$^{S51A/S51A}$ MEFs to α-Toxin

Results from Western blots (**Figure 4A**) raised the question whether α-toxin impacts translation differentially in eIF2α$^{S51A/S51A}$ vs. wild-type MEFs. Paradoxically, treatment with α-toxin caused dose-dependent attenuation of translation in eIF2α$^{S51A/S51A}$ MEFs, but had no such effect on wild-type MEFs (**Figure 4B**). This suggested that translation was inhibited in response to α-toxin through a mechanism that was independent of eIF2α-phosphorylation. Together with results from the foregoing ATP-assays, this also revealed that attenuation of translation *per se* is insufficient to maintain

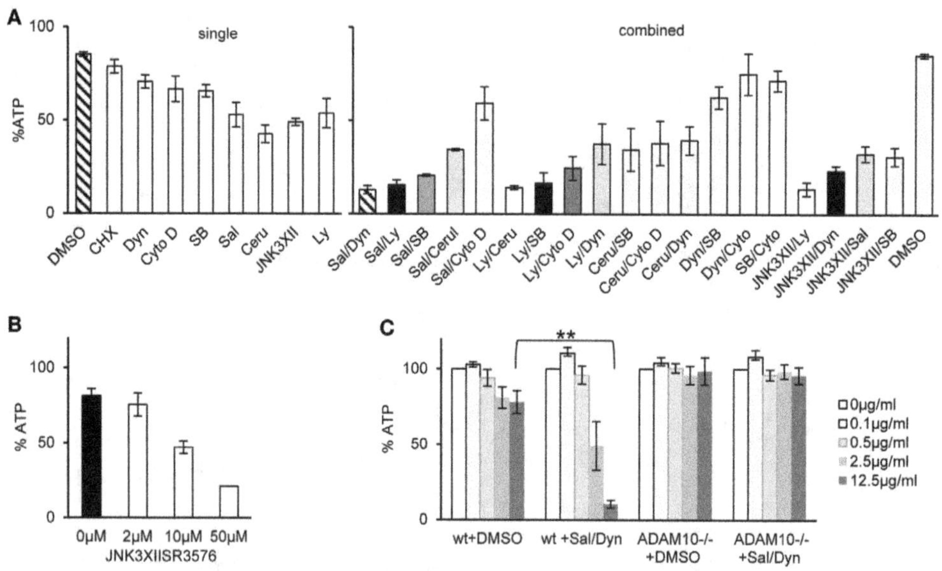

FIGURE 2 | Inhibitors of various pathways sensitize MEFs to α-toxin. **(A)** wt cells were treated as indicated either with single inhibitors or combinations of two inhibitors listed below and treated with 10 μg/ml α-toxin. Cellular ATP levels were determined after 2 h; data are percent of untreated controls; shown are mean values for single inhibitors ±SE with $n \geq 6$, or mean values of $n \geq 2$ for combinations of inhibitors, SE was <30% throughout. Inhibitor concentrations: Salubrinal (40 μM), Dynasore (80 μM), Ly29400L (100 μM), SB203580 (20 μM), Cerulenin (20 μM), Cytochalasin D (20 μM), and JNK3XIISR3576 (10 μM). **(B)** MEFs wt were treated with indicated doses of JNK3XIISR3576 and incubated with 10 μg/ml α-toxin. Cellular ATP levels were determined after 2 h; data are percent of controls treated with similar concentrations of JNK3XIISR3576 but not treated with α-toxin; mean values ±SE; $n \geq 3$. **(C)** MEFs wt or MEFs ADAM10$^{-/-}$ (mouse embryonal fibroblasts) were pretreated with 40 μM Salubrinal and 80 μM Dynasore or solvent alone for 30 min and treated or not with indicated doses of α-toxin. ATP levels were determined after 2 h; shown are percent of untreated controls; mean values ±SE; $n = 3$; two asterisks denote p-values ≤0.001.

metabolic homeostasis upon attack by α-toxin. Conversely, toxin-dependent ATP-loss in eIF2α$^{S51A/S51A}$ MEFs is not a consequence of translational arrest, because treatment of wild-type cells with CHX stops translation (**Figure 4B**), but does not hyper-sensitize MEFs for α-toxin (**Figure 4C**).

Wild-Type eIF2α Modulates Binding and Action of α-Toxin

That inhibition of translation did not protect eIF2α$^{S51A/S51A}$ MEFs provoked the question how phosphorylation of eIF2α tolerizes MEF. So, we investigated a primary event underlying many of the rapid molecular changes induced by PFT, i.e., disturbance of natural ion gradients (4). Loss of potassium appears to be one major trigger (11, 20, 24, 49). We measured loss of intracellular potassium by flame photometry. In line with results from ATP-assays, net loss of potassium was enhanced in GCN2$^{-/-}$ or Ppp1r15B$^{-/-}$ cells and even more so in eIF2α$^{S51A/S51A}$ MEFs (**Figure 5A**), suggesting that membrane damage was more severe in cells with deficiencies in regulation of eIF2α-phosphorylation.

Pore formation by α-toxin depends on oligomerization and insertion into the plasma membrane. Because oligomers resist SDS, they can be detected by SDS-PAGE. Using cell surface labeling after incubation with internally radio-labeled α-toxin, we compared the amount of α-toxin on the surface of wild-type MEFs and eIF2α$^{S51A/S51A}$ MEFs. The fluorographic analysis shown in **Figure 5B** reveals that more α-toxin is present at the surface of eIF2α$^{S51A/S51A}$ MEFs as compared to wild-type cells, providing a straightforward explanation for enhanced loss of potassium and ATP from these cells.

Wild-Type eIF2α Modulates Expression of ADAM10

Increased amounts of toxin associated with eIF2α$^{S51A/S51A}$ MEFs could be due to higher abundance of α-toxin receptors. Therefore, we compared ADAM10 expression in eIF2α$^{S51A/S51A}$ MEFs and wild-type MEFs. Western-blot analysis revealed that ADAM10 is over-expressed in eIF2α$^{S51A/S51A}$ MEFs; and more ADAM10 is exposed on the cell surface of eIF2α$^{S51A/S51A}$ MEFs as compared to wild-type cells (**Figure 5C**). No significant differences between wild-type cells and eIF2α$^{S51A/S51A}$ MEFs were found in a lipidomics analysis (data not shown). Treatment with α-toxin led to down-regulation of ADAM10 at the cell surface of both wild-type and mutant cells. Because basal eIF2α-phosphorylation in MEFs depends on GCN2 (**Figure 4A**), the collective data indicate that nutrient stress or basal activity of GCN2 modulates levels of ADAM10 in MEFs.

Role of GCN2 for Tolerance to α-Toxin is Not Conserved in BMDM

Like MEFs, BMDM proved to be highly tolerant to α-toxin although they were susceptible to *Vibrio cholera* cytolysin (VCC) (**Figure 6A**). However, lack of GCN2-expression in BMDM did not significantly alter susceptibility to α-toxin (**Figures 6A,D**); α-toxin-dependent phosphorylation of eIF2α appeared slightly reduced (**Figure 6B**). Further, infection of BMDM by *S. aureus*

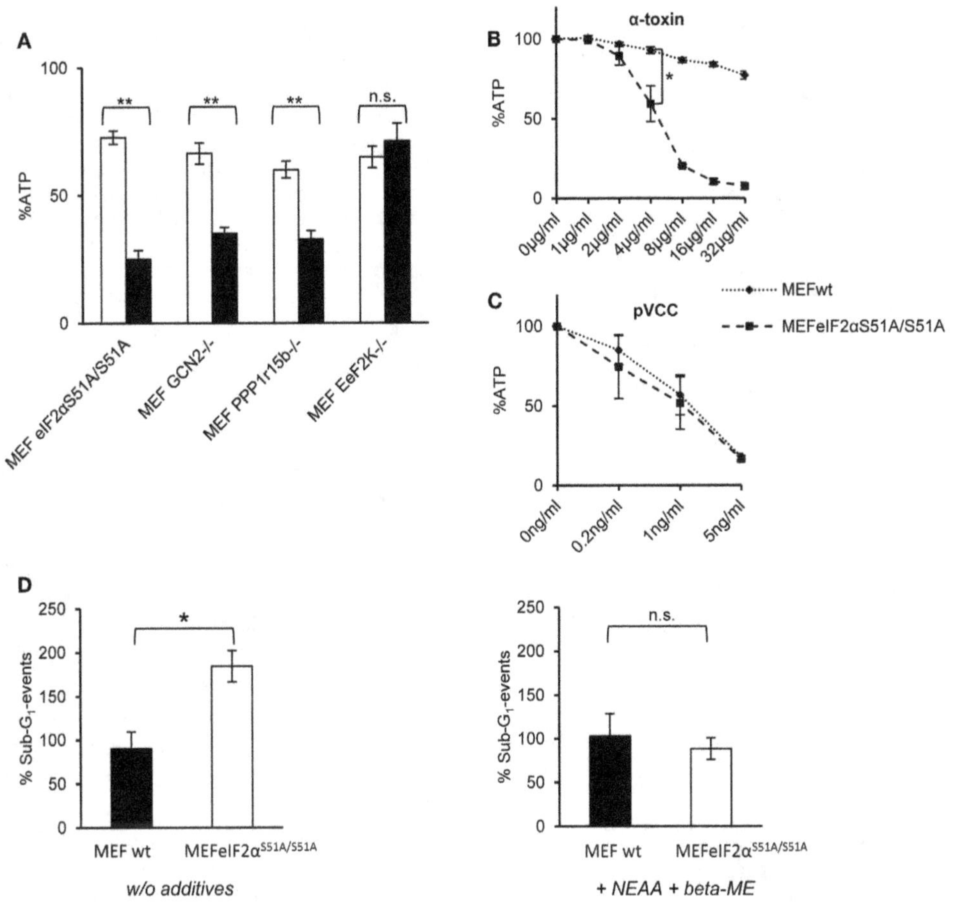

FIGURE 3 | Dysregulation of eIF2α-phosphorylation increases sensitivity for α-toxin. **(A)** MEFs eIF2$\alpha^{S51A/S51A}$, MEFs GCN2$^{-/-}$, MEFs Ppp1r15b$^{-/-}$, or MEFs EeF2K$^{-/-}$ were treated or not with 10 µg/ml α-toxin. ATP levels were determined after 2 h; shown are percent of untreated controls; mean values ±SE; $n \geq 4$. Black bars show data with MEFs cell variants as indicated; white bars corresponding control cells; two asterisks denote p-values $\leq$0.001. **(B)** MEFs wt or eIF2$\alpha^{S51A/S51A}$ were treated, or not with indicated doses of α-toxin. Cellular ATP levels were determined after 2 h; data are percent of untreated control; mean values ±SE; $n = 5$; asterisk: $p = 0.026$. **(C)** MEFs wt or MEFs eIF2$\alpha^{S51A/S51A}$ treated or not with indicated doses of pVCC. Cellular ATP levels were determined after 2 h; data are percent of untreated control; mean values ±SE; $n = 3$. **(D)** MEFs wt or MEFs eIF2$\alpha^{S51A/S51A}$ were cultured in standard media or in presence of β-ME and additional non-essential amino acids and treated with 10 µg/ml α-toxin for 48 h, stained with Propidium iodide and subsequently frequency of Sub-G$_1$-DNA was determined; data show percent of untreated controls; mean values ±SE; $n = 3$; asterisk denotes $p \leq 0,05$.

was equally efficient with both strains whether or not bacteria produced toxin or not (**Figure 6C**). Therefore, GCN2 seemed to play no major role for resistance or tolerance to α-toxin of BMDMs.

Small MW inhibitors of JNK and dynamin, a cocktail, which efficiently breaks tolerance to α-toxin in MEFs, moderately sensitized BMDM to purified α-toxin (**Figure 6D**). Inhibitors did not enhance ADAM10 expression (**Figure 6E**). Thus, MAPK enhance natural tolerance of BMDM to α-toxin, but modulation of ADAM10 expression appears not to be involved.

Discussion

One conclusion of this work is that regulators of translation initiation (eIF2α, GCN2, and Ppp1r15B) render murine embryonic fibroblasts tolerant to *S. aureus* α-toxin. This is consistent with our previous finding in human epithelial cells that these proteins promote recovery from successful attack (13). Although it remains to be investigated whether *a priori* tolerance of MEFs to α-toxin is likewise based on efficient endocytic removal of α-toxin pores from the cell surface, the present results document that the tolerogenic effect of eIF2α, GCN2, and Ppp1r15B is conserved in mice and man, and that it is observable in fibroblasts. That both lack of an eIF2α-kinase and of constitutive eIF2α-phosphatase reduce tolerance to α-toxin could be interpreted to show that balanced phosphorylation of eIF2α, or cycling of eIF2α between phosphorylated and unphosphorylated state is required for the tolerogenic effect.

A role of translational regulation for various aspects of innate immunity, and a protective function against PFT has been discussed in the recent literature (34, 50, 51). One established function of eIF2α-phosphorylation and the integrated stress–response is to reprogram expression of genes, including of genes that regulate immunity. Translational attenuation in host cells might

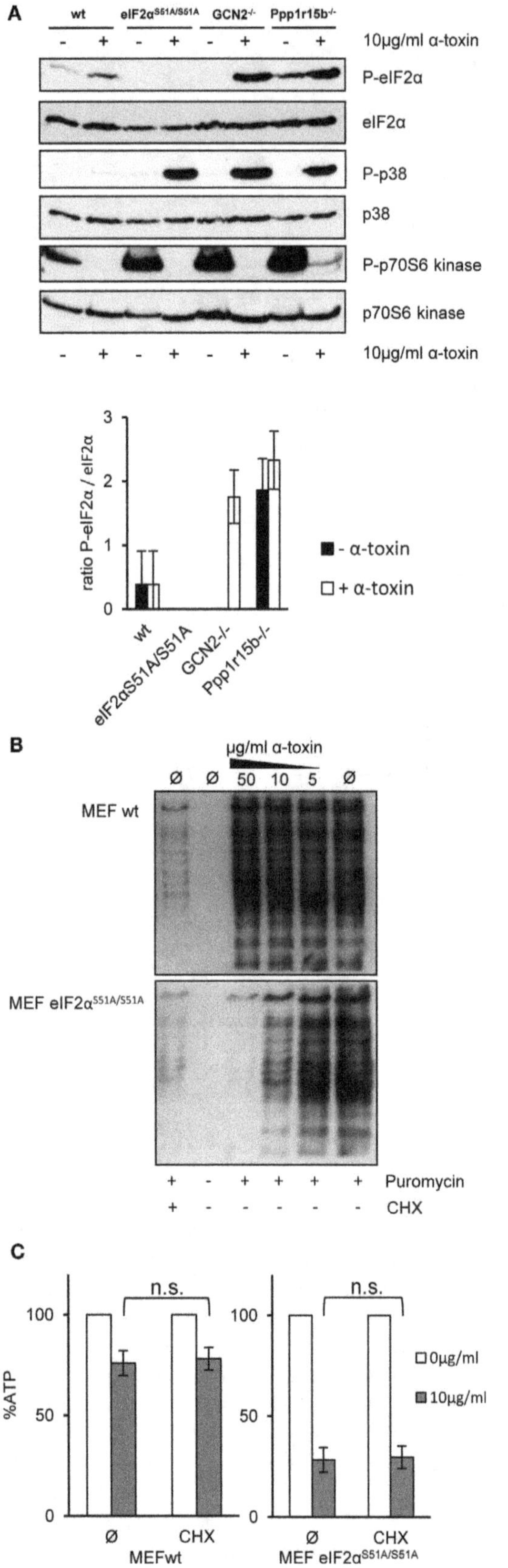

FIGURE 4 | α-toxin causes translational arrest in MEFs eIF2α$^{S51A/S51A}$.

(Continued)

FIGURE 4 | Continued
Cycloheximide does not sensitize wild-type MEFs **(A)** MEFs wt, MEFs eIF2α$^{S51A/S51A}$, MEFs GCN2$^{-/-}$, or MEFs Ppp1r15b$^{-/-}$ were treated or not with 10 μg/ml α-toxin for 2 h. Cell lysates were analyzed by Western blot for P-eIF2α, eIF2α, P-p70S6 kinase, p70S6 kinase P-p38, and p-38. One of three similar blots is shown; lower panel summarizes densitometric data for (P)-eIF2α, mean ± SE; $n = 3$ **(B)** MEFs wt or MEFs eIF2α$^{S51A/S51A}$ cells were treated or not with indicated concentrations of α-toxin and incubated for 1 h at 37°C. Treatment with CHX served as positive control for translational arrest. Subsequently, the cells were incubated for 1 h at 37°C with 10 μg/ml puromycin, which incorporates during ongoing synthesis into nascent proteins. Eventually, cells were analyzed by Western blot for puromycin. **(C)** MEFs wt or MEFs eIF2α$^{S51A/S51A}$ cells were treated or not with 10 μg/ml α-toxin and incubated with or without CHX at 37°C. Cellular ATP levels were determined after 2 h; data are percent of untreated controls; mean values ±SE; $n = 5$.

also help them to conserve energy, as proposed by some authors (20). However, (P)-eIF2α-dependent tolerance of MEF could not be explained by toxin-dependent translational attenuation *per se*: actually, α-toxin caused inhibition of translation in eIF2α$^{S51A/S51A}$ MEFs, but not in wild-type MEFs, and CHX did neither protect nor hyper-sensitize wild-type cells from/for α-toxin. Yet, eIF2α$^{S51A/S51A}$ MEFs proved to be significantly less tolerant to α-toxin. Probably as a consequence, protein synthesis was halted through alternative pathways. This could occur, for instance, by deactivation of mTOR, as indeed suggested by α-toxin-dependent dephosphorylation of p70S6K, a target of mTOR. The data may reflect a hierarchy of stress responses: Activation of eIF2α via nutritional sensor GCN2 contains α-toxin-dependent damage and stress, which would otherwise lead to exaggerated hyper-phosphorylation of eIF2α by (an)other eIF2α-kinase(s), possibly PERK and PKR. If eIF2α-phosphorylation fails, translation would be halted via robust deactivation of mTOR. Sustained inhibition of translation is obviously not tolerated by cells (35), but marked α-toxin-dependent inhibition of translation, observed in eIF2α$^{S51A/S51A}$ MEFs, does not explain the increased loss of ATP and potassium from these cells, because CHX did not affect these parameters in wild-type cells.

The fact that lack of phosphorylatable eIF2α was associated with increased toxin-dependent loss of potassium and ATP prompted us to measure the binding of α-toxin and expression of ADAM10, the proposed high-affinity receptor for α-toxin (52). This led to the unexpected finding that ADAM10 levels were significantly higher in eIF2α$^{S51A/S51A}$ cells. Therefore, tolerance to α-toxin in MEFs may be well due to GCN2/P-eIF2α-dependent modulation of its receptor, ADAM10. Whether or not the apparent link between basal nutrient stress and expression of ADAM10 was shaped by co-evolution of *S. aureus* and humans, it may also bear on functions of ADAM10 that are not related to infection.

How basal phosphorylation of eIF2α-levels modulates ADAM10-expression remains to be elucidated. Because P-eIF2α is required for starvation-dependent autophagy (53), a possible role of eIF2α in this context could be to maintain basal autophagic flux, which in turn could impact ADAM10 levels. Recently, Atg16L1, a protein essential for autophagy, has been shown to confer tolerance to α-toxin; the authors proposed that autophagy constitutively dampens ADAM10 levels in a cell-type selective manner (54). Similarly, eIF2α-dependent modulation

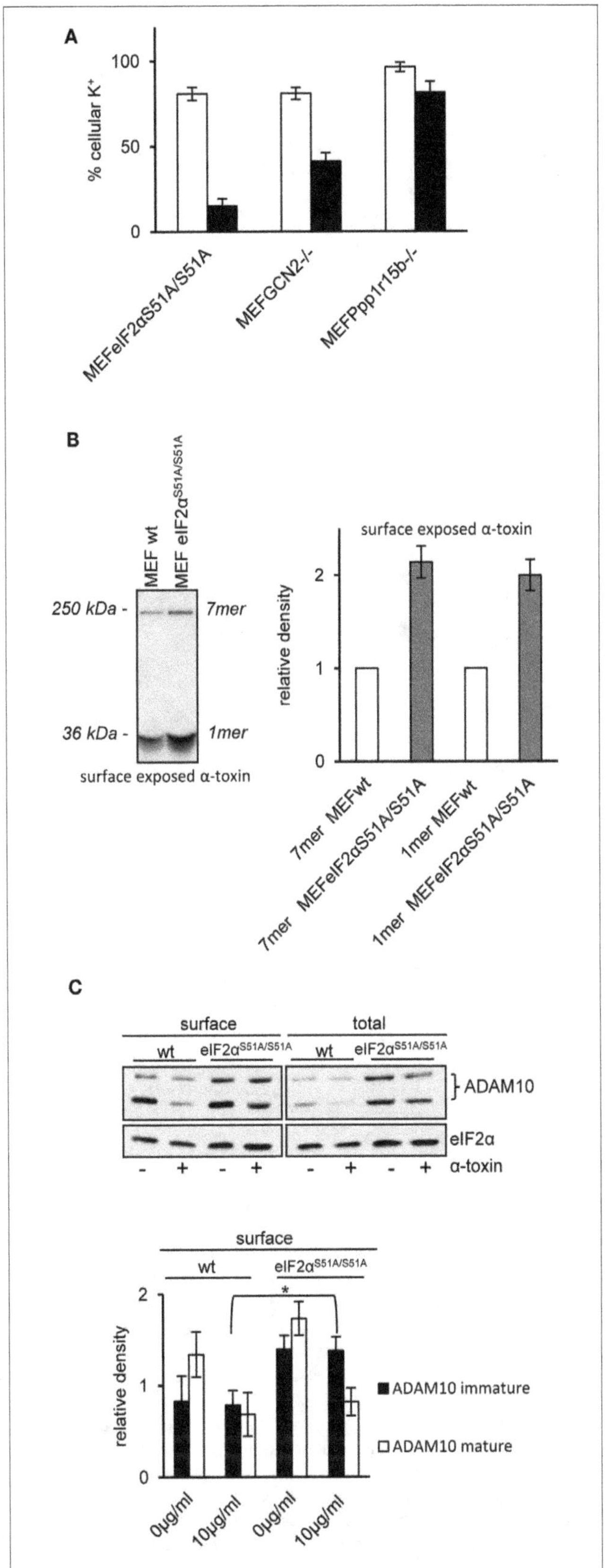

FIGURE 5 | MEFs eIF2α$^{S51A/S51A}$ over-express ADAM10 and bind higher amounts of α-toxin.

(Continued)

FIGURE 5 | Continued

(A) MEFs eIF2α$^{S51A/S51A}$, MEFs GCN2$^{-/-}$, MEFs Ppp1r15b$^{-/-}$, or corresponding control cells were treated or not with 10 µg/ml α-toxin. Potassium levels were determined after 2 h, shown are percent of untreated controls; mean values ±SE; $n \geq 4$. Black bars show data of MEFs cell variants as indicated; white bars corresponding control cells. **(B)** Right: MEFs wt or MEFs eIF2α$^{S51A/S51A}$ cells were incubated with radio-labeled α-toxin 2 and 8 µg/ml α-toxin at 37°C for 15 min. Subsequently, cells were surface biotinylated, lysates were obtained and subjected to sequential neutravidin-pulldown (NP) and immunoprecipitation (IP) followed by PAGE/fluography, as described in Kloft et al. (13). Left: band intensities were measured by densitometry using ImageJ software. Shown are mean values ±SE; $n = 4$. Variations of loading with toxin and plating of cells were <1 and <10%, respectively. **(C)** MEFs wt or MEFs eIF2α$^{S51A/S51A}$ were treated or not with 10 µg/ml α-toxin for 2 h. Subsequently, cells were surface biotinylated or not and lysed. Lysates were subjected to NP; both lysates and precipitation were analyzed by Western blot for ADAM10 and eIF2α (loading control); upper panel: one of four similar blots; lower panel: bar chart summarizing data (mean ± SE; $n = 4$).

of ADAM10 shown in the present work is constitutive and cell-type selective. Together this seems to suggest that basal eIF2α-phosphorylation functions upstream of autophagy to mediate tolerance. Alternatively, P-eIF2α, eIF2α-kinases, and -phosphatases could function through mechanisms acting in parallel to autophagy, for instance, by regulating endocytosis of membrane pores, as has been shown in epithelial cells (13). Whatever the effector mechanism(s) downstream of eIF2α are, our data reveal that GCN2-dependent basal phophorylation of eIF2α in MEF modulates ADAM10 levels as well as binding and action of α-toxin. Because GCN2 is activated by low levels of amino acids in cells, basal nutrient stress might be the driving force; the potential links discussed here are summarized in a model (**Figure 7**).

Inhibitors of various pathways hyper-sensitized MEFs to *S. aureus* α-toxin, which supports the notion that multiple signaling pathways are required to confer cellular tolerance to α-toxin. This raises the possibility that drugs used in pharmaco-therapy may have potential tolerance-modulating effects, an issue that warrants further investigation. The present data support a tolerogenic role of JNKs in both MEFs and BMDM. More work is required to understand the underlying mechanisms, but they seem to be distinct from eIF2α-dependent cellular tolerance to α-toxin. Conserved stress responses may fail to protect against some PFT, as exemplified here by *V. cholerae* cytolysin. Whether cells of the immune system tolerate attack of a given PFT will co-determine the ability of an organism to mount an effective "classic" immune response to infection with corresponding bacteria.

Materials and Methods

Antibodies

Antibodies against P-eIF2α (phosphorylated at Ser51), eIF2α, P-p38 (phosphorylated at Thr180/Tyr182), p-38, P-p70S6K (phosphorylated at Thr389), p70S6K, ADAM10, and α-tubulin were from Cell Signaling Technology. Antibodies against dynamin II were purchased from Santa-Cruz Biotechnology. Anti-Puromycin-antibody was from Merck Millipore. Antibodies against LC3 were bought from Sigma. HRP-conjugated secondary

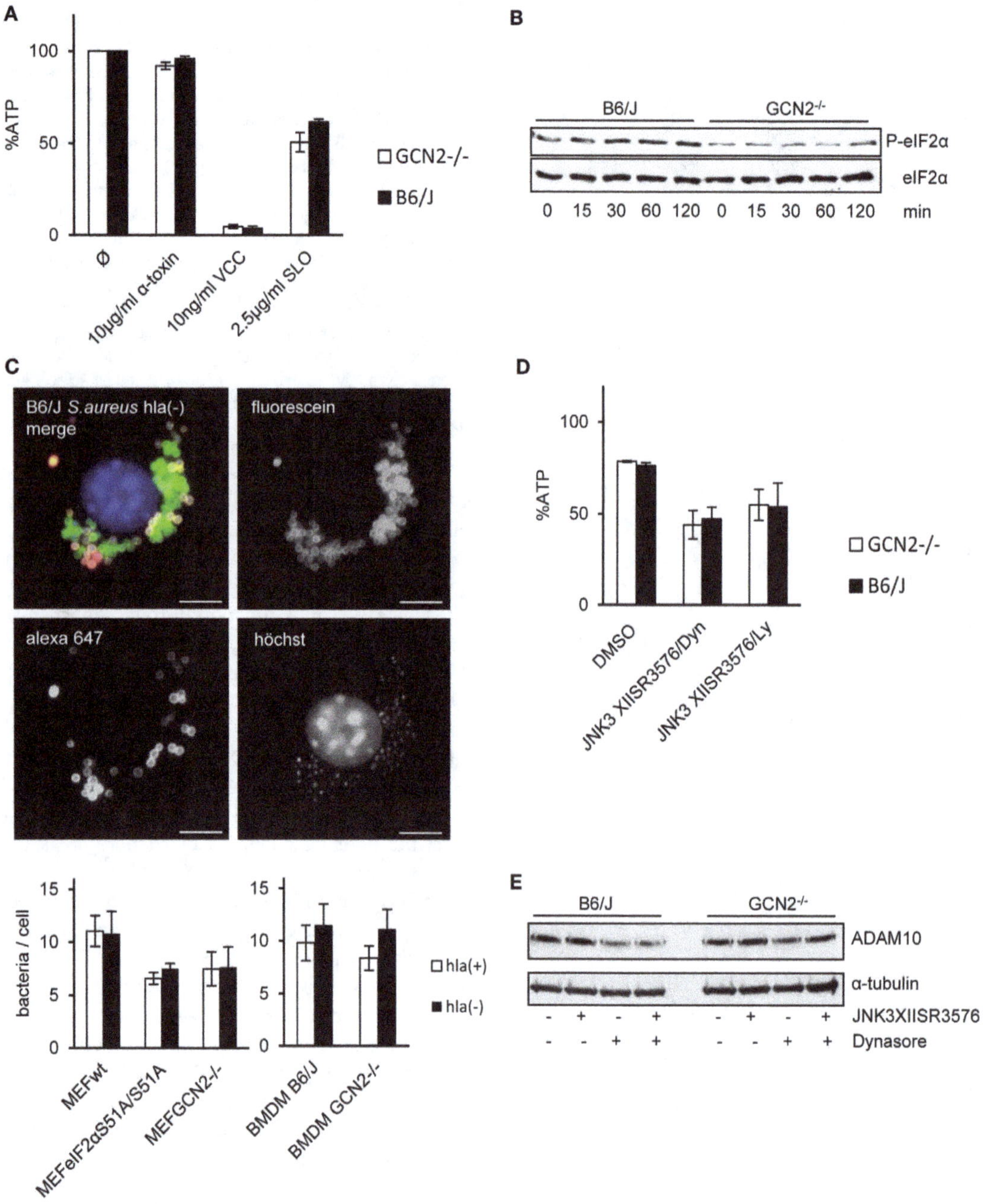

FIGURE 6 | GCN2 does not render BMDMs tolerant to α-toxin. **(A)** BMDMs isolated from GCN2$^{-/-}$ mice or control mice (B6/J) were treated or not with indicated doses of α-toxin, VCC, or SLO. Cellular ATP levels were determined after 2 h. White bars show BMDMs from GCN2$^{-/-}$ mice and black bars show BMDMs from control mice; mean values ±SE, $n \geq 3$ **(B)** BMDMs of GCN2$^{-/-}$ mice or control mice were incubated with 10 µg/ml α-toxin for indicated times. Cell lysates were analyzed by Western blot for P-eIF2α and eIF2α. **(C)** MEFs variants (right graph) or BMDMs of GCN2$^{-/-}$ and control mice (left graph) were incubated with fluorescein/biotin labeled *S. aureus* hla(−) or hla(+) strains (MOI 1:30) for 1 h, washed and incubated for an additional hour at 37°C. After fixation, extracellular bacteria were stained with streptavidin-coupled Alexa 647. Upper: exemplary picture of solely fluorescein-stained *S. aureus*, representing intracellular bacteria, and extracellular *S. aureus* that were accessible for Alexa 647-Streptavidin. Lower: graphs show counts of solely fluorescein-stained (intracellular) bacteria per cell; mean values ±SE; BMDMs $n = 2$; MEFs $n = 3$. **(D)** BMDMs of GCN2$^{-/-}$ mice or control mice were incubated with combinations of JNK3XIISR3576 (10 µM), Dynasore (80 µM), JNK3 XIISR3576 (10 µM), and Ly29400L (100 µM) or solvent alone and treated with 10 µg/ml α-toxin. Cellular ATP levels were determined after 2 h (percentage of controls). (Mean values ±SE; $n = 3$). **(E)** BMDMs of GCN2$^{-/-}$ and control mice were treated with the combination of Dynasore (80 µM) and JNK3XIISR3576 (10 µM), Dynasore (80 µM) alone, JNK3XIISR3576 (10 µM) alone, or solvent alone (DMSO) for 2.5 h. Cell lysates were analyzed by Western blot for ADAM10 and α-tubulin (loading control).

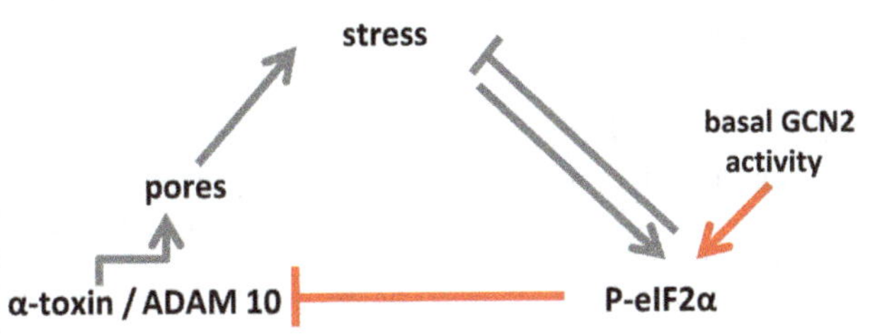

FIGURE 7 | Model of eIF2α-dependent cellular tolerance to α-toxin. Disruption of membrane integrity by insertion of membrane pores causes various forms of stresses in target cells (e.g., loss of potassium, starvation), which trigger an array of conserved responses, including phosphorylation of MAPK and of eIF2α. Not only may these pathways feed back to alleviate stress but also eIF2α may modulate formation of *S. aureus* α-toxin membrane pores (this work; highlighted in red in the scheme), or persistence of lesions (13), root causes of α-toxin-dependent stress. Notably, basal activity of GCN2 maintains low levels of ADAM10, resulting in bated binding of α-toxin. Thus, basal nutrient stress in cells could serve as a preemptive stimulus of cellular tolerance to *S. aureus* α-toxin. This link, which might have evolved from mutual adaptation of *S. aureus* and humans, is selective, as suggested by the fact that *Vibrio cholerae* cytolysin breaks cell autonomous defense.

antibodies were from Santa-Cruz Biotechnology (mouse) and Cell Signaling Technology (rabbit).

Inhibitors

Salubrinal (SAL), Cycloheximid (CHX), JNK3XIISR3576 (JNK3XII), and SB203580 (SB) were obtained from Calbiochem. Ly29400L (Ly) was from Cell Signaling. Bafilomycin, Cytochalasin D (Cyto D), and Cerulenin (Ceru) were from Sigma and Dynasore (Dyn) was from Tocris bioscience.

Chemicals

RNase, propidium iodide, and puromycin were purchased from Sigma. NHS-fluorescein and EZ-Link Sulfo-NHS-LC-biotin were from Thermo Fisher Scientific and Streptavidin-Alexa 647 was obtained from Molecular Probes. Rapamycin was from Calbiochem.

Toxins

α-toxin, internally radio-labeled α-toxin, streptolysin (SLO), and VCC were made as published elsewhere (10, 12).

S. aureus

In this study, *S. aureus* strain DU1090 (55) and α-toxin producing *S. aureus* strain, plasmid transformed derivative of DU1090 (55), were employed and referred to as hla(−) and hla(+), respectively.

Cells and Culture

Mouse embryonal fibroblasts GCN2$^{-/-}$ (56) were purchased from ATCC and PDV from CLS Cell Lines Service GmbH. MEFs eIF2α$^{S51A/S51A}$ and corresponding control cells MEFs eIF2α$^{S51/S51}$(57), MEFs Ppp1r15b$^{-/-}$ (58), MEFs EeF2K$^{-/-}$ (59), and MEFs ADAM10$^{-/-}$ (60) were kindly provided by Heather Harding and David Ron, Randal Kaufman, Alexey Ryazanov, and Paul Saftig, respectively. All MEFs cell lines were cultured in DMEM GlutaMAX™-I medium with 10% fetal calf serum, 1% HEPES buffer, 1% penicillin/streptomycin, 1% MEM NEAA, and 55 mM 2-mercaptoethanol. Under these conditions, we did not note significant differences in morphology, viability, or growth rate. PDV (murine keratinocyte cell line) were grown in DMEM GlutaMAX™-I medium with 10% fetal calf serum, 1% HEPES buffer, 1% penicillin/streptomycin without MEM NEAA, and 2-mercaptoethanol. HaCaT (non-virally transformed HaCaT) (61) were cultured in DMEM/F-12 GlutaMAX™-I medium with 10% fetal calf serum, 1% HEPES buffer, and 1% penicillin/streptomycin. All media and medium additives were obtained from Gibco by life technologies™. BMDMs were isolated from C57BL/6J (B6/J) or B6.129S6-EIF2αk4$^{tm1.2Dron/j}$ (GCN2$^{-/-}$) mice (Charles River laboratories) and cultured in DMEM GlutaMAX™-I supplemented with 20% fetal calf serum and M-CSF by adding 10% supernatant of L929 cells.

Western Blot

For Western blots, cell lysates were mixed with 2× SDS-loading buffer [65 mM Tris, 10% (v/v) glycerol, 5% (v/v) 2-mercaptoethanol, 2% (w/v) SDS, and bromphenol blue] and boiled for 5 min at 95°C. Proteins were separated by SDS-PAGE (10%) and electroblotted onto nitrocellulose membrane. After blocking for 1 h at room temperature in BSA or skim milk in TBST [Tris 50 μM, NaCl 0.15M, Tween 0.1% (v/v)], membrane was incubated with a primal antibody in BSA or skim milk in TBST, washed three times in TBST and incubated with HRP-conjugated second antibody for 1 h at room temperature. After three washing steps, bound antibody was detected by ECL (Roche Applied Science).

Puromycin Incorporation

The assay has been described elsewhere (62). MEFs eIF2α$^{S51A/S51A}$ cells were seeded at a density of 2×10^5 cells/well into six-well plates. Cells were incubated as indicated with several concentrations of pVCC or α-toxin for 1 h at 37°C. Thereafter, 10 μg/ml puromycin was added and cells were incubated for an additional hour at 37°C. Subsequently, medium was removed; cells were washed with PBS and analyzed by Western blot using anti-Puromycin-antibody in 2.5% (w/v) skim milk in TBST.

ATP-Measurements

Measurements of intracellular ATP were performed, as described elsewhere (22).

Potassium Efflux Measurements

For measurements of intracellular potassium, cells were seeded at a density of 3×10^5 cells/well into six-well plates. Cells were incubated or not with 8 μM Rapamycin for 3 h before adding 10 μg/ml α-toxin. After 2 h incubation at 37°C, cells were washed three times on ice with cold Choline-chlorid-buffer and solubilized in 2 ml Choline-chlorid-buffer containing 0.5% Triton X-100 (63) for 30 min at room temperature under constant shaking. Potassium content of supernatants of centrifuged cell lysates was determined using Sherwood single channel flame photometer M401.

Quantitation of Surface-Exposed or Internalized α-Toxin/ADAM10

Surface labeling, neutravidin pull-down, and fluorometric detection of α-toxin have been described previously (13). Variation of input (labeled S35-Met-α-toxin) was <1%; equal amounts of total protein were loaded.

Quantification of Internalized Bacteria

The assay has been performed, as described elsewhere (64).

Sub-G1 Analysis

Cells were seeded at a density of 1×10^5 cells/well into six-well plates and treated or not with 10 µg/ml α-toxin permanently for 48 h at 37°C. Cells were harvested including the detached cells in the supernatant, washed in PBS and centrifuged. Cells were resuspended and fixed in cold 70% ethanol in PBS/EDTA for 2 h at 4°C. Subsequently, the cells were washed and treated with RNase for 10 min at 37°C. Then, propidium iodide was added and cells were incubated for 5 min at room temperature. Eventually, cells were analyzed using a FACScan flow cytometer (BD Biosciences) and CellQuest software.

Statistical Analysis

Two-sided, unpaired Student's *t*-test was employed to assess statistical significance of differences between mean values. Significance was assumed at $p < 0.05$.

Acknowledgments

Research reported in this publication was supported by the National Institutes of Health NIH/NIFMS under award number 7-R0I-GM071603-09 to RA and MH (subrecipient); DK042394, DK088227, and HL052173 to RK; MH and GH received support by University Medical Center Mainz. We like to thank Dr. H. Harding and Dr. D. Ron for generously providing Ppp1r15b$^{-/-}$ MEFs, Dr. P. Saftig for ADAM10$^{-/-}$ MEFs, and Dr. B. Bruegger for a large scale lipid analysis. Further, we gratefully acknowledge support by Dr. A. Diefenbach.

References

1. Bhakdi S, Tranum-Jensen J. Membrane damage by pore-forming bacterial cytolysins. *Microb Pathog* (1986) **1**:5–14. doi:10.1016/0882-4010(86)90027-6
2. Iacovache I, Van Der Goot FG, Pernot L. Pore formation: an ancient yet complex form of attack. *Biochim Biophys Acta* (2008) **1778**:1611–23. doi:10.1016/j.bbamem.2008.01.026
3. Los FC, Randis TM, Aroian RV, Ratner AJ. Role of pore-forming toxins in bacterial infectious diseases. *Microbiol Mol Biol Rev* (2013) **77**:173–207. doi:10.1128/MMBR.00052-12
4. Bischofberger M, Iacovache I, Van Der Goot FG. Pathogenic pore-forming proteins: function and host response. *Cell Host Microbe* (2012) **12**:266–75. doi:10.1016/j.chom.2012.08.005
5. Thelestam M, Mollby R. Survival of cultured-cells after functional and structural disorganization of plasma-membrane by bacterial hemolysins and phospholipases. *Toxicon* (1983) **21**:805–15. doi:10.1016/0041-0101(83)90069-7
6. Walev I, Palmer M, Martin E, Jonas D, Weller U, Hohn-Bentz H, et al. Recovery of human fibroblasts from attack by the pore-forming alpha-toxin of *Staphylococcus aureus*. *Microb Pathog* (1994) **17**:187–201. doi:10.1006/mpat.1994.1065
7. Huffman DL, Abrami L, Sasik R, Corbeil J, Van Der Goot FG, Aroian RV. Mitogen-activated protein kinase pathways defend against bacterial pore-forming toxins. *Proc Natl Acad Sci U S A* (2004) **101**:10995–1000. doi:10.1073/pnas.0404073101
8. Husmann M, Dersch K, Bobkiewicz W, Beckmann E, Veerachato G, Bhakdi S. Differential role of p38 mitogen activated protein kinase for cellular recovery from attack by pore-forming *S. aureus* alpha-toxin or streptolysin O. *Biochem Biophys Res Commun* (2006) **344**:1128–34. doi:10.1016/j.bbrc.2006.03.241
9. Aroian R, Van Der Goot FG. Pore-forming toxins and cellular non-immune defenses (CNIDs). *Curr Opin Microbiol* (2007) **10**:57–61. doi:10.1016/j.mib.2006.12.008
10. Husmann M, Beckmann E, Boller K, Kloft N, Tenzer S, Bobkiewicz W, et al. Elimination of a bacterial pore-forming toxin by sequential endocytosis and exocytosis. *FEBS Lett* (2009) **583**:337–44. doi:10.1016/j.febslet.2008.12.028
11. Kloft N, Busch T, Neukirch C, Weis S, Boukhallouk F, Bobkiewicz W, et al. Pore-forming toxins activate MAPK p38 by causing loss of cellular potassium. *Biochem Biophys Res Commun* (2009) **385**:503–6. doi:10.1016/j.bbrc.2009.05.121
12. Kloft N, Neukirch C, Bobkiewicz W, Veerachato G, Busch T, Von Hoven G, et al. Pro-autophagic signal induction by bacterial pore-forming toxins. *Med Microbiol Immunol* (2010) **199**:299–309. doi:10.1007/s00430-010-0163-0
13. Kloft N, Neukirch C, Von Hoven G, Bobkiewicz W, Weis S, Boller K, et al. A subunit of eukaryotic translation initiation factor 2alpha-phosphatase (CreP/PPP1R15B) regulates membrane traffic. *J Biol Chem* (2012) **287**:35299–317. doi:10.1074/jbc.M112.379883
14. Stuart LM, Paquette N, Boyer L. Effector-triggered versus pattern-triggered immunity: how animals sense pathogens. *Nat Rev Immunol* (2013) **13**:199–206. doi:10.1038/nri3398
15. Kao CY, Los FC, Huffman DL, Wachi S, Kloft N, Husmann M, et al. Global functional analyses of cellular responses to pore-forming toxins. *PLoS Pathog* (2011) **7**:e1001314. doi:10.1371/journal.ppat.1001314
16. Gonzalez MR, Bischofberger M, Pernot L, Van Der Goot FG, Freche B. Bacterial pore-forming toxins: the (w)hole story? *Cell Mol Life Sci* (2008) **65**:493–507. doi:10.1007/s00018-007-7434-y
17. Porta H, Cancino-Rodezno A, Soberon M, Bravo A. Role of MAPK p38 in the cellular responses to pore-forming toxins. *Peptides* (2011) **32**:601–6. doi:10.1016/j.peptides.2010.06.012
18. Von Hoven G, Kloft N, Neukirch C, Ebinger S, Bobkiewicz W, Weis S, et al. Modulation of translation and induction of autophagy by bacterial exoproducts. *Med Microbiol Immunol* (2012) **201**:409–18. doi:10.1007/s00430-012-0271-0
19. Stassen M, Muller C, Richter C, Neudorfl C, Hultner L, Bhakdi S, et al. The streptococcal exotoxin streptolysin O activates mast cells to produce tumor necrosis factor alpha by p38 mitogen-activated protein kinase- and protein kinase C-dependent pathways. *Infect Immun* (2003) **71**:6171–7. doi:10.1128/IAI.71.11.6171-6177.2003
20. Gonzalez MR, Bischofberger M, Freche B, Ho S, Parton RG, Van Der Goot FG. Pore-forming toxins induce multiple cellular responses promoting survival. *Cell Microbiol* (2011) **13**(7):1026–43. doi:10.1111/j.1462-5822.2011.01600.x
21. Nagahama M, Shibutani M, Seike S, Yonezaki M, Takagishi T, Oda M, et al. The p38 MAPK and JNK pathways protect host cells against *Clostridium perfringens* beta-toxin. *Infect Immun* (2013) **81**:3703–8. doi:10.1128/IAI.00579-13
22. Haugwitz U, Bobkiewicz W, Han SR, Beckmann E, Veerachato G, Shaid S, et al. Pore-forming *Staphylococcus aureus* alpha-toxin triggers epidermal growth factor receptor-dependent proliferation. *Cell Microbiol* (2006) **8**:1591–600. doi:10.1111/j.1462-5822.2006.00733.x
23. Richter E, Harms M, Ventz K, Gierok P, Chilukoti RK, Hildebrandt JP, et al. A multi-omics approach identifies key hubs associated with cell type-specific responses of airway epithelial cells to staphylococcal alpha-toxin. *PLoS One* (2015) **10**:e0122089. doi:10.1371/journal.pone.0122089
24. Hamon MA, Cossart P. K+ efflux, is required for histone-H3 dephosphorylation by *Listeria* LLO and other pore forming toxins. *Infect Immun* (2011) **79**(7):2839–46. doi:10.1128/IAI.01243-10
25. Imre G, Heering J, Takeda AN, Husmann M, Thiede B, Zu Heringdorf DM, et al. Caspase-2 is an initiator caspase responsible for pore-forming toxin-mediated apoptosis. *EMBO J* (2012) **31**(11):2615–28. doi:10.1038/emboj.2012.93

26. Idone V, Tam C, Goss JW, Toomre D, Pypaert M, Andrews NW. Repair of injured plasma membrane by rapid Ca2+-dependent endocytosis. *J Cell Biol* (2008) **180**:905–14. doi:10.1083/jcb.200708010
27. Tam C, Idone V, Devlin C, Fernandes MC, Flannery A, He X, et al. Exocytosis of acid sphingomyelinase by wounded cells promotes endocytosis and plasma membrane repair. *J Cell Biol* (2010) **189**:1027–38. doi:10.1083/jcb.201003053
28. Babiychuk EB, Monastyrskaya K, Potez S, Draeger A. Blebbing confers resistance against cell lysis. *Cell Death Differ* (2011) **18**:80–9. doi:10.1038/cdd.2010.81
29. Draeger A, Monastyrskaya K, Babiychuk EB. Plasma membrane repair and cellular damage control: the annexin survival kit. *Biochem Pharmacol* (2011) **81**:703–12. doi:10.1016/j.bcp.2010.12.027
30. Corrotte M, Fernandes MC, Tam C, Andrews NW. Toxin pores endocytosed during plasma membrane repair traffic into the lumen of MVBs for degradation. *Traffic* (2012) **13**:483–94. doi:10.1111/j.1600-0854.2011.01323.x
31. Corrotte M, Almeida PE, Tam C, Castro-Gomes T, Fernandes MC, Millis BA, et al. Caveolae internalization repairs wounded cells and muscle fibers. *Elife* (2013) **2**:e00926. doi:10.7554/eLife.00926
32. Castro-Gomes T, Koushik AB, Andrews NW. ESCRT: nipping the wound in the bud? *Trends Biochem Sci* (2014) **39**:307–9. doi:10.1016/j.tibs.2014.06.001
33. Tattoli I, Sorbara MT, Vuckovic D, Ling A, Soares F, Carneiro LA, et al. Amino acid starvation induced by invasive bacterial pathogens triggers an innate host defense program. *Cell Host Microbe* (2012) **11**:563–75. doi:10.1016/j.chom.2012.04.012
34. Lemaitre B, Girardin SE. Translation inhibition and metabolic stress pathways in the host response to bacterial pathogens. *Nat Rev Microbiol* (2013) **11**:365–9. doi:10.1038/nrmicro3029
35. Chakrabarti S, Liehl P, Buchon N, Lemaitre B. Infection-induced host translational blockage inhibits immune responses and epithelial renewal in the *Drosophila* gut. *Cell Host Microbe* (2012) **12**:60–70. doi:10.1016/j.chom.2012.06.001
36. Wek RC, Jiang HY, Anthony TG. Coping with stress: eIF2 kinases and translational control. *Biochem Soc Trans* (2006) **34**:7–11. doi:10.1042/BST0340007
37. Dever TE, Feng L, Wek RC, Cigan AM, Donahue TF, Hinnebusch AG. Phosphorylation of initiation factor 2 alpha by protein kinase GCN2 mediates gene-specific translational control of GCN4 in yeast. *Cell* (1992) **68**:585–96. doi:10.1016/0092-8674(92)90193-G
38. Dong J, Qiu H, Garcia-Barrio M, Anderson J, Hinnebusch AG. Uncharged tRNA activates GCN2 by displacing the protein kinase moiety from a bipartite tRNA-binding domain. *Mol Cell* (2000) **6**:269–79. doi:10.1016/S1097-2765(00)00028-9
39. Deloche O, De La Cruz J, Kressler D, Doere M, Linder P. A membrane transport defect leads to a rapid attenuation of translation initiation in *Saccharomyces cerevisiae*. *Mol Cell* (2004) **13**:357–66. doi:10.1016/S1097-2765(04)00008-5
40. De Filippi L, Fournier M, Cameroni E, Linder P, De Virgilio C, Foti M, et al. Membrane stress is coupled to a rapid translational control of gene expression in chlorpromazine-treated cells. *Curr Genet* (2007) **52**:171–85. doi:10.1007/s00294-007-0151-0
41. Schneider DS, Ayres JS. Two ways to survive infection: what resistance and tolerance can teach us about treating infectious diseases. *Nat Rev Immunol* (2008) **8**:889–95. doi:10.1038/nri2432
42. Raberg L, Graham AL, Read AF. Decomposing health: tolerance and resistance to parasites in animals. *Philos Trans R Soc Lond B Biol Sci* (2009) **364**:37–49. doi:10.1098/rstb.2008.0184
43. Medzhitov R, Schneider DS, Soares MP. Disease tolerance as a defense strategy. *Science* (2012) **335**:936–41. doi:10.1126/science.1214935
44. De S, Olson R. Crystal structure of the *Vibrio cholerae* cytolysin heptamer reveals common features among disparate pore-forming toxins. *Proc Natl Acad Sci U S A* (2011) **108**:7385–90. doi:10.1073/pnas.1017442108
45. Tournier C, Hess P, Yang DD, Xu J, Turner TK, Nimnual A, et al. Requirement of JNK for stress-induced activation of the cytochrome c-mediated death pathway. *Science* (2000) **288**:870–4. doi:10.1126/science.288.5467.870
46. Boyce M, Bryant KF, Jousse C, Long K, Harding HP, Scheuner D, et al. A selective inhibitor of eIF2alpha dephosphorylation protects cells from ER stress. *Science* (2005) **307**:935–9. doi:10.1126/science.1101902
47. Jousse C, Oyadomari S, Novoa I, Lu P, Zhang Y, Harding HP, et al. Inhibition of a constitutive translation initiation factor 2alpha phosphatase, CReP, promotes survival of stressed cells. *J Cell Biol* (2003) **163**:767–75. doi:10.1083/jcb.200308075
48. Ryazanov AG, Shestakova EA, Natapov PG. Phosphorylation of elongation factor 2 by EF-2 kinase affects rate of translation. *Nature* (1988) **334**:170–3. doi:10.1038/334170a0
49. Vadia S, Seveau S. Fluxes of Ca2+ and K+ are required for the listeriolysin O-dependent internalization pathway of *Listeria monocytogenes*. *Infect Immun* (2014) **82**:1084–91. doi:10.1128/IAI.01067-13
50. Dunbar TL, Yan Z, Balla KM, Smelkinson MG, Troemel ER. *C. elegans* detects pathogen-induced translational inhibition to activate immune signaling. *Cell Host Microbe* (2012) **11**:375–86. doi:10.1016/j.chom.2012.02.008
51. Arguello RJ, Rodriguez Rodrigues C, Gatti E, Pierre P. Protein synthesis regulation, a pillar of strength for innate immunity? *Curr Opin Immunol* (2015) **32**:28–35. doi:10.1016/j.coi.2014.12.001
52. Wilke GA, Bubeck Wardenburg J. Role of a disintegrin and metalloprotease 10 in *Staphylococcus aureus* alpha-hemolysin-mediated cellular injury. *Proc Natl Acad Sci U S A* (2010) **107**:13473–8. doi:10.1073/pnas.1001815107
53. Talloczy Z, Jiang W, Virgin HWT, Leib DA, Scheuner D, Kaufman RJ, et al. Regulation of starvation- and virus-induced autophagy by the eIF2alpha kinase signaling pathway. *Proc Natl Acad Sci U S A* (2002) **99**:190–5. doi:10.1073/pnas.012485299
54. Maurer K, Reyes-Robles T, Alonzo F III, Durbin J, Torres VJ, Cadwell K. Autophagy mediates tolerance to *Staphylococcus aureus* alpha-toxin. *Cell Host Microbe* (2015) **17**:429–40. doi:10.1016/j.chom.2015.03.001
55. Jursch R, Hildebrand A, Hobom G, Tranum-Jensen J, Ward R, Kehoe M, et al. Histidine residues near the N terminus of staphylococcal alpha-toxin as reporters of regions that are critical for oligomerization and pore formation. *Infect Immun* (1994) **62**:2249–56.
56. Jiang HY, Wek SA, McGrath BC, Scheuner D, Kaufman RJ, Cavener DR, et al. Phosphorylation of the alpha subunit of eukaryotic initiation factor 2 is required for activation of NF-kappaB in response to diverse cellular stresses. *Mol Cell Biol* (2003) **23**:5651–63. doi:10.1128/MCB.23.16.5651-5663.2003
57. Scheuner D, Song B, McEwen E, Liu C, Laybutt R, Gillespie P, et al. Translational control is required for the unfolded protein response and in vivo glucose homeostasis. *Mol Cell* (2001) **7**:1165–76. doi:10.1016/S1097-2765(01)00265-9
58. Harding HP, Zhang Y, Scheuner D, Chen JJ, Kaufman RJ, Ron D. Ppp1r15 gene knockout reveals an essential role for translation initiation factor 2 alpha (eIF2alpha) dephosphorylation in mammalian development. *Proc Natl Acad Sci U S A* (2009) **106**:1832–7. doi:10.1073/pnas.0809632106
59. Chu HP, Liao Y, Novak JS, Hu Z, Merkin JJ, Shymkiv Y, et al. Germline quality control: eEF2K stands guard to eliminate defective oocytes. *Dev Cell* (2014) **28**:561–72. doi:10.1016/j.devcel.2014.01.027
60. Hartmann D, De Strooper B, Serneels L, Craessaerts K, Herreman A, Annaert W, et al. The disintegrin/metalloprotease ADAM 10 is essential for Notch signalling but not for alpha-secretase activity in fibroblasts. *Hum Mol Genet* (2002) **11**:2615–24. doi:10.1093/hmg/11.21.2615
61. Boukamp P, Petrussevska RT, Breitkreutz D, Hornung J, Markham A, Fusenig NE. Normal keratinization in a spontaneously immortalized aneuploid human keratinocyte cell line. *J Cell Biol* (1988) **106**:761–71. doi:10.1083/jcb.106.3.761
62. Schmidt EK, Clavarino G, Ceppi M, Pierre P. SUnSET, a nonradioactive method to monitor protein synthesis. *Nat Methods* (2009) **6**:275–7. doi:10.1038/nmeth.1314
63. Abrami L, Fivaz M, Glauser PE, Parton RG, Van Der Goot FG. A pore-forming toxin interacts with a GPI-anchored protein and causes vacuolation of the endoplasmic reticulum. *J Cell Biol* (1998) **140**:525–40. doi:10.1083/jcb.140.3.525
64. Agerer F, Waeckerle S, Hauck CR. Microscopic quantification of bacterial invasion by a novel antibody-independent staining method. *J Microbiol Methods* (2004) **59**:23–32. doi:10.1016/j.mimet.2004.05.008

Conflict of Interest Statement: The authors declare that the research was conducted in the absence of any commercial or financial relationships that could be construed as a potential conflict of interest.

9

Shigella Manipulates Host Immune Responses by Delivering Effector Proteins with Specific Roles

Hiroshi Ashida [1], Hitomi Mimuro [2] and Chihiro Sasakawa [1,3,4]*

[1] Division of Bacterial Infection Biology, Institute of Medical Science, University of Tokyo, Tokyo, Japan, [2] Division of Bacteriology, Department of Infectious Diseases Control, International Research Center for Infectious Diseases, The Institute of Medical Science, University of Tokyo, Tokyo, Japan, [3] Nippon Institute for Biological Science, Tokyo, Japan, [4] Medical Mycology Research Center, Chiba University, Chiba, Japan

Edited by:
Katharina F. Kubatzky,
University Hospital Heidelberg, Germany

Reviewed by:
Herve Agaisse,
Yale University, USA
Wendy Picking,
University of Kansas, USA

****Correspondence:***
Hiroshi Ashida,
Division of Bacterial Infection Biology, Institute of Medical Science, University of Tokyo, 4-6-1, Shirokanedai, Minato-ku, Tokyo 108-8639, Japan
hashida@ims.u-tokyo.ac.jp

The intestinal epithelium deploys multiple defense systems against microbial infection to sense bacterial components and danger alarms, as well as to induce intracellular signal transduction cascades that trigger both the innate and the adaptive immune systems, which are pivotal for bacterial elimination. However, many enteric bacterial pathogens, including *Shigella*, deliver a subset of virulence proteins (effectors) via the type III secretion system (T3SS) that enable bacterial evasion from host immune systems; consequently, these pathogens are able to efficiently colonize the intestinal epithelium. In this review, we present and select recently discovered examples of interactions between *Shigella* and host immune responses, with particular emphasis on strategies that bacteria use to manipulate inflammatory outputs of host-cell responses such as cell death, membrane trafficking, and innate and adaptive immune responses.

Keywords: *Shigella*, effector, inflammation, innate immunity

Introduction

Shigella is a causative agent of bacillary dysentery, which ultimately leads to severe bloody and mucous diarrhea (shigellosis). Most cases of shigellosis occur in developing countries and affect children under 5 years old. Although antibiotics are the standard care for shigellosis patients, antibiotic-resistant bacterium is becoming common. Therefore, it is urgently necessary to develop a safe and effective *Shigella* vaccine. *Shigella* have neither adherence factor nor flagella, but they are capable of efficiently invading the intestinal epithelium. *Shigella* injects a subset of effectors (secreted virulence proteins) via a type III secretion system (T3SS) (protein delivery system) into host cells, allowing the bacterium to invade, multiply within the intestinal epithelium, and subvert cellular and immune functions during bacterial internalization (1, 2). When *Shigella* cells are ingested via the oral route, the bacteria move down to the colon and rectum, and then preferentially enter the M cells overlying the follicle-associated epithelium of the Peyer's patches (3, 4) (**Figure 1**). Once the bacteria are endocytosed by the M cells, they are transcytosed toward the M cell pocket, where resident macrophages receive the bacteria. However, *Shigella* can disrupt the vacuolar membranes, disseminate into the cytoplasm, and multiply therein (5). Bacterial multiplication within the macrophages results in massive inflammatory cell death (6) (**Figure 1**). Meanwhile, *Shigella* cells that are released from dying macrophages subsequently enter the surrounding epithelium via the basolateral surface. Upon epithelial cell contact, the bacteria deliver a subset of T3SS effector proteins that trigger actin rearrangement, promoting bacterial uptake (7). Next, the *Shigella* cells are surrounded by a vacuolar membrane, but the bacteria rapidly disrupt this membrane and

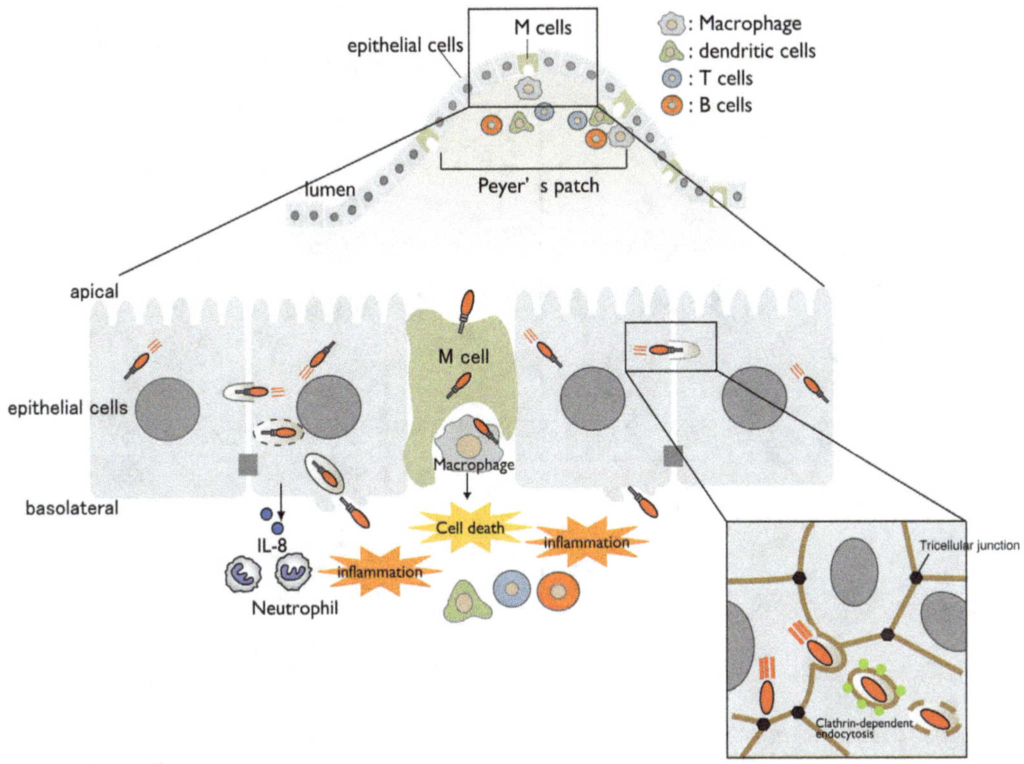

FIGURE 1 | A model for *Shigella* infection of the intestinal epithelium. A schematic representation of *Shigella* infection. Bacterial invasion and multiplication within macrophages and epithelial cells cause subsequent massive inflammatory colitis, termed as shigellosis. Cell-to-cell movement of *Shigella* occurs at tricellular junctions via clathrin-dependent endocytosis.

disseminate into the cytoplasm. As *Shigella* proliferates within the cytoplasm, it moves by inducing actin polymerization at one pole of the bacterium, providing the propulsive force required for inter- or intracellular movement (8–10). Intriguingly, *Shigella* cell-to-cell movement preferentially occurs at epithelial tricellular junctions, where three cells meet (**Figure 1**). At these positions, bacteria-containing pseudopodia are engulfed by neighboring cells via a clathrin-dependent endocytic pathway, resulting in dissemination of *Shigella* (11). By repeating these processes, the bacteria efficiently multiply by constantly renewing their replicative compartment. Thus, multiple infectious events during *Shigella* infection, including macrophage cell death, invasion of and multiplication within epithelial cells, cell-to-cell spreading, demise of the host epithelium, and alteration of the host inflammatory response, are major pathogenic events that lead to shigellosis (2) (**Figure 1**).

Cell Death

Host-cell death in response to microbial infection is an intrinsic immune defense stratagem against microbial intrusion. The sacrifice of infected cells plays a pivotal role in clearance of damaged cells, elimination of pathogens, local confinement of tissue damage and inflammation, and presentation of bacteria-derived antigens to the adaptive immune system (12). Cell death induced by bacterial infection can be classified into at least three types, depending on the type of cell and stage of infection: apoptosis, necrosis, and pyroptosis. Apoptosis is a non-inflammatory programed cell death triggered by the mitochondria-mediated pathway and receptor-mediated pathway, which eventually induce caspase activation (caspase-2, -3, -6, -7, -8, -9, and -10), chromatin condensation, cell shrinkage, plasma membrane blebbing, and cytoplasm retained in apoptotic bodies. On the other hand, necrosis is an inflammatory form of cell death characterized by cell swelling, membrane rupture, and intracellular content leakage. Pyroptosis is pro-inflammatory, lytic, and programed cell death that is accompanied by activation of caspase-1 or caspase-11 (human homologs: caspase-4/5) inflammasomes, leading to the production of IL-1β and IL-18.

When bacteria invade and multiply within host cells, they release bacterial components [e.g., lipopolysaccharide (LPS) and peptidoglycan (PGN)] and T3SS components, and further cause infection-associated cellular damage. These are recognized as pathogen-associated molecular patterns (PAMPs) and danger-associated molecular patterns (DAMPs) by recognition receptors, such as toll-like receptors (TLRs), nucleotide-binding oligomerization domain-like receptors (NLRs), AIM2-like receptors (ALRs), and RIG-like receptors, and trigger host immune responses against bacterial infection. Upon recognition of these PAMPs and DAMPs, a subset of NLRs (e.g., NLRP1, NLRP3, and NLRC4) and ALRs (e.g., AIM2) form inflammasomes, which are multi-protein signaling complexes composed of NLR/ALR, the adaptor protein ASC, and inflammatory caspase, such as caspase-1 (canonical inflammasomes) and caspase-11 (non-canonical

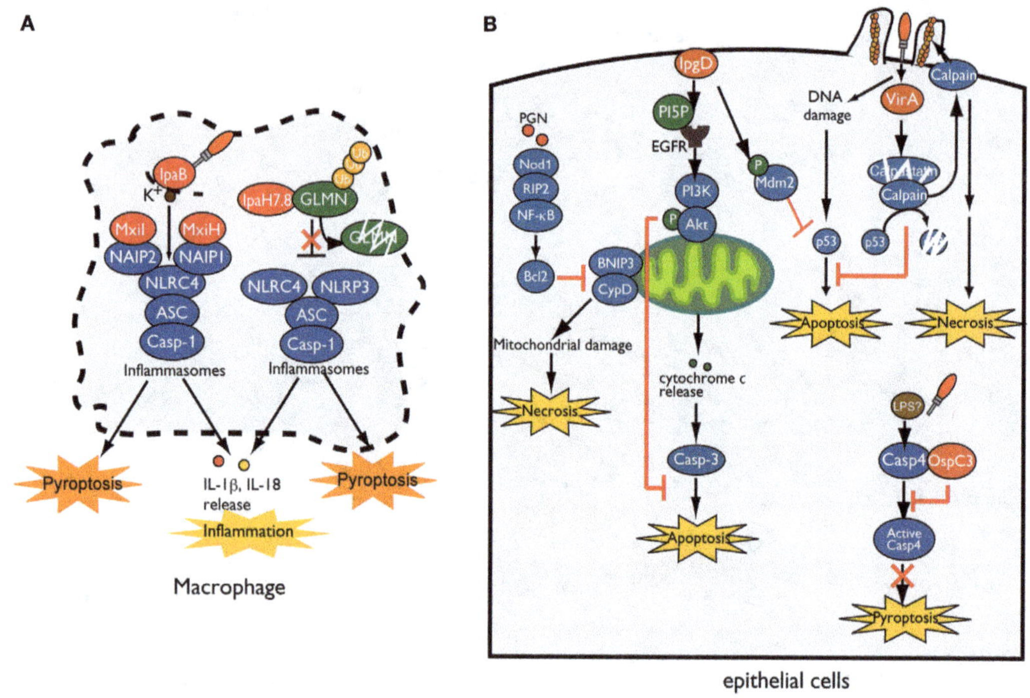

FIGURE 2 | *Shigella* manipulates host-cell death. **(A)** *Shigella* invasion and replication in macrophages trigger NLRC4/NLRP3-inflammasomes activation and pyroptotic cell death. The *Shigella* T3SS rod component MxiI and needle protein MxiH trigger NAIP2- and NAIP1-dependent NLRC4-inflammasome activation, respectively, whereas T3SS effector IpaB-mediated potassium influx triggers NLRC4-inflammasome activation and ultimately induces pyroptosis. IpaH7.8 ubiquitinates GLMN, and undergoes proteasome-dependent degradation, resulting in NLRP3/NLRC4-inflammasome activation and pyroptosis. **(B)** *Shigella* counteracts mitochondrial damage-dependent necrosis through BNIP3 and CypD by activating the PGN/Nod1/RIP2/NF-κB/Bcl2 pro-survival pathway. Furthermore, *Shigella* prevents epithelial cell death by delivering an array of T3SS effectors. PI5P, generated by IpgD, promotes EGFR activation, which contributes to sustained activation of the PI3K–Akt survival pathway. IpgD and VirA target Mdm2 and calpain, respectively, to prevent p53-dependent apoptosis. OspC3 binds to the caspase-4 subunit and inhibits its activation, thereby blocking caspase-4-dependent inflammatory cell death.

inflammasomes) (13, 14). Inflammasome activation ultimately results in release of pro-inflammatory cytokines (IL-1β and IL-18) and induction of pro-inflammatory lytic cell death (pyroptosis).

When *Shigella* invade and multiply within macrophages, they rapidly induce pyroptotic cell death accompanied by NLRP3- or NLRC4-inflammasome activation, leading to IL-1β and IL-18 secretion (15–18). T3SS needle or rod components indirectly activate NLRC4 inflammasomes through members of the NAIP subfamily of NLRs, which act as direct pathogen recognition sensors and determine the specificity of the NLRC4-inflammasome for various bacterial ligands. Recent reports have revealed that human NAIP and mouse NAIP1 bind to *Shigella* T3SS needle protein MxiH (19, 20). In addition, NAIP2 binds to the T3SS inner rod component MxiI (16, 21) (**Figure 2A**). Although *Shigella* lacks flagella, NAIP5 and NAIP6 specifically activate the NLRC4-inflammasome in response to other bacterial flagellins (22–25). Once these NAIP proteins bind to their ligands, they bind to NLRC4 and induce NLRC4-inflammasome activation and pyroptosis. In addition to bacterial T3SS components, cellular damage caused by *Shigella* also triggers inflammasome activation and pyroptosis. As *Shigella* cells multiply within macrophages, the T3SS effector IpaB assembles an ion channel within cell membrane to allow for potassium influx, which is recognized by the NLRC4-inflammasome and ultimately triggers pyroptosis (26) (**Figure 2A**).

As described above, because pyroptosis is accompanied by inflammation that would limit bacterial infection, it had not been unclear whether pyroptosis is beneficial for bacterial infection (27). However, a recent study showed that induction of inflammasome activation and pyroptosis in infected macrophages is a *Shigella* strategy that can promote bacterial survival and dissemination. Suzuki et al. showed that *Shigella* induces rapid macrophage pyroptosis via IpaH7.8, an IpaH family effector, mediated NLRP3- and NLRC4-dependent inflammasome activation (28) (**Figure 2A**). IpaH family effectors, which have a novel E3 ubiquitin ligase activity, are widely conserved among Gram-negative bacterial pathogens, including *Shigella*, *Salmonella*, *Yersinia*, and *Pseudomonas* spp. (29). IpaH7.8 targets GLMN (glomulin/flagellar-associated protein 68), a Cullin-RING E3 ligase inhibitor, for ubiquitination and undergoes proteasome-dependent degradation (**Figure 2A**) (28). Because GLMN acts as a negative regulator of NLR inflammasomes and pyroptosis, degradation of this factor induces inflammasome activation and pyroptosis. Consistent with results obtained *in vitro*, mice intranasally infected with *Shigella* WT or Δ*ipaH7.8/WT* complement strains induce more severe inflammatory responses and elevated numbers of colonized bacteria relative to Δ*ipaH7.8* or

ΔipaH7.8/CA E3 ligase-deficient mutant complemented strains (28). Therefore, IpaH7.8-mediated macrophage cell death is prerequisite for allowing bacteria to escape from macrophages, further enter surrounding epithelial cells, and spread to neighboring cells.

Epithelial Cell Death

When *Shigella* invade and multiply within epithelial cells, the cells generate an early genotoxic stress, mitochondrial damage, oxidative stress, and recognize PAMPs and DAMPs, which could induce several types of cell death as part of the host defense system aimed at terminating bacterial infection. However, in contrast to macrophage infection, *Shigella* seems to prevent epithelial cell death until the bacteria have fully multiplied, because it prefers these cells as replicative niche, spread to neighboring cells, and evasion of immune cells (30) (**Figure 2B**). To support this notion, *Shigella* has several countermeasures that inhibit epithelial cell death: (i) prevention of mitochondrial damage, (ii) activation of cell survival signaling [e.g., phosphoinositide-3 kinase (PI3K)-Akt and transcription factor nuclear factor κB (NF-κB)], and (iii) prevention of caspase activation. For example, *Shigella* prevents necrotic cell death mediated by mitochondrial damage (through BNIP3 and CypD) by activating the Nod1–RIP2–NF-κB–Bcl-2 pro-survival pathway (31) (**Figure 2B**). When *Shigella* invades epithelial cells, it delivers the T3SS effector IpgD (a homolog of *Salmonella* SopB), a phosphoinositide phosphatase that converts phosphatidylinositol 4,5-bisphosphate (PIP2) into phosphatidylinositol 5-phosphate (PI5P), at the bacterial entry site (32). Because PI plays pivotal roles in actin cytoskeleton rearrangement, elevated levels of PI5P at the plasma membrane promotes bacterial invasion. Furthermore, IpgD-mediated PI5P is an important factor involved in cell survival (33). At this step, the PI5P generated by IpgD contributes to epidermal growth factor receptor (EGFR) activation, which sustains the PI3K/Akt pro-survival pathway and thereby contributes indirectly to augmentation of pro-survival signaling (33, 34) (**Figure 2B**). A recent report showed that the early stage of *Shigella* infection induces genotoxic stress in epithelial cells, followed by p53 pro-apoptotic signaling activation and induction of apoptosis; however, *Shigella* promotes p53 degradation and antagonizes the early stage of cell death by delivering two T3SS effectors, IpgD and VirA (**Figure 2B**) (35). As described above, IpgD promotes Akt activation, which in turn phosphorylates and stabilizes the downstream E3 ubiquitin ligase Mdm2. Activated Mdm2 targets p53 for ubiquitination and leads to proteasome-dependent degradation, thereby inhibiting pro-apoptosis signaling by p53. Furthermore, VirA promotes further degradation of p53 by activating calpain protease, which is also important for bacterial invasion. VirA binds to the calpain inhibitor calpastatin and promotes its degradation, resulting in degradation of p53 by activated calpain and blocking the p53 pro-apoptotic signaling pathway. Although VirA-mediated calpain activation promotes bacterial entry and prevents the early stage of apoptosis, sustained calpain activation ultimately induces necrosis and restricts bacterial proliferation (35). In addition, *Shigella* exploits calpain activation and mitochondrial function to suppress the innate immune response without inducing host-cell death (see below) (36). These results indicate that *Shigella* deploys a sophisticated strategy that controls the delicate balance of host-cell death until *Shigella* has succeeded in primary colonization and proliferation within epithelial cells.

Recent reports have shown that non-canonical caspase-4/-11 inflammasome activation is an essential host defense mechanism of epithelial cells against enteric bacterial pathogens such as *Shigella*, *Salmonella*, and enteropathogenic *Escherichia coli* (EPEC). Caspase-4/-11 directly binds to cytoplasmic LPS of Gram-negative bacterial pathogens and activates caspase-4/-11-mediated inflammasomes and pyroptosis, ultimately inducing epithelial cell shedding to eliminate infected cells (37–39). However, *Shigella* antagonizes caspase-4-dependent inflammatory cell death by delivering the T3SS effector OspC3 and promoting epithelial infection (37). OspC3 deploys a unique mechanism that specifically targets and inactivates caspase-4, but not caspase-1 or mouse caspase-11. The C-terminal ankyrin-repeat (ANK) region of OspC3 interacts with the p19 subunit of caspase-4 and prevents its activation by inhibiting p19 and p10 dimerization (**Figure 2B**). In human epithelial cell lines, the *Shigella ΔospC3* mutant induces early caspase-4-dependent pyroptotic cell death and increased cytokine production relative to that of WT *Shigella*. Of note, the *Shigella ΔospC3* mutant also exhibited severe mucosal cell death and a reduction in the number of colonizing bacteria in a guinea pig rectal infection model, indicating the importance of OspC3-mediated cell death inhibition for bacterial infection (37). Thus, *Shigella* delivers a subset of T3SS effector proteins that delay epithelial cell death until the bacterial cells have fully replicated and further disseminated into surrounding cells.

Autophagy

Autophagy is an essential cellular catabolic process, which targets proteins, organelles, and large protein aggregates by sequestering deleterious cargos within a double-membrane compartment, the autophagosome. Autophagy also plays a pivotal role as a part of the innate immune system, by acting as a cytosolic sensor to recognize DAMPs and PAMPs, and as an "executioner" that engulfs bacteria in autophagosomes that fuse with lysosomes, ultimately destroying bacteria within lysosomal compartments.

Shigella enter epithelial cells, disseminate into the cytosol by disrupting the surrounding phagosomal membrane, and move into adjacent cells by inducing actin polymerization at one bacterial pole. In *Shigella* invasion of epithelial cells, Nod1 and Atg16L1 are recruited to the plasma membrane beneath the *Shigella* entry site and subsequently trigger autophagy (40). Furthermore, host-cell vacuolar membrane remnants generated by *Shigella* are recognized as DAMPs by galectin-8, and these remnants are also polyubiquitinated, followed by the recruitment of p62 (ubiquitin adaptor protein) and LC3 (autophagosome marker), resulting in autophagic activation (41, 42) (**Figure 3**). During multiplication within the cytosol, *Shigella* outer membrane protein VirG (IcsA) accumulates at one pole of the bacterial surface. There, VirG recruits and activates N-WASP, which subsequently recruits and activates the Arp2/3 complex, thereby inducing actin polymerization and bacterial motility within the cell (8–10). At this stage, VirG is recognized by the host autophagy protein Atg5, a protein essential for autophagosome maturation, resulting in *Shigella* uptake by autophagosomes. However, *Shigella* prevents

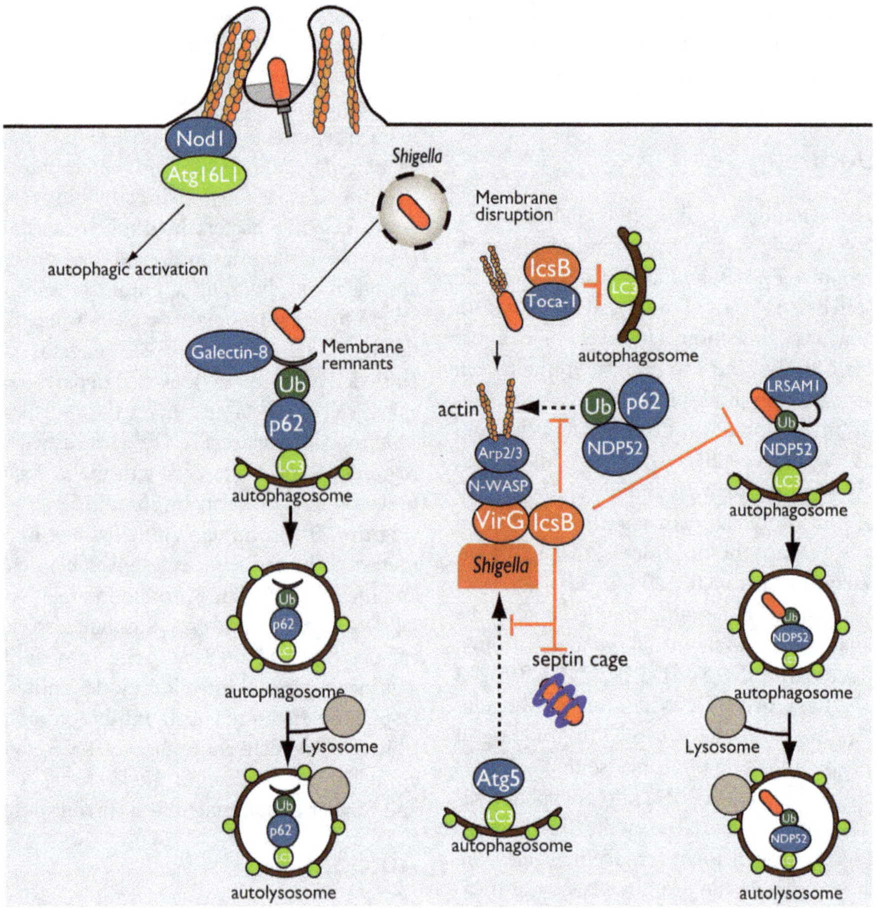

FIGURE 3 | *Shigella* prevents autophagic recognition. *Shigella* prevents autophagic clearance by evading autophagic recognition. *Shigella* delivers the T3SS effector IcsB, which binds to the bacterial outer membrane protein VirG and prevents Atg5–VirG interaction and ubiquitin recruitment, thereby evading autophagic recognition. IcsB also recruits Toca-1, and prevents the recruitment of LC3 to around intracellular bacteria.

autophagic recognition by delivering the IcsB T3SS effector. At an early stage of infection, IcsB binds to and recruits Toca-1, which is required for efficient formation of actin polymerization, around intracellular bacteria, and IcsB–Toca-1 prevents the recruitment of LC3 (43, 44). Furthermore, at the later stage of infection, IcsB plays a pivotal role in camouflage against autophagic recognition (45). IcsB and Atg5 interact with the same region on VirG, but the affinity of IcsB for VirG is stronger than that of Atg5; therefore, IcsB competitively inhibits the VirG–Atg5 interaction, masking the target VirG protein from autophagic recognition (45) (**Figure 3**). Consistent with this, the *Shigella* Δ*icsB* mutant is trapped in the autophagosome and delivered for lysosomal degradation (45). Intriguingly, IcsB has cholesterol binding region, and its ability is involved in evading autophagic recognition without affecting IcsB–VirG binding (46). In *Shigella* infection, ubiquitin-dependent selective autophagy is also triggered, but the bacteria evade ubiquitin recognition. Although *Shigella* WT is not targeted by ubiquitin, LRSAM1 (a mammalian LRR-containing RING E3 ligase), which itself acts as a bacterial recognition molecule, localizes to the *Shigella* Δ*icsB* mutant, and LRSAM1-mediated bacterial ubiquitylation has been observed *in vitro* (44). LRSAM1 recognition can trigger ubiquitin-dependent selective autophagy, resulting in restriction of bacterial replication (47). In addition, another host factor septin, which is GTP-binding protein, assembles at sites of VirG-induced actin polymerization and forms cages that surrounding bacteria, and prevents inter- and intracellular movement, thereby targeting by autophagy and restricting bacterial proliferation (48). A recent report showed that the ubiquitin adaptors p62 and NDP52 target *Shigella* for autophagy in an actin polymerization- and septin-dependent manner, and that the *Shigella* Δ*virG* mutant reduces the recruitment of p62 and NDP52 (49). The *Shigella* Δ*icsB* mutant increases septin-cage formation and the recruitment of ubiquitin, p62, and NDP52. Therefore, although VirG-mediated actin polymerization is targeted by septin- and ubiquitin-dependent selective autophagy, the IcsB–VirG interaction prevents the recruitment of ubiquitin, p62, and NDP52 around bacteria (49) (**Figure 3**).

Membrane Trafficking

Intracellular trafficking of membranes and proteins is essential for maintenance of epithelial homeostasis and barrier function, and also acts as a host defense system against bacterial pathogens. In eukaryotic cells, intracellular trafficking system can be divided into two pathways, endocytosis and the secretory pathway. During endocytosis, which internalizes extracellular molecules or delivers plasma membrane proteins to specialized sites, the cargo is sorted in the early endosome and further transported to Golgi, endoplasmic reticulum (ER), or late endosomes and lysosome for degradation, whereas some cargo proteins are transported back to the plasma membrane via recycling endosomes. On the other hand, the secretory pathway, which produces molecules such as antimicrobial peptides, growth factors, cell surface receptors, and cytokine secretion, transport molecules from the ER to the plasma membrane through the Golgi.

Exocytosis, endocytosis, phagocytosis, and cytokine secretion play essential roles as host defense systems for eliminating invading bacterial pathogens. However, many intracellular bacterial pathogens, such as *Shigella*, *Salmonella*, and *Legionella*, hijack and exploit host intracellular trafficking system to enter host cells, evade subsequent phagocytic destruction, and establish their safety replicative niche for their survival and proliferation (7, 50, 51). *Shigella* enter non-phagocytic cells by delivering a coordinated set of T3SS effector proteins that trigger actin cytoskeleton or plasma membrane remodeling, and promote subsequent bacterial uptake by host cells. Soon after internalization, the bacteria are engulfed within a membrane-bound vacuole that is derived from the host plasma membrane. After that, *Shigella* disrupt and escape from this vacuole and replicate within the cytoplasm (7). Vacuolar disruption or modulation by intracellular bacteria is related to the host membrane trafficking system, which is tightly regulated by small GTPases of the Rab and ARF families. The small GTPases act as molecular switches that cycle between the GTP-bound active form and GDP-bound inactive form, catalyzed by two classes of regulatory proteins, guanine nucleotide exchange factors (GEFs) and GTPase-activating proteins (GAPs). GEFs exchange GDP for GTP to activate GTPases, whereas GAPs inactivate GTPases by promoting hydrolysis of GTP to GDP (52). Rab and ARF GTPases play an important role in resisting bacterial infection, such as secretion of cytokines and antimicrobial peptides, maintenance of epithelial barrier integrity, and maturation of phagosomes into lysosomes to degrade entrapped bacteria. Therefore, *Shigella* target Rab GTPases and interfere with host membrane trafficking by delivering T3SS effector proteins (**Figure 4**).

In *Shigella* infection, the endocytic pathway is essential event for bacterial invasion, vacuolar rupture, and spread to neighboring cells. Using a high-content siRNA screening, Mellouk et al. recently demonstrated that *Shigella* targets and recruits Rab11, a component of host recycling endosomes, to disrupt the vacuole and efficiently escape from it by delivering the IpgD effector (53) (**Figure 4**). As described before, IpgD is a phosphoinositide phosphatase that produces PI5P, which is important for vesicular trafficking. In cells infected with *Shigella ΔipgD* or phosphatase-inactive *ΔipgD/ipgD* C438S complemented mutant, Rab11 is not recruited to the site of bacterial invasion, and vacuolar rupture is delayed; however, this deficiency is rescued by the complementation of *ΔipgD/ipgD* WT or WT *Shigella* infection, highlighting the requirement for the phosphoinositide phosphatase activity of IpgD. Furthermore, vacuole disruption by *Shigella* is delayed in Rab11 knockdown epithelial cells without affecting bacterial

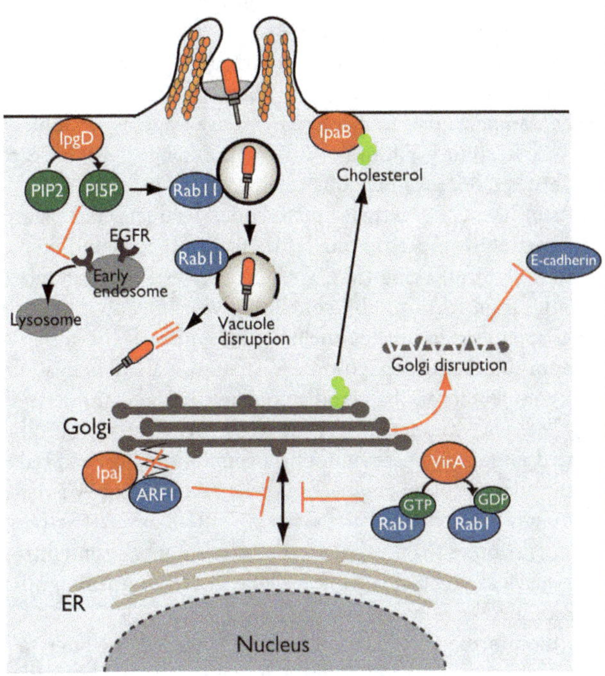

FIGURE 4 | *Shigella* alters host intracellular trafficking. *Shigella* exploits and alters intracellular trafficking to manipulate the host defense system. PI5P produced by IpgD recruits Rab11 to the bacterial entry site and promotes vacuolar membrane disruption. Production of PI5P by IpgD further relocates EGFR in early endosomes, resulting in blocking lysosomal degradation of EGFR and sustaining PI3K–Akt survival pathway. VirA and IpaJ target and inactivate Rab1 and ARF1, respectively, thereby preventing intracellular trafficking and Golgi disruption. IpaB binds to and relocates cholesterol from Golgi to bacterial entry sites, resulting in Golgi fragmentation.

invasion (53). Therefore, *Shigella* targets Rab11 to disrupt the vacuole and allow bacterial escape into the cytoplasm and promote further bacterial replication and dissemination.

In addition, as mentioned above briefly, IpgD subverts host intracellular trafficking to promote host-cell survival for bacterial prolonged colonization (34). The production of PI5P by IpgD at bacterial entry sites recruits active EGFR, which is required for the PI3K–Akt survival signaling pathway. An elevated level of PI5P at the plasma membrane, due to the activity of IpgD, alters EGFR trafficking and relocates it to early endosomes from late endosomes or lysosomes, thereby blocking lysosomal degradation of EGFR and sustaining EGFR–PI3K–Akt survival signaling activation (34) (**Figure 4**).

The Golgi apparatus is a central organelle involved in protein or lipid transport in eukaryotic cells. The Golgi processes and sorts proteins made by the ER, and transport them to various destinations, such as the plasma membrane and lysosome. Therefore, defects in Golgi function result in the loss of intracellular trafficking, including secretion of antimicrobial peptides and cytokines, and transport of epithelial junction proteins that are essential for epithelial barrier function against bacterial infection. Intriguingly, *Shigella* disrupts the Golgi apparatus by inhibiting vesicle trafficking in infected epithelial cells in a T3SS-dependent manner. Of note, three *Shigella* effectors, VirA, IpaJ, and IpaB, play important roles in disrupting the Golgi apparatus by targeting host intracellular trafficking (**Figure 4**).

Recent reports have shown that VirA and IpaJ, which have specific enzymatic activities targeting Rab and ARF GTPases, respectively, dampen host antibacterial defense systems by inhibiting the host secretory pathway. VirA (a homolog of EPEC EspG), which has TBC (homologous catalytic domain of GAP for Rab GTPase)-like GAP activity, preferentially targets and catalyzes GTP hydrolysis in Rab1, which localizes in the ER and mediates ER-to-Golgi trafficking (54). VirA inactivates Rab1, inhibits ER-to-Golgi transport of Rab1-containing vesicles, and disrupts the Golgi apparatus in host epithelial cells (**Figure 4**). Because Rab1 is important for autophagosome formation, Rab1 inactivation by VirA counteracts antibacterial autophagy. Of note, infection by a *Shigella* Δ*virA* mutant or GAP-inactive Δ*virA/virA-RQ* complemented mutant increases autophagosome formation and reduces the number of colonizing bacteria relative to that of WT *Shigella* in human epithelial cell lines. Thus, TBC-like GAP activity of VirA contributes to bacterial escape from autophagy and intracellular survival by blocking intracellular trafficking through Rab1 GTPase inactivation (54).

A bioinformatics approach revealed that IpaJ belongs to the cysteine protease family, whose members have Cys–His–Asp catalytic triad residues (55, 56). Further yeast genetic screening and mass spectrometry analysis demonstrated that IpaJ specifically cleaves the *N*-myristoylated glycine from ARF1 (**Figure 4**). *N*-myristoylation is an essential fatty-acid modification that is involved in protein localization, signal transduction, autophagosome maturation, and organelle function. Because *N*-myristoylation of ARF1 is essential for binding to the Golgi membrane, IpaJ-mediated cleavage induces the release of ARF1 from Golgi and its disruption, thereby inhibiting intracellular trafficking, including cytokine secretion and signal transduction (55, 56).

Another effector, IpaB, also disrupts the Golgi apparatus by targeting lipid trafficking, and thereby reduces epithelial barrier function. IpaB binds to and redirects cholesterol at the bacterial entry site from the Golgi apparatus, which depletes the lipid of Golgi and induces its fragmentation (**Figure 4**). The Golgi apparatus fragmentation further induces tubulation of the Rab11-positive compartment and reorganizes the recycling endosome, which is involved in trafficking of E-cadherin to adherence junction, resulting in epithelial junction disruption (57).

As described above, membrane and protein trafficking, such as the phagosome component, epithelial barrier components, antimicrobial peptides, cytokines, and cell surface receptors, is a major host defense system against bacterial infection. Therefore, *Shigella* blocks and alters host intracellular trafficking by delivering subsets of effectors that promote infection.

Manipulation of Host Innate Immunity

Once *Shigella* invades and replicates within host cells, the innate immune system quickly senses PAMPs or DAMPs, and transmits various alarm signals to the rest of the immune system, and ultimately triggers inflammation. Inflammation, which is accompanied by inflammatory cytokine secretion, neutrophil recruitment, and massive tissue destruction, is the hallmark of the host innate immune response that eventually restricts and eliminates bacterial infection. However, many bacterial pathogens, including *Shigella*, deliver a subset of T3SS effectors that manipulate host innate immune responses, thereby promoting bacterial colonization and survival (2) (**Figure 5**).

ATP Release

During bacterial infection, various molecules are released from damaged or stressed cells as DAMPs (e.g., ATP, uric acid, and membrane remnants), which trigger inflammatory immune responses. For example, infection of intestinal epithelial cells with enteric bacterial pathogens, such as *Shigella*, *Salmonella*, and EPEC, induce connexin hemichannel-dependent ATP release, which acts as an endogenous danger alarm against bacterial infection and triggers inflammatory responses (58). Hemichannel-dependent ATP release is a common host defense system against enteric bacterial infection. To counteract ATP-dependent inflammation, *Shigella* blocks ATP release by delivering IpgD to epithelial cells. IpgD blocks hemichannels and prevents ATP release through production of PI5P, which regulates hemichannel opening, thereby dampening ATP-dependent inflammation (**Figure 5**). Consistent with this, in both *in vitro* and *in vivo* infections, the *Shigella* Δ*ipgD* mutant causes elevated ATP release, severe intestinal inflammation, and mucosal damage, illustrating the key role of IpgD in preventing ATP-dependent inflammation (58).

Manipulation of NF-κB Signaling

During bacterial infection, the transcription factor NF-κB, the master regulator of pro-inflammatory cytokines, plays a major role in mediating an inflammatory signaling pathway that triggers a wide range of host inflammatory responses. In response to bacterial infection, several intracellular and extracellular stimuli activate signal transduction cascades, and NF-κB is translocated into the nucleus, where it promotes the transcription of target genes (59). In particular, ubiquitination of signaling factors is prerequisite for regulation of NF-κB activity; therefore, many bacterial pathogens, including *Shigella*, inhibit NF-κB activation by targeting signaling factors and altering signal transduction, thereby dampening inflammation, in order to promote infection (60).

Invasion of epithelial cells by *Shigella* produces membrane ruffles by remodeling the actin cytoskeleton around the bacterial entry site. Aberrant membrane ruffles, which protrude from the bacterial entry site and are accompanied by diacylglycerol (DAG) production, are sensed as DAMPs by the host innate immune system and trigger the activation of the DAG-CBM (CARMA-BCL10-MALT1)-TRAF6-NF-κB pathway. However, *Shigella* delivers OspI via the T3SS into the host cells; this factor targets and deamidates UBC13 (converts Gln-100 to Glu-100), an E2 required for TRAF6 E3 activity, resulting in abolition of its E2 activity and thereby interfering with the DAG–CBM–TRAF6–NF-κB pathway (61) (**Figure 5**).

Following membrane ruffling, *Shigella*-mediated vacuolar membrane ruptures are also recognized as DAMPs, triggering an additional alarm-signaling pathway via recruitment and activation of PKC, ultimately leading to the activation of the PKC–NF-κB pathway. To counteract PKC–NF-κB activation,

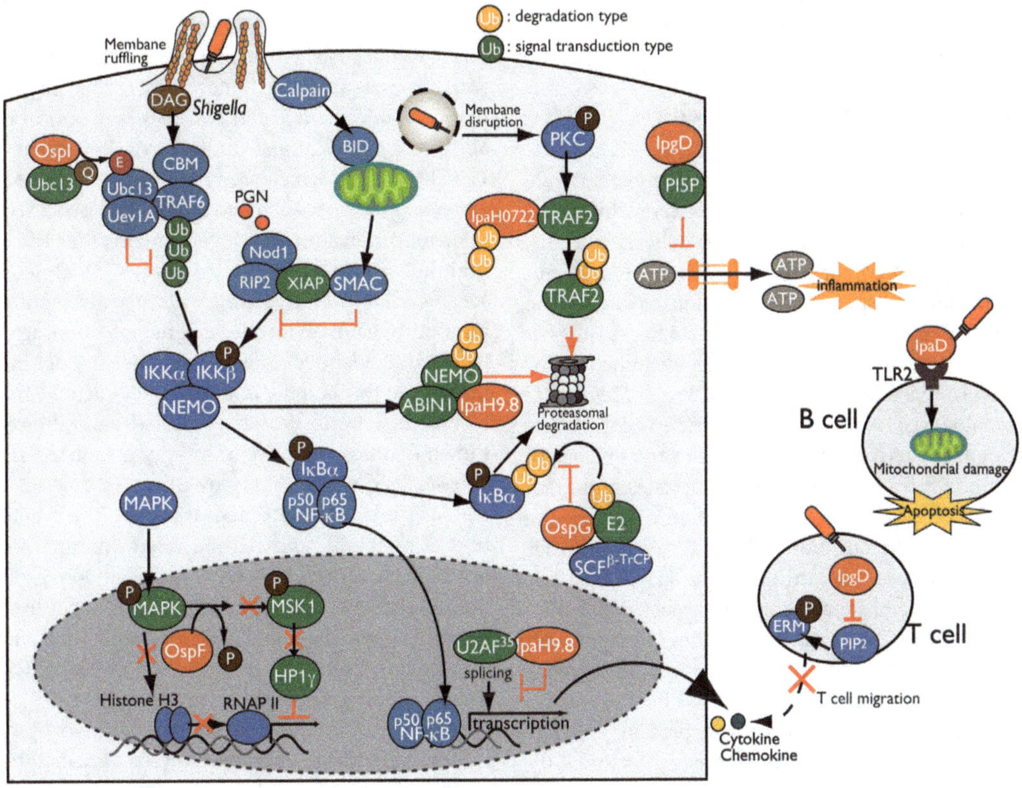

FIGURE 5 | *Shigella* manipulates the host innate and adaptive immune response. *Shigella* manipulates host inflammatory responses by delivering a subset of T3SS effectors, including IpgD, OspI, OspG, OspF, and IpaH. PI5P, generated by IpgD, prevents ATP release-dependent inflammation through hemichannels. OspG binds to ubiquitin and ubiquitinated E2 proteins and prevents IκBα ubiquitylation, which is required for NF-κB activation. OspI deamidates and inactivates Ubc13, resulting in inhibition of TRAF6 ubiquitination. IpaH0722 and IpaH9.8 target TRAF2 and NEMO, respectively, for ubiquitination, and undergo proteasome degradation. In addition, IpaH9.8 further targets U2AF35, preventing its participation in the splicing reaction via an unidentified bacterial factor, *Shigella* induces calpain-dependent BID activation, which, in turn, releases mitochondrial SMAC to antagonize XIAP-mediated inflammation. IpgD-mediated hydrolysis of PIP2 inactivates ERMs of T cells, which crosslink actin filaments with the plasma membrane, thereby preventing T cell migration. *Shigella* induces B cell death in both invaded and non-invaded cells. In non-invaded cells, the T3SS needle-tip protein IpaD interacts with TLR2 on B cells and triggers mitochondrial damage, which eventually induces apoptosis.

Shigella delivers IpaH0722, one of the IpaH family of E3 ubiquitin ligase effectors. IpaH0722 dampens the acute inflammatory response by ubiquitination of TRAF2, a molecule downstream of PKC, thereby preferentially inhibiting PKC-mediated NF-κB activation (62) (**Figure 5**).

During *Shigella* invasion of epithelial cells, Nod1 recognizes PGNs released from *Shigella* as PAMPs, triggering the Nod1–RIP2 pathway and activating the downstream mitogen-activated protein kinase (MAPK) and NF-κB signaling pathways. *Shigella* delivers IpaH9.8, an IpaH family E3 ubiquitin ligase effector, which preferentially prevents the Nod1-dependent NF-κB activation. IpaH9.8 interacts with NEMO/IKKγ, an essential component of the IKK kinase complex, and targets NEMO for ubiquitination. IpaH9.8 also interacts with ABIN-1, an ubiquitin-binding adaptor protein, to further promote polyubiquitylation of NEMO. Ubiquitinated NEMO by IpaH9.8 subsequently leads to proteasomal degradation, thereby diminishing NF-κB activation (63) (**Figure 5**). The activity of this E3 ligase effector during *Shigella* infection contributes to bacterial colonization in a mouse lung infection models (63).

Upon Nod1 activation, the X-linked inhibitor of apoptosis protein (XIAP) plays a crucial role in activating Nod1-dependent inflammatory responses. XIAP interacts with RIP2 and facilitates NF-κB activation, ultimately leading to pro-inflammatory gene transcription. A recent report showed that *Shigella* evades XIAP-mediated inflammation by actively releasing mitochondrial SMAC (36). *Shigella* infection triggers calpain activation and processes and activates BH3-only protein BID, which then translocates to the mitochondria and induces the release of SMAC. Released SMAC binds to XIAP, and inhibits XIAP-mediated NF-κB activation without inducing mitochondrial damage or cell death (**Figure 5**). Although it remains unknown whether BID-mediated release of SMAC by *Shigella* depends on a T3SS effector, *Shigella* uniquely exploits mitochondrial function to block the host innate immune response while avoiding cell death (36).

In addition, *Shigella* delivers another effector, OspG, which shares sequence similarity with mammalian serine/threonine kinases and inhibits NF-κB activation (64). OspG binds to ubiquitin and ubiquitinated E2s, which are required for phospho-IκBα ubiquitination by an E3 ligase such as $SCF^{\beta\text{-TrCP}}$, and formation of this OspG–E2-ubiquitin complex promotes OspG kinase activity and increases the prevention of NF-κB activation (64–67) (**Figure 5**).

Epigenetic Regulation of Immune Gene Expression

Because the host transcriptional program plays a pivotal role in triggering inflammation and eliminating bacterial infection, *Shigella* overcomes this defense mechanism by delivering a subset of T3SS effector proteins that regulate gene expression and counteract host immune responses. In addition to interfering with NF-κB signaling, as described above, *Shigella* hacks the host immune responses by reprograming the host epigenome to promote infection. The T3SS effector OspF (homolog of *Salmonella* SpvC and *Pseudomonas syringae* HopAl1), which has a unique phosphothreonine lyase activity, translocates into the nucleus of epithelial cells, where it irreversibly dephosphorylates and inactivates MAPKs (Erk and p38) through beta-elimination of the phosphate group (68, 69). Inactivation of MAPK by OspF further blocks downstream phosphorylation of histone H3 at Ser10 at the promoters of a subset of innate immune genes, such as IL-8, and promotes chromatin condensation, resulting in repression of transcription by masking NF-κB binding sites (68) (**Figure 5**). In addition to histone modification, OspF also alters the activity of the chromatin reader heterochromatin protein 1 (HP1) and represses host gene expression during *Shigella* infection (70). Although phosphorylated HP1γ at Ser83 proteins positively regulate euchromatin (a transcriptionally active state) and activate transcription, OspF inactivates Erk and consequently reduces the activity of the downstream kinase MSK1, a kinase for HP1γ at Ser83. As a result of HP1γ dephosphorylation, HP1γ dissociates from sites of transcriptional activation at OspF-target genes such as IL-8 (**Figure 5**). Consistent with this, the *Shigella ΔospF* mutant increased the level of HP1γS83 phosphorylation relative to WT *Shigella*-infected cells (70). Therefore, OspF manipulates host transcriptional response via two epigenetic modifications: (i) decreasing the level of phosphorylated histone H3, and (ii) altering the activity of HP1 through dephosphorylation of MAPK, which contributes to downregulation of host inflammatory responses.

In addition to OspF, *Shigella* delivers IpaH9.8 to regulate pro-inflammatory gene expression. IpaH9.8 is translocated into the nucleus of epithelial cells, where it binds to U2AF35, an mRNA splicing factor, and inhibits the U2AF35-dependent splicing reaction, enabling the bacterium to dampen expression of numerous genes, including some that encode pro-inflammatory cytokines and chemokines (71) (**Figure 5**).

Manipulation of Host Adaptive Immunity

Manipulation of the host innate immune response by *Shigella* is a pivotal survival strategy for promoting infection; however, the interactions between *Shigella* and adaptive immune response, such as T and B lymphocytes, have not been thoroughly investigated. One prominent issue is the lack of appropriate animal infection models that mimic human intestinal infection. Until now, however, several studies have proven the importance of adaptive immunity against *Shigella* infection using a mouse pulmonary infection model that mimics the acute inflammation that occurs during shigellosis. Mice that are genetically deficient in B, T, and NK cells are much more susceptible to *Shigella* infection than WT mice, and T and NK cells play critical roles in clearing *Shigella* (72). Additional studies have shown that CD4$^+$ T helper 17 (Th 17) cells, which are predominantly primed in response to *Shigella*, produce IL-17A and eventually restrict secondary *Shigella* infection, indicating the involvement of Th17 cells in adaptive immunity against *Shigella* infection (73). By contrast, antigen-specific CD8$^+$ T cells, which are usually required for adaptive immunity against cytosolic bacterial infection, are not primed and involved in adaptive immunity against *Shigella* infection (74). As the precise role and the importance of adaptive immunity against *Shigella* infection have been revealed, several recent studies have shown that host adaptive immunity is targeted and subverted by *Shigella* T3SS effectors (**Figure 5**).

T cell migration and activation are key events in the induction of antibody- and cell-mediated immune responses against bacterial infection. Recently, the Phalipon group investigated the interaction between T cells and *Shigella* using *in vitro* and *in vivo* approaches, and demonstrated that *Shigella* interferes with adaptive immune responses by targeting T cells. They also found that *Shigella* invades activated CD4$^+$ T cells and inhibit chemoattractant-mediated T cell migration by delivering IpgD (75). T-lymphocyte migration toward a chemoattractant depends on the membrane cytoskeleton crosslinkers proteins ERM (ezrin, radixin, and moesin), which are tightly converted between the active and inactive conformation by the concentration of PIP2 at the plasma membrane. Because IpgD is phosphoinositide phosphatase, it subsequently hydrolyzes and decreases the concentration of PIP2 at the plasma membrane, thereby inactivating ERM (**Figure 5**). In addition, *Shigella* impairs CD4$^+$ T cell dynamics within lymph nodes, where adaptive immunity is initiated, in the mouse infection model (76). Therefore, *Shigella* targets T-lymphocytes and inhibits their migration by delivering the T3S effector IpgD, thereby interfering with adaptive immunity.

B lymphocytes are key players in antibody-mediated immunity, and they produce and secrete cytokines that contribute to the antibody-independent immune response against bacterial infection. However, massive T- and B-cell deaths are observed in rectal biopsies of *Shigella*-infected humans, indicating that *Shigella* impairs B cell-mediated immunity (77). To support this notion, a recent report showed that *Shigella* targets B cells and induces cell death in both *Shigella*-invaded and non-invaded cells (78). *Shigella* invades B cells and replicates intracellularly, resulting in B cell death. Furthermore, *Shigella* induces B cell apoptosis via the T3SS needle-tip protein IpaD. IpaD binds to TLR2 on B cells and triggers the loss of mitochondrial dysfunction and apoptotic cell death signaling in non-invaded B cells, eventually helping the bacterium to avoid antibody-mediated immune responses (78) (**Figure 5**). Therefore, *Shigella* targets T and B cells and manipulates adaptive immunity against *Shigella* infection, thereby preventing antibody-mediated lasting immunity and promoting bacterial infection.

Conclusion

Although bacterial infection elicits host defense systems that would restrict and eliminate bacteria, many bacterial pathogens have evolved excellent strategies to manipulate host immune responses and enable bacterial colonization. Here, we review the current understanding of how *Shigella* manipulates host immune responses, focusing specifically on T3SS effector-mediated interference with host signaling transduction cascades, alteration of membrane trafficking, and modulation of host-cell death. Our understanding of the molecular basis of the interaction between bacterial effectors and host immune systems has advanced greatly during the past decade. Although accumulating evidence has revealed the importance of manipulation of the host immune system during bacterial infection, our knowledge of bacterial strategy is still in its infancy. Although approximately 50 T3SS effectors of *Shigella* are currently recognized, we have only elucidated the molecular function of one-third of them. As has been shown for other bacterial pathogens, the effector activity and strategies described above are not unique to *Shigella*, but are instead shared as common strategies by many bacterial pathogens. Therefore, the discovery of new *Shigella* strategies that manipulate the host immune system will not only provide a new understanding of how bacteria subvert the host immune system but also facilitate development of new types of antibacterial drugs that target bacterial effectors and bacterial live vaccines to overcome bacterial infections.

Acknowledgments

This work was supported by a Grant-in-Aid for Specially Promoted Research (23000012, to CS) and a Grant-in-Aid for Scientific Research (C) (25460527 to HA). Part of this work was supported by grants from the Naito Foundation (HA) and Mochida Memorial Foundation for Medical and Pharmaceutical Research (HA).

References

1. Parsot C. *Shigella* type III secretion effectors: how, where, when, for what purposes? *Curr Opin Microbiol* (2009) **12**:110–6. doi:10.1016/j.mib.2008.12.002
2. Ashida H, Ogawa M, Mimuro H, Kobayashi T, Sanada T, Sasakawa C. *Shigella* are versatile mucosal pathogens that circumvent the host innate immune system. *Curr Opin Immunol* (2011) **23**:448–55. doi:10.1016/j.coi.2011.06.001
3. Wassef JS, Keren DF, Mailloux JL. Role of M cells in initial antigen uptake and in ulcer formation in the rabbit intestinal loop model of shigellosis. *Infect Immun* (1989) **57**:858–63.
4. Perdomo OJ, Cavaillon JM, Huerre M, Ohayon H, Gounon P, Sansonetti PJ. Acute inflammation causes epithelial invasion and mucosal destruction in experimental shigellosis. *J Exp Med* (1994) **180**:1307–19. doi:10.1084/jem.180.4.1307
5. High N, Mounier J, Prévost MC, Sansonetti PJ. IpaB of *Shigella flexneri* causes entry into epithelial cells and escape from the phagocytic vacuole. *EMBO J* (1992) **11**:1991–9.
6. Zychlinsky A, Prevost MC, Sansonetti PJ. *Shigella flexneri* induces apoptosis in infected macrophages. *Nature* (1992) **358**:167–9. doi:10.1038/358167a0
7. Carayol N, Tran Van Nhieu G. Tips and tricks about *Shigella* invasion of epithelial cells. *Current Opin Microbiol* (2013) **16**:1–6. doi:10.1016/j.mib.2012.11.010
8. Suzuki T, Miki H, Takenawa T, Sasakawa C. Neural Wiskott-Aldrich syndrome protein is implicated in the actin-based motility of *Shigella flexneri*. *EMBO J* (1998) **17**:2767–76. doi:10.1093/emboj/17.10.2767
9. Suzuki T, Mimuro H, Miki H, Takenawa T, Sasaki T, Nakanishi H, et al. Rho family GTPase Cdc42 is essential for the actin-based motility of *Shigella* in mammalian cells. *J Exp Med* (2000) **191**:1905–20. doi:10.1084/jem.191.11.1905
10. Egile C, Loisel TP, Laurent V, Li R, Pantaloni D, Sansonetti PJ, et al. Activation of the CDC42 effector N-WASP by the *Shigella flexneri* IcsA protein promotes actin nucleation by Arp2/3 complex and bacterial actin-based motility. *J Cell Biol* (1999) **146**:1319–32. doi:10.1083/jcb.146.6.1319
11. Fukumatsu M, Ogawa M, Arakawa S, Suzuki M, Nakayama K, Shimizu S, et al. *Shigella* targets epithelial tricellular junctions and uses a noncanonical clathrin-dependent endocytic pathway to spread between cells. *Cell Host Microbe* (2012) **11**:325–36. doi:10.1016/j.chom.2012.03.001
12. Ashida H, Mimuro H, Ogawa M, Kobayashi T, Sanada T, Kim M, et al. Cell death and infection: a double-edged sword for host and pathogen survival. *J Cell Biol* (2011) **195**:931–42. doi:10.1083/jcb.201108081
13. Kayagaki N, Warming S, Lamkanfi M, Vande Walle L, Louie S, Dong J, et al. Non-canonical inflammasome activation targets caspase-11. *Nature* (2011) **479**:117–21. doi:10.1038/nature10558
14. Kayagaki N, Wong MT, Stowe IB, Ramani SR, Gonzalez LC, Akashi-Takamura S, et al. Noncanonical inflammasome activation by intracellular LPS independent of TLR4. *Science* (2013) **341**:1246–9. doi:10.1126/science.1240248
15. Willingham SB, Bergstralh DT, O'Connor W, Morrison AC, Taxman DJ, Duncan JA, et al. Microbial pathogen-induced necrotic cell death mediated by the inflammasome components CIAS1/cryopyrin/NLRP3 and ASC. *Cell Host Microbe* (2007) **2**:147–59. doi:10.1016/j.chom.2007.07.009
16. Miao EA, Mao DP, Yudkovsky N, Bonneau R, Lorang CG, Warren SE, et al. Innate immune detection of the type III secretion apparatus through the NLRC4 inflammasome. *Proc Natl Acad Sci U S A* (2010) **107**:3076–80. doi:10.1073/pnas.0913087107
17. Suzuki T, Franchi L, Toma C, Ashida H, Ogawa M, Yoshikawa Y, et al. Differential regulation of caspase-1 activation, pyroptosis, and autophagy via Ipaf and ASC in *Shigella*-infected macrophages. *PLoS Pathog* (2007) **3**:e111. doi:10.1371/journal.ppat.0030111
18. Davis BK, Roberts RA, Huang MT, Willingham SB, Conti BJ, Brickey WJ, et al. Cutting edge: NLRC5-dependent activation of the inflammasome. *J Immunol* (2011) **186**:1333–7. doi:10.4049/jimmunol.1003111
19. Rayamajhi M, Zak DE, Chavarria-Smith J, Vance RE, Miao EA. Cutting edge: mouse NAIP1 detects the type III secretion system needle protein. *J Immunol* (2013) **191**:3986–9. doi:10.4049/jimmunol.1301549
20. Yang J, Zhao Y, Shi J, Shao F. Human NAIP and mouse NAIP1 recognize bacterial type III secretion needle protein for inflammasome activation. *Proc Natl Acad Sci U S A* (2013) **110**:14408–13. doi:10.1073/pnas.1306376110
21. Suzuki S, Franchi L, He Y, Muñoz-Planillo R, Mimuro H, Suzuki T, et al. *Shigella* type III secretion protein MxiI is recognized by Naip2 to induce Nlrc4 inflammasome activation independently of Pkcδ. *PLoS Pathog* (2014) **10**:e1003926. doi:10.1371/journal.ppat.1003926
22. Lightfield KL, Persson J, Brubaker SW, Witte CE, von Moltke J, Dunipace EA, et al. Critical function for Naip5 in inflammasome activation by a conserved carboxy-terminal domain of flagellin. *Nat Immunol* (2008) **9**:1171–8. doi:10.1038/ni.1646
23. Kofoed EM, Vance RE. Innate immune recognition of bacterial ligands by NAIPs determines inflammasome specificity. *Nature* (2011) **477**:592–5. doi:10.1038/nature10394
24. Zhao Y, Yang J, Shi J, Gong YN, Lu Q, Xu H, et al. The NLRC4 inflammasome receptors for bacterial flagellin and type III secretion apparatus. *Nature* (2011) **477**:596–600. doi:10.1038/nature10510
25. Tenthorey JL, Kofoed EM, Daugherty MD, Malik HS, Vance RE. Molecular basis for specific recognition of bacterial ligands by NAIP/NLRC4 inflammasomes. *Mol Cell* (2014) **54**:17–29. doi:10.1016/j.molcel.2014.02.018
26. Senerovic L, Tsunoda SP, Goosmann C, Brinkmann V, Zychlinsky A, Meissner F, et al. Spontaneous formation of IpaB ion channels in host cell membranes reveals how *Shigella* induces pyroptosis in macrophages. *Cell Death Dis* (2012) **3**:e384. doi:10.1038/cddis.2012.124

27. Miao EA, Leaf IA, Treuting PM, Mao DP, Dors M, Sarkar A, et al. Caspase-1-induced pyroptosis is an innate immune effector mechanism against intracellular bacteria. *Nat Immunol* (2010) **11**:1136–42. doi:10.1038/ni.1960
28. Suzuki S, Mimuro H, Kim M, Ogawa M, Ashida H, Toyotome T, et al. *Shigella* IpaH7.8 E3 ubiquitin ligase targets glomulin and activates inflammasomes to demolish macrophages. *Proc Natl Acad Sci U S A* (2014) **111**:E4254–63. doi:10.1073/pnas.1324021111
29. Rohde JR, Breitkreutz A, Chenal A, Sansonetti PJ, Parsot C. Type III secretion effectors of the IpaH family are E3 ubiquitin ligase. *Cell Host Microbe* (2007) **1**:77–83. doi:10.1016/j.chom.2007.02.002
30. Ashida H, Kim M, Sasakawa C. Manipulation of the host cell death pathway by *Shigella*. *Cell Microbiol* (2014) **16**:1757–66. doi:10.1111/cmi.12367
31. Carneiro LA, Travassos LH, Soares F, Tattoli I, Magalhaes JG, Bozza MT, et al. *Shigella* induces mitochondrial dysfunction and cell death in nonmyleoid cells. *Cell Host Microbe* (2009) **5**:123–36. doi:10.1016/j.chom.2008.12.011
32. Niebuhr K, Giuriato S, Pedron T, Philpott DJ, Gaits F, Sable J, et al. Conversion of PtdIns(4,5)P(2) into PtdIns(5)P by the *S. flexneri* effector IpgD reorganizes host cell morphology. *EMBO J* (2002) **21**:5069–78. doi:10.1093/emboj/cdf522
33. Pendaries C, Tronchère H, Arbibe L, Mounier J, Gozani O, Cantley L, et al. PtdIns5P activates the host cell PI3-kinase/Akt pathway during *Shigella flexneri* infection. *EMBO J* (2006) **25**:1024–34. doi:10.1038/sj.emboj.7601001
34. Ramel D, Lagarrigue F, Pons V, Mounier J, Dupuis-Coronas S, Chicanne G, et al. *Shigella flexneri* infection generates the lipid PI5P to alter endocytosis and prevent termination of EGFR signaling. *Sci Signal* (2011) **4**:ra61. doi:10.1126/scisignal.2001619
35. Bergounioux J, Elisee R, Prunier AL, Donnadieu F, Sperandio B, Sansonetti P, et al. Calpain activation by the *Shigella flexneri* effector VirA regulates key steps in the formation and life of the bacterium's epithelial niche. *Cell Host Microbe* (2012) **11**:240–52. doi:10.1016/j.chom.2012.01.013
36. Andree M, Seeger JM, Schüll S, Coutelle O, Wagner-Stippich D, Wiegmann K, et al. BID-dependent release of mitochondrial SMAC dampens XIAP-mediated immunity against *Shigella*. *EMBO J* (2014) **33**:2171–87. doi:10.15252/embj.201387244
37. Kobayashi T, Ogawa M, Sanada T, Mimuro H, Kim M, Ashida H, et al. The *Shigella* OspC3 effector inhibits caspase-4, antagonizes inflammatory cell death, and promotes epithelial infection. *Cell Host Microbe* (2013) **13**:570–83. doi:10.1016/j.chom.2013.04.012
38. Knodler LA, Crowley SM, Sham HP, Yang H, Wrande M, Ma C, et al. Noncanonical inflammasome activation of caspase-4/caspase-11 mediates epithelial defenses against enteric bacterial pathogens. *Cell Host Microbe* (2014) **16**:249–56. doi:10.1016/j.chom.2014.07.002
39. Shi J, Zhao Y, Wang Y, Gao W, Ding J, Li P, et al. Inflammatory caspases are innate immune receptors for intracellular LPS. *Nature* (2014) **514**:187–92. doi:10.1038/nature13683
40. Travassos LH, Carneiro LA, Ramjeet M, Hussey S, Kim YG, Magalhães JG, et al. Nod1 and Nod2 direct autophagy by recruiting ATG16L1 to the plasma membrane at the site of bacterial entry. *Nat Immunol* (2010) **11**:55–62. doi:10.1038/ni.1823
41. Thurston TL, Wandel MP, von Muhlinen N, Foeglein A, Randow F. Galectin 8 targets damaged vesicles for autophagy to defend cells against bacterial invasion. *Nature* (2012) **482**:414–8. doi:10.1038/nature10744
42. Dupont N, Lacas-Gervais S, Bertout J, Paz I, Freche B, Van Nhieu GT, et al. *Shigella* phagocytic vacuolar membrane remnants participate in the cellular response to pathogen invasion and are regulated by autophagy. *Cell Host Microbe* (2009) **6**:137–49. doi:10.1016/j.chom.2009.07.005
43. Leung Y, Ally S, Goldberg MB. Bacterial actin assembly requires toca-1 to relieve N-wasp autoinhibition. *Cell Host Microbe* (2008) **3**:39–47. doi:10.1016/j.chom.2007.10.011
44. Baxt LA, Goldberg MB. Host and bacterial proteins that repress recruitment of LC3 to *Shigella* early during infection. *PLoS One* (2014) **9**:e94653. doi:10.1371/journal.pone.0094653
45. Ogawa M, Yoshimori T, Suzuki T, Sagara H, Mizushima N, Sasakawa C. Escape of intracellular *Shigella* from autophagy. *Science* (2005) **307**:727–31. doi:10.1126/science.1106036
46. Kayath CA, Hussey S, El Hajjami N, Nagra K, Philpott D, Allaoui A. Escape of intracellular *Shigella* from autophagy requires binding to cholesterol through the type III effector, IcsB. *Microbes Infect* (2010) **12**:956–66. doi:10.1016/j.micinf.2010.06.006
47. Huett A, Heath RJ, Begun J, Sassi SO, Baxt LA, Vyas JM, et al. The LRR and RING domain protein LRSAM1 is an E3 ligase crucial for ubiquitin-dependent autophagy of intracellular *Salmonella typhimurium*. *Cell Host Microbe* (2012) **12**:778–90. doi:10.1016/j.chom.2012.10.019
48. Mostowy S, Bonazzi M, Hamon MA, Tham TN, Mallet A, Lelek M, et al. Entrapment of intracytosolic bacteria by septin cage-like structures. *Cell Host Microbe* (2010) **8**:433–44. doi:10.1016/j.chom.2010.10.009
49. Mostowy S, Sancho-Shimizu V, Hamon MA, Simeone R, Brosch R, Johansen T, et al. p62 and NDP52 proteins target intracytosolic *Shigella* and *Listeria* to different autophagy pathways. *J Biol Chem* (2011) **286**:26987–95. doi:10.1074/jbc.M111.223610
50. McGhie EJ, Brawn LC, Hume PJ, Humphreys D, Koronakis V. *Salmonella* takes control: effector-driven manipulation of the host. *Curr Opin Microbiol* (2009) **12**:117–24. doi:10.1016/j.mib.2008.12.001
51. Sherwood RK, Roy CR. A Rab-centric perspective of bacterial pathogen-occupied vacuoles. *Cell Host Microbe* (2013) **14**:256–68. doi:10.1016/j.chom.2013.08.010
52. Stenmark H. Rab GTPases as coordinators of vesicle traffic. *Nat Rev Mol Cell Biol* (2009) **10**:513–25. doi:10.1038/nrm2728
53. Mellouk N, Weiner A, Aulner N, Schmitt C, Elbaum M, Shorte SL, et al. *Shigella* subverts the host recycling compartment to rupture its vacuole. *Cell Host Microbe* (2014) **16**:517–30. doi:10.1016/j.chom.2014.09.005
54. Dong N, Zhu Y, Lu Q, Hu L, Zheng Y, Shao F. Structurally distinct bacterial TBC-like GAPs link Arf GTPase to Rab1 inactivation to counteract host defenses. *Cell* (2012) **150**:1029–41. doi:10.1016/j.cell.2012.06.050
55. Burnaevskiy N, Fox TG, Plymire DA, Ertelt JM, Weigele BA, Selyunin AS, et al. Proteolytic elimination of *N*-myristoyl modifications by the *Shigella* virulence factor IpaJ. *Nature* (2013) **496**:106–9. doi:10.1038/nature12004
56. Burnaevskiy N, Peng T, Reddick LE, Hang HC, Alto NM. Myristoylome profiling reveals a concerted mechanism of ARF GTPase deacylation by the bacterial protease IpaJ. *Mol Cell* (2015) **58**:110–22. doi:10.1016/j.molcel.2015.01.040
57. Mounier J, Boncompain G, Senerovic L, Lagache T, Chrétien F, Perez F, et al. *Shigella* effector IpaB-induced cholesterol relocation disrupts the Golgi complex and recycling network to inhibit host cell secretion. *Cell Host Microbe* (2012) **12**:381–9. doi:10.1016/j.chom.2012.07.010
58. Puhar A, Tronchère H, Payrastre B, Nhieu GT, Sansonetti PJ. A *Shigella* effector dampens inflammation by regulating epithelial release of danger signal ATP through production of the lipid mediator PtdIns5P. *Immunity* (2013) **39**:1121–31. doi:10.1016/j.immuni.2013.11.013
59. Hayden MS, Ghosh S. NF-κB, the first quarter-century: remarkable progress and outstanding questions. *Genes Dev* (2012) **26**:203–34. doi:10.1101/gad.183434.111
60. Ashida H, Kim M, Sasakawa C. Exploitation of the host ubiquitin system by human bacterial pathogens. *Nat Rev Microbiol* (2014) **12**:399–413. doi:10.1038/nrmicro3259
61. Sanada T, Kim M, Mimuro H, Suzuki M, Ogawa M, Oyama A, et al. The *Shigella flexneri* effector OspI deamidates UBC13 to dampen the inflammatory response. *Nature* (2012) **483**:623–6. doi:10.1038/nature10894
62. Ashida H, Nakano H, Sasakawa C. *Shigella* IpaH0722 E3 ubiquitin ligase effector targets TRAF2 to inhibit PKC-NF-κB activity in invaded epithelial cells. *PLoS Pathog* (2013) **9**:e1003409. doi:10.1371/journal.ppat.1003409
63. Ashida H, Kim M, Schmidt-Supprian M, Ma A, Ogawa M, Sasakawa C. A bacterial E3 ubiquitin ligase IpaH9.8 targets NEMO/IKKγ to dampen the host NF-κB-mediated inflammatory response. *Nat Cell Biol* (2010) **12**:66–73. doi:10.1038/ncb2006
64. Kim DW, Lenzen G, Page AL, Legrain P, Sansonetti PJ, Parsot C. The *Shigella flexneri* effector OspG interferes with innate immune responses by targeting ubiquitin-conjugating enzymes. *Proc Natl Acad Sci U S A* (2005) **102**:14046–51. doi:10.1073/pnas.0504466102
65. Zhou Y, Dong N, Hu L, Shao F. The *Shigella* type three secretion system effector OspG directly and specifically binds to host ubiquitin for activation. *PLoS One* (2013) **8**:e57558. doi:10.1371/journal.pone.0057558
66. Pruneda JN, Smith FD, Daurie A, Swaney DL, Villén J, Scott JD, et al. E2~Ub conjugates regulate the kinase activity of *Shigella* effector OspG during pathogenesis. *EMBO J* (2014) **33**:437–49. doi:10.1002/embj.201386386
67. Grishin AM, Condos TE, Barber KR, Campbell-Valois FX, Parsot C, Shaw GS, et al. Structural basis for the inhibition of host protein ubiquitination by *Shigella* effector kinase OspG. *Structure* (2014) **22**:878–88. doi:10.1016/j.str.2014.04.010

68. Arbibe L, Kim DW, Batsche E, Pedron T, Mateescu B, Muchardt C, et al. An injected bacterial effector targets chromatin access for transcription factor NF-κB to alter transcription of host genes involved in immune responses. *Nat Immunol* (2007) **8**:47–56. doi:10.1038/ni1423
69. Li H, Xu H, Zhou Y, Zhang J, Long C, Li S, et al. The phosphothreonine lyase activity of a bacterial type III effector family. *Science* (2007) **315**:1000–3. doi:10.1126/science.1138960
70. Harouz H, Rachez C, Meijer BM, Marteyn B, Donnadieu F, Cammas F, et al. *Shigella flexneri* targets the HP1γ subcode through the phosphothreonine lyase OspF. *EMBO J* (2014) **33**:2606–22. doi:10.15252/embj.201489244
71. Okuda J, Toyotome T, Kataoka N, Ohno M, Abe H, Shimura Y, et al. *Shigella* effector IpaH9.8 binds to a splicing factor U2AF(35) to modulate host immune responses. *Biochem Biophys Res Commun* (2005) **333**:531–9. doi:10.1016/j.bbrc.2005.05.145
72. Le-Barillec K, Magalhaes JG, Corcuff E, Thuizat A, Sansonetti PJ, Phalipon A, et al. Roles for T and NK cells in the innate immune response to *Shigella flexneri*. *J Immunol* (2005) **175**:1735–40. doi:10.4049/jimmunol.175.3.1735
73. Sellge G, Magalhaes JG, Konradt C, Fritz JH, Salgado-Pabon W, Eberl G, et al. Th17 cells are the dominant T cell subtype primed by *Shigella flexneri* mediating protective immunity. *J Immunol* (2010) **184**:2076–85. doi:10.4049/jimmunol.0900978
74. Jehl SP, Doling AM, Giddings KS, Phalipon A, Sansonetti PJ, Goldberg MB, et al. Antigen-specific CD8(+) T cells fail to respond to *Shigella flexneri*. *Infect Immun* (2011) **79**:2021–30. doi:10.1128/IAI.00939-10
75. Konradt C, Frigimelica E, Nothelfer K, Puhar A, Salgado-Pabon W, di Bartolo V, et al. The *Shigella flexneri* type three secretion system effector IpgD inhibits T cell migration by manipulating host phosphoinositide metabolism. *Cell Host Microbe* (2011) **9**:263–72. doi:10.1016/j.chom.2011.03.010
76. Salgado-Pabón W, Celli S, Arena ET, Nothelfer K, Roux P, Sellge G, et al. *Shigella* impairs T lymphocyte dynamics in vivo. *Proc Natl Acad Sci U S A* (2013) **110**:4458–63. doi:10.1073/pnas.1300981110
77. Raqib R, Ekberg C, Sharkar P, Bardhan PK, Zychlinsky A, Sansonetti PJ, et al. Apoptosis in acute shigellosis is associated with increased production of Fas/Fas ligand, perforin, caspase-1, and caspase-3 but reduced production of Bcl-2 and interleukin-2. *Infect Immun* (2002) **70**:3199–207. doi:10.1128/IAI.70.6.3199-3207.2002
78. Nothelfer K, Arena ET, Pinaud L, Neunlist M, Mozeleski B, Belotserkovsky I, et al. B lymphocytes undergo TLR2-dependent apoptosis upon *Shigella* infection. *J Exp Med* (2014) **211**:1215–29. doi:10.1084/jem.20130914

Conflict of Interest Statement: The authors declare that the research was conducted in the absence of any commercial or financial relationships that could be construed as a potential conflict of interest.

10

Pasteurella multocida Toxin Manipulates T Cell Differentiation

Dagmar Hildebrand, Klaus Heeg and Katharina F. Kubatzky*

Zentrum für Infektiologie, Medizinische Mikrobiologie und Hygiene, Universitätsklinikum Heidelberg, Heidelberg, Germany

Edited by:
Amy Rasley,
Lawrence Livermore National Laboratory, USA

Reviewed by:
Alexis DeHaven Dunkle,
Genentech, USA
Brenda Anne Wilson,
University of Illinois at Urbana–Champaign, USA

Correspondence:
Katharina F. Kubatzky
kubatzky@uni-heidelberg.de

Pasteurella multocida causes various diseases in a broad range of wild and domestic animals. Toxigenic strains of the serotypes A and D produce an AB protein toxin named *Pasteurella multocida* toxin (PMT). PMT constitutively activates the heterotrimeric G protein subunits $G\alpha_q$, $G\alpha_{13}$, and $G\alpha_i$ through deamidation of a glutamine residue, which results in cytoskeletal rearrangements as well as increased proliferation and survival of the host cell. In human monocytes, PMT alters the lipopolysaccharide (LPS)-induced activation toward a phenotype that suppresses T cell activation. Here we describe that the toxin also modulates CD4-positive T helper (Th) cells directly. PMT amplifies the expansion of Th cells through enhanced cell cycle progression and suppression of apoptosis and manipulates the differentiation of Th subclasses through activation of Signal Transducers and Activators of Transcription (STAT) family members and induction of subtype-specific master transcription factors. A large population of toxin-treated T cells is double-positive for Foxp3 and RORγt, the transcription factors expressed by Treg and Th17 cells, respectively. This suggests that these cells could have the potential to turn into Th17 cells or suppressive Treg cells. However, in terms of function, the PMT-differentiated cells behave as inflammatory Th17 cells that produce IL-17 and trigger T cell proliferation.

Keywords: Th17 cells, regulatory T cell, STAT proteins, *Pasteurella multocida* toxin, T cell proliferation, T helper cell differentiation, Foxp3, RORγt

INTRODUCTION

An effective T cell-driven immune response against microbial pathogens depends on the T cell receptor (TCR)-mediated expansion of antigen-specific T cells as well as the differentiation of specialized T cell subsets. The nature of the invading pathogen determines the resulting CD4-positive Th subtype that is generated. Microbial components are recognized by distinct pattern recognition receptors (PRRs) on innate immune cells. As a consequence, professional antigen-presenting cells (APCs) perform phagocytosis and present foreign antigens on major histocompatibility complexes (MHC) to T cells. Binding of presented antigens to the TCR transmits the activation signal to intracellular molecules, which trigger cellular proliferation. In addition, APCs can express T cell-activating surface molecules that bind the co-receptor CD28. This ligand-receptor binding is required for the full activation and expansion of T cells. Depending on the activated PRR, APCs produce a specific set of cytokines that defines the direction of Th differentiation. The released cytokines bind to their responding receptors on the Th cell and induce signaling cascades that are transmitted through Signal Transducers and Activators of Transcription (STAT) proteins. Depending on the cytokine STAT-3, STAT-4, STAT-5, or STAT-6 then induce the expression of Th subtype-specific master transcription factors. Together with the STAT proteins, they finally determine the differentiation of effector cells by triggering gene expression of lineage-characteristic cytokines and surface molecules (O'Shea et al., 2011). In this way, pathogen-specific Th effector cells develop to help provide an immune response tailored to recognize and destroy the microorganism. Initially, only two resulting Th subtypes, Th1 and Th2, were known. While Th1 cells that release IFN-γ and TNF-α, stimulate innate and T cell-induced immunity to recognize intracellular bacteria, Th2 cells boost the response against extracellular pathogens in the humoral and mucosal immunity. Today, a much higher variety of effector Th cells such as Th3, Th9, TR1, T follicular helper cells, Th17 and the suppressive regulatory T cells (Tregs) are known (Zhu and Paul, 2010). They can be seen as separate types or as a specific state of a certain main lineage. The plasticity of T cell differentiation is remarkable and allows a quick adaption to the invading microbe. Furthermore, this plasticity allows the control of the sensitive balance of defense activation and suppression, which is a prerequisite for a successful and moderate immune response. Lately it has become clear that the interplay between Th17 cells and Tregs is particularly important to maintain homeostasis (Astry et al., 2015; Chen et al., 2015; Talaat et al., 2015) as these two T cell subtypes have opposite functions in the regulation of the immune system. Th17 cells are named after the IL-17 family of cytokines and activate a broad range of immune cells (Park et al., 2005), hence Th17 cells are considered potent inflammatory cells with a role in autoimmune disorders (reviewed in (Korn et al., 2009). In contrast, induced Tregs (iTregs) mediate immune suppression and protect from an overactive immune response (Shevach and Thornton, 2014), whereas natural Tregs (nTregs) develop from autoreactive thymocytes in the medulla of the thymus and sustain tolerance to self-antigens (Bettini and Vignali, 2010). The precise division between nTreg and iTreg-mediated modes of suppression however, is still under investigation (Curotto de Lafaille and Lafaille, 2009). Although Th17 cells and Tregs have opposite functions, the differentiation of both lineages is closely connected. Th17 cell development is mediated by TGF-β and IL-6, the activation of STAT-3 and the following induction of RORγt (Ivanov et al., 2006; Tanaka et al., 2014). Induced Tregs can be differentiated from peripheral $CD4^{+}CD25^{-}$ T cells through activation of the transcription factors STAT-5 and Foxp3 in the presence of TGF-β and IL-2 (Burchill et al., 2007; Schmitt and Williams, 2013). Thus Th17 differentiation, as well as Treg formation, are dependent on TGF-β. In addition, the transcription factors Foxp3 and RORγt can influence each other and generate intermediate T cell subtypes such as IL-17-releasing Foxp3-positive cells (Voo et al., 2009; Kryczek et al., 2011). Finally, Th17 cells can turn into iTregs and nTregs can be converted into Th17 cells, respectively, under inflammatory conditions (Kong et al., 2012). However, this plasticity of CD4-positive T cells also has disadvantages for the host as it represents a welcome target for pathogens to modulate the immune response. T cell differentiation can either be indirectly modulated through the manipulation of APCs or directly through induction of signaling cascades in the lymphocyte (Miethke et al., 1994; Smith et al., 2004; Ivars, 2007; Sakai et al., 2010; Rodriguez-Garcia et al., 2011). *Pasteurella multocida* toxin (PMT) from *P. multocida* manipulates the host's signaling cascades through constitutive activation of the alpha subunits of the heterotrimeric G proteins Gq, G13, and Gi through deamidation of glutamine residue (Orth and Aktories, 2010). Downstream of the activated G protein, signaling events such as calcium mobilization, phospholipase C activation, activation of the small GTPase RhoA or mitogen-activated protein (MAP) kinase pathways have been reported (Wilson and Ho, 2011). Furthermore, we could show that PMT mediates long-term activation of the Janus kinase (JAK)-STAT pathway through constitutive activation of Gq. This results in constitutive STAT-3 activation and aberrant expression of its target gene, the kinase Pim-1 which in turn phosphorylates and thus inhibits the negative feedback regulator supressor of cytokine signaling (SOCS)-1 (Orth et al., 2007; Hildebrand et al., 2010). Eventually this causes an enhanced expression of JAK2 and an increase in STAT-3 activity. As a consequence of PMT-induced signaling, the toxin stimulates proliferation of a variety of cells types such as fibroblasts, bladder epithelial cells or osteoclasts, and protects cells efficiently from apoptosis (Preuss et al., 2010; Wilson and Ho, 2012). Recently we have shown that PMT alters the lipopolysaccharide (LPS)-induced activation of human monocytes toward a phenotype which shows decreased T cell activation (Hildebrand et al., 2012). The study presented here aimed to answer the question whether PMT can also directly modulate CD4-positive T cells responses. Our experiments on human blood-derived Th cells demonstrate that the toxin amplifies the CD3/CD28-induced expansion of T cells through enhanced cell cycle progression and survival. Moreover, PMT modulates the differentiation of Th cells into specific subtypes. The toxin mediates activation of STAT family members and the expression of the subtype determining transcription factors RORγt and Foxp3. Our experiments show that the PMT-generated T cell phenotype is able to induce proliferation of naïve T cells as well as the differentiation of Th17 cells.

MATERIALS AND METHODS

Expression of Recombinant Protein

Recombinant PMT and PMT^{C1165S} were kindly provided by J. Orth and K. Aktories (Freiburg). The toxin was expressed and purified as described before. Possible endotoxin contaminations were removed by an endotoxin removing gel (Pierce; Hildebrand et al., 2014).

T Cell Isolation

Peripheral blood mononuclear cells were isolated from fresh blood or buffy coats by density gradient centrifugation (Pancoll 1.077 g/mL; PAN Biotech) and washed three times with PBS. CD4-positive T cells were negatively selected (untouched) by magnetic-associated cell sorting ($CD4^{+}$ T Cell Isolation Kit, Miltenyi Biotec) and AutoMACS technology twice. Sorted cells (93–97% purity) were cultured in RPMI 1640 medium (Biochrom AG) supplemented with 10% FBS and 1% penicillin and streptomycin (PAA laboratories) at 37°C in a humidified atmosphere in the presence of 5% CO_2.

Activation and Stimulation

Cells were seeded at a concentration of 2.5 million per ml and activated with anti-CD3 and anti-CD28-coated beads (T Cell Activation/Expansion Kit, Miltenyi Biotec) at a ratio of 1:2. Cells were stimulated with PMT, PMT^{C1165S} or heat-inactivated dT PMT (65°C, 1 h) at a concentration of 1 nM. Cholera toxin (CT) and nocodazole (Sigma–Aldrich) were used at 100 ng/ml.

T Cell Differentiation *In Vitro*

For the *in vitro* differentiation of Th17 cells, CD3/CD28-activated Th cells were stimulated 5–7 days with 100 ng/ml of recombinant IL-1β, IL-6, IL-23, and TGF-β (Immuno Tools GmbH). Regulatory T cells were generated by treatment with 100 ng/ml rapamycin (Calbiochem) and 200 ng/ml IL-2 (Immuno Tools GmbH).

Cell Viability Assay

T cells were stimulated for 5 days in a 96-well plate. Then the CellTiter-Glo® Luminescent Cell Viability Assay (Promega) was performed by following the manufacturer's protocol. By adding the CellTiter-Glo reagent, cells are lysed and a luminescent signal (measured with a microplate reader, Hidex) proportional to the amount of ATP is generated.

Carboxyfluorescein Succinimidyl Ester (CFSE) Staining

Isolated T cells were stained with carboxyfluorescein succinimidyl ester (CFSE; 2.5 million cells/ml, 5 μM CFSE, Biolegend) for 10 min at room temperature (r.t.) in the dark, before the reaction was stopped with cold medium containing 10% FCS. After washing the cells with medium, they were activated and stimulated as described above. Fluorescence intensity of the cells was quantified with a FACSCanto on day one (basis measurement) and day five.

ELISA

Supernatants of cultured T cells were collected at the indicated time points for the subsequent quantification of cytokines. IL-2, IL-17A, and IL-10 ELISAs (Biolegend) were performed following the manufacturer's instructions. Measurements were performed on a Tecan GENios Pro plate reader and analyzed with Magellan5 software.

Cell Cycle Stage Analysis

T cells were synchronized in G0 phase by culturing them overnight in the absence of an activation stimulus. Then the cells (250,000 cells per condition) were activated with anti-CD3/CD28 beads and stimulated with PMT, CT (G0/1 phase), or nocodazole (G2 phase). After 24 h cells were washed and the cell pellet was resolved in 70% ethanol and stored on ice for 15 min for fixation. Washed cells were then resolved in staining-solution (50 μg/ml propidium iodide, 100 μg/ml RNAse A, 0.02% Tween 20 in PBS) with constant vortexing. After incubation for 40 min at 37°C cells were analyzed with a FACSCanto (*x*-axis: PE-A, *y*-axis: PE-W).

Immunoblot Analysis

Cells were washed twice and lysed as described before (Hildebrand et al., 2010) in RIPA buffer and analyzed by western blotting. Samples were boiled in Laemmli buffer and separated on a 4–20% gradient polyacrylamide gel (Anamed GmbH) by SDS-PAGE, transferred to nitrocellulose, and immunoblotted. To determine STAT activation, antibodies against p(Y705)STAT-3, p(Y693)STAT-4, STAT-4, p(Y694)STAT-5, STAT-5, p(Y641)STAT-6, STAT-6, actin, and anti-rabbit horseradish peroxidase (HRP) from Cell Signaling Technology and against STAT-3 from Santa Cruz Biotechnology were used. To determine cell cycle-relevant proteins, nuclear extracts were produced as described before (Hildebrand et al., 2010). Antibodies were purchased from Cell Signaling Technology (anti-CDK6, -CDK4, -cyclin D1, -p27, -cyclin B, p(Y15) CDK1), Santa Cruz Biotechnology (anti-cyclin E, C-19), and Sigma–Aldrich (anti-B23). Primary antibodies were used 1:1000, HRP-coupled secondary antibodies 1:5000.

Caspase 3, 7, and 8 Activity Assay

T cells were seeded in 96-well plates, activated with anti-CD3/CD28-coated beads and treated for 1 or 5 days with PMT. Caspase 3, 7, and 8 activities were measured using the Caspase-3/7Glo and Caspase-8Glo Assay (Promega) in accordance to the manufacturer's instructions. Luminescence was measured using the Plate CHAMELEON™V microplate reader (Hidex).

Reverse Transcriptase-PCR

RNA was extracted from T cells with the "High pure RNA Isolation Kit" (Roche), following the manufacturers protocol. Total RNA was quantified with a NanoDrop and cDNA was prepared using "Reversed Aid First Strand cDNA Synthesis Kit" (Fermentas Life Science). The cDNA was then used for quantitative PCR analysis with the "SYBR Green Rox mix" (Thermo Scientific) and sequence-specific primers: *actin*, sense, 5′- AGA GCT ACG AGC TGC CTG AC -3′, antisense, 5′- AGC ACT GTG TTG GCG TAC AG-3′, human *FOXP3*: sense 5′- GAA ACA GCA CAT TCC CAG AGT TC-3′, antisense 5′- ATG GCC CAG CGG ATG AG -3′, human *TBET*: sense, 5′- AAG TTT AAT CAG CAC CAG ACA GAG ATG ATC -3′, antisense, 5′- AAC AGA TGT GTA CAT GGA CTC AAA GTT CTC -3′, human *GATA3*: sense 5′- GCT GTC TGA AGC CAG GAG AGC -3′, antisense 5′- ATG CAT CAA ACA ACT GTG GCC AA -3′, human *RORC*: sense 5′- CCG CTG AGA GGG CTT CAC -3′, antisense 5′- TGC AGG AGT AGG CCA CAT TAC A -3′. The results were analyzed using the Fast Real-Time PCR System (Applied Biosystems). The mean values of the results (mean ± SEM; $n = 2$) were normalized to *actin*.

FACS Staining

T cells were fixed with 4% paraformaldehyde/PBS at r.t. for 15 min. Subsequently, the cells were permeabilized in 0.1% Triton X-100/PBS for 5 min (r.t.). Afterward, one million cells were incubated with 2 μg of antibodies against CD4 (Miltenyi Biotec), RORγt, Foxp3, IL-10, or IL-17 (eBioscience) for 1 h at 4°C. After washing three times, cells were incubated for 30 min at 4°C with

1 μg of the secondary antibody per one million cells (anti-mouse FITC or anti-rat PE, eBioscience). The antibody against CD4 was directly labeled with an APC fluorophore. Washed cells were then analyzed on a FACSCanto.

Co-culture Experiment

Isolated, CD3/CD28-activated T cells were differentiated into Th17 cells or Tregs or stimulated with PMT for 7 days. Then untouched T cells from the same donor were isolated and stained with CFSE (5 μM CFSE/DMSO for 10 min at 37°C, before the staining reaction was stopped with ice cold medium/10% FCS). CFSE-stained cells were activated with CD3/CD28-beads and co-cultured with washed Tregs, Th17 cells, or PMT-generated T cells (ratio 1:1). After 5 days of co-culture the fluorescent signal of the CFSE-labeled cells was quantified on the FACSCanto (FITC channel). Histogram overlays were performed using the Weasel.jar software.

Statistics

Statistical significance was assessed using Prism6 software using a two-sided ANOVA multiple comparison test with $^{*}p \leq 0.05$, $^{**}p \leq 0.01$, $^{***}p \leq 0.001$, $^{****}p \leq 0.0001$.

RESULTS

PMT Augments CD3/CD28-activated Th Cell Proliferation

Our investigation aimed to identify whether Th cell differentiation is targeted by the bacterial toxin PMT. In the experiments, we treated blood-derived CD4-positive T cells with anti-CD3- and anti-CD28-coupled beads to achieve full activation of the cells in the absence of an APC. This activation induces a network of downstream signaling pathways that eventually leads to cell proliferation and subsequently, depending on the cytokine environment, to differentiation into specific effector cells (Coquet et al., 2015). To investigate whether PMT is toxic for T cells, we initially checked its influence on the viability of CD3/CD28-activated T cells. To test this, we quantified the amount of cellular ATP in an assay where the ATP is used to generate a luminescence signal. **Figure 1A** shows that the ATP level of CD3/CD28-activated cells was increased considerably after 7 days in culture which confirmed the activation of cells and suggested an increase of the cell number. The additional stimulation with PMT remarkably enhanced lymphocyte viability. A PMT mutant, where a point mutation within the catalytic domain renders the toxin inactive, as well as heat-inactivated PMT did not alter the CD3/CD28-mediated activation. Therefore we concluded that the observed effect is mediated by the enzymatic function of PMT and that a potential endotoxin contamination can be excluded. To investigate whether the increase in cell viability is due to an increase in Th cell proliferation, stainings with fluorescent CFSE that binds covalently to proteins and other amino groups were performed. During cell division the fluorescent dye is passed on to the dividing cells and thus the fluorescence intensity decreases with each division. **Figure 1B** shows that PMT treatment leads to an enhanced number of cells that underwent cell division. Additionally, the PMT-stimulated cells divided more frequently than the CD3/CD28-activated cells (mean fluorescence of 4222 for PMT-treated cells vs. 6562 for untreated cells).

The proliferation rate of T cells can be altered by various signals. IL-2 is a potent growth factor that mediates cell cycle entry, increases survival of dividing cells and also guides the differentiation of T cells (Oppenheim, 2007). To investigate whether PMT enhances T cell proliferation through an increased secretion of IL-2, we quantified the production of this cytokine by ELISA. The kinetics revealed that the IL-2 concentration of CD3/CD28-activated cells increased from day one to day seven, as expected. Addition of PMT effectively enhanced the release of the cytokine compared to CD3/CD28-activated cells, at least during the first 3 days of culture, which might be a reason for the observed increase in proliferation (**Figure 1C**).

PMT Mediates Cell Cycle Progression and Survival of Th Cells

In the following experiments we aimed to determine the influence of PMT on cellular proliferation in more detail. Therefore we analyzed the cell cycle stages of Th cells. The cell cycle can be divided into four major phases. In the G0 phase the cell rests. When it enters G1 phase it increases its size and produces proteins as well as other cellular components. DNA replication takes place in the synthesis (S) phase, which is followed by the preparation (G2 phase) for cell division. Finally the division of the nucleus (M Phase, mitosis) and the formation of two daughter cells (cytokinesis) occur. An effective T cell cycle run requires TCR activation to trigger transition from the G0 to G1 phase and subsequently activation of the IL-2 receptor, which mediates the entry into S phase (Cantrell and Smith, 1984; Meuer et al., 1984). To examine the effect of the toxin on the cell cycle, we quantified the DNA content of the cells. Cells were stained with propidium iodide (PI) to determine the amount of cellular DNA by FACS. As a control for G2/M, cells were treated with nocodazole, a substance that interferes with the polymerization of microtubules and thus inhibits division into daughter cells so that nocodazole-treated cells contain the maximum amount of DNA during cell cycle. As a control for G0/1 phase and therefore the minimal amount of DNA, cells were treated with CT from *Vibrio cholerae* which is known to induce cell cycle arrest in the early G1 phase (Zheng et al., 2014). **Figures 2A,B** show that non-activated Th cells paused in G1/0 phase. CD3/CD28-activation stimulated the increase of DNA per cell which is characteristic for the S phase of the cycle. Stimulation of CD3/CD28-activated cells with PMT induced a further increase of the PI signal and revealed a population with a DNA amount that is typical for the G2/M phase. We therefore conclude that the toxin mediates progression of the cell cycle and DNA synthesis. To link this to molecules known to be involved in cell cycle progression, we checked the impact of PMT on the expression and activation level of cyclin-dependent kinases (CDKs), the regulatory cyclins and CDK inhibitors (**Figure 2C**). CDK activity is regulated by its phosphorylation status, association with cyclins

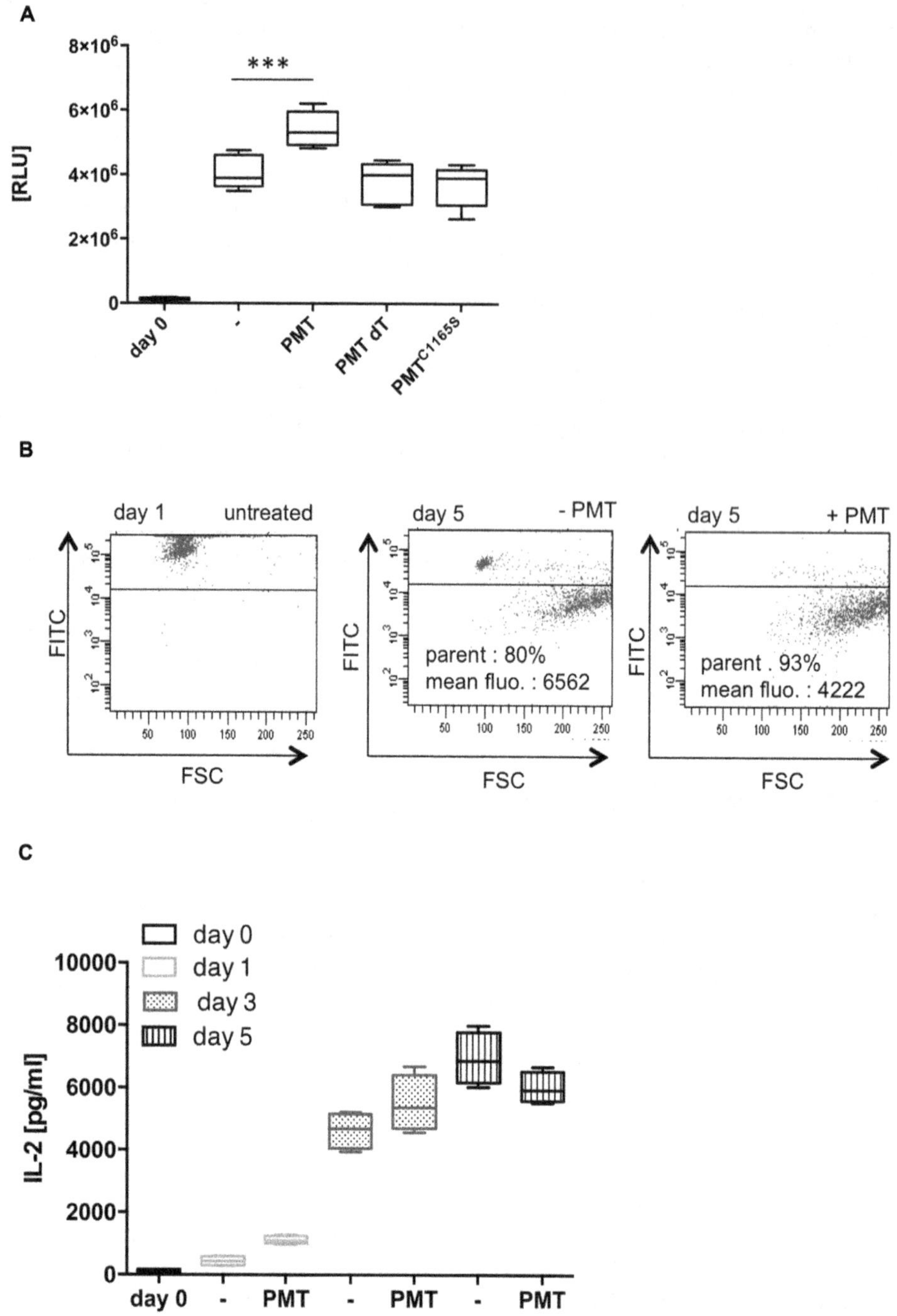

FIGURE 1 | *Pasteurella multocida* toxin (PMT) stimulates CD3/CD28-activated T cell proliferation. (A) CD4-positive T cells were activated with anti-CD3/CD28-coated beads (T Cell Activation/Expansion Kit, Miltenyi) and stimulated for 5 days with 1 nM PMT wild-type (wt), heat-inactivated PMT (dT), or a catalytically inactive mutant (PMT^{C1165S}). Viability of cells was determined by quantification of ATP using the CellTiter-Glo® Luminescent Cell Viability Assay (Promega). Shown is the mean ± SD, $n = 6$ (six donors, each experiment performed in duplicates). **(B)** Th cells were CFSE-labeled, CD3/CD28-activated and stimulated for 5 days with 1 nM PMT. Proliferation cycles were assayed by measuring the decreasing fluorescence by flow cytometry. The mean fluorescence and percentages refer to the gated (dividing) cells. Shown is the result of one representative donor out of three. **(C)** CD3/CD28-activated Th cells were stimulated with PMT for 1, 3, or 5 days. IL-2 ELISAs were performed (mean ± SD, $n = 4$; four donors, each experiment performed in duplicates). Statistical analysis was performed using two-sided ANOVA for multiple comparisons (***$p \leq 0.001$).

and binding of inhibitory molecules. Entry into S phase is mediated by CDK6/cyclin D and CDK4/cyclin D complexes and their activation requires the degradation of the inhibitor p27. The active complex then enables transcription factors to activate G1/S-phase gene expression (Bendris et al., 2015). The performed western blot analysis showed that PMT indeed mediates an

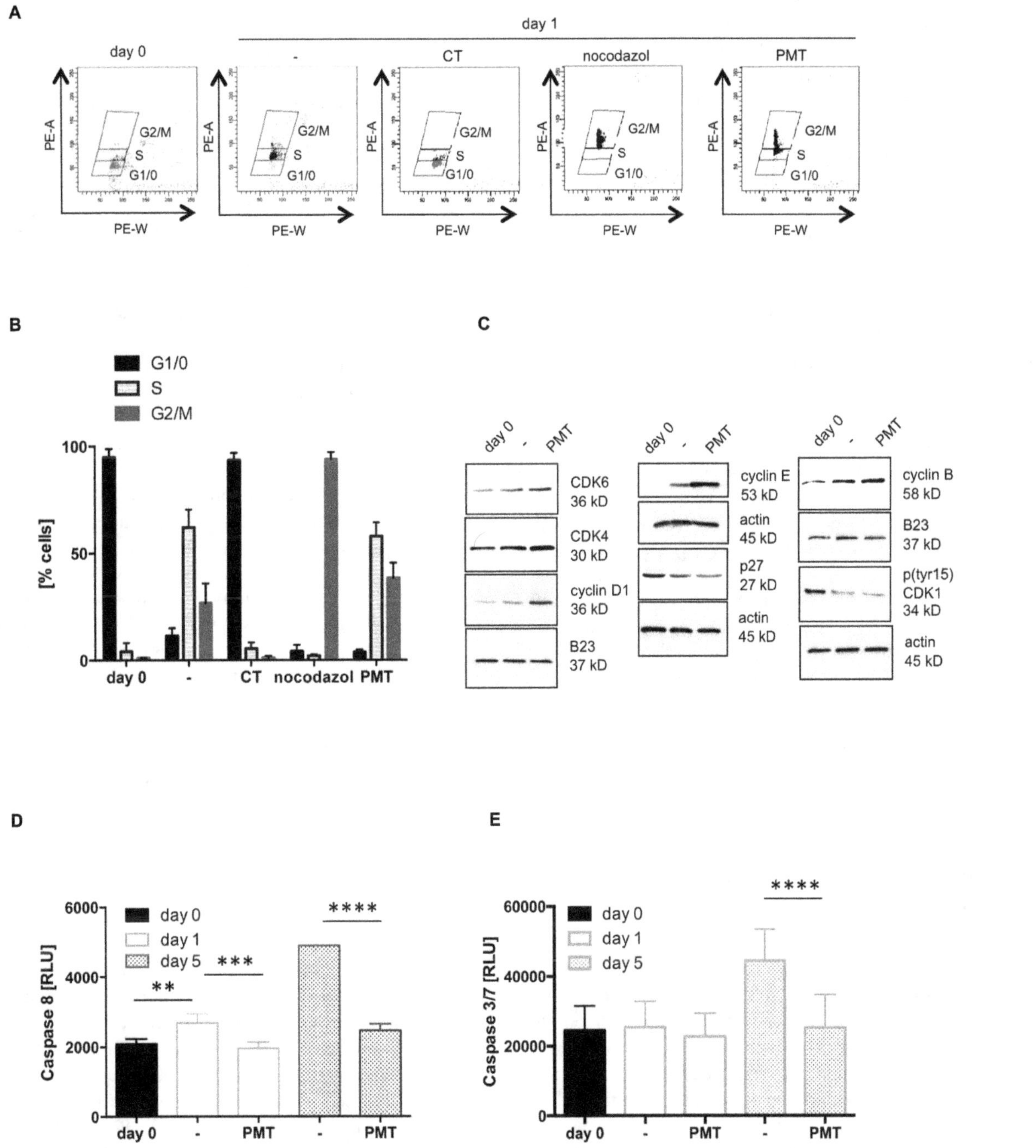

FIGURE 2 | PMT stimulates cell cycle progression. (A) CD4-positive T cells were cultured overnight without activation to synchronize cells in G0 phase. Then cells were CD3/CD28-activated and stimulated with PMT for 24 h. As controls, cells were treated with 100 ng/ml CT (G0/1 phase) or 100 ng/ml nocodazole (G2 phase). Nuclei were stained with propidium iodide (PI) and intensity of the PI signal was quantified by flow cytometry (*x*-axis: PE-A, *y*-axis: PE-W). **(B)** Quantification of the percentage of cells in G1/0, S and G2/M phase (mean ± SD, $n = 3$). **(C)** Cell lysates were used for western blot analysis with specific antibodies against cyclin B, pCDC2 (CDK1), cyclin D1, cyclin E, CDK4, or p27. Equal loading was verified by detection of actin (whole cell lysates) or B23 (nuclear extracts). The results were corroborated with two more donors. **(D,E)** T cells were stimulated with PMT for 1, 3, or 5 days, lysed and the activation of caspase-8 **(D)** or caspases 3 and 7 **(E)** was measured with a Caspase-8Glo® Assay or a Caspase 3/7-Glo® Assay, respectively (Promega). Shown is the mean ± SD, $n = 3$ donors. Statistical analysis for this figure **(B,D,E)** was performed using two-sided ANOVA for multiple comparisons (**$p \leq 0.01$; ***$p \leq 0.001$; ****$p \leq 0.0001$).

increased degradation of p27, an increase in cyclin E expression and additionally induces a stronger accumulation of CDK4, CDK6, and cyclin D in the nucleus than CD3/CD28. Furthermore PMT increases the expression of cyclin E which in complex with CDK2 promotes degradation of p27 (Bendris et al., 2015). Hence, toxin treatment seems to promote the entry into the S phase of the cell cycle. The entry of all eukaryotic cells into mitosis is regulated by activation of CDK1 (cdc2) at the G2/M transition. A critical regulatory step is the dephosphorylation of CDK1 at Thr14 and Tyr15 (Atherton-Fessler et al., 1994) and the phosphorylation-regulated translocation of cyclin B1 to the nucleus. Our results show reproducibly that PMT slightly enhances the translocation of cyclin B to the nucleus and a dephosphorylation of Tyr15 of CDK1. These results corroborate earlier findings reporting the upregulation of cell cycle progression by PMT using a fibroblast cell line (Wilson et al., 2000).

From our previous work in HEK 293 cells we knew that PMT is not only able to trigger proliferation, but also protects cells from staurosporine-induced apoptosis through the PI3K-dependent phosphorylation of Akt and the constitutive expression of the survival kinase Pim-1, respectively (Preuss et al., 2010). Programmed cell death is an integral part of the immune response, as the removal of T cells after clearance of the infection restores immune cell homeostasis (Pinkoski and Green, 2005). Taking this into consideration, the increased T cell number under PMT treatment could additionally be caused by the inhibition of apoptosis of activated T cells. To check this possibility, we analyzed caspase-8 and caspase-3/7 activation as central markers of the apoptotic response. The luminescence-based assays show that the activity of caspase-8 (**Figure 2D**) and caspase-3 and -7 (**Figure 2E**) increased considerably after 5 days of CD3/CD28-activation. Stimulation with PMT, however, kept the activity of the death caspases at the basal level of freshly isolated, naïve T cells. This supports our hypothesis that the toxin does not only trigger proliferation but also inhibits apoptosis of receptor-activated Th cells. In summary these data demonstrate that PMT expands CD3/CD28-activated Th cells by triggering proliferation and by suppressing apoptosis.

PMT Induces Foxp3 and RORγt-Positive T Cells

As we had observed that PMT supports the CD3/CD28-mediated proliferation of Th cells, we next aimed to investigate whether the toxin also modulates the differentiation of naïve T cells into specific T cell subtypes. T cell differentiation is dependent on the cytokine environment. Downstream of cytokine-induced receptor activation, STAT transcription factors play a pivotal role in T cell differentiation through the induction of subtype-specific genes. Hence our first question addressed the PMT-mediated phosphorylation and activation of STAT family members. Western blot analysis of toxin-stimulated cells (**Figure 3A**) shows a pronounced phosphorylation of STAT-3 in comparison to CD3/CD28-induced cells after 24 h of toxin treatment. As expected, STAT-5 phosphorylation was induced by CD3/CD28 activation, presumably via the production of IL-2, but this activation level could still be enhanced by PMT treatment. Phosphorylation of STAT-4 was not affected by antibody-activation and only slightly induced by PMT. Finally STAT-6 phosphorylation could not be observed under antibody or toxin conditions. As pSTAT-3 mediates differentiation of Th17 cells and pSTAT-5 is important in Treg formation, we generated these T cell subtypes *in vitro* and compared the STAT activation level of cytokine-differentiated control cells with PMT-treated cells. On day one, PMT treatment caused an intense phosphorylation of both transcription factors, however, on day seven only pSTAT-3 was detectable. Whereas fully differentiated (day 7) Tregs contained activated STAT-5 and only basal level of phosphorylated STAT-3, Th17 cells showed high STAT-3 but no STAT-5 activation. The data show that PMT-stimulated T cells display the same STAT activation pattern like Th17 cells (**Figure 3B**). When we analyzed the master transcription factors of subtype development after 24 h of treatment (**Figure 3C**), we found that in line with our previous results, toxin-stimulated CD4-positive cells contained higher copy numbers of *RORC* (STAT-3 target) and *FOXP3* (STAT-5 target) mRNA compared to *GATA3* (induced by STAT-6) and *TBET* (induced by STAT-4). This was confirmed on the protein level by intracellular FACS analysis (**Figure 3D**) where high expression of Foxp3 and RORγt could be found in cells that had been treated with PMT for 7 days. The CD3/CD28-activated population already contained a considerable amount of Foxp3, which is probably due to IL-2-mediated STAT-5 activation. These data raised the question whether PMT stimulates the differentiation of two different T cell subtypes or whether one subtype with two simultaneously expressed master transcription factors was produced. Double stainings for Foxp3 and RORγt (**Figure 3E**) revealed that the antibody-activated T cell population did not contain any double positive cells. In contrast to that the PMT-treated population contained around 30% RORγt-positive, 20% Foxp3-positive and almost 46% double positive T cells, even after 10 days of culture.

PMT Mediates Differentiation of an Inflammatory T Cell Subtype

Our results pointed to two different sets of T cell responses toward the bacterial toxin. Th17 cells have an autoinflammatory potential and support the immune response against extracellular bacteria, whereas Foxp3-positive Tregs have a suppressive role and protect the organism from autoimmunity. The different effector T cell types are ultimately defined by the cytokines that they release and by their function. Therefore we investigated which cytokines are actually released from PMT-stimulated cells and whether they support or suppress a T cell-driven immune response. ELISA analysis revealed that PMT-modulated cells release high amounts of IL-17A that is almost comparable to that of Th17 cells (**Figure 4A**). However, these cells produced nearly no immunosuppressive IL-10 (**Figure 4B**). FACS data corroborated these findings and additionally showed that there is no defect in the release of cytokines, but that IL-10 is hardly expressed by these lymphocytes (**Figure 4C**). Some of the CD3/CD28-activated T cells produced IL-10 (**Figures 4B,C**), which confirmed the differentiation of a few functional Foxp3-positive cells. To investigate whether the PMT-generated T cell subtype would rather have an activated or suppressive phenotype, we measured the ability of PMT-modulated T cells to increase or inhibit proliferation of CD3/CD28-activated CFSE-labeled CD4-positive T cells. Our data show that the CFSE signal of CD4-positive responder cells co-cultured with the PMT-differentiated effector subtype was lower than the signal obtained by CD3/CD28-activated responders. Additionally, the influence

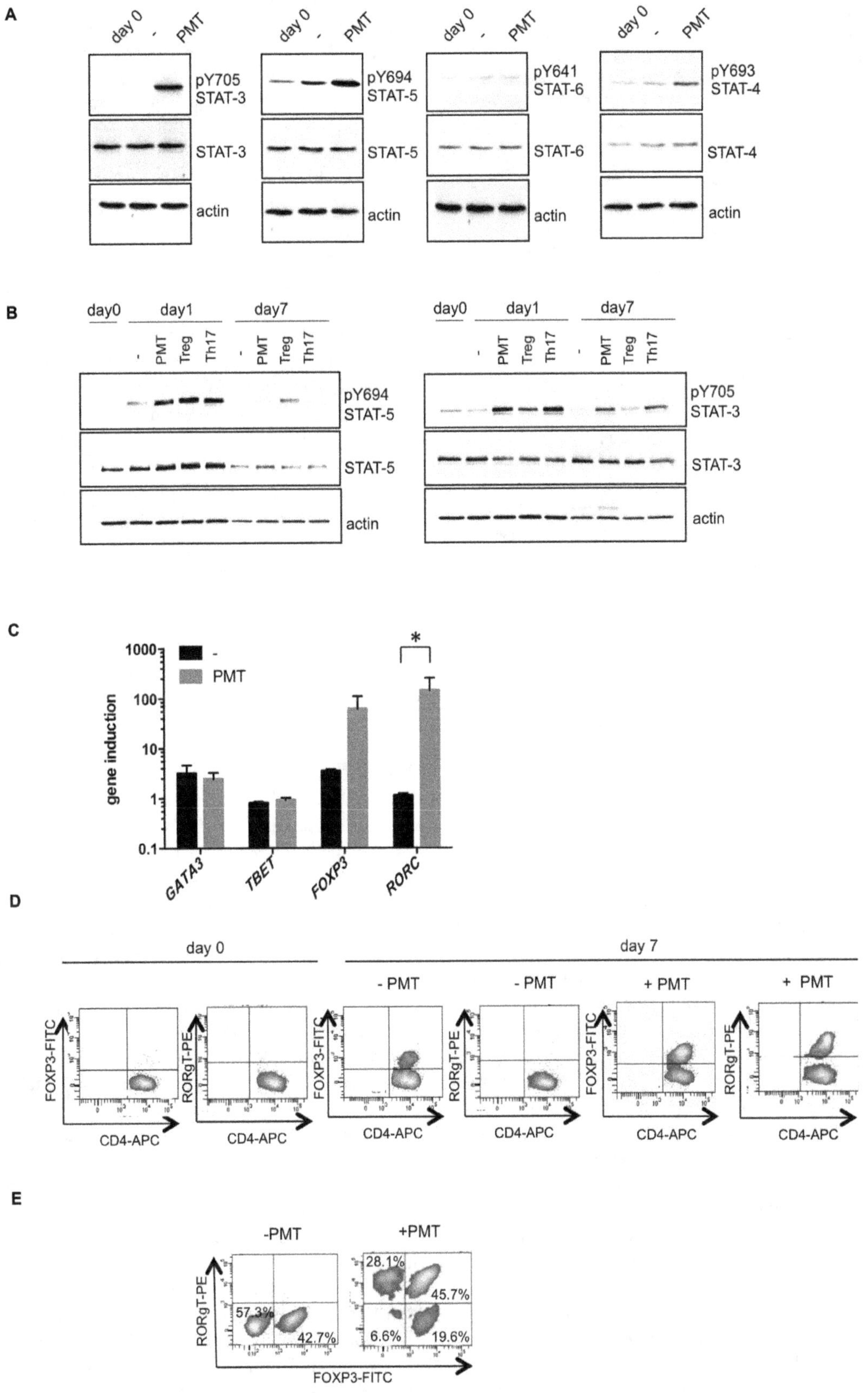

FIGURE 3 | Continued

FIGURE 3 | Continued

PMT induces Foxp3 and RORγt-positive T cells. CD3/CD28-activated T cells were stimulated with PMT overnight or for 7 days, respectively. **(A)** Whole cell lysates were produced and antibodies against p(Y705)STAT-3, STAT-3, p(Y694)STAT-5, STAT-5, p(Y641)STAT-6, STAT-6, and p(Y693)STAT-4, STAT-4 were used for immunoblotting. Equal loading of protein was verified by actin detection. **(B)** As a positive control, T cells were treated with 100 ng/ml IL-1β, IL-23, IL-6, TGF-β (Th17 cells), or 100 ng/ml IL-2 and rapamycin (Tregs). **(C)** Quantitative RT-PCRs were performed using SYBR Green and sequence specific primers for *FOXP3*, *TBET*, *GATA3*, and *RORC*. Results were normalized against the housekeeping gene *actin*. Shown is the mean of induction compared to non-activated cells. SD, $n = 2$ donors, experiment performed in duplicates. Statistical analysis was performed using two-sided ANOVA for multiple comparisons ($^*p \leq 0.05$). **(D)** Intracellular FACS analysis. Fixed and permeabilized cells were stained with anti-CD4-APC, purified anti-Foxp3 or anti-RORγt and FITC-/PE-labeled secondary antibodies. **(E)** Intracellular double stain of Foxp3 or RORγt of cells on day ten. Data shown in **(A,C,D)** are representative for three, and **(E)** for two experiments.

of PMT effector cells on proliferation was similar to the impact of Th17 cells. In contrast to this, Tregs showed an inhibitory impact on T cell proliferation (**Figure 4D**). This experiment suggests that the toxin modulates lineage decision of CD3/CD28-activated T helper cells toward an inflammatory subtype.

DISCUSSION

The relationship between the host and microbial pathogens can be quite complex. On the one hand the higher organism uses a highly specialized immune system for the detection and control of invading microorganisms, whereas pathogens have developed various strategies to avoid their recognition and elimination. The host's innate immune system can sense foreign structures through PRRs. In consequence of this recognition, pathogen-induced phagocytosis and presentation on MHC takes place. Presentation of the antigens to the TCR and the resulting association of CD3 with the TCR-peptide-MHC complex transmits the activation signal to intracellular molecules, which trigger a cascade of biochemical events that result in the expansion of antigen-specific T lymphocytes. In addition, the differentiation into effector types is crucial for the course of the immune response. Depending on the microbe and the activated PRRs, a specific set of cytokines is produced by the APC, which mediates the differentiation of appropriate effector types. The plasticity of Th cell differentiation during infection allows the development of pathogen-specific effector T cells. However, also the invading microorganism may profit from this plasticity by directing the differentiation to an ineffective or a suppressive effector type. One example for such a manipulation has been described for *Chlamydia muridarum*. Chlamydia-infected APCs are modulated in terms of the phenotype, cytokine secretion profile and antigen presentation. This eventually shifts the immune response from a protective Th1 to a non-protective Th2 response (Kaiko et al., 2008). Also extracellular bacteria such as *Staphylococcus aureus* and *Streptococcus pyogenes* disturb the balance between activation and inhibition of immune cells that protects the host from an over-reaction and re-establishes homeostasis after clearing the infection. Superantigens (SAgs) produced by these bacteria can activate up to 20% of all T cells in an individual, mediate a vast T cell expansion and induce a strong cytokine storm that causes serious harm to the host. Eventually, the majority of activated T cells die and the remaining cells will be anergic or even induce SAg-specific Treg-mediated tolerance to the microbe (Papiernik, 2001).

An excellent target to modulate the differentiation toward an ineffective T cell subtype are STAT transcription factors that transmit the cytokine-encoded differentiation signal. PMT activates the JAK-STAT pathway very efficiently in many types of cells (Kubatzky et al., 2013) and we found that the toxin also induces STAT activity in human T cells. After 1 day of PMT stimulation, the phosphorylation of both, STAT-3 and STAT-5 was very pronounced, whereas after 7 days of stimulation only phosphorylation of STAT-3 was detectable. Compatible with these results, RT-PCRs revealed a high expression of STAT-5-dependent *FOXP3* and STAT-3-promoted *RORC* in PMT-treated cells. This suggested that differentiation of Tregs and Th17 cells had taken place. Although the parallel development of two opposing subtypes appears paradox at first glance, it has been hypothesized that the interplay between iTregs and Th17 cells actually fine-tunes the activation to prevent hyperactivation and to facilitate the return to the steady state. This is supported by the fact that the development of these two subtypes is closely connected. Both subtypes need TGF-β for differentiation and the final direction of differentiation is induced by additional cytokines. TGF-β, in cooperation with IL-2-signaling, promotes STAT-5 activation and induction and maintenance of *FOXP3* via the induction of the Smad pathway. Nevertheless, TGF-β also mediates the maintenance of a functional IL-6 receptor on activated T cells. Under inflammatory conditions, with high concentrations of IL-6, TGF-β enables IL-6/STAT-3 signaling and eventually the differentiation of Th17 cells (Lee et al., 2009). The host's benefit from a high plasticity of Treg and Th17 cells is probably a fast reaction to the course of an infection. After the initial recognition of the pathogen, the inflammatory conditions change from the tolerogenic phenotype of nTregs toward the response-activating Th17 cells. In the process of clearance the cells are able to transdifferentiate again into iTregs to protect the body from hyperactivation (Gagliani et al., 2015). In consideration of this close connection of Th17 cells and Tregs and the complexity of the cytokine environment during infection it is not surprising that naïve, TGF-β-activated T cells can co-express RORγt and Foxp3. Our experiments show that after 1 week of culture several CD3/CD28-activated T cells are Foxp3 positive. This is probably due to the CD28-triggered and IL-2 mediated promotion of Foxp3 expression (Burchill et al., 2007; Guo et al., 2008) and may be due to an increased survival of nTregs (Weiss et al., 2011), (**Figure 3D**). Interestingly, 45% of the PMT-stimulated activated T lymphocytes are positive for both transcription factors. This could be caused by the strong activation of STAT-3 and STAT-5 during the first hours of stimulation. Furthermore it suggests that the observed toxin-mediated protection from apoptosis expands primarily Foxp3-positive cells, which later additionally express RORγt. The parallel production of RORγt and Foxp3 can lead to the

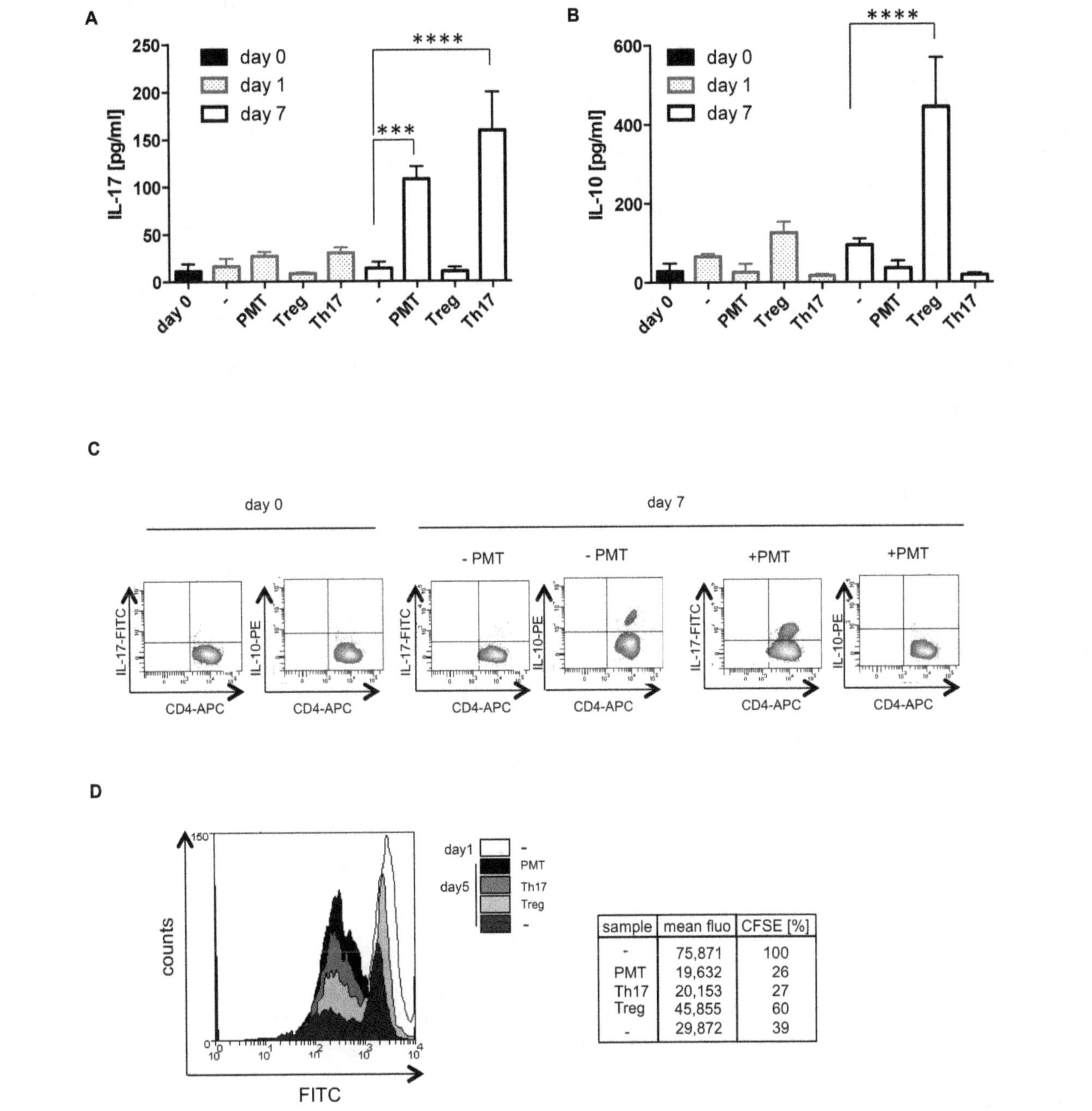

sample	mean fluo	CFSE [%]
-	75,871	100
PMT	19,632	26
Th17	20,153	27
Treg	45,855	60
-	29,872	39

FIGURE 4 | PMT mediates a Th17 phenotype. CD3/CD28-activated Th cells were stimulated with PMT for the indicated time points. In **(A)** IL-17A and **(B)** IL-10 ELISAs were performed (mean ± SD, $n = 3$). Statistical significance of the produced cytokine was compared to the value obtained on day 0 using two-sided ANOVA for multiple comparisons (***$p \leq 0.001$; ****$p \leq 0.0001$). **(C)** For intracellular FACS analysis, fixed and permeabilized cells were double-stained with anti-CD4-APC and purified anti-IL-10 or anti-IL-17 antibodies and the according FITC-/PE-labeled secondary antibodies. **(D)** Th cells were stimulated with PMT for 7 days. These PMT-generated T cells were then co-cultured with freshly isolated, CFSE-labeled Th cells (ratio 1:1) from the same donor. The fluorescent signal of the FITC-positive cells was quantified on day one and on day five of co-culture. Data shown in **(C,D)** are representative for three experiments (three donors).

induction of both IL-10 and IL-17 or just one of them (McGeachy et al., 2007; Ichiyama et al., 2008; Zhou et al., 2008) and Foxp3 can directly bind to RORγt and prevent its action. Only the additional stimulation with inflammatory cytokines such as IL-6 relieves Foxp3-mediated inhibition through a STAT-3-dependent suppression of *FOXP3* expression (Yao et al., 2007), and promotes differentiation of Th17 effector cells (Ichiyama et al., 2008; Zhou et al., 2008). Our experiments indicate that the release of IL-17 increases from day one to day seven of PMT stimulation whereas the IL-10 concentration did not. This suggests that in the beginning of PMT-stimulation the activation of STAT-3 and STAT-5 mediates the development of some IL-17 releasing Th17-cells, some IL-10 releasing Tregs, and some RORγt, Foxp3 double positive cells, which probably do not produce IL-17 due to an inhibitory interaction. At later time points, when only STAT-3 but no STAT-5 activation is observable, RORγt is released from Foxp3 inhibition and effectively induces IL-17 expression. An additional explanation for the low IL-10 production of toxin-treated cells could be the expression of the kinase Pim-1. While the expression of Pim-1 is inhibited by TCR activation, inflammatory conditions with high level of IL-6 lead to the induction of Pim-1 in human Tregs (Li et al., 2014). The kinase then phosphorylates Foxp3 at serine 422, which inhibits Foxp3 chromatin binding activity and eventually inhibits IL-10 production. Interestingly, our studies in HEK 293 cells show that PMT is a much more potent inducer of Pim-1 than IL-6 (Hildebrand et al., 2010). Therefore PMT stimulation could overcome the TCR-mediated inhibition of Pim-1 in the Foxp3-positive cells and the following mediated phosphorylation would suppress IL-10 production.

An interesting connection between Th17 cells and the bacterium is the interaction between the bone system and the immune system. IL-17 is involved in the excessive inflammation and activation of osteoclasts that strongly contributes to bone destruction associated with rheumatoid arthritis (Lubberts, 2015). Here, Th17 cells trigger the expression of receptor activator of nuclear factor-κB ligand (RANKL) in synovial cells, which then, in cooperation with other inflammatory cytokines, stimulate the differentiation and activation of monocytes into osteoclasts (Sato et al., 2006). PMT is known to be a mitogen for osteoclasts (Martineau-Doize et al., 1993; Jutras and Martineau-Doize, 1996) and pigs infected with *P. multocida* suffer from atrophic rhinitis, a pathologic condition characterized by PMT-stimulated osteoclastic bone resorption at the nasal turbinates and an inflammation of the nasal mucosa (Wilkie et al., 2012). We have previously shown that the PMT-triggered differentiation of bone marrow-derived mouse macrophages into osteoclasts is strongly dependent on cytokines released by B cells (Hildebrand et al., 2011). Cytokines produced by Th17-cells could additionally support the PMT-mediated osteoclast formation and the destruction of bone and thereby support the pathogenesis of toxigenic *P. multocida* strains. Nevertheless, the direction toward a Th17-mediated response is the normal course of an immune response against pulmonary and gastrointestinal pathogens like *P. multocida* (Garrido-Mesa et al., 2013). However, it has been suggested that pathogens that trigger a persistent Th17 response because they have not been not sufficiently cleared through the host's immune system, may be more beneficial for the microorganism than the host (van de Veerdonk et al., 2009).

In summary, we show that the bacterial toxin PMT enhances the proliferation and survival of CD3/CD28-activated Th cells and that the toxin manipulates the differentiation toward a Th17-subtype through the activation of STAT transcription factors. This once more illustrates how bacteria are able to target the high plasticity of T cell subtypes to reinforce their pathogenicity and possibly to gain an advantage in terms of survival and reproduction.

AUTHOR CONTRIBUTIONS

KK, KH, and DH designed the study. DH and KK wrote the final manuscript. DH performed the experiments. All authors read the manuscript and discussed the results.

FUNDING

This work was supported by Deutsche Forschungsgemeinschaft, a research grant (SPP1468 IMMUNOBONE, HI1747/4-2) to KK and a postdoctoral fellowship to DH (Medizinische Fakultät Heidelberg).

ACKNOWLEDGMENTS

We thank Ricarda Voigt and Eric Stricker for excellent technical assistance, as well as Sushmita Chakaraborty for the discussion of results. We thank Joachim Orth and Klaus Aktories (Freiburg) for providing us with recombinant PMT and Hilka Rauert-Wunderlich (Würzburg) for the gift of a cyclin E antibody (Santa Cruz).

REFERENCES

Astry, B., Venkatesha, S. H., and Moudgil, K. D. (2015). Involvement of the IL-23/IL-17 axis and the Th17/Treg balance in the pathogenesis and control of autoimmune arthritis. *Cytokine* 74, 54–61. doi: 10.1016/j.cyto.2014.11.020

Atherton-Fessler, S., Liu, F., Gabrielli, B., Lee, M. S., Peng, C. Y., and Piwnica-Worms, H. (1994). Cell cycle regulation of the p34cdc2 inhibitory kinases. *Mol. Biol. Cell* 5, 989–1001. doi: 10.1091/mbc.5.9.989

Bendris, N., Lemmers, B., and Blanchard, J. M. (2015). Cell cycle, cytoskeleton dynamics and beyond: the many functions of cyclins and CDK inhibitors. *Cell Cycle* 14, 1786–1798. doi: 10.1080/15384101.2014.998085

Bettini, M. L., and Vignali, D. A. (2010). Development of thymically derived natural regulatory T cells. *Ann. N. Y. Acad. Sci.* 1183, 1–12. doi: 10.1111/j.1749-6632.2009.05129.x

Burchill, M. A., Yang, J., Vogtenhuber, C., Blazar, B. R., and Farrar, M. A. (2007). IL-2 receptor beta-dependent STAT5 activation is required for the development of Foxp3+ regulatory T cells. *J. Immunol.* 178, 280–290. doi: 10.4049/jimmunol.178.1.280

Cantrell, D. A., and Smith, K. A. (1984). The interleukin-2 T-cell system: a new cell growth model. *Science* 224, 1312–1316. doi: 10.1126/science.6427923

Chen, Y., Fang, J., Chen, X., Pan, C., Liu, X., and Liu, J. (2015). Effects of the Treg/Th17 cell balance and their associated cytokines in patients with hepatitis B infection. *Exp. Ther. Med.* 9, 573–578. doi: 10.3892/etm.2014.2104

Coquet, J. M., Rausch, L., and Borst, J. (2015). The importance of co-stimulation in the orchestration of T helper cell differentiation. *Immunol. Cell Biol.* 93, 780–788. doi: 10.1038/icb.2015.45

Curotto de Lafaille, M. A., and Lafaille, J. J. (2009). Natural and adaptive foxp3+ regulatory T cells: more of the same or a division of labor? *Immunity* 30, 626–635. doi: 10.1016/j.immuni.2009.05.002

Gagliani, N., Vesely, M. C., Iseppon, A., Brockmann, L., Xu, H., Palm, N. W., et al. (2015). Th17 cells transdifferentiate into regulatory T cells during resolution of inflammation. *Nature* 523, 221–225. doi: 10.1038/nature14452

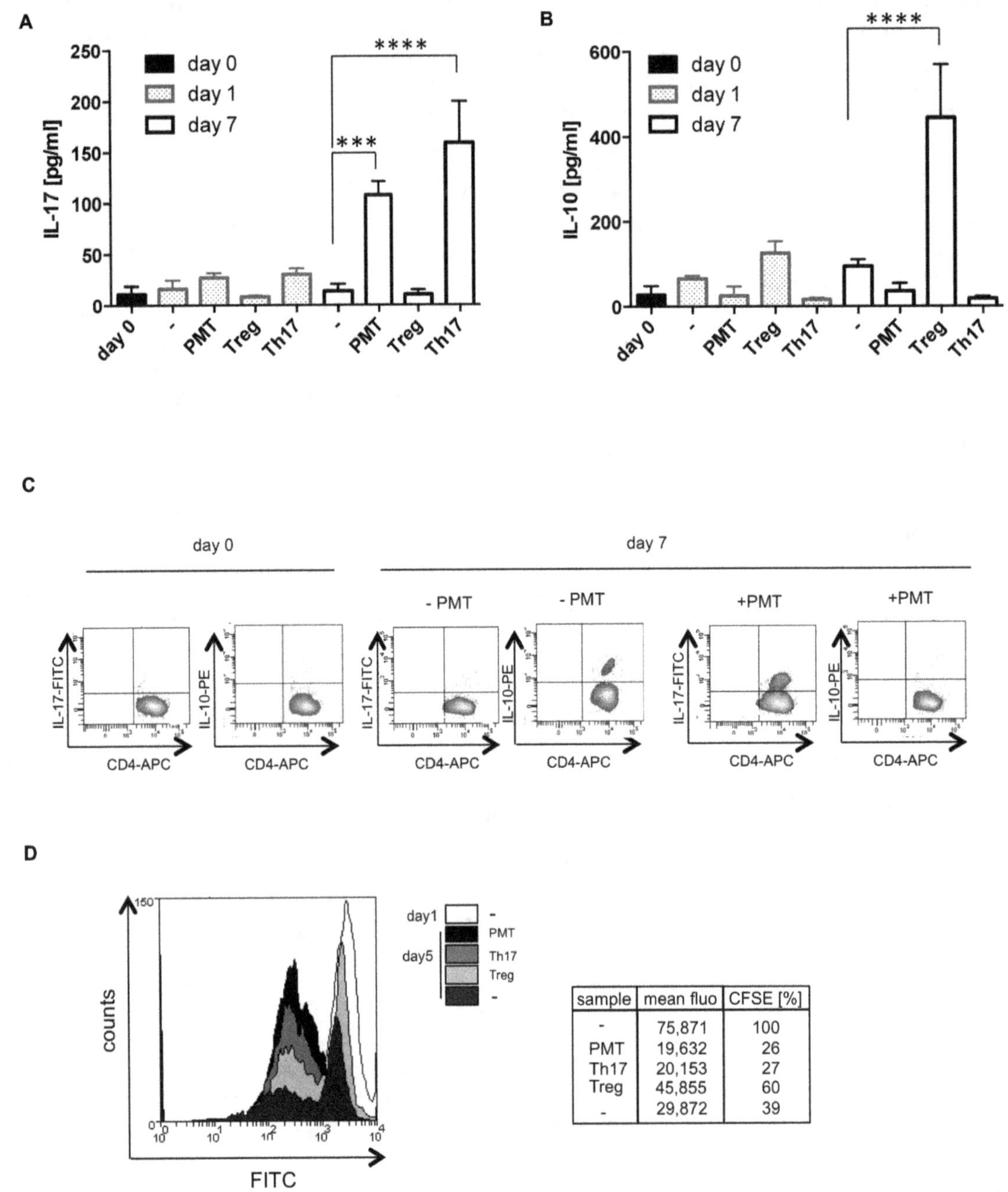

sample	mean fluo	CFSE [%]
-	75,871	100
PMT	19,632	26
Th17	20,153	27
Treg	45,855	60
-	29,872	39

FIGURE 4 | PMT mediates a Th17 phenotype. CD3/CD28-activated Th cells were stimulated with PMT for the indicated time points. In **(A)** IL-17A and **(B)** IL-10 ELISAs were performed (mean ± SD, $n = 3$). Statistical significance of the produced cytokine was compared to the value obtained on day 0 using two-sided ANOVA for multiple comparisons (***$p \leq 0.001$; ****$p \leq 0.0001$). **(C)** For intracellular FACS analysis, fixed and permeabilized cells were double-stained with anti-CD4-APC and purified anti-IL-10 or anti-IL-17 antibodies and the according FITC-/PE-labeled secondary antibodies. **(D)** Th cells were stimulated with PMT for 7 days. These PMT-generated T cells were then co-cultured with freshly isolated, CFSE-labeled Th cells (ratio 1:1) from the same donor. The fluorescent signal of the FITC-positive cells was quantified on day one and on day five of co-culture. Data shown in **(C,D)** are representative for three experiments (three donors).

induction of both IL-10 and IL-17 or just one of them (McGeachy et al., 2007; Ichiyama et al., 2008; Zhou et al., 2008) and Foxp3 can directly bind to RORγt and prevent its action. Only the additional stimulation with inflammatory cytokines such as IL-6 relieves Foxp3-mediated inhibition through a STAT-3-dependent suppression of *FOXP3* expression (Yao et al., 2007), and promotes differentiation of Th17 effector cells (Ichiyama et al., 2008; Zhou et al., 2008). Our experiments indicate that the release of IL-17 increases from day one to day seven of PMT stimulation whereas the IL-10 concentration did not. This suggests that in

Garrido-Mesa, N., Algieri, F., Rodriguez Nogales, A., and Galvez, J. (2013). Functional plasticity of Th17 cells: implications in gastrointestinal tract function. *Int. Rev. Immunol.* 32, 493–510. doi: 10.3109/08830185.2013.834899

Guo, F., Iclozan, C., Suh, W. K., Anasetti, C., and Yu, X. Z. (2008). CD28 controls differentiation of regulatory T cells from naive CD4 T cells. *J. Immunol.* 181, 2285–2291. doi: 10.4049/jimmunol.181.4.2285

Hildebrand, D., Bode, K. A., Riess, D., Cerny, D., Waldhuber, A., Rommler, F., et al. (2014). Granzyme A produces bioactive IL-1? through a nonapoptotic inflammasome-independent pathway. *Cell Rep.* 9, 910–917. doi: 10.1016/j.celrep.2014.10.003

Hildebrand, D., Heeg, K., and Kubatzky, K. F. (2011). *Pasteurella multocida* toxin-stimulated osteoclast differentiation is B cell dependent. *Infect. Immun.* 79, 220–228. doi: 10.1128/IAI.00565-510

Hildebrand, D., Sahr, A., Wolfle, S. J., Heeg, K., and Kubatzky, K. F. (2012). Regulation of Toll-like receptor 4-mediated immune responses through *Pasteurella multocida* toxin-induced G protein signalling. *Cell Commun. Signal.* 10, 22. doi: 10.1186/1478-811X-10-22

Hildebrand, D., Walker, P., Dalpke, A., Heeg, K., and Kubatzky, K. F. (2010). *Pasteurella multocida* Toxin-induced Pim-1 expression disrupts suppressor of cytokine signalling (SOCS)-1 activity. *Cell Microbiol.* 12, 1732–1745. doi: 10.1111/j.1462-5822.2010.01504.x

Ichiyama, K., Yoshida, H., Wakabayashi, Y., Chinen, T., Saeki, K., Nakaya, M., et al. (2008). Foxp3 inhibits RORgammat-mediated IL-17A mRNA transcription through direct interaction with RORgammat. *J. Biol. Chem.* 283, 17003–17008. doi: 10.1074/jbc.M801286200

Ivanov, I. I., Mckenzie, B. S., Zhou, L., Tadokoro, C. E., Lepelley, A., Lafaille, J. J., et al. (2006). The orphan nuclear receptor RORgammat directs the differentiation program of proinflammatory IL-17+ T helper cells. *Cell* 126, 1121–1133. doi: 10.1016/j.cell.2006.07.035

Ivars, F. (2007). Superantigen-induced regulatory T cells in vivo. *Chem. Immunol. Allergy* 93, 137–160. doi: 10.1159/0000100862

Jutras, I., and Martineau-Doize, B. (1996). Stimulation of osteoclast-like cell formation by *Pasteurella multocida* toxin from hemopoietic progenitor cells in mouse bone marrow cultures. *Can. J. Vet. Res.* 60, 34–39.

Kaiko, G. E., Phipps, S., Hickey, D. K., Lam, C. E., Hansbro, P. M., Foster, P. S., et al. (2008). Chlamydia muridarum infection subverts dendritic cell function to promote Th2 immunity and airways hyperreactivity. *J. Immunol.* 180, 2225–2232. doi: 10.4049/jimmunol.180.4.2225

Kong, N., Lan, Q., Chen, M., Wang, J., Shi, W., Horwitz, D. A., et al. (2012). Antigen-specific transforming growth factor beta-induced Treg cells, but not natural Treg cells, ameliorate autoimmune arthritis in mice by shifting the Th17/Treg cell balance from Th17 predominance to Treg cell predominance. *Arthritis Rheum.* 64, 2548–2558. doi: 10.1002/art.34513

Korn, T., Bettelli, E., Oukka, M., and Kuchroo, V. K. (2009). IL-17 and Th17 cells. *Annu. Rev. Immunol.* 27, 485–517. doi: 10.1146/annurev.immunol.021908.132710

Kryczek, I., Wu, K., Zhao, E., Wei, S., Vatan, L., Szeliga, W., et al. (2011). IL-17+ regulatory T cells in the microenvironments of chronic inflammation and cancer. *J. Immunol.* 186, 4388–4395. doi: 10.4049/jimmunol.1003251

Kubatzky, K. F., Kloos, B., and Hildebrand, D. (2013). Signaling cascades of *Pasteurella multocida* toxin in immune evasion. *Toxins (Basel)* 5, 1664–1681. doi: 10.3390/toxins5091664

Lee, Y. K., Turner, H., Maynard, C. L., Oliver, J. R., Chen, D., Elson, C. O., et al. (2009). Late developmental plasticity in the T helper 17 lineage. *Immunity* 30, 92–107. doi: 10.1016/j.immuni.2008.11.005

Li, Z., Lin, F., Zhuo, C., Deng, G., Chen, Z., Yin, S., et al. (2014). PIM1 kinase phosphorylates the human transcription factor FOXP3 at serine 422 to negatively regulate its activity under inflammation. *J. Biol. Chem.* 289, 26872–26881. doi: 10.1074/jbc.M114.586651

Lubberts, E. (2015). Role of T lymphocytes in the development of rheumatoid arthritis. Implications for treatment. *Curr. Pharm. Des.* 21, 142–146.

Martineau-Doize, B., Caya, I., Gagne, S., Jutras, I., and Dumas, G. (1993). Effects of *Pasteurella multocida* toxin on the osteoclast population of the rat. *J. Comp. Pathol.* 108, 81–91. doi: 10.1016/S0021-9975(08)80230-7

McGeachy, M. J., Bak-Jensen, K. S., Chen, Y., Tato, C. M., Blumenschein, W., Mcclanahan, T., et al. (2007). TGF-beta and IL-6 drive the production of IL-17 and IL-10 by T cells and restrain T(H)-17 cell-mediated pathology. *Nat. Immunol.* 8, 1390–1397. doi: 10.1038/ni1539

Meuer, S. C., Hussey, R. E., Cantrell, D. A., Hodgdon, J. C., Schlossman, S. F., Smith, K. A., et al. (1984). Triggering of the T3-Ti antigen-receptor complex results in clonal T-cell proliferation through an interleukin 2-dependent autocrine pathway. *Proc. Natl. Acad. Sci. U.S.A.* 81, 1509–1513. doi: 10.1073/pnas.81.5.1509

Miethke, T., Wahl, C., Gaus, H., Heeg, K., and Wagner, H. (1994). Exogenous superantigens acutely trigger distinct levels of peripheral T cell tolerance/immunosuppression: dose-response relationship. *Eur. J. Immunol.* 24, 1893–1902. doi: 10.1002/eji.1830240827

Oppenheim, J. J. (2007). IL-2: more than a T cell growth factor. *J. Immunol.* 179, 1413–1414. doi: 10.4049/jimmunol.179.3.1413

Orth, J. H., and Aktories, K. (2010). *Pasteurella multocida* toxin activates various heterotrimeric G proteins by deamidation. *Toxins (Basel)* 2, 205–214. doi: 10.3390/toxins2020205

Orth, J. H., Aktories, K., and Kubatzky, K. F. (2007). Modulation of host cell gene expression through activation of STAT transcription factors by *Pasteurella multocida* toxin. *J. Biol. Chem.* 282, 3050–3057. doi: 10.1074/jbc.M6090 18200

O'Shea, J. J., Lahesmaa, R., Vahedi, G., Laurence, A., and Kanno, Y. (2011). Genomic views of STAT function in CD4+ T helper cell differentiation. *Nat. Rev. Immunol.* 11, 239–250. doi: 10.1038/nri2958

Papiernik, M. (2001). Natural CD4+ CD25+ regulatory T cells. Their role in the control of superantigen responses. *Immunol. Rev.* 182, 180–189. doi: 10.1034/j.1600-065X.2001.1820114.x

Park, H., Li, Z., Yang, X. O., Chang, S. H., Nurieva, R., Wang, Y. H., et al. (2005). A distinct lineage of CD4 T cells regulates tissue inflammation by producing interleukin 17. *Nat. Immunol.* 6, 1133–1141. doi: 10.1038/ni1261

Pinkoski, M. J., and Green, D. R. (2005). Apoptosis in the regulation of immune responses. *J. Rheumatol. Suppl.* 74, 19–25.

Preuss, I., Hildebrand, D., Orth, J. H., Aktories, K., and Kubatzky, K. F. (2010). *Pasteurella multocida* toxin is a potent activator of anti-apoptotic signalling pathways. *Cell Microbiol.* 12, 1174–1185. doi: 10.1111/j.1462-5822.2010. 01462.x

Rodriguez-Garcia, M., Porichis, F., De Jong, O. G., Levi, K., Diefenbach, T. J., Lifson, J. D., et al. (2011). Expression of PD-L1 and PD-L2 on human macrophages is up-regulated by HIV-1 and differentially modulated by IL-10. *J. Leukoc. Biol.* 89, 507–515. doi: 10.1189/jlb.0610327

Sakai, S., Kawamura, I., Okazaki, T., Tsuchiya, K., Uchiyama, R., and Mitsuyama, M. (2010). PD-1-PD-L1 pathway impairs T(h)1 immune response in the late stage of infection with Mycobacterium bovis bacillus Calmette-Guerin. *Int. Immunol.* 22, 915–925. doi: 10.1093/intimm/dxq446

Sato, K., Suematsu, A., Okamoto, K., Yamaguchi, A., Morishita, Y., Kadono, Y., et al. (2006). Th17 functions as an osteoclastogenic helper T cell subset that links T cell activation and bone destruction. *J. Exp. Med.* 203, 2673–2682. doi: 10.1084/jem.20061775

Schmitt, E. G., and Williams, C. B. (2013). Generation and function of induced regulatory T cells. *Front. Immunol.* 4:152. doi: 10.3389/fimmu.2013. 00152

Shevach, E. M., and Thornton, A. M. (2014). tTregs, pTregs, and iTregs: similarities and differences. *Immunol. Rev.* 259, 88–102. doi: 10.1111/imr.12160

Smith, P., Walsh, C. M., Mangan, N. E., Fallon, R. E., Sayers, J. R., Mckenzie, A. N., et al. (2004). Schistosoma mansoni worms induce anergy of T cells via selective up-regulation of programmed death ligand 1 on macrophages. *J. Immunol.* 173, 1240–1248. doi: 10.4049/jimmunol.173.2.1240

Talaat, R. M., Mohamed, S. F., Bassyouni, I. H., and Raouf, A. A. (2015). Th1/Th2/Th17/Treg cytokine imbalance in systemic lupus erythematosus (SLE) patients: correlation with disease activity. *Cytokine* 72, 146–153. doi: 10.1016/j.cyto.2014.12.027

Tanaka, S., Suto, A., Iwamoto, T., Kashiwakuma, D., Kagami, S., Suzuki, K., et al. (2014). Sox5 and c-Maf cooperatively induce Th17 cell differentiation via RORgammat induction as downstream targets of Stat3. *J. Exp. Med.* 211, 1857–1874. doi: 10.1084/jem.20130791

van de Veerdonk, F. L., Gresnigt, M. S., Kullberg, B. J., Van Der Meer, J. W., Joosten, L. A., and Netea, M. G. (2009). Th17 responses and host defense against microorganisms: an overview. *BMB Rep.* 42, 776–787. doi: 10.5483/BMBRep.2009.42.12.776

Voo, K. S., Wang, Y. H., Santori, F. R., Boggiano, C., Wang, Y. H., Arima, K., et al. (2009). Identification of IL-17-producing FOXP3+ regulatory

T cells in humans. *Proc. Natl. Acad. Sci. U.S.A.* 106, 4793–4798. doi: 10.1073/pnas.0900408106

Weiss, E. M., Schmidt, A., Vobis, D., Garbi, N., Lahl, K., Mayer, C. T., et al. (2011). Foxp3-mediated suppression of CD95L expression confers resistance to activation-induced cell death in regulatory T cells. *J. Immunol.* 187, 1684–1691. doi: 10.4049/jimmunol.1002321

Wilkie, I. W., Harper, M., Boyce, J. D., and Adler, B. (2012). *Pasteurella multocida*: diseases and pathogenesis. *Curr. Top. Microbiol. Immunol.* 361, 1–22. doi: 10.1007/82_2012_216

Wilson, B. A., Aminova, L. R., Ponferrada, V. G., and Ho, M. (2000). Differential modulation and subsequent blockade of mitogenic signaling and cell cycle progression by *Pasteurella multocida* toxin. *Infect. Immun.* 68, 4531–4538. doi: 10.1128/IAI.68.8.4531-4538.2000

Wilson, B. A., and Ho, M. (2011). Cellular and molecular action of the mitogenic protein-deamidating toxin from *Pasteurella multocida*. *FEBS J.* 278, 4616–4632. doi: 10.1111/j.1742-4658.2011.08158.x

Wilson, B. A., and Ho, M. (2012). *Pasteurella multocida* toxin interaction with host cells: entry and cellular effects. *Curr. Top. Microbiol. Immunol.* 361, 93–111. doi: 10.1007/82_2012_219

Yao, Z., Kanno, Y., Kerenyi, M., Stephens, G., Durant, L., Watford, W.T., et al. (2007). Nonredundant roles for Stat5a/b in directly regulating *Foxp3*. *Blood* 109, 4368–4375. doi: 10.1182/blood-2006-11-055756

Zheng, X., Ou, Y., Shu, M., Wang, Y., Zhou, Y., Su, X., et al. (2014). Cholera toxin, a typical protein kinase A activator, induces G1 phase growth arrest in human bladder transitional cell carcinoma cells via inhibiting the c-Raf/MEK/ERK signaling pathway. *Mol. Med. Rep.* 9, 1773–1779. doi: 10.3892/mmr.2014.2054

Zhou, L., Lopes, J. E., Chong, M. M., Ivanov, I. I., Min, R., Victora, G. D., et al. (2008). TGF-beta-induced Foxp3 inhibits T(H)17 cell differentiation by antagonizing RORgammat function. *Nature* 453, 236–240. doi: 10.1038/nature06878

Zhu, J., and Paul, W. E. (2010). Peripheral CD4+ T-cell differentiation regulated by networks of cytokines and transcription factors. *Immunol. Rev.* 238, 247–262. doi: 10.1111/j.1600-065X.2010.00951.x

Conflict of Interest Statement: The authors declare that the research was conducted in the absence of any commercial or financial relationships that could be construed as a potential conflict of interest.

11

Lipoarabinomannan in Active and Passive Protection Against Tuberculosis

Margarida Correia-Neves[1,2,3], *Christopher Sundling*[3,4], *Andrea Cooper*[5] *and Gunilla Källenius*[3*]

[1] *Life and Health Sciences Research Institute (ICVS), School of Medicine, University of Minho, Braga, Portugal,* [2] *ICVS/3B's, PT Government Associate Laboratory, Braga, Guimarães, Portugal,* [3] *Division of Infectious Diseases, Department of Medicine Solna, Karolinska Institutet, Stockholm, Sweden,* [4] *Department of Infectious Diseases, Karolinska University Hospital, Stockholm, Sweden,* [5] *Leicester Tuberculosis Research Group (LTBRG), Department of Respiratory Sciences, University of Leicester, Leicester, United Kingdom*

Edited by:
Juraj Ivanyi,
King's College London,
United Kingdom

Reviewed by:
Angelo Izzo,
Colorado State University,
United States
Armando Acosta,
Universiti Sains Malaysia Health Campus, Malaysia
Ying Zhang,
Johns Hopkins University,
United States

***Correspondence:**
Gunilla Källenius
gunilla.kallenius@ki.se

Glycolipids of the cell wall of *Mycobacterium tuberculosis* (Mtb) are important immunomodulators in tuberculosis. In particular, lipoarabinomannan (LAM) has a profound effect on the innate immune response. LAM and its structural variants can be recognized by and activate human CD1b-restricted T cells, and emerging evidence indicates that B cells and antibodies against LAM can modulate the immune response to Mtb. Anti-LAM antibodies are induced during Mtb infection and after bacille Calmette–Guerin (BCG) vaccination, and monoclonal antibodies against LAM have been shown to confer protection by passive administration in mice and guinea pigs. In this review, we describe the immune response against LAM and the potential use of the mannose-capped arabinan moiety of LAM in the construction of vaccine candidates against tuberculosis.

Keywords: lipoarabinomannan, tuberculosis, Mycobacterium, glycolipid, antibodies, monoclonal, vaccine, immune response

INTRODUCTION

Tuberculosis (TB) causes significant mortality globally, accounting for 1.7 million deaths in 2016 (1). Morbidity is also significant, with 10.4 million people estimated to be affected by the disease in 2016, and the threat to global health is further amplified by the emergence of about 600,000 cases of drug-resistant TB per year (1). One-quarter of the world's population is estimated to be infected with the causative agent, *Mycobacterium tuberculosis* (Mtb), but the majority of those infected with the bacterium do not progress to active TB. Defining the essential components of the host response that stops disease development in the majority of infected individuals is critical to the ability to stop progression in those likely to develop disease. While clear progress has been made in defining the main elements that limit bacterial growth in infected individuals, we do not have a clear model of the immune mechanisms that determine infection control vs. disease progression (2, 3). Herein we discuss a less examined aspect of the immune response to Mtb infection—the role of the cell wall glycolipids of the bacterium in the generation and regulation of the immune response.

MYCOBACTERIAL GLYCOLIPIDS

There is currently no established immunological signal that differentiates those latently infected with Mtb from those developing active disease or indeed a signal that indicates whether a vaccinated individual will be protected from infection or disease. Despite this lack of signal, it is known that the absence of a T cell response and the inability to activate macrophages are associated with the development of active disease (3). Understandably, therefore, the focus of vaccine development has been on induction of antigen-specific T cells with the ability to produce macrophage-activating cytokines. While the vaccine trials associated with this approach have yielded significant understanding of immune responses, the trials themselves have not resulted in substantially improved protection. This inability to dramatically improve upon the current vaccine approach has led to other avenues, including investigation of the ability of carbohydrate antigens to drive humoral responses. Cell wall glycolipids from Mtb such as lipoarabinomannan (LAM), its precursors the phosphatidyl-inositol mannosides (PIMs) and lipomannan (LM), as well as phenolic glycolipids (4) trehalose dimycolate (cord factor) (5) and the sulfoglycolipids (6) have all been shown to have immunomodulatory activity, with both distinct activating and repressing activity.

LAM has been studied quite extensively for its immunomodulatory properties, and as a structurally unique glycolipid component of the envelope of all mycobacterial species, it provides an excellent model for discussing the role of glycolipids in immunity to Mtb (**Figure 1**). It is the main carbohydrate antigen of Mtb and accounts for up to 15% of the bacterial weight (7). LAM has a mannan core with a branched arabinan polymer and, in some species, capping motifs at the termini of the branched arabinan chains. The arabinan termini in Mtb are modified with caps consisting of oligomannosides, resulting in molecules designated ManLAM. LAM's mannan core is covalently linked to a mannosyl phosphatidyl-inositol (MPI) lipid moiety containing palmitic, tuberculostearic, and stearic acid residues (8–10). In addition to LAM, its precursors LM and the phosphatidyl-myo-inositol mannosides (PIMs) have been shown to have potent immunomodulatory effects on cells of the immune system (11). An extensively mannose-capped arabinomannan (ManAM) structurally related to the ManAM part of LAM is present in the capsule surrounding Mtb (12, 13).

INNATE IMMUNE RESPONSES TO LAM

LAM has a profound effect on the innate immune response, presenting varying degrees of activity in terms of cytokine production by human macrophages and dendritic cells (11, 14, 15), but little is known about which receptors mediate this signaling. Neither is it known in most cases which parts/epitopes of the LAM molecule are important for this activity. The mannose cap of mannosylated LAM (ManLAM) was long thought to be important in the pathogenesis of Mtb infection. However, it was shown that mannose caps do not affect Mtb virulence and have minimal impact on the *in vitro* inflammatory responses in macrophages in mice (16).

A large number of host cell receptors take part in the initial interaction between mycobacteria and cells of the innate immune system (17, 18). Several classes of pattern recognition receptors are involved in this interaction, including Toll-like receptors (TLRs) and C-type lectins. Most studies report little or no TLR2 or TLR4 agonist activity of ManLAM (14, 19–22). However, ManLAM has been shown to interact with several C-type lectin ligands on cells of the innate immune system (15, 23).

ManLAM is recognized by C-type lectins that bind to carbohydrate moieties, through carbohydrate recognition domains (CRDs) (24) on the surfaces of innate immune cells (macrophages, dendritic cells, and neutrophils) (25, 26). The responses mediated by these glycan-binding receptors can take several forms. CRDs target microorganisms for uptake by endocytosis and subsequent destruction (27). Endocytic receptors, including the mannose receptor, DC-SIGN, DC-SIGNR, and langerin, interact mainly with the mannose caps on LAM. Some of these receptors also interact with PIMs and mannose-containing phenolic glycolipids. Both the mannose receptor and DC-SIGN bind tightly to LAM found on the Mtb surface, and this binding can result in efficient internalization of Mtb (28, 29).

It has become clear that some glycan-binding receptors initiate signaling responses, often resulting in the activation of several antimicrobial mechanisms by macrophages and dendritic cells, including the secretion of pro-inflammatory cytokines (24, 30, 31). ManLAM was recently identified as a ligand of Dectin-2 (dendritic-associated C-type lectin-2) (19, 24, 32). Dectin-2 is a C-type lectin that is expressed on several cell types, including dendritic cells and Mφ (19). Purified Mtb ManLAM induces signaling via Dectin-2, an activity that requires ($\alpha 1 \rightarrow 2$)-linked dimannoside forming caps (33).

LAM AND T CELL–MEDIATED IMMUNITY

Traditionally, protective immunity to TB has primarily been ascribed to T cell–mediated immune responses, with $CD4^+$ T cells playing a crucial role. As Mtb is an intracellular pathogen, cell-mediated immunity characterized by interferon gamma IFN-γ-producing $CD4^+$ T helper type 1 (Th1) cells is regarded as essential for TB immunity (3). The role of vaccine-induced polyfunctional T cells, co-producing several pro-inflammatory cytokines such as IFN-γ, tumor necrosis factor (TNF), and interleukin-2 (IL-2), has also been extensively evaluated as a possible correlate of protection in both animal models and humans but has not been shown to be necessary to mediate protection (34).

$CD8^+$ T cells also produce IFN-γ and TNF upon Mtb infection (35–37). Moreover, $CD8^+$ T cells have cytotoxic activity and are thus able to eliminate Mtb-infected cells, using mechanisms such as the Fas- and TNF-mediated killing and granule exocytosis involving release of perforin and granzyme (in humans and mice) and granulysin (in humans) (38, 39).

Although glycolipids were long considered to not induce T cell–mediated immunity, several glycolipids, including LAM and its structural variants, are known to be recognized by and

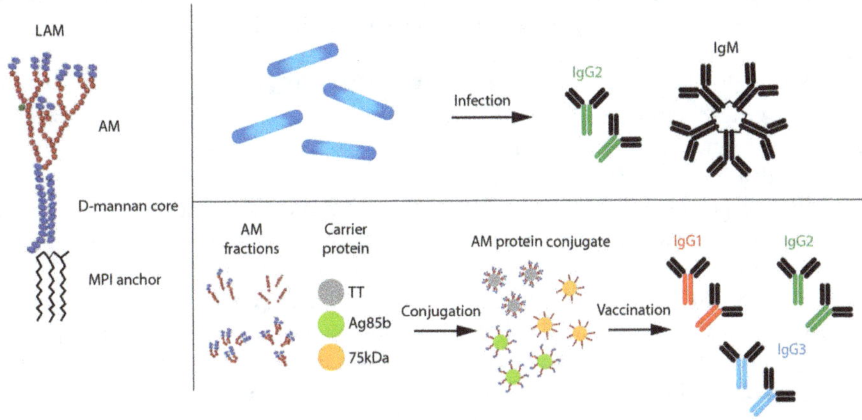

FIGURE 1 | Types of LAM antibodies produced after natural Mtb infection and vaccination with AM–protein conjugate vaccine, respectively. AM, arabinomannan; MPI, mannosyl phosphatidyl-inositol.

activate human CD1b-restricted T cells (40–43). In addition, liposomal delivery of LAM triggers LAM-specific T cells (44). The phosphatidylinositol moiety plays a central role in the binding of LAM to CD1b, although the exact epitopes involved in this binding have not yet been defined (45). CD1 molecules are expressed on dendritic cells, certain B and T lymphocytes, and epithelial cells (43, 46, 47). Upon recognition of mycobacterial lipids, CD1-restricted T cells can become activated and proliferate, secreting pro-inflammatory cytokines such as IFN-γ, and express cytolytic and bactericidal activity (41–43). CD1-restricted T cells have further been shown to support plasma cell formation and survival in mouse models, indicating a direct role in B cell responses (48).

An interesting strategy to search for immunological mechanisms/markers that predict protection against TB is to study the immune response in individuals with latent TB infection (LTBI) in whom the infection is controlled (49–52). Using this strategy, a correlation between the frequency of LAM-specific cytotoxic $CD8^+$ T cells and the ability to control infection was found in subjects latently infected with Mtb (52). These IFN-γ-producing CD1b-restricted $CD8^+$ T cells coexpressed perforin, granulysin, and granzyme B and limited bacterial growth *in vitro*, suggesting a possible mechanism of protection against TB (52). In contrast, a study using a whole blood assay stimulated with LAM showed a reduced induction of TNF and IFN-γ and a higher production of IL-10 in TB patients and contacts compared to healthy controls (53), indicating a complicated interplay between LAM and the immune system. It is now clear that LAM and other glycolipids play an important role in the immune response to infection by Mtb. Understanding to what extent T cell responses to these molecules are of relevance for the establishment of a protective immune response needs to be definitively addressed.

HUMORAL IMMUNITY INDUCED BY LAM AND THE ROLE OF B CELLS

As Mtb is a facultative intracellular pathogen, it has been assumed that antibodies (Abs) play no relevant role in the protection against TB and may even contribute to immunopathology (54). However, emerging evidence suggests that B cells and humoral immunity can modulate the immune response to various intracellular pathogens, including Mtb (55–59). Abs have been shown to confer protection against a number of intracellular pathogens by modulating immunity via Fc receptor–mediated phagocytosis, antibody-dependent cellular cytotoxicity (ADCC), and immunostimulation (57).

It has been shown that B cells can regulate the level of the granulomatous reaction, cytokine production, and the T cell response during Mtb infection (55, 56, 58–60). However, the regulation is complex due to the impact of both secreted products and direct cellular interactions. For example regulatory B cells, which produce IL-10 or transforming growth factor β, participate in the immunomodulation of immune responses, and it was recently shown that ManLAM induced IL-10 production by B cells, both *in vitro* and *in vivo*, predominantly through TLR2 (61).

It was recently shown that some B cells are functionally altered in active and latent TB (62). These "atypical" B cells displayed impaired proliferation, cytokine production, and differentiation to Ab-secreting cells. Similar atypical B cells have been described in several other infectious diseases and were shown to be rapidly induced and then accumulate, especially during chronic or recurrent infections, such as in malaria and infection by hepatitis B and C virus infection and HIV (63–65). The role of atypical B cells in immunity remains unclear. Due to a reduced capacity to respond to B cell receptor stimulation and differentiation to Ab-secreting cells *in vitro* (66–68), high levels of atypical B cells have been suggested to be detrimental for the pathogen-specific immune response (63). It remains to be determined if important protective Mtb-specific B cell reactivities are sequestered away within the atypical B cell compartment, where their contribution to the circulating Ab pool might be limited.

Anti-LAM Antibodies

It has long been known that Abs against mycobacterial antigens are formed during natural infection (57, 69). In a recent study, IgG levels against 15 Mtb protein antigens were shown to be higher in TB patients than in endemic and non-endemic healthy IGRA-negative controls, with one exception, Abs against the latency antigen Rv1733c that were preferentially present in endemic controls (70). During active TB, disease production of Abs is prominent (71, 72), but such Abs have been reported to have low avidity and low IgG/IgM ratio (73), potentially indicating low contribution of germinal center–derived B cell responses. However, there is evidence to suggest that Abs may limit the dissemination of Mtb and potentially also play a role in prevention of infection via mucosal immunity (74). In a few animal studies, Abs against Mtb have been shown to be protective against Mtb infection (75, 76). In addition, a human IgA monoclonal Ab (mAb) to 16 kDa α-crystallin (Acr) from Mtb reduced bacterial load in mice transgenic for the human FcαRI receptor (75). Coating bacille Calmette–Guerin (BCG) with mAbs to the heparin-binding haemagglutinin adhesin (HBHA) was also shown to impair dissemination after intranasal infection (77).

Anti-LAM Abs are induced both during Mtb infection (73, 78, 79) and after BCG vaccination (80–82). The titers of Abs against LAM and capsular arabinomannan (AM), were significantly lower in HIV-infected than in HIV-non-infected TB individuals, which might be due to impaired B cell function in HIV-positive patients (83–85). In children, an inverse correlation was detected between the presence of anti-LAM IgG Abs and disseminated TB, suggesting that LAM Abs might be important in limiting TB dissemination (86).

Anti-LAM Abs induced during infection have been associated with bacterial opsonization and restriction of intracellular growth (81, 82). Opsonization with anti-LAM Abs from latently infected individuals was shown to enhance uptake and killing of Mtb (87) and BCG (81) by donor macrophages. Key factors of macrophage microbicidal action, such as increased production of pro-inflammatory cytokines like IFN-γ and IL-6, phagosome acidification, and nitric oxide production, through expression of inducible nitric oxide synthase, were also significantly enhanced following Ab opsonization (87).

Anti-Mtb Ig Isotypes and Immune Response to Mtb

The activity of anti-Mtb Abs is strongly affected by their isotype (88). Distinct Ab glycosylation patterns have gained increasing attention as regulators of immune responses (49, 89). In patients with TB, the main Ig isotype response to polysaccharides (PSs), including LAM, is IgG of predominantly the IgG2 subclass (83, 85). This is in accordance with the general observation that Abs to carbohydrates tend to be IgG2 (90–92).

Vaccination of cattle with BCG induces both IgG1 and IgG2 Abs to ManLAM, which are transferred in colostrum to neonatal calves (93). LAM-specific IgA Abs can also be induced through oral vaccination with BCG (80). This is important as IgA, but not IgG, Abs specific for different Mtb surface antigens including ManLAM can block Mtb uptake by lung epithelial cells, independent of the expression of IgA Fc receptors (88).

Human levels of LAM-specific IgG2, but not IgM, correlate positively with classical complement activation induced by BCG (94). The binding of complement to BCG incubated with purified rabbit anti-LAM Abs followed by factor B-depleted serum indicates that anti-LAM IgG bound to the bacilli can mediate classical complement activation and deposition on the mycobacteria (94).

Monoclonal Antibodies Against LAM

Monoclonal Abs (mAbs) are increasingly used for therapeutic purposes. Yet, despite the clear role of Abs in many infections, only two monoclonal therapeutic Abs against infectious agents (human respiratory syncytial virus and *Bacillus anthracis*) have been licensed (95). There are, however, a number of therapeutic Abs against infectious diseases under development, mainly against viral diseases but also some bacterial agents (both Gram positive and Gram negative) and fungi (95).

Over the years, several mAbs have been raised against ManLAM [(76, 96–98); **Table 1**]. Some of these were shown to confer protection by passive administration in mice [(76, 96, 97, 99); **Table 1**]. However, these studies provided very little information about possible mechanisms of protection. Two mAbs, SMITB14 and 9d8, were shown to prolong survival of mice. Given intravenously to mice previously infected with Mtb (strain Harlingen), mAb SMITB14 reduced bacterial load and weight loss (76). The fact that the Fab2 fragments of SMITB14 showed a similar inhibitory effect as the full Abs indicates that the specific epitope targeted mediated the effect, rather than Fc-mediated functions. Incubation of Mtb (strain Erdman) with mAb 9d8, before intratracheal challenge, was shown not to reduce the bacterial load in the spleen, liver, or lungs (96). However, it profoundly altered the nature of the granulomas in the lungs, where bacilli localized within the center of well-organized granuloma. This suggests that the mAb conferred protection by enhancing the cellular immune response (96). Together, these observations suggest that targeting ManLAM can mediate protection via different mechanisms. In support of this, a single-stranded DNA aptamer that specifically binds to ManLAM, isolated from the virulent Mtb strain H37Rv, reduced Mtb infection in mice and rhesus monkeys (100).

While the majority of natural Abs against carbohydrate antigens are IgG2, most therapeutic Abs, licensed or in development, are of the human IgG1 subclass (95). Interestingly, none of the LAM-specific mAbs so far reported to induce protection against Mtb are of the IgG2 isotype (**Table 1**); one (SMITB14) was of the IgG1 type (76) and one (9d8) of the IgG3 type (96), indicating that these latter isotypes might be of importance in the design of protective Abs.

The affinity of IgG subclasses for varying FcγRs results in differing effector responses and has been made use of in the development of therapeutic Abs and vaccines (101). Signaling through FcγRs can modulate immunity upon infection with Mtb and significantly affect outcome (102). As a further complicating factor, in a murine model of progressive TB, the protective effect of the passive transfer of Abs depended on the glycosylation pattern of the Fc region of IgG Abs (103). Thus, the design of LAM-specific Abs with particular Fc characteristics should be explored for protective effect, including immunization strategies designed to augment cellular immune responses. Interestingly, it was recently reported that individuals with LTBI and active TB show different Mtb-specific humoral responses, where LTBI is associated with Ab Fc functional profiles associated with

TABLE 1 | *In vivo* effect of anti-LAM/AM monoclonal antibodies.

Antibody	Isotype	Epitopes	Administration route	Mouse strain	Mtb challenge strain	Challenge route	Biological effect	Comments	References
9d8	IgG3	AM	Together with challenge bacteria	BALB/c C57BL/c	Mtb Erdman	Respiratory	Prolonged survival	Bacteria were localized in center of granuloma.	(96)
9d8	IgG3	AM	NA	NA	NA	NA	NA	Binding of human Abs to AM inhibited by mAb 9d8 in patients with TB.	(13)
Sc11	IgM	AM, LAM, AG	NA	NA	NA	NA	LAM cleared from circulation when given i.v.	Deposition in spleen was reduced, while LAM deposition in the liver increased.	(97)
SMITB14	IgG1	AM	Intravenously	BALB/c	Mtb Harlingen	Intravenous	Increased long-term survival	Reduction in bacterial load in spleen and lung, decreased body weight.	(76)

Abs, antibodies; AG, antigen; AM, arabinomannan; Ig, immunoglobulin; i.v., intravenously; LAM, lipoarabinomannan; mAb, monoclonal Ab; Mtb, Mycobacterium tuberculosis; NA, not applicable; TB, tuberculosis.

enhanced binding to FcγRIII, and distinct Ab glycosylation patterns (49). In comparison to Abs from active TB patients, Abs from LTBI drive enhanced phagolysosomal maturation, inflammasome activation, and in particular, macrophage killing of intracellular Mtb (49). These findings indicate a potential role for Fc-mediated Ab effector functions in the control of Mtb, by differential glycosylation.

LAM Epitopes Recognized by Antibodies

The Abs against LAM/AM are heterogenous and differ in their specificity (13). Serum from mice immunized with AM conjugated to carrier proteins such as the Mtb antigen Ag85B or *Pseudomonas aeruginosa* exoprotein A (rEPA) (see ahead) recognizes a broad range of AM structural variants and specifically recognized arabinan fragments (104). The specificities of anti-LAM mAbs differ in their dependence on arabinofuranoside branching and capping motifs, and their specificities could be divided into two general groups: one dependent on the poly-Ara*f* backbone of LAM and the other dependent on specific mannose-capping motifs (98).

THE POTENTIAL ROLE OF LAM IN THE DESIGN OF TB VACCINES

The live attenuated BCG is still the only vaccine available against TB, but it has little effect on preventing pulmonary TB in adults (105). Since pulmonary TB is the most significant source of TB transmission, BCG is thought to have limited effect on the spread of TB. During the last decades, great efforts have been made to design new and better TB vaccines (106, 107). The vaccine candidates are intended to either replace BCG or to act as booster vaccines after infant BCG vaccination. To date, essentially all subcellular vaccines tested have been based on dominant antigens of a protein/peptide nature, such as the Mtb protein antigen 85A (Ag85A) or ESAT-6 (108), with induction of potent Th1 cell responses used as an immunological correlate of protection.

The first vaccine candidate to be tested in clinical trials was MVA85A, a recombinant strain of Modified Vaccinia virus Ankara expressing Ag85A. Several trials reported its efficacy in terms of protection in animal models (109) and safety and immunogenicity in humans. However, in a phase 2b trial conducted in South Africa with infants previously vaccinated with BCG, the vaccine was well-tolerated and immunogenic but failed to confer better protection against TB compared to BCG alone (110). In another study in South Africa, also with infants who were vaccinated with BCG, there was no correlation between the number of BCG-elicited T cells that produced IFN-γ or multiple cytokines (such as IFN-γ, IL-2, and TNF) and the development of active TB (111). An important message from these studies is that the immune markers that had been used in most previous preclinical studies as correlates of vaccine-induced protection show no correlation with actual protection. Thus, it is unknown which immunological parameters or biomarkers predict who is protected from infection or who will control the infection or develop clinical disease, both in natural immunity and in vaccine-induced immunity (112).

Carbohydrate Antigens as Vaccine Candidates

The possibility of using carbohydrate antigens as vaccine components in a TB vaccine is an interesting strategy (113). Some of the most successful vaccines against bacterial infections are based on carbohydrate antigens that are conjugated to a carrier protein. Glycoconjugate vaccines toward encapsulated *Haemophilus influenzae* type b (Hib), *Streptococcus pneumonia*, and *Neisseria meningitidis* have all been licensed (114). Additionally, a conjugate vaccine against *Salmonella typhi*, a facultative intracellular pathogen, is being developed and has shown >50% protection in a phase 2b challenge, leading to its licensing (115).

Pure oligosaccharides are poor immunogens, as they fail to recruit $CD4^+$ T cell help. They are therefore limited to T cell–independent B cell immune responses. Children under 2 years are not capable of producing effective Abs against PS antigens

(116). However, conjugating a bacterial PS to a carrier protein that provides T cell epitopes creates a T cell–dependent antigen, which can induce protective immunity in infants. Interestingly, during infections with PS-encapsulated extracellular bacteria, PS-specific IgG responses are largely $CD4^+$ T cell dependent (117), indicating that in the presence of whole bacilli, T cell–dependent B cell responses are favored over T cell–independent B cell responses.

Conjugate Vaccine Candidates Based on AM

AM isolated from LAM, from the bacterial cell wall (118, 119), or from the Mtb capsule (99, 104) conjugated to various carrier proteins has been used to create a number of conjugate vaccine candidates (**Table 2**). In a series of experiments, AM-derived oligosaccharides were covalently linked to Mtb proteins (Ag85B or a 75 kDa antigen) or tetanus toxoid (TT) and tested for their protective effect in mice (C57BL/6) and guinea pigs. Using a subcutaneous priming-intranasal boost regime, these conjugate vaccines showed good protective efficacy in mice after intranasal challenge with Mtb Harlingen and in guinea pigs after aerosol infection with H37Rv, with prolonged survival and reduced weight loss, similar to that observed after vaccination with BCG (118). Intranasal vaccination with AM–TT also had a slight effect in mice as a boost after BCG prime, with reduced bacterial load in the spleen and reduced granulomatous inflammation in the lungs (120). However, in a second set of experiments in guinea pigs, using the same infection model as previously, the vaccine candidate did not perform as well (121). The vaccine response only led to a modest reduction of bacterial load in spleens, and the guinea pigs showed poor survival compared to previous studies (118). It is possible that the route of immunization and choice of adjuvant could impact the type (IgG vs. IgA and cell mediated vs. Abs) and location (mucosal vs. systemic) of the immune response. In one study, guinea pigs were immunized subcutaneously on day 0 and boosted once intranasally on day 24 (118), while in the other, mice were immunized subcutaneously once and boosted twice intranasally 3 weeks apart (121). There was also a difference in the formulation of the adjuvant in terms of emulsion vs. suspension (see ahead).

Similarly, AM–rEPA conjugates were used for subcutaneous vaccination of mice, resulting in a moderate effect on the bacterial load in the lungs but not in the spleen early after challenge with Mtb Erdman (99). In a more recent publication, intraperitoneal immunization with either AM–Ag85b conjugate or AM conjugated to *Bacillus anthracis* protective antigen (PA) formulated in alum resulted in a reduced bacterial load in the lungs and spleen and modest increased survival after aerosol challenge with Mtb H37Rv (104). Thus, AM conjugated to various carrier proteins, using different vaccination protocols, showed protection against Mtb challenge. In some cases, the protection was modest, and it never surpassed the effect of vaccination with BCG.

Immunogenicity of AM Conjugate Vaccines

The mechanisms of the (partial) protection induced by AM conjugate vaccines reported so far are not clear. Most conjugate vaccines licensed to date are thought to work by inducing protective Abs. In accordance, neutralizing Ab titers are often used as correlates of protection (122). AM oligosaccharide conjugates were highly immunogenic in rabbits (119), and various AM oligosaccharide–protein conjugates were shown to induce robust LAM-specific IgG in mice (123), all inducing a boostable T helper cell–dependent IgG response.

Ab responses were not investigated in the few early protection studies with AM conjugates. More recently however, immunization with either AM–Ag85b or AM–PA conjugates was found to elicit an AM-specific Ab response in mice that were partially protected against Mtb challenge (104),

Abs raised after vaccination with conjugate vaccines are mostly of the IgG1 subclass. For example, the dominant subclass was IgG1 in subjects vaccinated with pneumococcal TT conjugate, whereas, as expected, Abs induced by pneumococcal PS had a higher proportion of IgG2 (124). A mix of IgM, IgG1, and IgG2b was observed in the AM–Ag85b immunized mice vs. an exclusive IgG2b Ab response in AM–PA immunized mice, in both cases with alum as an adjuvant (104).

FACTORS TO CONSIDER FOR THE DESIGN OF AM CONJUGATE VACCINES

The immunogenicity of conjugate vaccines depends on the nature and size of the PS, the nature of the carrier protein, the conjugation method and the adjuvant (125).

Size of Carbohydrates

AM oligosaccharides with an apparent molecular weight of 5.2 kDa determined by gel filtration coupled to carrier proteins were recognized by LAM-specific mAbs, irrespective of coupling procedure or carrier protein used (119). Conjugates with AM oligosaccharides with an apparent molecular weight of 28 kDa reacted with the mAbs. However, none of the conjugates containing smaller-sized (apparent molecular weight 1.0 kDa) AM fragments reacted with the mAbs, indicating that the epitopes were absent or destroyed during the conjugation process.

The Role of Carrier Protein

In the choice of carrier proteins, several considerations should be taken into account, in particular, potential pre-exposure or co-exposure to a given carrier, which may lead to immune interference and reduction of the anti-carbohydrate immune response, and the dual role of proteins as carrier and protective antigen (126). Combining AM with the mycobacterial antigen Ag85B or the 75 kDa antigen induces a higher protection than AM conjugated to a non-mycobacterial antigen, such as TT (118) or rEPA (104). This could indicate that the mycobacterial protein antigens serve not only as carrier proteins but also as efficient immunogens on their own. On the other hand, it could also indicate that the mycobacterial antigens are more efficient carrier proteins compared to TT.

TABLE 2 | AM conjugate vaccines.

Vaccine	Animal model	Immunization route	Adjuvant	Challenge route	Vaccine-challenge interval	Prolonged survival	Reduced cfu spleen	Reduced cfu liver	Reduced cfu lung	Other	References
AMOs–TT	Mouse	s.c., boosted nasally after 6 wks	L3 (in emulsion or suspension)	i.n. Mtb Harlingen 10^5	24 day	Yes (more so for L3 in emulsion)	ND	ND	ND	Reduced weight loss	(118)
AMOs–Ag85B	Mouse	s.c. × 2 (27 days apart)	Alum	i.v. H37Rv 10^5	32 day	Yes	ND	ND	ND	Reduced weight loss	(118)
AMOs–Ag85B	Guinea pig	s.c., boosted nasally after 24 days	L3	Aerosol H37Rv 10 cfu per lung	6 weeks	No	Yes	ND	No	Reduced pathology in lung and spleen	(118)
AM–rEPA	Mouse	s.c., boosted s.c. × 2 at 2 and 4 weeks	Freund's incomplete	i.v. Mtb Erdman 10^6	6 weeks?	No	ND	No	Yes	Early moderate reduction of cfu	(99)
AM–rEPA	Mouse	s.c., boosted s.c. × 2 at 2 and 4 weeks	Freund's incomplete	i.v. BCG 2.5 × 10^5	6 weeks?	No	Yes (p = 0.025)	Yes (p = 0.01)	No		(99)
AM–TT	Mouse	i.n. 6 months after BCG	L3	i.v. H37Rv 3 × 10^5	10 weeks	ND	Yes	ND	No	Reduced granulomatous inflammation in lungs	(120)
AMOs–Ag85B	Guinea pig	s.c. boosted nasally 3 wks later	L3	Aerosol H37Rv 10–50 cfu/lung	6 weeks	No	Yes	No	No		(121)
AMOs–TT	Guinea pig	s.c. and boosted × 2 nasally 3 weeks apart	L3	Aerosol H37Rv 500 cfu	8 weeks	No	No	No	No		(121)
AM–Ag85B	Mouse	i.p. × 3 (2 weeks apart)	Alum	Aerosol H37Rv 100 cfu	4 weeks	Yes, similar to BCG	Yes	ND	Yes	Reduction in diseased lung tissue	(104)
AM–PA	Mouse	i.p. × 3 (2 weeks apart)	Alum	Aerosol H37Rv 100 cfu	4 weeks	Slight increase in survival	Yes	ND	Yes	Reduction in diseased lung tissue	(104)

Ag, antigen; AM, arabinomannan; AMOs, AM oligosaccharide; BCG, Bacillus Calmette-Guérin; cfu, colony forming units; i.n., intranasally; i.p., intraperitoneally; i.v., intravenously; LAM, lipoarabinomannan; ND, not done; rEPA, Pseudomonas aeruginosa exoprotein A; s.c., sub-cutaneously; TT, tetanus toxoid.

Conjugation Method

In the construction of AM–protein conjugate vaccines, two different methods for conjugation of AM oligosaccharides to carrier proteins were evaluated: direct reductive amination (127) and a thioether linkage method (128). Using oligosaccharide fragments with an apparent size range of 1.0–5.2 kDa, both methods yielded oligosaccharide–protein conjugates with a similar degree of substitution. The thioether linkage method was also efficient for conjugation of AM oligosaccharides of higher apparent molecular mass (28 kDa) to both TT and cross-reacting material 197 (CRM197), while direct reductive amination was not effective in conjugating oligosaccharide fragments of apparent molecular mass of 28 kDa. This is most probably due to the linkage arm used in the thioether linkage method, circumventing steric hindrance during conjugation. The choice of the carbohydrate site for conjugation to carrier proteins heavily impacts the immune response. Upon conjugation to semi-synthetic LAM derivatives, starting from mannose and two disaccharide analogs, Ag85B maintains its ability to activate B cells after glycosylation while showing a significant reduction in stimulating T cell response (129).

The Role of Adjuvants

Adjuvant use impacts on the type and magnitude of the immune response induced. Alum compounds classically induce strong Ab responses, while emulsion-based adjuvants generate a more balanced humoral and cell-mediated immune response (130). In a study on the efficacy of various adjuvants using H56 (Ag85B–ESAT-6–Rv2660c) as an antigen, the highest level of protection was seen upon vaccination with H56 in combination with emulsion-based adjuvants generating Th1 responses, while H56 combined with alum was not protective (130). This could indicate that vaccine-induced Abs only play a minor role in the vaccine-induced protection against Mtb, but it is also possible that qualitative features induced by the different adjuvants have an important effect on the protective response. In support of this hypothesis, it was recently shown that Abs with improved binding to FcγRIII are associated with better control of Mtb infection in humans (49). The Ab isotypes primarily interacting with FcγRIII are IgG1 and IgG3, which are normally induced in response to intracellular pathogens via the secretion of Th1 cytokines. However, since LAM-specific Abs are commonly of the IgG2 subtype (83, 85), characteristic of a Th2-derived immune response, a novel vaccine would likely benefit from directing the Ab response toward a more balanced Th1/Th2 profile. Consistent with this, a TB vaccine using alum as an adjuvant did not perform as well-compared to when given with the liposomal L3 adjuvant (118). Using L3 in emulsion as an adjuvant was also shown to be more effective compared to L3 in suspension (131).

Liposomes are adjuvants based on different types of lipid formulations, which can either be given together with the antigen in solution or formulated within lipid nano-structure (132). Their mechanism of action remains unclear, although liposomes stimulate a strong pro-inflammatory immune environment at the site of injection, with efficient uptake of antigen by primarily monocytes and dendritic cells (133, 134). Although liposomal adjuvant MF59 is more efficient compared to alum at this process, the addition of a TLR ligand to alum can enhance the response to levels similar to MF59 (134).

Purified TLR ligands can be given together with vaccines or other adjuvants to modulate the immune response toward a certain response pattern or increase the immunogenicity of an antigen (135). The addition of TLR ligands to alum was shown to greatly enhance immune responses in both non-human primates and humans (136, 137). By combining liposomal adjuvants with TLR ligands, the immunostimulatory effect can be improved even further, as shown by responses to the hepatitis B surface antigen and malaria RTS,S vaccine (138, 139). This allows for reduction of each vaccine dose but also modulation of the immune response toward a more balanced immune profile, with improved functionality of both T cells and B cells, which is likely an important factor for an improved TB vaccine.

Route of Immunization and Immunization Regimens

Most TB vaccine candidates have been developed for parenteral use. However, several studies have used mucosal delivery (140) such as intranasal immunization (141), in particular to boost parenteral BCG immunization (142, 143). Intranasal administration of an AM–TT conjugate after BCG vaccination enhanced protection in the spleen of mice, indicating a systemic protective effect, but did not reduce the bacterial load in the lung, compared to non-boosted BCG vaccinated mice [(120); **Table 2**]. However, the extent of granulomatous inflammation in the lungs was significantly reduced.

The number of immunizations as well as the intervals between them may also be of importance. Various methods for slow vaccine release have been explored. Interestingly, two methods of slow delivery immunization were recently shown to enhance neutralizing HIV Abs and germinal center responses in rhesus monkeys, compared to conventional immunization strategies (144), indicating the importance of how the vaccine is delivered in the resulting immune response.

A few studies in guinea pigs have explored the potential use of mycobacterial glycolipids (145, 146) as vaccine candidates, including LAM (147). A LAM extract from *Mycobacterium paratuberculosis* was shown to induce specific Abs in cattle that, after challenge, reduce *M. paratuberculosis* excretion at least for 100 days (147). We found that native purified LAM does not induce IgG immune responses in mice that can be further boosted, indicating a lack of antigen-specific memory B or T cells. This is likely due to mice only expressing CD1d, which has not been found to present mycobacterial glycolipid antigens.

CONCLUDING REMARKS

Our increasing understanding of the immune response to Mtb glycolipids indicates that these responses play an important role in protection against TB. LAM-specific T cells have a direct role on infected cells and support the differentiation of LAM-specific B cells into plasma cells. LAM-specific Abs were shown to mediate moderate protection against Mtb infection in mouse models and correlate with better disease control in humans. However, the mechanism of protection induced by AM–protein conjugates should be further studied for the development of novel vaccination strategies, alone or in combination with mycobacterial protein antigens. Increasing knowledge on how to modulate the immune response toward a more optimal T and B cell response is necessary for the development of such a vaccine. Special effort should address the impact of adjuvants and immunization strategies but also the precise mechanisms by which LAM-specific Abs mediate protection, exploring not just the specific epitopes but also Ab Fc characteristics.

AUTHOR CONTRIBUTIONS

GK and MC-N decided on the structure and general content of the manuscript. GK, MC-N, and CS collected the data, summarized the information, and wrote the different chapters. AC performed an overall revision and provided important contributions that improved the manuscript. All authors discussed and commented on the structure and the content of the manuscript.

FUNDING

This work was supported by funds from the Swedish Research Council (2016-05683) and the Swedish Heart-Lung Foundation (20180386).

REFERENCES

1. World Health Organization. *Global Tuberculosis Report 2018*. Available online at: https://www.who.int/tb/publications/global_report/en/
2. O'Garra A, Redford PS, McNab FW, Bloom CI, Wilkinson RJ, Berry MP. The immune response in tuberculosis. *Annu Rev Immunol.* (2013) 31:475–527. doi: 10.1146/annurev-immunol-032712-095939
3. Cooper AM. Cell-mediated immune responses in tuberculosis. *Annu Rev Immunol.* (2009) 27:393–422. doi: 10.1146/annurev.immunol.021908.132703
4. Oldenburg R, Demangel C. Pathogenic and immunosuppressive properties of mycobacterial phenolic glycolipids. *Biochimie.* (2017) 141:3–8. doi: 10.1016/j.biochi.2017.03.012
5. Noll H, Bloch H, Asselineau J, Lederer E. The chemical structure of the cord factor of *Mycobacterium tuberculosis. Biochim Biophys Acta.* (1956) 20:299–309. doi: 10.1016/0006-3002(56)90289-X
6. Blanc L, Gilleron M, Prandi J, Song OR, Jang MS, Gicquel B, et al. *Mycobacterium tuberculosis* inhibits human innate immune responses *via the* production of TLR2 antagonist glycolipids. *Proc Natl Acad Sci USA.* (2017) 114:11205–10. doi: 10.1073/pnas.1707840114
7. Hunter SW, Gaylord H, Brennan PJ. Structure and antigenicity of the phosphorylated lipopolysaccharide antigens from the leprosy and tubercle bacilli. *J Biol Chem.* (1986) 261:12345–51.
8. Chatterjee D, Bozic CM, McNeil M, Brennan PJ. Structural features of the arabinan component of the lipoarabinomannan of *Mycobacterium tuberculosis. J Biol Chem.* (1991) 266:9652–60.
9. Hunter SW, Brennan PJ. Evidence for the presence of a phosphatidylinositol anchor on the lipoarabinomannan and lipomannan of *Mycobacterium tuberculosis. J Biol Chem.* (1990) 265:9272–9.
10. Chatterjee D, Bozic C, Aspinall GO, Brennan PJ. Glucuronic acid- and branched sugar-containing glycolipid antigens of *Mycobacterium avium. J Biol Chem.* (1988) 263:4092–7.
11. Kallenius G, Correia-Neves M, Buteme H, Hamasur B, Svenson SB. Lipoarabinomannan, and its related glycolipids, induce divergent and opposing immune responses to *Mycobacterium tuberculosis* depending on structural diversity and experimental variations. *Tuberculosis.* (2016) 96:120–30. doi: 10.1016/j.tube.2015.09.005
12. Ortalo-Magne A, Andersen AB, Daffe M. The outermost capsular arabinomannans and other mannoconjugates of virulent and avirulent tubercle bacilli. *Microbiology.* (1996) 142(Pt 4):927–35. doi: 10.1099/00221287-142-4-927
13. Navoa JA, Laal S, Pirofski LA, McLean GR, Dai Z, Robbins JB, et al. Specificity and diversity of antibodies to *Mycobacterium tuberculosis* arabinomannan. *Clin Diagn Lab Immunol.* (2003) 10:88–94. doi: 10.1128/CDLI.10.1.88-94.2003
14. Mazurek J, Ignatowicz L, Kallenius G, Svenson SB, Pawlowski A, Hamasur B. Divergent effects of mycobacterial cell wall glycolipids on maturation and function of human monocyte-derived dendritic cells. *PLoS ONE.* (2012) 7:e42515. doi: 10.1371/journal.pone.0042515
15. Vergne I, Gilleron M, Nigou J. Manipulation of the endocytic pathway and phagocyte functions by *Mycobacterium tuberculosis* lipoarabinomannan. *Front Cell Infect Microbiol.* (2014) 4:187. doi: 10.3389/fcimb.2014.00187
16. Afonso-Barroso A, Clark SO, Williams A, Rosa GT, Nobrega C, Silva-Gomes S, et al. Lipoarabinomannan mannose caps do not affect mycobacterial virulence or the induction of protective immunity in experimental animal models of infection and have minimal impact on *in vitro* inflammatory responses. *Cell Microbiol.* (2013) 15:660–74. doi: 10.1111/cmi.12065
17. Ernst JD. Macrophage receptors for *Mycobacterium tuberculosis. Infect Immun.* (1998) 66:1277–81.
18. Liu CH, Liu H, Ge B. Innate immunity in tuberculosis: host defense vs pathogen evasion. *Cell Mol Immunol.* (2017) 14:963–75. doi: 10.1038/cmi.2017.88
19. Yonekawa A, Saijo S, Hoshino Y, Miyake Y, Ishikawa E, Suzukawa M, et al. Dectin-2 is a direct receptor for mannose-capped lipoarabinomannan of mycobacteria. *Immunity.* (2014) 41:402–13. doi: 10.1016/j.immuni.2014.08.005
20. Puissegur MP, Lay G, Gilleron M, Botella L, Nigou J, Marrakchi H, et al. Mycobacterial lipomannan induces granuloma macrophage fusion *via a* TLR2-dependent, ADAM9- and beta1 integrin-mediated pathway. *J Immunol.* (2007) 178:3161–9. doi: 10.4049/jimmunol.178.5.3161
21. Dao DN, Kremer L, Guerardel Y, Molano A, Jacobs WR Jr, Porcelli SA, et al. *Mycobacterium tuberculosis* lipomannan induces apoptosis and interleukin-12 production in macrophages. *Infect Immun.* (2004) 72:2067–74. doi: 10.1128/IAI.72.4.2067-2074.2004
22. Nigou J, Vasselon T, Ray A, Constant P, Gilleron M, Besra GS, et al. Mannan chain length controls lipoglycans signaling *via* and binding to TLR2. *J Immunol.* (2008) 180:6696–702. doi: 10.4049/jimmunol.180.10.6696
23. Turner J, Torrelles JB. Mannose-capped lipoarabinomannan in *Mycobacterium tuberculosis* pathogenesis. *Pathog Dis.* (2018) 76:fty026. doi: 10.1093/femspd/fty026
24. Drickamer K, Taylor ME. Recent insights into structures and functions of C-type lectins in the immune system. *Curr Opin Struct Biol.* (2015) 34:26–34. doi: 10.1016/j.sbi.2015.06.003
25. Dambuza IM, Brown GD. C-type lectins in immunity: recent developments. *Curr Opin Immunol.* (2015) 32:21–7. doi: 10.1016/j.coi.2014.12.002
26. Schnaar RL. Glycans and glycan-binding proteins in immune regulation: a concise introduction to glycobiology for the allergist. *J Allergy Clin Immunol.* (2015) 135:609–15. doi: 10.1016/j.jaci.2014.10.057
27. McGreal EP, Miller JL, Gordon S. Ligand recognition by antigen-presenting cell C-type lectin receptors. *Curr Opin Immunol.* (2005) 17:18–24. doi: 10.1016/j.coi.2004.12.001
28. Kang PB, Azad AK, Torrelles JB, Kaufman TM, Beharka A, Tibesar E, et al. The human macrophage mannose receptor directs *Mycobacterium tuberculosis* lipoarabinomannan-mediated phagosome biogenesis. *J Exp Med.* (2005) 202:987–99. doi: 10.1084/jem.20051239
29. Maeda N, Nigou J, Herrmann JL, Jackson M, Amara A, Lagrange PH, et al. The cell surface receptor DC-SIGN discriminates between *Mycobacterium* species through selective recognition of the mannose caps on lipoarabinomannan. *J Biol Chem.* (2003) 278:5513–6. doi: 10.1074/jbc.C200586200
30. Drummond RA, Brown GD. Signalling C-type lectins in antimicrobial immunity. *PLoS Pathog.* (2013) 9:e1003417. doi: 10.1371/journal.ppat.1003417

31. Macauley MS, Crocker PR, Paulson JC. Siglec-mediated regulation of immune cell function in disease. *Nat Rev Immunol.* (2014) 14:653–66. doi: 10.1038/nri3737
32. Zheng RB, Jegouzo SAF, Joe M, Bai Y, Tran HA, Shen K, et al. Insights into interactions of mycobacteria with the host innate immune system from a novel array of synthetic mycobacterial glycans. *ACS Chem Biol.* (2017) 12:2990–3002. doi: 10.1021/acschembio.7b00797
33. Decout A, Silva-Gomes S, Drocourt D, Blattes E, Riviere M, Prandi J, et al. Deciphering the molecular basis of mycobacteria and lipoglycan recognition by the C-type lectin Dectin-2. *Sci Rep.* (2018) 8:16840. doi: 10.1038/s41598-018-35393-5
34. Lewinsohn DA, Lewinsohn DM, Scriba TJ. Polyfunctional CD4+ T cells as targets for tuberculosis vaccination. *Front Immunol.* (2017) 8:1262. doi: 10.3389/fimmu.2017.01262
35. Woodworth JS, Behar SM. *Mycobacterium tuberculosis*-specific CD8+ T cells and their role in immunity. *Crit Rev Immunol.* (2006) 26:317–52. doi: 10.1615/CritRevImmunol.v26.i4.30
36. Lewinsohn DM, Zhu L, Madison VJ, Dillon DC, Fling SP, Reed SG, et al. Classically restricted human CD8+ T lymphocytes derived from *Mycobacterium tuberculosis*-infected cells: definition of antigenic specificity. *J Immunol.* (2001) 166:439–46. doi: 10.4049/jimmunol.166.1.439
37. Tully G, Kortsik C, Hohn H, Zehbe I, Hitzler WE, Neukirch C, et al. Highly focused T cell responses in latent human pulmonary *Mycobacterium tuberculosis* infection. *J Immunol.* (2005) 174:2174–84. doi: 10.4049/jimmunol.174.4.2174
38. Woodworth JS, Wu Y, Behar SM. *Mycobacterium tuberculosis*-specific CD8+ T cells require perforin to kill target cells and provide protection *in vivo*. *J Immunol.* (2008) 181:8595–603. doi: 10.4049/jimmunol.181.12.8595
39. Stenger S, Hanson DA, Teitelbaum R, Dewan P, Niazi KR, Froelich CJ, et al. An antimicrobial activity of cytolytic T cells mediated by granulysin. *Science.* (1998) 282:121–5. doi: 10.1126/science.282.5386.121
40. Sieling PA, Chatterjee D, Porcelli SA, Prigozy TI, Mazzaccaro RJ, Soriano T, et al. CD1-restricted T cell recognition of microbial lipoglycan antigens. *Science.* (1995) 269:227–30. doi: 10.1126/science.7542404
41. Gilleron M, Stenger S, Mazorra Z, Wittke F, Mariotti S, Bohmer G, et al. Diacylated sulfoglycolipids are novel mycobacterial antigens stimulating CD1-restricted T cells during infection with *Mycobacterium tuberculosis*. *J Exp Med.* (2004) 199:649–59. doi: 10.1084/jem.20031097
42. Hiromatsu K, Dascher CC, LeClair KP, Sugita M, Furlong ST, Brenner MB, et al. Induction of CD1-restricted immune responses in guinea pigs by immunization with mycobacterial lipid antigens. *J Immunol.* (2002) 169:330–9. doi: 10.4049/jimmunol.169.1.330
43. Van Rhijn I, Moody DB. CD1 and mycobacterial lipids activate human T cells. *Immunol Rev.* (2015) 264:138–53. doi: 10.1111/imr.12253
44. Kallert S, Zenk SF, Walther P, Grieshober M, Weil T, Stenger S. Liposomal delivery of lipoarabinomannan triggers *Mycobacterium tuberculosis* specific T-cells. *Tuberculosis.* (2015) 95:452–62. doi: 10.1016/j.tube.2015.04.001
45. Torrelles JB, Sieling PA, Zhang N, Keen MA, McNeil MR, Belisle JT, et al. Isolation of a distinct *Mycobacterium tuberculosis* mannose-capped lipoarabinomannan isoform responsible for recognition by CD1b-restricted T cells. *Glycobiology.* (2012) 22:1118–27. doi: 10.1093/glycob/cws078
46. De Libero G, Mori L. The T-cell response to lipid antigens of *Mycobacterium tuberculosis*. *Front Immunol.* (2014) 5:219. doi: 10.3389/fimmu.2014.00219
47. Porcelli SA, Modlin RL. The CD1 system: antigen-presenting molecules for T cell recognition of lipids and glycolipids. *Annu Rev Immunol.* (1999) 17:297–329. doi: 10.1146/annurev.immunol.17.1.297
48. Lang ML, Glatman-Freedman A. Do CD1-restricted T cells contribute to antibody-mediated immunity against *Mycobacterium tuberculosis*? *Infect Immun.* (2006) 74:803–9. doi: 10.1128/IAI.74.2.803-809.2006
49. Lu LL, Chung AW, Rosebrock TR, Ghebremichael M, Yu WH, Grace PS, et al. A functional role for antibodies in tuberculosis. *Cell.* (2016) 167:433–43.e14. doi: 10.1016/j.cell.2016.08.072
50. van Rensburg IC, Loxton AG. Killer (FASL regulatory) B cells are present during latent TB and are induced by BCG stimulation in participants with and without latent tuberculosis. *Tuberculosis.* (2018) 108:114–7. doi: 10.1016/j.tube.2017.11.010
51. Roy Chowdhury R, Vallania F, Yang Q, Lopez Angel CJ, Darboe F, Penn-Nicholson A, et al. A multi-cohort study of the immune factors associated with *M. tuberculosis* infection outcomes. *Nature.* (2018) 560:644–8. doi: 10.1038/s41586-018-0439-x
52. Busch M, Herzmann C, Kallert S, Zimmermann A, Hofer C, Mayer D, et al. Lipoarabinomannan-responsive polycytotoxic t cells are associated with protection in human tuberculosis. *Am J Respir Crit Care Med.* (2016) 194:345–55. doi: 10.1164/rccm.201509-1746OC
53. Belay M, Legesse M, Mihret A, Bekele Y, Bjune G, Abebe F. Lipoarabinomannan-specific TNF-alpha and IFN-gamma as markers of protective immunity against tuberculosis: a cohort study in an endemic setting. *APMIS.* (2015) 123:851–7. doi: 10.1111/apm.12423
54. Orme IM. Vaccines to prevent tuberculosis infection rather than disease: physiological and immunological aspects. *Tuberculosis.* (2016) 101:210–6. doi: 10.1016/j.tube.2014.10.008
55. Kozakiewicz L, Phuah J, Flynn J, Chan J. The role of B cells and humoral immunity in *Mycobacterium tuberculosis* infection. *Adv Exp Med Biol.* (2013) 783:225–50. doi: 10.1007/978-1-4614-6111-1_12
56. Achkar JM, Chan J, Casadevall A. B cells and antibodies in the defense against *Mycobacterium tuberculosis* infection. *Immunol Rev.* (2015) 264:167–81. doi: 10.1111/imr.12276
57. Jacobs AJ, Mongkolsapaya J, Screaton GR, McShane H, Wilkinson RJ. Antibodies and tuberculosis. *Tuberculosis.* (2016) 101:102–13. doi: 10.1016/j.tube.2016.08.001
58. Maglione PJ, Xu J, Chan J. B cells moderate inflammatory progression and enhance bacterial containment upon pulmonary challenge with *Mycobacterium tuberculosis*. *J Immunol.* (2007) 178:7222–34. doi: 10.4049/jimmunol.178.11.7222
59. Hoff ST, Salman AM, Ruhwald M, Ravn P, Brock I, Elsheikh N, et al. Human B cells produce chemokine CXCL10 in the presence of *Mycobacterium tuberculosis* specific T cells. *Tuberculosis.* (2015) 95:40–7. doi: 10.1016/j.tube.2014.10.005
60. Torrado E, Fountain JJ, Robinson RT, Martino CA, Pearl JE, Rangel-Moreno J, et al. Differential and site specific impact of B cells in the protective immune response to *Mycobacterium tuberculosis* in the mouse. *PLoS ONE.* (2013) 8:e61681. doi: 10.1371/journal.pone.0061681
61. Yuan C, Qu ZL, Tang XL, Liu Q, Luo W, Huang C, et al. *Mycobacterium tuberculosis* mannose-capped lipoarabinomannan induces IL-10–producing B cells and hinders CD4(+)Th1 immunity. *iScience.* (2019) 11:13–30. doi: 10.1016/j.isci.2018.11.039
62. Joosten SA, van Meijgaarden KE, Del Nonno F, Baiocchini A, Petrone L, Vanini V, et al. Patients with tuberculosis have a dysfunctional circulating B-cell compartment, which normalizes following successful treatment. *PLoS Pathog.* (2016) 12:e1005687. doi: 10.1371/journal.ppat.1005687
63. Portugal S, Obeng-Adjei N, Moir S, Crompton PD, Pierce SK. Atypical memory B cells in human chronic infectious diseases: an interim report. *Cell Immunol.* (2017) 321:18–25. doi: 10.1016/j.cellimm.2017.07.003
64. Burton AR, Pallett LJ, McCoy LE, Suveizdyte K, Amin OE, Swadling L, et al. Circulating and intrahepatic antiviral B cells are defective in hepatitis B. *J Clin Invest.* (2018) 128:4588–603. doi: 10.1172/JCI121960
65. Sundling C, Ronnberg C, Yman V, Asghar M, Jahnmatz P, Lakshmikanth T, et al. B cell profiling in malaria reveals expansion and remodelling of CD11c+ B cell subsets. *JCI Insight.* (2019) 5:126492. doi: 10.1172/jci.insight.126492
66. Portugal S, Tipton CM, Sohn H, Kone Y, Wang J, Li S, et al. Malaria-associated atypical memory B cells exhibit markedly reduced B cell receptor signaling and effector function. *Elife.* (2015) 4:e07218. doi: 10.7554/eLife.07218
67. Sullivan RT, Kim CC, Fontana MF, Feeney ME, Jagannathan P, Boyle MJ, et al. FCRL5 delineates functionally impaired memory B cells associated with *Plasmodium falciparum* exposure. *PLoS Pathog.* (2015) 11:e1004894. doi: 10.1371/journal.ppat.1004894
68. Moir S, Ho J, Malaspina A, Wang W, DiPoto AC, O'Shea MA, et al. Evidence for HIV-associated B cell exhaustion in a dysfunctional memory B cell compartment in HIV-infected viremic individuals. *J Exp Med.* (2008) 205:1797–805. doi: 10.1084/jem.20072683
69. Steingart KR, Flores LL, Dendukuri N, Schiller I, Laal S, Ramsay A, et al. Commercial serological tests for the diagnosis of active pulmonary and extrapulmonary tuberculosis: an updated systematic review and meta-analysis. *PLoS Med.* (2011) 8:e1001062. doi: 10.1371/journal.pmed.1001062

70. Coppola M, Arroyo L, van Meijgaarden KE, Franken KL, Geluk A, Barrera LF, et al. Differences in IgG responses against infection phase related *Mycobacterium tuberculosis* (Mtb) specific antigens in individuals exposed or not to Mtb correlate with control of TB infection and progression. *Tuberculosis.* (2017) 106:25–32. doi: 10.1016/j.tube.2017.06.001
71. Gennaro ML, Affouf M, Kanaujia GV, Brusasca PN, Mangura B, Reichman L. Antibody markers of incident tuberculosis among HIV-infected adults in the USA: a historical prospective study. *Int J Tuberc Lung Dis.* (2007) 11:624–31.
72. Laal S, Samanich KM, Sonnenberg MG, Zolla-Pazner S, Phadtare JM, Belisle JT. Human humoral responses to antigens of *Mycobacterium tuberculosis*: immunodominance of high-molecular-mass antigens. *Clin Diagn Lab Immunol.* (1997) 4:49–56.
73. Perley CC, Frahm M, Click EM, Dobos KM, Ferrari G, Stout JE, et al. The human antibody response to the surface of *Mycobacterium tuberculosis. PLoS ONE.* (2014) 9:e98938. doi: 10.1371/journal.pone.0098938
74. Alvarez N, Otero O, Camacho F, Borrero R, Tirado Y, Puig A, et al. Passive administration of purified secretory IgA from human colostrum induces protection against *Mycobacterium tuberculosis* in a murine model of progressive pulmonary infection. *BMC Immunol.* (2013) 14(Suppl. 1):S3. doi: 10.1186/1471-2172-14-S1-S3
75. Balu S, Reljic R, Lewis MJ, Pleass RJ, McIntosh R, van Kooten C, et al. A novel human IgA monoclonal antibody protects against tuberculosis. *J Immunol.* (2011) 186:3113–9. doi: 10.4049/jimmunol.1003189
76. Hamasur B, Haile M, Pawlowski A, Schroder U, Kallenius G, Svenson SB. A mycobacterial lipoarabinomannan specific monoclonal antibody and its F(ab') fragment prolong survival of mice infected with *Mycobacterium tuberculosis. Clin Exp Immunol.* (2004) 138:30–8. doi: 10.1111/j.1365-2249.2004.02593.x
77. Pethe K, Alonso S, Biet F, Delogu G, Brennan MJ, Locht C, et al. The heparin-binding haemagglutinin of *M. tuberculosis* is required for extrapulmonary dissemination. *Nature.* (2001) 412:190–4. doi: 10.1038/35084083
78. Tessema TA, Bjune G, Hamasur B, Svenson S, Syre H, Bjorvatn B. Circulating antibodies to lipoarabinomannan in relation to sputum microscopy, clinical features and urinary anti-lipoarabinomannan detection in pulmonary tuberculosis. *Scand J Infect Dis.* (2002) 34:97–103. doi: 10.1080/00365540110077263
79. Shete PB, Ravindran R, Chang E, Worodria W, Chaisson LH, Andama A, et al. Evaluation of antibody responses to panels of *M. tuberculosis* antigens as a screening tool for active tuberculosis in Uganda. *PLoS ONE.* (2017) 12:e0180122. doi: 10.1371/journal.pone.0180122
80. Brown RM, Cruz O, Brennan M, Gennaro ML, Schlesinger L, Skeiky YA, et al. Lipoarabinomannan-reactive human secretory immunoglobulin A responses induced by mucosal bacille Calmette–Guerin vaccination. *J Infect Dis.* (2003) 187:513–7. doi: 10.1086/368096
81. de Valliere S, Abate G, Blazevic A, Heuertz RM, Hoft DF. Enhancement of innate and cell-mediated immunity by antimycobacterial antibodies. *Infect Immun.* (2005) 73:6711–20. doi: 10.1128/IAI.73.10.6711-6720.2005
82. Chen T, Blanc C, Eder AZ, Prados-Rosales R, Souza AC, Kim RS, et al. Association of human antibodies to arabinomannan with enhanced mycobacterial opsonophagocytosis and intracellular growth reduction. *J Infect Dis.* (2016) 214:300–10. doi: 10.1093/infdis/jiw141
83. Yu X, Prados-Rosales R, Jenny-Avital ER, Sosa K, Casadevall A, Achkar JM. Comparative evaluation of profiles of antibodies to mycobacterial capsular polysaccharides in tuberculosis patients and controls stratified by HIV status. *Clin Vaccine Immunol.* (2012) 19:198–208. doi: 10.1128/CVI.05550-11
84. Somi GR, O'Brien RJ, Mfinanga GS, Ipuge YA. Evaluation of the MycoDot test in patients with suspected tuberculosis in a field setting in Tanzania. *Int J Tuberc Lung Dis.* (1999) 3:231–8.
85. Da Costa CT, Khanolkar-Young S, Elliott AM, Wasunna KM, McAdam KP. Immunoglobulin G subclass responses to mycobacterial lipoarabinomannan in HIV-infected and non-infected patients with tuberculosis. *Clin Exp Immunol.* (1993) 91:25–9. doi: 10.1111/j.1365-2249.1993.tb03348.x
86. Costello AM, Kumar A, Narayan V, Akbar MS, Ahmed S, Abou-Zeid C, et al. Does antibody to mycobacterial antigens, including lipoarabinomannan, limit dissemination in childhood tuberculosis? *Trans R Soc Trop Med Hyg.* (1992) 86:686–92. doi: 10.1016/0035-9203(92)90192-F
87. Kumar SK, Singh P, Sinha S. Naturally produced opsonizing antibodies restrict the survival of *Mycobacterium tuberculosis* in human macrophages by augmenting phagosome maturation. *Open Biol.* (2015) 5:150171. doi: 10.1098/rsob.150171
88. Zimmermann N, Thormann V, Hu B, Kohler AB, Imai-Matsushima A, Locht C, et al. Human isotype-dependent inhibitory antibody responses against *Mycobacterium tuberculosis. EMBO Mol Med.* (2016) 8:1325–39. doi: 10.15252/emmm.201606330
89. Jennewein MF, Alter G. The immunoregulatory roles of antibody glycosylation. *Trends Immunol.* (2017) 38:358–72. doi: 10.1016/j.it.2017.02.004
90. Siber GR, Schur PH, Aisenberg AC, Weitzman SA, Schiffman G. Correlation between serum IgG-2 concentrations and the antibody response to bacterial polysaccharide antigens. *N Engl J Med.* (1980) 303:178–82. doi: 10.1056/NEJM198007243030402
91. Ferrante A, Beard LJ, Feldman RG. IgG subclass distribution of antibodies to bacterial and viral antigens. *Pediatr Infect Dis J.* (1990) 9(Suppl. 8):S16–24. doi: 10.1097/00006454-199008001-00004
92. Perlmutter RM, Hansburg D, Briles DE, Nicolotti RA, Davie JM. Subclass restriction of murine anti-carbohydrate antibodies. *J Immunol.* (1978) 121:566–72.
93. Foote MR, Nonnecke BJ, Beitz DC, Waters WR. Antigen-specific B-cell responses by neonatal calves after early vaccination. *J Dairy Sci.* (2007) 90:5208–17. doi: 10.3168/jds.2007-0285
94. Hetland G, Wiker HG, Hogasen K, Hamasur B, Svenson SB, Harboe M. Involvement of antilipoarabinomannan antibodies in classical complement activation in tuberculosis. *Clin Diagn Lab Immunol.* (1998) 5:211–8.
95. Irani V, Guy AJ, Andrew D, Beeson JG, Ramsland PA, Richards JS. Molecular properties of human IgG subclasses and their implications for designing therapeutic monoclonal antibodies against infectious diseases. *Mol Immunol.* (2015) 67(2 Pt A):171–82. doi: 10.1016/j.molimm.2015.03.255
96. Teitelbaum R, Glatman-Freedman A, Chen B, Robbins JB, Unanue E, Casadevall A, et al. A mAb recognizing a surface antigen of *Mycobacterium tuberculosis* enhances host survival. *Proc Natl Acad Sci USA.* (1998) 95:15688–93. doi: 10.1073/pnas.95.26.15688
97. Glatman-Freedman A, Mednick AJ, Lendvai N, Casadevall A. Clearance and organ distribution of *Mycobacterium tuberculosis* lipoarabinomannan (LAM) in the presence and absence of LAM-binding immunoglobulin M. *Infect Immun.* (2000) 68:335–41. doi: 10.1128/IAI.68.1.335-341.2000
98. Choudhary A, Patel D, Honnen W, Lai Z, Prattipati RS, Zheng RB, et al. Characterization of the antigenic heterogeneity of lipoarabinomannan, the major surface glycolipid of *Mycobacterium tuberculosis*, and complexity of antibody specificities toward this antigen. *J Immunol.* (2018) 200:3053–66. doi: 10.4049/jimmunol.1701673
99. Glatman-Freedman A, Casadevall A, Dai Z, Jacobs WR Jr, Li A, Morris SL, et al. Antigenic evidence of prevalence and diversity of *Mycobacterium tuberculosis* arabinomannan. *J Clin Microbiol.* (2004) 42:3225–31. doi: 10.1128/JCM.42.7.3225-3231.2004
100. Pan Q, Wang Q, Sun X, Xia X, Wu S, Luo F, et al. Aptamer against mannose-capped lipoarabinomannan inhibits virulent *Mycobacterium tuberculosis* infection in mice and rhesus monkeys. *Mol Ther.* (2014) 22:940–51. doi: 10.1038/mt.2014.31
101. Nimmerjahn F, Ravetch JV. Divergent immunoglobulin g subclass activity through selective Fc receptor binding. *Science.* (2005) 310:1510–2. doi: 10.1126/science.1118948
102. Maglione PJ, Xu J, Casadevall A, Chan J. Fc gamma receptors regulate immune activation and susceptibility during *Mycobacterium tuberculosis* infection. *J Immunol.* (2008) 180:3329–38. doi: 10.4049/jimmunol.180.5.3329
103. Olivares N, Marquina B, Mata-Espinoza D, Zatarain-Barron ZL, Pinzon CE, Estrada I, et al. The protective effect of immunoglobulin in murine tuberculosis is dependent on IgG glycosylation. *Pathog Dis.* (2013) 69:176–83. doi: 10.1111/2049-632X.12069
104. Prados-Rosales R, Carreno L, Cheng T, Blanc C, Weinrick B, Malek A, et al. Enhanced control of *Mycobacterium tuberculosis* extrapulmonary dissemination in mice by an arabinomannan–protein conjugate vaccine. *PLoS Pathog.* (2017) 13:e1006250. doi: 10.1371/journal.ppat.1006250

105. Mangtani P, Abubakar I, Ariti C, Beynon R, Pimpin L, Fine PE, et al. Protection by BCG vaccine against tuberculosis: a systematic review of randomized controlled trials. *Clin Infect Dis.* (2014) 58:470–80. doi: 10.1093/cid/cit790
106. Kaufmann SHE, Dockrell HM, Drager N, Ho MM, McShane H, Neyrolles O, et al. TBVAC2020: advancing tuberculosis vaccines from discovery to clinical development. *Front Immunol.* (2017) 8:1203. doi: 10.3389/fimmu.2017.01203
107. Kaufmann SH, Lange C, Rao M, Balaji KN, Lotze M, Schito M, et al. Progress in tuberculosis vaccine development and host-directed therapies—a state of the art review. *Lancet Respir Med.* (2014) 2:301–20. doi: 10.1016/S2213-2600(14)70033-5
108. Zhu B, Dockrell HM, Ottenhoff THM, Evans TG, Zhang Y. Tuberculosis vaccines: opportunities and challenges. *Respirology.* (2018) 23:359–68. doi: 10.1111/resp.13245
109. Kashangura R, Sena ES, Young T, Garner P. Effects of MVA85A vaccine on tuberculosis challenge in animals: systematic review. *Int J Epidemiol.* (2015) 44:1970–81. doi: 10.1093/ije/dyv142
110. Tameris MD, Hatherill M, Landry BS, Scriba TJ, Snowden MA, Lockhart S, et al. Safety and efficacy of MVA85A, a new tuberculosis vaccine, in infants previously vaccinated with BCG: a randomised, placebo-controlled phase 2b trial. *Lancet.* (2013) 381:1021–8. doi: 10.1016/S0140-6736(13)60177-4
111. Kagina BM, Abel B, Scriba TJ, Hughes EJ, Keyser A, Soares A, et al. Specific T cell frequency and cytokine expression profile do not correlate with protection against tuberculosis after bacillus Calmette–Guerin vaccination of newborns. *Am J Respir Crit Care Med.* (2010) 182:1073–9. doi: 10.1164/rccm.201003-0334OC
112. Nunes-Alves C, Booty MG, Carpenter SM, Jayaraman P, Rothchild AC, Behar SM. In search of a new paradigm for protective immunity to TB. *Nat Rev Microbiol.* (2014) 12:289–99. doi: 10.1038/nrmicro3230
113. Kallenius G, Pawlowski A, Hamasur B, Svenson SB. Mycobacterial glycoconjugates as vaccine candidates against tuberculosis. *Trends Microbiol.* (2008) 16:456–62. doi: 10.1016/j.tim.2008.07.007
114. Vella M, Pace D. Glycoconjugate vaccines: an update. *Expert Opin Biol Ther.* (2015) 15:529–46. doi: 10.1517/14712598.2015.993375
115. Meiring JE, Gibani M, TyVAC Consortium Meeting Group. The Typhoid Vaccine Acceleration Consortium (TyVAC): vaccine effectiveness study designs: accelerating the introduction of typhoid conjugate vaccines and reducing the global burden of enteric fever. Report from a meeting held on 26-27 October 2016, Oxford, UK. *Vaccine.* (2017) 35:5081–8. doi: 10.1016/j.vaccine.2017.08.001
116. Weintraub A. Immunology of bacterial polysaccharide antigens. *Carbohydr Res.* (2003) 338:2539–47. doi: 10.1016/j.carres.2003.07.008
117. Snapper CM. Mechanisms underlying *in vivo* polysaccharide-specific immunoglobulin responses to intact extracellular bacteria. *Ann N Y Acad Sci.* (2012) 1253:92–101. doi: 10.1111/j.1749-6632.2011.06329.x
118. Hamasur B, Haile M, Pawlowski A, Schroder U, Williams A, Hatch G, et al. *Mycobacterium tuberculosis* arabinomannan–protein conjugates protect against tuberculosis. *Vaccine.* (2003) 21:4081–93. doi: 10.1016/S0264-410X(03)00274-3
119. Hamasur B, Kallenius G, Svenson SB. Synthesis and immunologic characterisation of *Mycobacterium tuberculosis* lipoarabinomannan specific oligosaccharide–protein conjugates. *Vaccine.* (1999) 17:2853–61. doi: 10.1016/S0264-410X(99)00124-3
120. Haile M, Hamasur B, Jaxmar T, Gavier-Widen D, Chambers MA, Sanchez B, et al. Nasal boost with adjuvanted heat-killed BCG or arabinomannan–protein conjugate improves primary BCG-induced protection in C57BL/6 mice. *Tuberculosis.* (2005) 85:107–14. doi: 10.1016/j.tube.2004.09.013
121. Williams A, Hatch GJ, Clark SO, Gooch KE, Hatch KA, Hall GA, et al. Evaluation of vaccines in the EU TB vaccine cluster using a guinea pig aerosol infection model of tuberculosis. *Tuberculosis.* (2005) 85:29–38. doi: 10.1016/j.tube.2004.09.009
122. Plotkin SA. Correlates of protection induced by vaccination. *Clin Vaccine Immunol.* (2010) 17:1055–65. doi: 10.1128/CVI.00131-10
123. Wang L, Feng S, An L, Gu G, Guo Z. Synthetic and immunological studies of mycobacterial lipoarabinomannan oligosaccharides and their protein conjugates. *J Org Chem.* (2015) 80:10060–75. doi: 10.1021/acs.joc.5b01686
124. Eskola J. Immunogenicity of pneumococcal conjugate vaccines. *Pediatr Infect Dis J.* (2000) 19:388–93. doi: 10.1097/00006454-200004000-00035
125. Rinaldi F, Tengattini S, Piubelli L, Bernardini R, Mangione F, Bavaro T, et al. Rational design, preparation and characterization of recombinant Ag85B variants and their glycoconjugates with T-cell antigenic activity against *Mycobacterium tuberculosis. RSC Adv.* (2018) 8:23171–80. doi: 10.1039/C8RA03535K
126. Micoli F, Adamo R, Costantino P. Protein carriers for glycoconjugate vaccines: history, selection criteria, characterization and new trends. *Molecules.* (2018) 23:E1451. doi: 10.3390/molecules23061451
127. Jennings HJ, Lugowski C. Immunochemistry of groups A, B, and C meningococcal polysaccharide–tetanus toxoid conjugates. *J Immunol.* (1981) 127:1011–8.
128. Pawlowski A, Kallenius G, Svenson SB. A new method of non-cross-linking conjugation of polysaccharides to proteins *via* thioether bonds for the preparation of saccharide–protein conjugate vaccines. *Vaccine.* (1999) 17:1474–83. doi: 10.1016/S0264-410X(98)00385-5
129. Bavaro T, Tengattini S, Piubelli L, Mangione F, Bernardini R, Monzillo V, et al. Glycosylation of recombinant antigenic proteins from *Mycobacterium tuberculosis*: *in silico* prediction of protein epitopes and *ex vivo* biological evaluation of new semi-synthetic glycoconjugates. *Molecules.* (2017) 22:E1081. doi: 10.3390/molecules22071081
130. Knudsen NP, Olsen A, Buonsanti C, Follmann F, Zhang Y, Coler RN, et al. Different human vaccine adjuvants promote distinct antigen-independent immunological signatures tailored to different pathogens. *Sci Rep.* (2016) 6:19570. doi: 10.1038/srep19570
131. Haile M, Schroder U, Hamasur B, Pawlowski A, Jaxmar T, Kallenius G, et al. Immunization with heat-killed *Mycobacterium bovis* bacille Calmette–Guerin (BCG) in Eurocine L3 adjuvant protects against tuberculosis. *Vaccine.* (2004) 22:1498–508. doi: 10.1016/j.vaccine.2003.10.016
132. Alving CR, Beck Z, Matyas GR, Rao M. Liposomal adjuvants for human vaccines. *Expert Opin Drug Deliv.* (2016) 13:807–16. doi: 10.1517/17425247.2016.1151871
133. Awate S, Babiuk LA, Mutwiri G. Mechanisms of action of adjuvants. *Front Immunol.* (2013) 4:114. doi: 10.3389/fimmu.2013.00114
134. Liang F, Lindgren G, Sandgren KJ, Thompson EA, Francica JR, Seubert A, et al. Vaccine priming is restricted to draining lymph nodes and controlled by adjuvant-mediated antigen uptake. *Sci Transl Med.* (2017) 9:eaal2094. doi: 10.1126/scitranslmed.aal2094
135. Lore K, Karlsson Hedestam GB. Novel adjuvants for B cell immune responses. *Curr Opin HIV AIDS.* (2009) 4:441–6. doi: 10.1097/COH.0b013e32832da082
136. Thompson EA, Lore K. Non-human primates as a model for understanding the mechanism of action of toll-like receptor–based vaccine adjuvants. *Curr Opin Immunol.* (2017) 47:1–7. doi: 10.1016/j.coi.2017.06.006
137. Garcon N, Van Mechelen M. Recent clinical experience with vaccines using MPL- and QS-21-containing adjuvant systems. *Expert Rev Vaccines.* (2011) 10:471–86. doi: 10.1586/erv.11.29
138. Leroux-Roels G, Marchant A, Levy J, Van Damme P, Schwarz TF, Horsmans Y, et al. Impact of adjuvants on CD4(+) T cell and B cell responses to a protein antigen vaccine: results from a phase II, randomized, multicenter trial. *Clin Immunol.* (2016) 169:16–27. doi: 10.1016/j.clim.2016.05.007
139. Didierlaurent AM, Laupeze B, Di Pasquale A, Hergli N, Collignon C, Garcon N. Adjuvant system AS01: helping to overcome the challenges of modern vaccines. *Expert Rev Vaccines.* (2017) 16:55–63. doi: 10.1080/14760584.2016.1213632
140. Caetano LA, Almeida AJ, Goncalves LM. Approaches to tuberculosis mucosal vaccine development using nanoparticles and microparticles: a review. *J Biomed Nanotechnol.* (2014) 10:2295–316. doi: 10.1166/jbn.2014.1984
141. Kallenius G, Pawlowski A, Brandtzaeg P, Svenson S. Should a new tuberculosis vaccine be administered intranasally? *Tuberculosis.* (2007) 87:257–66. doi: 10.1016/j.tube.2006.12.006
142. Dietrich J, Andersen C, Rappuoli R, Doherty TM, Jensen CG, Andersen P. Mucosal administration of Ag85B-ESAT-6 protects against infection with *Mycobacterium tuberculosis* and boosts prior bacillus Calmette–Guerin immunity. *J Immunol.* (2006) 177:6353–60. doi: 10.4049/jimmunol.177.9.6353

143. Andersen CS, Dietrich J, Agger EM, Lycke NY, Lovgren K, Andersen P. The combined CTA1-DD/ISCOMs vector is an effective intranasal adjuvant for boosting prior *Mycobacterium bovis* BCG immunity to *Mycobacterium tuberculosis*. *Infect Immun.* (2007) 75:408–16. doi: 10.1128/IAI.01290-06
144. Cirelli KM, Carnathan DG, Nogal B, Martin JT, Rodriguez OL, Upadhyay AA, et al. Slow delivery immunization enhances HIV neutralizing antibody and germinal center responses *via* modulation of immunodominance. *Cell.* (2019) 177:1153–71 e28. doi: 10.1016/j.cell.2019.04.012
145. Larrouy-Maumus G, Layre E, Clark S, Prandi J, Rayner E, Lepore M, et al. Protective efficacy of a lipid antigen vaccine in a guinea pig model of tuberculosis. *Vaccine.* (2017) 35:1395–402. doi: 10.1016/j.vaccine.2017.01.079
146. Dascher CC, Hiromatsu K, Xiong X, Morehouse C, Watts G, Liu G, et al. Immunization with a mycobacterial lipid vaccine improves pulmonary pathology in the guinea pig model of tuberculosis. *Int Immunol.* (2003) 15:915–25. doi: 10.1093/intimm/dxg091
147. Jolly A, Morsella C, Bass L, Fiorentino MA, Paolicchi FA, Mundo SL. Bovine response to lipoarabinomannan vaccination and challenge with *Mycobacterium* paratuberculosis. *Braz J Microbiol.* (2013) 44:511–4. doi: 10.1590/S1517-83822013000200029

12

Interaction of Bacterial Exotoxins with Neutrophil Extracellular Traps: Impact for the Infected Host

Maren von Köckritz-Blickwede[1,2], Stefanie Blodkamp[1] and Victor Nizet[3]*

[1] Department of Physiological Chemistry, University of Veterinary Medicine Hannover, Hannover, Germany, [2] Research Center for Emerging Infections and Zoonoses, University of Veterinary Medicine Hannover, Hannover, Germany, [3] Department of Pediatric Pharmacology and Drug Discovery, San Diego School of Medicine, University of California, La Jolla, CA, USA

Edited by:
Katharina F. Kubatzky,
University Hospital Heidelberg,
Germany

Reviewed by:
Jos A. G. Van Strijp,
University Medical Center Utrecht,
Netherlands
Victor J. Torres,
New York University School
of Medicine, USA
Sonja Oehmcke-Hecht,
University Rostock, Germany

***Correspondence:**
Maren von Köckritz-Blickwede
mkoeckbl@tiho-hannover.de

Since their discovery in 2004, neutrophil extracellular traps (NETs) have been characterized as a fundamental host innate immune defense against various pathogens. Released in response to infectious and pro-inflammatory stimuli, NETs can immobilize invading pathogens within a fibrous matrix consisting of DNA, histones, and antimicrobial peptides. Conversely, excessive or dysregulated NET release may hold a variety of detrimental consequences for the host. A fine balance between NET formation and elimination is necessary to sustain a protective effect during infectious challenge. In recent years, a number of microbial virulence factors have been shown to modulate formation of NETs, thereby facilitating colonization or spread within the host. In this mini-review we summarize the contemporary research on the interaction of bacterial exotoxins with neutrophils that modulate NET production, focusing particular attention on consequences for the host. Understanding host–pathogen dynamics in this extracellular battlefield of innate immunity may provide novel therapeutic approaches for infectious and inflammatory disorders.

Keywords: phagocyte extracellular traps, neutrophils, macrophages, leukocidin, cell death

NEUTROPHIL EXTRACELLULAR TRAPS

The formation of neutrophil extracellular DNA traps (so called NETs) was first recognized as a host innate immune defense mechanism against infections by Brinkmann et al. (2004). This discovery altered the fundamental conception of the innate immune function of phagocytes against pathogenic microbes in a most fascinating way. Whereas it was previously believed that neutrophils kill invading pathogens by intracellular uptake (phagocytosis) and subsequent killing, the discovery of NETs revealed an additional phagocytosis-independent mechanism. The released nuclear material including histones and DNA within the extracellular trap (ET) immobilize and occasionally kill several medically important bacteria, viruses, or parasites (Brinkmann et al., 2004).

Regarding the cellular pathways that lead to NET formation, most studies show that the cell in question undergoes NETosis, a process of death that differs morphologically from necrosis and apoptosis (Fuchs et al., 2007). In contrast to apoptotic or necrotic cells, cells that undergo NETosis show a prelytic decondensation of the chromatin associated with the disruption of the nuclear membrane. This disintegration of the nuclear membrane allows mixing of DNA and histones with granule components, which are then extracellularly released as web-like fibers with a diameter of approximately 15 to 17 nm, associated with globular proteins or antimicrobial peptides.

NET-based antimicrobial activity remains active when cells are treated with the actin microfilament inhibitor cytochalasin D, indicating that this phenomenon is independent of phagocytosis. However, the antimicrobial activity of NETs can be abolished when NETs are treated with DNase, confirming that DNA comprises the functional backbone of the released fibers (Brinkmann et al., 2004; Fuchs et al., 2007).

The formation of NETs was initially thought to be an antimicrobial defense strategy and cell death pathway specific for neutrophils (Brinkmann et al., 2004). However, von Köckritz-Blickwede et al. (2008) mast cells were shown to also deploy mast cell extracellular traps (MCETs) in defense against bacterial pathogens. During recent years, further evidence has accumulated demonstrating ET formation occurs in eosinophils (Yousefi et al., 2008), basophils (Schorn et al., 2012), fibrocytes (Kisseleva et al., 2011), macrophages (Bartneck et al., 2010; Chow et al., 2010), and monocytes (Bartneck et al., 2010; Chow et al., 2010).

It is important to mention that some authors have demonstrated experimentally that eosinophils and neutrophils can release antimicrobial ETs in response to infection while remaining in a viable status (Yousefi et al., 2008, 2009). NET release by viable cells was confirmed *in vivo* in a murine model of *Staphylococcus aureus* skin infection (Yipp et al., 2012). Here, NET formation was seen as a dynamic process, which occurs by vesicular release of nuclear material during migration of neutrophils through the tissue. It is still unclear, however, how such budding and final release of nuclear material is initiated. Thus, based on current knowledge it seems that at least two different mechanisms can lead to NET formation: a viable form involving vesicular release and the more well-understood NETosis form associated with rupture of the nuclear membrane. Research has begun to explore how certain drugs, including statins (Chow et al., 2010) and tamoxifen (Corriden et al., 2015), modulate neutrophil functions to accentuate NET formation, which could be beneficial as an adjunctive therapy for extracellular pathogens such as *S. aureus* that efficiently avoid phagocytic killing by neutrophils.

In contrast to their protective effect against several infections, there is increasing evidence that a dysregulated NET release can provoke autoimmune reactions, tissue damage and impaired cellular functions (Villanueva et al., 2011; Saffarzadeh et al., 2012). Unchecked, aberrant NET formation can result in pathological damage as vascular thrombosis (Fuchs et al., 2010) or chronic lung inflammation in cystic fibrosis (Papayannopoulos et al., 2011). Countermeasures to the excessive NET formation may be of therapeutic utility in such cases as for example DNAse treatment in patients with cystic fibrosis (Fuchs et al., 1994) or the usage of anti-histone antibodies to alleviate vascular thrombosis (Semeraro et al., 2011). Since sustained blockade of NET formation may carry a risk of increased susceptibility to certain infections, anti-NET therapy might be targeted to severe autoimmune or inflammatory diseases, when pro-inflammatory activities of NETs outweigh their protective benefits (Saffarzadeh and Preissner, 2013).

THE ROLE OF NETs AGAINST INFECTION

In several studies, NETs were found to play a protective role against various infecting organisms, often employing mutagenesis of key evasion factors of the respective organisms. It is thought that NETs can act in two ways: (1) pathogen immobilization and (2) growth inhibition or killing of the microbe. The relative importance of killing is a point of some contention in the literature (Menegazzi et al., 2012). However, based on vital immunofluorescent staining of bacteria with DNA intercalating dyes, microbial killing by NETs has been demonstrated definitively by several authors (Lauth et al., 2009; Berends et al., 2010; Halverson et al., 2015). However, since the overall killing potential of NETs is physically restricted, the cumulative effect of NETs may be functionally bacteriostatic (Baums and von Köckritz-Blickwede, 2015).

Because some bacteria have evolved highly efficient resistance strategies against antimicrobial activity of NETs, in some specific circumstances, NETs can help to establish an infectious niche. For example, in the case of a primary influenza A infection of the middle ear which boosts formation of NETs by infiltrating neutrophils, resistant *Streptococcus pneumoniae* can use those NETs to multiply and persist in the middle ear cavity (Short et al., 2014). Additionally, some studies show that *S. pneumoniae* and *Haemophilus influenzae* can incorporate NETs into biofilms that promote their persistence in the host (Hong et al., 2009; Reid et al., 2009). In the end, it depends on the pathogen, its array of immune resistance factors and the anatomical site of infection, as to whether NETs can serve a protective function for the host during an infection (**Figure 1**).

CATEGORIES OF BACTERIAL NET EVASION FACTORS

Neutrophil extracellular trap evasion factors can be classified in three different phenotypes: (1) NET degradation, (2) resistance to the intrinsic antimicrobial effectors within NETs, or (3) the suppression of NET production (von Köckritz-Blickwede and Nizet, 2009). The best-studied NET evasion factor is the activity of microbial nucleases that degrade NETs deployed by innate immune cells to escape entrapment. Membrane-bound or released nucleases have been shown to be expressed by the following microbes as NET evasion factors: *Aeromonas hydrophila* (Brogden et al., 2012), *Escherichia coli* (Möllerherm et al., 2015), *Leptospira* sp. (Scharrig et al., 2015), *Neisseria gonorhoeae* (Juneau et al., 2015), *S. aureus* (Berends et al., 2010), *S. agalactiae* (Derré-Bobillot et al., 2013), *S. pneumoniae* (Beiter et al., 2006), *S. pyogenes* (Buchanan et al., 2006; Chang et al., 2011), *Streptococcus synguinis* (Morita et al., 2014), *S. suis* (de Buhr et al., 2014), *Vibrio cholera* (Seper et al., 2013), and *Yersinia enterocolitica* (Möllerherm et al., 2015). Whereas, resistance to antimicrobial effectors in ETs is mainly mediated by surface-bound virulence factors, e.g., exopolysaccharide capsules in case of *S. pyogenes* (Cole et al., 2010) or *S. pneumonia* (Wartha et al., 2007), the suppression of NET formation is mainly carried

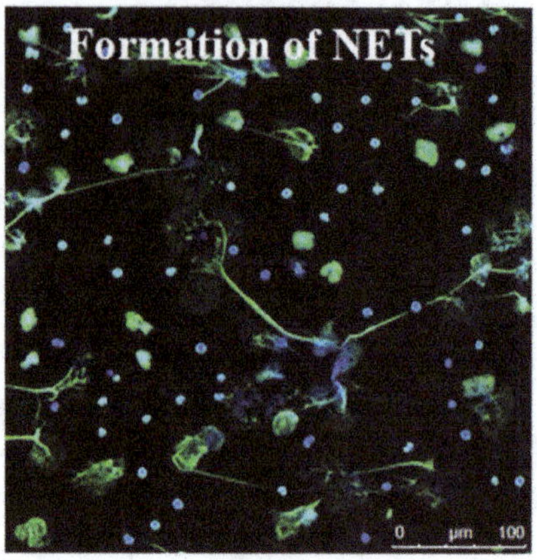

Protective immune defense of NETs against infections with:

- *Aeromonas hydrophila* (Brogden et al., 2014)
- *Aspergillus fumigatus* (Röhm et al., 2014)
- *Borrelia burgdorferi* (Menten-Dedoyart et al., 2012)
- *Candida albicans* (Urban et al., 2009)
- *Eimeria bovis* (Behrendt et al., 2010)
- *Escherichia coli* (Grinberg et al., 2008)
- *Group B Streptococcus* (Carlin et al., 2009)
- *Human immunodeficiency virus* (Saitoh et al., 2012)
- *Leishmania amazonensis* (Guimarães-Costa et al., 2009)
- *Leishmania donovani* (Gabriel et al., 2010)
- *Mannheimia haemolytica* (Aulik et al., 2010)
- *Mycobacterium tuberculosis* (Ramos-Kichik et al., 2009; Wong and Jacobs, 2013)
- *Myxoma virus (MYXV)* (Jenne et al., 2013)
- *Plasmodium falciparum* (Baker et al., 2008)
- *Pseudomonas aeruginosa* (Young et al., 2011)*
- *Salmonella typhimurium* (Brinkmann et al., 2004)
- *Shigella flexneri* (Brinkmann et al., 2004)
- *Staphylococcus aureus* (Berends et al., 2010)*
- *Streptococcus pneumoniae* (Beiter et al., 2006) *
- *Streptococcus pyogenes* (Lauth et al., 2009)
- *Streptococcus suis* (de Buhr et al., 2014)
- *Toxoplasma gondii* (Abi Abdallah et al., 2012)
- *Vibrio cholerae* (Seper et al., 2013)

Detrimental consequences of NETs for the host during infections with:

- *Feline leukemia virus* (Wardini et al., 2010)
- *Haemophilus influenzae* (Hong et al., 2009)
- *Influenza A virus* (Narasaraju et al., 2011)
- *Plasmodium falciparum* (Baker et al., 2008)
- *Pseudomonas aeruginosa* (Fuxman et al., 2010)*
- *Staphylococcus aureus* (Gonzalez et al., 2014)*
- *Streptococcus pneumoniae* (Reid et al., 2009) *

FIGURE 1 | Extracellular trap formation during infections: consequences for the host. The figure summarizes the protective versus detrimental role of NETs for the host during an infection with respective pathogens. In some cases (highlighted with an asterisk), there is benefit and harm depending on site, timing and magnitude of infection. The immunofluorescence micrograph is showing the release of NETs from human blood-derived neutrophils. NETs were stained with antibodies against histone-DNA complexes (green) and DAPI to stain the nuclei and nuclear DNA (blue).

out by released factors including proteases (Zinkernagel et al., 2008).

BACTERIAL EXOTOXINS THAT MODULATE NET FORMATION

During the last 10 years, there is increasing evidence in the literature that different exotoxins released by Gram-positive or Gram-negative bacteria are able to modulate the process of NET formation (**Table 1**): to date, at least 11 different exotoxins have been published to modulate NET formation, with nine of them reported to enhance NET formation while two suppress the process (**Table 1**). First we will set our attention on those toxins that induce NET formation.

The M1-protein of the Gram-positive bacterium *S. pyogenes*, also known as group A *Streptococcus* (GAS), was the first exotoxin described to induce formation of ETs by neutrophils and mast cells (Lauth et al., 2009). The surface-anchored M1 protein is a classical virulence factor that promotes resistance to phagocytosis. However, after proteolytic release from the *S. pyogenes* surface, the M1 protein forms a proinflammatory supramolecular network with fibrinogen that contributes to the pathophysiology of streptococcal toxic shock-like syndrome (Herwald et al., 2004). Two parallel studies showed in 2009, that the M1 protein is able to mediate formation of NETs as well as MCETs, and thereby contribute to entrapment and killing of the bacteria (Lauth et al., 2009; Oehmcke et al., 2009). Besides induction of ETs, M1 protein promotes resistance to the human cathelicidin antimicrobial peptide LL-37, an important effector of bacterial killing by neutrophils and mast cells (Lauth et al., 2009; LaRock et al., 2015).

Staphylococcus epidermidis is a common colonizer of healthy human skin and is hypothesized to play a mutually beneficial role in the cutaneous niche. Cogen et al. (2010) demonstrated that *S. epidermidis* phenol-soluble-modulin γ (δ-toxin) boosts NET production and further colocalizes with endogenous host antimicrobial peptides within the architecture of the expressed NETs. In antimicrobial assays against the pathogen *S. pyogenes*, δ-toxin cooperated with host-derived antimicrobial peptides cathelicidins (human LL-37 and murine CRAMP) and human beta-defensins (hBD2 and hBD3) to

TABLE 1 | Effect of bacterial exotoxins on formation of extracellular traps.

Microbe	Exoprotein	Effect on cells	Mechanism	Cell death	Putative consequences for the host	Reference
Bordetella pertussis	Adenylate cyclase toxin (ACT)	Suppression of NET formation and apoptosis	Generation of cAMP and inhibition of oxidative burst	Impairment of cell lysis	ACT promotes neutrophil infiltration by inhibiting neutrophil death	Eby et al., 2014
Escherichia coli	Hemolysin	Induction of ETs in murine and human macrophages	Not known	Not known	Not known	Aulik et al., 2012
Mannheimia haemolytica	Leukotoxin	Induction of ETs in bovine neutrophils, as well as bovine, murine, and human macrophages	CD18- and NADPH-oxidase dependent	Delayed LDH release, no necrosis, no aptoptosis	Trapping and killing of *M. haemolytica*	Aulik et al., 2010, 2012
Mycobacterium tuberculosis	Early secretory antigen-6 protein	Induction of secondary necrosis and ET formation of phosphatidylserine externalized neutrophils	Intracellular Ca^{2+} overload	Cell lysis	Necrotic granulomas during tuberculosis	Francis et al., 2014
Pseudomonas aeruginosa	Pyocyanin	Induction of NETs	NADPH- oxidase- and Jun N-terminal kinase-dependent	Cell lysis	Inflammatory condition during cystic fibrosis	Rada et al., 2013
Staphylococcus aureus	N-terminal ArgD peptides	Induction of NETs	Unknown	Cell lysis	Aggravation of skin lesions	Gonzalez et al., 2014
Staphylococcus aureus	Leukotoxin GH	Induction of NETs	Non-specific cytolysis	Non-specific cytolysis	Entrapment of *S. aureus*	Malachowa et al., 2013
Staphylococcus aureus	Panton-Valentin-leukocidin	Induction of rapid nuclear NET formation	Vesicular release of nuclear DNA, independent of NADPH- oxidase	No cell lysis	Entrapment of *S. aureus*	Pilsczek et al., 2010
Staphylococcus epidermidis	Phenol-soluble-modulin γ (δ-toxin)	Induces NET formation and colocalizes with NETs and host antimicrobial peptides	Physically binding to host derived antimicrobial peptides and DNA	Cell lysis	Cooperates with host antimicrobial peptides against bacterial pathogens	Cogen et al., 2010
Streptococcus pyogenes	M1 protein	Induces formation of NETs and MCETs	Unknown	Cell lysis	Trapping and killing of *S. pyogenes*	Lauth et al., 2009; Oehmcke et al., 2009
Streptococcus pyogenes	Streptolysin O	Suppression of NET formation	Impairment of oxidative burst	Sublytic con-centrations	Severe infection	Uchiyama et al., 2015

kill the pathogenic streptococci. Coimmunoprecipitation and tryptophan spectroscopy demonstrated direct binding of δ-toxin to LL-37, CRAMP, hBD2, and hBD3, as well as DNA. These data suggest that commensal *S. epidermidis*-derived δ-toxin cooperates with the host-derived antimicrobial peptides in the innate immune system to reduce survival of an important human bacterial pathogen. This could be corroborated in a mouse wound model where *S. pyogenes* survival was reduced when the wounds were pretreated with δ-toxin. These data described a novel therapeutic application of a bacterial toxin against pathogenic bacteria to enhance formation of NETs and support the host innate immune system.

Mannheimia haemolytica leukotoxin (LKT) induces formation of extracellular traps in bovine neutrophils and macrophages (Aulik et al., 2010, 2012). LKT is a member of the repeats-in-toxin (RTX) family of exoproteins produced by a wider variety of Gram-negative bacteria. In addition to the native fully active LKT, its non-cytolytic pro-LKT precursor also stimulated macrophage extracellular traps (METs) formation. Formation of METs in response to LKT required NADPH oxidase activity, as previously demonstrated for NETosis. METs induced in response to LKT trapped and killed a portion of the toxin-producing *M. haemolytica* cells. In contrast to NETosis, which is reported to occur between 10 min to 4 h after pathogen exposure, LKT-induced MET formation was extremely rapid, with significant accumulation of extracellular DNA within 2 min after macrophage stimulation.

Interestingly, the LKT-mediated cytotoxic effects are specific to ruminants, since LKT binds to amino acids 5–17 of the signal sequence of CD18, which is not present on mature leukocytes from humans and other mammalian species (Shanthalingam and Srikumaran, 2009). CD18 was confirmed by Aulik et al. (2010) to play a role in LKT-dependent NET formation in bovine phagocytes. Similarly, to the LKT, the related RTX family toxin, uropathgenic *E. coli* (UPEC) hemolysin, induced NET formation in the mouse and human monocyte/macrophage cell lines indicating that NET induction may be a general phenomenon in response to active RTX toxins (Aulik et al., 2012). At this point it remains difficult to assess whether NET production in response to RTX toxins leads to a protective host defense against *M. haemolytica* or other Gram-negative bacteria. In the case of LKT, bacterial cells entrapped in NETs can continue to secrete LKT, an important virulence determinant, which could exacerbate lung inflammation (Aulik et al., 2010, 2012).

The *Pseudomonas aeruginosa* virulence factor pyocyanin stimulates the release of NETs in a NADPH oxidase-dependent manner (Rada et al., 2013). Pyocyanin is a redox-active pigment associated with diminished lung function in cystic fibrosis. In cystic fibrosis airways, *P. aeruginosa* resides in biofilms protected from neutrophil phagocytic activity or from entrapment by NETs (Rada et al., 2013). The authors speculate that enhanced ROS-dependent NET formation by *P. aeruginosa* pyocyanin contributes to inflammatory conditions observed in chronically infected cystic fibrosis airways. Parenthetically, chronic granulomatous disease (CGD) patients, whose neutrophils are unable to make NETs, are not recognized to be disproportionately susceptible to infections with *Pseudomonas* species (Rada and Leto, 2008).

A specific activity of an exotoxin was shown for the *Mycobacterium tuberculosis* leukocidin ESAT-6, which induced NET formation only in a subpopulation of neutrophils with externalized phosphatidylserine as a marker for apoptosis (Francis et al., 2014). Thus, ESAT-6 induced NET formation was similar to secondary necrosis of pre-activated neutrophils and dependent on Ca^{2+}. In this case, increased NET formation was also speculated to contribute to virulence of tuberculosis, since it was recently observed that the development of necrotic granulomas in a mouse model of progressive tuberculosis were associated with the presence of extracellular bacteria, neutrophil necrosis and NET-like structures (Francis et al., 2014).

In the case of the leading human pathogen *S. aureus*, three different exotoxins enhance the release of NETs: N-terminal AgrD peptides (Gonzalez et al., 2014), leukotoxin GH (Malachowa et al., 2013), also known as LukAB (Dumont et al., 2011), thus named here LukGH/AB, and Panton-Valentin leukocidin (PVL; Pilsczek et al., 2010). AgrD is the precursor for the auto-inducing peptide in a well-known quorum sensing system regulating virulence phenotypes of *S. aureus*. A recent mass spectrometry-based study identified formylated and non-formylated peptide variants derived from AgrD N-terminal leader domain in *S. aureus* cell-free culture supernatant to act cytotoxic, modulate neutrophil chemotaxis and induce formation of NETs. As a consequence a detrimental effect for the host was hypothesized as indicated by aggravation of skin lesions *in vivo* in a murine model. The cellular pathways mediating this process of AgrD induced NET formation remain unclear (Gonzalez et al., 2014).

Staphylococcus aureus LukGH/AB, also promotes formation of NETs in association with death of the neutrophil (Malachowa et al., 2013). LukGH/AB is a pore-forming cytolytic toxin with proinflammatory properties, similar to those established for PVL. But unlike PVL, LukGH/AB did not prime human neutrophils for increased production of reactive oxygen species nor did it enhance binding and/or uptake of *S. aureus*. LukGH/AB promoted the release of NETs, which in turn ensnared but did not achieve killing of *S. aureus*. These authors found that electropermeabilization of human neutrophils, used to create pores in the neutrophil plasma membrane, induced NET formation in a similar way. This finding indicated that NETs can be generated during non-specific cytolysis.

The above phenomemon stands in contrast to NET formation initiated by PVL production in *S. aureus*. Pilsczek et al. (2010) demonstrated a novel rapid (5–60 min) process of NET formation that did not require cell lysis or even breach of the plasma membrane. The authors show that neutrophils treated with *S. aureus* supernatant show rounded and condensed nuclei, followed by the separation of the inner and outer nuclear membrane and budding of vesicles filled with nuclear DNA. The vesicles are extruded intact into extracellular space where they rupture, release the chromatin and form NETs. PVL was identified as one key effector of NET formation present in the *S. aureus* supernatant, but other yet unidentified molecules aside from PVL cannot be excluded as contributors to NET production. This rapid process of NET formation against *S. aureus* was dubbed "dynamic NETosis" and was recently proven to serve as an efficient host defense strategy during *S. aureus* skin infections in man and mice (Yipp et al., 2012). Since, PVL targets a different receptor on the neutrophil compared to the LukGH/AB, namely the C5aR receptor (Spaan et al., 2013) and the CD11b receptor (Dumont et al., 2013), respectively, it still remains to be determined if the receptor-mediated signaling leads to the different NET phenotypes.

Two toxins have been described so far to suppress NET formation, perhaps representing an immune evasion strategy by the respective pathogens: *Bordetella pertussis* adenylate cyclase toxin (Eby et al., 2014) and *S. pyogenes* streptolysin O (SLO; Uchiyama et al., 2015). Both mechanisms are coupled to suppression of neutrophil oxidative burst. In case of *S. pyogenes* SLO, this mechanism plays a key role of SLO in the pathogen's resistance to immediate neutrophil killing. In case of *B. pertussis*, the relevance of NET inhibition by the toxin is still unclear, since it is not known whether or not NETs are able to entrap and clear *B. pertussis*. This study highlighted how convalescent phase serum from humans following clinical pertussis blocked the ACT-mediated suppression of NET formation (Eby et al., 2014). Based on this finding, the authors mention that their data should alert investigators to the ability for ACT, and antibodies to ACT, to dysregulate neutrophil death mechanisms and potentially influence local tissue damage during infection with *B. pertussis*.

CONCLUDING COMMENTS

With increasing publications in the NET field it is well-known that NETs are on one hand protective against several infections, but may also result in detrimental effects when released in excessive amount. The complexity may be compared with the release of cytokines during infection and inflammation, which serve an essential role in an efficient immune response of the host against infections, but in which overwhelming cytokine storm can lead to septic shock and accelerate death of the host. The cumulative balance between NET formation and NET degradation – similar to cytokine release – defines the protective versus detrimental effects on the organism. Currently, it is under discussion in the literature if and how NETs can act as a novel therapeutic or prophylactic target for boosting immunity to bacterial infections or mitigating inflammatory diseases associated with detrimental NET formation. To rationally and effectively pharmacologically interfere with the process of NET formation, the cellular processes mediating this phenomenon need to be understood more in detail. Identification of bacterial factors, especially exotoxins, with specific roles in NET modulation, may serve as probes to understand the molecular basis of NET generation and antimicrobial activity. Furthermore, exotoxins might themselves harbor therapeutic potential when their effects against the host cells are fully characterized.

AUTHOR CONTRIBUTIONS

SB collected the literature, MK-B and VN wrote the manuscript. All authors have proofread the manuscript.

ACKNOWLEDGMENTS

SB was funded by a fellowship of the Ministry of Science and Culture of Lower Saxony (Georg-Christoph-Lichtenberg Scholarship) within the framework of the Ph.D. program 'EWI-Zoonoses' of the Hannover Graduate School for Veterinary Pathobiology, Neuroinfectiology, and Translational Medicine (HGNI).

REFERENCES

Abi Abdallah, D. S., Lin, C., Ball, C. J., King, M. R., Duhamel, G. E., and Denkers, E. Y. (2012). Toxoplasma gondii triggers release of human and mouse neutrophil extracellular traps. *Infect. Immun.* 80, 768–777. doi: 10.1128/IAI.05730-5711

Aulik, N. A., Hellenbrand, K. M., and Czuprynski, C. J. (2012). *Mannheimia haemolytica* and its leukotoxin cause macrophage extracellular trap formation by bovine macrophages. *Infect. Immun.* 80, 1923–1933. doi: 10.1128/IAI.06120-11

Aulik, N. A., Hellenbrand, K. M., Klos, H., and Czuprynski, C. J. (2010). *Mannheimia haemolytica* and its leukotoxin cause neutrophil extracellular trap formation by bovine neutrophils. *Infect. Immun.* 78, 4454–4466. doi: 10.1128/IAI.00840-10

Baker, V. S., Imade, G. E., Molta, N. B., Tawde, P., Pam, S. D., Obadofin, M. O., et al. (2008). Cytokine-associated neutrophil extracellular traps and antinuclear antibodies in *Plasmodium falciparum* infected children under six years of age. *Malar. J.* 7:41. doi: 10.1186/1475-2875-7-41

Bartneck, M., Keul, H. A., Zwadlo-Klarwasser, G., and Groll, J. (2010). Phagocytosis independent extracellular nanoparticle clearance by human immune cells. *Nano Lett.* 10, 59–63. doi: 10.1021/nl902830x

Baums, C. G., and von Köckritz-Blickwede, M. (2015). Novel role of DNA in neutrophil extracellular traps. *Trends Microbiol.* 23, 330–331. doi: 10.1016/j.tim.2015.04.003

Behrendt, J. H., Ruiz, A., Zahner, H., Taubert, A., and Hermosilla, C. (2010). Neutrophil extracellular trap formation as innate immune reactions against the apicomplexan parasite Eimeria bovis. *Vet. Immunol. Immunopathol.* 133, 1–8. doi: 10.1016/j.vetimm.2009.06.012

Beiter, K., Wartha, F., Albiger, B., Normark, S., Zychlinsky, A., and Henriques-Normark, B. (2006). An endonuclease allows *Streptococcus pneumoniae* to escape from neutrophil extracellular traps. *Curr. Biol.* 16, 401–407. doi: 10.1016/j.cub.2006.01.056

Berends, E. T., Horswill, A. R., Haste, N. M., Monestier, M., Nizet, V., and von Köckritz-Blickwede, M. (2010). Nuclease expression by *Staphylococcus aureus* facilitates escape from neutrophil extracellular traps. *J. Innate Immun.* 2, 576–586. doi: 10.1159/000319909

Brinkmann, V., Reichard, U., Goosmann, C., Fauler, B., Uhlemann, Y., Weiss, D. S., et al. (2004). Neutrophil extracellular traps kill bacteria. *Science* 303, 1532–1535. doi: 10.1126/science.1092385

Brogden, G., Krimmling, T., Adamek, M., Naim, H. Y., Steinhagen, D., and von Köckritz-Blickwede, M. (2014). The effect of β-glucan on formation and functionality of neutrophil extracellular traps in carp (*Cyprinus carpio* L.). *Dev. Comp. Immunol.* 44, 280–285. doi: 10.1016/j.dci.2014.01.003

Brogden, G., von Köckritz-Blickwede, M., Adamek, M., Reuner, F., Jung-Schroers, V., Naim, H. Y., et al. (2012). β-Glucan protects neutrophil extracellular traps against degradation by *Aeromonas hydrophila* in carp (*Cyprinus carpio*). *Fish Shellfish Immunol.* 33, 1060–1064. doi: 10.1016/j.fsi.2012.08.009

Buchanan, J. T., Simpson, A. J., Aziz, R. K., Liu, G. Y., Kristian, S. A., Kotb, M., et al. (2006). DNase expression allows the pathogen group a Streptococcus to escape killing in neutrophil extracellular traps. *Curr. Biol.* 16, 396–400. doi: 10.1016/j.cub.2005.12.039

Carlin, A. F., Uchiyama, S., Chang, Y. C., Lewis, A. L., Nizet, V., and Varki, A. (2009). Molecular mimicry of host sialylated glycans allows a bacterial pathogen to engage neutrophil Siglec-9 and dampen the innate immune response. *Blood* 113, 3333–3336. doi: 10.1182/blood-2008-11-187302

Chang, A., Khemlani, A., Kang, H., and Proft, T. (2011). Functional analysis of *Streptococcus pyogenes* nuclease A (SpnA), a novel group A streptococcal virulence factor. *Mol. Microbiol.* 79, 1629–1642. doi: 10.1111/j.1365-2958.2011.07550.x

Chow, O. A., Köckritz-Blickwede, M., von Bright, A. T., Hensler, M. E., Zinkernagel, A. S., Cogen, A. L., et al. (2010). Statins enhance formation of phagocyte extracellular traps. *Cell Host Microbe* 8, 445–454. doi: 10.1016/j.chom.2010.10.005

Cogen, A. L., Yamasaki, K., Muto, J., Sanchez, K. M., Crotty Alexander, L., Tanios, J., et al. (2010). Staphylococcus epidermidis antimicrobial delta-toxin (phenol-soluble modulin-gamma) cooperates with host antimicrobial peptides to kill group a Streptococcus. *PLoS ONE* 5:e8557. doi: 10.1371/journal.pone.0008557

Cole, J. N., Pence, M. A., von Köckritz-Blickwede, M., Hollands, A., Gallo, R. L., Walker, M. J., et al. (2010). M protein and hyaluronic acid capsule are essential for in vivo selection of covRS mutations characteristic of invasive serotype M1T1 group A Streptococcus. *MBio* 1:e191. doi: 10.1128/mBio.00191-10

Corriden, R., Hollands, A., Olson, J., Derieux, J., Lopez, J., Chang, J. T., et al. (2015). Tamoxifen augments the innate immune function of neutrophils through modulation of intracellular ceramide. *Nat. Commun.* 13:8369. doi: 10.1038/ncomms9369

de Buhr, N., Neumann, A., Jerjomiceva, N., Köckritz-Blickwede, M., and Baums, C. G. (2014). *Streptococcus suis* DNase SsnA contributes to degradation of neutrophil extracellular traps (NETs) and evasion of NET-mediated antimicrobial activity. *Microbiology* 160, 385–395. doi: 10.1099/mic.0.072199-0

Derré-Bobillot, A., Cortes-Perez, N. G., Yamamoto, Y., Kharrat, P., Couvé, E., Da Cunha, V., et al. (2013). Nuclease A (Gbs0661), an extracellular nuclease of *Streptococcus agalactiae*, attacks the neutrophil extracellular traps and is needed for full virulence. *Mol. Microbiol.* 89, 518–531. doi: 10.1111/mmi.12295

Dumont, A. L., Nygaard, T. K., Watkins, R. L., Smith, A., Kozhaya, L., Kreiswirth, B., et al. (2011). Characterization of a new cytotoxin that contributes to *Staphylococcus aureus* pathogenesis. *Mol. Microbiol.* 79, 814–825. doi: 10.1111/j.1365-2958.2010.07490.x

Dumont, A. L., Yoong, P., Day, C. J., Alonzo, F. III, McDonald, W. H., Jennings, M. P., et al. (2013). *Staphylococcus aureus* LukAB cytotoxin kills human neutrophils by targeting the CD11b subunit of the integrin Mac-1. *Proc. Natl. Acad. Sci. U.S.A.* 110, 10794–10799. doi: 10.1073/pnas.1305121110

Eby, J. C., Gray, M. C., and Hewlett, E. L. (2014). Cyclic AMP-mediated suppression of neutrophil extracellular trap formation and apoptosis by the *Bordetella pertussis* adenylate cyclase toxin. *Infect. Immun.* 82, 5256–5269. doi: 10.1128/IAI.02487-14

Francis, R. J., Butler, R. E., and Stewart, G. R. (2014). *Mycobacterium tuberculosis* ESAT-6 is a leukocidin causing Ca^{2+} influx, necrosis and neutrophil extracellular trap formation. *Cell Death Dis.* 5:e1474. doi: 10.1038/cddis.2014.394

Fuchs, H. J., Borowitz, D. S., Christiansen, D. H., Morris, E. M., Nash, M. L., Ramsey, B. W., et al. (1994). Effect of aerosolized recombinant human DNase on exacerbations of respiratory symptoms and on pulmonary function in patients with cystic fibrosis. The pulmozyme study group. *N. Engl. J. Med.* 331, 637–642. doi: 10.1056/NEJM199409083311003

Fuchs, T. A., Abed, U., Goosmann, C., Hurwitz, R., Schulze, I., Wahn, V., et al. (2007). Novel cell death program leads to neutrophil extracellular traps. *J. Cell Biol.* 176, 231–241. doi: 10.1083/jcb.200606027

Fuchs, T. A., Brill, A., Duerschmied, D., Schatzberg, D., Monestier, M., Myers, D. D. Jr., et al. (2010). Extracellular DNA traps promote thrombosis. *Proc. Natl. Acad. Sci. U.S.A.* 107, 15880–15885. doi: 10.1073/pnas.1005743107

Fuxman Bass, J. I., Russo, D. M., Gabelloni, M. L., Geffner, J. R., Giordano, M., Catalano, M., et al. (2010). Extracellular DNA: a major proinflammatory component of *Pseudomonas aeruginosa* biofilms. *J. Immunol.* 184, 6386–6395. doi: 10.4049/jimmunol.0901640

Gabriel, C., McMaster, W. R., Girard, D., and Descoteaux, A. (2010). *Leishmania donovani* promastigotes evade the antimicrobial activity of neutrophil extracellular traps. *J. Immunol.* 185, 4319–4327. doi: 10.4049/jimmunol.1000893

Gonzalez, D. J., Corriden, R., Akong-Moore, K., Olson, J., Dorrestein, P. C., and Nizet, V. (2014). N-terminal ArgD peptides from the classical *Staphylococcus aureus* Agr system have cytotoxic and proinflammatory activities. *Chem. Biol.* 21, 1457–1462. doi: 10.1016/j.chembiol.2014.09.015

Grinberg, N., Elazar, S., Rosenshine, I., and Shpigel, N. Y. (2008). Beta-hydroxybutyrate abrogates formation of bovine neutrophil extracellular traps and bactericidal activity against mammary pathogenic *Escherichia coli*. *Infect. Immun.* 76, 2802–2807. doi: 10.1128/IAI.00051-08

Guimarães-Costa, A. B., Nascimento, M. T., Froment, G. S., Soares, R. P., Morgado, F. N., Conceição-Silva, F., et al. (2009). *Leishmania amazonensis* promastigotes induce and are killed by neutrophil extracellular traps. *Proc. Natl. Acad. Sci. U.S.A.* 106, 6748–6753. doi: 10.1073/pnas.0900226106

Halverson, T. W., Wilton, M., Poon, K. K., Petri, B., and Lewenza, S. (2015). DNA is an antimicrobial component of neutrophil extracellular traps. *PLoS Pathog.* 11:e1004593. doi: 10.1371/journal.ppat.1004593

Herwald, H., Cramer, H., Mörgelin, M., Russell, W., Sollenberg, U., Norrby-Teglund, A., et al. (2004). M protein, a classical bacterial virulence determinant, forms complexes with fibrinogen that induce vascular leakage. *Cell* 116, 367–379. doi: 10.1016/S0092-8674(04)00057-1

Hong, W., Juneau, R. A., Pang, B., and Swords, W. E. (2009). Survival of bacterial biofilms within neutrophil extracellular traps promotes nontypeable *Haemophilus influenzae* persistence in the chinchilla model for otitis media. *J. Innate Immun.* 1, 215–224. doi: 10.1159/000205937

Jenne, C. N., Wong, C. H., Zemp, F. J., McDonald, B., Rahman, M. M., Forsyth, P. A., et al. (2013). Neutrophils recruited to sites of infection protect from virus challenge by releasing neutrophil extracellular traps. *Cell Host Microbe.* 13, 169–180. doi: 10.1016/j.chom.2013.01.005

Juneau, R. A., Stevens, J. S., Apicella, M. A., and Criss, A. K. (2015). A thermonuclease of *Neisseria gonorrhoeae* enhances bacterial escape from killing by neutrophil extracellular traps. *J. Infect. Dis.* 212, 316–324. doi: 10.1093/infdis/jiv031

Kisseleva, T., Köckritz-Blickwede, M., Reichart, D., McGillvray, S. M., Wingender, G., Kronenberg, M., et al. (2011). Fibrocyte-like cells recruited to the spleen support innate and adaptive immune responses to acute injury or infection. *J. Mol. Med.* 89, 997–1013. doi: 10.1007/s00109-011-0756-0

LaRock, C. N., Döhrmann, S., Todd, J., Corriden, R., Olson, J., Johannssen, T., et al. (2015). Group a streptococcal M1 protein sequesters cathelicidin to evade innate immune killing. *Cell Host Microbe* 18, 471–477. doi: 10.1016/j.chom.2015.09.004

Lauth, X., von Köckritz-Blickwede, M., McNamara, C. W., Myskowski, S., Zinkernagel, A. S., Beall, B., et al. (2009). M1 protein allows Group A streptococcal survival in phagocyte extracellular traps through cathelicidin inhibition. *J. Innate Immun.* 1, 202–214. doi: 10.1159/000203645

Malachowa, N., Kobayashi, S. D., Freedman, B., Dorward, D. W., and DeLeo, F. R. (2013). *Staphylococcus aureus* leukotoxin GH promotes formation of neutrophil extracellular traps. *J. Immunol.* 191, 6022–6029. doi: 10.4049/jimmunol.1301821

Menegazzi, R., Decleva, E., and Dri, P. (2012). Killing by neutrophil extracellular traps: fact or folklore? *Blood* 119, 1214–1216. doi: 10.1182/blood-2011-07-364604

Menten-Dedoyart, C., Faccinetto, C., Golovchenko, M., Dupiereux, I., Van Lerberghe, P. B., Dubois, S., et al. (2012). Neutrophil extracellular traps entrap and kill *Borrelia burgdorferi* sensu stricto spirochetes and are not affected by *Ixodes ricinus* tick saliva. *J. Immunol.* 189, 5393–5501. doi: 10.4049/jimmunol.1103771

Möllerherm, H., Neumann, A., Schilcher, K., Blodkamp, S., Zeitouni, N. E., Dersch, P., et al. (2015). *Yersinia enterocolitica*-mediated degradation of neutrophil extracellular traps (NETs). *FEMS Microbiol. Lett.* 362:fnv192. doi: 10.1093/femsle/fnv192

Morita, C., Sumioka, R., Nakata, M., Okahashi, N., Wada, S., Yamashiro, T., et al. (2014). Cell wall-anchored nuclease of *Streptococcus sanguinis* contributes to escape from neutrophil extracellular trap-mediated bacteriocidal activity. *PLoS ONE* 9:e103125. doi: 10.1371/journal.pone.0103125

Narasaraju, T., Yang, E., Samy, R. P., Ng, H. H., Poh, W. P., Liew, A. A., et al. (2011). Excessive neutrophils and neutrophil extracellular traps contribute to acute lung injury of influenza pneumonitis. *Am. J. Pathol.* 179, 199–210. doi: 10.1016/j.ajpath.2011.03.013

Oehmcke, S., Mörgelin, M., and Herwald, H. (2009). Activation of the human contact system on neutrophil extracellular traps. *J. Innate. Immun.* 1, 225–230. doi: 10.1159/000203700

Papayannopoulos, V., Staab, D., and Zychlinsky, A. (2011). Neutrophil elastase enhances sputum solubilization in cystic fibrosis patients receiving DNase therapy. *PLoS ONE.* 6:e28526. doi: 10.1371/journal.pone.0028526

Pilsczek, F. H., Salina, D., Poon, K. K., Fahey, C., Yipp, B. G., Sibley, C. D., et al. (2010). A novel mechanism of rapid nuclear neutrophil extracellular trap formation in response to Staphylococcus aureus. *J. Immunol.* 185, 7413–7425. doi: 10.4049/jimmunol.1000675

Rada, B., Jendrysik, M. A., Pang, L., Hayes, C. P., Yoo, D. G., Park, J. J., et al. (2013). Pyocyanin-enhanced neutrophil extracellular trap formation requires the NADPH oxidase. *PLoS ONE* 8:e54205. doi: 10.1371/journal.pone.0054205

Rada, B., and Leto, T. L. (2008). Oxidative innate immune defenses by Nox/Duox family NADPH oxidases. *Contrib. Microbiol.* 15, 164–187. doi: 10.1159/000136357

Ramos-Kichik, V., Mondragón-Flores, R., Mondragón-Castelán, M., Gonzalez-Pozos, S., Muñiz-Hernandez, S., Rojas-Espinosa, O., et al. (2009). Neutrophil extracellular traps are induced by *Mycobacterium tuberculosis*. *Tuberculosis* 89, 29–37. doi: 10.1016/j.tube.2008.09.009

Reid, S. D., Hong, W., Dew, K. E., Winn, D. R., Pang, B., Watt, J., et al. (2009). *Streptococcus pneumoniae* forms surface-attached communities in the middle ear of experimentally infected chinchillas. *J. Infect. Dis.* 199, 786–794. doi: 10.1086/597042

Röhm, M., Grimm, M. J., D'Auria, A. C., Almyroudis, N. G., Segal, B. H., and Urban, C. F. (2014). NADPH oxidase promotes neutrophil extracellular trap formation in pulmonary aspergillosis. *Infect. Immun.* 82, 1766–1777. doi: 10.1128/IAI.00096-14

Saffarzadeh, M., Juenemann, C., Queisser, M. A., Lochnit, G., Barreto, G., Galuska, S. P., et al. (2012). Neutrophil extracellular traps directly induce epithelial and endothelial cell death: a predominant role of histones. *PLoS ONE* 7:e32366. doi: 10.1371/journal.pone.0032366

Saffarzadeh, M., and Preissner, K. T. (2013). Fighting against the dark side of neutrophil extracellular traps in disease: manoeuvres for host protection. *Curr. Opin. Hematol.* 20, 3–9. doi: 10.1097/MOH.0b013e32835a0025

Saitoh, T., Komano, J., Saitoh, Y., Misawa, T., Takahama, M., Kozaki, T., et al. (2012). Neutrophil extracellular traps mediate a host defense response to human immunodeficiency virus-1. *Cell Host Microbe* 12, 109–116. doi: 10.1016/j.chom.2012.05.015

Scharrig, E., Carestia, A., Ferrer, M. F., Cédola, M., Pretre, G., Drut, R., et al. (2015). Neutrophil Extracellular Traps are involved in the innate immune response to infection with *Leptospira*. *PLoS Negl. Trop Dis.* 9:e0003927. doi: 10.1371/journal.pntd.0003927

Schorn, C., Janko, C., Latzko, M., Chaurio, R., Schett, G., and Herrmann, M. (2012). Monosodium urate crystals induce extracellular DNA traps in neutrophils, eosinophils, and basophils but not in mononuclear cells. *Front. Immunol.* 3:27. doi: 10.3389/fimmu.2012.0027

Semeraro, F., Ammollo, C. T., Morrissey, J. H., Dale, G. L., Friese, P., Esmon, N. L., et al. (2011). Extracellular histones promote thrombin generation through platelet-dependent mechanisms: involvement of platelet TLR2 and TLR4. *Blood* 118, 1952–1961. doi: 10.1182/blood-2011-03-343061

Seper, A., Hosseinzadeh, A., Gorkiewicz, G., Lichtenegger, S., Roier, S., Leitner, D. R., et al. (2013). *Vibrio cholerae* evades neutrophil extracellular traps by the activity of two extracellular nucleases. *PLoS Pathog.* 9:e1003614. doi: 10.1371/journal.ppat.1003614

Shanthalingam, S., and Srikumaran, S. (2009). Intact signal peptide of CD18, the beta-subunit of beta2-integrins, renders ruminants susceptible to *Mannheimia haemolytica* leukotoxin. *Proc. Natl. Acad. Sci. U.S.A.* 106, 15448–15453. doi: 10.1073/pnas.0906775106

Short, K. R., von Köckritz-Blickwede, M., Langereis, J. D., Chew, K. Y., Job, E. R., Armitage, C. W., et al. (2014). Antibodies mediate formation of neutrophil extracellular traps in the middle ear and facilitate secondary pneumococcal otitis media. *Infect. Immun.* 82, 364–370. doi: 10.1128/IAI.01104-13

Spaan, A. N., Henry, T., van Rooijen, W. J., Perret, M., Badiou, C., Aerts, P. C., et al. (2013). The staphylococcal toxin Panton-Valentine Leukocidin targets human C5a receptors. *Cell Host Microbe* 13, 584–594. doi: 10.1016/j.chom.2013.04.006

Uchiyama, S., Döhrmann, S., Timmer, A. M., Dixit, N., Ghochani, M., Bhandari, T., et al. (2015). Streptolysin o rapidly impairs neutrophil oxidative burst and antibacterial responses to group a *Streptococcus*. *Front. Immunol.* 6:581. doi: 10.3389/fimmu.2015.00581

Urban, C. F., Ermert, D., Schmid, M., Abu-Abed, U., Goosmann, C., Nacken, W., et al. (2009). Neutrophil extracellular traps contain calprotectin, a cytosolic protein complex involved in host defense against *Candida albicans*. *PLoS Pathog.* 5:e1000639. doi: 10.1371/journal.ppat.1000639

Villanueva, E., Yalavarthi, S., Berthier, C. C., Hodgin, J. B., Khandpur, R., Lin, A. M., et al. (2011). Netting neutrophils induce endothelial damage, infiltrate tissues, and expose immunostimulatory molecules in systemic lupus erythematosus. *J. Immunol.* 187, 538–552. doi: 10.4049/jimmunol.1100450

von Köckritz-Blickwede, M., and Nizet, V. (2009). Innate immunity turned inside-out: antimicrobial defense by phagocyte extracellular traps. *J. Mol. Med.* 87, 775–783. doi: 10.1007/s00109-009-0481-0

von Köckritz-Blickwede, M., von Goldmann, O., Thulin, P., Heinemann, K., Norrby-Teglund, A., Rohde, M., et al. (2008). Phagocytosis-independent antimicrobial activity of mast cells by means of extracellular trap formation. *Blood* 111, 3070–3080. doi: 10.1182/blood-2007-07-104018

Wardini, A. B., Guimarães-Costa, A. B., Nascimento, M. T., Nadaes, N. R., Danelli, M. G., Mazur, C., et al. (2010). Characterization of neutrophil extracellular traps in cats naturally infected with feline leukemia virus. *J. Gen. Virol.* 91, 259–264. doi: 10.1099/vir.0.014613-0

Wartha, F., Beiter, K., Albiger, B., Fernebro, J., Zychlinsky, A., Normark, S., et al. (2007). Capsule and D-alanylated lipoteichoic acids protect *Streptococcus pneumoniae* against neutrophil extracellular traps. *Cell Microbiol.* 9, 1162–1171. doi: 10.1111/j.1462-5822.2006.00857.x

Wong, K. W., and Jacobs, W. R. Jr. (2013). *Mycobacterium tuberculosis* exploits human interferon γ to stimulate macrophage extracellular trap formation and necrosis. *J. Infect. Dis.* 208, 109–119. doi: 10.1093/infdis/jit097

Yipp, B. G., Petri, B., Salina, D., Jenne, C. N., Scott, B. N., Zbytnuik, L. D., et al. (2012). Infection-induced NETosis is a dynamic process involving neutrophil multitasking in vivo. *Nat. Med.* 18, 1386–1393. doi: 10.1038/nm.2847

Young, R. L., Malcolm, K. C., Kret, J. E., Caceres, S. M., Poch, K. R., Nichols, D. P., et al. (2011). Neutrophil extracellular trap (NET)-mediated killing of *Pseudomonas aeruginosa*: evidence of acquired resistance within the CF airway, independent of CFTR. *PLoS ONE* 6:e23637. doi: 10.1371/journal.pone.0023637

Yousefi, S., Gold, J. A., Andina, N., Lee, J. J., Kelly, A. M., Kozlowski, E., et al. (2008). Catapult-like release of mitochondrial DNA by eosinophils contributes to antibacterial defense. *Nat. Med.* 14, 949–953. doi: 10.1038/nm.1855

Yousefi, S., Mihalache, C., Kozlowski, E., Schmid, I., and Simon, H. U. (2009). Viable neutrophils release mitochondrial DNA to form neutrophil extracellular traps. *Cell Death Differ.* 16, 1438–1444. doi: 10.1038/cdd.2009.96

Zinkernagel, A. S., Timmer, A. M., Pence, M. A., Locke, J. B., Buchanan, J. T., Turner, C. E., et al. (2008). The IL-8 protease SpyCEP/ScpC of group A *Streptococcus* promotes resistance to neutrophil killing. *Cell Host Microbe* 4, 170–178. doi: 10.1016/j.chom.2008.07.002

Conflict of Interest Statement: The authors declare that the research was conducted in the absence of any commercial or financial relationships that could be construed as a potential conflict of interest.

13

A Novel Pro-inflammatory Protein of *Streptococcus suis* 2 Induces the Toll-like Receptor 2-dependent Expression of Pro-inflammatory Cytokines in RAW 264.7 Macrophages via Activation of ERK1/2 Pathway

Qiang Zhang[1†], Yujie Yang[1†], Shuxian Yan[1], Jiantao Liu[1], Zhongmin Xu[1], Junping Yu[1], Yajing Song[1], Anding Zhang[1] and Meilin Jin[1,2]*

[1] *Unit of Animal Infectious Diseases, National Key Laboratory of Agricultural Microbiology, College of Veterinary Medicine, Huazhong Agricultural University, Wuhan, China,* [2] *Key Laboratory of Development of Veterinary Diagnostic Products, Ministry of Agriculture, Wuhan, China*

Edited by:
Denise Monack, Stanford University School of Medicine, USA

Reviewed by:
David Weiss, Emory University, USA
Heinrich Korner, Menzies Research Institute Tasmania, Australia

***Correspondence:**
Meilin Jin, Unit of Animal Infectious Diseases, National Key Laboratory of Agricultural Microbiology, College of Veterinary Medicine, Huazhong Agricultural University, No.1 Shizishan Street, Wuhan, Hongshan District, Hubei 430070, China
jinmeilin@mail.hzau.edu.cn

† These authors have contributed equally to this work.

Streptococcus suis 2 is an important swine pathogen and an emergent zoonotic pathogen. Excessive inflammation caused by *S. suis* is responsible for the high levels of early mortality observed in septic shock-like syndrome cases. However, the mechanisms through which *S. suis* 2 (SS2) causes excessive inflammation remain unclear. Thus, this study aimed to identify novel pro-inflammatory mediators that play important roles in the development of therapies against SS2 infection. In this study, the novel pro-inflammatory protein HP0459, which was encoded by the SSUSC84_0459 gene, was discovered. The stimulation of RAW 264.7 macrophages with recombinant HP0459 protein induced the expression of pro-inflammatory cytokines (IL-1β, MCP-1 and TNF-α). Compared with the wild-type (WT) strain, the isogenic knockout of HP0459 in SS2 led to reduced production of pro-inflammatory cytokines in RAW264.7 macrophages and *in vivo*. The pro-inflammatory activity of HP0459 was significantly reduced by an antibody against Toll-like receptor 2 (TLR2) in RAW264.7 macrophages and was lower in TLR2-deficient (TLR2−/−) macrophages than in WT macrophages. Furthermore, specific inhibitors of the extracellular signal-regulated kinase 1/2 (ERK1/2) pathways significantly decreased the HP0459-induced pro-inflammatory cytokine production, and a western blot assay showed that HP0459 stimulation induced the activation of the ERK1/2 pathway. Taken together, our data indicate that HP0459 is a novel pro-inflammatory mediator of SS2 and induces TLR2-dependent pro-inflammatory activity in RAW264.7 macrophages through the ERK1/2 pathway.

Keywords: *Streptococcus suis* 2, inflammation, pro-inflammatory protein, Toll-like receptor 2, ERK1/2 pathway

Introduction

Streptococcus suis is a major pathogen responsible for important economic losses to the swine industry worldwide, particularly during the past 20 years (Gottschalk et al., 2010). It causes meningitis, septicemia, pneumonia, endocarditis, arthritis, and other diseases in swine (Wei et al., 2009). Among the 33 serotypes based on capsular antigens that have been described, serotype 2 is the most frequently isolated from diseased pigs, particularly in Europe and Asia (Wisselink et al., 2000). Since the first human case was reported in Denmark in Perch et al. (1968), more *S. suis* infections in humans have been documented in several European and Asian countries as well as in North and South America, Australia, and New Zealand (Wertheim et al., 2009; Gottschalk et al., 2010). For a long time, it has been considered to be a fact that *S. suis* only infects people working with pigs or pork-derived products (Arends and Zanen, 1988); however, *S. suis* infections in the general population were recently reported in Southeast and East Asia (Gottschalk et al., 2010). Although most reports concern sporadic cases of infection, a large series of 151 *S. suis* meningitis cases was recently reported in southern Vietnam (Mai et al., 2008). Furthermore, an important outbreak, which involved 215 cases and 38 deaths, occurred in China in during 2005 (Yu et al., 2006). In addition, *S. suis* is considered one of the most important causes of meningitis in humans in various locations, including Vietnam, Thailand, and Hong Kong (Suankratay et al., 2004; Hui et al., 2005; Ip et al., 2007; Mai et al., 2008). These findings emphasize the importance of *S. suis* as an emerging zoonosis and indicate that *S. suis* represents a significant public health concern (Fittipaldi et al., 2012). The increased severity of *S. suis* infection in humans underscores the critical need to better understand the factors associated with the pathogenesis of *S. suis* infection (Gottschalk and Segura, 2000).

Although several virulence-related molecules have been proposed, only the capsular polysaccharide (CPS) has been proven to play a critical role in the virulence of *S. suis* (Charland et al., 1998; Smith et al., 1999; Segura et al., 2004). Some putative virulence factors have also been reported in *S. suis*, such as suilysin, muramidase-released protein (MRP), subtilisin-like protease (SspA), and LTA D-alanylation (DltA; Smith et al., 1992; Lun et al., 2003; Fittipaldi et al., 2008; Bonifait et al., 2011); however, the current knowledge on the pathogenesis of *S. suis* infection remains limited (Baums and Valentin-Weigand, 2009). To cause disease, *S. suis* must breach epithelial barriers, reach and survive in the bloodstream, invade different organs and cause exaggerated inflammation (Fittipaldi et al., 2012). The upregulated expression of several pro-inflammatory cytokines and chemokines, such as tumor necrosis factor-α (TNF-α), interleukin-1β (IL-1β), IL-6 and monocyte chemotactic protein-1 (MCP-1), has been reported in *S. suis* infection (Gottschalk and Segura, 2000; Segura et al., 2002; Al-Numani et al., 2003). Furthermore, inflammation is thought to be responsible for most clinical signs of meningitis, septicemia and sudden death (Segura et al., 2006). *S. suis* sequence type (ST) 7 was recognized as the causative agent for the Sichuan outbreak, and streptococcal toxic shock-like syndrome (STSLS) was observed for the first time in this large outbreak (Ye et al., 2006). A previous study showed that the increased virulence of *S. suis* ST7 is associated with an increased ability to stimulate excessive pro-inflammatory cytokines that may be responsible for the shock syndrome (Zheng et al., 2008). In addition, the most important clinical feature associated with *S. suis* is meningitis in pigs (Gottschalk and Segura, 2000); however, the mechanisms of *S. suis* crossing the blood–brain barrier (BBB) to cause meningitis are poorly understood. Even so, some mechanisms, such as the up-regulation of pro-inflammatory cytokines and increased leukocyte trafficking, have been proposed to contribute to the breakdown of the BBB (Vadeboncoeur et al., 2003; Adam et al., 2004; Jobin et al., 2005; Tenenbaum et al., 2005). The activation of the innate immune response depends on the recognition of pathogen-associated molecular patterns (PAMPs). Toll-like receptors (TLRs) are critical sensors that activate the innate immune response (Beutler, 2009; Kawai and Akira, 2010). For example, TLR2 can form heterodimers with TLR1 or TLR6 to recognize bacterial lipoprotein, lipoteichoic acid (LTA), peptidoglycans (PGNs) and zymosan and induce the release of many cytokines and chemokines responsible for inflammation (Akira and Hemmi, 2003; Beutler, 2004; Lachance et al., 2013). Many previous studies have reported that TLR2 is the major (but not exclusive) immune receptor involved in *S. suis* recognition (Graveline et al., 2007; Li et al., 2010; Lecours et al., 2012).

As mentioned above, inflammation has been thought to be a hallmark of *S. suis* infection (Gottschalk et al., 2007). However, the research on inflammation induced by *S. suis* remains limited. Therefore, it is important to identify novel pro-inflammatory mediators of *S. suis* in order to improve our understanding of the mechanism of inflammation induced by this pathogen. In our previous study, more than 50 extracellular proteins of *S. suis* were expressed, including membrane proteins, secreted proteins and cell wall proteins. And several pro-inflammatory proteins were identified, of which a novel protein HP0459 displayed rather robust pro-inflammatory activity (data not shown). In this study, through measuring the IL-1β and MCP-1 levels by relative quantitative polymerase chain reaction (qPCR) and enzyme-linked immunosorbent assay (ELISA), the pro-inflammatory ability of HP0459 was examined. We investigated the recognition receptor and signal transduction pathway through which HP0459 induces IL-1β, TNF-α and MCP-1 in RAW264.7 macrophages. As a result, we elucidated the mechanism through which HP0459 stimulation induces pro-inflammatory cytokine production.

Materials and Methods

Bacterial Strains, Plasmids and Growth Conditions

Streptococcus suis serotype 2 strain SC-19, which was isolated from the brain of a dead pig during the epidemic outbreak in Sichuan province of China in 2005, was selected as the wild-type (WT) strain. SC-19 was grown in Tryptic Soy Broth (TSB) or on Tryptic Soy Agar (TSA) plates (Difco, MI, USA) with 5% newborn bovine serum (Sijiqing Biological Engineering

Materials Co., Ltd., Hangzhou, China) at 37°C (Li et al., 2013). A temperature-sensitive *S. suis–Escherichia coli* shuttle vector (pSET4s) was used to construct the Δ*hp0459* mutant, which carries a spectinomycin resistance gene (*spc*^{r}). pSET2, a *S. suis–E. coli* shuttle vector carrying *spc*^{r}, was used in the construction of the complementary bacterium (Takamatsu et al., 2001).

Cell Culture

RAW 264.7 macrophages were grown in Dulbecco's modified Eagle's medium (DMEM) supplemented with 10% fetal bovine serum in a 5% CO_2 atmosphere at 37°C (Kang et al., 2009). The primary mouse macrophages were isolated from TLR2-deficient (TLR2−/−; JAX® Mice) and WT mice. The mice were injected intraperitoneally (i.p.) with 4% thioglycolate, and peritoneal exudate cells were harvested 4 days later (van Lint et al., 2010). More than 90% of the exudate cells were identified as macrophages by microscopic analysis and non-specific esterase staining (Sodhi et al., 2005). The macrophages were plated at a density of 10^6 cells per well in 12-well plates.

Cloning, Expression, Purification and Endotoxin Removal of HP0459 Protein

The HP0459 protein, which was reported as a secreted protein in a previous study, was encoded by SSUSC84_0459 (Liu et al., 2009). The HP0459 protein was cloned and purified according to published methods (Liu et al., 2012). Briefly, the *hp0459* gene was amplified from the chromosomal DNA of SC-19 by PCR using the primers listed in **Table 1**. The purified PCR product was inserted into pET-28a and harbored in *E. coli* BL21 (DE3) cells. HP0459 was induced with 0.5 mM isopropyl-b-D-thiogalactopyranoside (IPTG) and purified by ultrasonication and Ni-NTA agarose chromatography. The endotoxin in the purified recombinant protein was removed using an Endotoxin Removal Kit (Genmed Scientifics Inc. USA), and the endotoxin level was tested using a Quantitative Chromogenic Tachypleus Amebocyte Lysate For Endotoxin Detection kit (Chinese Horseshoe Crab Reagent Manufactory Co., Ltd., Xiamen, China; Liu et al., 2011). The protein was then treated using a 0.22-μm filter. After the above-mentioned treatment, the HP0459 protein was stored at −80°C.

RNA Extraction and qPCR Assay

After the treatment of RAW 264.7 cells with HP0459 at 10 μg ml^{-1} for 10 h, the total RNA of the cells was extracted with the TRIzol® reagent (Invitrogen, Paisley, UK), according to the manufacturer's guidelines. The RNA pellets were suspended in RNase-Free water, and the DNA contamination of the RNA was removed by DNase treatment (Promega, Madison, WI, USA). cDNA was obtained from 4 μg of RNA by reverse transcription using AMV reverse transcriptase (TAKARA, Japan) and the oligo-dT primer (300 pmol) in a total reaction volume of 40 μl (Moore et al., 2005). Relative quantitative PCR (qPCR) was performed to measure the mRNA levels of pro-inflammatory cytokines (IL-1β, MCP-1 and TNF-α) using a SYBR green PCR Kit (Roche) and the ABI ViiA7 instrument. Glyceraldehyde-3-phosphate dehydrogenase (GAPDH) was used as the reference gene, and all of the primers used in the

TABLE 1 | Oligonucleotide primers used in this study.

Primers	Primers sequence (5′–3′)[a]	Functions
hp0459-F	CCCGAATTCACGGAGGTAGCTAACGAAC	For amplification of the *hp0459* ORF gene
hp0459-R	CCCCTCGAGTTATTCGGGTGTTGTAAATAG	
hp0459-L1	AATGAATTCTTCCTGATAAGAAGGTGGCTAAC	Upstream border of *hp0459*
hp0459-L2	AACGTCGACCCTGAAGGATCCTAATGAAGTTT	
hp0459-R1	TTAGTCGACGCAATGTAATAACTCGAACTAG	Downstream border of *hp0459*
hp0459-R2	CTCAAGCTTGACTAAACCATTAAGCCA	
chp0459-1	CGCGCATGCCCTTTATTATGTCAAGTTCAGAT	To complement *hp0459* in the PCR assays
chp0459-2	CGCGAATTC TTATTCGGGTGTTGTAAATAG	
MCP1-F	TGGGTCCAGACATACATTA	For qPCR assay
MCP1-R	TCAGATTTACGGGTCAACT	
TNFα-F	CGATGAGGTCAATCTGCCCA	For qPCR assay
TNFα-R	CCAGGTCACTGTCCCAGCATC	
IL1β-F	CACCTGGTACATCAGCACCTCAC	For qPCR assay
IL1β-R	CATCAGAAACAGTCCAGCCCATAC	
GAPDH-F	CGTCGGTGCTGAGTATGTCGT	For qPCR assay
GAPDH-R	CAGTCTTCTGGGTGGCAGTGAT	
P1	TAGTTTCTGATAAACTTCATTAGGA	To identify the *hp0459* gene by PCR
P2	AAATGCGCTCGAAATGA	
P3	TGGAAATGTTCAAGTCAACC	To identify the *gdh* gene by PCR
P4	CGTTTTTCTTTGATGTCCAC	
P5	GCACAGATGCGTAAGGAG	To identify the pSET4s by PCR
P6	ACTCTGTAGCACCGCCTA	

[a] *The underlined nucleotides denote enzyme restriction sites.*

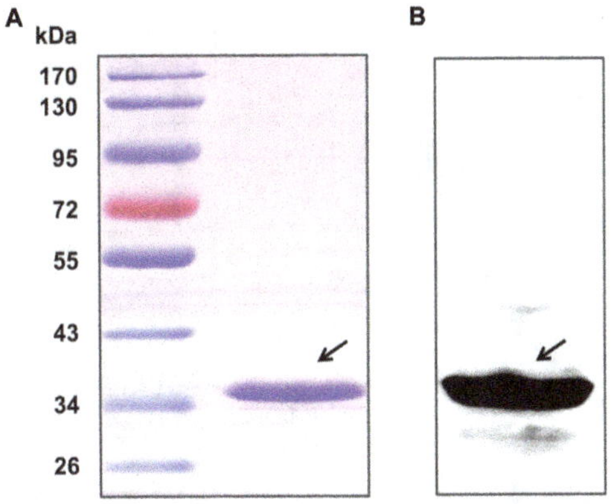

FIGURE 1 | Purification of the recombinant Hp0459 protein.
(A) SDS-PAGE. **(B)** Western blot analysis. The blot was probed with his tag monoclonal antibody (Cali-Bio).

qPCR assay are listed in **Table 1**. The data were analyzed using the ViiA7 software (Applied Biosystems; Zhao et al., 2014).

Enzyme-Linked Immunosorbent Assay for Cytokines

After treatment of RAW 264.7 cells with HP0459 at 10 μg ml^{-1} for 10 h, the protein levels of IL-1β, MCP-1 and TNF-α in the cell culture supernatants were determined using commercially available ELISA kits (Biolegend) according to the manufacturer's instructions.

Generation of an Isogenic *hp0459* Deletion Mutant and Complemented Strains

The construction of the Δ*hp0459*-knockout mutant was performed using a previously described procedure (Takamatsu et al., 2001). Briefly, DNA fragments were amplified from the genomic DNA of SC-19 by PCR using two pairs of specific primers, namely *hp0459*-L1/*hp0459*-L2 and *hp0459*-R1/*hp0459*-R2 (**Table 1**), which carry *EcoRI/SalI* and *SalI/HindI* restriction enzyme sites, respectively. The fragments were digested with the corresponding restriction enzymes and sequentially ligated into the temperature-sensitive *S. suis–E. coli* shuttle vector pSET4s to generate the *hp0459*-knockout vector pSET4sΔ*hp0459.* To obtain the isogenic mutant Δ*hp0459,* competent cells of SC-19

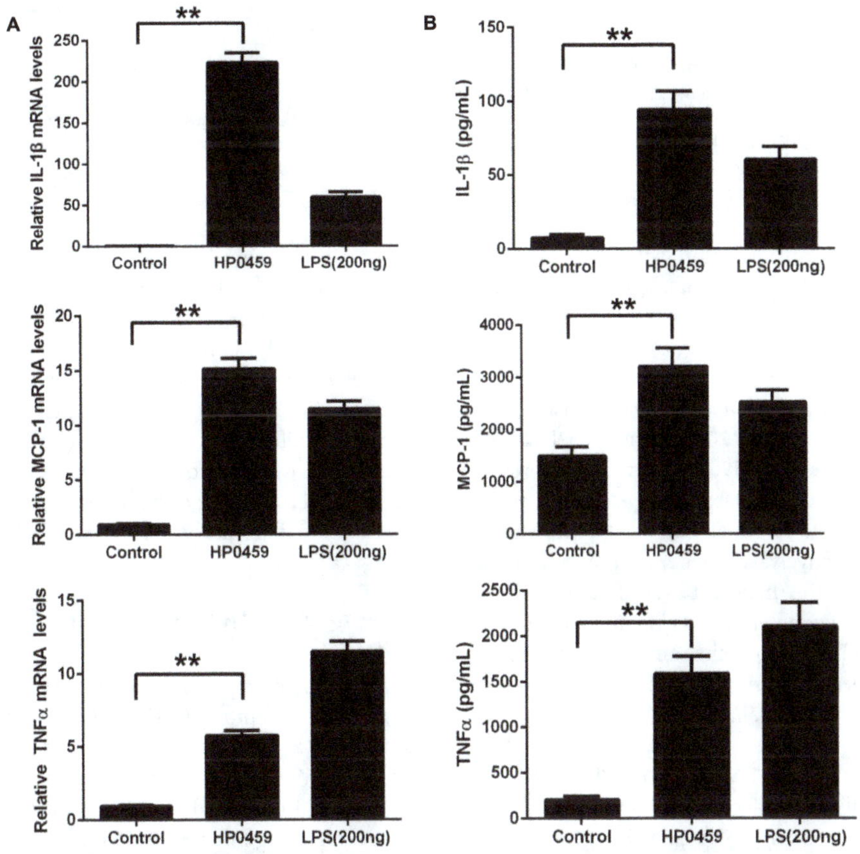

FIGURE 2 | Induction of cytokine mRNA and protein in RAW 264.7 macrophages by recombinant HP0459 stimulation. RAW 264.7 macrophages were incubated with 10 μg ml^{-1} HP0459 and 200 ng ml^{-1} LPS (positive control) for 10 h, as well as single culture media (negative control), **(A)** the cytokine mRNA levels were then determined by qPCR, **(B)** and the protein levels of IL-1β, MCP-1 and TNF-α in the culture supernatants were determined by ELISA. The bars represent the standard errors of the means, based on three independent experiments. $^{**}P < 0.01$.

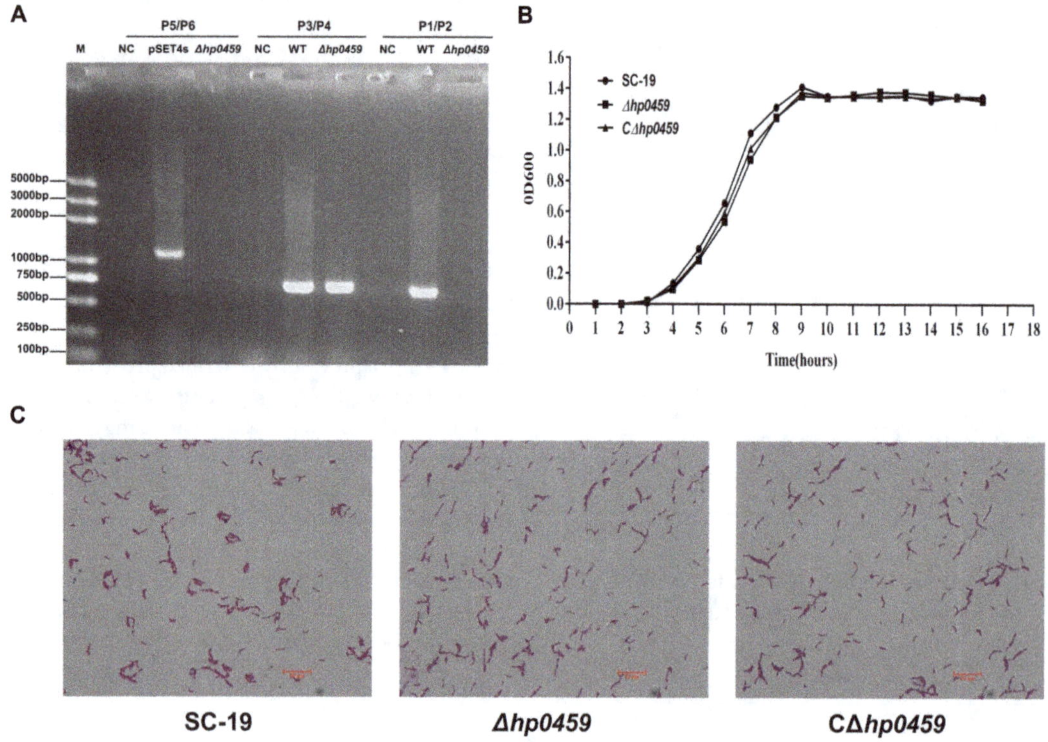

FIGURE 3 | Construction and confirmation of Δ*hp0459* and the complementary strain CΔ*hp0459*. (A) Confirmation of the Δ*hp0459* mutant by PCR using the primers pairs P1/P2 (to identify *hp0459*), P3/P4 (to identify *gdh*), and P5/P6 (to identify the pSET4s). **(B)** Growth curves of the SC-19, Δ*hp0459* and CΔ*hp0459* strains. The bacteria were cultured in TSB containing 5% newborn bovine serum at 37°C. The absorbance at 600 nm was measured at intervals of 1 h. Results shown are representative of three independent experiments. **(C)** Light microscope morphology of the SC-19, Δ*hp0459* and CΔ*hp0459* strains using Gram staining. The bar indicates the magnification size.

were subjected to electrotransformation with pSET4sΔ*hp0459* as described previously (Takamatsu et al., 2001). The suspected mutant was verified by PCR using three pairs of primers: P1/P2 (to identify *hp0459*), P3/P4 (to identify *gdh*) and P5/P6 (to identify the pSET4s).

The complemented strain of *hp0459* was constructed as described previously (Zhang et al., 2012). Briefly, a DNA fragment that contained the *hp0459* gene and its predicted upstream promoter was amplified by PCR using the primers *chp0459-1*/*chp0459-2* (**Table 1**), which carry *SphI*/*EcoRI* restriction enzyme sites, respectively. To generate the recombinant plasmid pSET2C*hp0459*, the fragment was digested with the appropriate restriction enzymes and was cloned into the *E. coli–S. suis* shuttle vector pSET2 carrying the same cohesive terminus. The plasmid was then electrotransformed into Δ*hp0459* to obtain the complemented CΔ*hp0459* strain.

Experimental Infections *In Vitro* and *In Vivo*

In vitro, RAW 264.7 cells were infected with 5×10^6 CFU of the WT (SC-19), Δ*hp0459* or CΔ*hp0459* strains in the logarithmic phase of growth. The supernatants were collected for western blot analysis after 10 h at 37°C.

All of the animal studies were performed according to the experimental protocols approved by the Laboratory Animal Monitoring Committee of Hubei Province, China. A total of 125 6-weeks-old female C57BL/6 mice were randomly divided into three groups with 40 mice per group, and the remaining five mice were used as controls. The three groups of mice were challenged intraperitoneally (i.p.) with 5×10^8 CFU log-phase WT (SC-19), Δ*hp0459* or CΔ*hp0459* strains, respectively. At certain times post infection (3, 6, 9, and 12 h), an equal number of mice in each group were killed. Bacteriological isolation from the blood or part of the spleen was performed essentially as described previously (Zhang et al., 2011), and the lungs or another part of the spleen were used to determine the IL-1β and TNF-α levels by qPCR.

Investigating the Recognition Receptor of HP0459

Antibody blocking assays were performed to investigate the recognition receptor of HP0459 in RAW264.7 cells using the anti-TLR2 (eBioscience) and anti-TLR4 (BioLegend) antibodies. Briefly, RAW264.7 cells were pretreated using 8 μg of anti-TLR2 and anti-TLR4 antibody for 30 min respectively and then incubated with 10 $\mu g\ ml^{-1}$ HP0459 for 10 h. The expression levels of various cytokines were determined by ELISA. According to the conditions of cytokine activation, the recognition receptor of the HP0459 was analyzed. In addition, TLR2−/− macrophages were isolated from TLR2−/− mice to verify the results of the blocking assays.

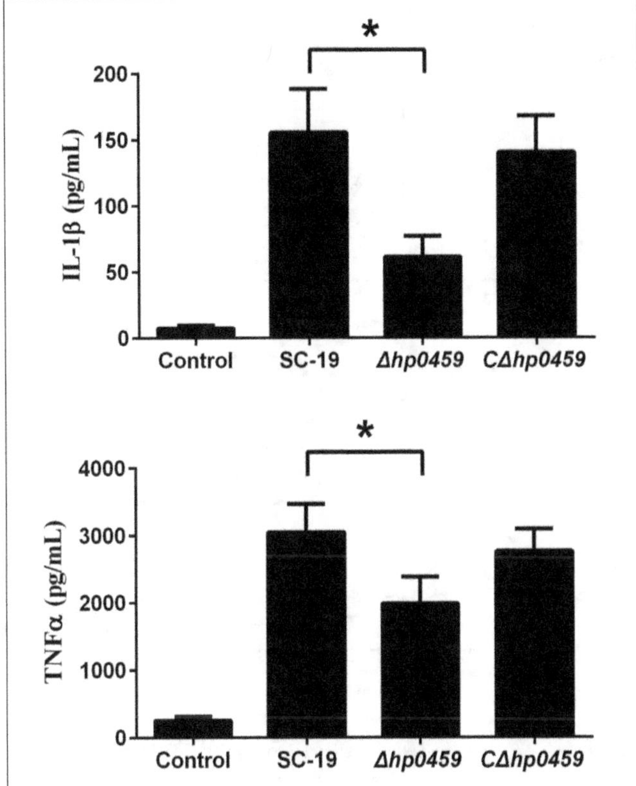

FIGURE 4 | Induction of IL-1β and TNF-α by SS2 strains in RAW 264.7 macrophages. RAW 264.7 macrophages were incubated with the SC-19, Δ*hp0459* and CΔ*hp0459* strains for 10 h. The levels of IL-1β and TNF-α in the supernatants were measured by ELISA. The bars represent the SEs of the means, based on three independent experiments. **P* < 0.05.

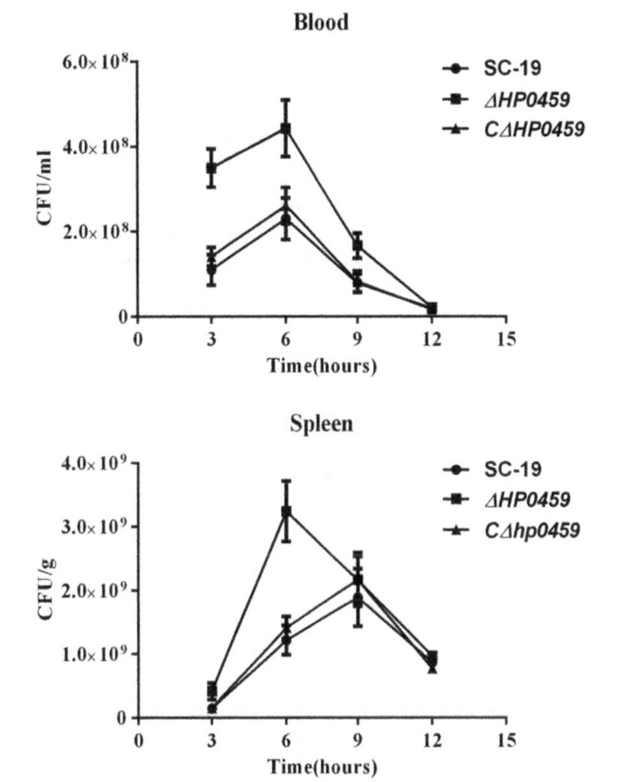

FIGURE 5 | Bacteriological isolation from blood and spleen. C57BL/6 mice were challenged with 5 × 10^8 CFU of the log-phase SC-19, Δ*hp0459* or CΔ*hp0459* strains. After infection for 3, 6, 9, and 12 h, bacteriological isolation from the blood and spleens was performed. Results shown are representative of three independent experiments.

Analysis of HP0459-Induced Cell Signal Transduction Pathways

RAW264.7 macrophages (1 × 10^6 cells ml^{-1}) were seeded into 12-well tissue culture plates. The cells were pretreated with the following specific inhibitors for 30 min prior to the addition of HP0459 (Liu et al., 2008): U0126 (for ERK1/2; 10 μM), SP600125 (for JNK; 10 μM), pyrrolidine dithiocarbamate (PDTC; for NF-κB; 20 μM) and LY294002 (for PI3K; 20 μM). All inhibitors were purchased from Cayman Chemical. Culture supernatants were collected at the indicated times and stored at −80°C until assayed.

SDS-PAGE and Western Blot Analysis

To confirm the HP0459-induced phosphorylation of signal transduction molecules, a western blot analysis was performed. To extract the cytosolic protein, after stimulation with HP0459 (10 μg ml^{-1}) for 10 h, RAW264.7 cells were washed with cold PBS and harvested by centrifugation. The pellets were then suspended in RIPA lysis buffer with phosphatase inhibitor (Roche) for 15 min on ice. The protein concentrations in the lysates were quantified with the Bradford protein assay, and 40 μg of proteins were subjected to 12% SDS-PAGE and transferred onto a 0.22-μm nitrocellulose membrane. Subsequently, these proteins were probed with specific Abs against the phosphorylated forms of ERK1/2 and NF-κB p65 (Cell Signaling Technology, Beverly, MA, USA), and β-actin was assessed as an internal control using anti-β-actin antibody (Wuhan PMK Biotechnology Co., Ltd.). The detection of the bands was performed using HRP-conjugated secondary antibody and an enhanced chemiluminescence (ECL) system (Amersham Life Science, Arlington Heights, IL, USA).

Statistical Analysis

The statistical significance of the data was determined using Student's *t* test with GraphPad Prism software (San Diego, CA, USA), and all the assays were repeated at least three times. For all tests, a value of $P < 0.05$ was considered as the threshold for significance.

Results

Cytokine Secretion from RAW264.7 Cells Stimulated with HP0459

After purification by Ni-NTA agarose chromatography, the SDS-PAGE (**Figure 1A**) and western blot analysis (**Figure 1B**) of HP0459 revealed that the HP0459 protein was successfully purified. The average endotoxin level in HP0459 was ~0.05 endotoxin

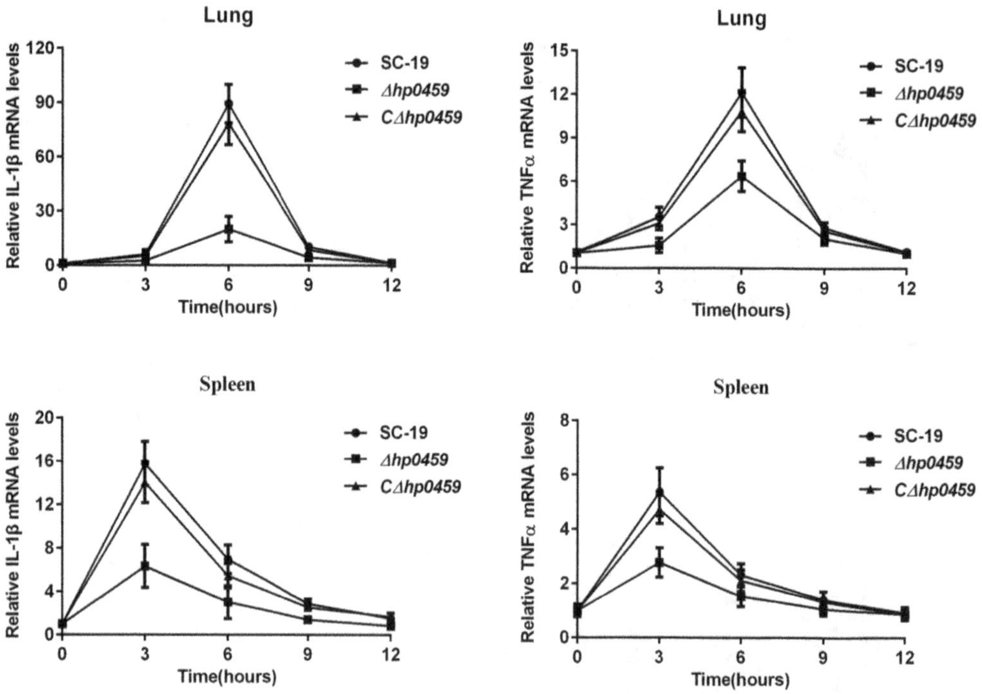

FIGURE 6 | Induction of IL-1β and TNF-α mRNA *in vivo* by stimulation with SS2 strains. C57BL/6 mice were challenged with 5 × 10^8 CFU of the log-phase SC-19, *Δhp0459* or *CΔhp0459* strains. After infection for 3, 6, 9, and 12 h, the mRNA levels of IL-1β and TNF-α in the lungs and spleens were measured by qPCR. Results shown are representative of three independent experiments.

units per milliliter via endotoxin removal. After the above-mentioned treatment, to determine the pro-inflammatory role of HP0459, RAW264.7 cells were stimulated with HP0459 at a concentration of 10 μg ml^{-1} for 10 h and analyzed by qPCR and ELISA respectively. As shown in **Figure 2**, it was confirmed that HP0459 stimulation significantly increased the expression levels of IL-1β, MCP-1, and TNF-α ($P < 0.01$) by qPCR (**Figure 2A**) and ELISA (**Figure 2B**) analyses.

Construction and Characterization of the Mutant Strain Δ*hp0459*

To study the role of HP0459 in the *S. suis*-induced pro-inflammatory response, the HP0459-knockout mutant Δ*hp0459* was constructed by homologous recombination, and the double-crossover event was confirmed by PCR (**Figure 3A**). To examine the growth characteristics of the mutants *in vitro*, the OD600 values of cultures of the SC-19, Δ*hp0459* and CΔ*hp0459* strains in TSB containing 10% newborn bovine serum at 37°C were determined (**Figure 3B**), and Gram-staining of these three strains was also performed (**Figure 3C**). No significant difference in growth was found between the SC-19, Δ*hp0459* and CΔ*hp0459* strains.

Pro-inflammatory Attenuation Induced by the Δ*hp0459* Strain *In Vitro* and *In Vivo*

After construction of the Δ*hp0459* strain, the role of HP0459 in the *S. suis*-induced pro-inflammatory response was assessed in RAW264.7 cells *in vitro*. The culture supernatants of RAW264.7 cells incubated with the SC-19, Δ*hp0459* and CΔ*hp0459* strains were harvested, and the levels of IL-1β and TNF-α in these supernatants were measured by ELISA. The results showed that the pro-inflammatory activity of the Δ*hp0459* strain was significantly lower than those of the SC-19 and CΔ*hp0459* strains (**Figure 4**). This finding suggested that HP0459 plays an important role in the *S. suis*-induced pro-inflammatory response *in vitro*.

The role of HP0459 *in vivo* in the *S. suis*-induced pro-inflammatory response was then assessed using an experimental infection model in C57BL/6 mice. Bacteriological isolation from the blood and part of the spleen showed that, compared with the SC-19 and CΔ*hp0459* strains, the bacterial content of the Δ*hp0459* strain was significantly increased (**Figure 5**). At the same time, the lungs and another part of the spleen from infected and control mice were separated to extract the total RNA, and the mRNA levels of IL-1β and TNF-α were measured by qPCR. The results indicated that the mRNA levels of IL-1β and TNF-α induced by the Δ*hp0459* strain *in vivo* were significantly lower than those induced by the SC-19 and CΔ*hp0459* strains (**Figure 6**). Thus, the HP0459 protein plays an important role in the *S. suis*-induced pro-inflammatory response.

HP0459 Protein Induced the Expression of IL-1β, MCP-1 and TNF-α by TLR2

To determine the recognition receptor responsible for the HP0459-mediated induction of cytokines, antibody blocking assays were performed. Compared with the positive control, anti-TLR2 antibody could significantly reduce the expression of

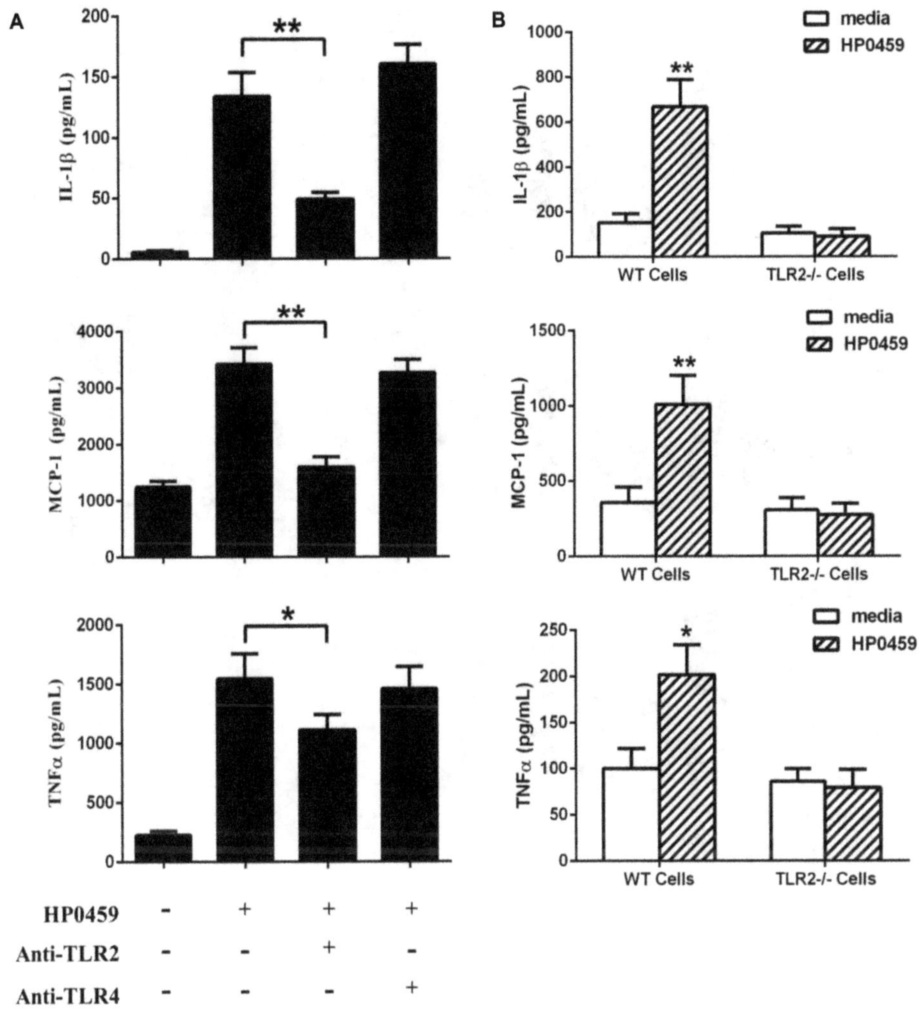

FIGURE 7 | Recognition receptor of the HP0459-stimulated pro-inflammatory response. (A) Antibody blocking assays. After pretreatment with 8 μg of anti-TLR2 and anti-TLR4 antibody for 30 min, RAW 264.7 macrophages were incubated with 10 μg ml^{-1} HP0459 for 10 h. The expression levels of various cytokines were then determined by ELISA. **(B)** Primary peritoneal macrophages were isolated from TLR2−/− and WT mice and incubated with 10 μg ml^{-1} HP0459 for 10 h. The induction of IL-1β, MCP-1 and TNF-α in the supernatants was then determined by ELISA. The bars represent the SEs of the means, based on three independent experiments. $^{*}P < 0.05$, $^{**}P < 0.01$.

IL-1β, MCP-1 and TNF-α induced by HP0459, whereas anti-TLR4 antibody could not (**Figure 7A**). The results showed that the HP0459-induced cytokine secretion may depend on TLR2. To verify this result, TLR2−/− and WT macrophages were isolated from TLR2−/− and WT mice, respectively, and were incubated with HP0459. The results show that HP0459 could significantly induce cytokine secretion in WT macrophages but not in TLR2−/− cells (**Figure 7B**), demonstrating that the HP0459 protein can induce a TLR2-dependent pro-inflammatory response.

HP0459-Induced Cell Signal Transduction Pathways in RAW264.7 Cells

Next, to further elucidate the mechanisms through which the HP0459 protein induced cytokine secretion, we investigated the cell signal transduction pathways in HP0459-stimulated RAW264.7 cells. RAW264.7 cells were pretreated with specific inhibitors of several cell signal pathways for 30 min and were then were incubated with HP0459 for 10 h. The levels of IL-1β, MCP-1, and TNF-α in the supernatants were quantified by ELISA. As shown in **Figure 8A**, the ERK 1/2 MAPK inhibitor (U0126) significantly decreased the HP0459-induced cytokine production, and the NF-κB inhibitor (PDTC) induced a lower degree of reduction. This result suggested that the HP0459-induced cytokine production likely primarily depends on the phosphorylation of ERK 1/2 MAPK. To verify this hypothesis, we performed a western blotting analysis to measure the phosphorylation of ERK 1/2 MAPK and NF-κB in RAW264.7 cells induced by HP0459 stimulation (**Figure 8B**). The results showed that the phosphorylation of ERK 1/2 MAPK was significantly enhanced, whereas the phosphorylation of NF-κB was slight. In addition, a western blot analysis with anti-actin antibody was

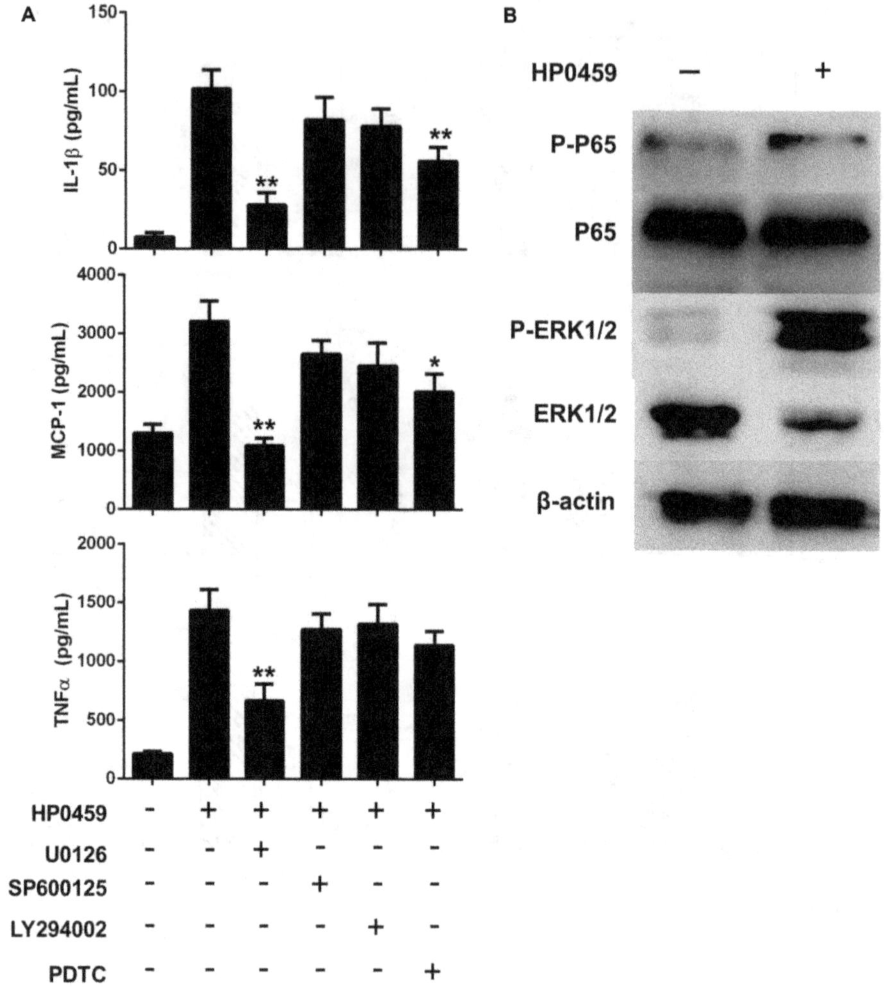

FIGURE 8 | Signal transduction pathways of the HP0459-stimulated pro-inflammatory response in RAW 264.7 macrophages. (A) Effect of ERK 1/2 (U0126), JNK (SP600125), PI3K (LY294002) and NF-κB (PDTC) inhibitors on cytokine production during HP0459 stimulation. After incubation with the inhibitors for 30 min, RAW 264.7 macrophages were stimulated with 10 μg ml^{-1} HP0459 for 10 h. The cytokine levels were then determined by ELISA. Data are expressed as the mean ± SD of three independent experiments. **(B)** HP0459-induced phosphorylation of ERK 1/2 MAPK in RAW264.7 macrophages. RAW264.7 macrophages were stimulated with HP0459 (10 μg ml^{-1}) for 10 h. The cell lysates were analyzed by western blotting using specific antibodies against ERK 1/2 MAPK, phospho-ERK 1/2 MAPK, NF-κB and phospho-NF-κB. At the same time, β-actin was assessed as an internal control using anti-β-actin antibody. Results shown are representative of three independent experiments. $^{*}P < 0.05$, $^{**}P < 0.01$.

used to confirm that equivalent amounts of the samples were loaded into the gels. These results suggested that signal transduction pathway ERK 1/2 MAPK play a primary role in the pro-inflammatory response induced by HP0459 stimulation in RAW264.7 cells.

Discussion

Due to the high prevalence of *S. suis*-induced diseases in humans in Southeast and East Asia, pathogen has been increasingly investigated (Fittipaldi et al., 2012). Although the pathogenesis of *S. suis* infection is not entirely known at present, several viewpoints have been generally recognized, and one of these is that inflammation plays an important role in *S. suis* infection (Segura et al., 2006). After pathogen invasion into a host, the innate immune system of the host will recognize key molecular signatures borne by PAMPs to activate the inflammatory response in order to clear pathogens (Martinon and Tschopp, 2005). However, during *S. suis* infection, the inflammatory response is excessively activated and thus plays an important role in most clinical signs of *S. suis* disease, including meningitis, septicemia and sudden death (Dominguez-Punaro Mde et al., 2008). Thus, it is significant to identify the pro-inflammatory molecules of *S. suis* in order to understand its pathogenesis. In this study, it was found that HP0459 could significantly activate inflammatory response in RAW264.7 cells, but heat-killed HP0459 couldn't (data not shown). Further, we elucidated the mechanism of HP0459 inducing inflammation. This result contributes to the understanding of the excessive inflammation induced by *S. suis*.

During *S. suis* infection, this pathogen can be recognized by some receptors, including CD14, TLR2, TLR6, and TLR9. CD14 has been considered one of the recognition receptors of LPS, the major component of the outer membrane of Gram-negative bacteria (Wright et al., 1990), and CD14 has been shown to be important in the recognition of cell wall constituents of Gram-positive bacteria (Wright, 1995; Moreillon and Majcherczyk, 2003). A previous study proved that *S. suis* can mediate CD14-dependent cytokine and chemokine production by human monocytes (Segura et al., 2002). Because CD14 lacks transmembrane and intracellular domains, it is not able to transduce the signal by itself. Thus, to activate the CD14-dependent inflammatory response, additional transmembrane receptors, such as TLR2, are required (Manukyan et al., 2005). TLR2, as a major pattern recognition receptor for ligands derived from Gram-positive bacteria (Kawai and Akira, 2005), has been shown to play an important role during *S. suis* ST1 strain interactions with mouse, swine and human cells (Graveline et al., 2007; Zheng et al., 2011, 2012). In addition, it has been indicated that not only TLR2 but also TLR6 and TLR9 play an important role on cell activation through *in vitro* studies carried out with the whole cells of the epidemic ST7 strain and human peripheral blood cells (Zheng et al., 2012). In the present study, using antibody blocking and TLR2−/− mouse macrophages assays, we found that TLR2 plays an important role in the inflammatory response activated by HP0459. This is consistent with *S. suis*-induced pro-inflammatory response, which is primary TLR2-dependent. However, LipoP prediction of HP0459 displayed that the signal peptide of HP0459 belonged to SpI (lipoprotein signal peptide belonged to SpII; Hutchings et al., 2009). This result suggested that the recognition of HP0459 may be different from bacterial lipoproteins. So, the mechanisms of HP0459 recognized by TLR2 need further research. We have proved that HP0459 contributes to pro-inflammatory response during *S. suis* infection. And it is known that there is a close link between excessive inflammation and the development of *Streptococcus* toxic shock syndrome (STSS; Zhao et al., 2011). This implies that HP0459 may be responsible for the pathogenesis of STSS caused by *S. suis* 2.

Compared with the SC-19 strains, the pro-inflammatory activation of the Δ*hp0459* strain *in vivo* was found to be significantly reduced in the present study. However, the mouse experiment showed that the lethality of the SC-19 and Δ*hp0459* strains were not significant different after challenge with the same CFUs (data not shown). To explain this result, we performed bacteriological isolation from the blood and spleen, and compared with the SC-19 and CΔ*hp0459* strains, the bacteria content of the Δ*hp0459* strain was significantly increased (**Figure 5**). It is known that higher bacterial counts of pathogenic bacteria *in vivo* may be an important cause leading to disease worsening (Sullivan et al., 1982). Thus, this finding suggests that inflammation is mainly but not exclusively responsible for the pathology of *S. suis* and that the bacteria content also plays an important role.

Finally, our data identified a novel pro-inflammatory protein denoted HP0459 from *S. suis*. Further, we demonstrated that HP0459 induces a TLR2-dependent pro-inflammatory response in RAW 264.7 macrophages via activation of the ERK1/2 pathway. These findings could be important for improving our understanding of the excessive inflammation induced by *S. suis* and may aid the development of therapies against SS2 infection.

Acknowledgments

The authors are grateful to the State Key Laboratory of Agricultural Microbiology (Wuhan, China) for supporting this study. This work was supported by the High-Tech Research and Development Program (863) of China (2011AA10A210), the National Basic Research Program (973) of China (2012CB518805), the Chinese Major Special Science and Technology Project (2012ZX10004214-005) the Special Fund for Public Welfare Industry of Chinese Ministry of Agriculture (201303041), and the Fundamental Research Funds for the Central Universities (2011PY006).

References

Adam, R. A., Tenenbaum, T., Valentin-Weigand, P., Laryea, M., Schwahn, B., Angelow, S., et al. (2004). Porcine choroid plexus epithelial cells induce *Streptococcus suis* bacteriostasis in vitro. *Infect. Immun.* 72, 3084–3087. doi: 10.1128/IAI.72.5.3084-3087.2004

Akira, S., and Hemmi, H. (2003). Recognition of pathogen-associated molecular patterns by TLR family. *Immunol. Lett.* 85, 85–95. doi: 10.1016/S0165-2478(02)00228-6

Al-Numani, D., Segura, M., Dore, M., and Gottschalk, M. (2003). Up-regulation of ICAM-1, CD11a/CD18 and CD11c/CD18 on human THP-1 monocytes stimulated by *Streptococcus suis* serotype 2. *Clin. Exp. Immunol.* 133, 67–77. doi: 10.1046/j.1365-2249.2003.02189.x

Arends, J. P., and Zanen, H. C. (1988). Meningitis caused by *Streptococcus suis* in humans. *Rev. Infect. Dis.* 10, 131–137. doi: 10.1093/clinids/10.1.131

Baums, C. G., and Valentin-Weigand, P. (2009). Surface-associated and secreted factors of *Streptococcus suis* in epidemiology, pathogenesis and vaccine development. *Anim. Health Res. Rev.* 10, 65–83. doi: 10.1017/S146625230999003X

Beutler, B. (2004). Inferences, questions and possibilities in Toll-like receptor signalling. *Nature* 430, 257–263. doi: 10.1038/nature02761

Beutler, B. (2009). Microbe sensing, positive feedback loops, and the pathogenesis of inflammatory diseases. *Immunol. Rev.* 227, 248–263. doi: 10.1111/j.1600-065X.2008.00733.x

Bonifait, L., Vaillancourt, K., Gottschalk, M., Frenette, M., and Grenier, D. (2011). Purification and characterization of the subtilisin-like protease of Streptococcus suis that contributes to its virulence. *Vet. Microbiol.* 148, 333–340. doi: 10.1016/j.vetmic.2010.09.024

Charland, N., Harel, J., Kobisch, M., Lacasse, S., and Gottschalk, M. (1998). *Streptococcus suis* serotype 2 mutants deficient in capsular expression. *Microbiology* 144(Pt 2), 325–332. doi: 10.1099/00221287-144-2-325

Dominguez-Punaro Mde, L., Segura, M., Radzioch, D., Rivest, S., and Gottschalk, M. (2008). Comparison of the susceptibilities of C57BL/6 and A/J mouse strains to *Streptococcus suis* serotype 2 infection. *Infect. Immun.* 76, 3901–3910. doi: 10.1128/IAI.00350-08.

Fittipaldi, N., Segura, M., Grenier, D., and Gottschalk, M. (2012). Virulence factors involved in the pathogenesis of the infection caused by the swine pathogen and zoonotic agent *Streptococcus suis*. *Future Microbiol.* 7, 259–279. doi: 10.2217/fmb.11.149

Fittipaldi, N., Sekizaki, T., Takamatsu, D., Harel, J., Dominguez-Punaro Mde, L., Von Aulock, S., et al. (2008). D-alanylation of lipoteichoic acid contributes to the virulence of *Streptococcus suis*. *Infect. Immun.* 76, 3587–3594. doi: 10.1128/IAI.01568-07.

Gottschalk, M., and Segura, M. (2000). The pathogenesis of the meningitis caused by *Streptococcus suis*: the unresolved questions. *Vet. Microbiol.* 76, 259–272. doi: 10.1016/S0378-1135(00)00250-9

Gottschalk, M., Segura, M., and Xu, J. (2007). *Streptococcus suis* infections in humans: the Chinese experience and the situation in North America. *Anim. Health Res. Rev.* 8, 29–45. doi: 10.1017/S1466252307001247.

Gottschalk, M., Xu, J., Calzas, C., and Segura, M. (2010). *Streptococcus suis*: a new emerging or an old neglected zoonotic pathogen? *Future Microbiol.* 5, 371–391. doi: 10.2217/fmb.10.2

Graveline, R., Segura, M., Radzioch, D., and Gottschalk, M. (2007). TLR2-dependent recognition of *Streptococcus suis* is modulated by the presence of capsular polysaccharide which modifies macrophage responsiveness. *Int. Immunol.* 19, 375–389. doi: 10.1093/intimm/dxm003

Hui, A. C., Ng, K. C., Tong, P. Y., Mok, V., Chow, K. M., Wu, A., et al. (2005). Bacterial meningitis in Hong Kong: 10-years' experience. *Clin. Neurol. Neurosurg.* 107, 366–370. doi: 10.1016/j.clineuro.2004.10.006

Hutchings, M. I., Palmer, T., Harrington, D. J., and Sutcliffe, I. C. (2009). Lipoprotein biogenesis in Gram-positive bacteria: knowing when to hold 'em, knowing when to fold 'em. *Trends Microbiol.* 17, 13–21. doi: 10.1016/j.tim.2008.10.001

Ip, M., Fung, K. S., Chi, F., Cheuk, E. S., Chau, S. S., Wong, B. W., et al. (2007). *Streptococcus suis* in Hong Kong. *Diagn. Microbiol. Infect. Dis* 57, 15–20. doi: 10.1016/j.diagmicrobio.2006.05.011

Jobin, M. C., Fortin, J., Willson, P. J., Gottschalk, M., and Grenier, D. (2005). Acquisition of plasmin activity and induction of arachidonic acid release by Streptococcus suis in contact with human brain microvascular endothelial cells. *FEMS Microbiol. Lett.* 252, 105–111. doi: 10.1016/j.femsle.2005.08.044

Kang, E. H., Gebru, E., Kim, M. H., Cheng, H., and Park, S. C. (2009). EstA protein, a novel virulence factor of Streptococcus pneumoniae, induces nitric oxide and pro-inflammatory cytokine production in RAW 264.7 macrophages through NF-kappaB/MAPK. *Microb. Pathog.* 47, 196–201. doi: 10.1016/j.micpath.2009.07.002

Kawai, T., and Akira, S. (2005). Pathogen recognition with Toll-like receptors. *Curr. Opin. Immunol* 17, 338–344. doi: 10.1016/j.coi.2005.02.007

Kawai, T., and Akira, S. (2010). The role of pattern-recognition receptors in innate immunity: update on Toll-like receptors. *Nat. Immunol.* 11, 373–384. doi: 10.1038/ni.1863

Lachance, C., Segura, M., Gerber, P. P., Xu, J., and Gottschalk, M. (2013). Toll-like receptor 2-independent host innate immune response against an epidemic strain of *Streptococcus suis* that causes a toxic shock-like syndrome in humans. *PLoS ONE* 8:e65031. doi: 10.1371/journal.pone.0065031

Lecours, M. P., Segura, M., Fittipaldi, N., Rivest, S., and Gottschalk, M. (2012). Immune receptors involved in *Streptococcus suis* recognition by dendritic cells. *PLoS ONE* 7:e44746. doi: 10.1371/journal.pone.0044746

Li, R., Zhang, A., Chen, B., Teng, L., Wang, Y., Chen, H., et al. (2010). Response of swine spleen to *Streptococcus suis* infection revealed by transcription analysis. *BMC Genomics* 11:556. doi: 10.1186/1471-2164-11-556

Li, W., Wan, Y., Tao, Z., Chen, H., and Zhou, R. (2013). A novel fibronectin-binding protein of *Streptococcus suis* serotype 2 contributes to epithelial cell invasion and in vivo dissemination. *Vet. Microbiol.* 162, 186–194. doi: 10.1016/j.vetmic.2012.09.004

Liu, B., Teng, D., Wang, X., Yang, Y., and Wang, J. (2012). Expression of the soybean allergenic protein P34 in *Escherichia coli* and its indirect ELISA detection method. *Appl. Microbiol. Biotechnol.* 94, 1337–1345. doi: 10.1007/s00253-012-4006-3.

Liu, D., Yumoto, H., Hirota, K., Murakami, K., Takahashi, K., Hirao, K., et al. (2008). Histone-like DNA binding protein of *Streptococcus* intermedius induces the expression of pro-inflammatory cytokines in human monocytes via activation of ERK1/2 and JNK pathways. *Cell Microbiol* 10, 262–276. doi: 10.1111/j.1462-5822.2007.01040.x

Liu, L., Cheng, G., Wang, C., Pan, X., Cong, Y., Pan, Q., et al. (2009). Identification and experimental verification of protective antigens against *Streptococcus suis* serotype 2 based on genome sequence analysis. *Curr. Microbiol.* 58, 11–17. doi: 10.1007/s00284-008-9258-x

Liu, N., Liu, J. T., Ji, Y. Y., and Lu, P. P. (2011). Rosiglitazone regulates c-reactive protein-induced inflammatory responses via glucocorticoid receptor-mediated inhibition of p38 mitogen-activated protein kinase-toll-like receptor 4 signal pathway in vascular smooth muscle cells. *J. Cardiovasc. Pharmacol.* 57, 348–356. doi: 10.1097/FJC.0b013e31820a0e67

Lun, S., Perez-Casal, J., Connor, W., and Willson, P. J. (2003). Role of suilysin in pathogenesis of *Streptococcus suis* capsular serotype 2. *Microb. Pathog.* 34, 27–37. doi: 10.1016/S0882-4010(02)00192-4

Mai, N. T., Hoa, N. T., Nga, T. V., Linh Le, D., Chau, T. T., Sinh, D. X., et al. (2008). *Streptococcus suis* meningitis in adults in Vietnam. *Clin. Infect. Dis.* 46, 659–667. doi: 10.1086/527385

Manukyan, M., Triantafilou, K., Triantafilou, M., Mackie, A., Nilsen, N., Espevik, T., et al. (2005). Binding of lipopeptide to CD14 induces physical proximity of CD14, TLR2 and TLR1. *Eur. J. Immunol.* 35, 911–921. doi: 10.1002/eji.200425336

Martinon, F., and Tschopp, J. (2005). NLRs join TLRs as innate sensors of pathogens. *Trends Immunol.* 26, 447–454. doi: 10.1016/j.it.2005.06.004

Moore, L. J., Pridmore, A. C., Lee, M. E., and Read, R. C. (2005). Induction of pro-inflammatory cytokine release by human macrophages during exposure of *Streptococcus pneumoniae* to penicillin is influenced by minimum inhibitory concentration ratio. *Int. J. Antimicrob. Agents* 26, 188–196. doi: 10.1016/j.ijantimicag.2005.06.006

Moreillon, P., and Majcherczyk, P. A. (2003). Proinflammatory activity of cell-wall constituents from gram-positive bacteria. *Scand. J. Infect. Dis.* 35, 632–641. doi: 10.1080/00365540310016259

Perch, B., Kristjansen, P., and Skadhauge, K. (1968). Group R streptococci pathogenic for man. Two cases of meningitis and one fatal case of sepsis. *Acta Pathol. Microbiol. Scand.* 74, 69–76. doi: 10.1111/j.1699-0463.1968.tb03456.x

Segura, M., Gottschalk, M., and Olivier, M. (2004). Encapsulated *Streptococcus suis* inhibits activation of signaling pathways involved in phagocytosis. *Infect. Immun.* 72, 5322–5330. doi: 10.1128/IAI.72.9.5322-5330.2004

Segura, M., Vadeboncoeur, N., and Gottschalk, M. (2002). CD14-dependent and -independent cytokine and chemokine production by human THP-1 monocytes stimulated by *Streptococcus suis* capsular type 2. *Clin. Exp. Immunol* 127, 243–254. doi: 10.1046/j.1365-2249.2002.01768.x

Segura, M., Vanier, G., Al-Numani, D., Lacouture, S., Olivier, M., and Gottschalk, M. (2006). Proinflammatory cytokine and chemokine modulation by *Streptococcus suis* in a whole-blood culture system. *FEMS Immunol. Med. Microbiol.* 47, 92–106. doi: 10.1111/j.1574-695X.2006.00067.x

Smith, H. E., Damman, M., Van Der Velde, J., Wagenaar, F., Wisselink, H. J., Stockhofe-Zurwieden, N., et al. (1999). Identification and characterization of the cps locus of *Streptococcus suis* serotype 2: the capsule protects against phagocytosis and is an important virulence factor. *Infect. Immun.* 67, 1750–1756.

Smith, H. E., Vecht, U., Gielkens, A. L., and Smits, M. A. (1992). Cloning and nucleotide sequence of the gene encoding the 136-kilodalton surface protein (muramidase-released protein) of *Streptococcus suis* type 2. *Infect. Immun.* 60, 2361–2367.

Sodhi, A., Sharma, R. K., and Batra, H. V. (2005). Yersinia rLcrV and rYopB inhibits the activation of murine peritoneal macrophages in vitro. *Immunol. Lett.* 99, 146–152. doi: 10.1016/j.imlet.2005.02.009

Suankratay, C., Intalapaporn, P., Nunthapisud, P., Arunyingmongkol, K., and Wilde, H. (2004). *Streptococcus suis* meningitis in Thailand. *Southeast Asian J. Trop. Med. Public Health* 35, 868–876.

Sullivan, T. D., Lascolea, L. J. Jr., and Neter, E. (1982). Relationship between the magnitude of bacteremia in children and the clinical disease. *Pediatrics* 69, 699–702.

Takamatsu, D., Osaki, M., and Sekizaki, T. (2001). Thermosensitive suicide vectors for gene replacement in *Streptococcus suis*. *Plasmid* 46, 140–148. doi: 10.1006/plas.2001.1532

Tenenbaum, T., Adam, R., Eggelnpohler, I., Matalon, D., Seibt, A., Ge, K. N., et al. (2005). Strain-dependent disruption of blood-cerebrospinal fluid barrier by *Streptoccocus suis* in vitro. *FEMS Immunol. Med. Microbiol.* 44, 25–34. doi: 10.1016/j.femsim.2004.12.006

Vadeboncoeur, N., Segura, M., Al-Numani, D., Vanier, G., and Gottschalk, M. (2003). Pro-inflammatory cytokine and chemokine release by human brain microvascular endothelial cells stimulated by *Streptococcus suis* serotype 2. *FEMS Immunol. Med. Microbiol.* 35, 49–58. doi: 10.1111/j.1574-695X.2003.tb00648.x

van Lint, A. L., Murawski, M. R., Goodbody, R. E., Severa, M., Fitzgerald, K. A., Finberg, R. W., et al. (2010). Herpes simplex virus immediate-early ICP0 protein inhibits Toll-like receptor 2-dependent inflammatory responses and NF-kappaB signaling. *J. Virol.* 84, 10802–10811. doi: 10.1128/JVI.00063-10

Wei, Z., Li, R., Zhang, A., He, H., Hua, Y., Xia, J., et al. (2009). Characterization of *Streptococcus suis* isolates from the diseased pigs in China between 2003 and 2007. *Vet. Microbiol.* 137, 196–201. doi: 10.1016/j.vetmic.2008.12.015

Wertheim, H. F., Nghia, H. D., Taylor, W., and Schultsz, C. (2009). *Streptococcus suis*: an emerging human pathogen. *Clin. Infect. Dis.* 48, 617–625. doi: 10.1086/596763

Wisselink, H. J., Smith, H. E., Stockhofe-Zurwieden, N., Peperkamp, K., and Vecht, U. (2000). Distribution of capsular types and production of muramidase-released protein (MRP) and extracellular factor (EF) of *Streptococcus suis* strains isolated from diseased pigs in seven European countries. *Vet. Microbiol.* 74, 237–248. doi: 10.1016/S0378-1135(00)00188-7

Wright, S. D. (1995). CD14 and innate recognition of bacteria. *J. Immunol.* 155, 6–8.

Wright, S. D., Ramos, R. A., Tobias, P. S., Ulevitch, R. J., and Mathison, J. C. (1990). CD14, a receptor for complexes of lipopolysaccharide (LPS) and LPS binding protein. *Science* 249, 1431–1433. doi: 10.1126/science.1698311

Ye, C., Zhu, X., Jing, H., Du, H., Segura, M., Zheng, H., et al. (2006). *Streptococcus suis* sequence type 7 outbreak, Sichuan, China. *Emerg. Infect. Dis* 12, 1203–1208. doi: 10.3201/eid1208.060232

Yu, H., Jing, H., Chen, Z., Zheng, H., Zhu, X., Wang, H., et al. (2006). Human *Streptococcus suis* outbreak, Sichuan, China. *Emerg. Infect. Dis.* 12, 914–920. doi: 10.3201/eid1206.051194

Zhang, A., Chen, B., Yuan, Z., Li, R., Liu, C., Zhou, H., et al. (2012). HP0197 contributes to CPS synthesis and the virulence of *Streptococcus suis* via CcpA. *PLoS ONE* 7:e50987. doi: 10.1371/journal.pone.0050987

Zhang, A., Mu, X., Chen, B., Han, L., Chen, H., and Jin, M. (2011). IgA1 protease contributes to the virulence of *Streptococcus suis. Vet. Microbiol.* 148, 436–439. doi: 10.1016/j.vetmic.2010.09.027

Zhao, Y., Liu, G., Li, S., Wang, M., Song, J., Wang, J., et al. (2011). Role of a type IV-like secretion system of *Streptococcus suis* 2 in the development of streptococcal toxic shock syndrome. *J. Infect. Dis.* 204, 274–281. doi: 10.1093/infdis/jir261

Zhao, Z., Yi, C., Zhao, L., Wang, S., Zhou, L., Hu, Y., et al. (2014). PB2-588I enhances 2009 H1N1 pandemic influenza virus virulence by increasing viral replication and exacerbating PB2 inhibition of beta interferon expression. *J. Virol.* 88, 2260–2267. doi: 10.1128/JVI.03024-13

Zheng, H., Luo, X., Segura, M., Sun, H., Ye, C., Gottschalk, M., et al. (2012). The role of toll-like receptors in the pathogenesis of *Streptococcus suis. Vet. Microbiol.* 156, 147–156. doi: 10.1016/j.vetmic.2011.10.014

Zheng, H., Punaro, M. C., Segura, M., Lachance, C., Rivest, S., Xu, J., et al. (2011). Toll-like receptor 2 is partially involved in the activation of murine astrocytes by *Streptococcus suis*, an important zoonotic agent of meningitis. *J. Neuroimmunol.* 234, 71–83. doi: 10.1016/j.jneuroim.2011.02.005

Zheng, H., Ye, C., Segura, M., Gottschalk, M., and Xu, J. (2008). Mitogenic effect contributes to increased virulence of *Streptococcus suis* sequence type 7 to cause streptococcal toxic shock-like syndrome. *Clin. Exp. Immunol* 153, 385–391. doi: 10.1111/j.1365-2249.2008.03722.x

Conflict of Interest Statement: The authors declare that the research was conducted in the absence of any commercial or financial relationships that could be construed as a potential conflict of interest.

14

Staphylococcal Superantigens Spark Host-Mediated Danger Signals

Teresa Krakauer[1], Kisha Pradhan[2] and Bradley G. Stiles[2]**

[1] Department of Immunology, Molecular Translational Sciences Division, United States Army Medical Research Institute of Infectious Diseases, Fort Detrick, Frederick, MD, USA, [2] Biology Department, Wilson College, Chambersburg, PA, USA

Edited by:
Katharina F. Kubatzky,
University Hospital Heidelberg, Germany

Reviewed by:
Howard M. Johnson,
University of Florida, USA
Klaus Heeg,
University of Heidelberg, Germany

***Correspondence:**
Teresa Krakauer
teresa.krakauer.civ@mail.mil;
Bradley G. Stiles
brad.stiles@wilson.edu

Staphylococcal enterotoxin B (SEB) of *Staphylococcus aureus*, and related superantigenic toxins produced by myriad microbes, are potent stimulators of the immune system causing a variety of human diseases from transient food poisoning to lethal toxic shock. These protein toxins bind directly to specific Vβ regions of T-cell receptors (TCR) and major histocompatibility complex (MHC) class II on antigen-presenting cells, resulting in hyperactivation of T lymphocytes and monocytes/macrophages. Activated host cells produce excessive amounts of proinflammatory cytokines and chemokines, especially tumor necrosis factor α, interleukin 1 (IL-1), IL-2, interferon γ (IFNγ), and macrophage chemoattractant protein 1 causing clinical symptoms of fever, hypotension, and shock. Because of superantigen-induced T cells skewed toward TH1 helper cells, and the induction of proinflammatory cytokines, superantigens can exacerbate autoimmune diseases. Upon TCR/MHC ligation, pathways induced by superantigens include the mitogen-activated protein kinase cascades and cytokine receptor signaling, resulting in activation of NFκB and the phosphoinositide 3-kinase/mammalian target of rapamycin pathways. Various mouse models exist to study SEB-induced shock including those with potentiating agents, transgenic mice and an "SEB-only" model. However, therapeutics to treat toxic shock remain elusive as host response genes central to pathogenesis of superantigens have only been identified recently. Gene profiling of a murine model for SEB-induced shock reveals novel molecules upregulated in multiple organs not previously associated with SEB-induced responses. The pivotal genes include intracellular DNA/RNA sensors, apoptosis/DNA damage-related molecules, immunoproteasome components, as well as antiviral and IFN-stimulated genes. The host-wide induction of these, and other, antimicrobial defense genes provide evidence that SEB elicits danger signals resulting in multi-organ damage and toxic shock. Ultimately, these discoveries might lead to novel therapeutics for various superantigen-based diseases.

Keywords: SEB, superantigens, toxic shock, damage response, therapy, animal models

THE BACTERIUM: *Staphylococcus aureus*

Staphylococcus aureus causes many human and animal diseases, some of them are life-threatening and found throughout the world (1–4). Humans are naturally colonized by *S. aureus* that often cause no problems and are considered a "harmless" commensal. However, when given an opportunity (i.e., weakened immune system, compromised epidermis, or mucosa, etc.), this bacterium can cause various diseases that become life threatening.

Staphylococcus aureus is a Gram-positive coccus and member of the Micrococcaceae family, which includes non-pathogenic genera found in soil, water, and on skin. Both planktonic- and biofilm-based versions of *S. aureus* can cause disease in an infected host, subsequently becoming (at times) quite difficult to clear from the body (5, 6). Characteristics of this genus called *Staphylococcus* ("staphyle" in Greek, meaning grape cluster and "coccus" meaning grain or berry) and species called *aureus* ("golden" in Latin, as per colony color on agar media), include Gram-positive, non-spore forming, facultative, β-hemolytic, catalase-positive, mannitol fermentation, and salt tolerant (7).

Staphylococcus aureus is known for producing many different protein toxins involved in pathogenesis. Clearly, the bacterium is very good at surviving harsh conditions in/on a host by using a vast array of virulence factors that promote the many diseases caused by infiltrating *S. aureus* and their toxins. One group of toxins, the staphylococcal enterotoxins (SEs), is the focus of this review. Antibiotic-resistant strains (community-, nursing home-, and hospital-acquired) are a particular problem in health care and its associated economics, requiring effective infection control plans that prevent *S. aureus* spread locally and globally (8).

Staphylococcus aureus was first described in the early 1880s by a Scotsman, Dr. Alexander Ogston, after microscopy analysis of over a hundred, pus-filled abscesses of human origins (9). *S. aureus* readily colonizes mammalian epidermis/dermis, mucosa, soft tissues, bone, and medical devices such as catheters. Furthermore, medically relevant strains of the bacterium have become variably resistant to many antibiotics, including methicillin and vancomycin, which negatively impacts psychological and economic aspects of human society (10–13). *S. aureus* strains that are methicillin resistant (MRSA) are commonly treated first with vancomycin. Both methicillin and vancomycin target bacterial cell-wall synthesis. Other antimicrobials that have been used for fighting *S. aureus* include (1) linezolid (inhibits protein synthesis at the 50S ribosome); (2) daptomycin (inserts into membranes); (3) fusidic acid (inhibits protein elongation at the ribosome); (4) teicoplanin (like vancomycin, inhibits cell wall synthesis but not used in the United States); and (5) tigecycline (inhibits protein synthesis at the 30S ribosome) (14, 15).

For various reasons, alternative methods for treating antibiotic-resistant strains of *S. aureus* (i.e., MRSA) are being explored by different groups around the world (16–23). Some of the exotic sources for these novel drugs include snake venom, tree bark/stems/leaves, and carnivorous plants. Risk factors for MRSA colonization include antibiotic exposure, admittance into an intensive care unit, surgery, long-term care residency, and exposure to others carrying MRSA (4, 24). In 2011, the United States Centers for Disease Control and Prevention reported >80,000 MRSA cases that were life threatening, with >11,000 fatalities (14%) (22, 25). This is remarkably a number of deaths greater than that attributed to acquired immunodeficiency syndrome (AIDS) for the same time period. Furthermore, the economic burden of MRSA in the United States is upwards to $13.8 billion/year and represents one of the most costly, acute infectious diseases requiring treatment (10). Invasive forms of MRSA include the following in order of prevalence: bacteremia, pneumonia, cellulitis, osteomyelitis, endocarditis, and septic shock (24). Black males older than 65 years are the most common patients suffering from invasive MRSA; however, the reasons for this are to date unknown (24). MRSA from hospital/health-care settings are perhaps the origins of community strains now seen throughout the human population (4). The latter types of *S. aureus* are often more susceptible to antibiotics than those from hospital/health-care settings.

From a historical perspective, MRSA was detected within a year after introducing this new beta-lactam antibiotic called methicillin in 1959, further highlighting the rapid adaptability of this microbe (26, 27). Methicillin was meant to overcome the increasingly prevalent, penicillin-resistant strains of *S. aureus* first detected during the mid-1940s in hospitals and throughout the community. These strains were isolated just 2 years after penicillin's introduction into clinical practice (27). This pattern of antibiotic resistance is evident with various pathogens, suggesting furthermore the importance of developing truly novel antimicrobials and embracing a paradigm shift for discovering/creating antimicrobials. Perhaps vaccines will play a bigger role in fighting various pathogens, in the future?

Transmission of antibiotic-resistant strains between livestock and humans working with and/or consuming the former is a real problem, poorly understood to date (28–31). Furthermore, aerosol transmission of MRSA from farm animals or human patients in hospital wards, to humans, is also a scary reality (32–34). As just one known example of *S. aureus* transmission between species, strain CC398 was most commonly associated with asymptomatic colonization of swine, but now causes invasive infections of humans (35). Evidently, farm animals are an important reservoir for disease-causing *S. aureus* that colonize humans. Humans (~30%) can be asymptomatic carriers of *S. aureus* strains that possess virulence factor genes encoding antibiotic resistance, toxins, and other proteins that promote some nasty, potentially fatal diseases (1, 3, 36–40) (**Table 1**). The risk factors are many for children carrying MRSA nasally, and include (1) <4 years of age; (2) being male; (3) family size >4; (4)

TABLE 1 | *Staphylococcus aureus*: some virulence factors and diseases.

Virulence factors
Antibiotic resistance (multiple mechanisms)
Biofilm/capsule
Coagulase
Exfoliative toxin
Microbial surface components recognizing adhesive matrix molecules (MSCRAMMs)
Plasminogen activator
Pore-forming toxins (hemolysins and leukocidins)
Quorum sensing mechanism
Superantigens (enterotoxins and toxic shock syndrome toxin-1)
Toll/interleukin 1 receptor (TIR)-like protein
Diseases
Bacteremia
Endocarditis
Osteomyelitis
Pneumonia
Pyoarthrosis
Skin and soft tissue infections (boils, cellulitis, impetigo, scalded-skin syndrome, sty)
Toxic shock syndrome

antibiotic use during previous 3 months; (5) smoking by parents; and (6) sleeping with parents (39). For adults, risk factors that enhance carriage of *S. aureus* include: (1) being male; (2) having diabetes; (3) use of hormonal contraception by women; (4) close physical contact with others (i.e., sports); and interestingly (5) being a non-smoker (40).

Clearly, *Homo sapiens* are not strangers to hosting *S. aureus* on, and in, themselves. This bacterium is rather adaptable and employs various virulence factors seemingly important in its survival. Are we, in the near and distant future, able to co-adapt and effectively counter this microbial medical challenge, which is just one of many? We now focus upon one remarkable virulence factor, a family of protein toxins that vary in structure and biological effects upon mammals: the SEs.

THE STAPHYLOCOCCAL ENTEROTOXINS

From a historical perspective, the SEs (and *S. aureus*) were first described as "ptomaine" (derived from Greek "ptoma," meaning corpse) in 1914 by Barber (41), after human consumption of cow's milk contaminated by staphylococci (from the cow) led to vomiting and diarrhea of unsuspecting visitors. Family members were remarkably resistant to this poisoning effect, perhaps due to repeated exposure and development of toxin-specific immunity. Barber included himself as a "human guinea pig" to prove his point that milk from this one farm, during the hot season in the Philippines, was indeed the medium for enteric illness. His final analysis revealed that short-term storage of the milk at ambient temperature enabled staphylococcal growth and toxin production (41).

Dack et al. reported in 1930 an enteric-acting toxin produced by a "yellow hemolytic *Staphylococcus*" isolated from sponge cake involved in food poisoning (42). Upon administering culture filtrate of the isolated bacterium grown in the laboratory to a rabbit (intravenous) or humans ($n = 3$; oral), there was respectively death and uniform nausea/diarrhea. Some of the SEs were first purified by various groups during the 1950s (43–45), further linked to a major cause of food poisoning evident around the world, and found to stimulate specific subsets of T cells leading to shock (1, 46, 47).

Regarding staphylococcal food poisoning and diagnosis, the latter can be done upon suspected food/drink using classic microbiology and agar plates (10^5 viable *S. aureus*/gram food), or by direct detection of toxin (47). SE detection in food is most commonly accomplished by employing immunological techniques (i.e., commercially available enzyme/fluorescent immune assays) that are to date limited in detecting relatively few toxin types. However, next generation assays involving mass spectrometry are now being developed for the future pending higher throughput and cost-effectiveness (47).

The SEs are characteristically stable proteins resistant to high heat, low pH, freezing, drying, and proteases. Partly because of their stable nature and powerful effects upon humans at very low doses, the SEs (particularly SEB) have been studied from a biodefense perspective spanning decades of research by various countries (48, 49). SEB is also considered a Category B select agent by various United States federal agencies. SEB, when inhaled, can induce several symptoms within 120 min involving an aching feeling (head and muscles), increased heartbeat, coughing, enteric dysfunction (i.e., nausea, vomiting, and diarrhea), as well as eye irritation (49). Nanogram levels of inhaled SEB are incapacitating while microgram levels can be fatal. SEB adversely affects the immune system, and it is plausible that opportunistic viruses and bacteria can cause further harm to the host exposed to any SE (50–52).

Upon ingestion, the SEs (A–V and counting) are responsible for a prevalent form of food poisoning globally (1, 38, 46, 47, 53). There are five, sequence-based groups of SEs generally recognized to date (1). These single-chain proteins (~25 kDa) are generally composed of two domains containing both alpha helices and beta sheets, separated by a groove, as evidenced by one of the first SEs to be crystallized: SEB (54). Since the mid-2000s, the literature has also adopted the designation of SE-like (SEl) (55). This defines a staphylococcal protein that shares amino acid sequence homology with a previously characterized SE, yet has not to date been tested (or at least reported in the literature) for enteric effects in primates (56). Regarding staphylococcal food poisoning, SEA is the most commonly detected culprit in the United States, followed by SED and SEB. Classic poisoning due to food-based SEs can occur after ingesting processed meats or dairy products contaminated by improper handling (i.e., *S. aureus* transfer from the skin of a food handler to food) and subsequent storage of food at an elevated temperature conducive to bacterial growth. Depending upon the *S. aureus* strain, there can be one (or more) SEs released into the tainted food. Following consumption of SEs in food/drink, one may still experience a general malaise 24–72 h later (56). Poisoning by the SEs *via* many different food types is rarely fatal for healthy individuals, and occurs around the world; however, as is often the case with many diseases, the very young and old represent higher risk groups for severe morbidity and possible death due to SE exposure (57).

Exactly how the SEs cause enteric illness is still remarkably unresolved, but prostaglandins and leukotrienes might mediate the effects (58, 59). Serotonin release in the intestines and vagal nerve stimulation may also be involved in SE-induced emesis (60, 61). The mode of action is seemingly more complex than many other enterotoxins that work directly upon gastric epithelial cells and/or supporting matrix.

The term "superantigen" commonly describes the *S. aureus* SEs and toxic shock syndrome toxin-1 (TSST-1), while related streptococcal pyrogenic exotoxins (SPEs), also known as superantigens, are produced by *Streptococcus pyogenes*. Superantigen was first used in late 1980s literature to describe microbial proteins that activate a large population of specific T-cells at very minute (picogram) concentrations (62, 63). Typical "conventional" antigens stimulate far fewer T-cells at higher concentrations. Superantigens and conventional counterparts also differ by: (1) superantigens bind to the exterior of the peptide-binding groove of major histocompatibility complex class II (MHC II); (2) superantigens bind to different MHC II types; (3) superantigenic effects occur without internalization and antigen processing; and (4) T-cell receptor (TCR) recognition of a superantigen: MHC II complex requires the variable region of a TCR β chain (Vβ), not the Vα–Vβ chain used by conventional antigens (62–67). From a virulence perspective, superantigens corrupt host immunity that then enables pathogen survival.

Microbial superantigens are reportedly produced by other bacterial genera (*Mycobacterium*, *Mycoplasma*, *Yersinia*), fungi (*Candida*), and even viruses (herpes, rabies), thus suggesting a conserved and successful strategy employed throughout Nature (68). Versus superantigens of *Staphylococcus* and *Streptococcus*, those from the other aforementioned microbes are poorly characterized to date.

RECEPTOR BINDING AND SIGNAL TRANSDUCTION

Superantigens interact with both CD4 and CD8 T-cells and the signaling, post-TCR binding, is similar to conventional antigen binding to TCR; however, binding is to specific Vβ regions of TCR unlike conventional antigens. In addition, the interaction of superantigen with MHC class II on antigen-presenting cells (APC) is also different, with superantigens binding outside the peptide-binding groove of MHC class II. Structural and binding studies indicate at least two different binding sites on MHC class II for SE and TSST-1. A common overlapping binding region exists on HLA-DR for these toxins, referred to as the generic MHC class II binding site involving the α chain. An additional binding site with higher affinity can be found within the C-terminus of SEA/SED/SEH, which binds to the HLA-DR β chain in a Zn^{2+}-dependent manner (69–71). The presence of two MHC binding sites allows SEA to cross-link MHC class II on APC and activate monocytes, inducing potent proinflammatory mediators. The binding of staphylococcal superantigens to MHC class II, and bridging to TCR, activates a high percentage of T cells. Two decades of intense investigations focused upon superantigen binding to cellular receptors reveal how superantigens activate cells of the immune system (46, 62, 72–76). Interaction of superantigen with TCR transmits the classical first signal for T-cell activation. The binding of superantigen to costimulatory receptors such as CD28 promotes supramolecular clusters, stabilizes cellular interaction, and optimizes protein kinase signal transduction (77). CD28 co-stimulation enhances mRNA stability of IL-2 and T-cell survival by increased expression of anti-apoptotic Bcl-xl (78). Other cell-surface molecules such as CD2, CD11a/ICAM-1, and ELAM facilitate optimal activation of endothelial and T cells by SEB (79). TCR and costimulatory receptors activate protein tyrosine kinases (PTKs), LCK and ZAP-70, resulting in phospholipase C gamma (PLCγ) activation, the release of intracellular second messengers, and subsequent protein kinase C (PKC) activation (67, 80, 81). Accompanying this T-cell activation is F-actin polymerization and increased intracellular Ca^{2+}. PTK and PKC activation lead to mitogen-activated protein kinase (MAPK), extracellular signal regulated kinase (ERK), and cJun N-terminal kinase (JNK) pathways that activate transcriptional factors NFκB, NF-AT, and AP-1 (67, 81, 82). PKCθ activation leads to CARMA1, Bcl10, and MALT1 (CBM) complexes that induce NFκB transcriptional activation and controls T-cell proliferation (83). Many proinflammatory cytokine genes contain NFκB-binding sites in their promotor region, and are induced by NFκB (84). The NFκB cascade is a major signal transduction pathway for many pattern recognition receptors (PRR), such as toll-like (TLR) and proinflammatory cytokine receptors (85). The cytokines IL-1, TNFα, IFNγ, IL-2, IL-6 and chemokines, specifically MCP-1, are induced by superantigens (74, 75, 86). IL-1 and TNFα also activate fibroblasts, epithelial and endothelial cells to induce NFκB and additional mediators, cell adhesion molecules, and tissue proteases (87). Mediators produced by superantigen-activated cells exert profound damaging effects on the immune and cardiovascular system, culminating in multi-organ dysfunction and lethal shock. PTKs and T-cell cytokines also activate phosphoinositide 3 kinase (PI3K), signaling protein kinase B (Akt), and mammalian target of rapamycin complex 1 (mTORC1) downstream (88). Two signaling molecules, NFκB and mTORC1, are key hubs mediating the major biological responses to SE from TCR, co-stimulator CD28, and cytokine signaling.

PI3K/MTORC1 PATHWAY IN T-CELL ACTIVATION

T-cell activation *via* the TCR-CD3 complex subsequently activates membrane proximal PTKs that phosphorylate TCR intracellular components, other cellular substrates, as well as adaptors (67, 80–82). PLCγ cleaves phosphatidylinositol 4,5-bisphosphate, generating second messengers to activate PKC and the proto-oncogene Ras. PTK also activates PI3K, generating several inositol phospholipids and ultimately activating Akt and PKCθ (89). A number of receptors besides TCR, including those for CD28, IL-2 (IL-2R), insulin, growth factor, and the G-protein-coupled receptor (GPCR), also transduce activation signals upon ligand binding *via* the PI3K pathway. The cytosolic signalsome formed by PKCθ and CBM is located at an immunological synapse formed after T-cell activation by anti-CD3 and -CD28 (89, 90). CBM complexes activate the inhibitor of κB (IκB) kinase complex (IKK) through ubiquitin ligases (91). IKK phosphorylates IκB, releasing NF-κB for nuclear translocation and gene activation. Nuclear NFκB binds DNA coding for proinflammatory cytokines and many other NFκB target genes, activating diverse proinflammatory mediators, as well as pro- and anti-apoptotic molecules.

Toxic shock syndrome toxin-1 induces inositol phospholipid turnover, PKC translocation, and calcium mobilization in human T-cells resembling responses from those of a mitogenic signal (92). SEE uses an alternative LCK-independent pathway by activating PLCβ signaling in T cells (93). This alternative pathway also triggers influx of Ca^{2+}, activates PKC, ERK1/2, and ultimately nuclear translocation of NFκB and NF-AT.

Downstream of PI3K is the serine/threonine kinase Akt, which mediates many diverse biological processes upon specific binding to TCR, growth factors, insulin receptor, tyrosine kinase receptor, and GPCR. Two potent cytokines from superantigen-stimulated T cells, IFNγ and IL-2, also activate the PI3K/Akt/mTORC1 pathway *via* the transducer Janus kinase 1 (JAK1) after binding IFNγ and IL-2 receptors, respectively (94, 95). Site-specific phosphorylation/dephosphorylation modulates the PI3K/Akt signaling to activate mTORC1. Activation of mTORC1 causes phosphorylation and activation of the ribosomal 40S protein p70S6 kinase (p70S6K) plus eukaryotic initiation factor binding protein 1 (4EBP1) (96–101). Thus mTORC1 controls protein translation essential for G1 to S phase transition (100). An mTORC1-specific inhibitor, rapamycin, blocks SEB-induced T cell proliferation as well as SEB-induced IL-2 and IFNγ *in vitro* and *in vivo* (102). Importantly, inhibition of the SEB-induced PI3K/Akt/mTORC1 pathway by rapamycin prevents lethal toxic shock in a mouse model (102).

INFLAMMATORY MEDIATORS, TCR/ COSTIMULATORS, AND MHC CLASS II CROSS LINKING ACTIVATE NFκB

In vitro and *in vivo* studies reveal that excessive release of proinflammatory cytokines (IL-1, TNFα, IL-6, and IFNγ) mediates the toxic effects of superantigens (49, 51, 52, 76, 79). Proinflammatory cytokines like IL-1 and TNFα activate the transcriptional factor NFκB in many cell types, thus perpetuating the inflammatory response. IL-1 and TNFα have tissue damaging effects, and together with SEB-induced metalloproteinases (MMP) and the pro-coagulant tissue factor (TF), promote inflammation and coagulation (87, 103). IL-2 from superantigen-activated T cells causes vasodilation, vascular leak, and edema.

Many cell types respond directly to staphylococcal superantigens besides T and APC/monocytes, including B, endothelial, synovial fibroblasts, intestinal epithelial, and mast (58, 79, 104–109). Cross-linking of TCR with MHC class II on B cells by superantigen triggers B cell proliferation and differentiation into plasma cells (104). Stimulation of synovial fibroblasts with superantigens induces chemokine gene expression, raising the possibility that superantigens can trigger chemotactic responses and initiate inflammatory arthritis (105). Human T84 colonic cells increase ion flow after incubation with SEB and PBMC, suggesting that superantigens indirectly affect gut mucosa *via* the immune system (106). SEB and TSST-1 transcytose intestinal epithelial cell barriers, unlike SEA (107), and TSST-1 can also induce chemotactic cytokines (IL-8 and MIP-3α) from human vaginal epithelial cells (109). *S. aureus* frequently colonizes vaginal mucosa, producing TSST-1, which penetrates the mucosa through chemokines from epithelial cells (110). Chemokines recruit neutrophils and other immune cells activated furthermore by superantigens. The potent activation of T cells by superantigen interaction of MHC class II and TCR induces T-cell proliferation *via* cell activation pathways in T cells and APC. The proliferative response as a result of IL-2 induction from superantigen-activated T cells is similar to that induced by mitogens. IL-2 increases vascular permeability, causing edema and multiple organ damage (111). TNFα synergizes with IL-2 to induce pulmonary vascular leak and lymphocyte accumulation (112). The combination of TNFα and IFNγ elaborated by SEB-activated cells disrupt the epithelial barrier, causing edema and vascular leak in gut mucosa (106).

Interleukin 1 interacts with IL-1 receptor 1 (IL-1R1) and an additional accessory protein, triggering downstream signaling molecules like the adaptor myeloid differentiation factor 88 (MyD88), IL-1R- associated protein kinase 1 (IRAK1), and TNF receptor-associated factor 6 (TRAF6) that activate NFκB (113). A set of structurally related receptors, the TLRs, signal with similar intracellular adaptors as those used for IL-1R1, but are not used for superantigen signaling (85, 114). The TLRs are conserved, type-1 transmembrane receptors that recognize pathogen-associated molecular patterns (PAMPs), such as lipoprotein, peptidoglycan, LPS, flagellin, dsRNA, and viral RNA, which stimulate host innate immunity and enhance adaptive immunity (85, 114). SEB reportedly increases cellular expression of TLR2 and TLR4, synergistically promoting lethal shock with LPS (115, 116). Activation of NFκB induces proinflammatory genes as well as pro- and anti-apoptotic genes. An auto-feedback loop exists to downregulate NFκB, which induces IκBα. Other cytosolic PRR sense intracellular damage-associated molecular patterns (DAMPs), triggering inflammasome activation and release of caspases plus IL-1β (117).

TNFα binds to TNF receptor 1 (TNFR1) or TNFR2, activating NFκB and inducing expression of other cytokines, adhesion, as well as co-stimulatory molecules (118, 119). The cytotoxic functions of TNFα are mostly mediated by binding TNFR1 *via* intracellular death domains that trigger apoptosis through caspase activation. SEB increases expression of CD95 (Fas), a receptor of the TNFR superfamily. Intracellular adaptors TNFR-associated death domain (TRADD), Fas-associated death domain (FADD), and receptor interacting protein kinase (RIP) are used by the TNFR superfamily to activate the caspase 8 cascade, JNK, and NFκB that subsequently elicit apoptosis, cell activation, coagulation, inflammation, and host defense (119). TNFα and IFNγ act synergistically on epithelial cells to increase ion transport, causing cell damage and epithelial leakage (120). The critical role of TNFα in mediating pathological effects of SEB and lethality was recognized early on as anti-TNFα antibodies confer protection from SEB-induced shock in a D-galactosamine (D-gal)-sensitized mouse model (121). Both IL-1 and TNFα enhance the procoagulant activity of TF and activate neutrophils, accounting for tissue damage and organ injury commonly seen in animal models of septic shock and SEB-induced lethal shock.

Type II interferon (IFNγ) binds to IFNγR, which belongs to the family of interferon receptors (IFNR), including the structurally different receptors for type I interferons (IFNα, IFNβ) (122, 123). The signal transducer and activator of transcription 1 (STAT1) is a common signal transducer for both types of IFNR, and phosphorylation of STAT1 by Janus kinases initiates signal transduction. IFNα and IFNβ induce indistinguishable signals, which include anti-proliferative and antiviral activities. Types I and II IFNs have many overlapping activities and stimulate many common interferon-stimulated genes (ISG) (123, 124). However, in addition to host-defense functions, IFNγ also induces immunoproteasomes and expression of MHC class II. IFNγ promotes cell-mediated immunological activities essential for antibacterial defense. The IFNγ-activated JAKs also activate PI3K in a STAT1-independent manner culminating in mTORC1 activation, promoting protein translation (125). IFNγ activates PKC leading to MAPK activation, which is also commonly activated by IL-1, TLR ligands, and TNFα. IFNγ has a critical role in host defense as the induction of immunoproteasome components and antigen-processing peptidases enhance cellular immune responses against pathogens. Both types of IFNs induce apoptosis and many

ISGs have antiviral, anti-angiogenic, and immunomodulatory functions. IFNs induce apoptosis by activating death receptors such as Fas (CD95), which then activates the adaptor FADD, leading to subsequent caspase-8 activation. Activation of the caspase 8 cascade causes cell death plus release of cytochrome c and mitochondrial DNA (mtDNA). SEA-induced hepatotoxicity is mediated by FasL and the hepatocellular damage is independent of leukocyte recruitment (126). IFNγ disrupts barrier function and ion transport in superantigen-activated epithelial cells. *In vitro*, disturbances in epithelial barrier function can be duplicated with IFNγ plus TNFα (106). Recent studies suggest that IFNγ downregulates regulatory T-cells (Treg), accounting for the potent polarizing effects of IFNγ on TH1/TH2 cell differentiation and an inflammatory environment for cell migration and activation (127).

IL-2 binds to IL-2R and activates PI3K and Ras (128). Activation of the PI3K/Akt/mTORC1 pathway and Ras controls proliferation, growth, and differentiation of many cell types. Ras activates the MAPK and ERK cascades, leading to transcriptional activation of AP-1, cJun/Fos, and NFAT. The MAPK cascade induces ER stress and NFκB (129). IL-2 induces vasodilation and increases microvascular permeability by suppressing endothelin-1, causing perivascular edema seen in SEB-induced acute lung injury (130, 131). IL-2-deficient mice are resistant to SEB-induced toxic shock, providing evidence for the critical roles of PI3K/mTORC1 and NFκB (132).

The chemokines IL-8, MCP-1, MIP-1α, and MIP-1β are induced directly by SEB or TSST-1. These chemoattractant chemokines activate leukocytes and influence migration of neutrophils, dendritic cells, and leukocytes (133, 134). Chemokines bind to seven-transmembrane GPCR, induce early Ca^{2+} flux, activate PLC and signal *via* the PI3K/mTORC1 pathway (133). Cytokine- and chemokine-activated neutrophils, recruited to sites of tissue injury and inflammation, produce reactive oxygen species (ROS) and activate MMPs contributing to organ dysfunction. MMPs cause tissue degradation and change chemokine interactions with the extracellular matrix (ECM), creating a chemokine gradient affecting cell recruitment (133). Exudates from superantigen-injected air pouches are predominantly neutrophils with some macrophages (135). Both systemic and intranasal administration of SEB cause acute lung injury characterized by increased: (1) expression of adhesion molecules ICAM-1 and VCAM; (2) neutrophil and mononuclear cell infiltrate; (3) endothelial cell injury; and (4) vascular permeability (135–137).

The PI3K/mTORC1 pathway plays a dominant role in superantigen-induced cell proliferation and migration as TCR, CD28, IL-2R, IFNγR, and chemokine receptors all signal through this pathway. IL-1 and TNFα independently activate NFκB *via* MyD88/TRAF6/IRAK and FADD/TRADD/RIP, respectively (87). TCR and CD28 *via* PKC also activate NFκB signaling. Other PRR, including surface and cytosolic TLRs, signal *via* MyD88 to activate NFκB *via* different adaptors. In addition, TLR3 and TLR4 activate TRIF (toll/IL-1 receptor homology domain-containing adaptor inducing interferon β) signaling, inducing interferon regulatory factors (IRFs) (114). The overlap of PRR signaling by NFκB and IRFs with TCR and co-stimulatory signals by superantigens to activate NFκB and mTORC1 cannot be understated. These pathways activate inflammatory genes, as well as antiviral, anti-apoptotic, and pro-apoptotic molecules. Thus the three initial signals provided by TCR, costimulatory receptors, and cytokines converge on NFκB and mTORC1 to promote host defense against superantigens.

IN VIVO EFFECTS OF SUPERANTIGEN

Injection of SEB into mice has been used to study activation-induced apoptosis and T-cell anergy *in vivo*. This effect may be linked to a rapid (within 1 h) loss of L-selectin on the surface of specific Vβ-bearing T cells, thus resulting in decreased signal transduction (138). *Via* endocytosis, surface levels of TCR-CD3 decrease ~50% among Vβ-reactive T cells within 30 min after SEB exposure (139). The rapid hyperactivation and proliferation of T cells in mice following an SEB injection is transient, as within 48 h the majority of proliferating T cells is eliminated by activation-induced cell death (140). CD95 mediates elimination of SEB-activated T cells and the residual Vβ-specific cells become anergic, or functionally unresponsive. However, controversy still exists regarding the functional ability and fate of these anergic T cells. After injection of SEB into mice, splenic $V\beta8^+$ T cells are deleted or no longer respond to SEB, and produce less IL-2 and IFNγ. In contrast, others report that these anergic cells synthesize less IL-2 but still secrete IFNγ that can mediate toxic shock following a subsequent dose of SEB (141). An evident paradox is that the anti-inflammatory cytokine IL-10, which protects against SE-induced shock (142), is also produced by SEB-primed T cells. Treg cells likely downregulate superantigen activation *in vitro* and *in vivo*.

Immune homeostasis is maintained by Treg cells through a number of mechanisms depending on the "immune environment" (143). Natural and induced $Foxp3^+$ Treg cells control excessive immune activation deleterious to the host. Treg cells also downregulate autoimmune responses, and superantigens can subvert the functional activity of Treg cells in atopic dermatitis (144). Although the immunosuppressive function of Treg cells is well known, the mechanisms of action and involvement of different cell types are still debated. The markers used in identifying different Treg cells are evolving, but the surface receptor cytotoxic T lymphocyte antigen 4 (CTLA4) likely plays a role particularly in superantigen-related counter regulation (145). CTLA4 interacts with costimulatory molecules CD80 and CD86, preventing costimulatory signals in superantigen-activated T cells (146). Systemic inflammatory responses elicited by SEB actually destabilize Treg cells and additional expansion of them *in vivo* does not protect transgenic mice from SEB-induced toxic shock (147).

SUPERANTIGEN-INDUCED AUTOIMMUNITY

The ability of superantigens to cross-link MHC class II and specific TCR Vβ enables these microbial toxins to stimulate the immune system and induce autoimmunity by activating APC and normally quiescent, autoreactive T- and B-cells. Activation results in cytokine and chemokine release, thus mediating a potently acute inflammatory response. Several experimental animal models show that staphylococcal superantigens are arthrogenic (148, 149). TSST-1 exacerbates bacterial cell wall-induced arthritis in rats, possibly linked to accumulating Vβ11$^+$ T cells and IFNγ production within arthritic joints (149). TSST-1 also plays a pivotal role in murine septic arthritis, as the frequency and severity of this disease are increased after intravenous administration of TSST-1-secreting *S. aureus* (150). SEA or SEB can also induce relapses of experimental autoimmune encephalomyelitis in a murine model for multiple sclerosis (151). How exogenously administered toxin triggers autoimmune processes like arthritis is unknown, but it is likely that proinflammatory cytokines and chemokines produced in response to superantigens facilitate specific recruitment and migration of autoreactive T-cells into synovial tissue and joints. Thus, proinflammatory cytokines and chemokines have acute toxic effects promoting cell activation and recruitment. Following minor tissue injury or inflammation due to superantigen exposure, an increased presence of immune cells might initiate a destructive autoimmune reaction.

Psoriasis and atopic dermatitis also represent autoimmune diseases linked to staphylococcal and streptococcal colonization of skin and subsequent production of superantigens (152). SEB on healthy human skin induces inflammatory reactions, possibly linked to degranulation of cutaneous mast cells (58). T cells from patients with severe atopic dermatitis are apoptotic, which may lead to chronic infections and subsequent worsening of disease (153). Bacterial density on skin can affect relative sensitivity toward these toxins and development of atopic dermatitis.

PATHWAY ANALYSIS YIELDS POTENTIAL DRUG TARGETS

There are currently no available therapeutics for treating superantigen-induced shock, except for the use of intravenous human immunoglobulin (154, 155). Targeting and neutralizing a superantigen directly is most suitable at early stages of exposure before cell activation and initiation of the proinflammatory cytokine cascade. Therapies targeting superantigens at the receptor level have been extensively reviewed recently (156) and include receptor-blocking peptides derived from toxins, chimeric inhibitors composed of Vβ and MHC class II domains, as well as synthetic blockers of co-stimulatory receptor CD28. However, preventing toxin-receptor interaction is ineffective post-toxin exposure and some inhibitors must be tailored to target individual toxins. Failed sepsis clinical trials of eritoran (anti-endotoxin), a drug that prevents the early steps of receptor interaction, suggest that blocking superantigen–receptor interactions will likely not protect against SEB-induced shock (157).

An important class of therapeutic compounds blocks signal transduction pathways activated by superantigens, and as these events are post-exposure, they are perhaps more amenable to suppressive manipulation. One example is the NFκB cascade containing many upstream activators. *In vitro* studies indicate numerous genes for cell adhesion molecules, cytokines, chemokines, acute phase proteins, and inducible nitric oxide synthase that are implicated in superantigen-induced lethal shock and contain NFκB binding sites in the promotor/enhancer region (84). Activation of NFκB leads to the inducible expression of many mediators involved in inflammation and tissue injury seen in SEB-induced lung injury and toxic shock models. Inhibition of NFκB is beneficial, or has no effect, in preventing SEB-induced shock depending upon the mouse model (136, 158, 159); however, NFκB inhibitor must be given early and for a long duration to afford any protection (136).

Pathway inhibitors are used for identifying molecules and signaling pathways crucial for cellular responses to a specific stimulus. The superantigenic properties of SEB make it an "ideal" toxin to study cell activation signals and molecular pathways. An obvious step in testing new therapeutic approaches for SEB-induced shock is finding relevant animal models that mimic important aspects of human disease. Mice are much less susceptible to SEB due to lower affinity of superantigen for mouse MHC class II (51, 62, 160). Potentiating agents such as LPS, viruses, D-gal, and actinomycin D are used together with SEB to sensitize mice (50–52). These synergistic agents alone activate similar host signaling pathways as superantigens, confounding the "pure" host response to SEB. An alternative model utilizes transgenic mice, with either human HLA-DR3 or -DQ8, that lethally respond to SEs without a potentiating agent (161, 162). Another recent, simplified model employs "double-hit," low dose SEB in C3H/HeJ mice, an LPS-resistant strain (131). Pathological lesions, cytokine response, and time to lethality in this "double-hit" model resemble findings in non-human primates (NHP) and staphylococcal toxic shock syndrome in patients (49). Host signatures arising from "SEB-only" exposure unexpectedly include many IFN-induced genes in multiple organs, not previously linked to SEB pathogenesis (163).

OLIGONUCLEOTIDE MICROARRAY REVEALS SEB-INDUCED DANGER SIGNALS

Oligonucleotide microarray analysis in the "double-hit" SEB mouse model reveals induction of danger signals bearing IFN signatures (163). These genes cover five important molecular hubs signaling danger similar to those activated by IFNs and pathogens. The upregulated transcripts present in PBMC, spleen, lung, liver, kidney, and heart include (1) innate mediators; (2) DNA/RNA sensing system; (3) ER stress; (4) metabolic/oxidative stress; and (5) the apoptosis pathway. Proinflammatory cytokines IL-1, IL-6, TNFα, IFNγ, and IFNγ-induced chemokines (CXCL9, CXCL10, CXCL11) are most prominent in SEB-stimulated PBMC, confirming previous observations *in vitro* and *in vivo* (51, 86, 135, 161, 162, 164). Activation of other innate host-defense genes include the Fc receptor for phagocytosis, MHC class I and II for increased APC function, as well as cell-surface receptors and adhesion molecules to promote cell recruitment. Z-DNA binding protein 1 (ZBP1), a DNA sensor triggering ISGs *via* DNA binding (165), is surprisingly upregulated in all tissues from the "double-hit" SEB mouse model (163). T-cell proliferation requires elevated protein translation and increased metabolism, resulting in misfolded proteins and oxidative stress. ER stress response genes such as ubiquitin ligases, immunoproteasome components, proteasome peptidases, and WARS (tryptophanyl-tRNA synthetase) are likely a result of Ca^{2+} flux, unfolded proteins, and activated PKC following cell activation and increased protein synthesis. Unresolved ER stress activates caspases and apoptosis (166, 167). Enhanced activity of the mitochondrial electron transport chain ultimately leads to oxidative stress, evidenced by activated NADPH oxidase in lung (163). MMP, cathepsins, and other proteases from lysosomes are also seen in this mouse model and *in vitro* stimulation of PBMC with SEA or SEB (168, 169). Increased ROS and protease levels are major factors in organ injury. Although the "double-hit" mouse model is not perfect, it mimics many *in vivo* responses of NHP to SEB and there are many similarities of gene expression by mouse and human PBMC incubated with SEB.

The cell death pathway triggered *in vitro* and *in vivo* includes genes associated with apoptosis such as FADD, death receptor ligand TRAIL, caspases, CARD, and phospholipid scramblase 1 (PLSCR1). These genes are activated in PBMC and major organs from the "double-hit" SEB model and human PBMC, following SEA or SEB exposure (163, 168, 169). PLSCR1 is implicated in moving phosphatidyl serine to the outside plasma membrane of apoptotic cells. Other danger signals include K^+ efflux, particulate ligands such as cholesterol crystals and lysosome destabilization that triggers inflammasome activation *via* NLRP3 which converts pro-IL-1β to IL-1β through caspase 1 (117). Catabolic enzymes cause destruction of cell matrix, accounting for the liver and lung injury seen in SEB-mediated shock.

Induction of ZBP1 leads to binding of DNA and endosomal TLR9, upregulating antiviral genes *via* the IRF3 and IRF7 (114, 170). TLR9 activation also promotes NFκB-mediated cytokine gene transcription and inflammasome activation (171). Mitochondria are the most likely source of cytosolic DNA induced by SEB, as ER stress and increased mitochondrial respiration due to T-cell proliferation and activation are known to induce ROS (172). Elevated ROS activates the intrinsic cell death pathway *via* caspase 9, leading to mitochondrial damage that releases cytochrome c and mtDNA (173). MtDNA acts as a direct inducer of ZBP1. In addition, mtDNA has motifs similar to bacterial DNA (CpG), binds cytosolic TLR9, and promotes activation of NFκB and IRFs. Mitochondrial ROS is also a potent inducer of the inflammasome NLRP3 (117). Both NLRP3 and CARD are identified in microarray studies *in vivo* and *in vitro* with superantigen-stimulated PBMC. Cathepsins, another category of "destroyer" molecules identified by microarray in SEB-activated PBMC (169), are indicative of lysosomal rupture which also activates inflammasomes.

Upon comparing the microarray data from an *in vivo* "SEB-only" mouse model (163), and human PBMC stimulated with SEA (168) or SEB (169), several commonly activated genes importantly induce the pathogenic effects of SEB (**Table 2**). These

TABLE 2 | Common differentially expressed genes induced by superantigens *in vitro* and *in vivo*.

Pathway/network	Gene	Major function
Innate response	IL6, TNFα, LTA, IL17A, IL22	Host defense, inflammation
	IFNγ	Host defense, antimicrobial
	CXCL11, CXXC5, CCL7, XCL1	Host defense, cell migration
	CISH, CIITA, GBP2, TRAF1, RGS16	Signal transduction
	PDE4DIP, PDE4B, PTGER3, P2RY14	Signal transduction
	NEDD9, GNAS, CSF1R	Signal transduction
	STAT1, STAT2, STAT3, IRF7	Transcription factor (TF)
	BATF, BATF2	IFN-inducible TF
	SOCS1, SOCS2, SOCS3	JAK/STAT counter-regulator
	CD69, CD74, ICAM	Immune regulation
	NRP2	Vascular signaling
	Rel A, Rel, NFκBia	NFκB regulator
DNA damage response	RIPK2	DNA sensor interactor
	CTPS, UPP1	Nucleic acid synthesis
	PIM1, PIM2	DNA repair/assembly
	GADD45G	DNA repair adaptor
ER stress/oxidative stress	SIAH2	Ubiquitin E3 ligase
	KCNE4	Membrane integrity
	JunB	Stress response TF
	MGST1	Cell protection
Metabolic stress	IL2, IL2RA, MACF1	Cell proliferation regulator
	FABP4, CD36	Fatty acid metabolism
	HK1, PDK4, PGS1	Cell metabolism
	TARS, NDST2	Synthetase
Apoptosis	PLSCR1, NR4A1	Membrane integrity
	CD40, TNFRSF9	TNFRSF, death receptor
	Casp 4, CFLAR	Caspase regulator
	VCAN, LMNB1	Cell matrix breakdown
	BCL2, BCL6	Anti-apoptotic regulator
	CCND2	Cell cycle regulator
	PLA2G7	Cardiovascular damage
Others	ARID5A, ZBTB32, NDST2	

ARID5A, AT-rich interactive domain 5A (MRF1-like); BATF2, basic leucine zipper transcription factor (TF); CCND2, cAMP specific cyclin D2; CFLAR, caspase 8 and FADD-like apoptosis regulator; CIITA, MHC class II transactivator; CISH, cytokine inducible SH2-containg protein; CTPS, cytidine 5′-triphosphate synthase; FABP4, fatty acid binding protein 4; GADD45G, growth arrest and DNA-damage-inducible 45 gamma; GBP2, IFN-inducible guanylate binding protein 2; GNAS, guanine nucleotide binding protein; HK1, hexokinase 1; KCNE3, potassium voltage-gated channel; LMNB1, lamin B1; MACF1, microtubule-actin crosslinking factor 1; MGST1, microsomal glutathione-S-transferase 1; NDST2, N-deacetylase/N-sulfotransferase; NEDD9 (HEF1), neural precursor cell expressed; NR4A1, nuclear receptor subfamily 4, group A, membrane 1; NRP2, neuropilin transmembrane protein receptor; P2RY14, purinergic receptor P2Y, G protein coupled; PDE4DIP, phosphodiesterase 4D interacting protein; PDE4B, cAMP specific phosphodiesterase 4B; PDK4, pyruvate dehydrogenase kinase isoenzyme 4; PGS1, phosphatidylglycerophosphate synthase 1; PIM1, pro-viral DNA integration; PLA2G7, phospholipase A2 group VII; PLSCR1, phospholipid scramblase; PIM1, proviral integration site 1; PTGER3, prostaglandin E receptor 3; RGS16, regulator of G protein signaling 16; RIPK2 (RIP2), TNFRSF-interacting protein kinase; SIAH2, seven in absentia 2; SLC30A1, solute carrier family 30 (zinc transporter); SOCS1, suppressor of cytokine signaling 1; TARS, threonyl-tRNA synthetase; UPP1, uridine phosphorylase 1; VCAN, verican; ZBTB32, zinc finger and BTB domain containing 32.

genes are activated by IFN and account for the therapeutic effectiveness of rapamycin in preventing SEB-induced shock (102).

DRUG TARGETS: WHAT WORKS AND WHAT DOES NOT

The early induction of three key proinflammatory cytokines, IL-1, TNFα, and IFNγ, work *via* individual receptor-mediated signaling molecules that, respectively, activate distinct pathways: (1) IL-1R/MyD88/NFκB; (2) TNFαR/FADD/RIP; and (3) IFNγR/JAK/IRF. IL-1, TNFα, and IFNγ have independent and synergistic effects signaling inflammation, caspase activation/cell death, and an antiviral response. IL-1 from inflammasome activation has pleiotropic effects, while TNFα has an established role that initiates cell death through the adaptor FADD, activating caspases 3 and 8. IFNγ triggers innate host defense responses, antiviral genes, apoptotic programs, immunoproteasomes, and has many immunomodulatory functions. Interruption of these concurrent cascades early after SEB exposure is effective in preventing SEB-induced lethal shock.

Decades of drug development against sepsis and septic shock point to the failure of using anti-inflammatory cytokines alone, and early interruption of cytokine release is perhaps a necessary but insufficient target. Drugs also act on multiple targets, some known and unknown, as exemplified by statins that inhibit 3-hydroxy-3-methylglutaryl-coenzyme A with anti-inflammatory effects (174). Statins, used therapeutically to reduce cholesterol, are under consideration for treating various inflammatory diseases (175). Superantigens trigger multiple pathways that cross-regulate each other positively and negatively, thus targeting downstream effectors might be more specific and perhaps interfere less with normal cell function. Knowledge of immunoregulation within activated pathways by SEB enables a better choice of inhibitors.

Apoptosis plays a critical role in sepsis-induced lethality (176). The two pathways leading to apoptosis are operative in SEB-induced lethal shock, as induction of genes for both pathways occurs in the "double-hit" SEB model and other mouse models employing potentiating agents such as D-gal. The death receptor pathway used by the TNFR superfamily with ligands like TNFα and FasL induces cell death following superantigen exposure (121, 126). A second apoptosis pathway, the intrinsic pathway, is dependent on mitochondria and the Bcl2 family of pro- and anti-apoptotic proteins. ROS and loss of mitochondrial transmembrane potential play important roles, as previous studies implicate both in SEB pathogenesis (177). Other cell-surface receptors like CD44 might also contribute to SEB-induced injury, as CD44-knockout mice have elevated liver damage in the D-gal-sensitized SEB model (178). Since D-gal is hepatotoxic, the induction of pro-apoptotic molecules by SEB likely act synergistically with D-gal to promote cell death.

Microarray gene analysis in an "SEB-only" mouse model implicates both extrinsic and intrinsic pathways in SEB-induced apoptosis. Apoptosis plays a role in down-regulating immune responses but simultaneously has devastating effects when apoptotic cells or associated molecules are not removed. Autophagy is a cellular mechanism that removes bacteria, protein aggregates, and damaged organelles to maintain homeostasis (176, 179). A recent study indicates that blocking autophagy augments T-cell activation (180). Degradation and removal of Bcl10, which is part of CBM and a critical component for TCR and costimulatory signaling, is dependent on autophagy (176). IFN induces many genes regulating NFκB and apoptosis. The damage response induced by superantigens likely starts with inflammatory cytokines and apoptotic programs activated by IFNγ and TNFα. DAMPs such as mitochondrial ROS and mtDNA trigger more apoptosis, activate inflammasomes, and induce transcription factors for ISGs. Increasing energy demand and mitochondrial respiratory-chain activity also lead to elevated ROS. Normally, mitochondria damaged by excessive membrane permeability and ROS are removed by a specialized form of autophagy called mitophagy. However, overactivation of PI3K/mTORC1 in superantigen-stimulated cells likely blocks autophagy, resulting in inflammasome activation and accumulation of damaged mitochondria. Rapamycin, a well-known inducer of autophagy, prevents SEB-induced shock by removing damage-inducing molecules and damaged mitochondria. Two other FDA-approved immunosuppressants for organ transplants, cyclosporine A and tacrolimus, do not protect against superantigen-induced shock in NHP and human HLA-DR3 transgenic mice, respectively (181, 182). The calcineurin inhibitor, cyclosporine A, protects D-gal-sensitized mice from SEB-induced shock (121) but does not protect NHP challenged with SEB (181). Tacrolimus suppresses SEB-induced T-cell proliferation *in vitro* but does not confer protection from toxic shock in transgenic mice (182). Tacrolimus also fails to protect mice from lethal pneumonia induced by superantigen-producing *S. aureus* (182).

There is good agreement between the genes significantly induced in mouse PBMC from this "double-hit" SEB model and SEA- or SEG-stimulated human PBMC (163, 168). Many genes of the apoptosis-related cell death pathway account for the damage response initiated by SEB. MtDNA is ancestrally related to bacterial DNA (CpG motifs), inducing a "foreign" DNA sensor (ZBP1) or alternatively binding endosomal TLR9 that triggers host defense including type 1 IFN-mediated responses *via* IRF3 (114). Important clues from animal models, old and new, reveal acute release of proinflammatory cytokines that culminate in damaged organs and lethal shock.

CONCLUSION

Staphylococcus aureus is a toxin-producing pathogen that causes various diseases found throughout the body. Increasingly, *S. aureus* becomes more resistant to various therapeutics (i.e., antibiotics) over time, thus our own immune systems must more effectively clear this pathogen. Further knowledge of our immune system will clearly enable us to better thwart *S. aureus* and other pathogens. Mammals sense invading microbes *via* conserved PRR for detecting molecular patterns on, or released by, various bacterial, viral, and fungal pathogens. This rapid innate immune response produces proinflammatory mediators, cell activation, and recruitment of inflammatory cells to infection sites. However, some sensors for detecting PAMPs also bind to host DAMPs, confusing the "stranger" versus "danger" signaling. Host response to staphylococcal superantigens typifies the generation of these danger signals as shown in a mouse model of SEB-induced shock. The induction of cell death through apoptotic proteins observed during sepsis or superantigen exposure may provide a common target for therapeutic intervention.

AUTHOR CONTRIBUTIONS

All authors listed have made substantial, direct and intellectual contributions to the work, and approved it for publication.

ACKNOWLEDGMENTS

We thank DTRA for funding. Brad Stiles acknowledges Wilson College for the continuing support of this, and many other, endeavors.

REFERENCES

1. Uchiyama T, Imanishi K, Miyoshi-Akiyama T, Kata H. Staphylococcal superantigens and the diseases they cause. 3rd ed. In: Alouf JE, Popoff MR, editors. *The Comprehensive Sourcebook of Bacterial Protein Toxins*. London: Academic Press (2006). p. 830–43.
2. Archer NK, Mazaitis MJ, Costerton JW, Leid JG, Powers ME, Shirtliff ME. *Staphylococcus aureus* biofilms: properties, regulation and roles in human disease. *Virulence* (2011) **2**(5):445–59. doi:10.4161/viru.2.5.17724
3. Thurlow LR, Joshi GS, Richardson AR. Virulence strategies of the dominant USA300 lineage of community-associated methicillin-resistant *Staphylococcus aureus* (CA-MRSA). *FEMS Immunol Med Microbiol* (2012) **65**:5–22. doi:10.1111/j.1574-695X.2012.00937.x
4. Chambers HF. The changing epidemiology of *Staphylococcus aureus*? *Emerg Infect Dis* (2001) **7**:178–82. doi:10.3201/eid0702.010204
5. Abdelhady W, Bayer AS, Seidl K, Nast CC, Kiedrowski MR, Horswill AR, et al. Reduced vancomycin susceptibility in an in vitro catheter-related biofilm model correlates with poor therapeutic outcomes in experimental endocarditis due to methicillin-resistant *Staphylococcus aureus*. *Antimicrob Agents Chemother* (2013) **57**:1447–54. doi:10.1128/AAC.02073-12
6. Windolf CD, Loegters T, Schoolz M, Windolf J, Flohe S. Lysostaphin-coated titan-implants preventing localized osteitis by *Staphylococcus aureus* in a mouse model. *PLoS One* (2014) **9**(12):e115940. doi:10.1371/journal.pone.0115940
7. Kloos WE, Smith PB. Staphylococci. In: Lennette EH, Balows A, Hausler WJ, Truant JP, editors. *Manual of Clinical Microbiology*. Washington, DC: American Society for Microbiology (1980). p. 83–7.
8. David MZ, Medvedev S, Hohmann SF, Ewigman B, Daum RS. Increasing burden of methicillin-resistant *Staphylococcus aureus* hospitalizations at US academic medical centers, 2003-2008. *Infect Control Hosp Epidemiol* (2012) **33**:782–9. doi:10.1086/666640
9. Ogston A. *Micrococcus* poisoning. *J Anat Physiol* (1882) **16**:526–67.
10. Lee BY, Singh A, David MZ, Bartsch SM, Slayton RB, Huang SS, et al. The economic burden of community-associated methicillin-resistant *Staphylococcus aureus* (CA-MRSA). *Clin Microbiol Infect* (2013) **19**:528–36. doi:10.1111/j.1469-0691.2012.03914.x
11. Antonanzas F, Lozano C, Torres C. Economic features of antibiotic resistance: the case of methicillin-resistant *Staphylococcus aureus*. *Pharmacoeconomics* (2015) **33**(4):285–325. doi:10.1007/s40273-014-0242-y
12. Cornejo-Juarez P, Vilar-Compte D, Perez-Jimenez C, Namendys-Silva SA, Sandoval-Hernandez S, Volkow-Fernandez P. The impact of hospital acquired infections with multidrug-resistant bacteria at an oncological intensive care unit. *Int J Infect Dis* (2015) **31**:31–4. doi:10.1016/j.ijid.2014.12.022
13. Mangen MJ, Bouwknegt M, Friesema IH, Haagsma JA, Kortbeek LM, Tariq L, et al. Cost-of-illness and disease burden of food-related pathogens in the Netherlands, 2011. *Int J Food Microbiol* (2015) **196**:84–93. doi:10.1016/j.ijfoodmicro.2014.11.022
14. Bouza E. New therapeutic choices for infections caused by methicillin-resistant *Staphylococcus aureus*. *Clin Microbiol Infect* (2009) **7**:44–52. doi:10.1111/j.1469-0691.2009.03091.x
15. Rivera AM, Boucher HW. Current concepts in antimicrobial therapy against select gram-positive organisms: methicillin-resistant *Staphylococcus aureus*, penicillin-resistant pneumococci, and vancomycin-resistant enterococci. *Mayo Clin Proc* (2011) **86**(12):1230–42. doi:10.4065/mcp.2011.0514
16. Samy RP, Stiles BG, Gopalakrishnakone P, Chow VT. Antimicrobial proteins from snake venoms: direct bacterial damage and activation of innate immunity against *Staphylococcus aureus* skin infection. *Curr Med Chem* (2011) **18**:5104–13. doi:10.2174/092986711797636108
17. Hiramatsu K, Igarashi M, Morimoto Y, Baba T, Umekita M, Akamatsu Y. Curing bacteria of antibiotic resistance: reverse antibiotics, a novel class of antibiotics in nature. *Int J Antimicrob Agents* (2012) **39**:478–85. doi:10.1016/j.ijantimicag.2012.02.007
18. Arunava K. Antibiotics and bioactive natural products in treatment of methicillin resistant *Staphylococcus aureus*: a brief review. *Pharmacogn Rev* (2015) **9**:29–34. doi:10.4103/0973-7847.156329
19. Coulibaly K, Zirihi G, Guessennd-Kouadio N, Oussou KR, Dosso M. Antibacterial properties of trunk barks of *Terminalia ivorensis*, a commercial and medicinal species on methicillin-resistant staphylococci species. *Afr Health Sci* (2014) **14**:753–6. doi:10.4314/ahs.v14i3.35
20. Krychowiak M, Grinholc M, Banasiuk R, Krauze-Baranowska M, Glod D, Kawiak A, et al. Combination of silver nanoparticles and *Drosera binata* extract as a possible alternative for antibiotic treatment of burn wound infections caused by resistant *Staphylococcus aureus*. *PLoS One* (2014) **9**:e115727. doi:10.1371/journal.pone.0115727
21. Rose T, Verbeken G, De Vos D, Merabishvilli M, Vaneechoutte M, Lavigne R, et al. Experimental phage therapy of burn wound infection: difficult first steps. *Int J Burns Trauma* (2014) **4**:66–73.
22. Rajamuthiah R, Fuchs BB, Conery AL, Kim W, Jayamani E, Kwon B, et al. Repurposing salicylanilide antihelmintic drugs to combat drug resistant *Staphylococcus aureus*. *PLoS One* (2015) **10**:e0124595. doi:10.1371/journal.pone.0124595
23. Rempe CS, Burris KP, Woo HL, Goodrich B, Gosnell DK, Tschaplinski TJ, et al. Computational ranking of yerba mate small molecules based on their predicted contribution to antibacterial activity against methicillin-resistant *Staphylococcus aureus*. *PLoS One* (2015) **10**:e0123925. doi:10.1371/journal.pone.0123925
24. Klevens RM, Morrison MA, Nadle J, Petit S, Gershman K, Ray S, et al. Invasive methicillin-resistant *Staphylococcus aureus* infections in the United States. *JAMA* (2007) **298**:1763–71. doi:10.1001/jama.298.15.1763
25. CDC. *Active Bacterial Core Surveillance (ABCs) Report Emerging Infections Program Network Methicillin-Resistant Staphylococcus aureus* (2011). Available from: http://www.cdc.gov/abcs/reports-findings/survreports/mrsa11.pdf
26. Jevons MP. Celbenin-resistant staphylococci. *Br Med J* (1961) **1**:124–5. doi:10.1136/bmj.1.5219.124-a
27. Lowy FD. Antimicrobial resistance: the example of *Staphylococcus aureus*. *J Clin Invest* (2003) **111**:1265–73. doi:10.1172/JCI18535
28. Attien P, Sina H, Moussaoui W, Zimmermann-Meisse G, Dadie T, Keller D, et al. Mass spectrometry and multiplex antigen assays to assess microbial quality and toxin production of *Staphylococcus aureus* strains isolated from clinical and food samples. *Biomed Res Int* (2014) **2014**:485620. doi:10.1155/2014/485620
29. Kadariya J, Smith TC, Thapaliya D. *Staphylococcus aureus* and staphylococcal food-borne disease: an ongoing challenge in public health. *Biomed Res Int* (2014) **2014**:827965. doi:10.1155/2014/827965
30. Nadimpalli M, Rinsky JL, Wing S, Hall D, Stewart J, Larsen J, et al. Persistence of livestock-associated antibiotic-resistant *Staphylococcus aureus* among

industrial hog operation workers in North Carolina over 14 days. *Occup Environ Med* (2015) **72**:90–9. doi:10.1136/oemed-2014-102095

31. Peton V, Le Loir Y. *Staphylococcus aureus* in veterinary medicine. *Infect Genet Evol* (2014) **21**:602–15. doi:10.1016/j.meegid.2013.08.011
32. Bos ME, Verstappen KM, van Cleef BA, Dohmen W, Dorado-Garcia A, Graveland H, et al. Transmission through air as a possible route of exposure for MRSA. *J Expo Sci Environ Epidemiol* (2014). doi:10.1038/jes.2014.85
33. Creamer E, Shore AC, Deasy EC, Galvin S, Dolan A, Walley N, et al. Air and surface contamination patterns of methicillin-resistant *Staphylococcus aureus* on eight acute hospital wards. *J Hosp Infect* (2014) **86**:201–8. doi:10.1016/j.jhin.2013.12.005
34. Patchanee P, Tadee P, Arjkumpa O, Love D, Chanachai K, Alter T, et al. Occurrence and characterization of livestock-associated methicillin-resistant *Staphylococcus aureus* in pig industries of Northern Thailand. *J Vet Sci* (2014) **15**:529–36. doi:10.4142/jvs.2014.15.4.529
35. Van der Mee-Marquet NL, Corvaglia A, Haenni M, Bertrand X, Franck J-B, Kluytmans J, et al. Emergence of a novel subpopulation of CC398 *Staphylococcus aureus* infecting animals is a serious hazard for humans. *Front Microbiol* (2014) **5**:00652. doi:10.3389/fmicb.2014.00652
36. Dinges MM, Orwin PM, Schlievert PM. Exotoxins of *Staphylococcus aureus*. *Clin Microbiol Rev* (2000) **13**:16–34. doi:10.1128/CMR.13.1.16-34.2000
37. Graham PL, Lin SX, Larson EL. A U.S. population-based survey of *Staphylococcus aureus* colonization. *Ann Intern Med* (2006) **144**:318–25. doi:10.7326/0003-4819-144-5-200603070-00006
38. Tang YW, Stratton CW. *Staphylococcus aureus*: an old pathogen with new weapons. *Clin Lab Med* (2010) **30**:179–208. doi:10.1016/j.cll.2010.01.005
39. Soltani B, Ardakani AT, Moravveji A, Erami M, Rezaei MH, Moniri R, et al. Risk factors for methicillin resistant *Staphylococcus aureus* nasal colonization of healthy children. *Jundishapur J Microbiol* (2014) **7**(9):e20025. doi:10.5812/jjm.20025
40. Sollid JUE, Furberg AS, Hanssen AM, Johannessen M. *Staphylococcus aureus*: determinants of human carriage. *Infect Genet Evol* (2014) **21**:531–41. doi:10.1016/j.meegid.2013.03.020
41. Barber MA. Milk poisoning due to a type of *Staphylococcus albus* occurring in the udder of a healthy cow. *Philippine J Sci* (1914) **9**:515–9.
42. Dack GM, Cary WE, Woolpert O, Wiggers H. An outbreak of food poisoning proved to be due to a yellow hemolytic staphylococcus. *J Prev Med* (1930) **4**:167–75.
43. Bergdoll MS, Kadavy JL, Surgalla MJ, Dack GM. Partial purification of staphylococcal enterotoxin. *Arch Biochem Biophys* (1951) **33**:259–62. doi:10.1016/0003-9861(51)90104-X
44. Bergdoll MS, Sugiyama H, Dack GM. Staphylococcal enterotoxin. 1. purification. *Arch Biochem Biophys* (1959) **85**:62–9. doi:10.1016/0003-9861(59)90447-3
45. Thatcher FS, Matheson BH, Simon WR. Studies with staphylococcal toxins: III. The application of paper ionophoresis to the resolution of components of toxic concentrates. *Can J Microbiol* (1955) **1**:401–14. doi:10.1139/m55-051
46. Fraser JD, Proft T. The bacterial superantigen and superantigen-like proteins. *Immunol Rev* (2008) **225**:226–43. doi:10.1111/j.1600-065X.2008.00681.x
47. Hennekinne J-A, de Buyser M-L, Dragacci S. *Staphylococcus aureus* and its food poisoning toxins: characterization and outbreak investigation. *FEMS Microbiol Rev* (2012) **36**:815–36. doi:10.1111/j.1574-6976.2011.00311.x
48. Madsen JM. Toxins as weapons of mass destruction. A comparison and contrast with biological-warfare and chemical-warfare agents. *Clin Lab Med* (2001) **21**:593–605.
49. Ulrich RG, Wilhelmsen CL, Krakauer T. Staphylococcal enterotoxin B and related toxins. In: Zygmund D, editor. *Textbook of Military Medicine: Medical Aspects of Biological Warfare*. Washington, DC: US Department of Army, Borden Institute (2007). p. 311–22.
50. Schlievert PM. Enhancement of host susceptibility to lethal endotoxin shock by staphylococcal pyrogenic exotoxin type C. *Infect Immun* (1982) **36**:123–8.
51. Stiles BG, Bavari S, Krakauer T, Ulrich RG. Toxicity of staphylococcal enterotoxin potentiated by lipopolysaccharide: major histocompatibility complex class II molecule dependency and cytokine release. *Infect Immun* (1993) **61**:5333–8.
52. Sarawar SR, Blackman MA, Doherty PC. Superantigen shock in mice with an inapparent viral infection. *J Infect Dis* (1994) **170**:1189–94. doi:10.1093/infdis/170.5.1189
53. Loir L, Baron F, Gautier M. *Staphylococcus aureus* and food poisoning. *Genet Mol Res* (2003) **2**:63–76.
54. Swaminathan S, Furey W, Pletcher J, Sax M. Crystal structure of staphylococcal enterotoxin B, a superantigen. *Nature* (1992) **359**:801–6. doi:10.1038/359801a0
55. Lina G, Bohach GA, Nair SP, Hiramatsu K, Jouvin-Marche E, Mariuzza R. Standard nomenclature for the superantigens expressed by *Staphylococcus*. *J Infect Dis* (2004) **189**:2334–6. doi:10.1086/420852
56. Bergdoll MS. Monkey feeding test for staphylococcal enterotoxin. *Methods Enzymol* (1988) **165**:324–33. doi:10.1016/S0076-6879(88)65048-8
57. Schelin J, Wallin-Carlquist N, Cohn MT, Lindqvist R, Barker GC, Radstrom P. The formation of *Staphylococcus aureus* enterotoxin in food environments and advances in risk assessment. *Virulence* (2011) **2**(6):580–92. doi:10.4161/viru.2.6.18122
58. Scheuber PH, Denzliner C, Wilker D, Beck G, Keppler D, Hammer DK. Staphylococcal enterotoxin B as a nonimmunological mast cell stimulus in primates: the role of endogenous cysteinyl leukotrienes. *Int Arch Allergy Appl Immunol* (1987) **82**:289–91. doi:10.1159/000234209
59. Jett M, Brinkley W, Neill R, Gemski P, Hunt R. *Staphylococcus aureus* enterotoxin B challenge of monkeys: correlation of plasma levels of arachidonic acid cascade products with occurrence of illness. *Infect Immun* (1999) **58**:3494–9.
60. Hu DL, Zhu G, Mori F, Omoe K, Okada M, Wakabayashi K, et al. Staphylococcal enterotoxin induces emesis through increasing serotonin release in intestine and it is downregulated by cannabinoid receptor 1. *Cell Microbiol* (2007) **9**:2267–77. doi:10.1111/j.1462-5822.2007.00957.x
61. Hu DL, Nakane A. Mechanisms of staphylococcal enterotoxin-induced emesis. *Eur J Pharmacol* (2014) **722**:95–107. doi:10.1016/j.ejphar.2013.08.050
62. Choi Y, Kotzin B, Herron L, Callahan J, Marrack P, Kappler J. Interaction of *Staphylococcus aureus* toxin "superantigens" with human T cells. *Proc Natl Acad Sci USA* (1989) **86**:8941–5. doi:10.1073/pnas.86.22.8941
63. Langley RJ, Fraser JD, Proft T. Bacterial superantigens and superantigen-like toxins. 4th ed. In: Alouf J, Ladant D, Popoff MR, editors. *The Comprehensive Sourcebook of Bacterial Protein Toxins*. London: Academic Press (2015). p. 911–74.
64. Pontzer CH, Russell JK, Johnson HM. Structural basis for differential binding of staphylococcal enterotoxin A and toxic shock syndrome toxin 1 to class II major histocompatibility molecules. *Proc Natl Acad Sci USA* (1991) **88**:125–8. doi:10.1073/pnas.88.1.125
65. Jablonsky MJ, Subramaniam PS, Johnson HM, Russell JK, Krishna NR. The solution structure of a class II major histocompatibility complex superantigen binding domain. *Biochem Biophys Res Commun* (1997) **234**:660–5. doi:10.1006/bbrc.1997.6692
66. Jardetzky TS, Brown JH, Gorga JC, Stern LJ, Urban RG, Chi YI, et al. Three-dimensional structure of a human class II histocompatibility molecule complexed with superantigen. *Nature* (1994) **368**:711–8. doi:10.1038/368711a0
67. Smith-Garvin JE, Koretzky GA, Jordan MS. T cell activation. *Annu Rev Immunol* (2009) **27**:591–619. doi:10.1146/annurev.immunol.021908.132706
68. Larkin EA, Carman RJ, Krakauer T, Stiles BG. *Staphylococcus aureus:* the toxic presence of a pathogen extraordinaire. *Curr Med Chem* (2009) **16**:4003–19. doi:10.2174/092986709789352321
69. Chintagumpala MM, Mollick JA, Rich RR. Staphylococcal toxins bind to different sites on HLA-DR. *J Immunol* (1991) **147**:3876–82.
70. Hudson KR, Tiedemann RE, Urban RG, Lowe SC, Strominger JL, Fraser JD. Staphylococcal enterotoxin A has two cooperative binding sites on major histocompatibility complex class II. *J Exp Med* (1995) **182**:711–20. doi:10.1084/jem.182.3.711
71. Mehindate K, Thibodeau J, Dohlsten M, Kalland T, Sékaly RP, Mourad W. Cross-linking MHC class II molecules by staphylococcal enterotoxin A superantigen is a requirement for inflammatory cytokine gene expression. *J Exp Med* (1995) **182**:1573–7. doi:10.1084/jem.182.5.1573
72. Fleischer B, Schrezenmeier H. T cell stimulation by staphylococcal enterotoxins. Clonally variable response and requirement for major histocompatibility complex class II molecules on accessory or target cells. *J Exp Med* (1988) **167**:1697–707. doi:10.1084/jem.167.5.1697
73. Chatila T, Geha RS. Signal transduction by microbial superantigens via MHC class II molecules. *Immunol Rev* (1993) **131**:43–59. doi:10.1111/j.1600-065X.1993.tb01529.x
74. Tiedemann RE, Fraser JD. Cross-linking of MHC class II molecules by staphylococcal enterotoxin A is essential for antigen-presenting cell and T cell activation. *J Immunol* (1996) **157**:3958–66.

75. Carlsson R, Fischer H, Sjogren HO. Binding of staphylococcal enterotoxin A to accessory cells is a requirement for its ability to activate human T cells. *J Immunol* (1998) **140**:2484–8.
76. Anderson MR, Tary-Lehmann M. Staphylococcal enterotoxin-B-induced lethal shock in mice is T-cell-dependent, but disease susceptibility is defined by the non-T-cell compartment. *Clin Immunol* (2001) **98**:85–94. doi:10.1006/clim.2000.4960
77. Cemerski S, Shaw A. Immune synapses in T-cell activation. *Curr Opin Immunol* (2006) **18**:298–304. doi:10.1016/j.coi.2006.03.011
78. Boise LH, Minn AJ, Noel PJ, June CH, Accavitti MA, Lindsten T, et al. CD28 costimulation can promote T cell survival by enhancing the expression of Bcl-xl. *Immunity* (1995) **3**:87–98. doi:10.1016/1074-7613(95)90161-2
79. Krakauer T. Co-stimulatory receptors for the superantigen staphyloccoccal enterotoxin B on human vascular endothelial cells and T cells. *J Leukoc Biol* (1994) **56**:458–63.
80. Weiss A. T lymphocyte activation. 4th ed. In: Paul W, editor. *Fundamental Immunology*. Philadelphia: Lippincott-Raven (1998). p. 411–47.
81. Van Leeuwen JE, Samelson LE. T cell-antigen receptor signal transduction. *Curr Opin Immunol* (1999) **11**:242–8. doi:10.1016/S0952-7915(99)80040-5
82. Isakov N, Altman A. PKC-theta-mediated signal delivery from the TCR/CD28 surface receptors. *Front Immunol* (2012) **3**:273. doi:10.3389/fimmu.2012.00273
83. Paul S, Schaefer BC. A new look at TCR signaling to NFκB. *Trends Immunol* (2013) **34**:269–81. doi:10.1016/j.it.2013.02.002
84. DiDonato JA, Mercurio F, Karin M. NFκB and the link between inflammation and cancer. *Immunol Rev* (2012) **246**:379–400. doi:10.1111/j.1600-065X.2012.01099.x
85. Takeuchi O, Akira S. Pattern recognition receptors and inflammation. *Cell* (2010) **140**:805–20. doi:10.1016/j.cell.2010.01.022
86. Krakauer T. The induction of CC chemokines in human peripheral blood mononuclear cells by staphylococcal exotoxins and its prevention by pentoxifylline. *J Leukoc Biol* (1999) **66**:158–64.
87. Krakauer T, Vilcek J, Oppenheim JJ. Proinflammatory cytokines: TNF and IL-1 families, chemokines, TGFß and others. 4th ed. In: Paul W, editor. *Fundamental Immunology*. Philadelphia: Lippincott-Raven (1998). p. 775–811.
88. Deane JA, Fruman DA. Phosphoinositide 3-kinase: diverse roles in immune cell activation. *Annu Rev Immunol* (2004) **22**:563–98. doi:10.1146/annurev.immunol.22.012703.104721
89. Park SG, Schulze-Luehrman J, Hayden MS, Hashimoto N, Ogawa W, Kasuga M, et al. Phosphoinositide-dependent kinase 1 integrates T cell receptor and CD28 co-receptor signaling to effect NFκB induction and T cell activation. *Nat Immunol* (2009) **10**:158–66. doi:10.1038/ni.1687
90. Cartwright NG, Kashyap AK, Schaefer BC. An active kinase domain is required for retention of PKCθ at the immunological synapse. *Mol Biol Cell* (2011) **22**:3491–7. doi:10.1091/mbc.E10-11-0916
91. Clark K, Nanda S, Cohen P. Molecular control of the MEMO family of ubiquitin-binding proteins. *Nat Rev Mol Cell Biol* (2013) **14**:673–85. doi:10.1038/nrm3644
92. Chatila T, Wood N, Parsonnet J, Geha RS. Toxic shock syndrome toxin-1 induces inositol phospholipid turnover, protein kinase C translocation, and calcium mobilization in human T cells. *J Immunol* (1988) **140**:1250–5.
93. Bueno C, Lemke CD, Criado G, Baroja ML, Ferguson SS, Rahman AK, et al. Bacterial superantigens bypass Lck-dependent T cell receptor signaling by activating a Galpha11-dependent, PLC-beta-mediated pathway. *Immunity* (2006) **25**:67–78. doi:10.1016/j.immuni.2006.04.012
94. Ghoreschi K, Laurence A, O'Shea JJ. Janus kinases in immune cell signaling. *Immunol Rev* (2009) **228**:273–87. doi:10.1111/j.1600-065X.2008.00754.x
95. Hulme MA, Wasserfall CH, Atkinson MA, Brusko TM. Central role for interleukin-2 in type 1 diabetes. *Diabetes* (2012) **61**:14–22. doi:10.2337/db11-1213
96. Thomson AW, Turnquist HR, Raimondi G. Immunoregulatory functions of mTOR inhibition. *Nat Rev Immunol* (2009) **9**:324–37. doi:10.1038/nri2546
97. Zoncu R, Sabatini DM, Efeyan A. mTOR: from growth signal integration to cancer, diabetes and ageing. *Nat Rev Mol Cell Biol* (2011) **12**:21–35. doi:10.1038/nrm3025
98. Kim SG, Buel GR, Blenis J. Nutrient regulation of the mTOR complex 1 signaling pathway. *Mol Cells* (2013) **35**:463–73. doi:10.1007/s10059-013-0138-2
99. Laplante M, Sabatini DM. mTOR signaling at a glance. *J Cell Sci* (2009) **122**:3389–94. doi:10.1242/jcs.051011
100. Abraham RT, Wiederrecht OJ. Immunopharmacology of rapamycin. *Annu Rev Immunol* (1996) **14**:483–510. doi:10.1146/annurev.immunol.14.1.483
101. Wullschleger S, Loewith R, Hall MN. TOR signaling in growth and metabolism. *Cell* (2006) **124**:471–84. doi:10.1016/j.cell.2006.01.016
102. Krakauer T, Buckley M, Issaq HJ, Fox SD. Rapamycin protects mice from staphylococcal enterotoxin B-induced toxic shock and blocks cytokine release *in vitro* and *in vivo*. *Antimicrob Agents Chemother* (2010) **54**:1125–31. doi:10.1128/AAC.01015-09
103. Mattsson E, Herwald H, Egsten A. Superantigens from *Staphylococcus aureus* induce procoagulant activity and monocyte tissue factor expression in whole blood and mononuclear cells via IL-1β. *J Thromb Haemost* (2003) **1**:2569–75. doi:10.1111/j.1538-7836.2003.00498.x
104. Stohl W, Elliott JE, Linsley PS. Human T cell-dependent B cell differentiation induced by staphylococcal superantigens. *J Immunol* (1994) **153**:117–27.
105. Mourad W, Mehindate K, Schall T, McColl S. Engagement of MHC class II molecules by superantigen induces inflammatory cytokine gene expression in human rheumatoid fibroblast-like synoviocytes. *J Exp Med* (1992) **175**:613–6. doi:10.1084/jem.175.2.613
106. Lu J, Philpott DJ, Saunders PR, Perdue MH, Yang PC, McKay DM. Epithelial ion transport and barrier abnormalities evoked by superantigen-activated immune cells are inhibited by interleukin-10 but not interleukin-4. *J Pharmacol Exp Ther* (1998) **287**:128–36.
107. Hamad AR, Marrack P, Kappler JW. Transcytosis of staphylococcal superantigen toxins. *J Exp Med* (1997) **185**:1447–54. doi:10.1084/jem.185.8.1447
108. Alber G, Scheuber PH, Reck B, Sailer-Krammer B, Hartmann A, Hammer DK. Role of substance P in immediate-type skin reactions induced by staphylococcal enterotoxin B in unsensitized monkeys. *J Allergy Clin Immunol* (1989) **84**:880–5. doi:10.1016/0091-6749(89)90383-7
109. Peterson M, Ault K, Kremer MJ, Klingelhutz AJ, Davis CC, Squier CA, et al. Innate immune system is activated by stimulation of vaginal epithelial cells with *Staphylococcus aureus* and toxic shock syndrome toxin 1. *Infect Immun* (2005) **73**:2164–74. doi:10.1128/IAI.73.4.2164-2174.2005
110. Brosnahan AJ, Schlievert PM. Gram positive bacterial superantigen outside-in signaling causes toxic shock syndrome. *FEBS J* (2011) **278**:4649–67. doi:10.1111/j.1742-4658.2011.08151.x
111. Vial T, Descotes J. Immune-mediated side-effects of cytokines in human. *Toxicol* (1995) **105**:31–57. doi:10.1016/0300-483X(95)03124-X
112. Dubinett SM, Huang M, Lichtenstein A, McBride WH, Wang J, Markovitz G, et al. Tumor necrosis factor-alpha plays a central role in interleukin-2 induced pulmonary vascular leak and lymphocyte accumulation. *Cell Immunol* (1994) **157**:170–80. doi:10.1006/cimm.1994.1214
113. Sims JE, Smith DE. The IL-1 family: regulators of immunity. *Nat Rev Immunol* (2010) **10**:89–102. doi:10.1038/nri2691
114. Kawai T, Akira S. Toll-like receptors and their crosstalk with other innate receptors in infection and immunity. *Immunity* (2011) **34**:637–50. doi:10.1016/j.immuni.2011.05.006
115. Hopkins PA, Fraser JD, Pridmore AC, Russell HH, Read RC, Sriskandan S. Superantigen recognition by HLA class II on monocytes up-regulates toll-like receptor 4 and enhances proinflammatory responses to endotoxin. *Blood* (2005) **105**:3655–62. doi:10.1182/blood-2004-07-2523
116. Hopkins PA, Pridmore AC, Ellmerich S, Fraser JD, Russell HH, Read RC, et al. Increased surface toll-like receptor 2 expression in superantigen shock. *Crit Care Med* (2008) **36**:1267–76. doi:10.1097/CCM.0b013e31816a0a78
117. Gross O, Thomas CJ, Guarda G, Tschopp J. The inflammasome: an integrated view. *Immunol Rev* (2011) **243**:136–51. doi:10.1111/j.1600-065X.2011.01046.x
118. Van Herreweghe F, Festjens N, Declercq W, Vandenabeele P. Tumor necrosis factor-mediated cell death: to break or to burst, that's the question. *Cell Mol Life Sci* (2010) **67**:1567–79. doi:10.1007/s00018-010-0283-0
119. Keystone EC, Ware CF. Tumor necrosis factor and anti-tumor necrosis factor therapies. *J Rheumatol* (2010) **85**:27–39. doi:10.3899/jrheum.091463
120. McKay DM. Bacterial superantigens: provocateurs of gut dysfunction and inflammation? *Trends Immunol* (2001) **22**:497–501. doi:10.1016/S1471-4906(01)02000-2
121. Miethke T, Wahl C, Heeg K, Echtenacher B, Krammer PH, Wagner H. T cell-mediated lethal shock triggered in mice by the superantigen staphylococcal enterotoxin B: critical role of tumor necrosis factor. *J Exp Med* (1992) **175**:91–8. doi:10.1084/jem.175.1.91

122. Dunn GP, Koebel CM, Schreiber RD. Interferons, immunity and cancer immunoediting. *Nat Rev Immunol* (2006) **6**:836–48. doi:10.1038/nri1961
123. Stark GR, Darnell JE. The JAK-STAT pathway at twenty. *Immunity* (2012) **36**:503–14. doi:10.1016/j.immuni.2012.03.013
124. McNab F, Mayer-Barber K, Sher A, Wack A, O'Garra A: Type 1 interferons in infectious disease. *Nat Rev Immunol* (2015) **15**:87–103. doi:10.1038/nri3787
125. Ramana CV, Gil MP, Schreiber RD, Stark GR. Stat-1-dependent and -independent pathways in IFN-dependent signaling. *Trends Immunol* (2002) **23**:96–101. doi:10.1016/S1471-4906(01)02118-4
126. Klintman D, Li X, Sato T, Wang Y, Jeppsson B, Thorlacius H. Staphylococcal enterotoxin A-induced hepatotoxicity is predominantly mediated by Fas ligand (CD95L). *Ann Surg* (2004) **240**(6):1065–72. doi:10.1097/01.sla.0000146159.88918.d2
127. Chang JH, Kim YJ, Han SH, Kang CY. IFN-γ-STAT1 signal regulates the differentiation of inducible Treg: potential role for ROS-mediated apoptosis. *Eur J Immunol* (2009) **39**:1241–51. doi:10.1002/eji.200838913
128. Malek TR, Castro I. Interleukin-2 receptor signaling: at the interface between tolerance and immunity. *Immunity* (2010) **33**:153–65. doi:10.1016/j.immuni.2010.08.004
129. Kyriakis JM, Avruch J. Mammalian MAPK signal transduction pathways activated by stress and inflammation. *Physiol Rev* (2012) **92**:689–737. doi:10.1152/physrev.00028.2011
130. Liu D, Zienkiewicz J, DiGiandomenico A, Hawiger J. Suppression of acute lung inflammation by intracellular peptide delivery of a nuclear import inhibitor. *Mol Ther* (2009) **17**:796–802. doi:10.1038/mt.2009.18
131. Huzella LM, Buckley MJ, Alves DA, Stiles BG, Krakauer T. Central roles for IL-2 and MCP-1 following intranasal exposure to SEB: a new mouse model. *Res Vet Sci* (2009) **86**:241–7. doi:10.1016/j.rvsc.2008.07.020
132. Khan AA, Priya S, Saha B. IL-2 regulates SEB induced toxic shock syndrome in BALB/c mice. *PLoS One* (2009) **4**:e8473. doi:10.1371/journal.pone.0008473
133. Sadik CD, Kim ND, Luster AD. Neutrophils cascading their way to inflammation. *Trends Immunol* (2011) **32**:452–60. doi:10.1016/j.it.2011.06.008
134. Zlotnik A, Yoshie D. The chemokine superfamily revisited. *Immunity* (2012) **36**:705–16. doi:10.1016/j.immuni.2012.05.008
135. Tessier PA, Naccache PH, Diener KR, Gladue RP, Neotem KS, Clark-Lewis I, et al. Induction of acute inflammation *in vivo* by staphylococcal superantigens. II. Critical role for chemokines, ICAM-1, and TNF-alpha. *J Immunol* (1998) **161**:1204–11.
136. Krakauer T, Buckley M, Huzella LM, Alves D. Critical timing, location and duration of glucocorticoid administration rescues mice from superantigen-induced shock and attenuates lung injury. *Int Immunopharmacol* (2009) **9**:1168–74. doi:10.1016/j.intimp.2009.06.004
137. Neumann B, Engelhardt B, Wagner H, Holzmann B. Induction of acute inflammatory lung injury by staphylococcal enterotoxin B. *J Immunol* (1997) **158**:1862–71.
138. Miethke T, Wahl C, Holzmann B, Heeg K, Wagner H. Bacterial superantigens induce rapid and T cell receptor V beta-selective down-regulation of L-selectin (gp90Mel-14) in vivo. *J Immunol* (1993) **151**:6777–82.
139. Niedergang F, Hemar A, Hewitt CR, Oven MJ, Dautry-Varset A, Alcover A. The *Staphylococcus aureus* enterotoxin B superantigen induces specific T cell receptor down-regulation by increasing its internalization. *J Biol Chem* (1995) **270**:12839–45. doi:10.1074/jbc.270.21.12839
140. MacDonald HR, Lees RK, Baschieri S, Herrmann T, Lussow AR. Peripheral T-cell reactivity to bacterial superantigens in vivo: the response/anergy paradox. *Immunol Rev* (1993) **133**:105–17. doi:10.1111/j.1600-065X.1993.tb01512.x
141. Florquin S, Amraoul Z, Goldman M. T cells made deficient in interleukin-2 production by exposure to staphylococcal enterotoxin B in vivo are primed for interferon-gamma and interleukin-10 secretion. *Eur J Immunol* (1995) **25**:1148–53. doi:10.1002/eji.1830250503
142. Bean AG, Freiberg RA, Andrade S, Menon S, Zlotnik A. Interleukin 10 protects mice against staphylococcal enterotoxin B-induced lethal shock. *Infect Immun* (1993) **61**:4937–9.
143. Shevach EM. Mechanisms of Fox3+ T regulatory cell-mediated suppression. *Immunity* (2009) **30**:636–45. doi:10.1016/j.immuni.2009.04.010
144. Ou L-S, Goleva E, Hall C, Leung DYM. T regulatory cells in atopic dermatitis and subversion of their activity by superantigens. *J Allergy Clin Immunol* (2004) **113**:756–63. doi:10.1016/j.jaci.2003.12.332
145. Saha B, Harlan DM, Lee KP, June CH, Abe R. Protection against lethal toxic shock by targeted disruption of the CD28 gene. *J Exp Med* (1996) **183**:2675–80. doi:10.1084/jem.183.6.2675
146. Linsley PS, Brady W, Urnes M, Grosmaire LS, Damle NK, Ledbetter JA. CTLA-4 is a second receptor for the B cell activation antigen B7. *J Exp Med* (1991) **174**:561–9. doi:10.1084/jem.174.3.561
147. Tilahun AY, Chowdhary VR, David CS, Rajagopalan G. Systemic inflammatory response elicited by superantigen destabilizes T regulatory cells, rendering them ineffective during toxic shock syndrome. *J Immunol* (2014) **193**:2919–30. doi:10.4049/jimmunol.1400980
148. Schwab JH, Brown RR, Anderle SK, Schlievert PM. Superantigen can reactivate bacterial cell wall-induced arthritis. *J Immunol* (1993) **150**:4151–9.
149. Zhao YX, Abdelnour A, Ljungdahl A, Olsson T, Tarkowski A. Patterns of interferon-gamma mRNA expression in toxic shock syndrome toxin-1 expanded Vβ11+ T lymphocytes. *Cell Immunol* (1995) **161**:28–33. doi:10.1006/cimm.1995.1005
150. Abdelnour AT, Bremell T, Tarkowski A. Toxic shock syndrome toxin 1 contributes to the arthritogenicity of *Staphylococcus aureus*. *J Infect Dis* (1994) **170**:94–9. doi:10.1093/infdis/170.1.94
151. Brocke S, Gaur A, Piercy C, Gautam A, Gijbels K, Fathman CG, et al. Induction of relapsing paralysis in experimental autoimmune encephalomyelitis by bacterial superantigen. *Nature* (1993) **365**:642–4. doi:10.1038/365642a0
152. Valdimarsson H, Baker BS, Jonsdottir I, Powles A, Fry L. Psoriasis: a T cell-mediated autoimmune disease induced by streptocococcal superantigens? *Immunol Today* (1995) **16**:145–9. doi:10.1016/0167-5699(95)80132-4
153. Yoshino T, Asada H, Sano S, Nakamura T, Itami S, Tamura M, et al. Impaired responses of peripheral blood mononuclear cells to staphylococcal superantigens in patients with severe atopic dermatitis: a role of T cell apoptosis. *J Invest Dermatol* (2000) **114**:281–8. doi:10.1046/j.1523-1747.2000.00878.x
154. Darenberg J, Soderquist B, Normark BH, Norrby-Teglund A. Differences in potency of intravenous polyspecific immunoglobulin G against streptococcal and staphylococcal superantigens: implications for therapy of toxic shock syndrome. *Clin Infect Dis* (2004) **38**:836–42. doi:10.1086/381979
155. Lappin E, Ferguson AJ. Gram-positive toxic shock syndromes. *Lancet Infect Dis* (2009) **9**:281–90. doi:10.1016/S1473-3099(09)70066-0
156. Krakauer T. Update on staphylococcal superantigen-induced signaling pathways and therapeutic interventions. *Toxins* (2013) **5**:1629–54. doi:10.3390/toxins5091629
157. Opal SM, Laterre PF, Francois B, LaRosa SP, Angus DC, Mira JP, et al. Effect of eritoran, an antagonist of MD2-TLR4, on mortality in patients with severe sepsis: the ACCESS randomized trial. *JAMA* (2013) **309**:1154–62. doi:10.1001/jama.2013.2194
158. Krakauer T, Buckley M. Dexamethasone attenuates staphylococcal enterotoxin B-induced hypothermic response and protects mice from superantigen-induced toxic shock. *Antimicrob Agents Chemother* (2006) **50**:391–5. doi:10.1128/AAC.50.1.391-395.2006
159. Tilahun AY, Theuer JE, Patel R, David CS, Rajagopalan G. Detrimental effect of the proteasome inhibitor, bortezomib in bacterial superantigen- and lipopolysaccharide-induced systemic inflammation. *Mol Ther* (2010) **18**:1143–54. doi:10.1038/mt.2010.53
160. Scholl P, Sekaly R, Diez A, Glimcher L, Geha R. Binding of toxic shock syndrome toxin-1 to murine major histocompatibility complex class II molecules. *Eur J Immunol* (1990) **20**:1911–6. doi:10.1002/eji.1830200907
161. Roy CJ, Warfield KL, Welcher BC, Gonzales RF, Larsen T, Hanson J, et al. Human leukocyte antigen-DQ8 transgenic mice: a model to examine the toxicity of aerosolized staphylococcal enterotoxin B. *Infect Immun* (2005) **73**:2452–60. doi:10.1128/IAI.73.4.2452-2460.2005
162. Tilahun AY, Marietta EV, Wu TT, Patel R, David CS, Rajagopalan G. Human leukocyte antigen class II transgenic mouse model unmasks the significant extrahepatic pathology in toxic shock syndrome. *Am J Pathol* (2011) **178**:2760–72. doi:10.1016/j.ajpath.2011.02.033
163. Ferreyra GA, Elinoff JM, Demirkale CY, Starost MF, Buckley M, Munson PJ, et al. Late multiple organ surge in interferon-regulated target genes characterizes staphylococcal enterotoxin B lethality. *PLoS One* (2014) **9**:e88756. doi:10.1371/journal.pone.0088756
164. Krakauer T, Buckley M, Fisher D. Proinflammatory mediators of toxic shock and their correlation to lethality. *Mediators Inflamm* (2010) **2010**:517594. doi:10.1155/2010/517594

165. Stetson DB, Medzhitov R. Recognition of cytosolic DNA activates an IRF3-dependent innate immune response. *Immunity* (2005) **24**:93–103. doi:10.1016/j.immuni.2005.12.003
166. Jimbo A, Fujita E, Kouroku Y, Ohnishi J, Inohara N, Kuida K, et al. ER stress induces caspase-8 activation, stimulating cytochrome c release and caspase-9 activation. *Exp Cell Res* (2003) **283**:156–66. doi:10.1016/S0014-4827(02)00033-2
167. Yoneda T, Imaizumi K, Oono K, Yui D, Gomi F, Katayama T, et al. Activation of caspase-12, an endoplastic reticulum (ER) resident caspase, through tumor necrosis factor receptor-associated factor 2-dependent mechanism in response to the ER stress. *J Biol Chem* (2001) **276**:13935–40. doi:10.1074/jbc.M010677200
168. Dauwalder O, Pachot A, Cazalis MA, Paye M, Faudot C, Badiou C, et al. Early kinetics of the transcriptional response of human leukocytes to staphylococcal superantigenic enterotoxins A and G. *Microb Pathog* (2009) **47**:171–6. doi:10.1016/j.micpath.2009.07.001
169. Mendis C, Das R, Hammamieh R, Royaee A, Yang D, Peel S, et al. Transcriptional response signature of human lymphoid cells to staphylococcal enterotoxin B. *Genes Immun* (2005) **6**:84–94. doi:10.1038/sj.gene.6364160
170. Takaoka A, Wang Z, Choi MK, Yanai H, Negishi H, Ban T, et al. DAI (DLM-1/ZBP1) is a cytosolic DNA sensor and an activator of innate immune response. *Nature* (2007) **448**:501–6. doi:10.1038/nature06013
171. Muruve DA, Pétrilli V, Zaiss AK, White LR, Clark SA, Ross PJ, et al. The inflammasome recognizes cytosolic microbial and host DNA and triggers an innate immune response. *Nature* (2008) **452**:103–7. doi:10.1038/nature06664
172. Zhang Q, Raoof M, Chen Y, Sumi Y, Sursai T, Junger W, et al. Circulating mitochondrial DAMPs cause inflammatory responses to injury. *Nature* (2010) **464**:104–7. doi:10.1038/nature08780
173. Helmerhorst HJ, Schultz MJ, van der Voort PH, de Jonge E, van Westerloo DJ. Bench-to-bedside review: the effects of hyperoxia during critical illness. *Crit Care* (2015) **19**:284–96. doi:10.1186/s13054-015-0996-4
174. Rosenson RS. Statins in atherosclerosis: lipid-lowering agents with antioxidant capabilities. *Atherosclerosis* (2004) **173**:1–12. doi:10.1016/S0021-9150(03)00239-9
175. Jacobson JR. Statins in endothelial signaling and activation. *Antioxid Redox Signal* (2009) **11**:811–21. doi:10.1089/ARS.2008.2284
176. Hotchkiss RS, Strasser A, McDunn JE, Swanson PE. Cell death in disease: mechanisms and emerging therapeutic concepts. *N Engl J Med* (2009) **361**:1570–83. doi:10.1056/NEJMra0901217
177. Franco-Penteado CF, Desouza I, Teixeira SA, Ribeiro-DaSilva G, De Nucci G, Antunes E. Role of nitric oxide on the increased vascular permeability and neutrophil accumulation induced by staphylococcal enterotoxin B into the mouse paw. *Biochem Pharmacol* (2001) **15**(61):1305–11. doi:10.1016/S0006-2952(01)00573-1
178. McKallip RJ, Fisher M, Gunthert U, Szakai AK, Nagarkatti PS, Nagarkatti M. Role of CD44 and its v7 isoform in staphylococcal enterotoxin B-induced toxic shock: CD44 deficiency on hepatic mononuclear cells leads to reduced activation-induced apoptosis that results in increased liver damage. *Infect Immun* (2005) **73**:50–61. doi:10.1128/IAI.73.1.50-61.2005
179. Martins JD, Liberal J, Silva A, Ferreira I, Neves BM, Cruz MT. Autophagy and inflammasome interplay. *DNA Cell Biol* (2015) **34**:274–81. doi:10.1089/dna.2014.2752
180. Paul S, Schaefer BC. Selective autophagy regulates T cell activation. *Autophagy* (2012) **8**(11):1690–2. doi:10.4161/auto.21581
181. Komisar JL, Weng CF, Oyejide A, Hunt RE, Briscoe C, Tseng J. Cellular and cytokine responses in the circulation and tissue reactions in the lung of rhesus monkeys (*Macaca mulatta*) pretreated with cyclosporine A and challenged with staphylococcal enterotoxin B. *Toxicol Pathol* (2001) **29**:369–78. doi:10.1080/019262301316905336
182. Tilahun AY, Darau MJ, Clar CR, Patel R, Rajagopalan G. The impact of tacrolimus on the immunopathogenesis of with staphylococcal enterotoxin-induced systemic inflammatory response syndrome and pneumonia. *Microbes Infect* (2012) **14**:528–36. doi:10.1016/j.micinf.2012.01.001

Disclaimer: The views expressed in this publication are those of the authors and do not reflect the official policy or position of the Department of the Army, the Department of Defense, or the U.S. Government.

Conflict of Interest Statement: The authors declare that the research was conducted in the absence of any commercial or financial relationships that could be construed as a potential conflict of interest.

15

Subversion of Cell-autonomous Immunity and Cell Migration by *Legionella pneumophila* Effectors

Sylvia Simon[1] *and Hubert Hilbi*[1,2*]

[1] *Institute of Medical Microbiology, University of Zürich, Zürich, Switzerland,* [2] *Max von Pettenkofer Institute, Ludwig-Maximilians University, Munich, Germany*

Edited by:
Katharina F. Kubatzky, University Hospital Heidelberg, Germany

Reviewed by:
Dario S. Zamboni, Universidade de São Paulo, Brazil
Amal O. Amer, The Ohio State University, USA

***Correspondence:**
Hubert Hilbi, Gloriastrasse 30, 8006 Zürich, Switzerland
hilbi@imm.uzh.ch

Bacteria trigger host defense and inflammatory processes, such as cytokine production, pyroptosis, and the chemotactic migration of immune cells toward the source of infection. However, a number of pathogens interfere with these immune functions by producing specific so-called "effector" proteins, which are delivered to host cells via dedicated secretion systems. Air-borne *Legionella pneumophila* bacteria trigger an acute and potential fatal inflammation in the lung termed Legionnaires' disease. The opportunistic pathogen *L. pneumophila* is a natural parasite of free-living amoebae, but also replicates in alveolar macrophages and accidentally infects humans. The bacteria employ the intracellular multiplication/defective for organelle trafficking (Icm/Dot) type IV secretion system and as many as 300 different effector proteins to govern host–cell interactions and establish in phagocytes an intracellular replication niche, the *Legionella*-containing vacuole. Some Icm/Dot-translocated effector proteins target cell-autonomous immunity or cell migration, i.e., they interfere with (i) endocytic, secretory, or retrograde vesicle trafficking pathways, (ii) organelle or cell motility, (iii) the inflammasome and programed cell death, or (iv) the transcription factor NF-κB. Here, we review recent mechanistic insights into the subversion of cellular immune functions by *L. pneumophila*.

Keywords: bacterial pathogenesis, *Dictyostelium*, inflammasome, *Legionella*, macrophage, pathogen vacuole, phosphoinositide, small GTPase

Introduction

Phagocytic cells of the innate immune system, such as macrophages, neutrophils, and dendritic cells (DC), produce pathogen recognition receptors (PRRs) comprising families of membrane-bound or cytosolic receptors (1–4). The membrane-bound receptors are toll-like receptors (TLRs) and C-type lectin receptors (CLRs), such as the mannose receptors. TLRs trigger cytokine production by signaling through the adaptor molecules MyD88 or TRIF and activate the transcription factor NF-κB or MAP kinase pathways. Cytosolic receptors are Nod-like receptors (NLRs) involved in inflammatory responses and cell death pathways through the activation of multiprotein complexes termed inflammasomes (5), as well as retinoic acid inducible gene-I (RIG-I)-like receptors (RLRs). Collectively, these receptors recognize pathogen-associated molecular patterns (PAMPs) and danger-associated molecular patterns (DAMPs), thus alerting the infected host cell, if material arising from cell infection or cell damage is present. Subsequent cytokine production promotes the clearance of invading microorganisms. The conserved process of cell-autonomous immunity includes not only inflammasome activation and interleukin (IL)-1β production, as well as autophagy, but also NF-κB- and type I interferon (IFN)-dependent cytokine production (6, 7).

Pathogenic bacteria target eukaryotic cells either as adversaries in the case of potentially bactericidal cells of the metazoan immune system and/or as a rich source of nutrients (8, 9). Accordingly, pathogens developed means to counteract the arsenal of the humoral and cellular components of the innate and acquired immune system and to survive and replicate within eukaryotic cells, including phagocytes (10). To this end, a number of pathogens produce specific so-called "effector" proteins, which are delivered via dedicated secretion systems into host cells, where they interfere with immune functions and cell migration (11).

Legionella pneumophila employs the intracellular multiplication/defective for organelle trafficking (Icm/Dot) type IV secretion system (T4SS) and as many as 300 different effector proteins to govern host–cell interactions. The role and molecular mode of action of effectors involved in pathogen vacuole formation has recently been reviewed in detail (12). Here, we will review mechanistic insights into the subversion of cell-autonomous immunity and cell migration by *L. pneumophila*.

Pathogenesis of *Legionella pneumophila*

Legionella pneumophila is an opportunistic pathogen and causes a severe pneumonia termed Legionnaires' disease. The Gram-negative genus *Legionella* comprises more than 55 species with several serogroups; yet, at least 85% of human infections are caused by *L. pneumophila* (13). Evolutionary adaptation allows *L. pneumophila* to persist in a variety of extra- and intracellular niches. The aerobic bacterium can not only replicate in biofilms but also resists degradation by free-living protozoa and replicates within, e.g., *Acanthamoeba*, *Hartmannella*, and *Tetrahymena* species, as well as in *Dictyostelium discoideum*, although the latter amoeba is likely not a natural host (14, 15). *L. pneumophila* is ubiquitously found in natural and technical water systems, including cooling towers, whirlpools, and showers.

Upon inhalation of contaminated aerosols, *L. pneumophila* resists degradation and replicates within alveolar macrophages, which is a precondition for the onset of disease (16). The acquisition of the pathogen from environmental sources is the only infection route; transmission between humans has never been observed. Since *L. pneumophila* probably mainly evolved as a parasite of free-living protozoa, the human host represents a dead-end for this "accidental" pathogen. Thus, *L. pneumophila* likely has not been exposed to a rigorous evolutionary selection to avoid recognition by mammalian PRRs, and accordingly, the bacteria trigger the activation of all PRR families (17).

Most humans and mice are able to clear a *Legionella* infection, and therefore, the development of a suitable small animal model was crucial. Initial studies using guinea pigs exposed to *L. pneumophila*-containing aerosols revealed a high susceptibility of these animals, which developed an illness reminiscent of typical human Legionnaires' disease (18). While most inbred mouse strains are resistant to *L. pneumophila* infection and disease progression, the A/J mouse strain was found to be susceptible and to present with acute pneumonia that resembled human disease (19). These results correspond to *in vitro* infections of peritoneal mouse macrophages, which indicated that cells from A/J mice were much more permissive for intracellular replication of *L. pneumophila* than macrophages from other mouse strains, such as C57BL/6 and BALB/c (20, 21). Macrophages from C57BL/6 and BALB/c mice restrict *L. pneumophila* by the activation of a programed cell death pathway as an ultimate line of defense against the intracellular pathogen (see below). Accordingly, mice lacking components of this pathway fail to restrict *L. pneumophila* and faithfully mimic *Legionella* pathology (17, 22).

In A/J mice, *L. pneumophila* elicits an acute inflammatory reaction, including production of the cytokines tumor necrosis factor (TNF)-α, IFN-γ, IL-12, and IL-18, which restrict pathogen replication (17, 22). IFN-γ is particularly important to inhibit bacterial growth in monocytes and alveolar macrophages, thus contributing to limiting the infection by *L. pneumophila* (23–25). These inflammatory cytokines recruit and activate polymorphonuclear neutrophil granulocytes (PMNs) (26–28). PMNs are central innate effector cells that not only resolve *L. pneumophila* infection but – in concert with IFN-γ-producing natural killer (NK) cells – also secrete cytokines, such as IL-18 (29, 30). In a feedback loop, IFN-γ triggers IL-12 production by DC, which activate NK cells. Therefore, DC are also essential to control *L. pneumophila* infection (31). Interestingly, DC restrict the intracellular growth of *L. pneumophila*, despite that the pathogen resides in an apparently non-bactericidal compartment derived from the endoplasmic reticulum (ER) (32).

Formation of the Intracellular Replication Niche

In permissive macrophages as well as in protozoa, *L. pneumophila* employs a complex and apparently evolutionarily conserved mechanism to establish its replication-permissive membrane-bound niche, the *Legionella*-containing vacuole (LCV) (**Figure 1**). LCV formation is governed by the Icm/Dot T4SS, which translocates approximately 300 different effector proteins into the host cell (12, 33–37). Since *L. pneumophila*–host interactions are defined to a large extent by the Icm/Dot apparatus, the T4SS represents a major virulence factor of *L. pneumophila*.

The Icm/Dot T4SS and translocated effectors control every step in the infection process of *L. pneumophila* (**Figure 1**), i.e., the uptake (38, 39), inhibition of fusion with lysosomes (40) and acidification (41), subversion of retrograde trafficking (42), as well as interception of secretory vesicle trafficking (43), coalescence with the ER (44–46), and finally, egress from the host cell (47). To interfere with host cell processes, many *L. pneumophila* effector proteins target pivotal components of eukaryotic membrane dynamics, such as phosphoinositide (PI) lipids and small GTPases (12, 33, 37). The PI lipids PtdIns(4)*P* and PtdIns(3)*P* control the secretory and the endosomal pathway, respectively, while the small GTPases Arf1 and Rab1 are pivotal regulators of ER-Golgi secretory trafficking.

Some Icm/Dot substrates subvert PI lipids by using PtdIns(4)*P* and PtdIns(3)*P* as membrane anchors, both of which are found on LCVs (33, 48–50). Accordingly, the effectors SidC, SdcA, SidM, LidA, Lpg1101, and Lpg2603 bind PtdIns(4)*P* (51–57), and

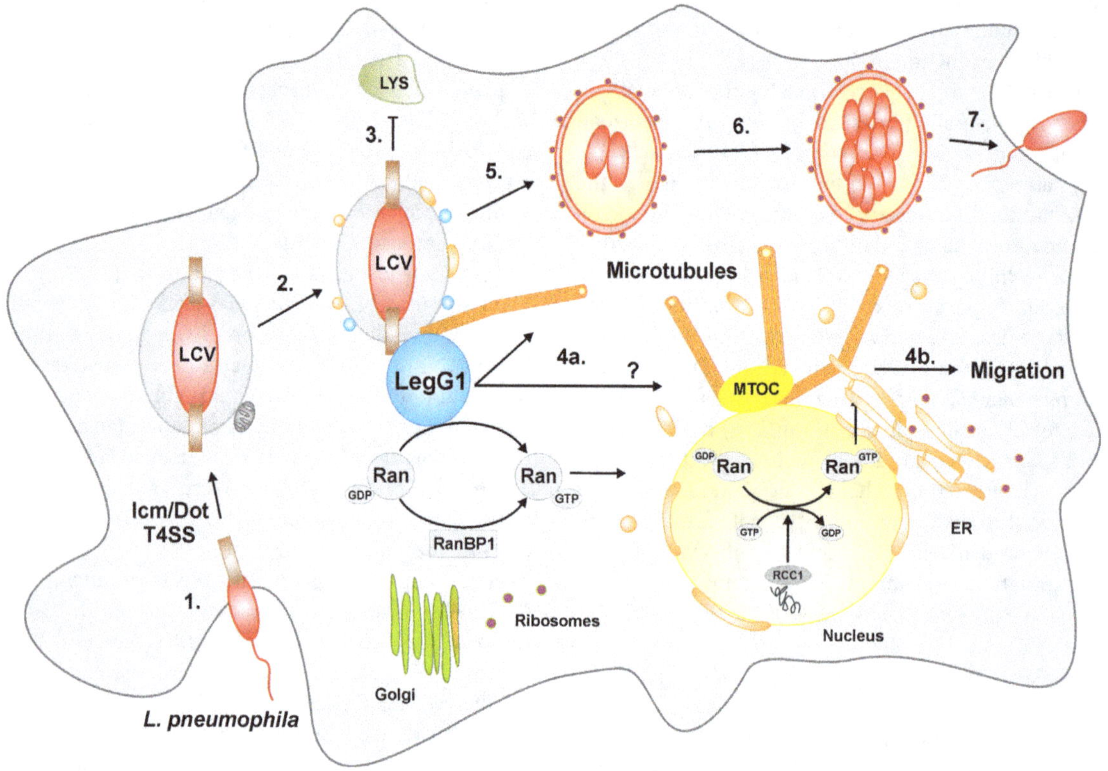

FIGURE 1 | LCV formation and LegG1-dependent modulation of cell migration. The intracellular phases of *L. pneumophila* can be divided in seven main steps: (1) Adhesion and entry into the host cell via macropinocytosis; (2) Formation of the LCV in an Icm/Dot T4SS-dependent manner and recruitment of vesicles from the ER as well as mitochondria; (3) Evasion from the lysosomal trafficking; (4a) The bacterial effector protein LegG1 activates the GTPase Ran, stabilizes microtubules in the vicinity of the LCV or possibly at a distance (?), and (4b) promotes cell migration. In the nucleus, Ran activation is triggered by the eukaryotic GEF RCC1. (5) The LCV communicates with host vesicle trafficking pathways, and acquires and eventually fuses with the ER; (6) The pathogen compartment turns in a rough-ER-like vacuole, wherein the bacteria replicate; (7) *L. pneumophila* is released and re-infects new host cells or is transmitted to other environmental niches. Abbreviations: ER, endoplasmic reticulum; LCV, *Legionella*-containing vacuole; MTOC, microtubule organizing center; Ran, Ras-related nuclear protein; RCC1, regulator of chromosome condensation 1; T4SS, type IV secretion system.

the effectors LidA, LptD, RidL, SetA, and LpnE bind PtdIns(3)*P* (42, 52, 58–60). Two *L. pneumophila* "CX$_5$R" domain PI phosphatases have been identified: SidF, a PI 3-phosphatase that hydrolyzes PtdIns(3,4)P_2 [and also PtdIns(3,4,5)P_3] *in vitro* (61), and SidP, a PI 3-phosphatase that hydrolyzes PtdIns(3)*P* as well as PtdIns(3,5)P_2 *in vitro* (62). Thus, these PI phosphatases might cause PtdIns(4)*P* to be formed on and PtdIns(3)*P* to be removed from LCV membranes. LppA is another Icm/Dot substrate that hydrolyzes polyphosphorylated PIs to mainly produce PtdIns(4)*P in vitro* (49). However, LppA does not affect the LCV PI pattern in infected cells, but rather functions as a translocated hexakisphosphate inositol phosphatase (phytase), which possibly promotes intracellular replication of *L. pneumophila* by removing the intracellular micronutrient chelator phytate.

Intracellular multiplication/defective for organelle trafficking T4SS substrates modulate the activity of the small GTPases Arf1 (RalF) (63) or Rab1 (SidM) (55, 64–67) by acting as guanine nucleotide exchange factors (GEFs), GTPase-activating proteins (GAPs; LepB) (68), AMPylase/deAMPylase (SidM, SidD) (69–72), or phosphocholinase/dephosphocholinase (AnkX, Lem3) (73–77). The covalent modifications of Rab1 by AMP or phosphocholine cause a prolonged activation of the GTPase, by preventing its inhibition by the GAP LepB. Similarly, the Icm/Dot substrate LidA binds to activated Rab1 and prevents inactivation by LepB, as well as covalent modifications by SidM or AnkX (64, 78–80). Finally, the PtdIns(4)*P*-binding effector SidC and its paralog SdcA promote the monoubiquitination of Rab1, in agreement with their function as an E3 ubiquitin ligase containing a catalytic Cys–His–Asp triad (81–83). In summary, after their translocation a number of *L. pneumophila* Icm/Dot substrates employ PI lipids on the cytoplasmic side of the LCV membrane as membrane anchors. Thus, PI lipids determine at least in part the subcellular localization of the effectors, which show different biochemical activities, including subversion of small GTPases, ubiquitinylation, or binding to and modulation of vesicle trafficking machinery.

Inhibition of Retrograde Trafficking and Autophagy

Retrograde trafficking in eukaryotes comprises the transport from early or late endosomes through the *trans*-Golgi network (TGN) to the ER (84). The cation-independent mannose 6-phosphate receptor (CI-MPR) is a well-documented substrate of the retrograde transport pathway. CI-MPR binds mannose present on hydrolases and transports the cargo destined for the endocytic pathway from the TGN to the endosomal system. After releasing the cargo in the lysosomal lumen, the receptor is recycled back through the retrograde pathway to the TGN (85). Proper recycling requires a protein complex termed the retromer, composed of a cargo recognition subcomplex (Vps26, Vps29, and Vps35) and a membrane-deforming subcomplex, which consists of a dimer of sorting nexins (SNXs) (86). The cargo recognition complex is recruited to the membrane by the activated small GTPases Rab5 and Rab7 (87).

Recent findings indicate that *L. pneumophila* communicates with the retrograde trafficking pathway. The small GTPases Rab5 and Rab7 (88–91), as well as the retromer subunits Vps26, Vps29, and Vps35 (42, 92) localize to the LCV. Interestingly, the Icm/Dot T4SS-translocated substrate RidL functions as a bacterial interactor of the eukaryotic retromer complex (42). RidL, which localizes on LCVs in *D. discoideum* cells and macrophages, binds to the Vps29 subunit of the cargo recognition subcomplex of the retromer. Subunits of the retromer cargo recognition complex localize to the LCV in an Icm/Dot-dependent but RidL-independent manner. RidL competes with SNX1 and SNX2 in its capacity for membrane binding specifically through PtdIns(3)*P*, thus possibly triggering the removal of SNXs. By altering the retrograde trafficking cascade, RidL might promote the formation of a non-lysosomal replicative vacuole and intracellular replication of *L. pneumophila* in protozoan and metazoan phagocytes (42). Earlier work already indicated that the retrograde pathway might control intracellular replication of *L. pneumophila.* The PtdIns(4,5)P_2 5-phosphatase OCRL1 and its *D. discoideum* homolog Dd5P4 are implicated in retrograde trafficking (93, 94) and localize to LCVs (60). Depletion of OCRL by RNA interference (42) or deletion of Dd5P4 (60) increased intracellular growth of *L. pneumophila*, indicating that the PI 5-phosphatase OCRL/Dd5P4 indeed restricts pathogen replication.

Autophagy is a major cell-autonomous defense mechanism used by infected cells against intracellular bacteria. The process of macroautophagy is responsible for the degradation of cytoplasmic constituents, such as bacteria or damaged organelles, which are engulfed by autophagosomes and subsequently fuse with lysosomes (95). In the course of microautophagy, the constituents are directly delivered to the lysosomes. Over 30 autophagy-related genes (Atg) have been discovered during the last years (96). An Atg protein essential for autophagy is the microtubule-associated protein light chain 3 (LC3, *alias* Atg8), which is conjugated to phosphatidylethanolamine and localized on autophagosomal membranes (97). Measurement of LC3/Atg8 during *L. pneumophila* infection revealed an Icm/Dot-dependent inhibition of autophagy by the bacteria. Intriguingly, the bacterial protein RavZ was identified as an effector required for autophagy inhibition, yet further effectors could be involved, since *L. pneumophila* lacking *ravZ* still evaded the autophagy cascade (98). RavZ is a cysteine protease, which irreversibly deconjugates phosphatidylethanolamine from LC3/Atg8, thus reducing its membrane accumulation and activity. While most autophagy factors are not necessary for *L. pneumophila* replication in *D. discoideum* (99), amoebae lacking Atg9 do not internalize the pathogen as efficient as wild-type cells, yet allow more effective intracellular growth (100). In summary, these findings suggest that *L. pneumophila* inhibits retrograde vesicle trafficking as well as autophagy to promote intracellular growth.

Modulation of Organelle Motility and Cell Migration

The Icm/Dot T4SS is crucially involved in the formation of the LCV, and many *L. pneumophila* effectors selectively decorate the pathogen vacuole membrane. Intact LCVs can be purified by a straight-forward two-step protocol involving immuno-affinity enrichment using an antibody against the *L. pneumophila* effector SidC selectively decorating the pathogen vacuole and a secondary antibody coupled to magnetic beads, followed by a conventional density centrifugation step. Proteomics analysis of purified preparations of LCVs from infected *D. discoideum* amoebae (91) and RAW 264.7 macrophages (90) revealed the presence of 670 and 1150 host proteins, respectively, including 13 small GTPases of the Rab family, as well as the small GTPase Ran and its effector Ran binding protein 1 (RanBP1).

Ran is a member of the Ras superfamily of small GTPases and is fundamental in numerous cellular processes, such as nuclear pore translocation (101), or mitotic spindle assembly and post-mitotic nuclear envelope formation (102, 103). Ran GTPase also plays an important role in cytoplasmic processes involving non-centrosomal microtubules, e.g., endocytic receptor trafficking and retrograde signaling along microtubules in nerve axons (104). Ran can be activated by a nuclear (or in mitotic cells: chromatin bound) Ran GEF termed regulator of chromosome condensation 1 (RCC1) (105). Ran(GTP) is inactivated by the cytoplasmic Ran GTPase-activating protein 1 (RanGAP1) together with RanBP1, which harbors a Ran(GTP)-binding domain (104).

Intriguingly, *L. pneumophila* produces an Icm/Dot-translocated effector termed LegG1 that harbors eukaryotic-like RCC1 domains (106, 107). The corresponding gene is conserved among the *L. pneumophila* genomes sequenced to date, including the strains Philadelphia-1, Paris, Lens, Corby, Alcoy, and Lorraine. LegG1 (Lpg1976/PieG) is encoded in the plasticity island of effectors (Pie) gene cluster (108). The C-terminal CAAX tetrapeptide motif of LegG1 is lipidated by the host prenylation machinery (109), thus promoting the membrane localization of the effector to small vesicular structures upon ectopic expression in eukaryotic cells (108).

Recent work revealed that LegG1 is indeed an *L. pneumophila* virulence factor, which promotes intracellular bacterial replication but is dispensable for uptake (110) (**Figure 1**). A Δ*legG1* mutant strain is not less cytotoxic than the parental strain, but outcompeted by wild-type *L. pneumophila* upon co-infection of *A. castellanii* amoebae. Moreover, Ran, RanBP1, and LegG1 accumulate in an Icm/Dot-dependent manner on the LCV, and the bacterial effector localizes to the cytosolic face of the LCV in *L. pneumophila*-infected phagocytes. *L. pneumophila* wild-type but not the Δ*legG1* mutant strain activates Ran on LCVs and in cell lysates; yet, while LegG1 promotes the accumulation of RanBP1 on LCVs, the effector is dispensable for the recruitment of the small GTPase. Several experimental approaches indicated that in infected phagocytes *L. pneumophila* triggers the polymerization of microtubules in a LegG1-dependent manner (110) (**Figure 1**). "Microbial microinjection" of LegG1 by *Yersinia enterocolitica* confirmed the positive effect of LegG1 on microtubule stabilization. Here, *Y. enterocolitica* harboring a T3SS, but lacking endogenous effectors, translocated LegG1 fused to a type III secretion signal into HeLa cells. Furthermore, while LCVs harboring wild-type *L. pneumophila* vividly move along microtubules in infected *D. discoideum*, the pathogen vacuole harboring Δ*legG1* mutant bacteria is stalled (110). In summary, the discovery and characterization of the *L. pneumophila* Ran activator LegG1 revealed an unexpected role of the small GTPase Ran in the formation of pathogen vacuoles.

Microtubule polarization and dynamics represent pivotal determinants of eukaryotic cell migration (111) (**Figure 2**). Given the prominent role of Ran and LegG1 on the dynamics of the microtubule cytoskeleton, we investigated the effect of *L. pneumophila* and LegG1 on host cell motility (112). Studies using *D. discoideum* amoebae or immune cells, such as RAW 264.7 macrophage-like cells or primary PMN, in different migration assays (under-agarose and Boyden chamber assays), revealed an Icm/Dot-dependent inhibition of migration. Phagocytes infected with wild-type *L. pneumophila* or *Legionella longbeachae* showed a substantially reduced migration when compared to cells lacking a functional Icm/Dot T4SS. Uptake and cytotoxicity assays demonstrated that the observed effect is not due to a defect in infection.

Based on these findings, the modulation of cell migration by LegG1 was assessed in *D. discoideum* as well as in RAW 264.7 macrophages or PMN (112) (**Figure 1**). Interestingly, the Δ*legG1* mutant strain hyper-inhibited the directed migration of phagocytes in the under-agarose assay, even to a larger extent than wild-type *L. pneumophila*. Overproduction of LegG1 in the Δ*legG1* mutant background re-established the migration range to an extent comparable to cells infected with a Δ*icmT* mutant strain. Single cell tracking revealed that the forward migration and the velocity of cells infected with wild-type or Δ*legG1 L. pneumophila* was impaired. Similarly, upon "microbial microinjection" by *Y. enterocolitica*, LegG1 was sufficient to stimulate migration of epithelial cells in scratch assays. Moreover, using RNA interference a role of Ran in the LegG1-dependent migration inhibition was demonstrated. Upon depletion of Ran, cells infected with the strain overproducing LegG1 were not able to migrate and to close scratch wounds over time (112). Taken together, the Ran activator LegG1 promotes cell motility by modulating microtubule dynamics and thus antagonizes Icm/Dot-dependent inhibition of cell migration. LegG1 might reverse the otherwise deleterious impact of other *L. pneumophila* effectors on the host cytoskeleton, and thereby

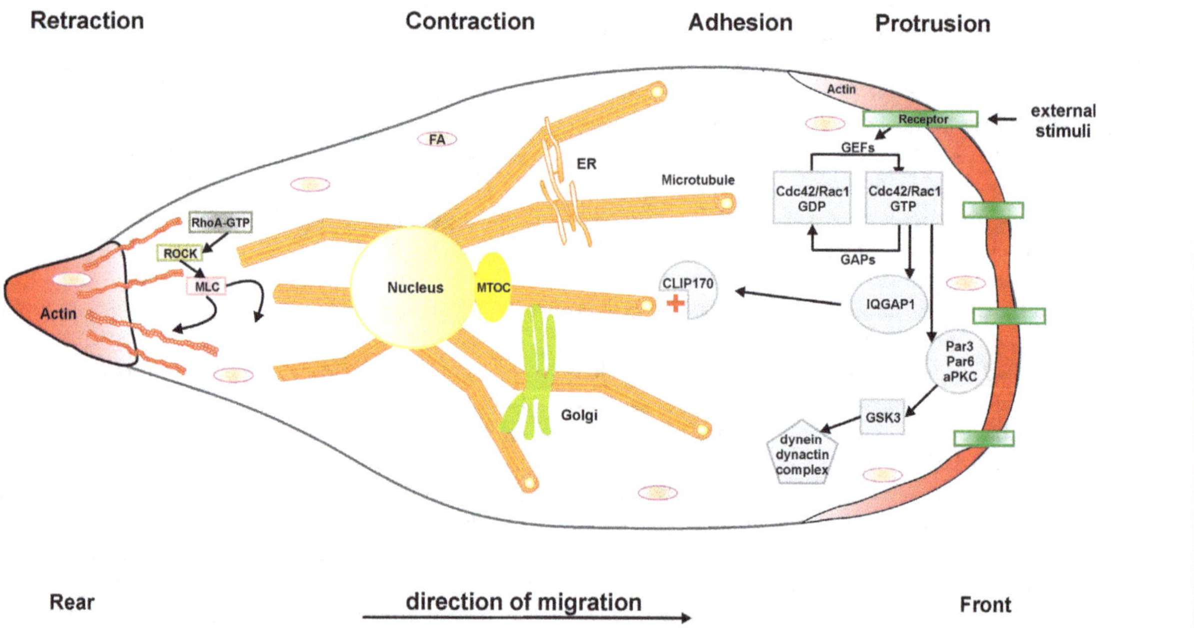

FIGURE 2 | Schematic representation of cell migration. Generally, the cellular migration cycle can be divided into four major steps: (1) protrusion at the leading edge, (2) cell adhesion, (3) contraction to generate the required forces, and (4) retraction at the rear edge. Microtubules and the actin cytoskeleton in concert with Rho GTPases are fundamental components of each phase. Abbreviations: aPKC, atypical protein kinase C; Cdc42, cell division control protein 42; CLIP170, cytoplasmic linker protein 170; FA, focal adhesion; GAP, GTPase-activating protein; GEF, guanine nucleotide exchange factor; GSK3, glycogen synthase kinase 3; IQGAP1, IQ motif-containing GTPase-activating protein 1; MLC, myosin light chain; MTOC, microtubule organizing center; Par3/6, partitioning defective 3/6 homolog; Rac1, Ras-related C3 botulinum toxin substrate 1; RhoA, Ras homolog gene family member A; ROCK, Rho-associated protein kinase. The *L. pneumophila* effector protein LegG1 promotes cell migration by activating the small GTPase Ran and stabilizing microtubules. Other *L. pneumophila* factors modulating cell migration are not known.

sustain vesicle trafficking and organelle motility required for the establishment and maintenance of LCVs.

Triggering Inflammasomes and Programed Cell Death

Maintenance of LCV integrity and regulation of programed cell death is critical for preserving the intracellular replication niche of *L. pneumophila* (113). Accordingly, impaired activation of programed cell death turned out to account for the failure of A/J mice to restrict *L. pneumophila*. The allele conferring sensitivity against *L. pneumophila* was mapped to the *NAIP5* gene (114, 115) within the *Lgn1* locus (116). The Naip5 (Birc1e)/Nlrc4 (Ipaf) inflammasome recognizes flagellin and triggers caspase-1 activation, pore formation, and pyroptosis (117–125). Naip5/Nlrc4 inflammasome activation represents a crucial mechanism of *L. pneumophila* restriction. In agreement with this notion, macrophages as well as DC restrict *L. pneumophila* replication through a cell death pathway mediated by Naip5, caspase-1, and caspase-3 (126). Yet, neither inflammasomes nor caspases are conserved in amoebae (127).

In addition to PAMPs, such as flagellin, a number of specific Icm/Dot substrates are implicated in the regulation of programed host cell death. The *L. pneumophila* PI 3-phosphatase SidF inactivates the pro-apoptotic factors BNIP3 and Bcl-rambo by an unknown mechanism, and thereby counteracts cell death induction (128). Furthermore, the effector SdhA plays a role in maintaining the LCV membrane integrity and also contributes to the prevention of cell death (129, 130). Cell death induction in absence of *sdhA* is suppressed by a secreted bacterial phospholipase A through an unknown mechanism. *L. pneumophila* lacking *sdhA* resides in the cytoplasm and triggers caspase-1 activation and IL-1β secretion, as well as macrophage pyroptosis through the DNA-sensing AIM2 inflammasome (131). The Icm/Dot substrate SdhA is also a key suppressor of the type I IFN (IFN-α/β) response to *L. pneumophila* through nucleic acid-sensing PRRs. Accordingly, RNA from *L. pneumophila* lacking *sdhA* triggers the RIG-I-dependent production of type I IFNs (132).

Legionella pneumophila also promotes programed cell death in an Icm/Dot-dependent manner. The phospholipase VipD destabilizes mitochondrial membranes by means of its phospholipase A activity, thus potentially contributing to the release of cytochrome *c* into the cytosol. This triggers pro-apoptotic caspase-3 activation and subsequently promotes host cell death (133). Analogously, the Icm/Dot-translocated phospholipase PlcC hydrolyzes several lipids, including phosphatidylcholine, phosphatidylglycerol, and phosphatidylinositol, which might destabilize target membranes and cause cell toxicity (134). Stimulation of apoptosis by an intracellular pathogen seems counterintuitive, but might reflect a tight spatial and temporal control of LCV maturation, followed by the release of the bacteria from the pathogen compartment and the host cell at the end of an infection cycle. Thus, the elaborate coordination of anti- and pro-apoptotic factors optimally supports intracellular bacterial proliferation.

Finally, caspases not only regulate cell death during *L. pneumophila* infection but also control vesicle trafficking pathways and thus contribute to the formation of LCVs. Dependent on the Nlrc4 inflammasome and Naip5, *L. pneumophila* activates caspase-7 downstream of caspase-1, and consequently, the pathogen is delivered to lysosomes (117, 135). In turn, upon deletion of the caspases, *L. pneumophila* forms an ER-derived replicative compartment. Similarly (but independently of caspase-1), active caspase-11 restricts the replication of *L. pneumophila* by promoting the fusion of the pathogen compartment with lysosomes (136).

Activation of the Transcription Factor NF-κB

The transcription factor NF-κB is a master regulator of the mammalian innate immune response and controls the production of anti-apoptotic and pro-survival factors as well as inflammatory mediators (137). Thus, the activation of NF-κB (which is not conserved in amoebae) represents another, more indirect way to prevent host cell death. NF-κB is composed of five Rel family proteins: RelA, RelB, c-Rel, and the precursors p100 and p105, which are processed to their mature forms, p52 and p50. The transcription factor is maintained in its inactive form in the cytoplasm by a family of inhibitors termed IκBs (IκBa, IκBb, IκBe, and Bcl-3).

Bacterial PAMPs, including LPS or flagellin, initiate the NF-κB cascade through TLR2 or TLR5 and the adaptor MyD88. *L. pneumophila* triggers the production of pro-inflammatory cytokines, such as IL-1β, and the activation of the inflammasome platform through these receptors (138–141). The deletion of single TLRs does not dramatically alter *L. pneumophila* infection; yet, mice lacking MyD88 fail to produce cytokines, such as NK cell-derived IFN-γ, and are highly susceptible to *L. pneumophila* infection (142–144, 29). Moreover, mice lacking the IL-1 receptor are impaired for PMN recruitment and bacterial clearance (145). Therefore, IL-1R/MyD88-dependant signaling is critical for host resistance to *L. pneumophila* infection.

The NF-κB pathway can be activated not only by bacterial PAMPs via a TLR-dependent pathway but also by distinct bacterial effectors (146). The Icm/Dot T4SS and its substrates LnaB and LegK1 strongly induce NF-κB (147–149). While the molecular mechanism of the novel effector LnaB is not understood, LegK1 is a Ser/Thr kinase that phosphorylates the NF-κB inhibitor IκBα, leading to a robust NF-κB activation by triggering the release and nuclear translocation of the transcription factor (150). Since *L. pneumophila* lacking *legK1* is not impaired for intracellular replication, other effectors might also modulate the NF-κB response. Among the five *L. pneumophila* Icm/Dot substrates that show protein kinase activity *in vitro* (LegK1–LegK5), LegK2 is a virulence factor, which promotes intracellular replication and efficient recruitment of the ER to LCVs (151).

Translation inhibition by the Icm/Dot substrates Lgt1-3, SidI, and SidL also specifically decreases the production of IκB, even though the effectors are overall cytotoxic (152, 153). To shut down translation, the UDP-glucosyltransferases Lgt1–3 modify the elongation factor eEF1A (154, 155), and SidI inactivates eEF1A/eEF1Bγ by an unknown mechanism (153). Finally, *L. pneumophila* also modulates host transcription more directly. The Icm/Dot substrate RomA/LegAS4 promotes intracellular replication as a SET domain-containing histone methyltransferase that modifies (immune) gene expression (156, 157). In summary, *L. pneumophila* modulates protein production and turnover through the activation of NF-κB and by altering the epigenetic pattern, as well as through the inhibition of translation.

Conclusion and Outlook

The ubiquitous environmental bacterium *L. pneumophila* triggers an acute and potential fatal pneumonia termed Legionnaires' disease. The opportunistic pathogen employs the Icm/Dot T4SS and as many as 300 different effector proteins to govern interactions with phagocytes and form an intracellular replication niche, the LCV. Some Icm/Dot-translocated effector proteins interfere with (i) endocytic, secretory, or retrograde vesicle trafficking pathways, (ii) organelle or cell motility, (iii) the inflammasome and programed cell death, or (iv) the transcription factor NF-κB. Future studies will address the molecular mechanisms of action of the many Icm/Dot substrates, which remain uncharacterized to date. The further analysis of *L. pneumophila* effector proteins will likely continue to provide novel insights into the elaborate pathogen–host interactions between a highly adapted opportunistic pathogen and its phagocytic host cells.

Acknowledgments

This work was supported by the Swiss National Science Foundation (SNF; 31003A_153200), the German Research Foundation (DFG; HI 1511/3-1, SPP 1580, SFB 914), and the Bundesministerium für Bildung und Forschung (BMBF; 031A410A; Infect-ERA project EUGENPATH).

References

1. Chow J, Franz KM, Kagan JC. PRRs are watching you: Localization of innate sensing and signaling regulators. *Virology* (2015) **479-480**:104–9. doi:10.1016/j.virol.2015.02.051
2. Creagh EM, O'Neill LA. TLRs, NLRs and RLRs: a trinity of pathogen sensors that co-operate in innate immunity. *Trends Immunol* (2006) **27**:352–57.
3. Janeway CA Jr, Medzhitov R. Innate immune recognition. *Ann Rev Immunol* (2002) **20**:197–216.
4. Sellge G, Kufer TA. PRR-signaling pathways - Learning from microbial tactics. *Semin Immunol* (2015) **27**:75–84. doi:10.1016/j.smim.2015.03.009
5. Schroder K, Tschopp J. The inflammasomes. *Cell* (2010) **140**:821–32. doi:10.1016/j.cell.2010.01.040
6. de Veer MJ, Holko M, Frevel M, Walker E, Der S, Paranjape JM, et al. Functional classification of interferon-stimulated genes identified using microarrays. *J Leukoc Biol* (2001) **69**:912–20.
7. Schoggins JW, Rice CM. Interferon-stimulated genes and their antiviral effector functions. *Curr Opin Virol* (2011) **1**:519–25. doi:10.1016/j.coviro.2011.10.008
8. Abu Kwaik Y, Bumann D. Microbial quest for food in vivo: 'nutritional virulence' as an emerging paradigm. *Cell Microbiol* (2013) **15**:882–90. doi:10.1111/cmi.12138
9. Manske C, Hilbi H. Metabolism of the vacuolar pathogen *Legionella* and implications for virulence. *Front Cell Infect Microbiol* (2014) **4**:125. doi:10.3389/fcimb.2014.00125
10. Flannagan RS, Cosio G, Grinstein S. Antimicrobial mechanisms of phagocytes and bacterial evasion strategies. *Nat Rev Microbiol* (2009) **7**:355–66. doi:10.1038/nrmicro2128
11. Asrat S, De Jesus DA, Hempstead AD, Ramabhadran V, Isberg RR. Bacterial pathogen manipulation of host membrane trafficking. *Annu Rev Cell Dev Biol* (2014) **30**:79–109. doi:10.1146/annurev-cellbio-100913-013439
12. Finsel I, Hilbi H. Formation of a pathogen vacuole according to *Legionella pneumophila*: how to kill one bird with many stones. *Cell Microbiol* (2015) **17**:935–50. doi:10.1111/cmi.12450
13. Bangsborg JM. Antigenic and genetic characterization of *Legionella* proteins: contributions to taxonomy, diagnosis and pathogenesis. *APMIS Suppl* (1997) **70**:1–53.
14. Fields BS. The molecular ecology of legionellae. *Trends Microbiol* (1996) **4**:286–90. doi:10.1016/0966-842X(96)10041-X
15. Steinert M, Heuner K. *Dictyostelium* as host model for pathogenesis. *Cell Microbiol* (2005) **7**:307–14. doi:10.1111/j.1462-5822.2005.00493.x
16. Hilbi H, Hoffmann C, Harrison CF. *Legionella* spp. outdoors: colonization, communication and persistence. *Environ Microbiol Rep* (2011) **3**:286–96. doi:10.1111/j.1758-2229.2011.00247.x
17. Massis LM, Zamboni DS. Innate immunity to *Legionella pneumophila*. *Front Microbiol* (2011) **2**:109. doi:10.3389/fmicb.2011.00109
18. Baskerville A, Fitzgeorge RB, Broster M, Hambleton P, Dennis PJ. Experimental transmission of Legionnaires' disease by exposure to aerosols of *Legionella pneumophila*. *Lancet* (1981) **2**:1389–90. doi:10.1016/S0140-6736(81)92803-8
19. Brieland J, Freeman P, Kunkel R, Chrisp C, Hurley M, Fantone J, et al. Replicative *Legionella pneumophila* lung infection in intratracheally inoculated A/J mice. A murine model of human Legionnaires' disease. *Am J Pathol* (1994) **145**:1537–46.
20. Yamamoto Y, Klein TW, Newton CA, Widen R, Friedman H. Growth of *Legionella pneumophila* in thioglycolate-elicited peritoneal macrophages from A/J mice. *Infect Immun* (1988) **56**:370–5.
21. Yoshida S, Goto Y, Mizuguchi Y, Nomoto K, Skamene E. Genetic control of natural resistance in mouse macrophages regulating intracellular *Legionella pneumophila* multiplication in vitro. *Infect Immun* (1991) **59**:428–32.
22. Newton HJ, Ang DK, Van Driel IR, Hartland EL. Molecular pathogenesis of infections caused by *Legionella pneumophila*. *Clin Microbiol Rev* (2010) **23**:274–98. doi:10.1128/CMR.00052-09
23. Bhardwaj N, Nash TW, Horwitz MA. Interferon-gamma-activated human monocytes inhibit the intracellular multiplication of *Legionella pneumophila*. *J Immunol* (1986) **137**:2662–9.
24. Byrd TF, Horwitz MA. Interferon gamma-activated human monocytes downregulate transferrin receptors and inhibit the intracellular multiplication of *Legionella pneumophila* by limiting the availability of iron. *J Clin Invest* (1989) **83**:1457–65. doi:10.1172/JCI114038
25. Nash TW, Libby DM, Horwitz MA. IFN-gamma-activated human alveolar macrophages inhibit the intracellular multiplication of *Legionella pneumophila*. *J Immunol* (1988) **140**:3978–81.
26. Blanchard DK, Djeu JY, Klein TW, Friedman H, Stewart WE. Protective effects of tumor necrosis factor in experimental *Legionella pneumophila* infections of mice via activation of PMN function. *J Leukoc Biol* (1988) **43**:429–35.
27. Blanchard DK, Friedman H, Klein TW, Djeu JY. Induction of interferon-gamma and tumor necrosis factor by *Legionella pneumophila*: augmentation of human neutrophil bactericidal activity. *J Leukoc Biol* (1989) **45**:538–45.
28. Tateda K, Moore TA, Newstead MW, Tsai WC, Zeng X, Deng JC, et al. Chemokine-dependent neutrophil recruitment in a murine model of *Legionella* pneumonia: potential role of neutrophils as immunoregulatory cells. *Infect Immun* (2001) **69**:2017–24. doi:10.1128/IAI.69.4.2017-2024.2001
29. Spörri R, Joller N, Albers U, Hilbi H, Oxenius A. MyD88-dependent IFN-gamma production by NK cells is key for control of *Legionella pneumophila* infection. *J Immunol* (2006) **176**:6162–71. doi:10.4049/jimmunol.176.10.6162
30. Spörri R, Joller N, Hilbi H, Oxenius A. A novel role for neutrophils as critical activators of NK cells. *J Immunol* (2008) **181**:7121–30. doi:10.4049/jimmunol.181.10.7121
31. Ang DK, Oates CV, Schuelein R, Kelly M, Sansom FM, Bourges D, et al. Cutting edge: pulmonary *Legionella pneumophila* is controlled by plasmacytoid dendritic cells but not type I IFN. *J Immunol* (2010) **184**:5429–33. doi:10.4049/jimmunol.1000128

32. Neild AL, Roy CR. *Legionella* reveal dendritic cell functions that facilitate selection of antigens for MHC class II presentation. *Immunity* (2003) **18**:813–23. doi:10.1016/S1074-7613(03)00140-7
33. Haneburger I, Hilbi H. Phosphoinositide lipids and the *Legionella* pathogen vacuole. *Curr Top Microbiol Immunol* (2013) **376**:155–73. doi:10.1007/82_2013_341
34. Hilbi H, Haas A. Secretive bacterial pathogens and the secretory pathway. *Traffic* (2012) **13**:1187–97. doi:10.1111/j.1600-0854.2012.01344.x
35. Hubber A, Roy CR. Modulation of host cell function by *Legionella pneumophila* type IV effectors. *Annu Rev Cell Dev Biol* (2010) **26**:261–83. doi:10.1146/annurev-cellbio-100109-104034
36. Isberg RR, O'Connor TJ, Heidtman M. The *Legionella pneumophila* replication vacuole: making a cosy niche inside host cells. *Nat Rev Microbiol* (2009) 7:13–24. doi:10.1038/nrmicro1967
37. Sherwood RK, Roy CR. A Rab-centric perspective of bacterial pathogen-occupied vacuoles. *Cell Host Microbe* (2013) **14**:256–68. doi:10.1016/j.chom.2013.08.010
38. Hilbi H, Segal G, Shuman HA. Icm/Dot-dependent upregulation of phagocytosis by *Legionella pneumophila*. *Mol Microbiol* (2001) **42**:603–17. doi:10.1046/j.1365-2958.2001.02645.x
39. Watarai M, Derre I, Kirby J, Growney JD, Dietrich WF, Isberg RR. *Legionella pneumophila* is internalized by a macropinocytotic uptake pathway controlled by the Dot/Icm system and the mouse lgn1 locus. *J Exp Med* (2001) **194**:1081–96. doi:10.1084/jem.194.8.1081
40. Roy CR, Berger KH, Isberg RR. *Legionella pneumophila* DotA protein is required for early phagosome trafficking decisions that occur within minutes of bacterial uptake. *Mol Microbiol* (1998) **28**:663–74. doi:10.1046/j.1365-2958.1998.00841.x
41. Xu L, Shen X, Bryan A, Banga S, Swanson MS, Luo ZQ. Inhibition of host vacuolar H^+-ATPase activity by a *Legionella pneumophila* effector. *PLoS Pathog* (2010) **6**:e1000822. doi:10.1371/journal.ppat.1000822
42. Finsel I, Ragaz C, Hoffmann C, Harrison CF, Weber S, Van Rahden VA, et al. The *Legionella* effector RidL inhibits retrograde trafficking to promote intracellular replication. *Cell Host Microbe* (2013) **14**:38–50. doi:10.1016/j.chom.2013.06.001
43. Kagan JC, Roy CR. *Legionella* phagosomes intercept vesicular traffic from endoplasmic reticulum exit sites. *Nat Cell Biol* (2002) **4**:945–54. doi:10.1038/ncb883
44. Lu H, Clarke M. Dynamic properties of *Legionella*-containing phagosomes in *Dictyostelium* amoebae. *Cell Microbiol* (2005) **7**:995–1007. doi:10.1111/j.1462-5822.2005.00528.x
45. Robinson CG, Roy CR. Attachment and fusion of endoplasmic reticulum with vacuoles containing *Legionella pneumophila*. *Cell Microbiol* (2006) **8**:793–805. doi:10.1111/j.1462-5822.2005.00666.x
46. Swanson MS, Isberg RR. Association of *Legionella pneumophila* with the macrophage endoplasmic reticulum. *Infect Immun* (1995) **63**:3609–20.
47. Chen J, De Felipe KS, Clarke M, Lu H, Anderson OR, Segal G, et al. *Legionella* effectors that promote nonlytic release from protozoa. *Science* (2004) **303**:1358–61. doi:10.1126/science.1094226
48. Hilbi H, Weber S, Finsel I. Anchors for effectors: subversion of phosphoinositide lipids by *Legionella*. *Front Microbiol* (2011) **2**:91. doi:10.3389/fmicb.2011.00091
49. Weber S, Stirnimann CU, Wieser M, Frey D, Meier R, Engelhardt S, et al. A type IV translocated *Legionella* cysteine phytase counteracts intracellular growth restriction by phytate. *J Biol Chem* (2014) **289**:34175–88. doi:10.1074/jbc.M114.592568
50. Weber S, Wagner M, Hilbi H. Live-cell imaging of phosphoinositide dynamics and membrane architecture during *Legionella* infection. *MBio* (2014) **5**:e839–813. doi:10.1128/mBio.00839-13
51. Weber SS, Ragaz C, Reus K, Nyfeler Y, Hilbi H. *Legionella pneumophila* exploits PI(4)P to anchor secreted effector proteins to the replicative vacuole. *PLoS Pathog* (2006) **2**:e46. doi:10.1371/journal.ppat.0020046
52. Brombacher E, Urwyler S, Ragaz C, Weber SS, Kami K, Overduin M, et al. Rab1 guanine nucleotide exchange factor SidM is a major phosphatidylinositol 4-phosphate-binding effector protein of *Legionella pneumophila*. *J Biol Chem* (2009) **284**:4846–56. doi:10.1074/jbc.M807505200
53. Del Campo CM, Mishra AK, Wang YH, Roy CR, Janmey PA, Lambright DG. Structural basis for PI(4)*P*-specific membrane recruitment of the *Legionella pneumophila* effector DrrA/SidM. *Structure* (2014) **22**:397–408. doi:10.1016/j.str.2013.12.018
54. Dolinsky S, Haneburger I, Cichy A, Hannemann M, Itzen A, Hilbi H. The *Legionella longbeachae* Icm/Dot substrate SidC selectively binds phosphatidylinositol 4-phosphate with nanomolar affinity and promotes pathogen vacuole-endoplasmic reticulum interactions. *Infect Immun* (2014) **82**:4021–33. doi:10.1128/IAI.01685-14
55. Hubber A, Arasaki K, Nakatsu F, Hardiman C, Lambright D, De Camilli P, et al. The machinery at endoplasmic reticulum-plasma membrane contact sites contributes to spatial regulation of multiple *Legionella* effector proteins. *PLoS Pathog* (2014) **10**:e1004222. doi:10.1371/journal.ppat.1004222
56. Ragaz C, Pietsch H, Urwyler S, Tiaden A, Weber SS, Hilbi H. The *Legionella pneumophila* phosphatidylinositol-4 phosphate-binding type IV substrate SidC recruits endoplasmic reticulum vesicles to a replication-permissive vacuole. *Cell Microbiol* (2008) **10**:2416–33. doi:10.1111/j.1462-5822.2008.01219.x
57. Schoebel S, Blankenfeldt W, Goody RS, Itzen A. High-affinity binding of phosphatidylinositol 4-phosphate by *Legionella pneumophila* DrrA. *EMBO Rep* (2010) **11**:598–604. doi:10.1038/embor.2010.97
58. Harding CR, Mattheis C, Mousnier A, Oates CV, Hartland EL, Frankel G, et al. LtpD is a novel *Legionella pneumophila* effector that binds phosphatidylinositol 3-phosphate and inositol monophosphatase IMPA1. *Infect Immun* (2013) **81**:4261–70. doi:10.1128/IAI.01054-13
59. Jank T, Bohmer KE, Tzivelekidis T, Schwan C, Belyi Y, Aktories K. Domain organization of *Legionella* effector SetA. *Cell Microbiol* (2012) **14**:852–68. doi:10.1111/j.1462-5822.2012.01761.x
60. Weber SS, Ragaz C, Hilbi H. The inositol polyphosphate 5-phosphatase OCRL1 restricts intracellular growth of *Legionella*, localizes to the replicative vacuole and binds to the bacterial effector LpnE. *Cell Microbiol* (2009) **11**:442–60. doi:10.1111/j.1462-5822.2008.01266.x
61. Hsu F, Zhu W, Brennan L, Tao L, Luo ZQ, Mao Y. Structural basis for substrate recognition by a unique *Legionella* phosphoinositide phosphatase. *Proc Natl Acad Sci U S A* (2012) **109**:13567–72. doi:10.1073/pnas.1207903109
62. Toulabi L, Wu X, Cheng Y, Mao Y. Identification and structural characterization of a *Legionella* phosphoinositide phosphatase. *J Biol Chem* (2013) **288**:24518–27. doi:10.1074/jbc.M113.474239
63. Nagai H, Kagan JC, Zhu X, Kahn RA, Roy CR. A bacterial guanine nucleotide exchange factor activates ARF on *Legionella* phagosomes. *Science* (2002) **295**:679–82. doi:10.1126/science.1067025
64. Machner MP, Isberg RR. Targeting of host Rab GTPase function by the intravacuolar pathogen *Legionella pneumophila*. *Dev Cell* (2006) **11**:47–56. doi:10.1016/j.devcel.2006.05.013
65. Machner MP, Isberg RR. A bifunctional bacterial protein links GDI displacement to Rab1 activation. *Science* (2007) **318**:974–7. doi:10.1126/science.1149121
66. Murata T, Delprato A, Ingmundson A, Toomre DK, Lambright DG, Roy CR. The *Legionella pneumophila* effector protein DrrA is a Rab1 guanine nucleotide-exchange factor. *Nat Cell Biol* (2006) **8**:971–7. doi:10.1038/ncb1463
67. Schoebel S, Oesterlin LK, Blankenfeldt W, Goody RS, Itzen A. RabGDI displacement by DrrA from *Legionella* is a consequence of its guanine nucleotide exchange activity. *Mol Cell* (2009) **36**:1060–72. doi:10.1016/j.molcel.2009.11.014
68. Ingmundson A, Delprato A, Lambright DG, Roy CR. *Legionella pneumophila* proteins that regulate Rab1 membrane cycling. *Nature* (2007) **450**:365–9. doi:10.1038/nature06336
69. Hardiman CA, Roy CR. AMPylation is critical for Rab1 localization to vacuoles containing *Legionella pneumophila*. *MBio* (2014) **5**:e1035–1013. doi:10.1128/mBio.01035-13
70. Müller MP, Peters H, Blumer J, Blankenfeldt W, Goody RS, Itzen A. The *Legionella* effector protein DrrA AMPylates the membrane traffic regulator Rab1b. *Science* (2010) **329**:946–9. doi:10.1126/science.1192276
71. Neunuebel MR, Chen Y, Gaspar AH, Backlund PS, Yergey A, Machner MP. De-AMPylation of the small GTPase Rab1 by the pathogen *Legionella pneumophila*. *Science* (2011) **333**:453–6. doi:10.1126/science.1207193

72. Tan Y, Luo ZQ. *Legionella pneumophila* SidD is a deAMPylase that modifies Rab1. *Nature* (2011) **475**:506–9. doi:10.1038/nature10307
73. Campanacci V, Mukherjee S, Roy CR, Cherfils J. Structure of the *Legionella* effector AnkX reveals the mechanism of phosphocholine transfer by the FIC domain. *EMBO J* (2013) **32**:1469–77. doi:10.1038/emboj.2013.82
74. Goody PR, Heller K, Oesterlin LK, Müller MP, Itzen A, Goody RS. Reversible phosphocholination of Rab proteins by *Legionella pneumophila* effector proteins. *EMBO J* (2012) **31**:1774–84. doi:10.1038/emboj.2012.16
75. Mukherjee S, Liu X, Arasaki K, Mcdonough J, Galan JE, Roy CR. Modulation of Rab GTPase function by a protein phosphocholine transferase. *Nature* (2011) **477**:103–6. doi:10.1038/nature10335
76. Pan X, Lührmann A, Satoh A, Laskowski-Arce MA, Roy CR. Ankyrin repeat proteins comprise a diverse family of bacterial type IV effectors. *Science* (2008) **320**:1651–4. doi:10.1126/science.1158160
77. Tan Y, Arnold RJ, Luo ZQ. *Legionella pneumophila* regulates the small GTPase Rab1 activity by reversible phosphorylcholination. *Proc Natl Acad Sci U S A* (2011) **108**:21212–7. doi:10.1073/pnas.1114023109
78. Cheng W, Yin K, Lu D, Li B, Zhu D, Chen Y, et al. Structural insights into a unique *Legionella pneumophila* effector LidA recognizing both GDP and GTP bound Rab1 in their active state. *PLoS Pathog* (2012) **8**:e1002528. doi:10.1371/journal.ppat.1002528
79. Neunuebel MR, Mohammadi S, Jarnik M, Machner MP. *Legionella pneumophila* LidA affects nucleotide binding and activity of the host GTPase Rab1. *J Bacteriol* (2012) **194**:1389–400. doi:10.1128/JB.06306-11
80. Schoebel S, Cichy AL, Goody RS, Itzen A. Protein LidA from *Legionella* is a rab GTPase supereffector. *Proc Natl Acad Sci U S A* (2011) **108**:17945–50. doi:10.1073/pnas.1113133108
81. Gazdag EM, Schobel S, Shkumatov AV, Goody RS, Itzen A. The structure of the N-terminal domain of the *Legionella* protein SidC. *J Struct Biol* (2014) **186**:188–94. doi:10.1016/j.jsb.2014.02.003
82. Horenkamp FA, Mukherjee S, Alix E, Schauder CM, Hubber AM, Roy CR, et al. *Legionella pneumophila* subversion of host vesicular transport by SidC effector proteins. *Traffic* (2014) **15**:488–99. doi:10.1111/tra.12158
83. Hsu F, Luo X, Qiu J, Teng YB, Jin J, Smolka MB, et al. The *Legionella* effector SidC defines a unique family of ubiquitin ligases important for bacterial phagosomal remodeling. *Proc Natl Acad Sci U S A* (2014) **111**:10538–43. doi:10.1073/pnas.1402605111
84. Bonifacino JS, Hurley JH. Retromer. *Curr Opin Cell Biol* (2008) **20**:427–36. doi:10.1016/j.ceb.2008.03.009
85. Arighi CN, Hartnell LM, Aguilar RC, Haft CR, Bonifacino JS. Role of the mammalian retromer in sorting of the cation-independent mannose 6-phosphate receptor. *J Cell Biol* (2004) **165**:123–33. doi:10.1083/jcb.200312055
86. Seaman MN. The retromer complex – endosomal protein recycling and beyond. *J Cell Sci* (2012) **125**:4693–702. doi:10.1242/jcs.103440
87. Rojas R, Van Vlijmen T, Mardones GA, Prabhu Y, Rojas AL, Mohammed S, et al. Regulation of retromer recruitment to endosomes by sequential action of Rab5 and Rab7. *J Cell Biol* (2008) **183**:513–26. doi:10.1083/jcb.200804048
88. Clemens DL, Lee BY, Horwitz MA. Deviant expression of Rab5 on phagosomes containing the intracellular pathogens *Mycobacterium tuberculosis* and *Legionella pneumophila* is associated with altered phagosomal fate. *Infect Immun* (2000) **68**:2671–84. doi:10.1128/IAI.68.5.2671-2684.2000
89. Clemens DL, Lee BY, Horwitz MA. *Mycobacterium tuberculosis* and *Legionella pneumophila* phagosomes exhibit arrested maturation despite acquisition of Rab7. *Infect Immun* (2000) **68**:5154–66. doi:10.1128/IAI.68.9.5154-5166.2000
90. Hoffmann C, Finsel I, Otto A, Pfaffinger G, Rothmeier E, Hecker M, et al. Functional analysis of novel Rab GTPases identified in the proteome of purified *Legionella*-containing vacuoles from macrophages. *Cell Microbiol* (2014) **16**:1034–52. doi:10.1111/cmi.12256
91. Urwyler S, Nyfeler Y, Ragaz C, Lee H, Mueller LN, Aebersold R, et al. Proteome analysis of *Legionella* vacuoles purified by magnetic immunoseparation reveals secretory and endosomal GTPases. *Traffic* (2009) **10**:76–87. doi:10.1111/j.1600-0854.2008.00851.x
92. Shevchuk O, Batzilla C, Hägele S, Kusch H, Engelmann S, Hecker M, et al. Proteomic analysis of *Legionella*-containing phagosomes isolated from *Dictyostelium*. *Int J Med Microbiol* (2009) **299**:489–508. doi:10.1016/j.ijmm.2009.03.006
93. Choudhury R, Diao A, Zhang F, Eisenberg E, Saint-Pol A, Williams C, et al. Lowe syndrome protein OCRL1 interacts with clathrin and regulates protein trafficking between endosomes and the trans-Golgi network. *Mol Biol Cell* (2005) **16**:3467–79. doi:10.1091/mbc.E05-02-0120
94. van Rahden VA, Brand K, Najm J, Heeren J, Pfeffer SR, Braulke T, et al. The 5-phosphatase OCRL mediates retrograde transport of the mannose 6-phosphate receptor by regulating a Rac1-cofilin signalling module. *Hum Mol Genet* (2012) **21**:5019–38. doi:10.1093/hmg/dds343
95. Glick D, Barth S, Macleod KF. Autophagy: cellular and molecular mechanisms. *J Pathol* (2010) **221**:3–12. doi:10.1002/path.2697
96. Nakatogawa H, Suzuki K, Kamada Y, Ohsumi Y. Dynamics and diversity in autophagy mechanisms: lessons from yeast. *Nat Rev Mol Cell Biol* (2009) **10**:458–67. doi:10.1038/nrm2708
97. Feng Y, Yao Z, Klionsky DJ. How to control self-digestion: transcriptional, post-transcriptional, and post-translational regulation of autophagy. *Trends Cell Biol* (2015) **25**:354–63. doi:10.1016/j.tcb.2015.02.002
98. Choy A, Dancourt J, Mugo B, O'Connor TJ, Isberg RR, Melia TJ, et al. The *Legionella* effector RavZ inhibits host autophagy through irreversible Atg8 deconjugation. *Science* (2012) **338**:1072–6. doi:10.1126/science.1227026
99. Otto GP, Wu MY, Clarke M, Lu H, Anderson OR, Hilbi H, et al. Macroautophagy is dispensable for intracellular replication of *Legionella pneumophila* in *Dictyostelium discoideum*. *Mol Microbiol* (2004) **51**:63–72. doi:10.1046/j.1365-2958.2003.03826.x
100. Tung SM, Unal C, Ley A, Pena C, Tunggal B, Noegel AA, et al. Loss of *Dictyostelium* ATG9 results in a pleiotropic phenotype affecting growth, development, phagocytosis and clearance and replication of *Legionella pneumophila*. *Cell Microbiol* (2010) **12**:765–80. doi:10.1111/j.1462-5822.2010.01432.x
101. Stewart M. Molecular mechanism of the nuclear protein import cycle. *Nat Rev Mol Cell Biol* (2007) **8**:195–208. doi:10.1038/nrm2114
102. Clarke PR, Zhang C. Spatial and temporal coordination of mitosis by Ran GTPase. *Nat Rev Mol Cell Biol* (2008) **9**:464–77. doi:10.1038/nrm2410
103. Goodman B, Zheng Y. Mitotic spindle morphogenesis: ran on the microtubule cytoskeleton and beyond. *Biochem Soc Trans* (2006) **34**:716–21. doi:10.1042/BST0340716
104. Yudin D, Fainzilber M. Ran on tracks – cytoplasmic roles for a nuclear regulator. *J Cell Sci* (2009) **122**:587–93. doi:10.1242/jcs.015289
105. Bischoff FR, Ponstingl H. Catalysis of guanine nucleotide exchange on Ran by the mitotic regulator RCC1. *Nature* (1991) **354**:80–2. doi:10.1038/354080a0
106. de Felipe KS, Glover RT, Charpentier X, Anderson OR, Reyes M, Pericone CD, et al. *Legionella* eukaryotic-like type IV substrates interfere with organelle trafficking. *PLoS Pathog* (2008) **4**:e1000117. doi:10.1371/journal.ppat.1000117
107. de Felipe KS, Pampou S, Jovanovic OS, Pericone CD, Ye SF, Kalachikov S, et al. Evidence for acquisition of *Legionella* type IV secretion substrates via interdomain horizontal gene transfer. *J Bacteriol* (2005) **187**:7716–26. doi:10.1128/JB.187.22.7716-7726.2005
108. Ninio S, Celli J, Roy CR. A *Legionella pneumophila* effector protein encoded in a region of genomic plasticity binds to Dot/Icm-modified vacuoles. *PLoS Pathog* (2009) **5**:e1000278. doi:10.1371/journal.ppat.1000278
109. Ivanov SS, Charron G, Hang HC, Roy CR. Lipidation by the host prenyltransferase machinery facilitates membrane localization of *Legionella pneumophila* effector proteins. *J Biol Chem* (2010) **285**:34686–98. doi:10.1074/jbc.M110.170746
110. Rothmeier E, Pfaffinger G, Hoffmann C, Harrison CF, Grabmayr H, Repnik U, et al. Activation of Ran GTPase by a *Legionella* effector promotes microtubule polymerization, pathogen vacuole motility and infection. *PLoS Pathog* (2013) **9**:e1003598. doi:10.1371/journal.ppat.1003598
111. Etienne-Manneville S. Microtubules in cell migration. *Annu Rev Cell Dev Biol* (2013) **29**:471–99. doi:10.1146/annurev-cellbio-101011-155711
112. Simon S, Wagner MA, Rothmeier E, Müller-Taubenberger A, Hilbi H. Icm/Dot-dependent inhibition of phagocyte migration by *Legionella* is antagonized by a translocated Ran GTPase activator. *Cell Microbiol* (2014) **16**:977–92. doi:10.1111/cmi.12258
113. Luo ZQ. Striking a balance: modulation of host cell death pathways by *Legionella pneumophila*. *Front Microbiol* (2011) **2**:36. doi:10.3389/fmicb.2011.00036

114. Diez E, Lee SH, Gauthier S, Yaraghi Z, Tremblay M, Vidal S, et al. Birc1e is the gene within the Lgn1 locus associated with resistance to *Legionella pneumophila*. *Nat Genet* (2003) **33**:55–60. doi:10.1038/ng1065
115. Wright EK, Goodart SA, Growney JD, Hadinoto V, Endrizzi MG, Long EM, et al. Naip5 affects host susceptibility to the intracellular pathogen *Legionella pneumophila*. *Curr Biol* (2003) **13**:27–36. doi:10.1016/S0960-9822(02)01359-3
116. Dietrich WF, Damron DM, Isberg RR, Lander ES, Swanson MS. Lgn1, a gene that determines susceptibility to *Legionella pneumophila*, maps to mouse chromosome 13. *Genomics* (1995) **26**:443–50. doi:10.1016/0888-7543(95)80161-E
117. Amer A, Franchi L, Kanneganti TD, Body-Malapel M, Ozoren N, Brady G, et al. Regulation of *Legionella* phagosome maturation and infection through flagellin and host Ipaf. *J Biol Chem* (2006) **281**:35217–23. doi:10.1074/jbc.M604933200
118. Coers J, Vance RE, Fontana MF, Dietrich WF. Restriction of *Legionella pneumophila* growth in macrophages requires the concerted action of cytokine and Naip5/Ipaf signalling pathways. *Cell Microbiol* (2007) **9**:2344–57. doi:10.1111/j.1462-5822.2007.00963.x
119. Fortier A, De Chastellier C, Balor S, Gros P. Birc1e/Naip5 rapidly antagonizes modulation of phagosome maturation by *Legionella pneumophila*. *Cell Microbiol* (2007) **9**:910–23. doi:10.1111/j.1462-5822.2006.00839.x
120. Lamkanfi M, Amer A, Kanneganti TD, Munoz-Planillo R, Chen G, Vandenabeele P, et al. The Nod-like receptor family member Naip5/Birc1e restricts *Legionella pneumophila* growth independently of caspase-1 activation. *J Immunol* (2007) **178**:8022–7. doi:10.4049/jimmunol.178.12.8022
121. Lightfield KL, Persson J, Brubaker SW, Witte CE, Von Moltke J, Dunipace EA, et al. Critical function for Naip5 in inflammasome activation by a conserved carboxy-terminal domain of flagellin. *Nat Immunol* (2008) **9**:1171–8. doi:10.1038/ni.1646
122. Molofsky AB, Byrne BG, Whitfield NN, Madigan CA, Fuse ET, Tateda K, et al. Cytosolic recognition of flagellin by mouse macrophages restricts *Legionella pneumophila* infection. *J Exp Med* (2006) **203**:1093–104. doi:10.1084/jem.20051659
123. Ren T, Zamboni DS, Roy CR, Dietrich WF, Vance RE. Flagellin-deficient *Legionella* mutants evade caspase-1- and Naip5-mediated macrophage immunity. *PLoS Pathog* (2006) **2**:e18. doi:10.1371/journal.ppat.0020018
124. Silveira TN, Zamboni DS. Pore formation triggered by *Legionella* spp. is an Nlrc4 inflammasome-dependent host cell response that precedes pyroptosis. *Infect Immun* (2010) **78**:1403–13. doi:10.1128/IAI.00905-09
125. Zamboni DS, Kobayashi KS, Kohlsdorf T, Ogura Y, Long EM, Vance RE, et al. The Birc1e cytosolic pattern-recognition receptor contributes to the detection and control of *Legionella pneumophila* infection. *Nat Immunol* (2006) **7**:318–25. doi:10.1038/ni1305
126. Nogueira CV, Lindsten T, Jamieson AM, Case CL, Shin S, Thompson CB, et al. Rapid pathogen-induced apoptosis: a mechanism used by dendritic cells to limit intracellular replication of *Legionella pneumophila*. *PLoS Pathog* (2009) **5**:e1000478. doi:10.1371/journal.ppat.1000478
127. Hoffmann C, Harrison CF, Hilbi H. The natural alternative: protozoa as cellular models for *Legionella* infection. *Cell Microbiol* (2014) **16**:15–26. doi:10.1111/cmi.12235
128. Banga S, Gao P, Shen X, Fiscus V, Zong WX, Chen L, et al. *Legionella pneumophila* inhibits macrophage apoptosis by targeting pro-death members of the Bcl2 protein family. *Proc Natl Acad Sci U S A* (2007) **104**:5121–6. doi:10.1073/pnas.0611030104
129. Creasey EA, Isberg RR. The protein SdhA maintains the integrity of the *Legionella*-containing vacuole. *Proc Natl Acad Sci U S A* (2012) **109**:3481–6. doi:10.1073/pnas.1121286109
130. Laguna RK, Creasey EA, Li Z, Valtz N, Isberg RRA. *Legionella pneumophila*-translocated substrate that is required for growth within macrophages and protection from host cell death. *Proc Natl Acad Sci U S A* (2006) **103**:18745–50. doi:10.1073/pnas.0609012103
131. Ge J, Gong YN, Xu Y, Shao F. Preventing bacterial DNA release and absent in melanoma 2 inflammasome activation by a *Legionella* effector functioning in membrane trafficking. *Proc Natl Acad Sci U S A* (2012) **109**:6193–8. doi:10.1073/pnas.1117490109
132. Monroe KM, Mcwhirter SM, Vance RE. Identification of host cytosolic sensors and bacterial factors regulating the type I interferon response to *Legionella pneumophila*. *PLoS Pathog* (2009) **5**:e1000665. doi:10.1371/journal.ppat.1000665
133. Zhu W, Hammad LA, Hsu F, Mao Y, Luo ZQ. Induction of caspase 3 activation by multiple *Legionella pneumophila* Dot/Icm substrates. *Cell Microbiol* (2013) **15**:1783–95. doi:10.1111/cmi.12157
134. Aurass P, Schlegel M, Metwally O, Harding CR, Schroeder GN, Frankel G, et al. The *Legionella pneumophila* Dot/Icm-secreted effector PlcC/CegC1 together with PlcA and PlcB promotes virulence and belongs to a novel zinc metallophospholipase C family present in bacteria and fungi. *J Biol Chem* (2013) **288**:11080–92. doi:10.1074/jbc.M112.426049
135. Akhter A, Gavrilin MA, Frantz L, Washington S, Ditty C, Limoli D, et al. Caspase-7 activation by the Nlrc4/Ipaf inflammasome restricts *Legionella pneumophila* infection. *PLoS Pathog* (2009) **5**:e1000361. doi:10.1371/journal.ppat.1000361
136. Akhter A, Caution K, Abu Khweek A, Tazi M, Abdulrahman BA, Abdelaziz DH, et al. Caspase-11 promotes the fusion of phagosomes harboring pathogenic bacteria with lysosomes by modulating actin polymerization. *Immunity* (2012) **37**:35–47. doi:10.1016/j.immuni.2012.05.001
137. Gilmore TD. Introduction to NF-kappaB: players, pathways, perspectives. *Oncogene* (2006) **25**:6680–4. doi:10.1038/sj.onc.1209954
138. Akamine M, Higa F, Arakaki N, Kawakami K, Takeda K, Akira S, et al. Differential roles of toll-like receptors 2 and 4 in in vitro responses of macrophages to *Legionella pneumophila*. *Infect Immun* (2005) **73**:352–61. doi:10.1128/IAI.73.1.352-361.2005
139. Hawn TR, Berrington WR, Smith IA, Uematsu S, Akira S, Aderem A, et al. Altered inflammatory responses in TLR5-deficient mice infected with *Legionella pneumophila*. *J Immunol* (2007) **179**:6981–7. doi:10.4049/jimmunol.179.10.6981
140. Hawn TR, Smith KD, Aderem A, Skerrett SJ. Myeloid differentiation primary response gene (88)- and toll-like receptor 2-deficient mice are susceptible to infection with aerosolized *Legionella pneumophila*. *J Infect Dis* (2006) **193**:1693–702. doi:10.1086/504525
141. Vinzing M, Eitel J, Lippmann J, Hocke AC, Zahlten J, Slevogt H, et al. NAIP and Ipaf control *Legionella pneumophila* replication in human cells. *J Immunol* (2008) **180**:6808–15. doi:10.4049/jimmunol.180.10.6808
142. Archer KA, Alexopoulou L, Flavell RA, Roy CR. Multiple MyD88-dependent responses contribute to pulmonary clearance of *Legionella pneumophila*. *Cell Microbiol* (2009) **11**:21–36. doi:10.1111/j.1462-5822.2008.01234.x
143. Archer KA, Roy CR. MyD88-dependent responses involving toll-like receptor 2 are important for protection and clearance of *Legionella pneumophila* in a mouse model of Legionnaires' disease. *Infect Immun* (2006) **74**:3325–33. doi:10.1128/IAI.02049-05
144. Neild AL, Shin S, Roy CR. Activated macrophages infected with *Legionella* inhibit T cells by means of MyD88-dependent production of prostaglandins. *J Immunol* (2005) **175**:8181–90. doi:10.4049/jimmunol.175.12.8181
145. Mascarenhas DP, Pereira MS, Manin GZ, Hori JI, Zamboni DS. Interleukin 1 receptor-driven neutrophil recruitment accounts to MyD88-dependent pulmonary clearance of *Legionella pneumophila* infection in vivo. *J Infect Dis* (2015) **211**:322–30. doi:10.1093/infdis/jiu430
146. Bartfeld S, Engels C, Bauer B, Aurass P, Flieger A, Brüggemann H, et al. Temporal resolution of two-tracked NF-kappaB activation by *Legionella pneumophila*. *Cell Microbiol* (2009) **11**:1638–51. doi:10.1111/j.1462-5822.2009.01354.x
147. Abu-Zant A, Jones S, Asare R, Suttles J, Price C, Graham J, et al. Anti-apoptotic signalling by the Dot/Icm secretion system of *L. pneumophila*. *Cell Microbiol* (2007) **9**:246–64. doi:10.1111/j.1462-5822.2006.00785.x
148. Losick VP, Haenssler E, Moy MY, Isberg RR. LnaB: a *Legionella pneumophila* activator of NF-kappaB. *Cell Microbiol* (2010) **12**:1083–97. doi:10.1111/j.1462-5822.2010.01452.x
149. Losick VP, Isberg RR. NF-kappaB translocation prevents host cell death after low-dose challenge by *Legionella pneumophila*. *J Exp Med* (2006) **203**:2177–89. doi:10.1084/jem.20060766
150. Ge J, Xu H, Li T, Zhou Y, Zhang Z, Li S, et al. *Legionella* type IV effector activates the NF-kappaB pathway by phosphorylating the IkappaB family of inhibitors. *Proc Natl Acad Sci U S A* (2009) **106**:13725–30. doi:10.1073/pnas.0907200106
151. Hervet E, Charpentier X, Vianney A, Lazzaroni JC, Gilbert C, Atlan D, et al. Protein kinase LegK2 is a type IV secretion system effector involved in endoplasmic reticulum recruitment and intracellular replication of *Legionella pneumophila*. *Infect Immun* (2011) **79**:1936–50. doi:10.1128/IAI.00805-10
152. Fontana MF, Banga S, Barry KC, Shen X, Tan Y, Luo ZQ, et al. Secreted bacterial effectors that inhibit host protein synthesis are critical for induction of

the innate immune response to virulent *Legionella pneumophila*. *PLoS Pathog* (2011) **7**:e1001289. doi:10.1371/journal.ppat.1001289
153. Shen X, Banga S, Liu Y, Xu L, Gao P, Shamovsky I, et al. Targeting eEF1A by a *Legionella pneumophila* effector leads to inhibition of protein synthesis and induction of host stress response. *Cell Microbiol* (2009) **11**:911–26. doi:10.1111/j.1462-5822.2009.01301.x
154. Belyi Y, Niggeweg R, Opitz B, Vogelsgesang M, Hippenstiel S, Wilm M, et al. *Legionella pneumophila* glucosyltransferase inhibits host elongation factor 1A. *Proc Natl Acad Sci U S A* (2006) **103**:16953–8. doi:10.1073/pnas.0601562103
155. Belyi Y, Tabakova I, Stahl M, Aktories K. Lgt: a family of cytotoxic glucosyl-transferases produced by *Legionella pneumophila*. *J Bacteriol* (2008) **190**:3026–35. doi:10.1128/JB.01798-07
156. Li T, Lu Q, Wang G, Xu H, Huang H, Cai T, et al. SET-domain bacterial effectors target heterochromatin protein 1 to activate host rDNA transcription. *EMBO Rep* (2013) **14**:733–40. doi:10.1038/embor.2013.86
157. Rolando M, Sanulli S, Rusniok C, Gomez-Valero L, Bertholet C, Sahr T, et al. *Legionella pneumophila* effector RomA uniquely modifies host chromatin to repress gene expression and promote intracellular bacterial replication. *Cell Host Microbe* (2013) **13**:395–405. doi:10.1016/j.chom.2013.03.004

Conflict of Interest Statement: The authors declare that the research was conducted in the absence of any commercial or financial relationships that could be construed as a potential conflict of interest.

16

Streptolysin O Rapidly Impairs Neutrophil Oxidative Burst and Antibacterial Responses to Group A *Streptococcus*

*Satoshi Uchiyama[1†], Simon Döhrmann[1†], Anjuli M. Timmer[1], Neha Dixit[2], Mariam Ghochani[3], Tamara Bhandari[1], John C. Timmer[4], Kimberly Sprague[1], Juliane Bubeck-Wardenburg[5], Scott I. Simon[2] and Victor Nizet[1,6]**

[1] Department of Pediatrics, University of California San Diego, La Jolla, CA, USA, [2] Department of Biomedical Engineering, University of California Davis, Davis, CA, USA, [3] Department of Biological Sciences, San Diego State University, San Diego, CA, USA, [4] Department of Pharmacology, University of California San Diego, La Jolla, CA, USA, [5] Department of Pediatrics, University of Chicago, Chicago, IL, USA, [6] Skaggs School of Pharmacy and Pharmaceutical Sciences, University of California San Diego, La Jolla, CA, USA

Edited by:
Inka Sastalla,
National Institutes of Health, USA

Reviewed by:
Francis Alonzo,
Loyola University Chicago, USA
Stephen B. Beres,
Houston Methodist Research Institute, USA

***Correspondence:**
Victor Nizet
vnizet@ucsd.edu

[†] Satoshi Uchiyama and Simon Döhrmann have contributed equally to this work.

Group A *Streptococcus* (GAS) causes a wide range of human infections, ranging from simple pharyngitis to life-threatening necrotizing fasciitis and toxic shock syndrome. A globally disseminated clone of M1T1 GAS has been associated with an increase in severe, invasive GAS infections in recent decades. The secreted GAS pore-forming toxin streptolysin O (SLO), which induces eukaryotic cell lysis in a cholesterol-dependent manner, is highly upregulated in the GAS M1T1 clone during bloodstream dissemination. SLO is known to promote GAS resistance to phagocytic clearance by neutrophils, a critical first element of host defense against invasive bacterial infection. Here, we examine the role of SLO in modulating specific neutrophil functions during their early interaction with GAS. We find that SLO at subcytotoxic concentrations and early time points is necessary and sufficient to suppress neutrophil oxidative burst, in a manner reversed by free cholesterol and anti-SLO blocking antibodies. In addition, SLO at subcytotoxic concentrations blocked neutrophil degranulation, interleukin-8 secretion and responsiveness, and elaboration of DNA-based neutrophil extracellular traps, cumulatively supporting a key role for SLO in GAS resistance to immediate neutrophil killing. A non-toxic SLO derivate elicits protective immunity against lethal GAS challenge in a murine infection model. We conclude that SLO exerts a novel cytotoxic-independent function at early stages of invasive infections (<30 min), contributing to GAS escape from neutrophil clearance.

Keywords: group A *Streptococcus*, *Streptococcus pyogenes*, pore-forming toxin, streptolysin O, neutrophils, oxidative burst, neutrophil extracellular traps, infection

INTRODUCTION

The Gram-positive bacterium group A *Streptococcus* (GAS; *Streptococcus pyogenes*) is a leading human pathogen ranked among the top 10 causes of infection-associated mortality worldwide (1). An estimate of 5–15% of all humans are colonized with GAS, which can cause a wide spectrum of infections ranging from self-limiting pharyngitis and impetigo to invasive and life-threatening diseases including streptococcal toxic shock syndrome and necrotizing fasciitis (2). Among 700 million GAS infections worldwide annually, an estimated 1.8 million severe infections occur with a mortality rate as high as 25% (1). A commercial vaccine to protect against GAS infections is not yet available (3).

The ability of GAS to produce serious disease, sometimes in previously healthy children and adults, defines a capacity of the organism to resist clearance by frontline elements of the host innate immune system, including neutrophils, the most abundant circulating leukocytes, which are rapidly recruited to sites of bacterial infection (4, 5). GAS express an array of virulence factors that allow the bacterium to counteract neutrophil processes and molecular effectors of bactericidal activity (6), including the anti-phagocytic surface hyaluronic acid capsule and M protein (7, 8), nuclease-mediated degradation of neutrophil extracellular traps (NETs) (9, 10), factors that neutralize reactive oxygen species (ROS) (11), and inhibitors of neutrophil-derived cationic antimicrobial peptides (12). A global increase in invasive GAS infections has been attributed to the rise of a single, globally disseminated GAS M1T1 clone (13, 14). GAS M1T1 strains can undergo a spontaneous mutation in the two component global regulatory system *covR/S* (also known as *csrR/S*) *in vivo* leading to upregulation of capsule, nuclease, and other GAS virulence factors (15, 16), and neutrophil resistance may be an important selection pressure for the survival and bloodstream dissemination of these mutants (17).

In addition to these "defensive" survival properties, GAS is remarkable for its potential to induce the apoptotic cell death of neutrophils (18), a phenotype that was subsequently linked to the bacterium's production of the potent pore-forming toxin, streptolysin O (SLO) (19). The *slo* gene has universally been found to be present in the genome of all GAS serotypes and strains, and encodes a toxin belonging to the cholesterol-dependent cytolysins (CDCs), a family comprising at least 28 members/bacterial species, including *Streptococcus pneumoniae* pneumolysin and *Listeria monocytogenes* listeriolysin O (20). SLO is a multi-functional protein with pore-dependent and -independent functions (21, 22). SLO is highly expressed by epidemic M1 GAS strains (23), and is further upregulated upon *covR/S* mutation in the GAS M1T1 background (15).

In a general sense, SLO toxicity to neutrophils has been hypothesized to contribute to GAS bloodstream survival and virulence (24, 25). Given the strong bioactivities and high expression of this toxin, as well as its prominence in invasive M1T1 GAS infection, we undertook a mechanistic analysis of its impact on specific neutrophil functions and bactericidal activity. We discovered that SLO has the capacity to suppress neutrophil oxidative burst and other key neutrophil functions including degranulation, directed migration, and formation of NETs, together promoting neutrophil resistance. Immunization with an inactivated form of SLO (rSLOmut) induced high titers of neutralizing IgG and IgM anti-SLO antibodies affording passive protection in a mouse model of systemic GAS infection, confirming its attractiveness as a component of a future multicomponent vaccine against this pathogen.

MATERIALS AND METHODS

Bacterial Strains

Group A *Streptococcus* wild-type (WT) strain M1T1 5448 was originally isolated from a patient with necrotizing fasciitis and toxic shock syndrome (26). The isogenic M1T1 5448 ΔSLO mutant and complemented (ΔSLO + pSLO) strain were described previously (19). The animal-passaged (AP) version of the M1T1 GAS parent strain contains a single inactivating adenine insertion at the 877-bp position of *covS* (16). A panel of 33 GAS isolates of various M serotypes and clinical associations was provided from collections of U.S. Centers for Disease Control and Prevention (CDC). Additional strains used were *S. pneumoniae* D39, *Staphylococcus aureus* Newman strain, and methicillin-resistant *S. aureus* (MRSA) TCH1516 WT and its isogenic ΔHla mutant. Bacteria were cultivated in Todd–Hewitt broth (THB) at 37°C.

Isolation of Human and Murine Neutrophils

Neutrophils were isolated from freshly collected whole blood of healthy donors under a protocol approved by the University of California San Diego (UCSD) Human Research Protections Program, using PolyMorphPrep Kit (Fresenius Kabi), as previously described (27). Mice were sacrificed and blood was collected via terminal cardiac puncture under a protocol approved by the UCSD Institutional Animal Care and Use Committee. Mouse blood was diluted fivefold in phosphate-buffered saline (PBS, Cellgro) and neutrophils were isolated by layers of Histopaque-1077 and Histopaque-119 (Sigma). Remaining red blood cells (RBCs) were eliminated in lysis buffer (0.15M NH_4Cl, 10 mM $KHCO_3$, 0.1 mM EDTA).

Oxidative Burst Assay

Oxidative burst assays were performed as previously described (28). Briefly, 2×10^6/mL human or 2×10^5/mL mouse neutrophils were loaded with 20 μM 2,7-dichlorofluorescein diacetate (DCFH-DA; Fisher) in Hank's balanced salt solution (HBSS, Cellgro) without Ca^{2+} and Mg^{2+} and incubated with rotation at 37°C for 20 min. Neutrophils were resuspended to a density of 1×10^7 cells/mL and 1×10^6 cells/well were infected at a multiplicity of infection (MOI) = 1 bacteria/cell and incubated for 20 min at 37°C/5% CO_2. Assays were performed using untreated control (control), 25 nM phorbol 12-myristate 13-acetate (PMA, Sigma) as positive control, 10 μM diphenyleneiodonium (DPI, Sigma) to block generation of ROS, or recombinant SLO protein (rSLO), anti-SLO antibody (Abnova), or water-soluble cholesterol (Sigma) at the stated concentrations. The fluorescence intensity was quantified on a SpectraMax M3 plate reader at 485 nm excitation/520 nm emission using SoftMax Pro software.

Neutrophil Killing Assays

Neutrophils were isolated as described above, and 1×10^6 cells in 100 μL were challenged with bacteria at MOI = 10 bacteria/cell at 37°C for 30 min with rotation. To pinpoint extracellular killing, cells were treated with 10 μg/mL cytochalasin D (cyt.

D, Sigma) for 10 min prior to infection to inhibit phagocytosis, in the presence or absence of 500 mU/mL desoxyribonuclease (DNase, Sigma) to evaluate the contribution of DNA-based NETs to bacterial killing. Surviving colony-forming units (CFU) for all killing assays were enumerated by Todd-Hewit agar (THA) dilution plating. Survival was calculated as the percentage of the initial inoculum.

Red Blood Cell Hemolysis Assay

Red blood cells were isolated from human blood and supernatants from bacterial cultures at various growth phases OD_{600} = 0.3–0.7, and tested for hemolysis activity of secreted SLO. In these assays, supernatant was incubated with 4 mM dithiothreitol (DTT) to stabilize SLO and 0.0004% trypan blue at room temperature (RT) for 10 min to inhibit streptolysin S activity. RBCs were incubated with bacterial culture supernatant with or without anti-SLO antibodies or water-soluble cholesterol (Sigma) as indicated for 30 min at 37°C, with PBS (0% hemolysis) or H_2O (100% hemolysis) as positive and negative controls. Supernatants were collected from assay wells after centrifugation at 3000 × *g* for 15 min, and hemolysis determined by absorbance with a SpectraMax M3 plate reader at 541 nm using SoftMax Pro software. Each titer was recorded as the point that the hemolysis reached half of the 100% RBC lysis (H_2O) control.

Neutrophil Elastase and IL-8 Release

Neutrophils were infected with bacteria at an MOI of 1 for 30 min as described earlier. The release of neutrophil elastase in response to bacteria was indirectly determined via the proteolytic activity of elastase. Cell-free supernatant was incubated with 20 µM of peptide substrate *N*-(Methoxysuccinyl)-Ala-Ala-Pro-Val 4-nitroanilide (Sigma) for 20 min at RT. The absorbance was measured with SpectraMax M3 plate reader at 405 nm for elastase using SoftMax Pro software. IL-8 release from neutrophils was quantified using the Quantikine ELISA kit (R&D Systems) following the manufacturer's manual. Briefly, 1×10^6 neutrophils were infected with live GAS WT and ΔSLO at an MOI of 1 for 60 min and the release of IL-8 was determined via absorbance with SpectraMax M3 plate reader at 450 nm using SoftMax Pro software.

Neutrophil Migration

Migration distance was measured by perfusing neutrophils with GAS WT, ΔSLO mutant bacteria at an MOI of 10 and 50. Five nM IL-8 was used as a chemotactic stimulus. Migration of individual neutrophils was quantified by measuring the distance in micrometer after 3 min compared to starting point by microscopy (Zeiss Axiolab).

Immunostaining of NETs and NET Quantification

Neutrophil extracellular trap visualization was performed as described previously (29). Briefly, 5×10^6 neutrophils were infected with GAS WT and ΔSLO mutant was added at MOI = 1 and incubated for 4 h in 37°C/5% CO_2. Cells were fixed with 4% paraformaldehyde and stained with anti-myeloperoxidase (MPO) antibody (1:300 diluted, Calbiochem) in PBS + 2% bovine serum albumin (BSA, Sigma) for 1 h at RT, followed by incubation with goat anti-rabbit Alexa 488 antibody (1:500 dilution, Life Technologies). Cells were counterstained with ProlongGold + 4′,6′-diamidino-2-phenylindole (DAPI, Invitrogen) and imaged on a fluorescent microscope. Representative, randomized images (n = 5) were taken for each condition and individual experiment. The ratio of NET-releasing cells to non-NET-releasing cells was determined as % of total cells releasing NETs.

Transmission Electron Microscopy

Neutrophils were infected with GAS WT and ΔSLO mutant for 4 h at an MOI of 1 and cells were prepared for transmission electron microscopy (TEM) as previously described (19, 30). Briefly, cells were fixed with 3% formaldehyde, 1.5% glutaraldehyde, 0.1M sodium cacodylate trihydrate, 5 mM $CaCl_2$ and 2.5% sucrose at pH 7.4 buffer for 1 h. Cells were washed three times in ice-cold 0.1M sodium cacodylate buffer containing 2.5% sucrose and fixed cells were incubated with Palade's OsO_4 by incubation with 1% osmium tetroxide in acetate/veronal solution for 1 h on ice. After three washing with acetate/veronal solution, the cells were stained and stabilized *en bloc* in 0.5% uranyl acetate with acetate/veronal solution overnight at RT in the dark. After one rinse with ddH_2O and one rinse with ice-cold 50% ethanol, the cells were dehydrated in an ice-cold dehydration series of 70, 95, and 100% ethanol for 15 min followed by three washes in 100% acetone at RT. Cells were then infiltrated in well-mixed Epon-ethanol resin series of three: 3% for 7 h, 66% for 7 h, followed by 100% at least overnight with agitation at RT. Next, the samples were allowed to polymerize in 100% Epon blocks at 60–80°C for a minimum of 48 h. Thin sections (70 nm) were examined with FEI Tecnai 12 transmission electron microscope and images taken with Tietz 214 CCD camera. Images were adjusted in brightness and contrast with Adobe Photoshop CS version 8.

Generation of Inactivated rSLOmut

Based on the SLO crystal structure (31), a recombinant SLO derivate (rSLOmut) with mutated tryptophan residue in the membrane-binding loop W535A was generated similarly as described (32). Loss of hemolytic activity of the rSLOmut protein was determined in J774 murine macrophages by TUNEL assay as described (19). The rSLOmut protein was cloned by overlapping PCR into a pET15b expression vector and expressed in *Escherichia coli* BL21 by induction at OD_{600} 0.6–1.0 with 1 mM isopropyl-beta-D-thiogalactopyranoside (IPTG) for 4 h. Protein purification and quantification by A_{280} was performed as described (19).

SLO Immunization Study

Eight-week-old male CD1 mice (Charles River Laboratories) were immunized three times, 2 weeks apart, with recombinant SLOmut protein or PBS control, plus adjuvant. The first immunization was done with Freund's complete adjuvant, and the following two immunizations were done with Freund's incomplete adjuvant. Protein or PBS were mixed with adjuvant 1:1 and emulsified, resulting in a final protein concentration of 500 µg/

mL. Each mouse was injected intraperitonealy (i.p.) with 100 μL of the protein/adjuvant emulsified mixture. Mice were infected i.p. with 8×10^7 CFU bacteria 2 weeks after the last immunization. Survival was monitored daily for 14 days. Anti-SLO IgG and IgM antibody titer was measured by ELISA using the serum collected 3 days before challenging the mice with GAS.

Statistical Analysis

All data were collected from at least three independent experiments in triplicate. Experiments using neutrophils were performed with a minimum of three different healthy volunteers. The data were combined and expressed as mean ± SEM except stated differently. All data were analyzed by unpaired Student's *t*-test using GraphPad Prism version 5.0 (GraphPad Software Inc.). *P* values <0.05 were considered statistically significant.

RESULTS

GAS Production of SLO Suppresses Neutrophil Oxidative Burst

A select group of Gram-positive human pathogens including GAS, *S. aureus*, and *S. pneumoniae* are leading causes of invasive, potentially life-threatening infections worldwide (1, 33, 34). Neutrophils are a critical first line of defense against such invading pathogens, and the rapid production of bactericidal ROS is key element of their effectiveness. Compared to *S. aureus* and *S. pneumoniae*, GAS of the M1 and M3 serotypes elicited a reduced level of oxidative burst from freshly isolated human neutrophils 20 min post infection (**Figure 1A**). As each bacterial species is replete with pathogen-associated molecular patterns (e.g., lipoteichoic acid, peptidoglycan, and formyl peptides) to stimulate neutrophil activation, we hypothesized that GAS produced factor(s) that suppressed the full host oxidative burst potential. By screening a panel of isogenic M1T1 GAS mutants lacking defined virulence factors, we identified a higher neutrophil oxidative burst in response to a GAS ΔSLO mutant, suggesting that the toxin could play a key role in oxidative burst suppression (**Figure 1B**).

Streptolysin O is a well-studied, secreted virulence factor of GAS that induces apoptosis in neutrophils and macrophages at later time points (~4 h) (19). We, thus, focused the aim of the current study to investigate the role of SLO in the early GAS–neutrophil interaction (<30 min) independent of cell death. Incubation of human neutrophils with GAS WT or ΔSLO mutant at MOI of 1, 10, or 50 bacteria/cell for 30 min did not produce cell death as assessed by LDH release (Figure S1A in Supplementary Material) or Live–Dead staining (Figures S1B,C in Supplementary Material). Complementation of the ΔSLO mutant with the *slo* gene expressed on a plasmid vector (ΔSLO + pSLO) restored the WT phenotype confirming that SLO dampens the neutrophil ROS response (**Figure 1C**). Additionally, purified recombinant SLO (rSLO) protein was sufficient to reduce the oxidative burst in neutrophils in a dose-dependent manner to the levels of the untreated control (**Figure 1D**). Oxidative burst suppression was not a universal property of bacterial pore-forming toxins, as live methicillin-resistant *S. aureus* (MRSA) USA300 producing the pore-forming virulence factor α-hemolysin (Hla) did not differ from an isogenic ΔHla mutant in elicitation of neutrophil ROS (Figure S2A in Supplementary Material), nor did recombinant Hla toxin at concentrations between 10 and 1000 ng/mL modulate oxidative burst (Figure S2B in Supplementary Material).

As ROS production is an important bactericidal effector of neutrophils, we confirmed the overall effect of SLO on neutrophil bacterial killing. The ΔSLO mutant was significantly more susceptible to rapid neutrophil killing after 30 min of infection than the M1T1 GAS parent strain, and WT resistance was restored in the complemented mutant (**Figure 1E**).

GAS M1 Serotype and *covRS* Mutation Are Associated with Higher SLO Activity

An increase in the production of SLO played a key role in the evolution of the globally disseminated M1T1 epidemic clone responsible for GAS infections of increased frequency and severity (13, 14, 23). Analysis of 33 clinical isolates obtained from the U.S. Center for Disease Control and Prevention (CDC) revealed a higher median hemolytic activity of M1 serotypes compared to non-M1 serotypes (**Figure 2A**). In this study, we utilized GAS strain 5448, a representative of the globally disseminated M1T1 clone, which is the most frequently isolated clone from patients suffering invasive infections (35, 36). During invasive infections in humans, M1T1 GAS (15); and isolates of certain additional serotypes (37) undergo spontaneous mutations in the CovRS regulator, leading to upregulation of virulence factors including SLO and hyaluronic acid capsule, coupled with loss of cysteine protease SpeB expression, which can degrade SLO during stationary phase (38). This "genetic switch" is recapitulated by AP of M1T1 GAS in mice (15, 38). Consistent with this model, we found that the SLO activity of WT M1T1 GAS is maximal in log-phase (OD_{600} 0.4–0.6) and declines in stationary phase, while its expression is markedly increased in the AP strain bearing a *covS* mutation, remaining high through late-log phase and stationary phase due to SpeB inactivation (**Figure 2B**).

Inhibitors of SLO Activity Increase the Oxidative Burst Response of Neutrophils to GAS

To confirm a key role of SLO in suppression of neutrophil oxidative burst, we utilized anti-SLO neutralizing antibodies and water-soluble cholesterol, a known inhibitor of SLO activity (39). Dose-dependent activity of each inhibitor against SLO was confirmed by complete ablation of hemolytic activity (**Figure 3A**). Both anti-SLO antibodies and cholesterol reversed the suppressive effect of SLO on neutrophil oxidative burst in dose-dependent manners (**Figures 3B,C**). Interestingly, delayed administration of cholesterol to neutrophils 30 min after exposure to WT GAS restored oxidative burst expression as measured at 60 min (**Figure 3D**), indicating the effects of the toxin are at least partially reversible upon SLO inhibition. When the pharmacological NADPH oxidase inhibitor, DPI was used to block neutrophil ROS generation (40), the survival difference between the WT and ΔSLO mutant GAS strains was no longer observed (**Figure 3E**); DPI treatment enhanced survival of both WT and

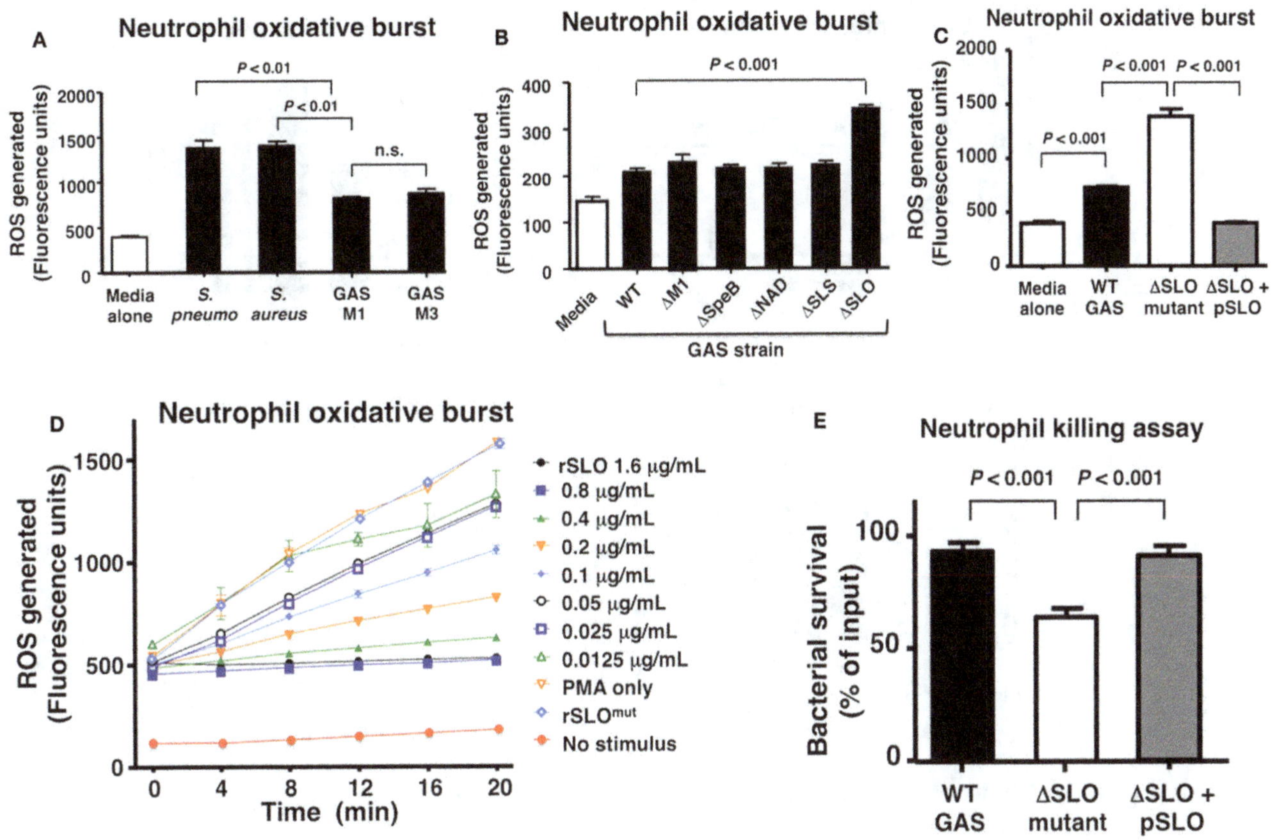

FIGURE 1 | Group A *Streptococcus* SLO suppresses neutrophil oxidative burst. (A) GAS WT serotype M1 and M3 suppress oxidative burst response in human neutrophils compared to representative WT strains of *Streptococcus pneumoniae* (D39) and *Staphylococcus aureus* (Newman) and media only control. **(B)** An isogenic GAS ΔSLO mutant fails to suppress oxidative burst in neutrophils compared to isogenic ΔM1, ΔSpeB, ΔNAD, and ΔSLS mutants and GAS WT. **(C)** The complemented GAS ΔSLO mutant (ΔSLO + pSLO) restores the suppression of oxidative burst. **(D)** Recombinant, active SLO protein suppresses oxidative burst in a dose-dependent manner in PMA-stimulated neutrophils, including media alone as a negative control. **(E)** SLO protects against killing by neutrophils at an MOI of 10 at 30 min post infection. Results are given in average ± SEM and analyzed by Student's *t*-test (n.s., not significant).

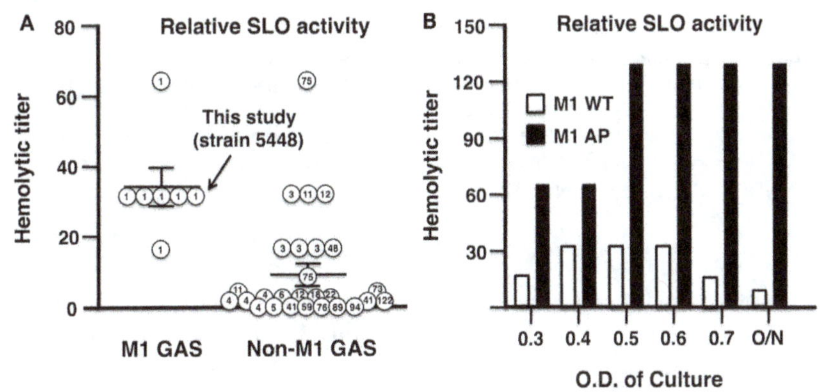

FIGURE 2 | Streptolysin O activity is enhanced in GAS M1 serotype and further enhanced after AP of GAS M1T1 clone. (A) Thirty-three clinical GAS isolates from the CDC were grouped into M1 serotype and non-M1 serotypes, where M1 serotypes including M1 5448 show increased hemolytic activity compared to non-M1 serotypes. **(B)** Hemolytic activity of SLO in GAS WT peaks in log-phase, whereas in an AP strain, M1 is elevated and reaches maximal hemolytic activity $OD_{600} \geq 0.5$.

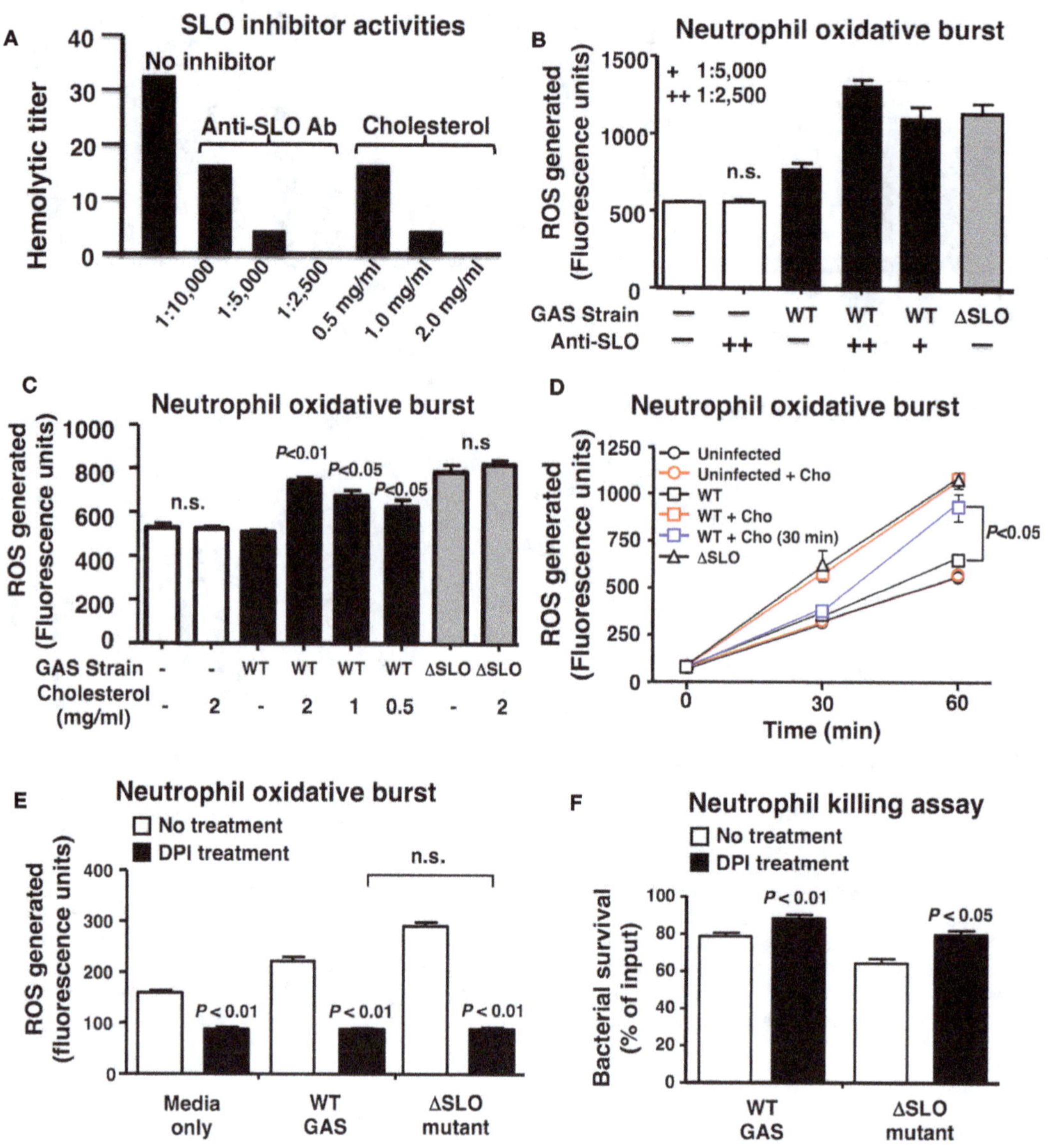

FIGURE 3 | Inhibition of SLO reverses the suppression of oxidative burst. (A) Anti-SLO antibodies and water-soluble cholesterol block hemolytic activity of SLO in a dose-dependent manner compared to control without inhibitor. **(B)** Inhibition of SLO with anti-SLO blocking antibodies or **(C)** cholesterol prevents suppression of oxidative burst in neutrophils in WT GAS with no effect on ΔSLO mutant or media only control 20 min post infection. **(D)** Oxidative burst in neutrophils infected with bacteria was monitored over a 60 min time course and cholesterol (cho) added after 30 min reversed the suppression by SLO. Effect of pharmacological inhibition of NADPH oxidase activity by DPI compared to untreated control in response to live GAS WT and ΔSLO mutant on **(E)** neutrophil oxidative burst and **(F)** neutrophil killing. Results are given in average ± SEM and analyzed by Student's *t*-test (n.s., not significant).

ΔSLO strains, demonstrating that ROS contributes to neutrophil killing of GAS (**Figure 3F**).

SLO Suppresses Neutrophil Degranulation and IL-8 Release/ Responsiveness

Reactive oxygen species have been implicated in the activation of neutrophil granule proteases including elastase (41, 42). GAS suppressed the degranulation of neutrophils in a SLO-dependent manner as quantified by elastase release into the supernatant (**Figure 4A**). Activated neutrophils themselves are an important source of chemokine interleukin-8 (IL-8), the release of which recruits additional neutrophils to the site of inflammation or infection (43). After 60 min bacterial coincubation, we observed a clear reduction in IL-8 secretion by neutrophils exposed to WT GAS compared to the ΔSLO mutant (**Figure 4B**). Finally, the mean distance of neutrophil migration over 3 min in response to IL-8 as stimulus was significantly reduced in neutrophils exposed to WT GAS compared to those exposed to the ΔSLO mutant

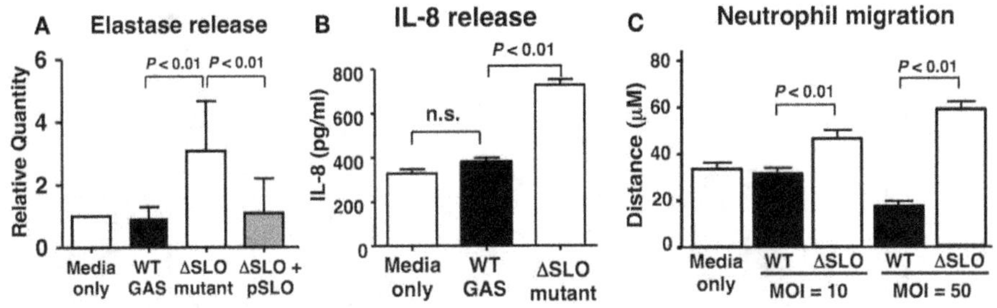

FIGURE 4 | Streptolysin O inhibits neutrophil degranulation and IL-8 release/responsiveness. **(A)** Neutrophil degranulation was determined by elastase release upon infection with live GAS WT, ΔSLO mutant, complemented bacteria and media only control, as quantified by ELISA. **(B)** SLO prevents IL-8 release from infected neutrophils compared to ΔSLO mutant bacteria and media only control as quantified by ELISA. **(C)** SLO inhibits neutrophil migration in MOI-dependent manner in contrast to ΔSLO mutant bacteria and media only control as measured by microscopic analysis. Results are given in average ± SEM and analyzed by Student's *t*-test (n.s., not significant).

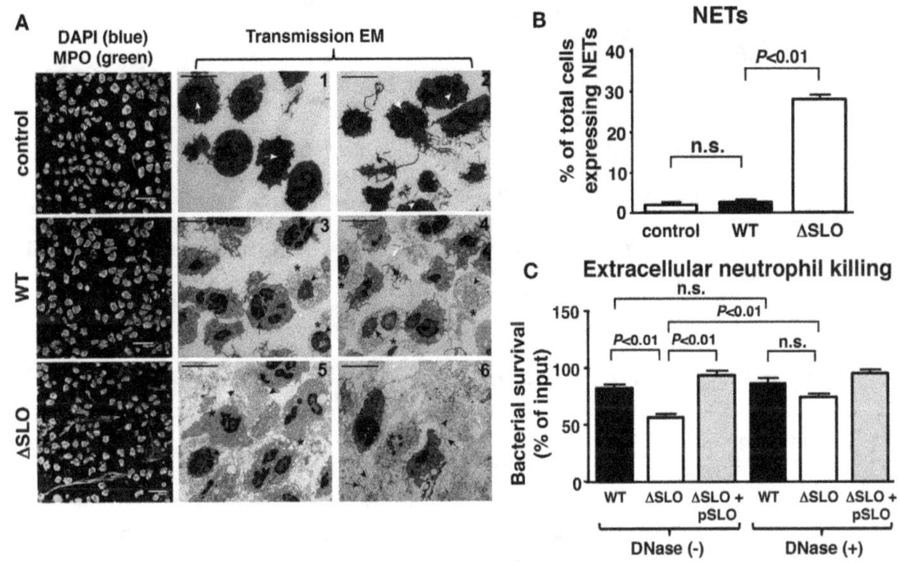

FIGURE 5 | Streptolysin O inhibits NET formation and extracellular killing. **(A)** Representative images from NET induction with GAS are shown by immunofluorescence (IF) and transmission electron microscopy (TEM) for untreated control [white arrows = multi-lobulated nuclei; white arrowheads = granules; white arrows = multi-lobulated nuclei], GAS WT [asterisk = decondensed chromatin; arrowhead = disintegrated nuclear envelope; arrow = disrupted nuclear membrane; asterisk = decondensation of chromatin; arrowhead = disintegrated nuclear envelope; arrow = disrupted cell membrane] and ΔSLO mutant [arrow = dilated space between inner and outer membrane; arrowhead = cytoplasmic protrusions and vacuolization; white arrowhead = phagocytosed bacteria; asterisk = dissolved membrane and DNA; arrowhead = vesicles of globular proteins; arrow = fibrous DNA strands; asterisk = dissolved membrane], bar = 5 μm TEM. **(B)** SLO suppresses NET formation as quantified for GAS WT, ΔSLO mutant and media only control by IF. **(C)** Extracellular killing in the presence of cytochalasin D of GAS WT, ΔSLO and complemented strain shows that SLO to protect against killing in the presence/absence of DNase control to degrade NETs. Results are given in average ± SEM and analyzed by Student's *t*-test (n.s., not significant).

(**Figure 4C**). Thus, SLO production by GAS impairs neutrophil IL-8 release and their activation and migratory responses to the chemokine.

SLO Inhibits Neutrophil Extracellular Trap Formation

Another critical innate immune function from neutrophils dependent on oxidative burst activity is NET formation, a specialized cell death process in which the nuclear DNA, embedded with antimicrobial histones and proteins, is released to ensnare and kill bacteria at tissue foci of infection (44, 45). Neutrophils were incubated with GAS WT and ΔSLO mutant at MOI of 1 bacteria/cell for 4 h then stained with antibodies against myeloperoxidase (MPO), the most abundant neutrophil granule protein that colocalizes within NETs (46). GAS SLO expression inhibited NETosis indicated by dissolved cell membranes and fibrous, extracellular DNA strands as determined by TEM and immunofluorescence (IF) microscopy (**Figure 5A**). The percentage of NET-releasing

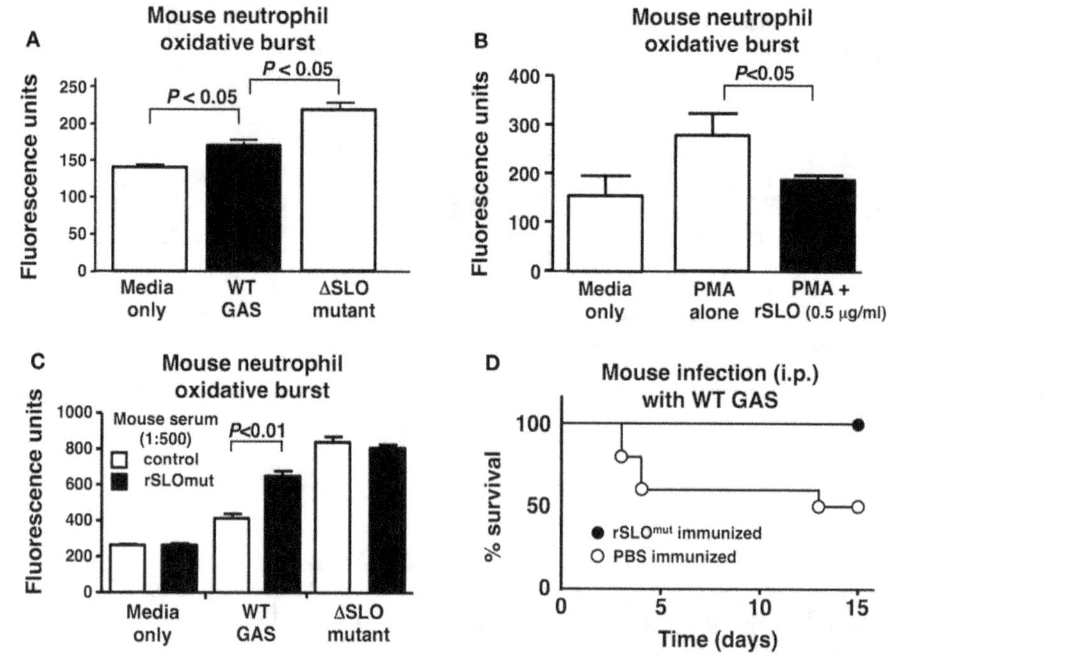

FIGURE 6 | Immunization with inactivated rSLOmut protects against GAS systemic infection. SLO suppresses the oxidative burst in mouse neutrophils with **(A)** live GAS WT compared to ΔSLO mutant or **(B)** rSLO protein after PMA-stimulation including media alone as control. Serum from rSLOmut-immunized mice inhibits SLO suppression of neutrophil oxidative burst from live bacteria **(C)**. Immunization of mice with rSLOmut protein Vs mock prevents mortality of mice challenged with GAS WT in systemic infection model **(D)**. Results are shown as ±SEM from a minimum of three independent experiments. Data were analyzed by Student's *t*-test.

neutrophils was quantified by counting random IF microscopy fields (**Figure 5B**). M1T1 GAS strains also produce a nuclease (Sda1) that degrades the DNA backbone of NETs (9, 10). We confirmed that the GAS WT and ΔSLO mutant did not differ in their nuclease activity (Figure S3 in Supplementary Material). When neutrophils were treated with cytochalasin D to block phagocytic uptake of bacteria, survival of the GAS WT and complemented strain was greater than the survival of the ΔSLO mutant, differences that were abolished by DNase treatment (**Figure 5C**). This indicates that a DNA-dependent, extracellular killing process contributes to the neutrophil resistance phenotype provided by SLO expression.

Inactivated SLO as a Protective Vaccine Antigen *In vivo*

Streptolysin O is a potent toxin that triggers cell apoptosis *in vivo* (47), and deletion of SLO from highly hemolytic M1 and M3 GAS strains (**Figure 2A**) has previously been associated with loss of virulence in murine models of systemic infection (19, 48). Potentially, due to GAS being a human-adapted pathogen, the oxidative burst response of human neutrophils was more robust than that of mouse neutrophils; nevertheless, we found a similar effect wherein murine neutrophil ROS production was significantly reduced upon exposure to WT GAS compared to the ΔSLO mutant (**Figure 6A**). Recombinant SLO also suppressed the oxidative burst response of PMA-stimulated murine neutrophils (**Figure 6B**). Native SLO may be a poor vaccine candidate because of its high cell toxicity, but strategies exist to mutate the cell-binding domain to create a toxoid SLO protein (rSLOmut) (32). Mice were immunized three times 2 weeks apart with rSLOmut, and a significant response in anti-SLO IgG and IgM antibodies was documented in serum by ELISA (Figure S4 in Supplementary Material). The post-immune antisera was found to reverse the inhibitory effect of SLO on the oxidative burst of mouse neutrophils (**Figure 6C**), demonstrating the capacity of the antibodies to block SLO neutrophil suppression. Mice immunized with rSLOmut were significantly protected against lethal systemic challenge with WT GAS M1 strain (**Figure 6D**), further supporting the inactivated SLO antigen as an attractive component for future multivalent GAS vaccines.

DISCUSSION

Group A *Streptococcus* remains one of the top 10 causes of infection-related morbidity and mortality worldwide (1). The aim of this study was to investigate the mechanisms GAS employs at early time points of infection (<30 min) to resist innate immune clearance by neutrophils. Generation of ROS upon bacterial encounter is an immediate neutrophil response to kill pathogens and to activate inflammatory signaling to recruit immune cells to the site of infection. As GAS is a non-pigmented and catalase-negative bacterium, it is dependent on alternative strategies to circumvent hostile ROS-rich host environments (11). Here, we identify SLO as the major bacterial factor responsible for suppressing the oxidative burst in neutrophils, facilitating GAS escape from innate immune killing. Previously, a study using the recombinant Mac-1/IdeS protease from GAS proposed that this virulence factor could be suppressing ROS production (49). However, physiological levels of Mac-1/IdeS from live bacteria had no effect on ROS response or virulence (50). To the best of our knowledge, SLO is the first GAS virulence factor shown to be necessary and sufficient for suppressing production of bactericidal ROS, thereby subverting neutrophil ROS-dependent killing. Our study shows that SLO produced by live bacteria as well as recombinant SLO protein is necessary and sufficient to modulate innate immune cell functions at early time points of infection. A specific suppression of oxidative burst from neutrophils by SLO was further demonstrated by using specific inhibitor of SLO activity, cholesterol, or anti-SLO blocking antibodies.

Streptolysin O, PLY, LLO, and PFO, all belong to the same family of CDCs (20). Interestingly, neither the pore-forming staphylococcal α-hemolysin nor its production from live *S. aureus* had a similar effect on oxidative burst in neutrophils, suggesting a specific unique role of SLO and related toxins that is not a general consequence of membrane perturbation and disruption. Recently, LLO and PFO have been shown to suppress oxidative burst in murine macrophages by preventing the localization of NADPH oxidase with the phagosome (51).

To extend the analysis of SLO in suppressing neutrophil functions, we investigated its role in additional ROS-dependent functions critical in clearance of bacterial pathogens. The production of ROS is linked to degranulation and the elaboration of extracellular traps, which is a critical function for entrapment and extracellular killing of bacteria (44). Our findings show that SLO prevents the release of IL-8 and elastase from neutrophils and blocks the formation of NETs. In our study, we confirmed that SLO contributes to resistance to an extracellular killing mechanism that is DNA-dependent. The hyperinvasive M1T1 GAS clone expresses a potent phage encoded nuclease (Sda1) that, like SLO, is upregulated on *covR*/S mutations arising *in vivo* under innate immune selection (15, 16). Thus, GAS has evolved a dual-pronged approach to both reduce the production of NETs, and to further dissolve NETs that are established, effectively neutralizing this arm of innate immunity.

In summary, we described a number of incapacitating effects of SLO on neutrophil function that act rapidly to suppress neutrophil function in advance of the toxin's ultimate and previously established role in triggering cell death pathways in the phagocyte. The importance of these virulence phenotypes is underscored by the protective effect of active immunization with an rSLOmut toxoid vaccine against invasive GAS infection. As SLO is present in all GAS serotypes and strains, our findings support proposals (3, 32) to include inactivated SLO toxoid as a component of a multicomponent vaccine against a leading human pathogen, GAS.

ACKNOWLEDGMENTS

Support for this work was provided by the National Institutes of Health Grants AI077780 and AI096837 to VN, AI047294 to SS, and a German Academic Exchange Service (DAAD) graduate fellowship to SD.

REFERENCES

1. Carapetis JR, Steer AC, Mulholland EK, Weber M. The global burden of group A streptococcal diseases. *Lancet Infect Dis* (2005) **5**:685–94. doi:10.1016/S1473-3099(05)70267-X
2. Walker MJ, Barnett TC, Mcarthur JD, Cole JN, Gillen CM, Henningham A, et al. Disease manifestations and pathogenic mechanisms of group A *Streptococcus*. *Clin Microbiol Rev* (2014) **27**:264–301. doi:10.1128/CMR.00101-13
3. Good MF, Batzloff MR, Pandey M. Strategies in the development of vaccines to prevent infections with group A *Streptococcus*. *Hum Vaccin Immunother* (2013) **9**:2393–7. doi:10.4161/hv.25506
4. Nathan C. Neutrophils and immunity: challenges and opportunities. *Nat Rev Immunol* (2006) **6**:173–82. doi:10.1038/nri1785
5. Jaillon S, Galdiero MR, Del Prete D, Cassatella MA, Garlanda C, Mantovani A. Neutrophils in innate and adaptive immunity. *Semin Immunopathol* (2013) **35**:377–94. doi:10.1007/s00281-013-0374-8
6. Kwinn LA, Nizet V. How group A *Streptococcus* circumvents host phagocyte defenses. *Future Microbiol* (2007) **2**:75–84. doi:10.2217/17460913.2.1.75
7. Dale JB, Washburn RG, Marques MB, Wessels MR. Hyaluronate capsule and surface M protein in resistance to opsonization of group A streptococci. *Infect Immun* (1996) **64**:1495–501.
8. Moses AE, Wessels MR, Zalcman K, Alberti S, Natanson-Yaron S, Menes T, et al. Relative contributions of hyaluronic acid capsule and M protein to virulence in a mucoid strain of the group A *Streptococcus*. *Infect Immun* (1997) **65**:64–71.
9. Sumby P, Barbian KD, Gardner DJ, Whitney AR, Welty DM, Long RD, et al. Extracellular deoxyribonuclease made by group A *Streptococcus* assists pathogenesis by enhancing evasion of the innate immune response. *Proc Natl Acad Sci U S A* (2005) **102**:1679–84. doi:10.1073/pnas.0406641102
10. Buchanan JT, Simpson AJ, Aziz RK, Liu GY, Kristian SA, Kotb M, et al. DNase expression allows the pathogen group A *Streptococcus* to escape killing in neutrophil extracellular traps. *Curr Biol* (2006) **16**:396–400. doi:10.1016/j.cub.2005.12.039
11. Henningham A, Dohrmann S, Nizet V, Cole JN. Mechanisms of group A *Streptococcus* resistance to reactive oxygen species. *FEMS Microbiol Rev* (2015) **39**:488–508. doi:10.1093/femsre/fuu009
12. Larock CN, Nizet V. Cationic antimicrobial peptide resistance mechanisms of streptococcal pathogens. *Biochim Biophys Acta* (2015) **1848**(11 Pt B):3047–54. doi:10.1016/j.bbamem.2015.02.010
13. Sumby P, Porcella SF, Madrigal AG, Barbian KD, Virtaneva K, Ricklefs SM, et al. Evolutionary origin and emergence of a highly successful clone of serotype M1 group A *Streptococcus* involved multiple horizontal gene transfer events. *J Infect Dis* (2005) **192**:771–82. doi:10.1086/432514
14. Nasser W, Beres SB, Olsen RJ, Dean MA, Rice KA, Long SW, et al. Evolutionary pathway to increased virulence and epidemic group A *Streptococcus* disease derived from 3,615 genome sequences. *Proc Natl Acad Sci U S A* (2014) **111**:E1768–76. doi:10.1073/pnas.1403138111
15. Sumby P, Whitney AR, Graviss EA, Deleo FR, Musser JM. Genome-wide analysis of group A streptococci reveals a mutation that modulates global phenotype and disease specificity. *PLoS Pathog* (2006) **2**:e5. doi:10.1371/journal.ppat.0020005
16. Walker MJ, Hollands A, Sanderson-Smith ML, Cole JN, Kirk JK, Henningham A, et al. DNase Sda1 provides selection pressure for a switch to invasive group A streptococcal infection. *Nat Med* (2007) **13**:981–5. doi:10.1038/nm1612
17. Li J, Liu G, Feng W, Zhou Y, Liu M, Wiley JA, et al. Neutrophils select hypervirulent CovRS mutants of M1T1 group A *Streptococcus* during subcutaneous infection of mice. *Infect Immun* (2014) **82**:1579–90. doi:10.1128/IAI.01458-13

18. Kobayashi SD, Braughton KR, Whitney AR, Voyich JM, Schwan TG, Musser JM, et al. Bacterial pathogens modulate an apoptosis differentiation program in human neutrophils. *Proc Natl Acad Sci U S A* (2003) **100**:10948–53. doi:10.1073/pnas.1833375100
19. Timmer AM, Timmer JC, Pence MA, Hsu LC, Ghochani M, Frey TG, et al. Streptolysin O promotes group A *Streptococcus* immune evasion by accelerated macrophage apoptosis. *J Biol Chem* (2009) **284**:862–71. doi:10.1074/jbc.M804632200
20. Tweten RK. Cholesterol-dependent cytolysins, a family of versatile pore-forming toxins. *Infect Immun* (2005) **73**:6199–209. doi:10.1128/IAI.73.10.6199-6209.2005
21. Magassa N, Chandrasekaran S, Caparon MG. *Streptococcus pyogenes* cytolysin-mediated translocation does not require pore formation by streptolysin O. *EMBO Rep* (2010) **11**:400–5. doi:10.1038/embor.2010.37
22. Logsdon LK, Hakansson AP, Cortes G, Wessels MR. Streptolysin O inhibits clathrin-dependent internalization of group A *Streptococcus*. *MBio* (2011) **2**:e310–32. doi:10.1128/mBio.00332-10
23. Zhu L, Olsen RJ, Nasser W, Beres SB, Vuopio J, Kristinsson KG, et al. A molecular trigger for intercontinental epidemics of group A *Streptococcus*. *J Clin Invest* (2015) **125**:3545–59. doi:10.1172/JCI82478
24. Sierig G, Cywes C, Wessels MR, Ashbaugh CD. Cytotoxic effects of streptolysin o and streptolysin s enhance the virulence of poorly encapsulated group A streptococci. *Infect Immun* (2003) **71**:446–55. doi:10.1128/IAI.71.1.446-455.2003
25. Ato M, Ikebe T, Kawabata H, Takemori T, Watanabe H. Incompetence of neutrophils to invasive group A *Streptococcus* is attributed to induction of plural virulence factors by dysfunction of a regulator. *PLoS One* (2008) **3**:e3455. doi:10.1371/journal.pone.0003455
26. Chatellier S, Ihendyane N, Kansal RG, Khambaty F, Basma H, Norrby-Teglund A, et al. Genetic relatedness and superantigen expression in group A *Streptococcus* serotype M1 isolates from patients with severe and nonsevere invasive diseases. *Infect Immun* (2000) **68**:3523–34. doi:10.1128/IAI.68.6.3523-3534.2000
27. Kristian SA, Datta V, Weidenmaier C, Kansal R, Fedtke I, Peschel A, et al. D-alanylation of teichoic acids promotes group A *Streptococcus* antimicrobial peptide resistance, neutrophil survival, and epithelial cell invasion. *J Bacteriol* (2005) **187**:6719–25. doi:10.1128/JB.187.19.6719-6725.2005
28. Chow OA, Von Köckritz-Blickwede M, Bright AT, Hensler ME, Zinkernagel AS, Cogen AL, et al. Statins enhance formation of phagocyte extracellular traps. *Cell Host Microbe* (2009) **8**:445–54. doi:10.1016/j.chom.2010.10.005
29. Von Köckritz-Blickwede M, Chow O, Ghochani M, Nizet V. Visualization and functional evaluation of phagocyte extracellular traps. *Methods Microbiol* (2010) **37**:139–60. doi:10.1016/S0580-9517(10)37007-3
30. Song Z, Ghochani M, Mccaffery JM, Frey TG, Chan DC. Mitofusins and OPA1 mediate sequential steps in mitochondrial membrane fusion. *Mol Biol Cell* (2009) **20**:3525–32. doi:10.1091/mbc.E09-03-0252
31. Feil SC, Ascher DB, Kuiper MJ, Tweten RK, Parker MW. Structural studies of *Streptococcus pyogenes* streptolysin O provide insights into the early steps of membrane penetration. *J Mol Biol* (2014) **426**:785–92. doi:10.1016/j.jmb.2013.11.020
32. Chiarot E, Faralla C, Chiappini N, Tuscano G, Falugi F, Gambellini G, et al. Targeted amino acid substitutions impair streptolysin O toxicity and group A *Streptococcus* virulence. *MBio* (2013) **4**:e312–87. doi:10.1128/mBio.00387-12
33. Lowy FD. *Staphylococcus aureus* infections. *N Engl J Med* (1998) **339**:520–32. doi:10.1056/NEJM199808203390806
34. O'brien KL, Wolfson LJ, Watt JP, Henkle E, Deloria-Knoll M, Mccall N, et al. Burden of disease caused by *Streptococcus pneumoniae* in children younger than 5 years: global estimates. *Lancet* (2009) **374**:893–902. doi:10.1016/S0140-6736(09)61204-6
35. Aziz RK, Kotb M. Rise and persistence of global M1T1 clone of *Streptococcus pyogenes*. *Emerg Infect Dis* (2008) **14**:1511–7. doi:10.3201/eid1410.071660
36. Steer AC, Law I, Matatolu L, Beall BW, Carapetis JR. Global emm type distribution of group A streptococci: systematic review and implications for vaccine development. *Lancet Infect Dis* (2009) **9**:611–6. doi:10.1016/S1473-3099(09)70178-1
37. Maamary PG, Sanderson-Smith ML, Aziz RK, Hollands A, Cole JN, Mckay FC, et al. Parameters governing invasive disease propensity of non-M1 serotype group A streptococci. *J Innate Immun* (2010) **2**:596–606. doi:10.1159/000317640
38. Cole JN, Barnett TC, Nizet V, Walker MJ. Molecular insight into invasive group A streptococcal disease. *Nat Rev Microbiol* (2011) **9**:724–36. doi:10.1038/nrmicro2648
39. Ofek I, Bergner-Rabinowitz S, Ginsburg I. Oxygen-stable hemolysins of group A streptococci. 8. Leukotoxic and antiphagocytic effects of streptolysins S and O. *Infect Immun* (1972) **6**:459–64.
40. Hancock JT, Jones OT. The inhibition by diphenyleneiodonium and its analogues of superoxide generation by macrophages. *Biochem J* (1987) **242**:103–7. doi:10.1042/bj2420103
41. Reeves EP, Lu H, Jacobs HL, Messina CG, Bolsover S, Gabella G, et al. Killing activity of neutrophils is mediated through activation of proteases by K+ flux. *Nature* (2002) **416**:291–7. doi:10.1038/416291a
42. Ahluwalia J, Tinker A, Clapp LH, Duchen MR, Abramov AY, Pope S, et al. The large-conductance Ca2+-activated K+ channel is essential for innate immunity. *Nature* (2004) **427**:853–8. doi:10.1038/nature02356
43. Baggiolini M, Walz A, Kunkel SL. Neutrophil-activating peptide-1/interleukin 8, a novel cytokine that activates neutrophils. *J Clin Invest* (1989) **84**:1045–9. doi:10.1172/JCI114265
44. Brinkmann V, Reichard U, Goosmann C, Fauler B, Uhlemann Y, Weiss DS, et al. Neutrophil extracellular traps kill bacteria. *Science* (2004) **303**:1532–5. doi:10.1126/science.1092385
45. Fuchs TA, Abed U, Goosmann C, Hurwitz R, Schulze I, Wahn V, et al. Novel cell death program leads to neutrophil extracellular traps. *J Cell Biol* (2007) **176**:231–41. doi:10.1083/jcb.200606027
46. Akong-Moore K, Chow OA, Von Kockritz-Blickwede M, Nizet V. Influences of chloride and hypochlorite on neutrophil extracellular trap formation. *PLoS One* (2012) **7**:e42984. doi:10.1371/journal.pone.0042984
47. Halbert SP, Bircher R, Dahle E. The analysis of streptococcal infections. V. Cardiotoxicity of streptolysin O for rabbits in vivo. *J Exp Med* (1961) **113**:759–84. doi:10.1084/jem.113.4.759
48. Limbago B, Penumalli V, Weinrick B, Scott JR. Role of streptolysin O in a mouse model of invasive group A streptococcal disease. *Infect Immun* (2000) **68**:6384–90. doi:10.1128/IAI.68.11.6384-6390.2000
49. Lei B, Deleo FR, Hoe NP, Graham MR, Mackie SM, Cole RL, et al. Evasion of human innate and acquired immunity by a bacterial homolog of CD11b that inhibits opsonophagocytosis. *Nat Med* (2001) **7**:1298–305. doi:10.1038/nm1201-1298
50. Okumura CY, Anderson EL, Dohrmann S, Tran DN, Olson J, Von Pawel-Rammingen U, et al. IgG protease Mac/IdeS is not essential for phagocyte resistance or mouse virulence of M1T1 group A *Streptococcus*. *MBio* (2013) **4**:e413–99. doi:10.1128/mBio.00499-13
51. Lam GY, Fattouh R, Muise AM, Grinstein S, Higgins DE, Brumell JH. Listeriolysin O suppresses phospholipase C-mediated activation of the microbicidal NADPH oxidase to promote *Listeria monocytogenes* infection. *Cell Host Microbe* (2011) **10**:627–34. doi:10.1016/j.chom.2011.11.005

Conflict of Interest Statement: The authors declare that the research was conducted in the absence of any commercial or financial relationships that could be construed as a potential conflict of interest.

17

Clostridial C3 Toxins Target Monocytes/ Macrophages and Modulate their Functions

Holger Barth[1], Stephan Fischer[1,2], Amelie Möglich[1] and Christina Förtsch[1]*

[1] Institute of Pharmacology and Toxicology, University of Ulm Medical Center, Ulm, Germany, [2] Institute of Organic Chemistry III, University of Ulm, Ulm, Germany

Edited by:
Katharina F. Kubatzky,
University Hospital Heidelberg,
Germany

Reviewed by:
Xingmin Sun,
Tufts University Cummings School of Veterinary Medicine, USA
Masataka Oda,
Niigata University, Japan

***Correspondence:**
Holger Barth,
Institute of Pharmacology and Toxicology, University of Ulm Medical Center, Albert-Einstein-Allee 11,
Ulm D-89081, Germany
holger.barth@uni-ulm.de

The C3 enzymes from *Clostridium (C.) botulinum* (C3bot) and *Clostridium limosum* (C3lim) are single chain protein toxins of about 25 kDa that mono-ADP-ribosylate Rho-A, -B, and -C in the cytosol of mammalian cells. We discovered that both C3 proteins are selectively internalized into the cytosol of monocytes and macrophages by an endocytotic mechanism, comparable to bacterial AB-type toxins, while they are not efficiently taken up into the cytosol of other cell types including epithelial cells and fibroblasts. C3-treatment results in disturbed macrophage functions, such as migration and phagocytosis, suggesting a novel function of clostridial C3 toxins as virulence factors, which selectively interfere with these immune cells. Moreover, enzymatic inactive C3 protein serves as a transport system to selectively deliver pharmacologically active molecules into the cytosol of monocytes/macrophages without damaging these cells. This review addresses also the generation of C3-based molecular tools for experimental macrophage pharmacology and cell biology as well as the exploitation of C3 for development of novel therapeutic strategies against monocyte/macrophage-associated diseases.

Keywords: C3, Rho, macrophage, cellular uptake, targeted drug delivery

Introduction: Bacterial C3 Proteins

Various Gram-positive bacteria produce and secrete C3 proteins (~25 kDa, pI > 9) that selectively mono-ADP-ribosylate the small GTPases Rho-A, -B, and -C. C3 ADP-ribosyltransferases catalyze the covalent transfer of the ADP-ribose group from the cosubstrate NAD onto Asn-41 of Rho (1, 2). In 1987, Aktories and co-workers described the first C3 protein (C3bot), which is produced from *Clostridium botulinum* C and D strains (3, 4). It became evident that there are two isoforms, C3bot1 and C3bot2 with 60% sequence identity. Further, C3 proteins were identified in *Clostridium limosum* (C3lim) (5), *Bacillus cereus* (C3cer) (6), as well as *Staphylococcus aureus* (C3stau) (7). C3stau *alias* epithelial differentiation inhibitor (EDIN) includes three isoforms (A, B, and C), which share ~35% sequence identity to C3bot1 and ADP-ribosylates RhoE in addition to Rho-A, -B, and -C (7).

The pathophysiological role of the clostridial C3 proteins is not clear so far, but it was suggested that C3stau might play a role as virulence factor, since more C3stau-producing *S. aureus* were isolated from patients with impetigo, diabetic foot ulcers, and other skin infections than from healthy carriers (8–11). In contrast to the clostridia, *S. aureus* enters mammalian cells and releases C3stau into the host cell cytosol, where the C3stau-catalyzed ADP-ribosylation of Rho results in reorganization of the actin cytoskeleton (12). In tissues, this mode of action causes disruption of cell–cell contacts and barrier functions, which allows the dissemination of *S. aureus* through the disturbed tissue barriers into deeper tissues, as already demonstrated in animal models (12, 13). Moreover, in endothelial cells, the C3stau-mediated reorganization of the actin cytoskeleton results in the formation of large transendothelial channels, so-called macroapertures (TEMs) (14–16), which increase the endothelial permeability and facilitate the dissemination of *S. aureus* from blood to deeper tissue layers.

C3 Enzymes Mono-ADP-Ribosylate Rho and Inhibit Rho Signaling

Up to date, C3 enzymes represent the only known specific Rho inhibitors and were widely used in biochemistry, cell biology, and experimental pharmacology as highly valuable molecular tools to investigate the role of Rho-signaling *in vitro* and in living cells (17, 18). However, it soon became evident that C3 proteins are not efficiently taken up into the cytosol of the tested cell types including epithelial cells and fibroblasts and it was suggested that C3 proteins might be exoenzymes rather than typical exotoxins, which become internalized only by non-specific mechanisms, such as pinocytosis, when cultured cells were treated for long incubation periods (>24 h) with high C3 concentrations (>10 μg/ml) in the medium (19). Therefore, C3 proteins were introduced into the cytosol of cells by artificial methods including microinjection, transfection, or by the use of molecular transporters like cell-penetrating peptides, viruses, or portions of bacterial AB-toxins (18, 20).

In line with the observation of their very limited and non-specific uptake in the tested cell types, no binding and translocation subunit was found in the C3 proteins, which could mediate their uptake into the host cell cytosol. Structure analysis of C3bot and C3stau revealed that these proteins consist of a single enzyme domain with a catalytic core containing the conserved NAD binding site and a catalytic pocket with an α-helix bent lying over two antiparallel β-sheets, which form a central cleft (21, 22) and show high structure similarity with the catalytic domains of binary actin-ADP-ribosylating toxins (18, 22, 23). Moreover, like the binary actin-ADP-ribosylating toxins, C3 enzymes contain the highly conserved amino acids, which were also identified in other ADP-ribosyltransferases: the STS-motif 174Ser-Thr-Ser176, which is flanked by Arg128 and Glu214 (22) and the "catalytic glutamate" Glu214, which plays a central role in the covalent transfer of ADP-ribose onto Asn41 of Rho. Furthermore, the binary actin ADP-ribosylating toxins and the C3 enzymes contain the so-called ADP-ribosylating toxin turn-turn (ARTT) motif, two adjacent protruding turns, which play a role for the toxin-specific recognition of actin or Rho: at position 212 in turn 2, the Rho-modifying C3 enzymes have a Gln residue, the actin-ADP-ribosylating toxins a Glu residue (22, 24). The precise molecular mechanisms of the ADP-ribosylation as well as the function of the individual amino acid residues of the ADP-ribosyltransferases is described in more detail in highly acknowledged reviews (18, 25).

Besides the Rho ADP-ribosylation, C3 acts on RalA, a member of the Ras GTPase family, in a non-enzymatic manner independent from its ADP-ribosyltransferase activity. C3 directly binds to RalA (6), which keeps RalA in its inactive GDP-bound conformation and prevents the activation of downstream RalA effectors (26, 27). However, the cellular consequences of this C3–RalA interaction are not known so far.

Molecular and Cellular Consequences of the C3-Catalyzed Rho ADP-Ribosylation

The C3 enzymes specifically mono-ADP-ribosylate the isoforms Rho A, -B, and -C at Asn-41 and prefer Rho–GDP as substrate (28, 29) because in this structure, Asn41 is accessible to C3 (30). As a consequence, the ADP-ribosylated Rho–GDP binds more efficiently to GDI, which traps Rho in the Rho–GDI complexes in the cytosol [Ref. (31), for review on Rho regulation see Ref. (32)]. This prevents the translocation of cytosolic Rho to the cytoplasmic membrane and consequently its activation to Rho–GTP and the subsequent activation of the various cellular Rho-effector molecules (33, 34). This disturbed Rho signaling is the reason underlying most of the cellular effects observed after treatment of cells with C3. However, it should be kept in mind that due to the very limited cellular uptake of C3 proteins, all the experiments with cultured cells or tissues summarized below, were performed either by incubating the cells for a long time with high concentrations of C3 protein, or by introducing C3 into the cytosol via artificial approaches.

In 1989, Chardin and co-workers described that C3-treatment of Vero cells results in a characteristic change of the cell morphology with cell-rounding, formation of long protrusions, and reorganization of the actin filaments (35). Importantly, this was the first evidence that Rho signaling might be associated with the structure of actin filaments in mammalian cells. Due to its substrate specificity, C3 proved to be a very useful tool to unravel the role of Rho for the organization of the actin cytoskeleton and for actin-dependent processes in various eukaryotic cell types. It was found that C3-treatment inhibited endocytosis, exocytosis, cytokinesis, cell cycle progression, modulated the neuronal plasticity, and induced apoptosis (25, 36, 37). In leukocytes, the C3-catalyzed Rho-inhibition inhibited migration (38–40), adhesion (41, 42), and phagocytosis (43), suggesting a pathophysiological role of C3 proteins toward such immune cells. However, the role of C3 proteins remained unclear, mainly because of their limited cellular uptake.

The increasing knowledge on the C3-mediated effects resulted in the pharmacological exploitation of C3 for novel therapeutic strategies. The C3-catalyzed inhibition of Rho signaling has protective effects on neurons of the central nervous system because it prevented the ephrin-A5-induced growth cone collapse of such cells *in vitro* (44). Moreover, independent of its ADP-ribosyltransferase activity, C3bot – but not C3lim – exhibits neurotrophic effects on cells of the central nerve system, as demonstrated for murine hippocampal neurons (45–48), but not on cells of the peripheral nerve system (49). The recombinant cell-permeable C3bot protein BA-210 (Cethrin™), which acts as Rho inhibitor, became a drug for treatment of acute spinal cord injury in men. This Rho-inhibitor showed promising effects after local application in pre-clinical animal studies (50) and is evaluated by the U.S. National Institutes of Health in a phase IIb clinical trial for its efficacy and safety.

Clostridial C3 Toxins Selectively Enter Monocytes/Macrophages and Allow for Targeted Pharmacological Modulation of Rho-Dependent Processes in these Cells

Based on reported inhibitory effects of C3 on basic leukocyte functions, a potential pathophysiological role of C3 toward immune cells was suggested and in 2010, Barth and co-workers identified monocytes/macrophages as target cells for C3bot and C3lim (19). It was demonstrated that within short incubation periods (e.g., 3 h), comparatively low concentrations of C3bot or C3lim (e.g., 0.5–1 μg/ml) were efficiently taken up into the cytosol of cultured monocytes and macrophages, such as murine (J774A.1, RAW 264.7) and human cell lines, and primary cultured human macrophages after their differentiation from blood monocytes, while under comparable experimental conditions, no relevant uptake of clostridial C3 toxins into the cytosol of cultured epithelial cells and fibroblasts was observed. Most likely, C3bot and C3lim enter the cytosol of monocytes/macrophages by a specific endocytotic mechanism because their uptake was decreased after pretreatment of cells with bafilomycin A1 (19, 51), which inhibits endosomal acidification. Taken together, these results suggest that the clostridial C3 proteins might act on monocytes/macrophages like fully functional bacterial exotoxins but the precise molecular mechanisms underlying their cellular uptake and intracellular transport are not known so far. However, Just and co-workers, who confirmed and extended the studies on the interaction of clostridial C3 proteins with macrophages, got experimental evidence that C3 binds to proteinaceous structures on J774A.1 macrophages, which exhibit significantly more C3-binding sites compared to other cell types (52). Moreover, vimentin might be involved in the uptake of C3 protein into macrophages (52).

The C3-treatment of macrophages resulted in characteristic morphological changes due to a reorganization of the actin cytoskeleton (19), as shown in **Figures 1A,B**. Importantly, this effect strictly depended on the C3-catalyzed ADP-ribosylation of Rho in the cytosol of the macrophages as enzymatically inactive C3botE174Q had no effect on cell morphology but was taken up into macrophages comparable to wild-type C3 (19). The C3-mediated impairment of Rho-signaling inhibits essential macrophage functions, such as phagocytosis (53) and migration [Ref. (54), **Figure 1C**], suggesting an immunosuppressive mode of action of the clostridial C3 proteins.

Moreover, treatment of RAW264.7 macrophages with recom-[illegible] C3bot and C3lim proteins prevented their differentiation [illegible]ent of already differentiated [illegible]hich can be considered as spec[illegible] their resorbing activity *in vitro* (55). Interestingly, a recom[illegible]t C3lim fusion protein (C2IN–C3lim), which contains C2IN, a[illegible] enzymatically inactive portion of C2I, the enzyme component o[illegible] the binary actin ADP-ribosylating *C. botulinum* C2 toxin (56), w[illegible]s much more efficient then C3bot and C3lim regarding the targeted pharmacological manipulation of osteoclast formation and activity *in vitro* (55); most likely, this fusion protein is more efficiently internalized into the macrophages then the wild-type C3 proteins. Taken together, the recombinant C3 proteins and fusion proteins represent attractive candidates for targeted pharmacological manipulation of osteoclast formation and activity with great potential for the development of novel anti-resorptive therapies in low bone mass diseases, such as osteoporosis.

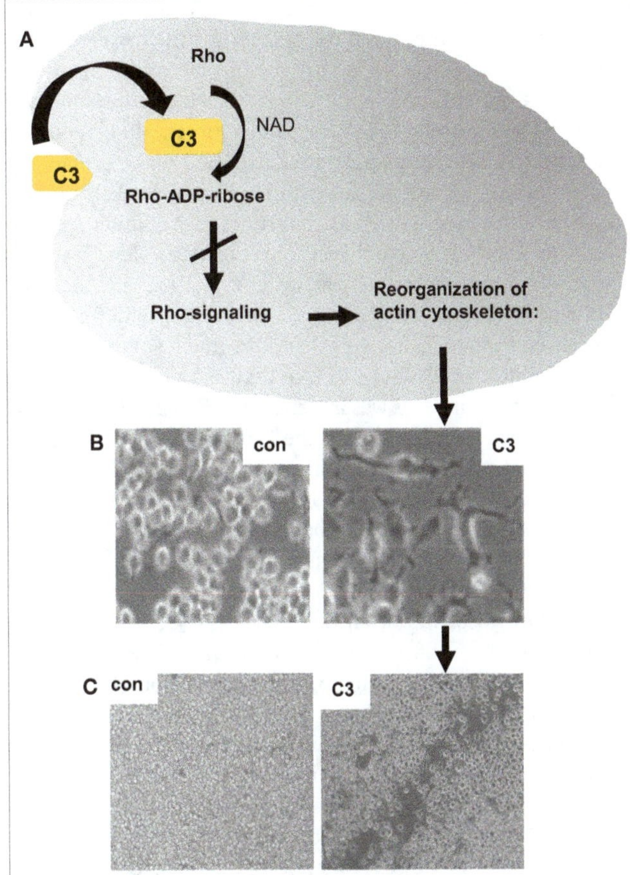

FIGURE 1 | Effects of clostridial C3 toxin on macrophages. The uptake of clostridial C3 toxin into the cytosol of macrophages causes the C3-catalyzed ADP-ribosylation of Rho and inhibition of Rho signaling, which results in a reorganization of the actin cytoskeleton [schematically depicted in **(A)**], a dramatic change of the cell morphology **(B)** and an inhibition of the migration of cultured macrophages, as shown by the scratch test experiment **(C)**. NAD, nicotinamide adenine dinucleotide; con, untreated J774A.1 macrophages, which were grown for 24 h after the removal of a portion of the cells by a scratch; C3, J774A.1 cells treated for 24 h with 300 nM C3bot1 after removal of some cells by a scratch. After treatment with C3bot1, the scratch remains open, indicating less macrophage migration of C3-treated macrophages compared to untreated control cells.

Clostridial C3 Proteins for Targeted Monocyte/ Macrophage-Selective Drug Delivery

Due to the fact that comparatively low concentrations of the clostridial C3 proteins are much more efficiently taken up into monocytes and macrophages then in other cell types, such as epithelial cells or fibroblasts, enzymatically inactive C3 protein should represent an optimal transporter for the targeted delivery of pharmacologically active (macro)molecules including therapeutic proteins and peptides into the cytosol of monocytes/macrophages. Therefore, Barth and co-workers developed and characterized novel transport systems based on enzymatically inactive C3bot1E174Q (57), which is efficiently taken up into monocytes/macrophages (19), but does not exhibit adverse effects or induces macrophage activation (51). For proof-of-concept, a pharmacologically active enzyme, namely, the actin ADP-ribosylating *C. botulinum* C2I, was genetically fused as a reporter enzyme to C3botE174Q and the uptake of the resulting recombinant C3botE174Q-C2I (see **Figure 2A**) into the cytosol of cultured macrophages was investigated. Indeed, C3bot1E174Q served for delivery of C2I into the cytosol of macrophage-like cell lines and primary human macrophages derived from blood monocytes. Incubation with C3bot1E174Q-C2I in the culture medium resulted in the ADP-ribosylation of actin in the cytosol of the macrophages and consequently in the depolymerization of F-actin (51), a clear indication that enzymatically active C2I reached the cytosol. Importantly, C3bot1E174Q-C2I had no effect on epithelial cells or fibroblasts, indicating the cell-type selectivity of this fusion toxin (51). Moreover, an application of C2I alone into the medium had no effect on the macrophages because C2I is not taken up into cells. In conclusion, enzymatically inactive C3 protein was established as a novel macrophage/monocyte-selective transport system. The generated recombinant actin-inhibitor C3bot1E174Q-C2I can serve for the targeted modulation of actin-dependent processes in monocytes and macrophages, such as the inhibition of migration and phagocytosis in the context of macrophage-associated diseases.

Prompted by this first successful exploitation of C3 for protein delivery, novel modular transporters based on biotin/streptavidin technology were developed (58, 59), as depicted in **Figure 2B**. Here, the C3botE174Q moiety mediates the specific transport of streptavidin into the cytosol of monocytes/macrophages. Streptavidin was either coupled to C3botE174Q by chemical crosslinking (59) or by genetic fusion (58) and serves as a delivery platform for biotin-labeled cargo molecules, which are then released in the host cell cytosol to elicit their pharmacological effects, as demonstrated for the enzymatic domain DTA of diphtheria toxin (58), which triggers cell death by inactivating the elongation factor EF-2, which is essential for protein biosynthesis (60) (see **Figure 2B**). In conclusion, enzymatically inactive C3 protein serves as a transport system for delivery of "foreign" proteins including pharmacologically active enzymes into the cytosol of cultured monocytes/macrophages.

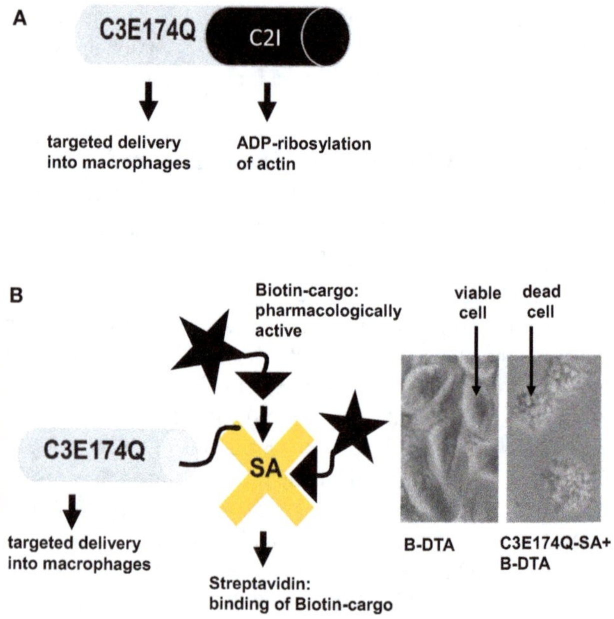

FIGURE 2 | Recombinant fusion proteins based on enzymatically inactive C3botE174Q for the targeted delivery of pharmacologically active molecules into the cytosol of macrophages and modulation of macrophage functions. (A) Schematic representation of the recombinant C3botE174Q-C2I fusion toxin, which specifically ADP-ribosylates G-actin in macrophages, resulting in depolymerization of the actin filaments and inhibition of actin-dependent cell functions. **(B)** Schematic representation of the recombinant C3botE174Q-Streptavidin (indicated as C3botE174Q-SA) fusion protein, which serves as transporter for the targeted delivery of biotinylated (indicated as B-) pharmacologically active cargo molecules into the cytosol of macrophages, which are alone not taken up, in order to specifically modulate macrophage functions (left panel). Right panel: the enzyme domain of diphtheria toxin (DTA), which ADP-ribosylates elongation factor 2 in cells and triggers cell death, was biotinylated and used as a model cargo for proof-of-concept. J774A.1 macrophages incubated with biotin-labeled DTA (B-DTA) or with B-DTA, which was previously bound to the transporter molecule C3botE174Q-SA. When B-DTA was delivered into the cytosol by C3botE174Q-SA the cells died, while cells treated with B-DTA alone stayed viable [modified from Christow et al. (58)].

Conclusion

Recombinant C3 toxins and C3 fusion toxins proved to represent valuable tools to modulate Rho- and actin-associated cell functions in monocytes and macrophages and the novel C3-based transport systems are attractive carriers for targeted drug delivery into these cells. The detailed characterization of the cellular uptake and the cellular consequences of the C3-based molecules in clinically relevant animal models, for example, after local application of C3 molecules coupled to tailored supramolecular nanocarriers (61, 62), will show whether the very promising results obtained *in vitro* and *ex vivo* can be exploited for the targeted pharmacological modulation of cellular functions in the context of monocyte/macrophage-associated diseases *in vivo*, such as inflammation, infection, or low bone mass diseases.

Acknowledgments

The research underlying parts of this review, which was performed in the laboratory of HB was supported by the Deutsche Forschungsgemeinschaft (SFB1149, project A04), the International Graduate School in Molecular Medicine Ulm (IGradU) and the Medical Faculty of the University of Ulm. SF and AM are members of IGradU.

References

1. Aktories K. Bacterial protein toxins that modify host regulatory GTPases. *Nat Rev Microbiol* (2011) **9**:487–98. doi:10.1038/nrmicro2592
2. Ménétrey J, Flatau G, Stura EA, Charbonnier JB, Gas F, Teulon JM, et al. NAD binding induces conformational changes in Rho ADP-ribosylating *Clostridium botulinum* C3 exoenzyme. *J Biol Chem* (2002) **277**:30950–7. doi:10.1074/jbc.M201844200
3. Aktories K, Weller U, Chatwal GS. *Clostridium botulinum* type C produces a novel ADP-ribosyltransferase distinct from botulinum C2 toxin. *FEBS Lett* (1987) **212**:109–13. doi:10.1016/0014-5793(87)81566-1
4. Rubin EJ, Gill DM, Boquet P, Popoff MR. Functional modification of a 21-kilodalton G protein when ADP-ribosylated by exoenzyme C3 of *Clostridium botulinum*. *Mol Cell Biol* (1988) **8**:418–26.
5. Just I, Mohr C, Schallehn G, Menard L, Didsbury JR, Vandekerckhove J, et al. Purification and characterization of an ADP-ribosyltransferase produced by *Clostridium limosum*. *J Biol Chem* (1992) **267**:10274–80.
6. Wilde C, Vogelsgesang M, Aktories K. Rho-specific *Bacillus cereus* ADP-ribosyltransferase C3cer cloning and characterization. *Biochemistry* (2003) **42**:9694–702. doi:10.1021/bi034583b
7. Wilde C, Chhatwal GS, Schmalzing G, Aktories K, Just I. A novel C3-like ADP-ribosyltransferase from *Staphylococcus aureus* modifying RhoE and Rnd3. *J Biol Chem* (2001) **276**:9537–42. doi:10.1074/jbc.M011035200
8. Czech A, Yamaguchi T, Bader L, Linder S, Kamisnki K, Sugai M, et al. Prevalence of Rho-inactivating epidermal cell differentiation inhibitor toxins in clinical *Staphylococcus aureus* isolates. *J Infect Dis* (2001) **184**:785–8. doi:10.1086/322983
9. O'Neill AJ, Larsen AR, Skov R, Henriksen AS, Chopra I. Characterization of the epidemic European fusidic acid-resistant impetigo clone of *Staphylococcus aureus*. *J Clin Microbiol* (2007) **45**:1505–10. doi:10.1128/JCM.01984-06
10. Messad N, Landraud L, Canivet B, Lina G, Richard JL, Sotto A, et al. Distribution of edin in *Staphylococcus aureus* isolated from diabetic foot ulcers. *Clin Microbiol Infect* (2013) **19**:875–80. doi:10.1111/1469-0691.12084
11. Munro P, Clement R, Lavigne JP, Pulcini C, Lemichez E, Landraud L. High prevalence of edin-C encoding RhoA-targeting toxin in clinical isolates of *Staphylococcus aureus*. *Eur J Clin Microbiol Infect Dis* (2011) **30**:965–72. doi:10.1007/s10096-011-1181-6
12. Molinari G, Rohde M, Wilde C, Just I, Aktories K, Chhatwal GS. Localization of the C3-like ADP-ribosyltransferase from *Staphylococcus aureus* during bacterial invasion of mammalian cells. *Infect Immun* (2006) **74**:3673–7. doi:10.1128/IAI.02013-05
13. Munro P, Benchetrit M, Nahori MA, Stefani C, Clement R, Michiels JF, et al. The *Staphylococcus aureus* epidermal cell differentiation inhibitor toxin promotes formation of infection foci in a mouse model of bacteremia. *Infect Immun* (2010) **78**:3404–11. doi:10.1128/IAI.00319-10
14. Boyer L, Doye A, Rolando M, Flatau G, Munro P, Gounon P, et al. Induction of transient macroapertures in endothelial cells through RhoA inhibition by *Staphylococcus aureus* factors. *J Cell Biol* (2006) **173**:809–19. doi:10.1083/jcb.200509009
15. Lemichez E, Aktories K. Hijacking of Rho GTPases during bacterial infection. *Exp Cell Res* (2013) **319**:2329–36. doi:10.1016/j.yexcr.2013.04.021
16. Maddugoda MP, Stefani C, Gonzalez-Rodriguez D, Saarikangas J, Torrino S, Janel S, et al. cAMP signaling by anthrax edema toxin induces transendothelial cell tunnels, which are resealed by MIM via Arp2/3-driven actin polymerization. *Cell Host Microbe* (2011) **10**:464–74. doi:10.1016/j.chom.2011.09.014
17. Nobes CD, Hall A. *Clostridium botulinum* C3 exoenzyme and studies on Rho proteins. In: Aktories K, editor. *Bacterial Toxins – Tools in Cell Biology and Pharmacology*. Weinheim: Chapman & Hall (1993). p. 71–83.

and specific internalization of clostridial C3 ADP-ribosyltransferases into the cytosol of macrophages. *Cell Microbiol* (2010) **12**:233–47. doi:10.1111/j.1462-5822.2009.01393.x

20. Barth H, Blöcker D, Aktories K. The uptake machinery of clostridial actin ADP-ribosylating toxins – a cell delivery system for fusion proteins and polypeptide drugs. *Naunyn Schmiedebergs Arch Pharmacol* (2002) **366**:501–12. doi:10.1007/s00210-002-0626-y
21. Evans HR, Sutton JM, Holloway DE, Ayriss J, Shone CC, Acharya KR. The crystal structure of C3stau2 from *Staphylococcus aureus* and its complex with NAD. *J Biol Chem* (2003) **278**:45924–30. doi:10.1074/jbc.M307719200
22. Han S, Arvai AS, Clancy SB, Tainer JA. Crystal structure and novel recognition motif of Rho-ADP-ribosylating exoenzyme from *Clostridium botulinum*: structural insights for recognition specificity and catalysis. *J Mol Biol* (2001) **305**:95–107. doi:10.1006/jmbi.2000.4292
23. Tsuge H, Nagahama M, Nishimura H, Hisatsune J, Sakaguchi Y, Itogawa Y, et al. Crystal structure and site-directed mutagenesis of enzymatic components from *Clostridium perfringens* iota-toxin. *J Mol Biol* (2003) **325**:471–83. doi:10.1016/S0022-2836(02)01247-0
24. Han S, Craig JA, Putnam CD, Carozzi NB, Tainer JA. Evolution and mechanism from structures of an ADP-ribosylating toxin and NAD complex. *Nat Struct Biol* (1999) **6**:932–6. doi:10.1038/13300
25. Aktories K, Wilde C, Vogelsgesang M. Rho-modifying C3-like ADP-ribosyltransferases. *Rev Physiol Biochem Pharmacol* (2004) **152**:1–22. doi:10.1007/s10254-004-0034-4
26. Holbourn KP, Sutton JM, Evans HR, Shone CC, Acharya KR. Molecular recognition of an ADP-ribosylating *Clostridium botulinum* C3 exoenzyme by RalA GTPase. *Proc Natl Acad Sci U S A* (2005) **102**:5357–62. doi:10.1073/pnas.0501525102
27. Pautsch A, Vogelsgesang M, Tränkle J, Herrmann C, Aktories K. Crystal structure of the C3bot-RalA complex reveals a novel type of action of a bacterial exoenzyme. *EMBO J* (2005) **24**:3670–80. doi:10.1038/sj.emboj.7600813
28. Aktories K, Frevert J. ADP-ribosylation of a 21-24 kDa eukaryotic protein(s) by C3, a novel botulinum ADP-ribosyltransferase, is regulated by guanine nucleotide. *Biochem J* (1987) **247**:363–8.
29. Habermann B, Mohr C, Just I, Aktories K. ADP-ribosylation and de-ADP-ribosylation of the Rho protein by *Clostridium botulinum* exoenzyme C3. Regulation by EDTA, guanine nucleotides and pH. *Biochim Biophys Acta* (1991) **1077**:253–8. doi:10.1016/0167-4838(91)90537-A
30. Wei Y, Zhang Y, Derewenda U, Liu X, Minor W, Nakamoto RK, et al. Crystal structure of RhoA-GDP and its functional implications. *Nat Struct Biol* (1997) **4**:699–703. doi:10.1038/nsb0997-699
31. Fujihara H, Walker LA, Gong MC, Lemichez E, Boquet P, Somlyo AV, et al. Inhibition of RhoA translocation and calcium sensitization by in vivo ADP-ribosylation with the chimeric toxin DC3B. *Mol Biol Cell* (1997) **8**:2437–47. doi:10.1091/mbc.8.12.2437
32. Etienne-Manneville S, Hall A. Rho GTPases in cell biology. *Nature* (2002) **420**:629–35. doi:10.1038/nature01148
33. Sehr P, Joseph G, Genth H, Just I, Pick E, Aktories K. Glucosylation and ADP ribosylation of rho proteins: effects on nucleotide binding, GTPase activity, and effector coupling. *Biochemistry* (1998) **37**:5296–304. doi:10.1021/bi972592c
34. Genth H, Gerhard R, Maeda A, Amano M, Kaibuchi K, Aktories K, et al. Entrapment of Rho ADP-ribosylated by *Clostridium botulinum* C3 exoenzyme in the Rho-guanine nucleotide dissociation inhibitor-1 complex. *J Biol Chem* (2003) **278**:28523–7. doi:10.1074/jbc.M301915200
35. Chardin P, Boquet P, Madaule P, Popoff MR, Rubin EJ, Gill DM. The mammalian G protein rho C is ADP-ribosylated by *Clostridium botulinum* exoenzyme C3 and affects actin microfilament in vero cells. *EMBO J* (1989) **8**:1087–92.
36. Aktories K, Barth H, Just I. *Clostridium botulinum* C3 exoenzyme and C3-like transferases. In: Aktories K, Just I, editors. *Bacterial Protein Toxins*. Berlin: Springer (2000). p. 207–33.
37. Nusrat A, Giry M, Turner JR, Colgan SP, Parkos CA, Carnes D, et al. Rho protein regulates tight junctions and perijunctional actin organization in polarized epithelia. *Proc Natl Acad Sci U S A* (1995) **92**:10629–33. doi:10.1073/pnas.92.23.10629
38. Worthylake RA, Lemoine S, Watson JM, Burridge K. RhoA is required for monocyte tail retraction during transendothelial migration. *J Cell Biol* (2001) **154**:147–60. doi:10.1083/jcb.200103048

Rho is a negative regulator

shape changes of T-lymphoma cells. *Eur J Cell Biol*

41. Laudanna C, Campbell JJ, Butcher EC. Role of Rho in chemoattractant-activated leukocyte adhesion through integrins. *Science* (1996) **271**:981–3. doi:10.1126/science.271.5251.981
42. Liu L, Schwartz BR, Lin N, Winn RK, Harlan JM. Requirement for RhoA kinase activation in leukocyte de-adhesion. *J Immunol* (2002) **169**:2330–6. doi:10.4049/jimmunol.169.5.2330

43. Caron E, Hall A. Identification of two distinct mechanisms of phagocytosis controlled by different Rho GTPases. *Science* (1998) **282**:1717–21. doi:10.1126/science.282.5394.1717
44. Wahl S, Barth H, Ciossek T, Aktories K, Müller BK. Ephrin-A5 induces collapse of growth cones by activating Rho and Rho kinase. *J Cell Biol* (2000) **149**:263–70. doi:10.1083/jcb.149.2.263
45. Ahnert-Hilger G, Höltje M, Grosse G, Pickert G, Mucke C, Nixdorf-Bergweiler B, et al. Differential effects of Rho GTPases on axonal and dendritic development in hippocampal neurons. *J Neurochem* (2004) **90**:9–18. doi:10.1111/j.1471-4159.2004.02475.x
46. Höltje M, Djalali S, Hofmann F, Münster-Wandowski A, Hendrix S, Boato F, et al. A 29-amino acid fragment of *Clostridium botulinum* C3 protein enhances neuronal outgrowth, connectivity, and reinnervation. *FASEB J* (2009) **23**:1115–26. doi:10.1096/fj.08-116855
47. Boato F, Hendrix S, Huelsenbeck SC, Hofmann F, Grosse G, Djalali S, et al. C3 peptide enhances recovery from spinal cord injury by improved regenerative growth of descending fiber tracts. *J Cell Sci* (2010) **123**:1652–62. doi:10.1242/jcs.066050
48. Loske P, Boato F, Hendrix S, Piepgras J, Just I, Ahnert-Hilger G, et al. Minimal essential length of *Clostridium botulinum* C3 peptides to enhance neuronal regenerative growth and connectivity in a non-enzymatic mode. *J Neurochem* (2012) **120**:1084–96. doi:10.1111/j.1471-4159.2012.07657.x
49. Auer M, Allodi I, Barham M, Udina E, Neiss WF, Navarro X, et al. C3 exoenzyme lacks effects on peripheral axon regeneration in vivo. *J Peripher Nerv Syst* (2013) **18**:30–6. doi:10.1111/jns5.12004
50. Fehlings MG, Theodore N, Harrop J, Maurais G, Kuntz C, Shaffrey CI, et al. A phase I/IIa clinical trial of a recombinant Rho protein antagonist in acute spinal cord injury. *J Neurotrauma* (2011) **28**:787–96. doi:10.1089/neu.2011.1765
51. Dmochewitz L, Förtsch C, Zwerger C, Väth M, Felder E, Huber-Lang M, et al. A recombinant fusion toxin based on enzymatic inactive C3bot1 selectively targets macrophages. *PLoS One* (2013) **8**:e54517. doi:10.1371/journal.pone.0054517
52. Rohrbeck A, von Elsner L, Hagemann S, Just I. Binding of *Clostridium botulinum* C3 exoenzyme to intact cells. *Naunyn Schmiedebergs Arch Pharmacol* (2014) **387**:523–32. doi:10.1007/s00210-014-0963-7
53. Park J, Kim JS, Jung KC, Lee HJ, Kim JI, Kim J, et al. Exoenzyme Tat-C3 inhibits association of zymosan particles, phagocytosis, adhesion, and complement binding in macrophage cells. *Mol Cells* (2003) **16**:216–23.
54. Rotsch J, Rohrbeck A, May M, Kolbe T, Hagemann S, Schelle I, et al. Inhibition of macrophage migration by *C. botulinum* exoenzyme C3. *Naunyn Schmiedebergs Arch Pharmacol* (2012) **385**:883–90. doi:10.1007/s00210-012-0764-9
55. Tautzenberger A, Förtsch C, Zwerger C, Dmochewitz L, Kreja L, Ignatius A, et al. C3 Rho-inhibitor for targeted pharmacological manipulation of osteoclast-like cells. *PLoS One* (2013) **8**:e85695. doi:10.1371/journal.pone.0085695
56. Barth H, Hofmann F, Olenik C, Just I, Aktories K. The N-terminal part of the enzyme component (C2I) of the binary *Clostridium botulinum* C2 toxin interacts with the binding component C2II and functions as a carrier system for a Rho ADP-ribosylating C3-like fusion toxin. *Infect Immun* (1998) **66**:1364–9.
57. Aktories K, Jung M, Böhmer J, Fritz G, Vandekerckhove J, Just I. Studies on the active site structure of C3-like exoenzymes: involvement of glutamic acid in catalysis of ADP-ribosylation. *Biochimie* (1995) **77**:326–32. doi:10.1016/0300-9084(96)88142-9
58. Christow H, Lillich M, Sold A, Fahrer J, Barth H. Recombinant streptavidin-C3bot for delivery of proteins into macrophages. *Toxicon* (2013) **75**:144–7. doi:10.1016/j.toxicon.2013.02.002
59. Lillich M, Chen X, Weil T, Barth H, Fahrer J. Streptavidin-conjugated C3 protein mediates the delivery of mono-biotinylated RNAse A into macrophages. *Bioconjug Chem* (2012) **23**:1426–36. doi:10.1021/bc300041z
60. Collier RJ. Understanding the mode of action of diphtheria toxin: a perspective on progress during the 20th century. *Toxicon* (2001) **39**:1793–803. doi:10.1016/S0041-0101(01)00165-9
61. Ng DYW, Fahrer J, Wu Y, Kuan SL, Barth H, Weil T. Supramolecular dendronized streptavidin nanocarriers: a versatile and synergistic strategy for the delivery of therapeutically active proteins into human cells. *Adv Healthc Mater* (2013) **2**:1620–9. doi:10.1002/adhm.201200419
62. Kuan S, Ng D, Wu Y, Förtsch C, Barth H, Doroshenko M, et al. pH responsive Janus-like supramolecular fusion proteins for functional protein delivery. *J Am Chem Soc* (2013) **135**:17254–7. doi:10.1021/ja4084122

Conflict of Interest Statement: The authors declare that the research was conducted in the absence of any commercial or financial relationships that could be construed as a potential conflict of interest.

Tasting *Pseudomonas aeruginosa* Biofilms: Human Neutrophils Express the Bitter Receptor T2R38 as Sensor for the Quorum Sensing Molecule *N*-(3-oxododecanoyl)-L-homoserine Lactone

*Susanne Maurer[1], Guido H. Wabnitz[1], Nadine A. Kahle[1], Sabine Stegmaier[1], Birgit Prior[1], Thomas Giese[1], Matthias Martin Gaida[1,2], Yvonne Samstag[1] and Gertrud Maria Hänsch[1]**

[1] *Institute for Immunology, University of Heidelberg, Heidelberg, Germany,* [2] *Institute of Pathology, University of Heidelberg, Heidelberg, Germany*

Edited by:
Inka Sastalla,
National Institutes of Health, USA

Reviewed by:
Hridayesh Prakash,
University of Hyderabad, India
Rudra Bhowmick,
Oklahoma State University, USA

***Correspondence:**
Gertrud Maria Hänsch,
Institute for Immunology, University of Heidelberg, Im Neuenheimer Feld 305, Heidelberg 69120, Germany
maria.haensch@urz.uni-heidelberg.de

Bacteria communicate with one another via specialized signaling molecules, known as quorum sensing molecules or autoinducers. The *Pseudomonas aeruginosa*-derived quorum sensing molecule *N*-(3-oxododecanoyl)-L-homoserine lactone (AHL-12), however, also activates mammalian cells. As shown previously, AHL-12-induced chemotaxis, up-regulated CD11b expression, and enhanced phagocytosis of polymorphonuclear neutrophils. Circumstantial evidence concurred with a receptor for AHL-12, which has been elusive so far. We now investigated the bitter receptor T2R38 as a potential candidate. Although identified as a taste receptor, extragustatory cells express T2R38, for example, epithelial cells in the lung. We now detected T2R38 in peripheral blood neutrophils, monocytes, and lymphocytes. T2R38 is not only found on the cell membrane but also intracellular. In neutrophils, T2R38 was located in vesicles with characteristics of lipid droplets, and super-resolution microscopy showed a co-localization with the lipid droplet membrane. Neutrophils take up AHL-12, and it co-localized with T2R38 as seen by laser scan microscopy. Binding of AHL-12 to T2R28 was confirmed by pull-down assays using biotin-coupled AHL-12 as bait. A commercially available antibody to T2R38 inhibited binding of AHL-12 to neutrophils, and this antibody by itself stimulated neutrophils, similarly to AHL-12. In conclusion, our data provide evidence for expression of functional T2R38 on neutrophils, and are compatible with the notion that T2R38 is the receptor for AHL-12.

Keywords: neutrophils, *N*-(3-oxododecanoyl)-L-homoserine lactone, quorum sensing, bitter receptor T2R38, lipid

Introduction

The quorum sensing molecule, *N*-(3-oxododecanoyl)-L-homoserine lactone (AHL-12), is produced by *Pseudomonas aeruginosa* and other Gram-negative bacteria. AHL-12 participates in the generation of virulence factors, and is particularly well studied in the context of biofilm formation (1–4). Of note, for AHL-12, a so-called "interkingdom-signaling" was reported, defined as activation of mammalian cells by bacterial signaling molecules (5, 6). Inhibition of various lymphocyte functions by AHL-12 was reported (7, 8), and was interpreted as a down-regulation of the immune response (9). On the other hand, we and others described activation of phagocytic cells, including enhancement of phagocytosis, increased expression of adhesion receptors, and induction of chemotaxis (10–12), leading to the presumption that recognition of AHL-12 might reinforce the local immune defense by preventing biofilm formation [reviewed in Ref. (13)].

AHL-12 is highly hydrophobic. For T cells and model lipid membranes, free diffusion through membranes was reported (14, 15). A nuclear receptor, PPARγ, was described in epithelial cells, although physical binding of AHL-12 to PPARγ has not yet been proven (16). In macrophages and neutrophils, the AHL-12 signaling pathway was compatible with that of a G-protein coupled receptor, whereas participation of toll-like receptors could be excluded (11, 13, 17–20), a receptor, however, has not yet been identified.

More recently, effects of AHL-12 on epithelial cells were described, such as enhanced ciliary movement and NO-production (21), indicating that AHL-12 induces activation of defense-related functions in these cells. On airway epithelial cells, the so-called bitter receptor T2R38 has been indicated as receptor for AHL-12, and there is evidence for a correlation between T2R38 receptor allotypes and susceptibility to airway infections (21–23).

The bitter receptor T2R38 belongs to a family of taste receptors. As the name implies, the receptor is sensing bitter tasting substances, a phenomenon that is thought to prevent the intake of bitter and potentially toxic substances [reviewed in Ref. (24–27)]. Bitter receptors were first described on cells of the taste buds in the oral cavity. Meanwhile, a wider distribution was reported, e.g., expression on epithelial cells of the airways, in the colon, in the brain, and in tumor cells as well (21, 28–30). The role of T2R38 in tissue outside the gustatory systems is still elusive, and so far, AHL-12 appears to be the first natural ligand for T2R38, which leads to the notion that T2R38 participates in "sensing of bacteria." In that context, we now addressed the question whether T2R38 could also be the receptor for AHL-12 on neutrophils.

Materials and Methods

Human Blood and Human Neutrophils

Blood from volunteers (mainly laboratory personal and students) was drawn into heparin-coated tubes (Sarstedt, Nümbrecht, Germany). Neutrophils were isolated by PolymorphPrep™(Axis-Shield, Oslo, Norway) and residual erythrocytes were removed by hypotonic lysis. The remaining cells were suspended in Hanks balanced salt solution (HBSS), containing 0.5% bovine serum albumin (BSA), at a final concentration of 5×10^6 cells/ml. This procedure yielded 85–95% neutrophils as judged by cytofluorometry using CD66b as marker for neutrophils. Written informed consent was obtained from the volunteers and the ethic committee of the University of Heidelberg approved the procedure.

Cell Lines

U937 and HL-60 were purchased from ATCC. The cell lines were propagated in RPMI, containing fetal calf serum (10%), L-glutamine (1%), and penicillin-streptomycin (1%) (all purchased from Gibco, Eggenstein, Germany). The cells were used in their exponential growth phase. Differentiation of U937 was induced by phorbol ester (100 ng/ml); HL-60 was differentiated with DMSO (1.5%).

Antibodies

The following antibodies to T2R38 were used: bs-86508-A488 [rabbit (rb)IgG, labeled with Alexa Fluor 488, Bioss, Woburn, MA, USA]; ab65509 (rb serum); ab 130503 (rb IgG) (both obtained from abcam, Cambridge, UK); sc-76108 (rb IgG), and sc-34294 (goat IgG) (Santa Cruz, Dallas, TX, USA). Isotype controls were rb IgG labeled with Alexa Fluor (bs-0295P-A488, Bioss, Woburn, MA, USA); rabbit IgG (b176094, abcam), or goat IgG (sc-3887, Santa Cruz). Rabbit serum was obtained from our animal facility. Secondary antibodies were FITC-labeled anti-rb IgG from goat (111-096-003, Dianova, Hamburg, Germany) or Alexa Flour 488-labeled anti-goat IgG from life technologies (A11055, Darmstadt, Germany). CD11b expression was determined using a FITC-labeled-antibody (IM0530, Beckman Coulter, Germany Krefeld). The antigen peptide, sc-34294P, used to raise the antibody to T2R38 (sc-34294), was purchased from Santa Cruz.

Cytofluorometry

To determine the expression of surface receptors on neutrophils, whole blood (100 µl) or isolated neutrophils (10^7/ml) were pre-incubated with an Fc-receptor blocking agent (Biolegend 422302, San Diego, CA, USA), then incubated with the respective antibodies or isotypic antibodies. In the whole blood assays, erythrocytes were lyzed using BD Facs Lysing solution (BD Biosciences, NJ, USA). Cells were subjected to cytofluorometry using FACSCalibur and CellQuest Pro software (Becton Dickinson, Heidelberg, Germany). For intracellular staining, the cells were treated with FACS Permeabilizing Solution 2 (BD Biosciences), pre-incubated with the Fc-receptor blocking agent and then incubated with anti-T2R38 (2 µg), as described above.

Uptake of AHL-12 and Inhibition Experiments

Isolated neutrophils (1×10^6/ml) in HBSS, containing 1% BSA and 0.1% sodium azide ("FACS buffer") were incubated with AHL-FITC (Cayman Chemical, Ann Arbor, MI, USA) (50–100 µM) for 30 min at 4°C, and then subjected to cytofluorometry. For inhibition experiments, isolated neutrophils were washed with FACS buffer and incubated with antibodies to T2R38 for 30 min at 4°C, or with the antigen peptide (2–4 µg). AHL-FITC (100 µM) was added and after for 30 min at 4°C, fluorescence associated with the cells was measured.

Detection of T2R38

A test kit (SEF811Hu) was purchased from Cloud Clone Corporation, Houston, TX, USA, and the test was carried out according to the protocol supplied.

Gene Expression Analysis

DNA was isolated from 10^6 purified neutrophils using the MagnaPure mRNA Isolation Kit I. mRNA was isolated with the

MagnaPure-LC device. An aliquot of mRNA was reversely transcribed using AMV-RT and oligo-dT as primer (First Strand cDNA synthesis kit, Roche) according to the manufactures protocol. For PCR analysis, primer sets optimized for the LightCycler® (RAS, Mannheim, Germany) were developed by and purchased from SEARCH-LC GmbH (www.Search-LC.com). The PCR was performed with the LightCycler® FastStart DNA Sybr GreenI kit (RAS). To correct for differences in the content of mRNA, the calculated transcript numbers were normalized according to the expression of the housekeeping gene peptidylprolyl isomerase B (PPIB). Values represent number of transcripts per 1000 transcripts of PPIB.

Western Blot and Pull-Down Assays

Isolated neutrophils or HL-60 (10^7 cells) were lysed with RIPA buffer (Tris-buffered saline containing 1% Non-idet P-40, 0.5% sodium deoxycholate, 0.1% sodium dodecyl sulfate, 0.0004% sodium azide, 0.2M orthovanadate, and 0.5M phenylmethylsulfonyl fluoride). Of each sample, 25 µl were mixed with 5 µl 5× Laemmli-buffer and applied to a SDS-Gel (9%). Following blotting, the membrane was incubated with 5% milk powder in TBS, containing 0.1% Tween20. Of the antibody to T2R38 (sc-34294) a 1:1000 dilution was used, of antigen peptide sc-34294P 0.5/10 µl antibody. Secondary antibody was a POX conjugated mouse anti-goat IgG (#205-035-108, Jackson Immuno Research, Suffolk, UK, 1:20,000). Blots were developed using Amersham Prime Western Blotting detection Reagent (GE Healthcare, Freiburg, Germany).

For the pull-down assay, neutrophils or HL-60 (2×10^7) were lysed in RIPA buffer (2 ml) and incubated overnight at 4°C with or without AHL-12-biotin (300 µM; Cayman Chemical). Then streptavidin sepharose beads (Cell signaling, Danvers, MA, USA) were added for 2 h at 4°C, then washed with PBS, and incubated with SDS loading buffer containing β-mercaptoethanol. The samples were applied to a 12% SDS-Gel. Silver staining was performed and Western blotting using sc-34294 and the respective antigenic peptide. Alternatively, to whole cell lysate, membrane proteins were extracted using the kit supplied by Calbiochem for transmembrane proteins (#71772-3, Merck, Darmstadt). Extracts of $4–5 \times 10^7$ cells were used (neutrophils or HL-60, as indicated in the respective experiments).

Isolation of Lipid Droplets

Essentially, the method described by Wan et al. (31) was used. In brief, 3×10^8 U937 cells were suspended in 3 ml ice-cold disruption buffer [25 mM Tris-HCl pH 7.4, 5 mM EDTA, 1 mM EGTA, 0.2 mM PMSF, 50 µg/ml *N*-α-*p*-tosyl-L-lysin-chloromethyl-ketone (TLCK), 1 µg/ml leupeptin, 1 µg/ml [illegible] for 30 min, 1000 g at 4°C, to eliminate nuclei and intact cells and then subjected to sucrose density centrifugation. After centrifugation (3.5 h, 34,000 rpm, 4°C), the lipid droplets were in the fraction with 0.54M sucrose. This fraction was diluted to 0.35M sucrose with 0.15M NaCl, overlaid with sucrose (0.27, 0.135 and 0M) for a second ultracentrifugation step (34,000 rpm, 4°C, 3.5 h). Lipid droplets were now in the fraction with 0.35M sucrose. Droplets could be recognized by fluorescence microscopy after having incorporated Nile red (see below). For Western blotting, the proteins were precipitated with methanol/chloroform, resuspended in Laemmli Buffer (95°C for 10 min) and applied to a 12% SDS-Gel. Silver staining and Western blotting was performed was described above.

Laser Scan Microscopy

To assess binding of AHL-12 to cells and co-localization of T2R38, neutrophils were incubated with AHL-12-FITC (50 µM, 20 min at 4°C), placed on cover slips, and fixed with 2% PFA for 15 min. Then, anti-T2R38 [ab65509 (diluted 1:300) or sc-34294 (4 µg)] with the respective secondary antibodies (Alexa Fluor 488-labeled anti-rabbit IgG from donkey (bs-0295D-A488, Bioss); Alexa Fluor 488 Donkey anti-goat IgG (life technologies, Carlsbad, CA, USA) were added, and for comparison rabbit serum or goat IgG. The samples were mounted with Moviol (Sigma-Aldrich, St. Louis, MO, USA) and viewed by Laser Scan microscopy (LSM) (Nikon) using a 40× objective. To detect lipid droplets and the association of T2R38 with the droplets, neutrophils cells were placed on slides by cytospin, fixed with ice-cold methanol for 5 min and washed with PBS. Nile red (Sigma-Aldrich) was added (10 µg/ml 20 min, room temperature), and after washing Hoechst 33342 nuclear stain was applied (H3570, life technologies, Darmstadt, Germany) 1:10,000 for 10 min. The samples were embedded with Moviol. To visualize T2R38, 5% goat serum was used as blocking agent and anti-T2R38 (ab65509) in a dilution of 1:300 (for comparison, normal rabbit serum was used). After 30 min, the secondary antibody was applied for 30 min, then Nile red was added (10 µg/ml, 20 min). In addition, experiments with anti-T2R38 (sc-34294) were performed. In parallel, the antibody was added to the cells together with the peptide sc-34294P. Droplets were visualized by LSM and structured illumination microscopy (SIM) (D-Structured Illumination Microscopy), respectively.

Phagocytosis

A commercially available kit (Glycotope Biotechnology, Heidelberg, Germany) with FITC-labeled bacteria was used according to the instructions supplied.

Migration Assays

Chemotaxis was determined using the BD Falcon HTS Fluoroblok 96-Multiwell Insert System with an insert PET membrane of 3 µm pore size. Isolated polymorphonuclear neutrophils (PMNs) were labeled with calcein. Cells (10^6 cells/ml) were incubated for 15 min at 37°C with calcein (life technologies; 1 µg/ml), then washed with HBSS and seeded in a concentration of 2×10^5/50 µl into the [illegible] determined by a Perkin Elmer Wallac Victor 2 Microplate Reader using Wallac Workstation software.

For the "under-agarose" migration assay, cover slips (1.5 × 4 cm) were pretreated with 3M HCl and ethanol (95%). Following washing with water, gelatine (0.5% in RPMI) was applied as a thin film. The cover slips were air dried, then agarose (1% in RPMI) was added (3 ml per cover slip). When the agarose was solid, two holes were punched in a distance of 5 mm. Cells were placed in one hole (5×10^4 in 50 µl), in the other the chemokine to be tested. After 2 h at 37°C, the cells were fixed with methanol for 30 min, the agarose was removed and the remaining cells were stained by Giemsa an examined microscopically.

Results

Expression of the Bitter Receptor T2R38 on Neutrophils and Monocytes

T2R38 was detected on neutrophils, monocytes, and to a lesser extent on lymphocytes in whole blood samples by cytofluorometry. Comparing samples of seven individuals, a considerably variation was seen (**Figures 1A,B**). Data obtained with the T2R38 antibody sc-34294 are shown, however, antibodies directed against other epitopes within the N-terminus of T2R38 (sc-67108; ab65509) gave essentially similar results (data not shown). For the antibody sc-34294, the antigenic peptide used to raise the antibody was available. This peptide inhibited binding of the antibody to the cells by 50% (**Figure 1C**). An antibody directed against an amino acid sequence of the C-terminus (ab130503) reacted only with permeabilized cells (**Figure 1D**). By Western blotting, T2R38 was detected in whole cell lysates and in cell membrane preparations. Specificity of the antibody binding was controlled by the antigenic peptide. In cell membranes, two bands with an apparent molecular weight of 90 and 50 kDa were detected, in the cell lysates multiple bands at 100, 90, 55, and 50 kDa, respectively (**Figure 1E**). The expected molecular weight of T2R38 is 38 kDa, but post-translational modifications and polymerization are described (for these experiments, sc-34294 was used).

T2R38 was also expressed by U937 and HL-60 cells (data not shown). Following differentiation of HL-60 by DMSO surface expression of T2R38 increased moderately (39.9% increase of the mean fluorescence intensity; average of three experiments).

The distribution of T2R38 in neutrophils was analyzed by confocal LSM. Accumulation at the membrane was seen and a speckled pattern in the cytoplasm (images with two different antibodies are shown in **Figure 2**). The latter finding was compatible with the notion that intracellular T2R38 was associated with membranous vesicles and/or granules. Stimuli known to induce degranulation, including IL8, f-MLP, C5a, or phorbol ester, did not up-regulate surface expression of T2R38 when applied under conditions causing up-regulation of CD11b. Moreover, T2R38 was not released into the cell supernatant as determined by ELISA

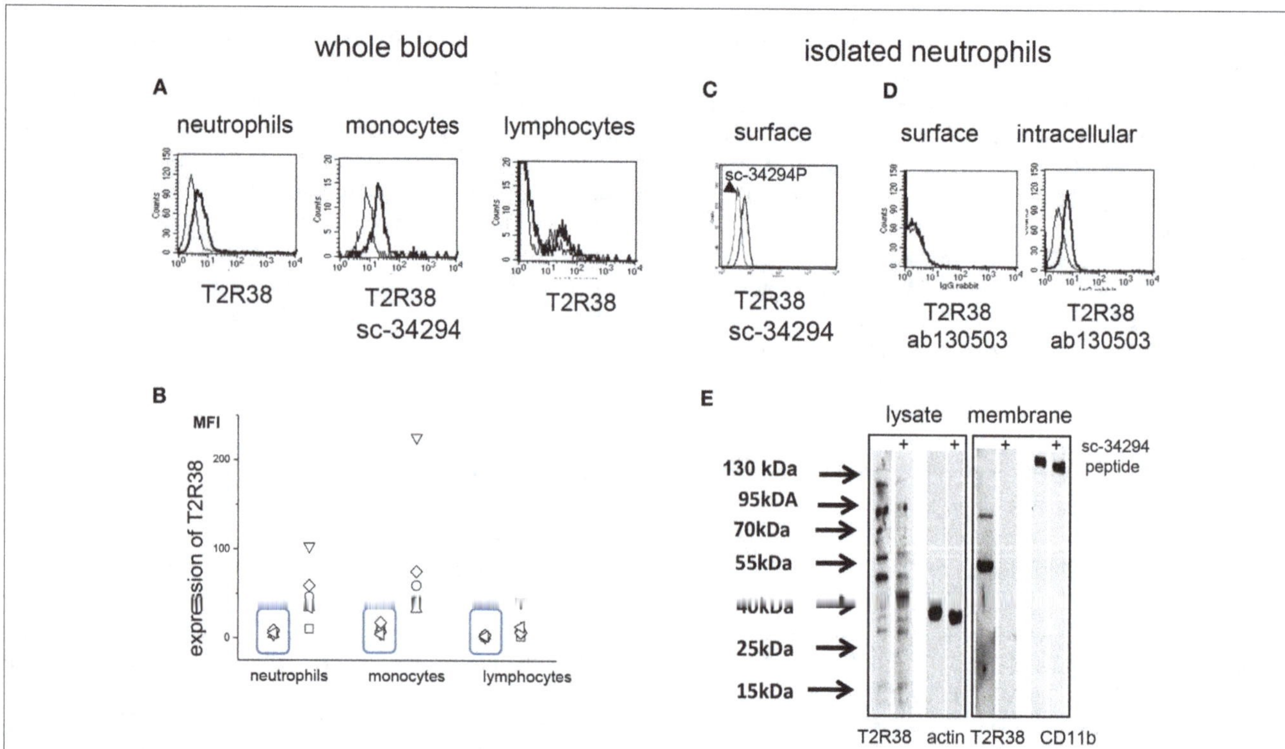

FIGURE 1 | Expression of T2R38 on peripheral blood cells. **(A)** Cytofluorometry using the antibody sc-34294 (thick line) showed expression of T2R38 [the gates were set according to the cell populations identified by expression of CD66b (neutrophils), CD14 (monocytes), and CD3 (lymphocytes)]. For comparison goat IgG (thin line) was used. **(B)** T2R38 was determined by cytofluorometry on peripheral blood cells of seven individuals (each represented by a symbol) and mean fluorescence intensity (MFI) is shown. The rectangles contain the data for the antibody isotype binding. **(C)** T2R38 was also seen on isolated neutrophils (thick line) and binding of the antibody could be inhibited by the specific antigen peptide (thin line) **(D)** The antibody to T2R38 that recognizes the C-terminal domain of the receptor (ab130503) reacts only following permeabilization of the cells. **(E)** Western blot of neutrophil lysates with anti-T2R38 (sc-34294) revealed multiple bands. Intensity of bands with an apparent molecular weight of 90–100 kDa was reduced by the antigen peptide sc-34294p, as was the intensity of bands at 50–55 kDa (actin was used as loading control). Isolated membranes showed only two protein bands at approximately 90 and 50 kDa (CD11b was used as loading control). (Lysates and membranes were run on different gels, and are aligned here according to the molecular weight markers) (one of four blot is shown).

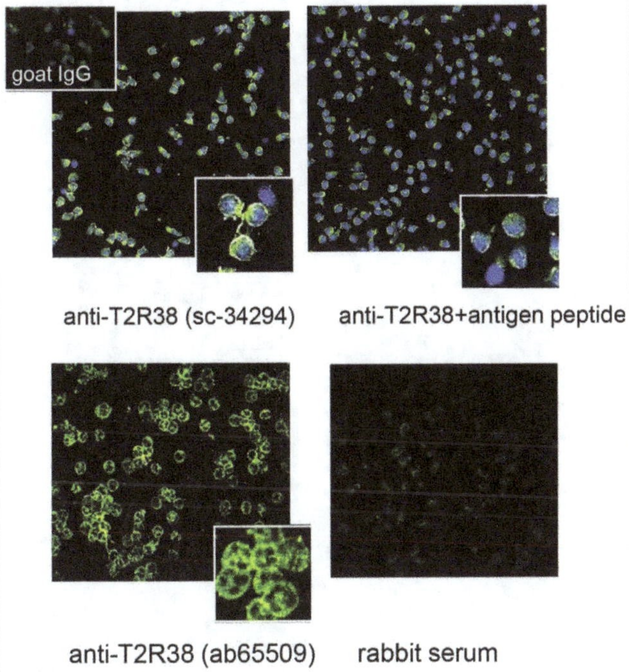

FIGURE 2 | Distribution of T2R38 on neutrophils: by laser scan microscopy, T2R38 was detected in neutrophils. The upper panel shows an image with anti-T2R38 sc-34294 (green) and nuclei are stained in blue. Binding of the antibody was partly inhibited by the antigen peptide. The antibody ab65509 (lower panel) also detected T2R38 (green), in association with the membrane, and with a speckled pattern intracellular. The lower right panel shows the image obtained with rabbit serum instead of the antibody (images were taken with 40× objective; digital zooms of representative areas are shown in the insets).

(data not shown). The failure to mobilize T2R38 makes its location in granules unlikely. Hence, we assessed association of T2R38 with lipid droplets. Neutrophils contain numerous lipid droplets visualized by incorporation of Nile red (**Figure 3A**). Confocal LSM indicated an association of T2R38 with droplets, particularly with their membrane (**Figure 3B**). To support this finding, we performed SIM. The circular morphology of the droplets could be resolved by this super-resolution microscopy, and T2R38 clearly was seen on the droplet membrane (**Figure 3C**). To verify the association of T2R38 with lipid droplets further, droplets were isolated by density centrifugation following the protocol by Wan et al. (31). Droplets of different sizes were seen, and again an association of T2R38 with the droplet membrane (**Figures 3D,E**). Lysates of the droplet fraction were positive for T2R38 as tested by blotting (**Figure 3F**).

Effect of Antibodies to T2R38 on AHL-12-Mediated Function

To answer the most pertinent question whether T2R38 might function as a receptor for AHL-12, the effect of antibodies to T2R38 on AHL-12 binding and AHL-12-induced neutrophil activation was tested. For these experiments, FITC-labeled AHL-12 was used. AHL-12-FITC binds to neutrophils, as seen by cytofluorometry and confocal LSM (**Figures 4A,B**). Binding of AHL-12-FITC could be inhibited by anti-T2R38 (sc-34294), on average by 35%, when 4 µg antibody were used per 1×10^6 neutrophils/ml (mean of three experiments with cells of different individuals). With higher antibody concentrations, also the isotype control was inhibitory, precluding the correct interpretation of the data. The other antibodies to T2R38 (sc-67108 and ab65509) did not inhibit AHL-12 binding.

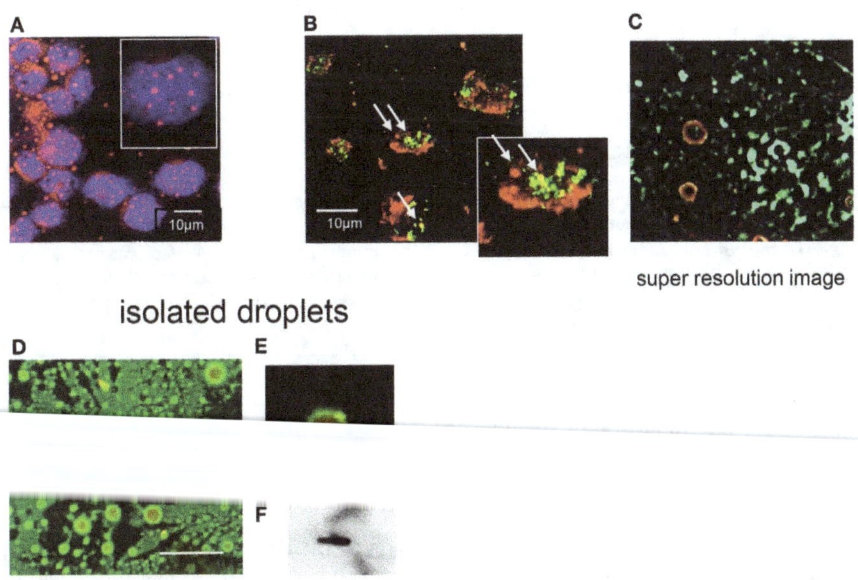

FIGURE 3 | Association of T2R38 with lipid droplets: (A) shows neutrophils stained with Nile red to visualize the lipid droplets; nuclei are in blue. (B) With anti-T2R38 (green) and staining with Nile red, droplets containing T2R38 can be recognized within the cell (marked by arrows). **(C)** A SIM image of the area in the rectangle shows two lipids droplets with a distinct yellow staining of their membranes, indicative of a close association of T2R38 with the droplet membrane. **(D)** The images show isolated lipid droplets stained with Nile red and anti-T2R38; localization of T2R38 on the droplets membrane is seen especially in the zoomed image **(E)**. **(F)** shows a blot of isolated droplets with ab65509.

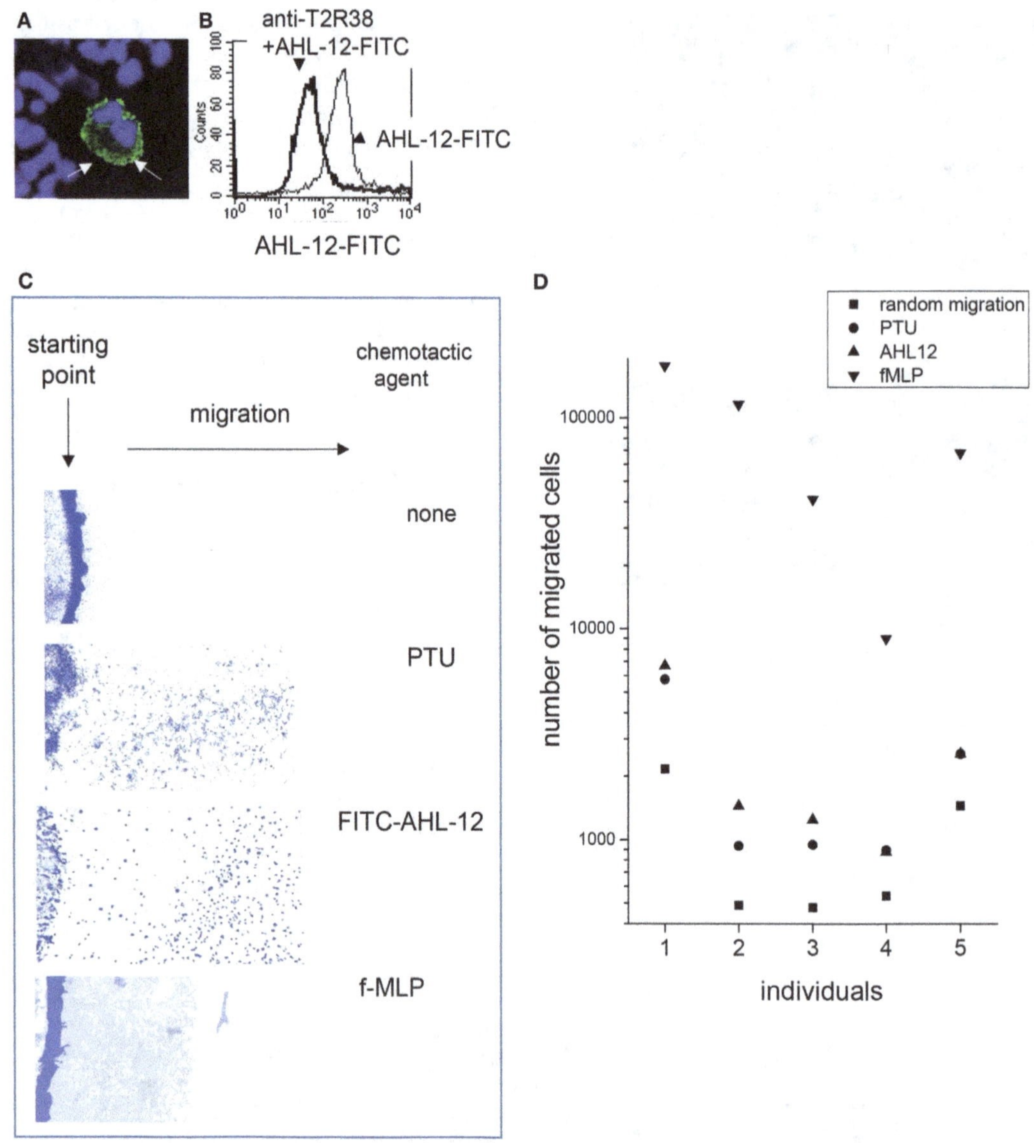

FIGURE 4 | Uptake of AHL-12-FITC by neutrophils and activation of cells: laser scan microscopy (A) and cytofluorometry (B) revealed binding of AHL-12-FITC to neutrophils (thin line). Binding could be inhibited by anti-T2R38 (sc-34294) (thick line) **(C)** AHL-12-FITC induced migration of neutrophils [illegible] migration (cells are stained in blue), as did PTU. In comparison to f-MLP, fewer cells had migrated but had covered a greater distance. **(D)** Chemotaxis (here migration through porous membrane is shown; data with neutrophils of seven individuals are summarized) was induced by AHL-12, PTU, f-MLP, and compared to the random migration. [illegible] measured, the number of cell migrating toward f-MLP was considerably higher.

Next, the effect of the antibody sc-34294 on AHL-12-mediated activation of neutrophils was assessed. Induction of chemotaxis was measured, enhancement of phagocytosis, and up-regulation of CD11b. Chemotaxis in response to AHL-12 was not inhibited, nor was migration toward PTU, which is widely used as ligand for T2R38, and which also induced chemotaxis of neutrophils (data based on two different methods, under-agarose migration and migration through a filter membrane, are shown in **Figures 4C,D**). Furthermore, the antibody did not inhibit the AHL-12 mediated up-regulation of CD11b nor the enhancement of phagocytosis. Rather, the antibody sc-34294 by itself activated neutrophils. Following antibody binding surface expression of CD11b was enhanced (on average by 1.7-fold; mean of three experiments), as was the oxidative burst elicited by phagocytosis of opsonized *E. coli* (on average by 3.5-fold) (examples in **Figures 5A,B**).

Association of AHL-12 with T2R38 on Neutrophils

When neutrophils were incubated with limited amounts of AHL-12-FITC (50 μM) at 4°C and then incubated with anti-T2R38, confocal LSM revealed a close proximity of AHL-12 and T2R38 (**Figures 6A,B**), suggesting a physical association of AHL-12 and T2R38. "Pull-down" assays using biotin-labeled AHL-12 as bait were performed with lysates of neutrophils or HL-60. Cell lysates

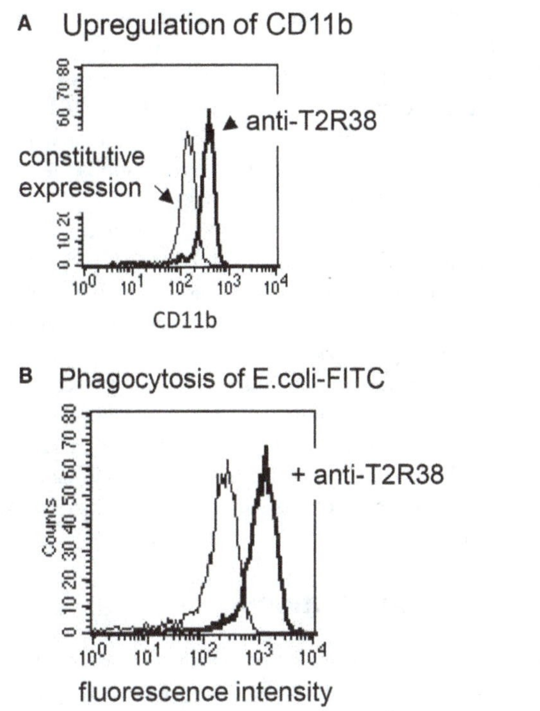

FIGURE 5 | Antibodies to T2R38 activate neutrophils. (A) The antibody sc-34294-induced up-regulation of CD11b on neutrophils (thick line; the thin shows constitutive CD11b expression) and **(B)** it enhanced the phagocytosis of FITC-labeled bacteria, measured as increase of cell-associated fluorescence intensity (thick line; the thin line shows phagocytosis in presence of the isotype IgG).

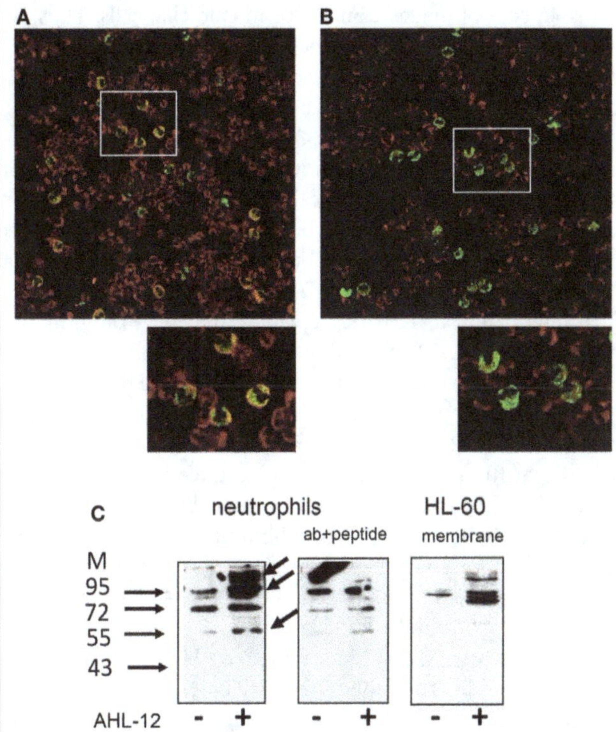

FIGURE 6 | Co-localization of AHL-12 and T2R38: neutrophils were incubated with AHL-12-FITC, placed on slides, fixed, and incubated with either (A) anti-T2R38 or (B) isotypic antibody. Laser scan microscopy shows co-expression of T2R38 and AHL-12-FITC as "yellow cells." The isotypic antibody shows some background staining, but the majority of cells remained "green." Selected areas (white rectangles) were zoomed to show details. **(C)** A pull-down assays showed binding of AHL-12 to T2R38: AHL-12-biotin was incubated with neutrophils; then lysates were prepared, and incubated with streptavidin beads, and the eluates were subjected to SDS-PAGE and Western blotting using anti-T2R38 sc-34294 (lanes marked with +). In comparison to neutrophils having not been exposed to AHL-12-biotin (lanes marked with −), major bands were seen, at 100–130 kDa, and 55 kDa, the intensity of which could be reduced with the specific antigen peptide. Experiments with HL-60 showed essentially similar results. Here, membrane preparations are shown.

were incubated with AHL-12-biotin, and then with streptavidin beads. Eluates from the beads yielded numerous proteins. By Western blotting T2R38 could be identified with an apparent molecular weight of 100 and 60 kDa (data for neutrophils and for HL-60 are shown **Figure 6C**). When the pull-down assay with AHL-12-biotin was carried with membrane protein fractions, T2R38 could be detected again by Western blotting (**Figure 6C**).

Discussion

Our data provide evidence for an extragustatory expression of the bitter taste receptor T2R38 on neutrophils, monocytes, and myeloid cell lines. By a variety of methods and with antibodies directed against different epitopes, intracellular expression as well as expression on the cell membrane was seen. At difference with the predicted molecular weight of 38 kDa, proteins with higher

receptor, as it had been described for T2R38 in epithelial cells (32).

Surface expression of T2R38 varied among individuals, as did the number of gene transcripts. We presume a genetic predisposition, because multiple testing revealed individuals expressing little T2R38 and others expressing moderate levels (own unpublished data). The histograms obtained by cytofluorometry revealed a more or less symmetric peak, indicating that T2R38 expression was not restricted to a subpopulation. The number of gene transcripts was rather low. Absent or marginal gene transcription – even for highly expressed proteins – is not unusual for neutrophils, because in these cells a great majority of proteins is produced during maturation only, and are then stored in intracellular compartments (33, 34).

Well-characterized storage compartments of neutrophils are granules, which are generated as the precursor cells mature, and which enclose proteins as they are synthesized [reviewed

fMLF, C5a, or phorbol ester (36). Because these stimuli failed to mobilize T2R38, we looked for other storage compartments and found T2R38 in association with vesicles having the properties of lipid droplets. Lipid droplets were originally described as lipid storage sites, but more recent studies described them as functional organelles that actively participate in lipid turnover, cellular signaling, and membrane trafficking (37, 38). As far as we know, storage in or association with vesicles – lipid droplets or others – has not been investigated for T2R38 in its usual habitat – that is, in cells of the taste buds – or in epithelial cells. Thus, the question remains whether or not storage of T2R38 in vesicles is typical for neutrophils or whether it is a general feature of T2R38 (our own not yet published data point to the latter).

Having established that T2R38 is present in neutrophils, the question was addressed whether it is the receptor for AHL-12 on these cells. LSM showed a co-localization of AHL-12 and T2R38, and the "pull-down-assay" with biotin-labeled AHL-12 indicated a physical interaction of AHL-12 with T2R38, which implies that T2R38 is indeed the receptor for AHL-12 on neutrophils.

To link T2R38 to AHL-12 signaling, the effect of antibodies to T2R38 on AHL-12-induced functions was tested. Only one of the antibodies (sc-34294) inhibited binding of AHL-12 to some extent, but not entirely. A likely explanation is the hydrophobic nature of AHL-12, which allows attachment to cells and probably also partition into the membrane in a receptor-independent manner. That the other antibodies – though binding to the cells – did not block potential AHL-12 binding sites is in line with data reported for ab65509 by others in another setting (32). As anticipated, these antibodies did not inhibit AHL-12-mediated activation of neutrophils.

The antibody sc-34294, which inhibited AHL-12 binding, stimulated neutrophils. The antibody by itself up-regulated CD11b cell surface expression, and it enhanced phagocytosis, two functions that are also induced by AHL-12. Thus, T2R38 is a functional receptor on neutrophils, and according to the co-localization data, AHL-12 is the most likely ligand.

Expression of T2R38 on neutrophils fits the concept that the cells "sense" bacterial infection, and up-regulate defense-relevant actions in consequence. In that, T2R38 qualifies as pathogen-associated molecular pattern recognition receptor. Because bacteria produce and release AHL-12 as an autoinducer, crucially required for biofilm formation, neutrophils will be attracted to the infectious site before or concurrent with biofilm formation. Because planktonic bacteria are in general more susceptible to host defense mechanisms compared to bacteria organized as biofilm, the infection can be cleared. Valid objections to this notion are that micromolar concentrations of AHL-12 are required to activate neutrophils – roughly 100- to 10,000-fold more than for established chemokines – and that the weak to moderate surface expression of T2R38 questions the effectiveness of an AHL-12–T2R38 interaction. However, according to our experience, AHL-12 is not very stable under cell culture conditions, meaning that the abundance of active molecules might be lower than the calculated input. In addition, local concentrations of AHL-12 could be rather high, as data using supernatants of biofilm-forming *P. aeruginosa* imply (20). Without question, the biological relevance of our finding is still under scrutiny.

Our data, however, establish for the first time extra gustatory expression of T2R38 on cells of the myeloid lineage, and are compatible with the notion that T2R38 functions as a receptor for AHL-12 on neutrophils as a novel means of recognizing of bacterial components by the host defense.

Author Contributions

All authors approved the final version of the manuscript, and contributed substantially to the study: SM did the cytofluorometry experiments, GW and SS the laser scan microscopy and SIM, NK the initial experiments on neutrophil activation and the uptake studies. BP performed the functional assays (chemotaxis and phagocytosis), TG did the quantitative PCRs, SM, MG, YS, GW, and GH did the data analysis and wrote the manuscript.

Acknowledgments

The discussion with the colleagues from the Karlsruhe Institut für Technology (KIT), Ursula Obst, Gerald Brenner-Weiss, Thomas Schwarz, and Jörg Overhage are gratefully acknowledged.

References

1. Whiteley M, Lee KM, Greenberg EP. Identification of genes controlled by quorum sensing in *Pseudomonas aeruginosa*. *Proc Natl Acad Sci U S A* (1999) **96**:13904–9. doi:10.1073/pnas.96.24.13904
2. Fuqua C, Parsek MR, Greenberg EP. Regulation of gene expression by cell-to-cell communication: acyl-homoserine lactone quorum sensing. *Annu Rev Genet* (2001) 35:439–68. doi:10.1146/annurev.genet.35.102401.090913
3. Karatan E, Watnick P. Signals, regulatory networks, and materials that build and break bacterial biofilms. *Microbiol Mol Biol Rev* (2009) **73**:310–47. doi:10.1128/MMBR.00041-08
4. Schuster M, Sexton DJ, Diggle SP, Greenberg EP. Acyl-homoserine lactone quorum sensing: from evolution to application. *Annu Rev Microbiol* (2013) **67**:43–63. doi:10.1146/annurev-micro-092412-155635
5. Shiner EK, Rumbaugh KP, Williams SC. Interkingdom signaling: deciphering the language of acyl homoserine lactones. *FEMS Microbiol Rev* (2005) **29**:935–47. doi:10.1016/j.femsre.2005.03.001
6. Hughes DT, Sperandio V. Inter-kingdom signalling: communication between bacteria and their hosts. *Nat Rev Microbiol* (2008) **6**:111–20. doi:10.1038/nrmicro1836
7. Telford G, Wheeler D, Williams P, Tomkins PT, Appleby P, Sewell H, et al. The *Pseudomonas aeruginosa* quorum-sensing signal molecule N-(3-oxododecanoyl)-L-homoserine lactone has immunomodulatory activity. *Infect Immun* (1998) **66**:36–42.
8. Ritchie AJ, Yam AO, Tanabe KM, Rice SA, Cooley MA. Modification of in vivo and in vitro T- and B-cell-mediated immune responses by the *Pseudomonas aeruginosa* quorum-sensing molecule N-(3-oxododecanoyl)-L-homoserine lactone. *Infect Immun* (2003) **71**:4421–31. doi:10.1128/IAI.71.8.4421-4431.2003
9. Cooley M, Chhabra SR, Williams P. N-acylhomoserine lactone-mediated quorum sensing: a twist in the tail and a blow for host immunity. *Chem Biol* (2008) 15:1141–7. doi:10.1016/j.chembiol.2008.10.010
10. Vikstrom E, Magnusson KE, Pivoriunas A. The *Pseudomonas aeruginosa* quorum-sensing molecule N-(3-oxododecanoyl)-L-homoserine lactone stimulates phagocytic activity in human macrophages through the p38 MAPK pathway. *Microbes Infect* (2005) **7**:1512–8. doi:10.1016/j.micinf.2005.05.012
11. Zimmermann S, Wagner C, Muller W, Brenner-Weiss G, Hug F, Prior B, et al. Induction of neutrophil chemotaxis by the quorum-sensing molecule N-(3-oxododecanoyl)-L-homoserine lactone. *Infect Immun* (2006) **74**:5687–92. doi:10.1128/IAI.01940-05
12. Wagner C, Zimmermann S, Brenner-Weiss G, Hug F, Prior B, Obst U, et al. The quorum-sensing molecule N-3-oxododecanoyl homoserine lactone (3OC12-HSL) enhances the host defence by activating human polymorphonuclear neutrophils (PMN). *Anal Bioanal Chem* (2007) **387**:481–7. doi:10.1007/s00216-006-0698-5
13. Hansch GM. Editorial: molecular eavesdropping: phagocytic cells spy on bacterial communication. *J Leukoc Biol* (2012) **91**:3–5. doi:10.1189/jlb.0711358

14. Ritchie AJ, Whittall C, Lazenby JJ, Chhabra SR, Pritchard DI, Cooley MA. The immunomodulatory *Pseudomonas aeruginosa* signalling molecule N-(3-oxododecanoyl)-L-homoserine lactone enters mammalian cells in an unregulated fashion. *Immunol Cell Biol* (2007) **85**:596–602. doi:10.1038/sj.icb.7100090
15. Barth C, Jakubczyk D, Kubas A, Anastassacos F, Brenner-Weiss G, Fink K, et al. Interkingdom signaling: integration, conformation, and orientation of N-acyl-l-homoserine lactones in supported lipid bilayers. *Langmuir* (2012) **28**:8456–62. doi:10.1021/la301241s
16. Jahoor A, Patel R, Bryan A, Do C, Krier J, Watters C, et al. Peroxisome proliferator-activated receptors mediate host cell proinflammatory responses to *Pseudomonas aeruginosa* autoinducer. *J Bacteriol* (2008) **190**:4408–15. doi:10.1128/JB.01444-07
17. Kravchenko VV, Kaufmann GF, Mathison JC, Scott DA, Katz AZ, Wood MR, et al. N-(3-oxo-acyl)homoserine lactones signal cell activation through a mechanism distinct from the canonical pathogen-associated molecular pattern recognition receptor pathways. *J Biol Chem* (2006) **281**:28822–30. doi:10.1074/jbc.M606613200
18. Karlsson T, Musse F, Magnusson KE, Vikstrom E. N-Acylhomoserine lactones are potent neutrophil chemoattractants that act via calcium mobilization and actin remodeling. *J Leukoc Biol* (2012) **91**:15–26. doi:10.1189/jlb.0111034
19. Karlsson T, Turkina MV, Yakymenko O, Magnusson KE, Vikstrom E. The *Pseudomonas aeruginosa* N-acylhomoserine lactone quorum sensing molecules target IQGAP1 and modulate epithelial cell migration. *PLoS Pathog* (2012) **8**:e1002953. doi:10.1371/journal.ppat.1002953
20. Kahle NA, Brenner-Weiss G, Overhage J, Obst U, Hansch GM. Bacterial quorum sensing molecule induces chemotaxis of human neutrophils via induction of p38 and leukocyte specific protein 1 (LSP1). *Immunobiology* (2013) **218**:145–51. doi:10.1016/j.imbio.2012.02.004
21. Lee RJ, Xiong G, Kofonow JM, Chen B, Lysenko A, Jiang P, et al. T2R38 taste receptor polymorphisms underlie susceptibility to upper respiratory infection. *J Clin Invest* (2012) **122**:4145–59. doi:10.1172/JCI64240
22. Shah AS, Ben-Shahar Y, Moninger TO, Kline JN, Welsh MJ. Motile cilia of human airway epithelia are chemosensory. *Science* (2009) **325**:1131–4. doi:10.1126/science.1173869
23. Tizzano M, Gulbransen BD, Vandenbeuch A, Clapp TR, Herman JP, Sibhatu HM, et al. Nasal chemosensory cells use bitter taste signaling to detect irritants and bacterial signals. *Proc Natl Acad Sci U S A* (2010) **107**:3210–5. doi:10.1073/pnas.0911934107
24. Adler E, Hoon MA, Mueller KL, Chandrashekar J, Ryba NJ, Zuker CS. A novel family of mammalian taste receptors. *Cell* (2000) **100**:693–702. doi:10.1016/S0092-8674(00)80705-9
25. Chandrashekar J, Mueller KL, Hoon MA, Adler E, Feng L, Guo W, et al. T2Rs function as bitter taste receptors. *Cell* (2000) **100**:703–11. doi:10.1016/S0092-8674(00)80706-0
26. Meyerhof W, Behrens M, Brockhoff A, Bufe B, Kuhn C. Human bitter taste perception. *Chem Senses* (2005) **30**(Suppl 1):i14–5. doi:10.1093/chemse/bjh089
27. Behrens M, Foerster S, Staehler F, Raguse JD, Meyerhof W. Gustatory expression pattern of the human TAS2R bitter receptor gene family reveals a heterogenous population of bitter responsive taste receptor cells. *J Neurosci* (2007) **27**:12630–40. doi:10.1523/JNEUROSCI.1168-07.2007
28. Rozengurt N, Wu SV, Chen MC, Huang C, Sternini C, Rozengurt E. Colocalization of the alpha-subunit of gustducin with PYY and GLP-1 in L cells of human colon. *Am J Physiol Gastrointest Liver Physiol* (2006) **291**:G792–802. doi:10.1152/ajpgi.00074.2006
29. Singh N, Vrontakis M, Parkinson F, Chelikani P. Functional bitter taste receptors are expressed in brain cells. *Biochem Biophys Res Commun* (2011) **406**:146–51. doi:10.1016/j.bbrc.2011.02.016
30. Viswanathan VK. Sensing bacteria, without bitterness? *Gut Microbes* (2013) **4**:91–3. doi:10.4161/gmic.23776
31. Wan HC, Melo RC, Jin Z, Dvorak AM, Weller PF. Roles and origins of leukocyte lipid bodies: proteomic and ultrastructural studies. *FASEB J* (2007) **21**:167–78. doi:10.1096/fj.06-6711com
32. Behrens M, Born S, Redel U, Voigt N, Schuh V, Raguse JD, et al. Immunohistochemical detection of TAS2R38 protein in human taste cells. *PLoS One* (2012) **7**:e40304. doi:10.1371/journal.pone.0040304
33. Theilgaard-Monch K, Jacobsen LC, Rasmussen T, Niemann CU, Udby L, Borup R, et al. Highly glycosylated alpha1-acid glycoprotein is synthesized in myelocytes, stored in secondary granules, and released by activated neutrophils. *J Leukoc Biol* (2005) **78**:462–70. doi:10.1189/jlb.0105042
34. Borregaard N, Sorensen OE, Theilgaard-Monch K. Neutrophil granules: a library of innate immunity proteins. *Trends Immunol* (2007) **28**:340–5. doi:10.1016/j.it.2007.06.002
35. Borregaard N, Cowland JB. Granules of the human neutrophilic polymorphonuclear leukocyte. *Blood* (1997) **89**:3503–21.
36. Witko-Sarsat V, Rieu P, Descamps-Latscha B, Lesavre P, Halbwachs-Mecarelli L. Neutrophils: molecules, functions and pathophysiological aspects. *Lab Invest* (2000) **80**:617–53. doi:10.1038/labinvest.3780067
37. Goodman SB. Wear particles, periprosthetic osteolysis and the immune system. *Biomaterials* (2007) **28**:5044–8. doi:10.1016/j.biomaterials.2007.06.035
38. Bozza PT, Viola JP. Lipid droplets in inflammation and cancer. *Prostaglandins Leukot Essent Fatty Acids* (2010) **82**:243–50. doi:10.1016/j.plefa.2010.02.005

Conflict of Interest Statement: The authors declare that the research was conducted in the absence of any commercial or financial relationships that could be construed as a potential conflict of interest.

19

T-cell Activation or Tolerization: The Yin and Yang of Bacterial Superantigens

*Aline Sähr†, Sandra Förmer†, Dagmar Hildebrand and Klaus Heeg**

Medical Microbiology and Hygiene, Department of Infectious Diseases, University Hospital Heidelberg, Heidelberg, Germany

Edited by:
Inka Sastalla,
National Institutes of Health, USA

Reviewed by:
Dane Parker,
Columbia University, USA
Hatice Karauzum,
National Institutes of Health, USA

***Correspondence:**
Klaus Heeg
klaus.heeg@med.uni-heidelberg.de

† These authors have contributed equally to this work.

Bacterial superantigens (SAg) are exotoxins from pathogens which interact with innate and adaptive immune cells. The paradox that SAgs cause activation and inactivation/anergy of T-cells was soon recognized. The structural and molecular events following SAg binding to antigen presenting cells (APCs) followed by crosslinking of T-cell receptors were characterized in detail. Activation, cytokine burst and T-cell anergy have been described *in vitro* and *in vivo*. Later it became clear that SAg-induced T-cell anergy is in part caused by SAg-dependent activation of T-regulatory cells (Tregs). Although the main focus of analyses was laid on T-cells, it was also shown that SAg binding to MHC class II molecules on APCs induces a signal, which leads to activation and secretion of pro-inflammatory cytokines. Accordingly APCs are mandatory for T-cell activation. So far it is not known, whether APCs play a role during SAg-triggered activation of Tregs. We therefore tested whether in SAg (Streptococcal pyrogenic exotoxin A) -treated APCs an anti-inflammatory program is triggered in addition. We show here that not only the anti-inflammatory cytokine IL-10 and the co-inhibitory surface molecule PD-L1 (CD274) but also inhibitory effector systems like indoleamine 2,3-dioxygenase (IDO) or intracellular negative feedback loops (suppressor of cytokine signaling molecules, SOCS) are induced by SAgs. Moreover, cyclosporine A completely prevented induction of this program. We therefore propose that APCs triggered by SAgs play a key role in T-cell activation as well as inactivation and induction of Treg cells.

Keywords: superantigen, anergy, Treg, class II signaling, co-inhibitory molecules, STAT3, IDO, PD-L1

INTRODUCTION

Superantigens (SAg) are bacterial exotoxins which share unique immunological properties. SAg released by staphylococci or streptococci during infection or even colonization induce a strong activation of the immune system. Major hallmark of this activation is the fulminant release of cytokines (Carlsson and Sjögren, 1985) leading to a disastrous cytokine storm (Miethke et al., 1992; Michie et al., 1994) which might lead to an uncontrolled systemic shock with high lethality. The toxic shock syndrome (TSS) was recognized to be mediated by a SAg, the TSS toxin 1, TSST-1 (Miethke et al., 1993). In the meantime, many bacterial exotoxins have been classified as SAgs (Fraser and Proft, 2008), including the erythrogenic toxins of *Streptococcus pyogenes* (SPEA, SPEC) and the enterotoxins from *Staphlococcus aureus* (Lina et al., 2004).

Superantigen behave like bifunctional agents: the bind to conserved regions of MHC class II molecules and to V-beta encoded regions of the T-cell receptor (TCR) (Choi et al., 1990; Dellabona et al., 1990). Crosslinking class II on APC with TCR induces T-cell activation with subsequent cytokine release. While all SAg bind to class II, the binding to the TCR is V-beta specific, thus single SAg activates a V-beta defined subfraction of the T-cell pool (Kappler et al., 1989). Nevertheless, in the human system a single SAg can activate 1 to 10% of the T-cell pool, leading to a strong oligoclonal T-cell response which exceeds the clone size activated by a protein antigen at least by the factor of 1000 (Herrmann and MacDonald, 1991).

The tracking of SAg reactive T-cells using the V-beta TCR expression has greatly facilitated the analyses of SAg-induced T-cell responses. After initial activation accompanied with secretion of cytokines including TNF, interleukin-2 (IL-2) and IFN-gamma a phase of clonal T-cell expansion follows which is then succeeded by apoptosis and clonal retraction (Herrmann et al., 1992; Lee and Vitetta, 1992; Huang and Crispe, 1993; Miethke et al., 1995). The apoptosis is not completely, roughly 50% of the initial numbers of V-beta bearing T-cells survive. However, these T-cells fail to respond to further stimulation, i.e., display an anergic phenotype (MacDonald et al., 1993; Wahl et al., 1993). Unfolding of unresponsiveness is prevented in the presence of the T-cell immunosupressive agent cyclosporine A (CsA) but not rapamycin (Vanier and Prud'homme, 1992; Prud'homme et al., 1995). This was taken as an indication that anergy induction is dependent on calcineurin and is triggered primarily in T-cells. Besides anergy induction it was also shown that after stimulation with SAg the fraction of $CD4^{++}CD25^{+}$ $foxp3^{+}$ Treg within the T-cell pool is significantly augmented (Wang et al., 1998; Papiernik, 2001; Feunou et al., 2003; Grundström et al., 2003; Ivars, 2007). The cellular interactions and signaling pathways leading to Treg induction after SAg stimulation are not well understood.

TABLE 1 | Primers used.

Gene	Forward – primer	Reverse – primer
ß-Actin	aga gct acg agc tgc ctg ac	agc act gtg ttg gcg tac ag
calcineurin	aaa cag tga ctg gcg cat c	ccg gct tac agc aaa aga ag
IDO	tta gag tca aat ccc tca gtc c	ttt gca gat ggt agc tcc tc
IL-1b	agc tga tgg ccc taa aca ga	gca tct tcc tca gct tgt cc
IRF-1	gct ggg aca tca aca agg at	tgg tct ttc acc tcc tcg at
JNK	gca tgg gct aca agg aaa ac	ttc agg aca tgg tgt tcc aa
p38	gac aca aaa acg ggg tta cg	tgg gtc acc aga tac aca tca
p44/42	agt aca tcc act ccg cca ac	cgt agc cac ata ctc cgt ca
CD274	tgc tgt ctt tat att cat gac cta c	tcc tcc att tcc caa tag aca
SOCS1	tcc ccc tca acc ccg t	cat ccg ctc cct cca acc
SOCS3	ggg agt ccc ccc aga aga g	ata gga gtc cag gtg gcc gt
STAT1	ccg ttt tca tga cct cct gt	ggc gtt ttc cag aat ttt cc
STAT3	cag gtt gct ggt caa att cc	tgt gtt tgt gcc cag aat gt
TDO	ggt tcc tca ggc tat cac tac c	cag tgt cgg gga atc agg t

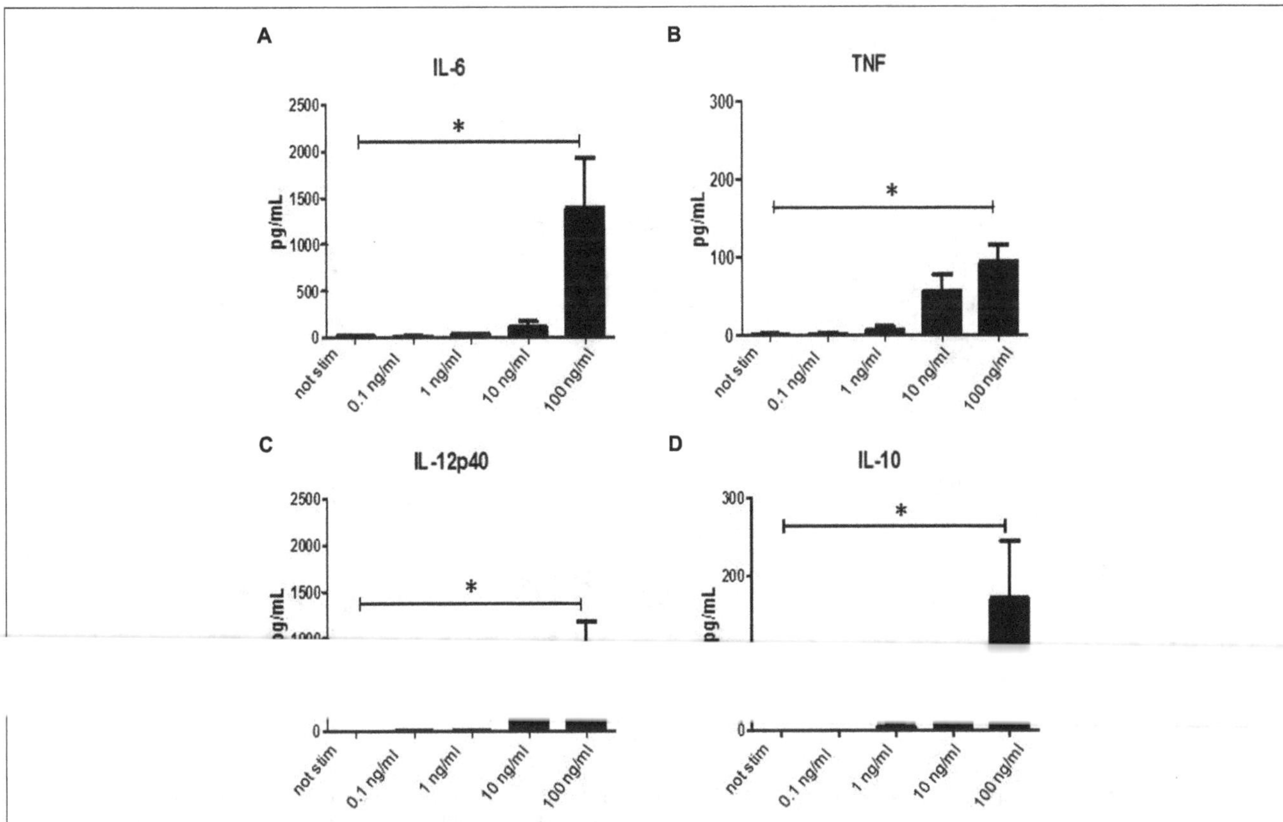

FIGURE 1 | Streptococcal pyrogenic exotoxin A (SPEA)- induced cytokines. $CD14^{+}$ monocytes were stimulated with increasing concentrations of SPEA. 24 h after stimulation supernatants were analyzed by ELISA for **(A)** IL-6, **(B)** TNF, **(C)** IL-12p40 and **(D)** IL-10 (mean ± SD, $n = 5$). Statistical analysis was performed using a Multiple Comparison of Means (Tukey Contrasts). Significance code: 0 '***'; 0.001 '**'; 0.01 '*'.

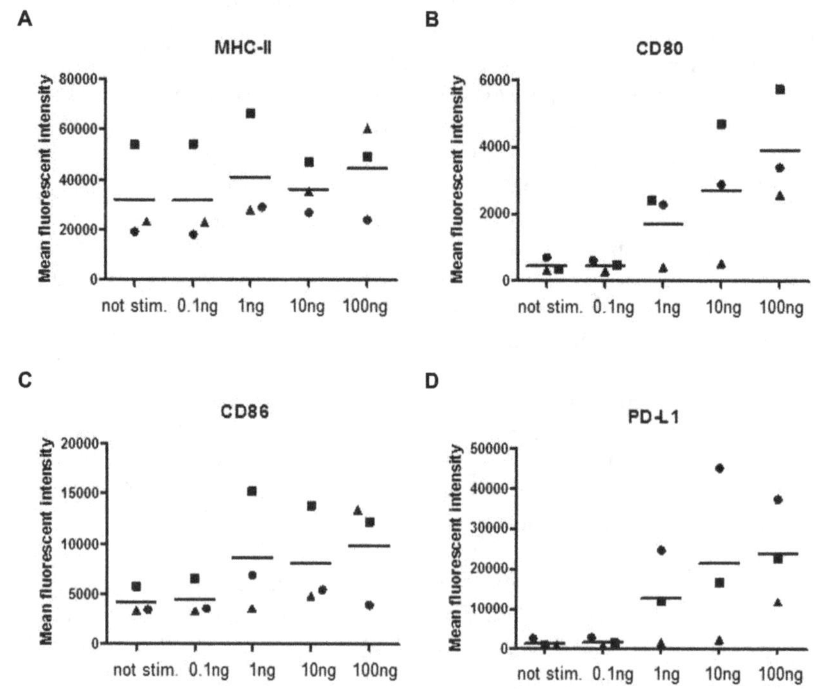

FIGURE 2 | Streptococcal pyrogenic exotoxin A- induced surface molecules. $CD14^+$ monocytes were stimulated with increasing concentrations of SPEA as indicated. 24 h after stimulation cells were harvested and analyzed for surface markers by flow cytometry **(A)** MHC-II, **(B)** CD80, **(C)** CD86 and **(D)** PD-L1. The mean fluorescent intensity of three different experiments is shown. The line represents the mean of the experiments.

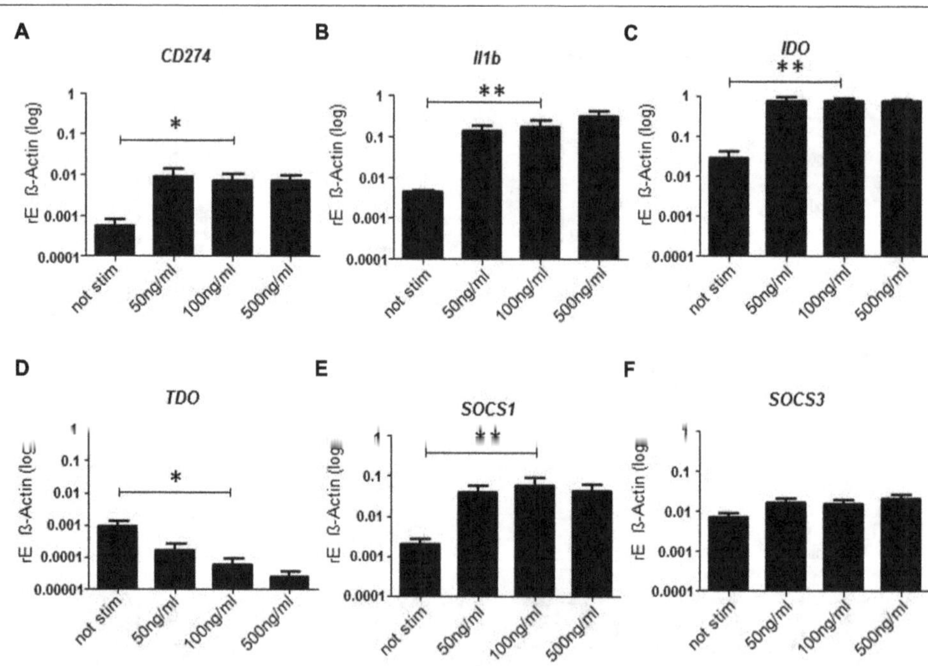

FIGURE 3 | Streptococcal pyrogenic exotoxin A- induced mRNA expression of inhibitory pathways. $CD14^+$ monocytes were stimulated 24 h with increasing concentrations of SPEA. Quantitative real-time PCR of cDNA was performed for **(A)** CD274, **(B)** Il1b, **(C)** IDO, **(D)** TDO, **(E)** SOCS1, and **(F)** SOCS3. Shown is the mean of induction compared to ß-Actin (±SD) of 5 donors. Statistical analysis was performed using a Multiple Comparison of Means (Tukey Contrasts). Significance code: 0 '***'; 0.001 '**'; 0.01 '*'.

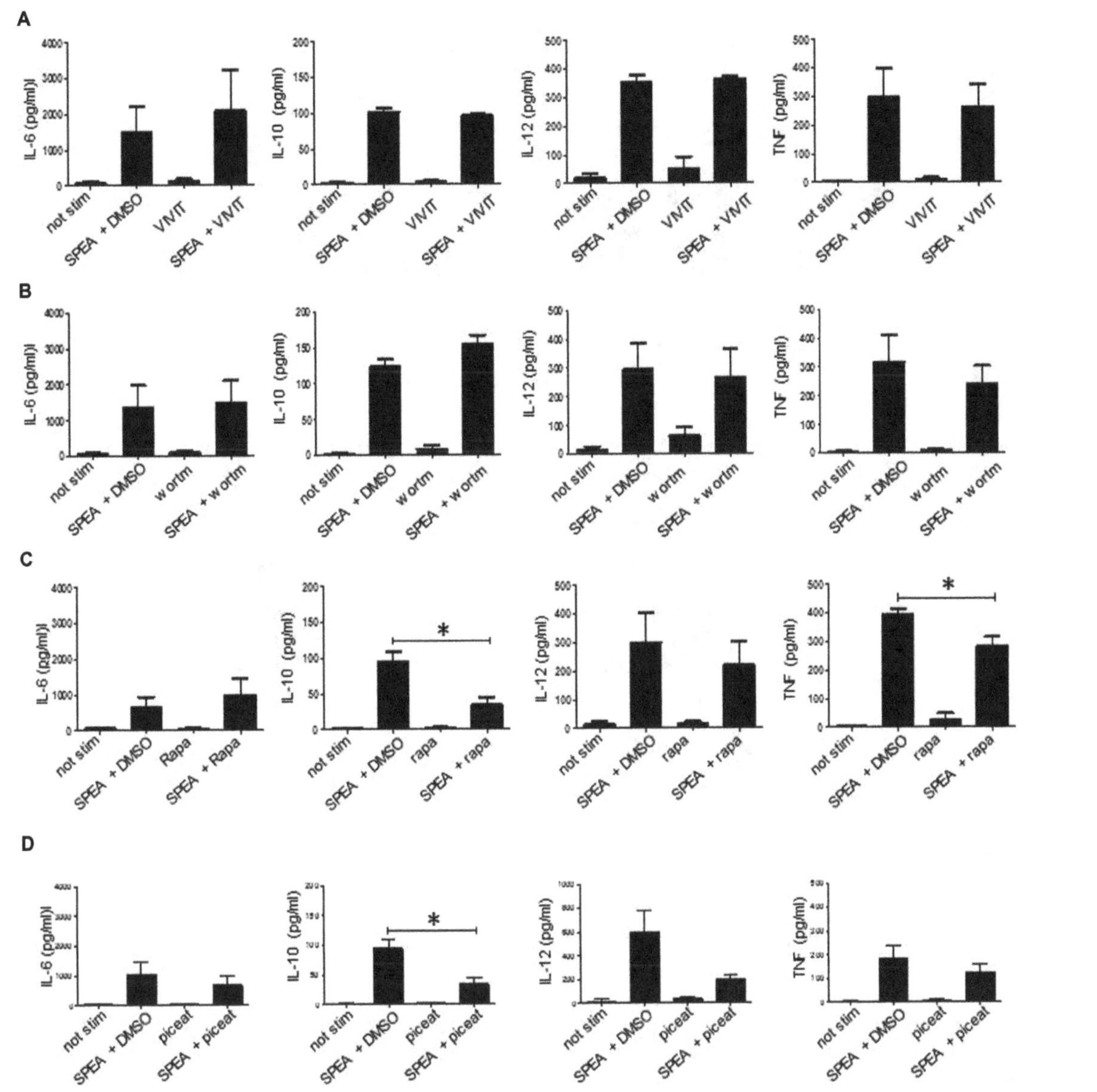

FIGURE 4 | Signaling pathways involved in SPEA- mediated activation of APC. $CD14^+$ monocytes were stimulated with SPEA (100ng/ml) in the presence or absence of **(A)** VIVIT (1 M), **(B)** Wortmannin (50 nM), **(C)** rapamycin (50nM) and **(D)** piceatannol (50 μM) or DMSO (equal volume to inhibitor) as solvent control. 24 h after stimulation supernatants were harvested and ELISAs were performed for IL-6, IL-10, IL-12p40 and TNF. Shown is the mean ±SD from 3 different donors, except VIVIT ($n = 2$). Statistical analysis was performed using a Multiple Comparison of Means (Tukey Contrasts). Significance codes: 0 '***'; 0.001 '**'; 0.01 '*'.

For T-cell activation, the presence of APC and the binding of SAg to their MHC class II molecules are mandatory (MacDonald [illegible] became evident that SAg not only binds to [illegible] the ability at different levels to crosslink MHC class II molecules (Hudson et al., 1995; Kozono et al., 1995). This suggested that SAg might confer a signal to the MHC class II expressing APC. Indeed MHC class II signaling after crosslinking has been observed in B-cells and stem cells which was characterized by activation of tyrosine kinases like Syk (Mooney et al., 1990, 1994; Scholl and Geha, 1994; Kanner et al., 1995; Yamaguchi et al., 1999). In monocytes binding of SAg and thus crosslinking of MHC class II molecules was followed by an intracellular [illegible] et al., 1995; Espel et al., 1996; Khan et al., 2008). Interestingly Treg seem to utilize class II signaling too. Mature Tregs express LAG-3 (CD223) which has been shown to bind to MHC class II and to crosslink subsequently the molecules (Hemon et al., 2011).

Since SAg-mediated stimulation of T-cells includes activation as well as tolerization processes we hypothesized that both

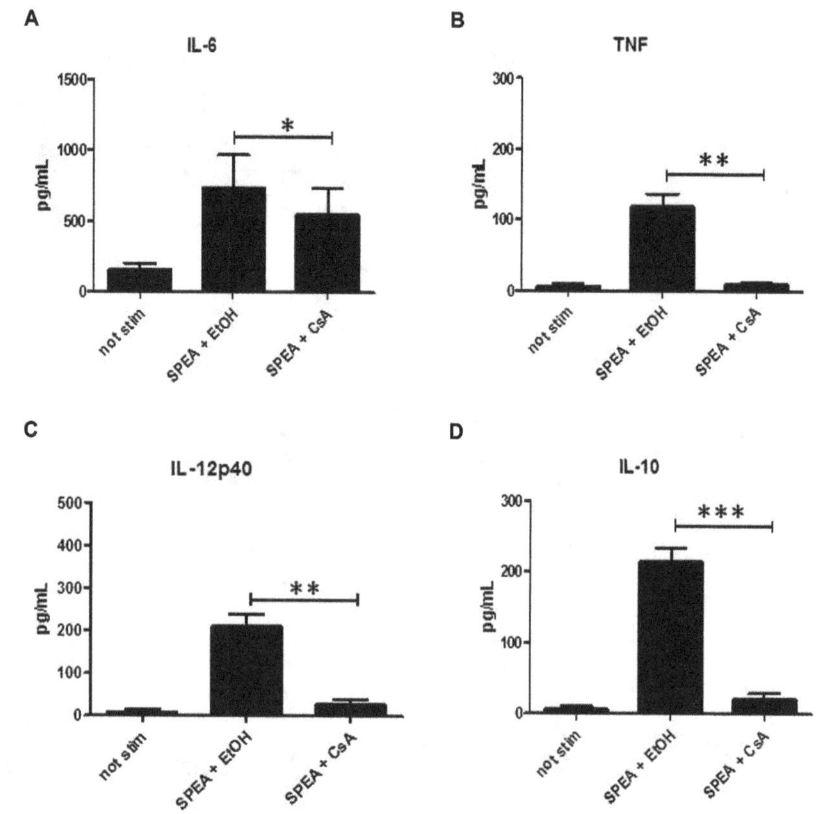

FIGURE 5 | Cyclosporine A (CsA) prevents secretion of cytokines. CD14^{+} monocytes were stimulated with SPEA (100 ng/ml) and CsA (1 μM) or Ethanol (equal volume to CsA) for 24 h. Supernatants were harvested and analyzed by ELISA for **(A)** IL-6, **(B)** TNF, **(C)** IL-12p40 and **(D)** IL-10. Shown is the mean concentration with standard deviation of 5 different donors. Significance codes: 0 '***'; 0.001 '**'; 0.01 '*' (Tukey Contrasts).

events are determined by respective cellular or humoral events triggered in APC by SAg. We therefore tested whether co-inhibitory molecules (such as PD-L1; Francisco et al., 2009, 2010), negative signaling circuits (such as suppressor of cytokine signaling molecules, SOCS; Alexander, 2002; Strebovsky et al., 2011), or inhibitory effector systems (such as IDO; Lestage et al., 2002; Hill et al., 2007) are induced by SAg in APC.

MATERIALS AND METHODS

Reagents

Streptococcal pyrogenic exotoxin A (SPEA) was purchased from Toxin Technology Inc. (Sarasota, FL, USA). The mTOR-inhibitor rapamycin (50 nM), NFAT-inhibitor VIVIT (1 M), PI3-kinase-inhibitor Wortmannin (50 nM) and piceatannol (50 μM), a Syk-inhibitor were acquired from Calbiochem (Schwalbach, Germany). CsA was purchased from R&D Systems (Wiesbaden, Germany).

Isolation of Primary Cells

Human peripheral blood mononuclear cells (PBMCs) were isolated from fresh blood by density gradient centrifugation (Pancoll 1.077 g/ml; PAN Biotech). Monocytes were isolated via CD14 MicroBeads (Miltenyi Biotech, Bergisch-Gladbach, Germany) with the autoMACS separator. 2×10^6 cells were cultured in RPMI 1640 (Sigma-Aldrich, Taufkirchen, Germany) supplemented with 100 IU/mL of penicillin, 100 μg/mL streptomycin containing 10% heat inactivated fetal calf serum (Promocell, Heidelberg, Germany) at 37°C in a humidified atmosphere in the presence of 5% CO_2 and stimulated for 24 h.

Flow Cytometry

Twenty four hours after stimulation monocytes were analyzed for the surface markers CD14 (clone TÜK4), CD80 (clone L307.4), CD86 (clone IT2.2), PD-L1 (clone MIH1) and MHC-II (clone Tu39). Analyses were performed on a FACS Canto I (BD Biosciences).

The antibodies were purchased from Becton Dickinson (Heidelberg, Germany), except PD-L1 (eBioscience, Frankfurt/Main, Germany).

Western Blotting

8×10^6 cells were lysed 24 h after stimulation in RIPA lysis buffer containing protease inhibitor cocktail and phosphatase inhibitor cocktail from Roche (Mannheim,

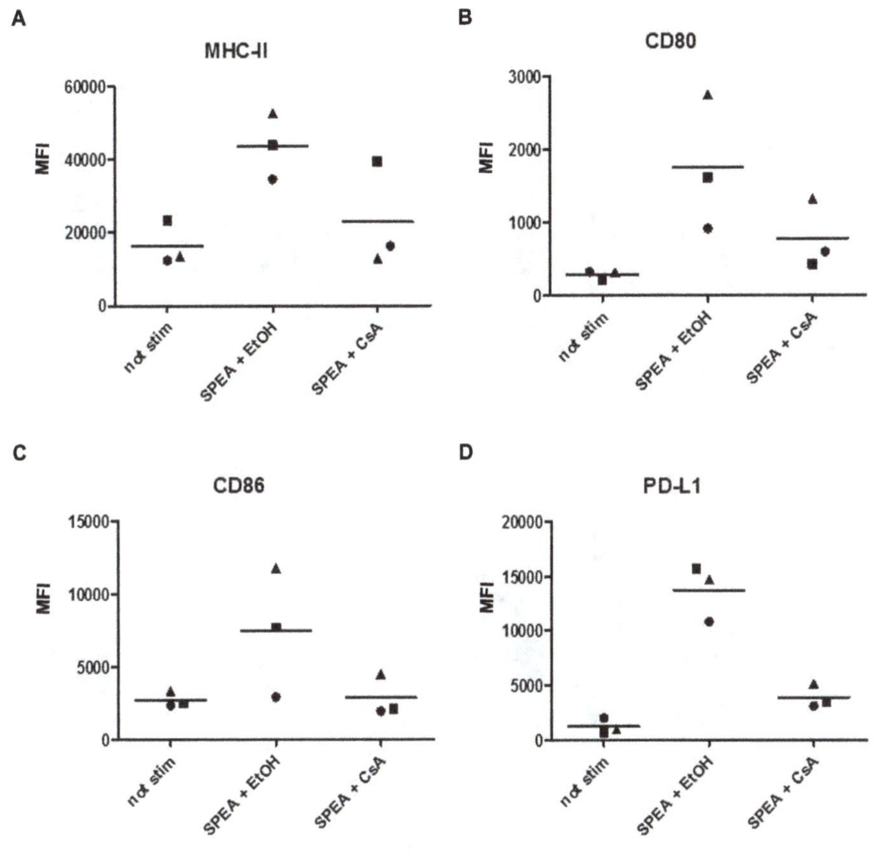

FIGURE 6 | Cyclosporine A prevents the expression of costimulatory and coinhibitory molecules. $CD14^+$ monocytes were stimulated with SPEA (100 ng/ml) and CsA (1 μM) or Ethanol (equal volume to CsA) for 24 h. Cells were harvested and analyzed for surface markers by flow cytometry **(A)** MHC-II, **(B)** CD80, **(C)** CD86 and **(D)** PD-L1. The mean fluorescence intensity of three different donors is shown.

Germany). Equal amounts of the lysates were fractionated by SDS-PAGE and electrotransferred to nitrocellulose membranes. After blocking and washing steps the indicated antibodies, purchased from Cell Signaling Technology (Danvers, MA, USA), were incubated for 24 h and detected via chemiluminescence (ECL; Perkin Elmer, Groningen, Netherlands).

ELISA

Cell-free supernatants were harvested 24 h after stimulation and analyzed for IL-6, IL-10, IL-12p40 and TNF by commercial available ELISA kits from Becton Dickinson (OptEIA; Becton Dickinson, Heidelberg, Germany) according to the

RNA Purification and Quantitative Real – Time PCR

Total RNA from 4×10^6 cells was isolated using the High Pure RNA isolation Kit (Roche, Mannheim, Germany), cDNA was synthesized from equal amounts of RNA using the first strand cDNA kit from Thermo Scientific (Waltham, MA USA). Quantitative real-time RT-PCR was performed using ABsoluteTM QPCR SYBR®Green Low ROX Mix (Thermo Scientific, Waltham, MA, USA). Relative expression was calculated by normalization to β-Actin mRNA expression levels as 2-ΔCt. All primers were synthesized by Eurofins MWG Operon (Ebersberg, Germany) (**Table 1**).

Statistics

Statistical significance was assessed using SPSS statistics software and paired student's *t*-test evaluation with *: $p \leq 0.05$, **: $p \leq 0.005$. Further on we confirmed the results with Multiple Comparisons of Means (Tukey Contrasts), performed with R. Significance codes: 0 '***' 0.001 '**' 0.01 '*' 0.05 '.' 0.1 ' ' 1. When the results of the methods differ, significance of Tukey's test is shown.

Superantigens Induce Cytokine Secretion in APC

It has been shown previously that binding of SAg to MHC class II molecules induces activation and secretion of cytokines. When we incubated monocytes for 24 h with graded doses of SPEA a dose dependent induction of cytokine secretion was found (**Figures 1A–C**). Not only pro-inflammatory cytokines

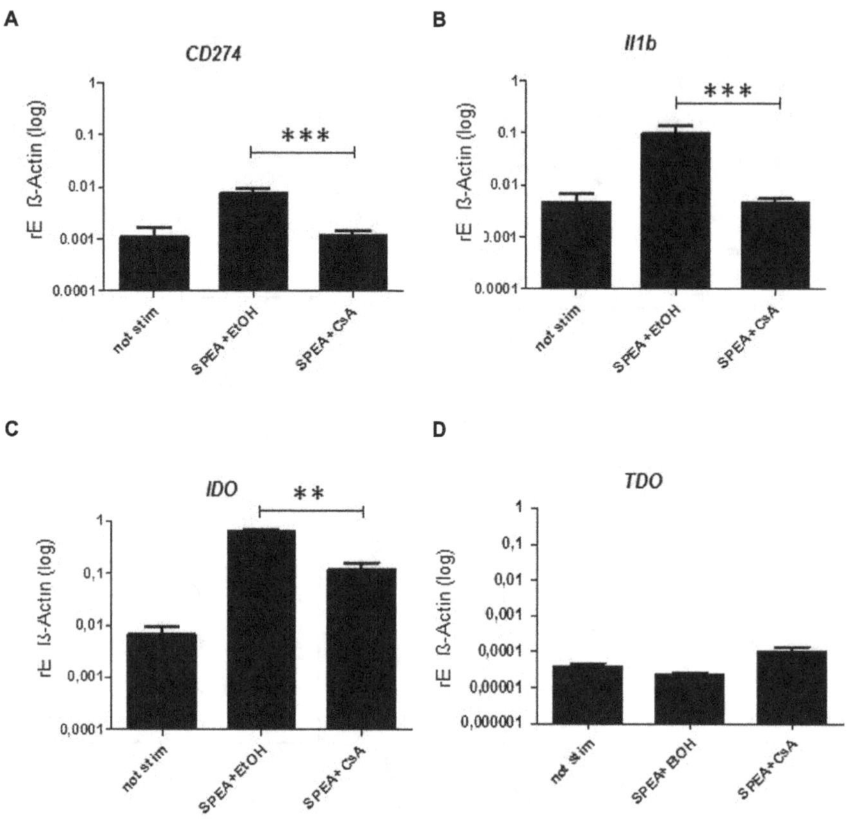

FIGURE 7 | Cyclosporine A – mediated expression of mRNA. CD14^{+} monocytes were stimulated 24 h with SPEA (100 ng/ml) and CsA (1 μM) or Ethanol (equal volume to CsA). Quantitative real-time PCR of cDNA was performed with sequence-specific primers for **(A)** CD274, **(B)** Il1b, **(C)** IDO, **(D)** TDO. Data were normalized to ß-Actin. Shown is the mean and standard deviation of 5 donors. Statistical analysis was performed using a Multiple Comparison of Means (Tukey Contrasts). Significance codes: 0 '***'; 0.001 '**'; 0.01 '*'.

like IL-6, TNF, or IL-12p40 were induced but also secretion of anti-inflammatory IL-10 could be recorded (**Figure 1D**).

Next we assayed for the expression of surface molecules involved in antigen presentation and T-cell interaction (**Figure 2**). The surface expression of MHC class II molecules showed no dependency on SPEA stimulation (**Figure 2A**), while a dose-dependent expression of the co-stimulatory molecules CD80 and CD86 was observed (**Figures 2B,C**). Interestingly the co-inhibitory molecule PD-L1 (CD274) was strongly induced (**Figure 2D**, Supplementary **Figure S1**).

Superantigens Induce mRNA Expression of Inhibitory Pathways

Since PD-L1 expression indicated that inhibitory pathways might be induced we analyzed the mRNA expression of molecules involved in negative regulatory circuits (**Figure 3**). As expected mRNA expression of PD-L1 was significantly enhanced after treatment with SPEA (**Figure 3A**). The same was true for IL-1beta (**Figure 3B**). As IL-1 beta is produced as inactive precursor we additionally confirmed the elevated release of the active cytokine (Supplementary Figure S2). Surprisingly expression of the inhibitory effector enzyme IDO was observed (**Figure 3C**). In contrast, expression of tryptophan 2,3-dioxygenase (TDO) was suppressed (**Figure 3D**) suggesting an inverse regulation. Within APCs negative feedback loops have been identified which regulate the response to cytokine stimulation. Molecules of the SOCS family play a pivotal role during these processes. We therefore analyzed the expression of SOCS1 and SOCS3 after

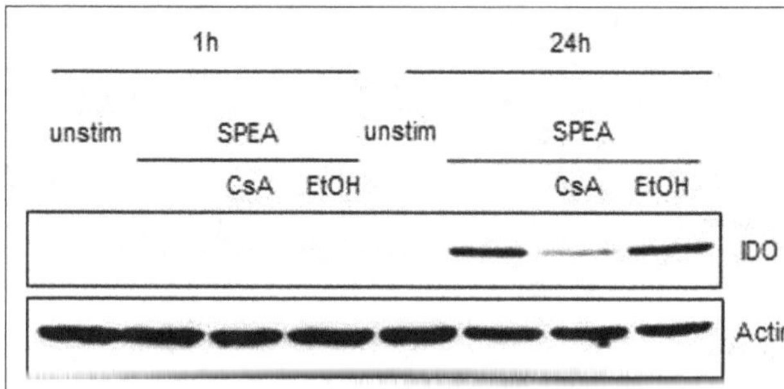

FIGURE 8 | Indoleamine 2,3-dioxygenase expression after SPEA stimulation. CD14^{+} monocytes were stimulated for 1 h or 24 h with 100 ng/ml SPEA and 1 μM CsA (Ethanol was used equal to the amount of CsA). Cells were lysed and analyzed by western blot for IDO. ß-Actin as loading control was displayed. Shown is one representative experiment out of three.

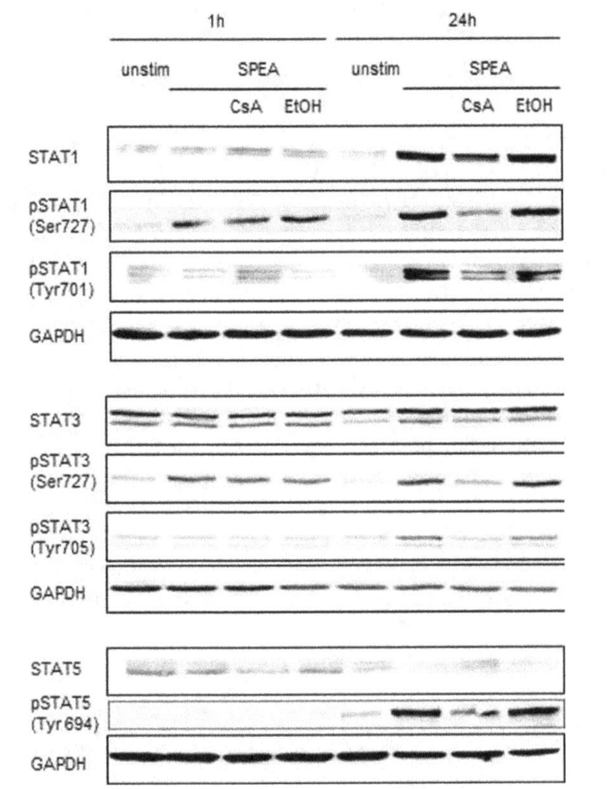

FIGURE 9 | STAT activation after SPEA stimulation. (A) CD14$^+$ monocytes were stimulated for 1 or 24 h with 100 ng/ml SPEA and 1 μM CsA (Ethanol was used equal to the amount of CsA). Cells were lysed and analyzed by western blotting with specific antibodies against STAT1, phospho-STAT1 (Ser727, Tyr701), STAT3, phospho-STAT3 (Ser727, Tyr 705), STAT5 and phospho-STAT5 (Tyr 694). GAPDH as loading control was displayed. Shown is one representative experiment out of three.

stimulation of monocytes with SAg. While SOCS1 showed a clear induction, SOCS3 was only slightly induced (**Figures 3E,F**). mRNA expression of other signaling molecules was unchanged except for STAT1 and IRF-1which showed a slight increased expression (Supplementary Figure S3).

Signaling Pathways Involved in SAg-Mediated Activation of APC

It was suggested that class II-signaling induces Ca-mobilization (Mooney et al., 1990) and subsequent activation. SAg-mediated signaling in APC could thus be mediated by non-classical

other non-canonical pathways (mTOR, PI3K) could mediate SAg stimulation of APC. To test this, we resorted to classical inhibition assays (**Figure 4**). When we tested for inhibition of the induction of pro-inflammatory and anti-inflammatory cytokines we failed to observe an effect with the inhibitors of NFAT (VIVIT) (**Figure 4A**) and PI3K (Wortmannin) (**Figure 4B**). Inhibition of mTOR pathway by rapamycin or Syk pathway by piceatannol had no effect on the induction of pro-inflammatory cytokines (TNF, IL-6, IL-12), however, rapamycin and piceatannol prevented release of IL-10 indicating a differential regulation of pro- and anti-inflammatory cytokines (**Figures 4C,D**).

Cyclosporine A interacts with cyclophilin and suppresses activation of T-cells and B-cells (Fischer et al., 1989). We thus hypothesized that CsA could interfere with SAg-mediated stimulation of APC. Indeed stimulation of monocytes with SPEA in the presence of CsA prevented the secretion of pro-inflammatory and anti-inflammatory cytokines significantly (**Figure 5**). Further on CsA modulated the SPEA-induced expression of co-stimulatory and co-inhibitory molecules, although not in a statistically significant mode (**Figure 6**). A similar pattern was observed for the induction of mRNA: CsA prevented the upregulation of mRNA coding for PD-L1 (**Figure 7A**), IL-1beta (**Figure 7B**), and partially reduced induction of IDO (**Figure 7C**). Accordingly, expression of TDO was not downregulated (**Figure 7D**).

Indoleamine 2,3-dioxygenase is a rate-limiting enzyme of tryptophan catabolism resulting in kynurenine production. Depletion of tryptophan causes halted growth of T-cells while kynurenine activates the aryl hydrocarbon (AH) receptor system leading to induction of Tregs (Opitz et al., 2011). We therefore tested for the expression of IDO and the subsequent depletion of tryptophan and production of kynurenine after SAg stimulation (**Figure 8**). In SPEA-stimulated APC IDO protein expression was detected (**Figure 8**) and the levels of kynurenine were enhanced (data not shown). Both IDO expression and kynurenine production was sensitive to inhibition by CsA corroborating the effects seen with mRNA induction (**Figure 7C**).

STAT Activation is a Hallmark of SAg-Mediated Stimulation of APCs

To further elucidate the signaling pathways in SAg-mediated stimulation of APC we analyzed the phosphorylation pattern of STAT molecules and MAPK after stimulation. While phosphorylation of MAPK (p38, JNK, P44/42) was not changed (Supplementary Figure S4), STAT1 and STAT3 were immediately phosphorylated, independent on CsA (**Figure 9**). STAT5 showed no phosphorylation. In contrast, after 24 h STAT1, STAT3, and STAT5 were strongly phosphorylated (**Figure 9**). The phosphorylation observed was sensitive to inhibition by CsA.

DISCUSSION

Superantigens are bacterial exotoxins that interact with immune

evasion strategy of the pathogens (MacDonald et al., 1993). It is unquestionable that acute release of pro-inflammatory cytokines leads to a dysregulation of the immune and other systemic responses eventually causing septic shock. Staphylococcal or streptococcal TSS are examples of this pathogenesis (Miethke et al., 1993). They resemble cytokine storm events comparable to those observed in humans after accidentally triggering T-cells with monoclonal antibodies (Suntharalingam et al., 2006). The induction of T-cells requires binding of SAg to MHC class II molecules of APC and the followed crosslinking of T-cells to APCs (Rink et al., 1997). Binding of SAg activates APC, resulting in production of co-stimulatory cytokines and co-stimulatory molecules (Ohnishi et al., 1995).

Concomitantly inhibitory circuits are induced that subsequently dominate. After initial deletion of SAg-reactive T-cells the remaining T-cells become unresponsive and display an anergic phenotype (Kisielow et al., 1991; Wahl et al., 1993). Moreover it was shown that anergy and unresponsiveness are accompanied by the induction of $CD4^{+}CD25^{+}$ T regulatory cells (Wang et al., 1998; Noël et al., 2001; Papiernik, 2001). It is quite obvious that cytokine storm, anergy and induction of Treg induce a milieu of dysregulation and suppression that precludes a coordinated immune response and thus allows the pathogen to subvert anti-infective strategies of the host.

We show here that APC function after SAg-binding not only includes the induction of pro-inflammatory responses in terms of cytokine release and costimulation but also the induction of co-inhibitory circuits including anti-inflammatory cytokines (IL-10), co-inhibitory molecules (PD-L1) and also an induction of inhibitory effector programs (IDO). While IDO might result in an unspecific immunosupression by depleting tryptophan (Taylor and Feng, 1991) PD-L1 and IL-10 suggest an at least bystanding activity to induce Tregs (Unger et al., 2009; Wölfle et al., 2011). Moreover kynurenine an intermediate of the tryptophan metabolism produced by IDO, can further screw T-cell differentiation in direction of Tregs (Quintana et al., 2008; Gandhi et al., 2010; Opitz et al., 2011). Altogether an immunosuppressive milieu is induced which clearly foster the generation of Tregs and thus prevents an active immune response. Indeed our experiments with SPEA-treated monocytes confirm their inhibitory influence on CD3-mediated T-cell proliferation (Supplementary Figure S5). Furthermore SPEA-treated co-culture experiments with APCs and T-cells reveal a $CD4^{+}CD25^{+}Foxp3^{+}$ Treg population (Supplementary Figure S6) that is functionally active and inhibits T-cell proliferation (Supplementary Figure S7).

Interestingly, negative feedback regulators for IFN signaling (SOCS-1) were also induced by SPEA (Song and Shuai, 1998; Dalpke et al., 2008; Masters et al., 2010). This could explain why APCs are refractory to stimulation with proinflammatory cytokines like IFN-gamma and thus maintain their inhibitory phenotype. We also observed a long lasting phosphorylation of STAT1 as well as STAT3 after SAg treatment of APC. Phosphorylated STAT3 induce expression of PD-L1 (Wölfle et al., 2011) and IDO (Litzenburger et al., 2014) and thus contribute to the immunosuppressive milieu. Constitutive STAT1 phosphorylation seems to contradict the immunosuppressive phenotype. Although we have not addressed this here in detail, it was shown that STAT1 action itself is under control of SOCS-1, which could explain why STAT1 phosphorylation does not necessarily induce proinflammatory cascades (Hildebrand et al., 2010).

A critical role during these processes of tolerance induction plays the APC. APC represent the scaffold to present SAg to T-cells (Dellabona et al., 1990), yet also provide other signals to the T-cells (Mehindate et al., 1995). Therefore class II-binding of SAg has to deliver a signal to the APC. Class II signaling has been described for B-cells (Mooney et al., 1990; Nabavi et al., 1992) and APCs (Chatila and Geha, 1993). Thereby different signal intermediates such as Ca^{2+} (Damaj et al., 1992), PKC (Palkama and Hurme, 1993), NFAT and MAPK (Haylett et al., 2009), and tyrosine kinases (Palkama and Hurme, 1993; Morio et al., 1994; Kanner et al., 1995) were shown. When we analyzed SAg-induced signaling using various inhibitors, we found no indication for an involvement of PI3K (Wortmannin), mTOR (rapamycin) or the tyrosine kinase Syk (piceatannol) (**Figure 4**). Since Ca^{2+} mobilization can be a consequence of SAg-activation (Damaj et al., 1992) we presumed that NFAT would be involved, yet this was not the case. In contrast all responses induced by SAg could be blocked by CsA. This was not an entirely surprising finding since we have reported recently that in an analogous stimulation model in B-cells signaling was also sensitive to inhibition with CsA but not dependent on NFAT (Ziegler et al., 2014). The sensitivity to CsA of the induction of an immunosuppressive milieu of APC fits well in older reports on the unfolding of tolerance and unresponsiveness. Sakaguchi reported that neonatal administration of CsA induced autoimmune disease in a model which was depended on Tregs (Sakaguchi and Sakaguchi, 1989). It was also shown that CsA inhibits Treg generation (Wang et al., 2006) and that immunosuppressive therapy with CsA might reduce the number of Treg after allogeneic renal transplantation (Korczak-Kowalska et al., 2007).

Cyclosporine A might interfere with the induction of Treg in two ways. Once it inhibits the Treg inducing milieu after class II binding of SAgs as described above. Secondly, CsA might affect the self-stabilizing feedback loop of Treg-APC interaction. Tregs express LAG-3 which interacts with class II molecules (Liang et al., 2008). Interaction could lead to APC activation and subsequent production of inhibitory cytokines and molecules as shown above. That would lead to a self-propagation of Tregs and thus stabilize and expand the Treg pool. Accordingly CsA would prevent this loop by inhibiting the response to class II crosslinking. It has been shown that blockade of PD-L1 and LAG-3 rapidly cleared infection with plasmodia (Butler et al., 2012), indicating that during infection this loop is operative and could be manipulated.

Taken together we show here that APC triggered by SAg are not only responsible for the initial induction of pro-inflammatory responses but are also crucial for the induction and maintenance of unresponsiveness and anergy. Paradoxically the T-cell immunosuppressive drug CsA prevents in SAg triggered APC the manifestation of the immunosuppressive program, indicating that CsA could be utilized to prevent SAg-induced anergy and unresponsiveness.

FUNDING

This work was supported by a project of the collaborative research center 938 (SFB938) to KH and a postdoctoral fellowship to DH (Medizinische Fakultät Heidelberg).

AUTHOR CONTRIBUTIONS

AS, SF, KH, and DH designed the study. DH and KH wrote the final manuscript. AS, SF performed the experiments. All authors read the manuscript and discussed the results.

ACKNOWLEDGMENTS

We thank Sabrina Klein, Nina Sinke, and Dennis Nurjadi for the fruitful discussion of results and Sebastien Boutin for statistical analysis. We are grateful to Christiane Opitz, German Cancer Research Center (DKFZ), for the discussions and technical advice (kynurenine assay).

REFERENCES

Alexander, W. S. (2002). Suppressors of cytokine signalling (SOCS) in the immune system. *Nat. Rev. Immunol.* 2, 410–416.

Butler, N. S., Moebius, J., Pewe, L. L., Traore, B., Doumbo, O. K., Tygrett, L. T., et al. (2012). Therapeutic blockade of PD-L1 and LAG-3 rapidly clears established blood-stage *Plasmodium* infection. *Nat. Immunol.* 13, 188–195. doi: 10.1038/ni.2180

Carlsson, R., and Sjögren, H. O. (1985). Kinetics of IL-2 and interferon-gamma production, expression of IL-2 receptors, and cell proliferation in human mononuclear cells exposed to staphylococcal enterotoxin A. *Cell. Immunol.* 96, 175–183. doi: 10.1016/0008-8749(85)90349-1

Chatila, T., and Geha, R. S. (1993). Signal transduction by microbial superantigens via MHC class II molecules. *Immunol. Rev.* 131, 43–59. doi: 10.1111/j.1600-065X.1993.tb01529.x

Choi, Y. W., Herman, A., DiGiusto, D., Wade, T., Marrack, P., and Kappler, J. W. (1990). Residues of the variable region of the T-cell-receptor beta- chain that interact with *S. aureus* toxin superantigens. *Nature* 346, 471–473. doi: 10.1038/346471a0

Dalpke, A., Heeg, K., Bartz, H., and Baetz, A. (2008). Regulation of innate immunity by suppressor of cytokine signaling (SOCS) proteins. *Immunobiology* 213, 225–235. doi: 10.1016/j.imbio.2007.10.008

Damaj, B., Mourad, W., and Naccache, P. H. (1992). Superantigen-mediated human monocyte-T lymphocyte interactions are associated with an MHC class II-, TCR/CD3-, and CD4-dependent mobilization of calcium in monocytes. *J. Immunol.* 149, 1497–1503.

Dellabona, P., Peccoud, J., Kappler, J. W., Marrack, P., Benoist, C., and Mathis, D. (1990). Superantigens interact with MHC class II molecules outside of the antigen groove. *Cell Camb.* 62, 1115–1121. doi: 10.1016/0092-8674(90)90388-U

Espel, E., Garcia-Sanz, J. A., Aubert, V., Menoud, V., Sperisen, P., Fernández, N., et al. (1996). Transcriptional and translational control of TNF-α gene expression in human monocytes by major histocompatibility complex class II ligands. *Eur. J. Immunol.* 26, 2417–2424. doi: 10.1002/eji.1830261023

Feunou, P., Poulin, L., Habran, C., Le Moine, A., Goldman, M., and Braun, M. Y. (2003). CD4+CD25+ and CD4+CD25- T cells act respectively as inducer and effector T suppressor cells in superantigen-induced tolerance. *J. Immunol.* 171, 3475–3484. doi: 10.4049/jimmunol.171.7.3475

Fischer, G., Wittmann-Liebold, B., Lang, K., Kiefhaber, T., and Schmid, F. X. (1989). Cyclophilin and peptidyl-prolyl cis-trans isomerase are probably identical proteins. *Nature* 337, 476–478. doi: 10.1038/337476a0

Francisco, L. M., Sage, P. T., and Sharpe, A. H. (2010). The PD-1 pathway in

Grundström, S., Cederbom, L., Sundstedt, A., Scheipers, P., and Ivars, F. (2003). Superantigen-induced regulatory T cells display different suppressive functions in the presence or absence of natural CD4+CD25+ regulatory T cells in vivo. *J. Immunol.* 170, 5008–5017. doi: 10.4049/jimmunol.170.10.5008

Haylett, R. S., Koch, N., and Rink, L. (2009). MHC class II molecules activate NFAT and the ERK group of MAPK through distinct signaling pathways in B cells. *Eur. J. Immunol.* 39, 1947–1955. doi: 10.1002/eji.200838992

Hemon, P., Jean-Louis, F., Ramgolam, K., Brignone, C., Viguier, M., Bachelez, H., et al. (2011). MHC class II engagement by its ligand LAG-3 (CD223) contributes to melanoma resistance to apoptosis. *J. Immunol.* 186, 5173–5183. doi: 10.4049/jimmunol.1002050

Herrmann, T., Baschieri, S., Lees, R. K., and MacDonald, H. R. (1992). In vivo responses of CD4+ and CD8+ cells to bacterial superantigens. *Eur. J. Immunol.* 22, 1935–1938. doi: 10.1002/eji.1830220739

Herrmann, T., and MacDonald, H. R. (1991). T cell recognition of superantigens. *Curr. Top. Microbiol. Immunol.* 174, 21–38.

Hildebrand, D., Walker, P., Dalpke, A., Heeg, K., and Kubatzky, K. F. (2010). *Pasteurella multocida* Toxin-induced Pim-1 expression disrupts suppressor of cytokine signalling (SOCS)-1 activity. *Cell. Microbiol.* 12, 1732–1745. doi: 10.1111/j.1462-5822.2010.01504.x

Hill, M., Tanguy-Royer, S., Royer, P., Chauveau, C., Asghar, K., Tesson, L., et al. (2007). IDO expands human CD4+CD25 high regulatory T cells by promoting maturation of LPS-treated dendritic cells. *Eur. J. Immunol.* 37, 3054–3062. doi: 10.1002/eji.200636704

Huang, L., and Crispe, I. N. (1993). Superantigen-driven peripheral deletion of T cells. Apoptosis occurs in cells that have lost the alpha/beta T cell receptor. *J. Immunol.* 151, 1844–1851.

Hudson, K. R., Tiedemann, R. E., Urban, R. G., Lowe, S. C., Strominger, J. L., and Fraser, J. D. (1995). Staphylocaoccal enterotoxin A has two cooperative binding sites on major histocompatibility complex class II. *J. Exp. Med.* 182, 711–720. doi: 10.1084/jem.182.3.711

Ivars, F. (2007). Superantigen-induced regulatory T cells in vivo. *Chem. Immunol. Allergy* 93, 137–160. doi: 10.1159/0000100862

Kanner, S. B., Grosmaire, L. S., Blake, J., Schieven, G. L., Masewicz, S., Odum, N., et al. (1995). ZAP-70 and p72syk are signaling response elements through MHC class II molecules. *Tissue Antigens* 46, 145–154. doi: 10.1111/j.1399-0039.1995.tb03113.x

Kappler, J. W., Kotzin, B. L., Herron, L., Gelfand, E. W., Bigler, R. D., Boylston, A., et al. (1989). V beta-specific stimulation of human T cells by staphylococcal toxins. *Sci. Wash.* 244, 811–813. doi: 10.1126/science.25

Kuchroo, V. K., et al. (2009). PD-L1 regulates the development, maintenance, and function of induced regulatory T cells. *J. Exp. Med.* 206, 3015–3029. doi: 10.1084/jem.20090847

Fraser, J. D., and Proft, T. (2008). The bacterial superantigen and superantigen-like proteins. *Immunol. Rev.* 225, 226–243. doi: 10.1111/j.1600-065X.2008.00681.x

Gandhi, R., Kumar, D., Burns, E. J., Nadeau, M., Dake, B., Laroni, A., et al. (2010). Activation of the aryl hydrocarbon receptor induces human type 1 regulatory T cell-like and Foxp3+ regulatory T cells. *Nat. Immunol.* 11, 846–853. doi: 10.1038/ni.1915

10.1189/jlb.1007686

Kisielow, P., Swat, W., Rocha, B., and von Boehmer, H. (1991). Induction of immunological unresponsiveness in vivo and in vitro by conventional and super-antigens in developing and mature T cells. *Immunol. Rev.* 122, 69–85. doi: 10.1111/j.1600-065X.1991.tb00597.x

Korczak-Kowalska, G., Wierzbicki, P., Bocian, K., Klosowska, D., Niemczyk, M., Wyzgal, J., et al. (2007). The influence of immuosuppressive therapy on the development of CD4+CD25+ T cells after renal transplantation. *Transplant. Proc.* 39, 2721–2723. doi: 10.1016/j.transproceed.2007.09.015

Kozono, H., Parker, D., White, J., Marrack, P., and Kappler, J. (1995). Multiple binding sites for bacterial superantigens on soluble class II MHC molecules. *Immunity* 3, 187–196. doi: 10.1016/1074-7613(95)90088-8

Lee, W. T., and Vitetta, E. S. (1992). Memory T cells are anergic to the superantigen staphylococcal enterotoxin B. *J. Exp. Med.* 176, 575–579. doi: 10.1084/jem.176.2.575

Lestage, J., Verrier, D., Palin, K., and Dantzer, R. (2002). The enzyme indoleamine 2,3-dioxygenase is induced in the mouse brain in response to peripheral administration of lipopolysaccharide and superantigen. *Brain Behav. Immun.* 16, 596–601. doi: 10.1016/S0889-1591(02)00014-4

Liang, B., Workman, C., Lee, J., Chew, C., Dale, B. M., Colonna, L., et al. (2008). Regulatory T cells inhibit dendritic cells by Lymphocyte activation gene-3 engagement of MHC class II. *J. Immunol.* 180, 5916–5926. doi: 10.4049/jimmunol.180.9.5916

Lina, G., Bohach, G. A., Nair, S. P., Hiramatsu, K., Jouvin-Marche, E., and Mariuzza, R. (2004). Standard nomenclature for the superantigens expressed by *Staphylococcus*. *J. Infect. Dis.* 189, 2334–2336. doi: 10.1086/420852

Litzenburger, U. M., Opitz, C. A., Sahm, F., Rauschenbach, K. J., Trump, S., Winter, M., et al. (2014). Constitutive IDO expression in human cancer is sustained by an autocrine signaling loop involving IL-6, STAT3 and the AHR. *Oncotarget* 5, 1038–1051. doi: 10.18632/oncotarget.1637

MacDonald, H. R., Lees, R. K., Baschieri, S., Herrmann, T., and Lussow, A. R. (1993). Peripheral T cell reactivity to bacterial superantigens in vivo: the response/anergy paradox. *Immunol. Rev.* 133, 105–117. doi: 10.1111/j.1600-065X.1993.tb01512.x

Masters, S. L., Mielke, L. A., Cornish, A. L., Sutton, C. E., O'Donnell, J., Cengia, L. H., et al. (2010). Regulation of interleukin-1[beta] by interferon-[gamma] is species specific, limited by suppressor of cytokine signalling 1 and influences interleukin-17 production. *EMBO Rep.* 11, 640–646. doi: 10.1038/embor.2010.93

Mehindate, K., Thibodeau, J., Dohlsten, M., Kalland, T., Sekaly, R. P., and Mourad, W. (1995). Cross-linking of major histocompatibility complex class Ii molecules by staphylococcal enterotoxin A superantigen is a requirement for inflammatory cytokine gene expression. *J. Exp. Med.* 182, 1573–1577. doi: 10.1084/jem.182.5.1573

Michie, C., Scott, A., Cheesbrough, J., Beverley, P., and Pasvol, G. (1994). Streptococcal toxic shock-like syndrome: evidence of superantigen activity and its effects on T lymphocyte subsets in vivo. *Clin. Exp. Immunol.* 98, 140–144. doi: 10.1111/j.1365-2249.1994.tb06620.x

Miethke, T., Duschek, K., Wahl, C., Heeg, K., and Wagner, H. (1993). Pathogenesis of the toxic shock syndrome: T cell mediated lethal shock caused by the superantigen TSST-1. *Eur. J. Immunol.* 23, 1494–1500. doi: 10.1002/eji.1830230715

Miethke, T., Wahl, C., Heeg, K., Echtenacher, B., Krammer, P. H., and Wagner, H. (1992). T cell-mediated lethal shock triggered in mice by the superantigen staphylococcal enterotoxin B: critical role of tumor necrosis factor. *J. Exp. Med.* 175, 91–98. doi: 10.1084/jem.175.1.91

Miethke, T., Wahl, C., Heeg, K., and Wagner, H. (1995). Superantigens: the paradox of T cell activation versus inactivation. *Int. Arch. Allergy Immunol.* 106, 3–7. doi: 10.1159/000236882

Mooney, N. A., Grillot-Courvalin, C., Hivroz, C., Ju, L. Y., and Charron, D. (1990). Early biochemical events after MHC class II-mediated signaling on human B lymphocytes. *J. Immunol.* 145, 2070–2076.

Mooney, N. A., Ju, L., Brick-Ghannam, C., and Charron, D. J. (1994). Bacterial superantigen signaling via HLA class II on human B lymphocytes. *Mol. Immunol.* 31, 675–681. doi: 10.1016/0161-5890(94)90177-5

Morio, T., Geha, R. S., and Chatila, T. A. (1994). Engagement of MHC class II molecules by staphylococcal superantigens activates src-type protein tyrosine kinases. *Eur. J. Immunol.* 24, 651–658. doi: 10.1002/eji.1830240325

Nabavi, N., Freeman, G. J., Gault, A., Godfrey, D., Nadler, L. M., and Glimcher, L. H. (1992). Signalling through the MHC class II cytoplasmic domain is required foe antigen presentation and induces B7 expression. *Nature* 360, 266–268. doi: 10.1038/360266a0

Noël, C., Florquin, S., Goldman, M., and Braun, M. Y. (2001). Chronic exposure to superantigen induces regulatory CD4+ T cells with IL-10-mediated suppressive activity. *Int. Immunol.* 13, 431–439. doi: 10.1093/intimm/13.4.431

Ohnishi, H., Ledbetter, J. A., Kanner, S. B., Linsley, P. S., Tanaka, T., Geller, A. M., et al. (1995). CD28 cross-linking augments TCR-mediated signals and costimulates superantigen responses. *J. Immunol.* 154, 3180–3193.

Opitz, C. A., Litzenburger, U. M., Sahm, F., Ott, M., Tritschler, I., Trump, S., et al. (2011). An endogenous tumour-promoting ligand of the human aryl hydrocarbon receptor. *Nature* 478, 197–203. doi: 10.1038/nature10491

Palkama, T., and Hurme, M. (1993). Signal transduction mechanisms of HLA-DR-mediated interleukin-1 beta production in human monocytes. Role of protein kinase C and tyrosine kinase activation. *Hum. Immunol.* 36, 259–267.

Papiernik, M. (2001). Natural CD4+ CD25+ regulatory T cells. Their role in the control of superantigen responses. *Immunol. Rev.* 182, 180–189. doi: 10.1034/j.1600-065X.2001.1820114.x

Prud'homme, G. J., Vanier, L. E., Bocarro, D. C., and Croix, H. (1995). Effects of cyclosporin A, rapamycin, and FK520 on peripheral T- cell deletion and anergy. *Cell. Immunol.* 164, 47–56. doi: 10.1006/cimm.1995.1141

Quintana, F. J., Basso, A. S., Iglesias, A. H., Korn, T., Farez, M. F., Bettelli, E., et al. (2008). Control of T(reg) and T(H)17 cell differentiation by the aryl hydrocarbon receptor. *Nature* 453, 65–71. doi: 10.1038/nature06880

Rink, L., Nicklas, W., Alvarez-Ossorio, L., Fagin, U., and Kirchner, H. (1997). Microbial superantigens stimulate T cells by the superantigen bridge and independently by a cytokine pathway. *J. Interferon Cytokine Res.* 17, 489–499. doi: 10.1089/jir.1997.17.489

Sakaguchi, S., and Sakaguchi, N. (1989). Organ-specific autoimmune disease induced in mice by elimination of T cell subsets. V. Neonatal administration of cyclosporin A causes autoimmune disease. *J. Immunol.* 142, 471–480.

Scholl, P. R., and Geha, R. S. (1994). MHC class II signaling in B-cell activation. *Immunol. Today* 15, 418–422. doi: 10.1016/0167-5699(94)90271-2

Scholl, P. R., Trede, N., Chatila, T. A., and Geha, R. S. (1992). Role of protein tyrosine phosphorylation in monokine induction by the staphylococcal superantigen toxic shock syndrome toxin-1. *J. Immunol.* 148, 2237–2241.

Song, M. M., and Shuai, K. (1998). The suppressor of cytokine signaling (SOCS) 1 and SOCS3 but not SOCS2 proteins inhibit interferon-mediated antiviral and antiproliferative activities. *J. Biol. Chem.* 273, 35056–35062. doi: 10.1074/jbc.273.52.35056

Strebovsky, J., Walker, P., Lang, R., and Dalpke, A. H. (2011). Suppressor of cytokine signaling 1 (SOCS1) limits NFκB signaling by decreasing p65 stability within the cell nucleus. *FASEB J.* 25, 863–874. doi: 10.1096/fj.10-170597

Suntharalingam, G., Perry, M. R., Ward, S., Brett, S. J., Castello-Cortes, A., Brunner, M. D., et al. (2006). Cytokine storm in a phase 1 trial of the anti-CD28 monoclonal antibody TGN1412. *N. Engl. J. Med.* 355, 1018–1028. doi: 10.1056/NEJMoa063842

Taylor, M. W., and Feng, G. S. (1991). Relationship between interferon-gamma, indoleamine 2,3-dioxygenase, and tryptophan catabolism. *FASEB J.* 5, 2516–2522.

Trede, N. S., Castigli, E., Geha, R. S., and Chatila, T. (1993a). Microbial superantigens induce NF-kB in the human monocytic cell line THP-1. *J. Immunol.* 150, 5604–5613.

Trede, N. S., Chatila, T., and Geha, R. S. (1993b). Activator protein-1 (AP-1) is stimulated by microbial superantigens in human monocytic cells. *Eur. J. Immunol.* 23, 2129–2135. doi: 10.1002/eji.1830230913

Unger, W. W. J., Laban, S., Kleijwegt, F. S., van der Slik, A. R., and Roep, B. O. (2009). Induction of Treg by monocyte-derived DC modulated by vitamin D3 or dexamethasone: differential role for PD-L1. *Eur. J. Immunol.* 39, 3147–3159. doi: 10.1002/eji.200839103

Vanier, L. E., and Prud'homme, G. J. (1992). Cyclosporin A markedly enhances superantigen-induced peripheral T cell deletion and inhibits anergy induction. *J. Exp. Med.* 176, 37–46. doi: 10.1084/jem.176.1.37

Wahl, C., Miethke, T., Heeg, K., and Wagner, H. (1993). Clonal deletion as direct consequence of an in vivo T cell response to bacterial superantigen. *Eur. J. Immunol.* 23, 1197–1200. doi: 10.1002/eji.1830230536

Wang, H., Zhao, L., Sun, Z., Sun, L., Zhang, B., and Zhao, Y. (2006). A potential side effect of cyclosporin A: inhibition of CD4(+)CD25(+) regulatory T cells in mice. *Transplantation* 82, 1484–1492. doi: 10.1097/01.tp.0000246312.89689.17

Wang, Z.-Q., Orlikowsky, T., Dudhane, A., Trejo, V., Dannecker, G. E., Pernis, B., et al. (1998). Staphylococcal enterotoxin B-induced T-cell anergy is mediated by regulatory T cells. *Immunology* 94, 331–339. doi: 10.1046/j.1365-2567.1998.00519.x

Wölfle, S. J., Strebovsky, J., Bartz, H., Sähr, A., Arnold, C., Kaiser, C., et al. (2011). PD-L1 expression on tolerogenic APCs is controlled by STAT-3. *Eur. J. Immunol.* 41, 413–424. doi: 10.1002/eji.201040979

Yamaguchi, M., Nadler, S., Lee, J.-W., and Deeg, H. J. (1999). Induction of negative regulators of haematopoiesis in human bone marrow cells by HLA-DR cross-linking. *Transpl. Immunol.* 7, 159–168. doi: 10.1016/S0966-3274(99)80035-5

Ziegler, S., Gartner, K., Scheuermann, U., Zoeller, T., Hantzschmann, J., Over, B., et al. (2014). Ca(2+) -related signaling events influence TLR9-induced IL-10 secretion in human B cells. *Eur. J. Immunol.* 44, 1285–1298. doi: 10.1002/eji.201343994

Conflict of Interest Statement: The authors declare that the research was conducted in the absence of any commercial or financial relationships that could be construed as a potential conflict of interest.

Permissions

All chapters in this book were first published in BEHBFIS, by Frontiers; hereby published with permission under the Creative Commons Attribution License or equivalent. Every chapter published in this book has been scrutinized by our experts. Their significance has been extensively debated. The topics covered herein carry significant findings which will fuel the growth of the discipline. They may even be implemented as practical applications or may be referred to as a beginning point for another development.

The contributors of this book come from diverse backgrounds, making this book a truly international effort. This book will bring forth new frontiers with its revolutionizing research information and detailed analysis of the nascent developments around the world.

We would like to thank all the contributing authors for lending their expertise to make the book truly unique. They have played a crucial role in the development of this book. Without their invaluable contributions this book wouldn't have been possible. They have made vital efforts to compile up to date information on the varied aspects of this subject to make this book a valuable addition to the collection of many professionals and students.

This book was conceptualized with the vision of imparting up-to-date information and advanced data in this field. To ensure the same, a matchless editorial board was set up. Every individual on the board went through rigorous rounds of assessment to prove their worth. After which they invested a large part of their time researching and compiling the most relevant data for our readers.

The editorial board has been involved in producing this book since its inception. They have spent rigorous hours researching and exploring the diverse topics which have resulted in the successful publishing of this book. They have passed on their knowledge of decades through this book. To expedite this challenging task, the publisher supported the team at every step. A small team of assistant editors was also appointed to further simplify the editing procedure and attain best results for the readers.

Apart from the editorial board, the designing team has also invested a significant amount of their time in understanding the subject and creating the most relevant covers. They scrutinized every image to scout for the most suitable representation of the subject and create an appropriate cover for the book.

The publishing team has been an ardent support to the editorial, designing and production team. Their endless efforts to recruit the best for this project, has resulted in the accomplishment of this book. They are a veteran in the field of academics and their pool of knowledge is as vast as their experience in printing. Their expertise and guidance has proved useful at every step. Their uncompromising quality standards have made this book an exceptional effort. Their encouragement from time to time has been an inspiration for everyone.

The publisher and the editorial board hope that this book will prove to be a valuable piece of knowledge for researchers, students, practitioners and scholars across the globe.

List of Contributors

Ana do Vale
Host Interaction and Response, Instituto de Investigação e Inovação em Saúde, Universidade do Porto, Porto, Portugal
Group of Fish Immunology and Vaccinology, Instituto de Biologia Molecular e Celular, Universidade do Porto, Porto, Portugal

Didier Cabanes and Sandra Sousa
Host Interaction and Response, Instituto de Investigação e Inovação em Saúde, Universidade do Porto, Porto, Portugal
Group of Molecular Microbiology, Instituto de Biologia Molecular e Celular, Universidade do Porto, Porto, Portugal

Andreu Garcia-Vilanova and Jordi B. Torrelles
Population Health Program, TB Group, Texas Biomedical Research Institute, San Antonio, TX, United States

John Chan
Department of Medicine (Infectious Diseases), Albert Einstein College of Medicine & Montefiore Medical Center, Bronx, NY, United States
Department of Microbiology and Immunology, Albert Einstein College of Medicine & Montefiore Medical Center, Bronx, NY, United States

Yi-Chia Liu
Division of Molecular Microbiology, School of Life Sciences, University of Dundee, Dundee, UK

Kok-Gan Chan
Division of Genetics and Molecular Biology, Institute of Biological Sciences, Faculty of Science, University of Malaya, Kuala Lumpur, Malaysia

Chien-Yi Chang
Centre for Bacterial Cell Biology, Medical School, Newcastle University, Newcastle upon Tyne, UK
Interdisciplinary Computing and Complex BioSystems [illegible]

Charlotte Lombardi, Luca Signor, Daphna Fenel, Jean-Marie Teulon, Juliane Brock and Jean-Luc Pellequer
Univ. Grenoble Alpes, CEA, CNRS, Institut de Biologie Structurale (IBS), Grenoble, France

James Tolchard, Birgit Habenstein and Antoine Loquet
Institute of Chemistry and Biology of Membranes and Nanoobjects, Institut Européen de Chimie et Biologie (CBMN), UMR5248 CNRS, University of Bordeaux, Pessac, France

Stephanie Bouillot, Caroline Gebus, David Liebl, Eric Faudry and Ina Attrée
Univ. Grenoble Alpes, Bacterial Pathogenesis and Cellular Responses Group, U1036 INSERM, ERL5261 CNRS, CEA, Grenoble, France

Viviana Job
Univ. Grenoble Alpes, CEA, CNRS, Institut de Biologie Structurale (IBS), Grenoble, France
Univ. Grenoble Alpes, Bacterial Pathogenesis and Cellular Responses Group, U1036 INSERM, ERL5261 CNRS, CEA, Grenoble, France

Andréa Dessen
Univ. Grenoble Alpes, CEA, CNRS, Institut de Biologie Structurale (IBS), Grenoble, France
Brazilian Biosciences National Laboratory (LNBio), Centro Nacional de Pesquisa em Energia e Materiais (CNPEM), Campinas, Brazil

Zhichao Xu, Yao Qin, Yongqiang Wang, Xiaoqi Li, Hong Cao and Shijun J. Zheng
State Key Laboratory of Agrobiotechnology, China Agricultural University, Beijing, China
Key Laboratory of Animal Epidemiology and Zoonosis, Ministry of Agriculture, China Agricultural University, Beijing, China
College of Veterinary Medicine, China Agricultural University, Beijing, China

Allison J. Greaney, Stephen H. Leppla and Mahtab Moayeri
Microbial Pathogenesis Section, Laboratory of Parasitic Diseases, National Institute of Allergy and Infectious Diseases, National Institutes of Health, Bethesda, MD, USA

[illegible]
Oxford, United Kingdom

Gisela von Hoven, Claudia Neukirch, Martina Meyenburg, Sabine Füser, Maria Bidna Petrivna, Amable J. Rivas and Matthias Husmann
University Medical Center, Institute of Medical Microbiology and Hygiene, Johannes Gutenberg-University, Mainz, Germany

Alexey Ryazanov
Department of Pharmacology, Rutgers Robert Wood Johnson Medical School, Piscataway, NJ, USA

Randal J. Kaufman
Degenerative Diseases Program, Sanford-Burnham Medical Research Institute, La Jolla, CA, USA

Raffi V. Aroian
University of Massachusetts Medical School, Worcester, MA, USA

Hiroshi Ashida
Division of Bacterial Infection Biology, Institute of Medical Science, University of Tokyo, Tokyo, Japan

Hitomi Mimuro
Division of Bacteriology, Department of Infectious Diseases Control, International Research Center for Infectious Diseases, The Institute of Medical Science, University of Tokyo, Tokyo, Japan

Chihiro Sasakawa
Division of Bacterial Infection Biology, Institute of Medical Science, University of Tokyo, Tokyo, Japan
Nippon Institute for Biological Science, Tokyo, Japan
Medical Mycology Research Center, Chiba University, Chiba, Japan

Dagmar Hildebrand, Klaus Heeg and Katharina F. Kubatzky
Zentrum für Infektiologie, Medizinische Mikrobiologie und Hygiene, Universitätsklinikum Heidelberg, Heidelberg, Germany

Margarida Correia-Neves
Life and Health Sciences Research Institute (ICVS), School of Medicine, University of Minho, Braga, Portugal
ICVS/3B's,PT Government Associate Laboratory, Braga, Guimarães, Portugal
Division of Infectious Diseases, Department of Medicine Solna, Karolinska Institutet, Stockholm, Sweden

Gunilla Källenius
Division of Infectious Diseases, Department of Medicine Solna, Karolinska Institutet, Stockholm, Sweden

Christopher Sundling
Division of Infectious Diseases, Department of Medicine Solna, Karolinska Institutet, Stockholm, Sweden
Department of Infectious Diseases, Karolinska University Hospital, Stockholm, Sweden

Andrea Cooper
Leicester Tuberculosis Research Group (LTBRG), Department of Respiratory Sciences, University of Leicester, Leicester, United Kingdom

Stefanie Blodkamp
Department of Physiological Chemistry, University of Veterinary Medicine Hannover, Hannover, Germany

Maren von Köckritz-Blickwede
Department of Physiological Chemistry, University of Veterinary Medicine Hannover, Hannover, Germany
Research Center for Emerging Infections and Zoonoses, University of Veterinary Medicine Hannover, Hannover, Germany

Victor Nizet
Department of Pediatric Pharmacology and Drug Discovery, San Diego School of Medicine, University of California, La Jolla, CA, USA

Qiang Zhang, Yujie Yang, Shuxian Yan, Jiantao Liu, Zhongmin Xu, Junping Yu, Yajing Song and Anding Zhang
Unit of Animal Infectious Diseases, National Key Laboratory of Agricultural Microbiology, College of Veterinary Medicine, Huazhong Agricultural University, Wuhan, China

Meilin Jin
Unit of Animal Infectious Diseases, National Key Laboratory of Agricultural Microbiology, College of Veterinary Medicine, Huazhong Agricultural University, Wuhan, China
Key Laboratory of Development of Veterinary Diagnostic Products, Ministry of Agriculture, Wuhan, China

Teresa Krakauer
Department of Immunology, Molecular Translational Sciences Division, United States Army Medical Research Institute of Infectious Diseases, Fort Detrick, Frederick, MD, USA

Kisha Pradhan and Bradley G. Stiles
Biology Department, Wilson College, Chambersburg, PA, USA

Sylvia Simon
Institute of Medical Microbiology, University of Zürich, Zürich, Switzerland

Hubert Hilbi
Institute of Medical Microbiology, University of Zürich, Zürich, Switzerland
Max von Pettenkofer Institute, Ludwig-Maximilians University, Munich, Germany

Satoshi Uchiyama, Simon Döhrmann, Anjuli M. Timmer, Tamara Bhandari and Kimberly Sprague
Department of Pediatrics, University of California San Diego, La Jolla, CA, USA

Neha Dixit and Scott I. Simon
Department of Biomedical Engineering,University of California Davis, Davis, CA, USA

Mariam Ghochani
Department of Biological Sciences, San Diego State University, San Diego, CA, USA

John C. Timmer
Department of Pharmacology, University of California San Diego, La Jolla, CA, USA

Juliane Bubeck-Wardenburg
Department of Pediatrics,University of Chicago, Chicago, IL, USA

Victor Nizet
Department of Pediatrics, University of California San Diego, La Jolla, CA, USA
Skaggs School of Pharmacy and Pharmaceutical Sciences, University of California San Diego, La Jolla, CA, US

Holger Barth, Amelie Möglich and Christina Förtsch
Institute of Pharmacology and Toxicology, University of Ulm Medical Center, Ulm, Germany

Stephan Fischer
Institute of Pharmacology and Toxicology, University of Ulm Medical Center, Ulm, Germany
Institute of Organic ChemistryIII, University of Ulm, Ulm, Germany

Susanne Maurer, Guido H. Wabnitz, Nadine A. Kahle, Sabine Stegmaier, Birgit Prior,Thomas Giese,Yvonne Samstag and Gertrud Maria Hänsch
Institute for Immunology, University of Heidelberg, Heidelberg, Germany

Matthias Martin Gaida
Institute for Immunology, University of Heidelberg, Heidelberg, Germany
Institute of Pathology, University of Heidelberg,Heidelberg, Germany

Aline Sähr, Sandra Förmer, Dagmar Hildebrand and Klaus Heeg
Medical Microbiology and Hygiene, Department of Infectious Diseases, University Hospital Heidelberg, Heidelberg, Germany

Index

www.ingramcontent.com/pod-product-compliance
Lightning Source LLC
LaVergne TN
LVHW060939220926
848485LV00015B/1116